	$f(t)$	$\mathcal{L}\{f(t)\} = F(s)$
39.	$\dfrac{e^{bt} - e^{at}}{t}$	$\ln\dfrac{s-a}{s-b}$
40.	$\dfrac{2(1 - \cos kt)}{t}$	$\ln\dfrac{s^2 + k^2}{s^2}$
41.	$\dfrac{2(1 - \cosh kt)}{t}$	$\ln\dfrac{s^2 - k^2}{s^2}$
42.	$\dfrac{\sin at}{t}$	$\arctan\left(\dfrac{a}{s}\right)$
43.	$\dfrac{\sin at \cos bt}{t}$	$\dfrac{1}{2}\arctan\dfrac{a+b}{s} + \dfrac{1}{2}\arctan\dfrac{a-b}{s}$
44.	$\dfrac{1}{\sqrt{\pi t}}\, e^{-a^2/4t}$	$\dfrac{e^{-a\sqrt{s}}}{\sqrt{s}}$
45.	$\dfrac{a}{2\sqrt{\pi t^3}}\, e^{-a^2/4t}$	$e^{-a\sqrt{s}}$
46.	$\operatorname{erfc}\left(\dfrac{a}{2\sqrt{t}}\right)$	$\dfrac{e^{-a\sqrt{s}}}{s}$
47.	$2\sqrt{\dfrac{t}{\pi}}\, e^{-a^2/4t} - a\operatorname{erfc}\left(\dfrac{a}{2\sqrt{t}}\right)$	$\dfrac{e^{-a\sqrt{s}}}{s\sqrt{s}}$
48.	$e^{ab}e^{b^2t}\operatorname{erfc}\left(b\sqrt{t} + \dfrac{a}{2\sqrt{t}}\right)$	$\dfrac{e^{-a\sqrt{s}}}{\sqrt{s}(\sqrt{s} + b)}$
49.	$-e^{ab}e^{b^2t}\operatorname{erfc}\left(b\sqrt{t} + \dfrac{a}{2\sqrt{t}}\right)$ $+ \operatorname{erfc}\left(\dfrac{a}{2\sqrt{t}}\right)$	$\dfrac{be^{-a\sqrt{s}}}{s(\sqrt{s} + b)}$
50.	$\delta(t)$	1
51.	$\delta(t - t_0)$	e^{-st_0}
52.	$e^{at}f(t)$	$F(s - a)$
53.	$f(t - a)\mathcal{U}(t - a)$	$e^{-as}F(s)$
54.	$\mathcal{U}(t - a)$	$\dfrac{e^{-as}}{s}$
55.	$f^{(n)}(t)$	$s^n F(s) - s^{(n-1)}f(0) - \cdots - f^{(n-1)}(0)$
56.	$t^n f(t)$	$(-1)^n \dfrac{d^n}{ds^n}\, F(s)$
57.	$\displaystyle\int_0^t f(\tau)g(t - \tau)\, d\tau$	$F(s)G(s)$

DIFFERENTIAL EQUATIONS WITH BOUNDARY-VALUE PROBLEMS

THE PRINDLE, WEBER & SCHMIDT SERIES IN MATHEMATICS

Althoen and Bumcrot, *Introduction to Discrete Mathematics*
Bean, Sharp, and Sharp, *Precalculus*
Boye, Kavanaugh, and Williams, *Elementary Algebra*
Boye, Kavanaugh, and Williams, *Intermediate Algebra*
Burden and Faires, *Numerical Analysis, Fifth Edition*
Cass and O'Connor, *Fundamentals with Elements of Algebra*
Cass and O'Connor, *Basic Mathematics: Prealgebra and Algebra*
Cullen, *Linear Algebra and Differential Equations, Second Edition*
Dick and Patton, *Calculus, Volumes I and II*
Dick and Patton, *Technology in Calculus: A Sourcebook of Activities*
Eves, *In Mathematical Circles*
Eves, *Mathematical Circles Squared*
Eves, *Return to Mathematical Circles*
Faires and Burden, *Numerical Methods*
Fletcher, Hoyle, and Patty, *Foundations of Discrete Mathematics*
Fletcher and Patty, *Foundations of Higher Mathematics, Second Edition*
Fraser, *Intermediate Algebra: An Early Functions Approach*
Gantner and Gantner, *Trigonometry*
Geltner and Peterson, *Geometry for College Students, Second Edition*
Gilbert and Gilbert, *Elements of Modern Algebra, Third Edition*
Gobran, *Beginning Algebra, Fifth Edition*
Gobran, *Intermediate Algebra, Fourth Edition*
Gordon, *Calculus and the Computer*
Hall, *Beginning Algebra*
Hall, *Intermediate Algebra*
Hall, *Algebra for College Students, Second Edition*
Hall, *College Algebra with Applications, Third Edition*
Hartfiel and Hobbs, *Elementary Linear Algebra*
Humi and Miller, *Boundary-Value Problems and Partial Differential Equations*
Kaufmann, *Elementary Algebra for College Students, Fourth Edition*
Kaufmann, *Intermediate Algebra for College Students, Fourth Edition*
Kaufmann, *Algebra for College Students, Fourth Edition*
Kaufmann, *Algebra with Trigonometry for College Students, Third Edition*
Kaufmann, *College Algebra, Second Edition*
Kaufmann, *Trigonometry*
Kaufmann, *Precalculus, Second Edition*
Kennedy and Green, *Prealgebra for College Students*
Laufer, *Discrete Mathematics and Applied Modern Algebra*
Lavoie, *Discovering Mathematics*
Nicholson, *Elementary Linear Algebra with Applications, Second Edition*
Nicholson, *Introduction to Abstract Algebra*
Pence, *Calculus Activities for Graphic Calculators*
Pence, *Calculus Activities for the TI-81 Graphic Calculator*
Plybon, *An Introduction to Applied Numerical Analysis*
Powers, *Elementary Differential Equations*
Powers, *Elementary Differential Equations with Boundary-Value Problems*
Proga, *Arithmetic and Algebra, Third Edition*
Proga, *Basic Mathematics, Third Edition*

Rice and Strange, *Plane Trigonometry, Sixth Edition*
Rogers, Haney, and Laird, *Fundamentals of Business Mathematics*
Schelin and Bange, *Mathematical Analysis for Business and Economics, Second Edition*
Sgroi and Sgroi, *Mathematics for Elementary School Teachers*
Swokowski and Cole, *Fundamentals of College Algebra, Eighth Edition*
Swokowski and Cole, *Fundamentals of Algebra and Trigonometry, Eighth Edition*
Swokowski and Cole, *Fundamentals of Trigonometry, Eighth Edition*
Swokowski and Cole, *Algebra and Trigonometry with Analytic Geometry, Eighth Edition*
Swokowski, *Precalculus: Functions and Graphs, Sixth Edition*
Swokowski, *Calculus, Fifth Edition*
Swokowski, *Calculus, Fifth Edition, Late Trigonometry Edition*
Swokowski, *Calculus of a Single Variable*
Tan, *Applied Finite Mathematics, Third Edition*
Tan, *Calculus for the Managerial, Life, and Social Sciences, Second Edition*
Tan, *Applied Calculus, Second Edition*
Tan, *College Mathematics, Second Edition*
Trim, *Applied Partial Differential Equations*
Venit and Bishop, *Elementary Linear Algebra, Alternate Second Edition*
Venit and Bishop, *Elementary Linear Algebra, Third Edition*
Wiggins, *Problem Solver for Finite Mathematics and Calculus*
Willard, *Calculus and Its Applications, Second Edition*
Wood and Capell, *Arithmetic*
Wood and Capell, *Intermediate Algebra*
Wood, Capell, and Hall, *Developmental Mathematics, Fourth Edition*
Zill, *Calculus, Third Edition*
Zill, *A First Course in Differential Equations, Fifth Edition*
Zill and Cullen, *Differential Equations with Boundary-Value Problems, Third Edition*
Zill and Cullen, *Advanced Engineering Mathematics*

THE PRINDLE, WEBER & SCHMIDT SERIES IN ADVANCED MATHEMATICS

Brabenec, *Introduction to Real Analysis*
Ehrlich, *Fundamental Concepts of Abstract Algebra*
Eves, *Foundations and Fundamental Concepts of Mathematics, Third Edition*
Kirkwood, *An Introduction to Real Analysis*
Patty, *Foundations of Topology*
Ruckle, *Modern Analysis: Measure Theory and Functional Analysis with Applications*
Sieradski, *An Introduction to Topology and Homotopy*

3rd EDITION

DIFFERENTIAL EQUATIONS WITH BOUNDARY-VALUE PROBLEMS

Dennis G. Zill
Michael R. Cullen
Loyola Marymount University

PWS–KENT Publishing Company
Boston

PWS-KENT
Publishing Company

20 Park Plaza
Boston, Massachusetts 02116

Sponsoring Editor: Steve Quigley
Production Editor: Patricia Adams
Developmental Editor: Barbara Lovenvirth
Manufacturing Coordinator: Lisa Flanagan
Cover and Interior Design: Patricia Adams
Cover Photo: Bernard Van Berg/The Image Bank
Interior Illustrations: Deborah Schneck and Network Graphics
Typesetting: Polyglot Pte Ltd
Cover and Insert Printing: New England Book Components, Inc.
Printing and Binding: Arcata Graphics/Halliday

PWS-KENT Publishing Company is a division of Wadsworth, Inc.

Portions of this book also appear in *A First Course in Differential Equations*, Fifth Edition, by Dennis G. Zill, copyright © 1993 by PWS-KENT Publishing Company.

Insert 1: p. i, top: Terje Rakke/The Image Bank; bottom: Andrea Pistolesi/The Image Bank; p. ii, top: Lawrence Hughes/The Image Bank; bottom: R. Hayo/The Image Bank; p. iii, top: NASA; bottom: The University Museum, University of Pennsylvania Neg. # T4-143c2; p. iv, top: Garry Gay/The Image Bank; bottom: Biophoto Associates, Science Source/Photo Researchers Inc. *Insert 2*: p. i, top: NASA; bottom: Michael Quackenbush/The Image Bank; p. ii, top: NASA; bottom: Ford Motor Company; p. ii, top: Lockheed Aeronautical Systems Co, A Div. of Lockheed Corporation; bottom: Wm. Thomas King/The Image Bank; p. iii, top: Tacoma-Pierce County Visitor & Convention Bureau; bottom: Burton McNeely/The Image Bank. *Insert 3*: p. i, top: Guido Alberto Rossi/The Image Bank; bottom: Axel Foerster/The Image Bank. p. ii, top: Sobel/Klonsky/The Image Bank; bottom: Santi Visalli/The Image Bank.

Library of Congress Cataloging-in-Publication Data

Zill, Dennis G.
    Differential equations with boundary-value problems / Dennis G.
Zill. and Michael R. Cullen. — 3rd ed.
      p.   cm.
    Includes index.
    ISBN 0-534-93158-8   1. Differential equations.   2. Boundary value problems.
    I. Cullen, Michael R.   II. Title.
    QA371.Z55 1992
    515'.35—dc20   92-17344
                                        CIP

Printed in the United States of America
93 94 95 96 97 — 10 9 8 7 6 5 4 3 2 1

CONTENTS

CHAPTER 1

INTRODUCTION TO DIFFERENTIAL EQUATIONS **1**

1.1   Basic Definitions and Terminology   **2**
[O] 1.2   Some Mathematical Models   **12**
      Chapter 1 Review   **29**   *Chapter 1 Review Exercises*   **30**

CHAPTER 2

FIRST-ORDER DIFFERENTIAL EQUATIONS   **33**

2.1   Preliminary Theory   **34**
2.2   Separable Variables   **38**
2.3   Homogeneous Equations   **46**
2.4   Exact Equations   **53**
2.5   Linear Equations   **62**
[O] 2.6   Equations of Bernoulli, Ricatti, and Clairaut   **71**
[O] 2.7   Substitutions   **76**
[O] 2.8   Picard's Method   **80**
      Chapter 2 Review   **83**   *Chapter 2 Review Exercises*   **84**

CHAPTER 3

APPLICATIONS OF FIRST-ORDER DIFFERENTIAL EQUATIONS   **87**

3.1   Orthogonal Trajectories   **88**
3.2   Applications of Linear Equations   **94**
3.3   Applications of Nonlinear Equations   **108**
      Chapter 3 Review   **120**   *Chapter 3 Review Exercises*   **121**
      Essay: Population Dynamics by Michael Olinick   **123**

CHAPTER 4

LINEAR DIFFERENTIAL EQUATIONS OF HIGHER-ORDER   **129**

4.1   Preliminary Theory   **130**
     4.1.1   Initial-Value and Boundary-Value Problems   **130**
     4.1.2   Linear Dependence and Linear Independence   **134**
     4.1.3   Solutions of Linear Equations   **139**

[O] = *optional*

4.2   Constructing a Second Solution from a Known Solution   **154**

4.3   Homogeneous Linear Equations with Constant Coefficients   **159**

4.4   Undetermined Coefficients—Superposition Approach   **169**

4.5   Differential Operators   **181**

4.6   Undetermined Coefficient—Annihilator Approach   **187**

4.7   Variation of Parameters   **194**
      Chapter 4 Review   **203**   *Chapter 4 Review Exercises*   **204**
      Essay: Chaos by John H. Hubbard   **207**

CHAPTER 5

APPLICATIONS OF SECOND-ORDER DIFFERENTIAL EQUATIONS: VIBRATIONAL MODELS   211

5.1   Simple Harmonic Motion   **212**

5.2   Damped Motion   **222**

5.3   Forced Motion   **232**

5.4   Electric Circuits and Other Analogous Systems   **242**
      Chapter 5 Review   **248**   *Chapter 5 Review Exercises*   **249**
      Essay: Tacoma Narrows Suspension Bridge Collapse
          by Gilbert N. Lewis   **253**

CHAPTER 6

DIFFERENTIAL EQUATIONS WITH VARIABLE COEFFICIENTS   257

6.1   Cauchy-Euler Equation   **258**

6.2   Review of Power Series; Power Series Solutions   **267**

6.3   Solutions About Ordinary Points   **277**

6.4   Solutions About Singular Points   **286**
      6.4.1   Regular Singular Points; Method of Frobenius—Case I   **286**
      6.4.2   Method of Frobenius—Cases II and III   **296**

6.5   Two Special Equations   **306**
      6.5.1   Solution of Bessel's Equation   **307**
      6.5.2   Solution of Legendre's Equation   **314**
      Chapter 6 Review   **321**   *Chapter 6 Review Exercises*   **322**

CHAPTER 7

LAPLACE TRANSFORM   323

7.1   Laplace Transform   **324**

7.2   Inverse Transform   **335**

7.3   Translation Theorems and Derivatives of a Transform   **343**

7.4   Transforms of Derivatives, Integrals, and Periodic Functions   **355**

7.5   Applications   **363**

[0] 7.6   Dirac Delta Function   **379**
      Chapter 7 Review   **384**   *Chapter 7 Review Exercises*   **385**

CHAPTER 8

SYSTEMS OF LINEAR DIFFERENTIAL EQUATIONS   389

8.1   Operator Method   **390**

8.2   Laplace Transform Method   **398**

8.3   Systems of Linear First-Order Equations   **407**

8.4   Introduction to Matrices   **413**
      8.4.1   Basic Definitions and Theory   **413**
      8.4.2   Gaussian and Gauss-Jordan Elimination Methods   **423**
      8.4.3   The Eigenvalue Problem   **427**

8.5   Matrices and Systems of Linear First-Order Equations   **436**
      8.5.1   Preliminary Theory   **436**
      8.5.2   A Fundamental Matrix   **446**

8.6   Homogeneous Linear Systems   **453**
      8.6.1   Distinct Real Eigenvalues   **453**
      8.6.2   Complex Eigenvalues   **457**
      8.6.3   Repeated Eigenvalues   **462**

[0] 8.7   Undetermined Coefficients   **469**

8.8   Variation of Parameters   **473**

[0] 8.9   Matrix Exponential   **478**
      Chapter 8 Review   **481**   *Chapter 8 Review Exercises*   **482**

CHAPTER 9

NUMERICAL METHODS FOR ORDINARY DIFFERENTIAL EQUATIONS   487

9.1   Direction Fields   **488**

9.2   The Euler Methods   **494**

9.3   The Three-Term Taylor Method   **503**

9.4   The Runge-Kutta Method   **507**

9.5   Multistep Methods   **513**

9.6   Errors and Stability   **517**

9.7   Higher-Order Equations and Systems   **523**

9.8   Second-Order Boundary-Value Problems   **528**
      Chapter 9 Review   **533**   *Chapter 9 Review Exercises*   **534**
      Essay: Nerve Impulse Models by C. J. Knickerbocker   **537**

CHAPTER 10

PLANE AUTONOMOUS SYSTEMS AND STABILITY   543

10.1   Autonomous Systems, Critical Points, and Periodic Solutions   544
10.2   Stability of Linear Systems   554
10.3   Linearization and Local Stability   565
10.4   Applications of Autonomous Systems   578
         Chapter 10 Review   593   Chapter 10 Review Exercises   594

CHAPTER 11

ORTHOGONAL FUNCTIONS AND FOURIER SERIES   597

11.1   Orthogonal Functions   598
11.2   Fourier Series   604
11.3   Fourier Cosine and Sine Series   610
11.4   Sturm-Liouville Problem   622
11.5   Bessel and Legendre Series   631
         Chapter 11 Review   639   Chapter 11 Review Exercises   640

CHAPTER 12

BOUNDARY-VALUE PROBLEMS IN RECTANGULAR COORDINATES   643

12.1   Separable Partial Differential Equations   644
12.2   Classical Equations and Boundary-Value Problems   649
12.3   Heat Equation   657
12.4   Wave Equation   660
12.5   Laplace's Equation   666
12.6   Nonhomogeneous Equations and Boundary Conditions   670
12.7   Use of Generalized Fourier Series   674
[O] 12.8   Boundary-Value Problems Involving Fourier Series in Two Variables   680
         Chapter 12 Review   683   Chapter 12 Review Exercises   683
         Essay: Where Is Middle C? by Ruth G. Favro   687

CHAPTER 13

BOUNDARY-VALUE PROBLEMS IN OTHER COORDINATE SYSTEMS   693

13.1   Problems Involving Laplace's Equation in Polar Coordinates   694

13.2   Problems in Polar and Cylindrical Coordinates   699

13.3   Problems in Spherical Coordinates   707
Chapter 13 Review   710   Chapter 13 Review Exercises   710

CHAPTER 14

INTEGRAL TRANSFORM METHOD   715

14.1   Error Function   716

14.2   Applications of the Laplace Transform   718

14.3   Fourier Integral   728

14.4   Fourier Transform   735
Chapter 14 Review   743   Chapter 14 Review Exercises   744

CHAPTER 15

NUMERICAL METHODS FOR PARTIAL DIFFERENTIAL EQUATIONS   747

15.1   Elliptic Equations   748

15.2   Parabolic Equations   757

15.3   Hyperbolic Equations   767
Chapter 15 Review   773   Chapter 15 Review Exercises   774

APPENDICES   APP–1

I   Gamma Function   APP–2

II   Laplace Transforms   APP–4

III   Review of Determinants   APP–7

IV   Complex Numbers   APP–12

ANSWERS TO ODD-NUMBERED PROBLEMS   A–1

INDEX   I–1

PREFACE

Our goal for this third edition was to achieve a balance between the concepts and presentation of material that appealed to users of the previous editions and the substantive changes made to strengthen and modernize some aspects of the text. We feel that we have achieved this balance, thus enabling the text to appeal to an even wider audience. Many of the additions and changes are the result of user and reviewer comments and suggestions. Moreover, these changes were made with the ultimate audience in mind—the students who will be using it. For this reason, solutions of every example have been read with an eye to improving their clarity. In various places we have either added further explanations where we thought they might be helpful, or have added "guidance boxes" at crucial points in the flow of the solution.

As before, this text is intended to satisfy the needs of an instructor who desires more than just an introduction to the subject. In addition to the basic material on ordinary differential equations, this text contains a chapter on nonlinear equations and stability and chapters on partial differential equations and boundary-value problems. It can be used for either a one- or two-semester course.

NEW FEATURES

Some new features, that we hope students will find both interesting and motivational, have been added to the text. Essays, written by mathematicians prominent in their specialty are included after Chapters 3, 4, 5, 9, and 12. Each essay reflects the thoughts, creativity, and opinions of the individual author and is intended to enhance the material found in the preceding chapter. It is our hope that the addition of these essays will spark the interest of the students, encourage them to read mathematics, and help them to gain a realization that differential equations is not simply a dry collection of methods, facts, and formulas, but a vibrant field in which people can, and do, work.

Color inserts also have been added at intervals in the text. These pages consist of illustrations matched with photographs relating to some of the applications found in the text. We feel that these pairings contribute to the visualization of the applications and thereby provide an added insight to students.

CHANGES IN THIS EDITION

- Section 1.2 is now devoted solely to the concept of a differential equation as a mathematical model.

- The material on the differential equation of a family of curves has been deleted. A brief discussion of this concept is now given in Section 3.1 (Orthogonal Trajectories).

- The method of undetermined coefficients is one of the more controversial topics in a course in differential equations. In the last three editions, this topic was developed from the viewpoint of using a differential annihilator as an aid in determining the correct form of a particular solution. While preparing this revision, a substantial number of reviewers indicated that the annihilator approach was too sophisticated for their students and requested a simpler rule-based approach. Other reviewers, however, desired no change. In order to satisfy each of these preferences, both approaches are presented in this edition. The instructor can now choose between undetermined coefficients based on the superposition principle for nonhomogeneous linear differential equations (Section 4.4) or those based on the concept of differential annihilators (Section 4.6). Moreover, in this edition the notion of a differential operator is now introduced in a separate section (Section 4.5) Thus, covering Section 4.4 does not preclude coverage of the otherwise useful concept of a differential operator.

- The review of power series in Section 6.2 has been greatly expanded. A discussion of the arithmetic of powers series (addition, multiplication, and division of series) has been added.

- A brief discussion of the "cover-up method" for determining coefficients in a partial fraction decomposition and a historical note on Oliver Heaviside have been added to Section 7.2.

- The discussion on the operational properties of the Laplace transform has now been divided into two sections: Section 7.3, Translation Theorems and Derivatives of Transforms, and Section 7.4, Transforms of Derivatives, Integrals, and Periodic Functions. This separation allows for a clearer, more comprehensive treatment of these topics.

- Gaussian elimination, in addition to Gauss-Jordan elimination, is now discussed in Section 8.4. The notation for indicating row operations on an augmented matrix has been improved.

- Chapter 9, "Numerical Methods for Ordinary Differential Equations," has been significantly expanded and partially rewritten. The Adams-Bashforth/Adams-Moulton multistep method has been added to Section 9.5. Section 9.6, Errors and Stability, and Section 9.8, Second-Order Boundary-Value Problems, are new to this edition.

- The BASIC programs previously in Chapter 9 have been deleted. Instead, we offer these programs, along with their FORTRAN and Pascal versions, on disk.

- Chapter 10, Plane Autonomous Systems and Stability," is new to this edition. These topics were added in response to favorable reactions from the users and reviewers of the previous edition.

- The material in Chapters 10, 11, and 12 has been expanded and reorganized as Chapters 11, 12, 13, and 14 in this edition.
- Chapter 15, "Numerical Methods for Partial Differential Equations," is new.
- New problems, applications, illustrations, remarks, and historical footnotes have been added throughout the text.

SUPPLEMENTS AVAILABLE

For Instructors

- *Complete Solutions Manual*, Warren S. Wright, Loyola Marymount University. This manual contains complete, worked-out solutions to every problem in the text.

For Students

- *Student Solutions Manual*, Warren S. Wright. This manual provides solutions to every third problem in each exercise set.

Software

- *Computer Programs*, C. J. Knickerbocker, St. Lawrence University. This disk contains a listing of computer programs for many of the numerical methods considered in this text. Each program is written in three languages: BASIC, FORTRAN, and Pascal. (For IBM or MAC.)
- *The Grapher*, Steve Scarborough, Loyola Marymount University. For the Macintosh, this flexible program can generate graphs of equations in rectangular and polar coordinates as well as curves defined parametrically. In addition, it has routines for graphing: series, interpolating polynomials, direction fields, and numerical solutions of differential equations. The updated version also contains numerical integration and numerical root finding.
- *Computer Simulations in DE Workbook and Software.*

Dennis G. Zill
Michael R. Cullen
Los Angeles

ACKNOWLEDGMENTS

It takes the teamwork of many people to compile a revision of a textbook. We are especially grateful to Barbara Lovenvirth, developmental editor, Patty Adams, production editor, Carol Reitz, copy editor, Warren and Carol Wright for their help in the manuscript preparation and for producing the excellent solutions manuals, John Ellison of Grove City College, Grove City, PA, for his valuable contribution to Chapter 9, and to the following reviewers for their counsel, comments, criticisms, and compliments:

Linda J. S. Allen, *Texas Tech University*

Stephen Breen, *California State University*

Dean R. Brown, *Youngstown State University*

Kalin N. Godev, *Penn State University*

Thomas G. Kudzuma, *University of Massachusetts at Lowell*

Gilbert N. Lewis, *Michigan Technological University*

Clarence A. McGuff, *Austin Community College*

We also want to recognize and extend our sincere appreciation to the following individuals for taking time out of their busy schedules to contribute the essays found in this edition:

Michael Olinick, Department of Mathematics and Computer Science, *Middlebury College*, **Population Dynamics**

John H. Hubbard and Beverly West, Department of Mathematics, *Cornell University*, **Chaos**

Gilbert N. Lewis, Department of Mathematical and Computer Sciences, *Michigan Technological University*, **Tacoma Narrows Suspension Bridge Collapse**

C. J. Knickerbocker, Department of Mathematics, *St. Lawrence University*, **Nerve Impulse Models**

Ruth Favro, Department of Mathematics and Computer Science, *Lawrence Technological University*, **Where Is Middle C?**

D.G.Z.
M.R.C.

INTRODUCTION TO DIFFERENTIAL EQUATIONS

1.1 Basic Definitions and Terminology

[O] **1.2** Some Mathematical Models

Chapter 1 Review

Chapter 1 Review Exercises

Important Concepts

Ordinary differential equation

Partial differential equation

Order of an equation

Linear equation

Nonlinear equation

Solution

Trivial solution

Explicit and implicit solutions

***n*-parameter family of solutions**

Particular solution

Singular solution

General solution

Mathematical model

The words *differential* and *equations* certainly suggest solving some kind of equation that contains derivatives. So it is; in fact, the preceding sentence tells the complete story about the course that you are about to begin. But before you start solving anything, you must learn some of the basic definitions and terminology of the subject. This is what Section 1.1 is all about. Section 1.2 is intended to be motivational. Why should you, an erstwhile scientist or engineer, study this subject? The answer is simple: Differential equations are the mathematical backbone of many areas of science and engineering. Hence, in Section 1.2 we examine, albeit briefly, how differential equations arise from attempts to formulate, or describe, certain physical systems in terms of mathematics.

1.1 BASIC DEFINITIONS AND TERMINOLOGY

In calculus you learned that given a function $y = f(x)$, the derivative

$$\frac{dy}{dx} = f'(x)$$

is itself a function of x and is found by some appropriate rule. For example, if $y = e^{x^2}$, then

$$\frac{dy}{dx} = 2xe^{x^2} \quad \text{or} \quad \frac{dy}{dx} = 2xy. \tag{1}$$

The problem that we face in this course is not: given a function $y = f(x)$, find its derivative. Rather, our problem is: if we are given an equation such as $dy/dx = 2xy$, somehow to find a function $y = f(x)$ that satisfies the equation. In a word, we wish to *solve* differential equations.

DEFINITION 1.1   Differential Equation

An equation containing the derivatives or differentials of one or more dependent variables, with respect to one or more independent variables, is said to be a **differential equation** (DE).

Differential equations are classified according to **type**, **order**, and **linearity**.

Classification by Type

If an equation contains only ordinary derivatives of one or more dependent variables, with respect to a single independent variable, it is then said to be an **ordinary differential equation** (ODE). For example,

$$\frac{dy}{dt} - 5y = 1$$

$$(y - x)\,dx + 4x\,dy = 0$$

$$\frac{du}{dx} - \frac{dv}{dx} = x$$

$$\frac{d^2y}{dx^2} - 2\frac{dy}{dx} + 6y = 0$$

are ordinary differential equations. An equation involving the partial derivatives of one or more dependent variables of two or more independent variables is called a **partial differential equation** (PDE). For example,

$$\frac{\partial u}{\partial y} = -\frac{\partial v}{\partial x}$$

$$x\frac{\partial u}{\partial x} + y\frac{\partial u}{\partial y} = u$$

$$\frac{\partial^2 u}{\partial x^2} = \frac{\partial^2 u}{\partial t^2} - 2\frac{\partial u}{\partial t}$$

are partial differential equations.

Classification by Order

The order of the highest-order derivative in a differential equation is called the **order of the equation**. For example,

$$\frac{d^2 y}{dx^2} + 5\left(\frac{dy}{dx}\right)^3 - 4y = e^x$$

is a second-order ordinary differential equation. Since the differential equation $(y - x)\,dx + 4x\,dy = 0$ can be put into the form

$$4x\frac{dy}{dx} + y = x$$

by dividing by the differential dx, it is a first-order ordinary differential equation. The equation

$$a^2\frac{\partial^4 u}{\partial x^4} + \frac{\partial^2 u}{\partial t^2} = 0$$

is a fourth-order partial differential equation.

Although partial differential equations are very important, their study demands a good foundation in the theory of ordinary differential equations. Consequently, in the discussion that follows we shall confine our attention to ordinary differential equations.

A general nth-order, ordinary differential equation is often represented by the symbolism

$$F\left(x, y, \frac{dy}{dx}, \ldots, \frac{d^n y}{dx^n}\right) = 0. \tag{2}$$

The following is a special case of (2).

Classification as Linear or Nonlinear

A differential equation is said to be **linear** if it can be written in the form

$$a_n(x)\frac{d^n y}{dx^n} + a_{n-1}(x)\frac{d^{n-1} y}{dx^{n-1}} + \cdots + a_1(x)\frac{dy}{dx} + a_0(x)y = g(x).$$

It should be observed that linear differential equations are characterized by two properties:

(i) The dependent variable y and all its derivatives are of the first degree; that is, the power of each term involving y is 1.

(ii) Each coefficient depends on only the independent variable x.

An equation that is not linear is said to be **nonlinear**.

The equations

$$x \, dy + y \, dx = 0$$

$$y'' - 2y' + y = 0$$

and

$$x^3 \frac{d^3y}{dx^3} - x^2 \frac{d^2y}{dx^2} + 3x \frac{dy}{dx} + 5y = e^x$$

are linear first-, second-, and third-order ordinary differential equations, respectively. On the other hand,

| coefficient depends on y | | power not 1 |

$$yy'' - 2y' = x \quad \text{and} \quad \frac{d^3y}{dx^3} + y^2 = 0$$

are nonlinear second- and third-order ordinary differential equations, respectively.

Solutions

As mentioned before, our goal in this course is to solve, or find solutions of, differential equations.

DEFINITION 1.2   Solution of a Differential Equation
Any function f defined on some interval I, which when substituted into a differential equation reduces the equation to an identity, is said to be a **solution** of the equation on the interval.

In other words, a solution of an ordinary differential equation

$$F(x, y, y', \ldots, y^{(n)}) = 0$$

is a function f that possesses at least n derivatives and *satisfies* the equation; that is,

$$F(x, f(x), f'(x), \ldots, f^{(n)}(x)) = 0$$

for every x in the interval I. The precise form of the interval I is purposely left vague in Definition 1.2. Depending on the context of the discussion, I could represent an open interval (a, b), a closed interval $[a, b]$, an infinite interval $(0, \infty)$, and so on.

EXAMPLE 1

Verify that $y = x^4/16$ is a solution of the nonlinear equation

$$\frac{dy}{dx} = xy^{1/2}$$

on the interval $(-\infty, \infty)$.

Solution   One way of verifying that the given function is a solution is to write the differential equation as $dy/dx - xy^{1/2} = 0$ and then see, after substituting, whether the sum $dy/dx - xy^{1/2}$ is zero for every x in the interval. Using

$$\frac{dy}{dx} = 4\frac{x^3}{16} = \frac{x^3}{4} \quad \text{and} \quad y^{1/2} = \left(\frac{x^4}{16}\right)^{1/2} = \frac{x^2}{4},$$

we see that

$$\frac{dy}{dx} - xy^{1/2} = \frac{x^3}{4} - x\left(\frac{x^4}{16}\right)^{1/2} = \frac{x^3}{4} - \frac{x^3}{4} = 0$$

for every real number. ∎

EXAMPLE 2

The function $y = xe^x$ is a solution of the linear equation

$$y'' - 2y' + y = 0$$

on $(-\infty, \infty)$. To see this, we compute

$$y' = xe^x + e^x \quad \text{and} \quad y'' = xe^x + 2e^x.$$

Observe $y'' - 2y' + y = (xe^x + 2e^x) - 2(xe^x + e^x) + xe^x = 0$

for every real number. ∎

Notice that in Examples 1 and 2 the constant function $y = 0$, for $-\infty < x < \infty$, also satisfies the given differential equation. A solution of a differential equation that is identically zero on an interval I is often referred to as a **trivial solution**.

Not every differential equation that we write necessarily has a solution.

EXAMPLE 3

(a) The first-order differential equations

$$\left(\frac{dy}{dx}\right)^2 + 1 = 0 \quad \text{and} \quad (y')^2 + y^2 + 4 = 0$$

possess no real solutions. Why?

(b) The second-order equation $(y'')^2 + 10y^4 = 0$ possesses only one real solution. What is it?   ∎

Explicit and Implicit Solutions

You should be familiar with the notions of explicit and implicit functions from your study of calculus. Similarly, solutions of differential equations are distinguished as either explicit solutions or implicit solutions. A solution of an ordinary differential equation (2) that can be written in the form $y = f(x)$ is said to be an **explicit solution**. We saw in our initial discussion that $y = e^{x^2}$ is an explicit solution of $dy/dx = 2xy$. In Examples 1 and 2, $y = x^4/16$ and $y = xe^x$ are explicit solutions of $dy/dx = xy^{1/2}$ and $y'' - 2y' + y = 0$, respectively. A relation $G(x, y) = 0$ is said to be an **implicit solution** of an ordinary differential equation (2) on an interval I provided it defines one or more explicit solutions on I.

EXAMPLE 4

For $-2 < x < 2$ the relation $x^2 + y^2 - 4 = 0$ is an implicit solution of the differential equation

$$\frac{dy}{dx} = -\frac{x}{y}.$$

By implicit differentiation it follows that

$$\frac{d}{dx}(x^2) + \frac{d}{dx}(y^2) - \frac{d}{dx}(4) = 0$$

$$2x + 2y\frac{dy}{dx} = 0 \quad \text{or} \quad \frac{dy}{dx} = -\frac{x}{y}.$$ ∎

The relation $x^2 + y^2 - 4 = 0$ in Example 4 defines two explicit differentiable functions: $y = \sqrt{4 - x^2}$ and $y = -\sqrt{4 - x^2}$ on the interval $(-2, 2)$. Also note that any relation of the form $x^2 + y^2 - c = 0$ *formally* satisfies $dy/dx = -x/y$ for any constant c. However, it is naturally understood that the relation should always make sense in the real number system; thus we cannot say that $x^2 + y^2 + 1 = 0$ determines a solution of the differential equation.

Since the distinction between an explicit and an implicit solution should be intuitively clear, we shall not belabor the issue by always saying "here is an explicit (implicit) solution."

Number of Solutions

You should become accustomed to the fact that a given differential equation usually possesses an *infinite* number of solutions. By direct substitution, we can prove that any curve—that is, function—in the one-parameter family $y = ce^{x^2}$, where c is any arbitrary constant, also satisfies (1). As indicated in Figure 1.1, the trivial solution is a member of this family of solutions corresponding to $c = 0$. In Example 2, tracing back through the work reveals that $y = cxe^x$ is a family of solutions of the given differential equation.

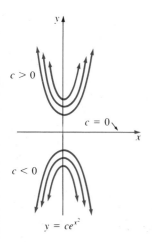

$c > 0$

$c = 0$

$c < 0$

$y = ce^{x^2}$

Figure 1.1

EXAMPLE 5

For any value of c the function $y = c/x + 1$ is a solution of the first-order differential equation

$$x \frac{dy}{dx} + y = 1$$

on the interval $(0, \infty)$. We have

$$\frac{dy}{dx} = c \frac{d}{dx}(x^{-1}) + \frac{d}{dx}(1) = -cx^{-2} = -\frac{c}{x^2}$$

so that

$$x \frac{dy}{dx} + y = x \left(-\frac{c}{x^2} \right) + \left(\frac{c}{x} + 1 \right) = 1.$$

By choosing c to be any real number, we can generate an infinite number of solutions. In particular, for $c = 0$ we obtain a constant solution $y = 1$. See Figure 1.2. ∎

In Example 5, $y = c/x + 1$ is a solution of the differential equation on any interval not containing the origin. The function is not differentiable at $x = 0$.

Under some circumstances when we add two solutions of a differential equation, we get another solution.

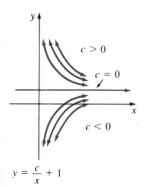

$c > 0$

$c = 0$

$c < 0$

$y = \dfrac{c}{x} + 1$

Figure 1.2

EXAMPLE 6

(a) The functions $y = c_1 \cos 4x$ and $y = c_2 \sin 4x$, where c_1 and c_2 are arbitrary constants, are solutions of the differential equation

$$y'' + 16y = 0.$$

For $y = c_1 \cos 4x$, the first and second derivatives are

$$y' = -4c_1 \sin 4x \quad \text{and} \quad y'' = -16c_1 \cos 4x,$$

and so $\quad y'' + 16y = -16c_1 \cos 4x + 16(c_1 \cos 4x) = 0$.

Similarly, for $y = c_2 \sin 4x$,

$$y'' + 16y = -16c_2 \sin 4x + 16(c_2 \sin 4x) = 0.$$

(b) The sum of the solutions in part (a), $y = c_1 \cos 4x + c_2 \sin 4x$, can also be shown to be a solution of $y'' + 16y = 0$. ∎

EXAMPLE 7

You should be able to show that

$$y = e^x, \quad y = e^{-x}, \quad y = c_1 e^x, \quad y = c_2 e^{-x}, \quad \text{and} \quad y = c_1 e^x + c_2 e^{-x}$$

are all solutions of the linear second-order differential equation

$$y'' - y = 0.$$

Note that $y = c_1 e^x$ is a solution for any choice of c_1, but $y = e^x + c_1, c_1 \neq 0$, does *not* satisfy the equation since, for this latter family of functions, we would get $y'' - y = -c_1$. ∎

The next example shows that a solution of a differential equation can be a piecewise-defined function.

EXAMPLE 8

Any function in the one-parameter family $y = cx^4$ is a solution of the differential equation

$$xy' - 4y = 0.$$

We have $xy' - 4y = x(4cx^3) - 4cx^4 = 0$. The piecewise-defined function

$$y = \begin{cases} -x^4, & x < 0 \\ x^4, & x \geq 0 \end{cases}$$

is also a solution. Observe that this function cannot be obtained from $y = cx^4$ by a single selection of the parameter c. See Figure 1.3(b). ∎

(a)

(b)

Figure 1.3

Further Terminology

The study of differential equations is similar to integral calculus. When evaluating an antiderivative or indefinite integral, we utilize a single constant of integration. In like manner, when solving a first-order differential equation

$F(x, y, y') = 0$, we usually obtain a family of curves or functions $G(x, y, c) = 0$ containing one arbitrary parameter such that each member of the family is a solution of the differential equation. In fact, when solving an nth-order equation $F(x, y, y', \ldots, y^{(n)}) = 0$, where $y^{(n)}$ means $d^n y/dx^n$, we expect an **n-parameter family of solutions** $G(x, y, c_1, \ldots, c_n) = 0$.

A solution of a differential equation that is free of arbitrary parameters is called a **particular solution**. One way of obtaining a particular solution is to choose specific values of the parameter(s) in a family of solutions. For example, it is readily seen that $y = ce^x$ is a one-parameter family of solutions of the simple first-order equation $y' = y$. For $c = 0, -2$, and 5, we get the particular solutions $y = 0$, $y = -2e^x$, and $y = 5e^x$, respectively.

Sometimes a differential equation possesses a solution that cannot be obtained by specializing the parameters in a family of solutions. Such a solution is called a **singular solution**.

EXAMPLE 9

In Section 2.1 we shall prove that a one-parameter family of solutions of $y' = xy^{1/2}$ is given by $y = (x^2/4 + c)^2$. When $c = 0$, the resulting particular solution is $y = x^4/16$. In this case the trivial solution $y = 0$ is a singular solution of the equation, since it cannot be obtained from the family for any choice of the parameter c.   ∎

If *every* solution of $F(x, y, y', \ldots, y^{(n)}) = 0$ on an interval I can be obtained from $G(x, y, c_1, \ldots, c_n) = 0$ by appropriate choices of the c_i, $i = 1, 2, \ldots, n$, we then say that the n-parameter family is the **general**, or **complete**, solution of the differential equation.

Remark   There are two schools of thought concerning the concept of a "general solution." An alternative viewpoint holds that a general solution of an nth-order differential equation is a family of solutions containing n essential* parameters. Period! In other words, the family is not required to contain all solutions of the differential equation on some interval. The difference in these opinions is really a distinction between the solutions to linear and nonlinear equations. In solving linear differential equations we shall impose relatively simple restrictions on the coefficients; with these restrictions one can always be assured not only that a solution does exist on an interval but also that a family of solutions indeed yields all possible solutions.

Another fact deserves mention at this time. Nonlinear equations, with the exception of some first-order equations, are usually difficult or *impossible*

---

* We won't try to define this concept. But roughly it means: Don't play games with the constants. Certainly $y = x + c_1 + c_2$ represents a family of solutions of $y' = 1$. By re-naming $c_1 + c_2$ as c, the family has *essentially* one constant: $y = x + c$. You should verify that $y = c_1 + \ln c_2 x$ is a solution of $x^2 y'' + xy' = 0$ on the interval $(0, \infty)$ for any choice of c_1 and $c_2 > 0$. Are c_1 and c_2 essential parameters?

to solve in terms of familiar elementary functions such as algebraic functions, exponential and logarithmic functions, and trigonometric and inverse trigonometric functions. Furthermore, if we happen to have a family of solutions for a nonlinear equation, it is not obvious when this family constitutes a general solution. On a practical level, then, the designation "general solution" is applied only to linear differential equations.

1.1 EXERCISES

Answers to odd-numbered problems begin on page A–1.

In Problems 1–10 state whether the given differential equations are linear or nonlinear. Give the order of each equation.

1. $(1 - x)y'' - 4xy' + 5y = \cos x$

2. $x\dfrac{d^3 y}{dx^3} - 2\left(\dfrac{dy}{dx}\right)^4 + y = 0$

3. $yy' + 2y = 1 + x^2$

4. $x^2\, dy + (y - xy - xe^x)\, dx = 0$

5. $x^3 y^{(4)} - x^2 y'' + 4xy' - 3y = 0$

6. $\dfrac{d^2 y}{dx^2} + 9y = \sin y$

7. $\dfrac{dy}{dx} = \sqrt{1 + \left(\dfrac{d^2 y}{dx^2}\right)^2}$

8. $\dfrac{d^2 r}{dt^2} = -\dfrac{k}{r^2}$

9. $(\sin x)y''' - (\cos x)y' = 2$

10. $(1 - y^2)\, dx + x\, dy = 0$

In Problems 11–40 verify that the indicated function is a solution of the given differential equation. Where appropriate, c_1 and c_2 denote constants.

11. $2y' + y = 0;\quad y = e^{-x/2}$

12. $y' + 4y = 32;\quad y = 8$

13. $\dfrac{dy}{dx} - 2y = e^{3x};\quad y = e^{3x} + 10e^{2x}$

14. $\dfrac{dy}{dt} + 20y = 24;\quad y = \tfrac{6}{5} - \tfrac{6}{5}e^{-20t}$

15. $y' = 25 + y^2;\quad y = 5\tan 5x$

16. $\dfrac{dy}{dx} = \sqrt{\dfrac{y}{x}};\quad y = (\sqrt{x} + c_1)^2,\ x > 0,\ c_1 > 0$

17. $y' + y = \sin x;\quad y = \tfrac{1}{2}\sin x - \tfrac{1}{2}\cos x + 10e^{-x}$

18. $2xy\, dx + (x^2 + 2y)\, dy = 0;\quad x^2 y + y^2 = c_1$

19. $x^2\, dy + 2xy\, dx = 0;\quad y = -\dfrac{1}{x^2}$

20. $(y')^3 + xy' = y;\quad y = x + 1$

21. $y = 2xy' + y(y')^2;\quad y^2 = c_1(x + \tfrac{1}{4}c_1)$

22. $y' = 2\sqrt{|y|};\quad y = x|x|$

23. $y' - \dfrac{1}{x}y = 1;\quad y = x\ln x,\ x > 0$

24. $\dfrac{dP}{dt} = P(a - bP);\quad P = \dfrac{ac_1 e^{at}}{1 + bc_1 e^{at}}$

25. $\dfrac{dX}{dt} = (2 - X)(1 - X); \quad \ln\dfrac{2 - X}{1 - X} = t$

26. $y' + 2xy = 1; \quad y = e^{-x^2}\displaystyle\int_0^x e^{t^2}\, dt + c_1 e^{-x^2}$

27. $(x^2 + y^2)\, dx + (x^2 - xy)\, dy = 0; \quad c_1(x + y)^2 = xe^{y/x}$

28. $y'' + y' - 12y = 0; \quad y = c_1 e^{3x} + c_2 e^{-4x}$

29. $y'' - 6y' + 13y = 0; \quad y = e^{3x}\cos 2x$

30. $\dfrac{d^2 y}{dx^2} - 4\dfrac{dy}{dx} + 4y = 0; \quad y = e^{2x} + xe^{2x}$

31. $y'' = y; \quad y = \cosh x + \sinh x$

32. $y'' + 25y = 0; \quad y = c_1\cos 5x$

33. $y'' + (y')^2 = 0; \quad y = \ln|x + c_1| + c_2$

34. $y'' + y = \tan x; \quad y = -\cos x\,\ln(\sec x + \tan x)$

35. $x\dfrac{d^2 y}{dx^2} + 2\dfrac{dy}{dx} = 0; \quad y = c_1 + c_2 x^{-1}$

36. $x^2 y'' - xy' + 2y = 0; \quad y = x\cos(\ln x),\ x > 0$

37. $x^2 y'' - 3xy' + 4y = 0; \quad y = x^2 + x^2\ln x,\ x > 0$

38. $y''' - y'' + 9y' - 9y = 0; \quad y = c_1\sin 3x + c_2\cos 3x + 4e^x$

39. $y''' - 3y'' + 3y' - y = 0; \quad y = x^2 e^x$

40. $x^3\dfrac{d^3 y}{dx^3} + 2x^2\dfrac{d^2 y}{dx^2} - x\dfrac{dy}{dx} + y = 12x^2; \quad y = c_1 x + c_2 x\ln x + 4x^2,\ x > 0$

In Problems 41 and 42 verify that the indicated piecewise-defined function is a solution of the given differential equation.

41. $xy' - 2y = 0; \quad y = \begin{cases} -x^2, & x < 0 \\ \ \ x^2, & x \ge 0 \end{cases}$

42. $(y')^2 = 9xy; \quad y = \begin{cases} 0, & x < 0 \\ x^3, & x \ge 0 \end{cases}$

43. Verify that a one-parameter family of solutions for

$$y = xy' + (y')^2 \quad \text{is} \quad y = cx + c^2.$$

Determine a value of k such that $y = kx^2$ is a singular solution of the differential equation.

44. Verify that a one-parameter family of solutions for

$$y = xy' + \sqrt{1 + (y')^2} \quad \text{is} \quad y = cx + \sqrt{1 + c^2}.$$

Show that the relation $x^2 + y^2 = 1$ defines a singular solution of the equation on the interval $(-1, 1)$.

45. A one-parameter family of solutions for

$$y' = y^2 - 1 \quad \text{is} \quad y = \frac{1 + ce^{2x}}{1 - ce^{2x}}.$$

By inspection,* determine a singular solution of the differential equation.

46. On page 6 we saw that $y = \sqrt{4 - x^2}$ and $y = -\sqrt{4 - x^2}$ are solutions of $dy/dx = -x/y$ on the interval $(-2, 2)$. Explain why

$$y = \begin{cases} \sqrt{4 - x^2}, & -2 < x < 0 \\ -\sqrt{4 - x^2}, & 0 \le x < 2 \end{cases}$$

is not a solution of the differential equation on the interval.

In Problems 47 and 48 find values of m so that $y = e^{mx}$ is a solution of each differential equation.

47. $y'' - 5y' + 6y = 0$ **48.** $y'' + 10y' + 25y = 0$

In Problems 49 and 50 find values of m so that $y = x^m$ is a solution of each differential equation.

49. $x^2 y'' - y = 0$ **50.** $x^2 y'' + 6xy' + 4y = 0$

51. Show that $y_1 = x^2$ and $y_2 = x^3$ are both solutions of

$$x^2 y'' - 4xy' + 6y = 0.$$

Are the constant multiples $c_1 y_1$ and $c_2 y_2$, with c_1 and c_2 arbitrary, also solutions? Is the sum $y_1 + y_2$ a solution?

52. Show that $y_1 = 2x + 2$ and $y_2 = -x^2/2$ are both solutions of

$$y = xy' + (y')^2/2.$$

Are the constant multiples $c_1 y_1$ and $c_2 y_2$, with c_1 and c_2 arbitrary, also solutions? Is the sum $y_1 + y_2$ a solution?

53. By inspection determine, if possible, a real solution of the given differential equation.

(a) $\left| \dfrac{dy}{dx} \right| + |y| = 0$ **(b)** $\left| \dfrac{dy}{dx} \right| + |y| + 1 = 0$ **(c)** $\left| \dfrac{dy}{dx} \right| + |y| = 1$

[o] 1.2 SOME MATHEMATICAL MODELS

In science, engineering, economics, and even psychology, we often wish to describe or **model** the behavior of some system or phenomenon in mathematical terms. This description starts with

_____

* Translated, this means take a good guess and see if it works.

(i) identifying the variables that are responsible for changing the system, and

(ii) a set of reasonable assumptions about the system.

These assumptions also include any empirical laws that are applicable to the system. The mathematical construct of all these assumptions, or the **mathematical model** of the system, is in many instances a differential equation or a system of differential equations. We expect a reasonable mathematical model of a system to have a solution that is consistent with the known behavior of the system.

A mathematical model of a *physical* system often involves the variable time. The solution of the model then gives the **state of the system**; in other words, for appropriate values of time t, the values of the dependent variable (or variables) describe the system in the past, present, and future.

Falling Body

The mathematical description of a body falling vertically under the influence of gravity leads to a simple second-order differential equation. The solution of this equation gives the position of the body relative to the ground.

Figure 1.4

EXAMPLE 1

It is well known that free-falling objects close to the surface of the earth accelerate at a constant rate g. Acceleration is the derivative of the velocity, and this, in turn, is the derivative of the distance s. Suppose a rock is tossed upward from the roof of a building as illustrated in Figure 1.4. If we assume that the upward direction is positive, then the mathematical statement

$$\frac{d^2 s}{dt^2} = -g$$

is the differential equation that governs the vertical distance that the body travels. The minus sign is used because the weight of the body is a force directed opposite to the positive direction.

If we suppose further that the height of the building is s_0 and the initial velocity of the rock is v_0, then we must find a solution of the differential equation

$$\frac{d^2 s}{dt^2} = -g, \qquad 0 < t < t_1,$$

that also satisfies the side conditions $s(0) = s_0$ and $s'(0) = v_0$. Here $t = 0$ is the initial time the rock leaves the roof of the building, and t_1 is the elapsed time when the rock hits the ground. Since the rock is thrown upward in the positive direction, it is naturally assumed that $v_0 > 0$.

Note that this formulation of the problem ignores other forces such as air resistance acting on the body.   ∎

Spring-Mass System

When Newton's second law of motion is combined with Hooke's law, we can derive a differential equation governing the motion of a mass attached to a spring.

EXAMPLE **2**

To find the vertical displacement $x(t)$ of a mass attached to a spring, we use two different empirical laws: Newton's second law of motion and Hooke's law. The former law states that the net force acting on the system in motion is $F = ma$, where m is the mass and a is acceleration. Hooke's law states that the restoring force of a stretched spring is proportional to the elongation $s + x$; that is, the restoring force is $k(s + x)$, where $k > 0$ is a constant. As shown in Figure 1.5(b), s is the elongation of the spring after the mass has been attached and the system hangs at rest in the *equilibrium position*. When the system is in motion, the variable x represents a directed distance of the mass beyond the equilibrium position. In Chapter 5 we shall prove that when the system is in motion, the *net force* acting on the mass is simply $F = -kx$. Thus, in the absence of damping and other external forces that might be impressed on the system, the differential equation of the vertical motion through the center of gravity of the mass can be obtained by equating:

$$m\frac{d^2x}{dt^2} = -kx.$$

Here the minus sign means that the restoring force of the spring acts opposite to the direction of motion, that is, toward the equilibrium position. In practice this second-order differential equation is often written as

$$\frac{d^2x}{dt^2} + \omega^2 x = 0, \tag{1}$$

where $\omega^2 = k/m$.

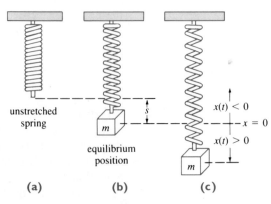

(a)   **(b)**   **(c)**

Figure 1.5

∎

Units

A word is in order regarding the system of units that is used in describing dynamic problems such as those illustrated in the last two examples. Three commonly used systems of units are summarized in the following table. In each system the basic unit of time is the second.

Quantity	Engineering System[*]	SI System[†]	cgs
Force	pound (lb)	newton (N)	dyne
Mass	slug	kilogram (kg)	gram (g)
Distance	foot (ft)	meter (m)	centimeter (cm)
Acceleration of gravity g (approximate)	32 ft/s^2	9.8 m/s^2	980 cm/s^2

The gravitational *force* exerted by the earth on a body of mass m is called its *weight W*. In the absence of air resistance, the only force acting on a freely falling body is its weight. Hence, from Newton's second law of motion, it follows that mass m and weight W are related by

$$W = mg.$$

For example, in the engineering system a mass of $1/4$ slug corresponds to an 8-lb weight. Since $m = W/g$, a 64-lb weight corresponds to a mass of $64/32 = 2$ slugs. In the cgs system a weight of 2450 dynes has a mass of $2450/980 = 2.5$ grams. In the SI system a weight of 50 newtons has a mass of $50/9.8 = 5.1$ kilograms. We note that

$$1 \text{ newton} = 10^5 \text{ dynes} = 0.2247 \text{ pound.}$$

In the next example we derive the differential equation that describes the motion of a *simple pendulum*.

Simple Pendulum

Any object that swings back and forth is called a **physical pendulum**. The simple pendulum is a special case of the physical pendulum and consists of a rod to which a mass is attached at one end. In describing the motion of a **simple pendulum**, we make the simplifying assumptions that the mass of the rod is negligible and that no external damping forces act on the system (such as air resistance).

---

[*] Also known as the English gravitational system or British engineering system.

[†] International system of units. SI is the abbreviation for *Systeme International*.

EXAMPLE 3

A mass m having weight W is suspended from the end of a rod of constant length l. For motion in a vertical plane, we would like to determine the displacement angle θ, measured from the vertical, as a function of time t (we consider $\theta > 0$ to the right of OP and $\theta < 0$ to the left of OP). Recall that an arc s of a circle of radius l is related to the central angle θ through the formula $s = l\theta$. Hence the angular acceleration is

$$a = \frac{d^2s}{dt^2} = l\frac{d^2\theta}{dt^2}.$$

From Newton's second law we then have

$$F = ma = ml\frac{d^2\theta}{dt^2}.$$

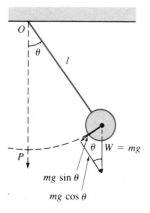

From Figure 1.6 we see that the tangential component of the force due to the weight W is $mg \sin\theta$. We equate the two different formulations of the tangential force to obtain

$$ml\frac{d^2\theta}{dt^2} = -mg\sin\theta \quad \text{or} \quad \frac{d^2\theta}{dt^2} + \frac{g}{l}\sin\theta = 0. \tag{2}$$

Figure 1.6

Because of the presence of $\sin\theta$, the differential equation (2) is nonlinear. It is also known that this equation cannot be solved in terms of elementary functions. Hence a further simplifying assumption is made. If the angular displacements θ are not too large, we can use the approximation $\sin\theta \approx \theta$,* so that (2) can be replaced with the linear second-order differential equation

$$\frac{d^2\theta}{dt^2} + \frac{g}{l}\theta = 0. \tag{3}$$

If we set $\omega^2 = g/l$, (3) has the exact same structure as the differential equation (1) governing the free vibrations of a weight on a spring. The fact that one basic differential equation can describe many diverse physical or even social/economic phenomena is a common occurrence in the study of applicable mathematics.

Rotating String

We encounter equation (1) again in the analysis of a rotating string.

---

* For small values of θ (in radians), powers θ^3 and higher can be ignored in the Maclaurin series $\sin\theta = \theta - \theta^3/3! + \cdots$, and so we get $\sin\theta \approx \theta$. Use a calculator and compare the values of $\sin(0.05)$ and $\sin(0.005)$ with 0.05 and 0.005.

ILLUSTRATED APPLICATIONS

The swinging bob in a grandfather's clock and a child's swing are examples of pendulums. The displacement angle θ of a simple plane pendulum of length l is determined from the nonlinear second-order differential equation $d^2\theta/dt^2 + (g/l)\sin\theta = 0$. When the displacements of the pendulum are not too large, we can use the replacement $\sin\theta \approx \theta$ and thus approximate θ by solving the linear equation $d^2\theta/dt^2 + (g/l)\theta = 0$. See pages 15–16 and 246.

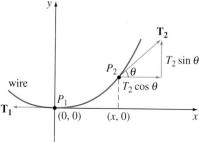

To determine the shape of a wire hanging under its own weight, such as a wire strung between two telephone poles, we must solve the nonlinear differential equation $d^2y/dx^2 + (w/T)\sqrt{1 + (dy/dx)^2} = 0$. It turns out that the wire has essentially the shape of the graph of a hyperbolic cosine. This graph of a hyperbolic cosine is called a *catenary*. The famous Gateway Arch in St. Louis has the shape of an inverted catenary. See pages 20–21.

Under some circumstances a body moving through the air encounters a resistance that is proportional to its velocity v. In general, air resistance is directly proportional to a positive power of the velocity of the body—the faster the body travels, the greater the resistance. For bodies moving at high speeds, such as projectiles or freely falling sky divers, air resistance is often taken to be proportional to v^2. See Problem 27, page 106 and Problem 8, page 121.

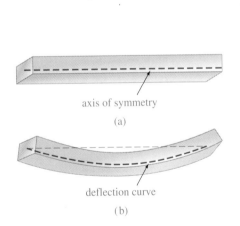

axis of symmetry

(a)

deflection curve

(b)

Forces acting on beams and girders cause them to distort or deform. This deformation, or deflection $y(x)$ is described by the fourth-order differential equation $EIy^{(4)} = w(x)$. A beam that is clamped or embedded at one end and free at the other is called a *cantilever beam*. A diving board, an outstretched arm, and an airplane wing are common examples of such beams; but even flagpoles, skyscrapers, and the Washington Monument act as cantilever beams. See pages 22–33, 373, 378, and 384.

Imagine that a shaft is drilled from one side of the earth to the opposite side through its center, and a heavy object, such as a bowling ball, is dropped into the shaft. What would be the motion of the object? Would it drop straight through? Would it stop at the center? In Problem 20 of Exercise 1.2 you are asked to construct a mathematical model for the motion of the object. You can then solve this differential equation with the techniques discussed in Section 4.2.

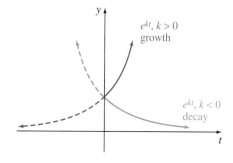

Any phenomenon modeled by the simple differential equation $dx/dt = kx$ grows ($k > 0$) or decays ($k < 0$) exponentially. The growth in the population P of bacteria, insects, or even humans can often be predicted over short time periods by the exponential solution $P(t) = ce^{kt}$. The study of substances decaying through radioactivity led to the discovery of carbon dating, which is a means of dating fossils or even a mummy. See Section 3.2 and the essay at the end of Chapter 3.

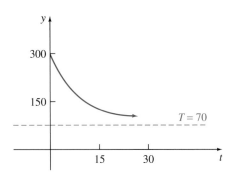

Before taking the first sip of coffee, one usually waits a short time for the liquid to cool. A forgotten cup of coffee is almost undrinkable when it has cooled to room temperature. An empirical law of cooling, attributed to Isaac Newton, asserts that the rate at which the temperature of a body cools is proportional to the difference between the temperature of the body and the temperature of the surrounding medium. The preceding sentence is a verbal description of a differential equation. See pages 19 and 98–99.

Suppose that a cell is suspended in a solution containing a solute, such as potassium, with a constant concentration and that the solute is absorbed into the cell through its permeable membrane. A mathematical model for the concentration $C(t)$ of the solute inside the cell at time t can be found using Fick's principle of passive diffusion. See Problem 30, Exercises 3.2.

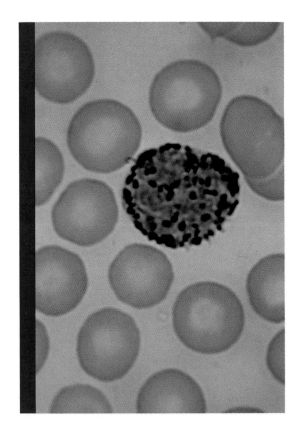

EXAMPLE 4

Suppose a string of length L with constant linear density ρ (mass per unit length) is stretched along the x-axis and fixed at $x = 0$ and $x = L$. Suppose the string is then rotated about that axis at a constant angular speed ω. This is analogous to two persons holding a jump rope and then twirling it in a synchronous manner. See Figure 1.7(a). We want to find the differential equation that defines the shape $y(x)$ of the string, or the **deflection curve** away from its initial position. See Figure 1.7(b). To do this, consider a portion of the string on the interval $[x, x + \Delta x]$, where Δx is small. If the magnitude T of the tension $\mathbf{T}$ acting tangential to the string is constant along the string, then the desired differential equation can be obtained by equating two different formulations of the net force acting on the string on the interval $[x, x + \Delta x]$. First, we see from Figure 1.7(c) that the net vertical force is

$$F = T \sin \theta_2 - T \sin \theta_1. \tag{4}$$

When the angles θ_1 and θ_2 (measured in radians) are small, we have

$$\sin \theta_2 \approx \tan \theta_2 \approx y'(x + \Delta x) \quad \text{and} \quad \sin \theta_1 \approx \tan \theta_1 \approx y'(x),$$

and so (4) becomes

$$F = T[y'(x + \Delta x) - y'(x)]. \tag{5}$$

Second, the net force is also given by Newton's second law $F = ma$. Here the mass of string on the interval is $m = \rho \, \Delta x$; the centripetal acceleration of a point rotating with angular speed ω in a circle of radius r is $a = r\omega^2$. With Δx small we take $r = y$. Thus another formulation of the net force is

$$F \approx -(\rho \, \Delta x)y\omega^2, \tag{6}$$

where the minus sign comes from the fact that the acceleration points in the direction opposite to the positive y direction. Now by equating (5) and (6), we have

$$T[y'(x + \Delta x) - y'(x)] \approx -(\rho \, \Delta x)y\omega^2 \quad \text{or} \quad T\frac{y'(x + \Delta x) - y'(x)}{\Delta x} \approx -\rho\omega^2 y. \tag{7}$$

(a)

(b)

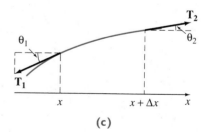

(c)

Figure 1.7

For Δx close to zero, $[y'(x + \Delta x) - y'(x)]/\Delta x \approx d^2y/dx^2$, so the last expression in (7) gives

$$T\frac{d^2y}{dx^2} = -\rho\omega^2 y \quad \text{or} \quad T\frac{d^2y}{dx^2} + \rho\omega^2 y = 0. \tag{8}$$

Since the string is anchored at $x = 0$ and $x = L$ we expect that the solution $y(x)$ of the last equation would also satisfy the side conditions $y(0) = 0$ and $y(L) = 0$. ∎

Dividing the last equation in (8) by T gives

$$\frac{d^2y}{dx^2} + \frac{\rho\omega^2}{T} y = 0,$$

which is analogous to (1) and (3). If the magnitude T of the tension is not constant throughout the interval $[0, L]$, then it can be shown that the differential equation for the deflection curve of the string is

$$\frac{d}{dx}\left[T(x)\frac{dy}{dx}\right] + \rho\omega^2 y = 0. \tag{9}$$

Series Circuits

According to Kirchhoff's second law, the impressed voltage $E(t)$ on a closed loop must equal the sum of the voltage drops in the loop. Figure 1.8 shows the symbols and the formulas for the respective voltage drops across an inductor, a capacitor, and a resistor. The current in a circuit after a switch is closed is denoted by $i(t)$; the charge on a capacitor at time t is denoted by $q(t)$. The letters L, C, and R are constants known as inductance, capacitance, and resistance, respectively.

inductor	capacitor	resistor
$L\frac{di}{dt}$	$\frac{1}{C}q$	iR
(a)	**(b)**	**(c)**

Figure 1.8

EXAMPLE 5

Consider the single-loop series circuit containing an inductor, resistor, and capacitor shown in Figure 1.9. A second-order differential equation for the charge $q(t)$ on the capacitor can be obtained by adding the voltage drops

Figure 1.9

$$\text{inductor} = L\frac{di}{dt} = L\frac{d^2q}{dt^2}$$

current i related
to charge q by $i = dq/dt$

$$\text{resistor} = iR = R\frac{dq}{dt}$$

$$\text{capacitor} = \frac{1}{C}q$$

and equating the sum to the impressed voltage $E(t)$:

$$L\frac{d^2q}{dt^2} + R\frac{dq}{dt} + \frac{1}{C}q = E(t). \tag{10}$$

■

In Example 5 the side conditions $q(0)$ and $q'(0)$ represent the charge on the capacitor and the current in the circuit, respectively, at time $t = 0$. Also, the impressed voltage $E(t)$ is said to be an **electromotive force**, or **emf**. An emf, as well as a charge on a capacitor, causes the current in a circuit to flow. The following table shows the basic units of measurement used in circuit analysis.

Quantity	Unit
Impressed voltage, or emf	volt (V)
Inductance L	henry (H)
Capacitance C	farad (F)
Resistance R	ohm (Ω)
Charge q	coulomb (C)
Current i	ampere (A)

Newton's Law of Cooling

According to Newton's empirical law of cooling, the rate at which a body cools is proportional to the difference between the temperature of the body and the temperature of the surrounding medium, the so-called ambient temperature.

EXAMPLE 6

Suppose $T(t)$ denotes the temperature of a body at any time t, and T_m is the constant temperature of the surrounding medium. If dT/dt represents the rate at which a body cools, then Newton's law of cooling translates into the mathematical statement

$$\frac{dT}{dt} \propto T - T_m \quad \text{or} \quad \frac{dT}{dt} = k(T - T_m), \tag{11}$$

where k is a constant of proportionality. Since the body is assumed to be cooling, we must have $T > T_m$ and so it stands to reason that $k < 0$. ∎

Suspended Wire

Suppose a suspended wire hangs under its own weight. As Figure 1.10(a) shows, a physical model for this could be a long telephone wire strung between two posts. As in Example 4, our goal in the example that follows is to determine the differential equation that describes the shape that the hanging wire assumes.

EXAMPLE 7

Let us examine only a portion of the wire between the lowest point P_1 and any arbitrary point P_2. See Figure 1.10 (b). Three forces are acting on the wire: the weight of the segment $P_1 P_2$ and the tensions $\mathbf{T}_1$ and $\mathbf{T}_2$ in the wire at P_1 and P_2, respectively. If w is the linear density (measured, say, in lb/ft) and s is the length of the segment $P_1 P_2$, its weight is necessarily ws.

Now the tension $\mathbf{T}_2$ resolves into horizontal and vertical components (scalar quantities) $T_2 \cos \theta$ and $T_2 \sin \theta$. Because of equilibrium we can write

$$|\mathbf{T}_1| = T_1 = T_2 \cos \theta \quad \text{and} \quad ws = T_2 \sin \theta.$$

Dividing the last two equations, we find

$$\tan \theta = \frac{ws}{T_1}$$

or

$$\frac{dy}{dx} = \frac{ws}{T_1}. \tag{12}$$

Now since the length of the arc between points P_1 and P_2 is

$$s = \int_0^x \sqrt{1 + \left(\frac{dy}{dx}\right)^2}\, dx,$$

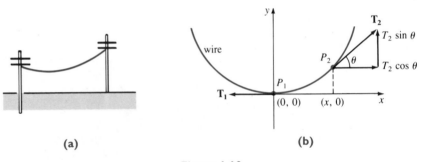

(a)   **(b)**

Figure 1.10

it follows from one form of the fundamental theorem of calculus that

$$\frac{ds}{dx} = \sqrt{1 + \left(\frac{dy}{dx}\right)^2}.$$ (13)

Differentiating (12) with respect to x and using (13) lead to

$$\frac{d^2y}{dx^2} = \frac{w}{T_1}\frac{ds}{dx} \quad \text{or} \quad \frac{d^2y}{dx^2} = \frac{w}{T_1}\sqrt{1 + \left(\frac{dy}{dx}\right)^2}.$$ (14)

■

One might conclude from Figure 1.10 that the shape the hanging wire assumes is parabolic. However, this is not the case; a wire or heavy rope hanging under only its own weight takes the shape of a hyperbolic cosine. See Problem 12, Exercises 3.3. Recall that the graph of the hyperbolic cosine is called a **catenary**, which stems from the Latin word *catena* meaning "chain." The Romans used the catena as a dog leash. Probably the most graphic example of the shape of a catenary is the 630-ft-high Gateway Arch in St. Louis, Missouri.

Discharge Through an Orifice

In hydrodynamics, Torricelli's theorem states that the speed v of efflux of water through a sharp-edged orifice at the bottom of a tank filled to a depth h is the same as the speed that a body (in this case a drop of water) would acquire in falling freely from a height h:

$$v = \sqrt{2gh},$$

where g is the acceleration due to gravity. This last expression comes from equating the kinetic energy $\frac{1}{2}mv^2$ with the potential energy mgh and solving for v.

EXAMPLE 8

Suppose a tank filled with water is allowed to discharge through an orifice under the influence of gravity. We would like to find the depth h of water remaining in the tank at any time t.

Consider the tank shown in Figure 1.11. If the area of the orifice is A_o (in ft^2) and the speed of the water leaving the tank is $v = \sqrt{2gh}$ (in ft/s), then the volume of water leaving the tank per second is $A_o\sqrt{2gh}$ (in ft^3/s). Thus, if $V(t)$ denotes the volume of water in the tank at any time t, then

$$\frac{dV}{dt} = -A_o\sqrt{2gh},$$ (15)

where the minus sign indicates that V is decreasing. Note here that we are ignoring the possibility of friction at the orifice, which might reduce the rate of flow there.

Figure 1.11

Now if the tank is such that the volume of water in it at any time t can be written as $V(t) = A_w h$, where A_w (in ft^2) is the *constant* area of the upper surface of the water (see Figure 1.11), then $dV/dt = A_w\,(dh/dt)$. Substituting this last expression into (15) gives us the desired differential equation for the height h of the water at any time:

$$\frac{dh}{dt} = -\frac{A_o}{A_w}\sqrt{2gh}.\tag{16}$$

∎

It is interesting to observe that (16) remains valid even when A_w is not constant. In this case we must express the upper surface area of the water as a function of h: $A_w = A(h)$. See Problem 9 in Exercises 1.2 and Problem 19 in the Chapter 1 Review Exercises.

Deflection of Beams

axis of symmetry

(a)

deflection curve

(b)

Figure 1.12

In engineering an important problem is to determine the static deflection of an elastic beam caused by its weight or by an external load. We assume that the beam is homogeneous and has uniform cross sections along its length. Let L denote the length of the beam. In the absence of any load on the beam (including its weight), a curve joining the centroids of all its cross sections is a straight line called the **axis of symmetry**. See Figure 1.12(a). If a load is applied to the beam in a vertical plane containing the axis of symmetry, then, as shown in Figure 1.12(b), the beam undergoes a distortion and the curve connecting the centroids of all cross sections is called the **deflection curve** or **elastic curve**. In the next example we derive the differential equation of the deflection curve. This derivation uses principles from elasticity and a concept from calculus called curvature.

EXAMPLE 9

Figure 1.13

For the sake of illustration let us consider a cantilever beam embedded at its left end and free at its right end. As shown in Figure 1.13, we let the embedded end of the beam coincide with $x = 0$ and its free end with $x = L$. The x-axis coincides with the axis of symmetry, and the deflection $y(x)$ is measured from this axis and is considered positive if downward. Now in the theory of elasticity it is shown that the bending moment $M(x)$ at a point x along the beam is related to the load per unit length $w(x)$ by the equation

$$\frac{d^2 M}{dx^2} = w(x).\tag{17}$$

In addition, the bending moment $M(x)$ is proportional to the curvature κ of the elastic curve:

$$M(x) = EI\kappa,\tag{18}$$

where E and I are constants; E is Young's modulus of elasticity of the material of the beam and I is the moment of inertia of a cross-section of the beam (about an axis known as the neutral axis). The product EI is called the flexural rigidity of the beam.

Now, from calculus, curvature is given by

$$\kappa = \frac{y''}{[1 + (y')^2]^{3/2}}.$$

When the deflection $y(x)$ is small, the slope $y' \approx 0$ and so $[1 + (y')^2]^{3/2} \approx 1$. If we let $\kappa = y''$, equation (18) becomes $M = EIy''$. The second derivative of this last expression is

$$\frac{d^2M}{dx^2} = EI\frac{d^2}{dx^2}y'' = EI\frac{d^4y}{dx^4}. \tag{19}$$

Using the given result in (17) to replace d^2M/dx^2 in (19), we see that the deflection $y(x)$ satisfies the fourth-order differential equation

$$EI\frac{d^4y}{dx^4} = w(x). \tag{20}$$

■

As we shall see later on, it is extremely important to note any side conditions that accompany a differential equation in the mathematical description of a physical phenomenon. For the cantilever beam in Example 9, in addition to satisfying (20) we would expect the deflection $y(x)$ to satisfy the following conditions at the ends of the beam:

- $y(0) = 0$ since there is no deflection at the embedded end.
- $y'(0) = 0$ since the deflection curve is tangent to the x-axis at the embedded end.
- $y''(L) = 0$ since the bending moment is zero at a free end.
- $y'''(L) = 0$ since the shear force is zero at a free end.

The function $F(x) = dM/dx = EI\,(d^3y/dx^3)$ is called the shear force.

Population Growth

In the next examples we examine some mathematical models of biological growth.

EXAMPLE 10

It seems plausible to expect that the rate at which a population P expands is proportional to the population that is present at any time. Roughly put, the more people there are, the more there are going to be. Thus one model for population growth is given by the differential equation

$$\frac{dP}{dt} = kP, \tag{21}$$

where k is a constant of proportionality. Since we also expect the population to expand, we must have $dP/dt > 0$ and thus $k > 0$. ∎

EXAMPLE 11

In the spread of a contagious disease, for example a flu virus, it is reasonable to assume that the rate, dx/dt, at which the disease spreads is proportional not only to the number of people, $x(t)$, who have contracted the disease, but also to the number of people, $y(t)$, who have not yet been exposed; that is,

$$\frac{dx}{dt} = kxy, \tag{22}$$

where k is the usual constant of proportionality. If one infected person is introduced into a fixed population of n people, then x and y are related by

$$x + y = n + 1. \tag{23}$$

Using (23) to eliminate y in (22), we get

$$\frac{dx}{dt} = kx(n + 1 - x). \tag{24}$$

The obvious side condition accompanying equation (24) is $x(0) = 1$. ∎

Logistic Equation

The nonlinear first-order equation (24) is a special case of a more general equation

$$\frac{dP}{dt} = P(a - bP), \qquad a \text{ and } b \text{ constants}, \tag{25}$$

known as the **logistic equation**. See Section 3.3. The solution of this equation is very important in ecological, sociological, and even managerial sciences.

Continuous Compounding of Interest

It is very common for savings institutions to advertise that interest is being compounded daily, by the hour, or even by the minute. Of course, there is no reason to stop there; interest could as well be compounded every second, every half-second, every microsecond, and so on. That is to say, interest could be compounded continuously.

Figure 1.14

EXAMPLE 12

When interest is compounded continuously, the rate at which an amount of money S grows is proportional to the amount of money present at any time; that is,

$$\frac{dS}{dt} = rS, \tag{26}$$

where r is the annual rate of interest. This mathematical description is analogous to the population growth of Example 10. The rate of growth is large when the amount of money present in the account is also large. Translated geometrically, this means the tangent line is steep when S is large. See Figure 1.14.

The definition of a derivative provides an interesting derivation of (26). Suppose $S(t)$ is the amount of money accrued in a savings account after t years when the annual rate of interest r is compounded continuously. If h denotes an increment in time, then the interest obtained in the time span $(t + h) - t$ is the difference in the amounts accrued:

$$S(t + h) - S(t). \tag{27}$$

Since interest is given by (rate) × (time) × (principal), we can approximate the interest earned in this same time period by either

$$rhS(t) \quad \text{or} \quad rhS(t + h).$$

Intuitively $rhS(t)$ and $rhS(t + h)$ are lower and upper bounds, respectively, for the actual interest (27); that is,

$$rhS(t) \le S(t + h) - S(t) \le rhS(t + h)$$

or $$rS(t) \le \frac{S(t + h) - S(t)}{h} \le rS(t + h). \tag{28}$$

Taking the limit of (28) as $h \to 0$, we get

$$rS(t) \le \lim_{h \to 0} \frac{S(t + h) - S(t)}{h} \le rS(t),$$

and so it must follow that

$$\lim_{h \to 0} \frac{S(t + h) - S(t)}{h} = rS(t) \quad \text{or} \quad \frac{dS}{dt} = rS. \qquad \blacksquare$$

1.2 EXERCISES

Answers to odd-numbered problems begin on page A–2.

Figure 1.15

Figure 1.16

Figure 1.17

Figure 1.18

In Problems 1–22 derive the differential equation(s) describing the given physical situation.

1. Under some circumstances a falling body B of mass m, such as the sky diver shown in Figure 1.15, encounters air resistance proportional to its instantaneous velocity v. Use Newton's second law to find the differential equation for the velocity v of the body at any time. Recall that acceleration is $a = dv/dt$. Assume in this case that the positive direction is downward.

2. What is the differential equation for the velocity v of a body of mass m falling vertically downward through a medium (such as water) that offers a resistance proportional to the square of its instantaneous velocity? Assume the positive direction is downward.

3. By Newton's universal law of gravitation the free-fall acceleration a of a body, such as the satellite shown in Figure 1.16, falling a great distance to the surface is *not* the constant g. Rather, the acceleration a is inversely proportional to the square of the distance r from the center of the earth, $a = k/r^2$, where k is the constant of proportionality.

   (a) Use the fact that at the surface of the earth $r = R$ and $a = g$ to determine the constant of proportionality k.

   (b) Use Newton's second law and part (a) to find a differential equation for the distance r.

   (c) Use the chain rule in the form

   $$\frac{d^2r}{dt^2} = \frac{dv}{dt} = \frac{dv}{dr}\frac{dr}{dt}$$

   to express the differential equation in part (b) as a differential equation involving v and dv/dr.

4. **(a)** Use part (b) of Problem 3 to find the differential equation for r if the resistance to the falling satellite is proportional to its instantaneous velocity.

   (b) Near the surface of the earth, use the approximation $R \approx r$ to show that the differential equation in part (a) reduces to the equation derived in Problem 1.

5. A series circuit contains a resistor and an inductor as shown in Figure 1.17. Determine the differential equation for the current $i(t)$ if the resistance is R, the inductance is L, and the impressed voltage is $E(t)$.

6. A series circuit contains a resistor and a capacitor as shown in Figure 1.18. Determine the differential equation for the charge $q(t)$ on the capacitor if the resistance is R, the capacitance is C, and the impressed voltage is $E(t)$.

7. Suppose a tank is discharging water through a circular orifice of cross-

Figure 1.19

sectional area A_o at its bottom. It has been shown experimentally that when friction at the orifice is taken into consideration, the volume of water leaving the tank per second is approximately $0.6 A_o \sqrt{2gh}$. Find the differential equation for the height h of water at any time t for the cubical tank in Figure 1.19. The radius of the orifice is 2 in. and $g = 32$ ft/s^2.

8. A tank in the form of a right circular cylinder of radius 2 ft and height 10 ft is standing on end. The tank is initially full of water, and water leaks from a circular hole of radius $1/2$ in. at its bottom. Use the information in Problem 7 to obtain the differential equation for the height h of the water at any time t.

9. A water tank has the shape of a hemisphere with radius 5 ft. Water leaks out of a circular hole of radius 1 in. at its flat bottom. Use the information in Problem 7 to obtain the differential equation for the height h of the water at any time t.

10. The rate at which a radioactive substance decays is proportional to the amount of the substance remaining at any time. Determine the differential equation for the amount $A(t)$.

11. A drug is infused into a patient's bloodstream at a constant rate of r grams per second. Simultaneously, the drug is removed at a rate proportional to the amount $x(t)$ of the drug present at any time. Determine the differential equation governing the amount $x(t)$.

12. A projectile shot from a gun has weight $w = mg$ and velocity $\mathbf{v}$ tangent to its path of motion. Ignoring air resistance and all other forces except its weight, find the system of differential equations that describes the motion. See Figure 1.20. [*Hint:* Use Newton's second law in the x and y direction.]

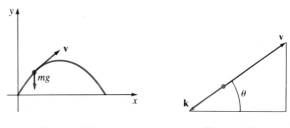

Figure 1.20   **Figure 1.21**

13. Determine the equations of motion if the projectile in Problem 12 encounters a retarding force $\mathbf{k}$ (of magnitude k) acting tangent to the path but opposite to the motion. See Figure 1.21. [*Hint:* $\mathbf{k}$ is a multiple of the velocity, say $c\mathbf{v}$.]

14. Two chemicals A and B react to form a new chemical C. Assuming that the concentrations of both A and B decrease by the amount of C formed, find the differential equation governing the concentration $x(t)$ of the

Figure 1.22

(a)

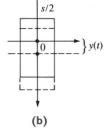

(b)

Figure 1.23

chemical C if the rate at which the chemical reaction takes place is proportional to the product of the remaining concentrations of A and B.

15. Light strikes a plane curve C in such a manner that all beams L parallel to the y-axis are reflected to a single point 0. Determine the differential equation for the function $y = f(x)$ describing the shape of the curve. (The fact that the angle of incidence is equal to the angle of reflection is a principle of optics.) [*Hint:* Inspection of Figure 1.22 shows that the inclination of the tangent line from the horizontal at $P(x, y)$ is $\pi/2 - \theta$ and that we can write $\phi = 2\theta$. (Why?) Also, don't be afraid to use a trigonometric identity.]

16. A cylindrical barrel s feet in diameter of weight w lb is floating in water. After an initial depression, the barrel exhibits an up-and-down bobbing motion along a vertical line. Using Figure 1.23(b), determine the differential equation for the vertical displacements $y(t)$ if the origin is taken to be on the vertical axis at the surface of the water when the barrel is at rest. Use Archimedes' principle that the buoyancy, or upward force of the water on the barrel, is equal to the weight of the water displaced, and the fact that the density of water is 62.4 lb/ft^3. Assume that the downward direction is positive. Ignore the resistance of the water.

17. A rocket is shot vertically upward from the surface of the earth. After all its fuel has been expended, the mass of the rocket is a constant m. Use Newton's second law of motion and the fact that the force of gravity varies inversely as the square of the distance to find the differential equation for distance y from the earth's center to the rocket at any time after burnout. State appropriate conditions at $t = 0$ associated with this differential equation.

18. Newton's second law $F = ma$ can be written $F = d/dt(mv)$. When the mass of an object is variable, this latter formulation is used. The mass $m(t)$ of a rocket launched upward changes as its fuel is consumed.* If $v(t)$ denotes its velocity at any time, it can be shown that

$$-mg = m\frac{dv}{dt} - V\frac{dm}{dt}, \tag{29}$$

where V is the constant velocity of the exhaust gases relative to the rocket. Use (29) to find the differential equation for v if it is known that $m(t) = m_0 - at$ and $V = -b$, where m_0, a, and b are constants.

19. A person P, starting at the origin, moves in the direction of the positive x-axis, pulling a weight along the curve C (called a **tractrix**) as shown in

---

* It is assumed that the *total* mass:

mass of vehicle + mass of fuel + mass of exhaust gases

is constant. In this case $m(t) = $ *mass of vehicle + mass of fuel.*

Figure 1.24

Figure 1.25

Figure 1.24. The weight, initially located on the y-axis at $(0, s)$, is pulled by a rope of constant length s, which is kept taut throughout the motion. Find the differential equation of the path of motion. [*Hint:* The rope is always tangent to C; consider the angle of inclination θ as shown in the figure.]

20. Suppose a hole is drilled through the center of the earth. A body with mass m is dropped into the hole. Let the distance from the center of the earth to the mass at any time t be denoted by r. See Figure 1.25.

(a) Let M denote the mass of the earth and M_r denote the mass of that portion of the earth within a sphere of radius r. The gravitational force on m is $F = -kM_r m/r^2$, where the minus sign indicates that the force is one of attraction. Use this fact to show that

$$F = -k\frac{mM}{R^3}r.$$

[*Hint:* Assume that the earth is homogeneous, that is, has a constant density δ. Use mass = density × volume.]

(b) Use Newton's second law and the result in part (a) to derive the differential equation

$$\frac{d^2r}{dt^2} + \omega^2 r = 0,$$

where $\omega^2 = kM/R^3 = g/R$.

21. In the theory of learning, the rate at which a subject is memorized is assumed to be proportional to the amount that is left to be memorized. If M denotes the total amount that is to be memorized and $A(t)$ the amount memorized in time t, find the differential equation for A.

22. In Problem 21 assume that the amount of material forgotten is proportional to the amount memorized in time t. What is the differential equation for A when forgetfulness is taken into account?

Chapter 1   *REVIEW*

We classify a differential equation by its type: **ordinary** or **partial**; by its **order**; and by whether it is **linear** or **nonlinear**.

A **solution** of a differential equation is any function having a sufficient number of derivatives that satisfies the equation identically on some interval.

When solving an nth-order ordinary differential equation, we expect to find an n-parameter family of solutions. A **particular solution** is any solution free of arbitrary parameters that satisfies the differential equation. A **singular solution** is any solution that cannot be obtained from an n-parameter family of solutions by assigning values to the parameters. When an n-parameter family of solutions gives every solution of a differential equation on some interval, it is called a **general**, or **complete**, solution.

In the analysis of physical problems, many differential equations can be obtained by equating two different empirical formulations of the same situation. For example, a differential equation of motion can sometimes be obtained by simply equating Newton's second law of motion with the net forces acting on a body.

Chapter I *REVIEW EXERCISES*

Answers to odd-numbered problems begin on page A–2.

In Problems 1–4 classify the given differential equation as to type and order. Classify the ordinary differential equations as to linearity.

1. $(2xy - y^2)\,dx + e^x\,dy = 0$

2. $(\sin xy)y''' + 4xy' = 0$

3. $\dfrac{\partial^2 u}{\partial x^2} + \dfrac{\partial^2 u}{\partial y^2} = u$

4. $x^2\dfrac{d^2y}{dx^2} - 3x\dfrac{dy}{dx} + y = x^2$

In Problems 5–8 verify that the indicated function is a solution of the given differential equation.

5. $y' + 2xy = 2 + x^2 + y^2;\quad y = x + \tan x$

6. $x^2y'' + xy' + y = 0;\quad y = c_1\cos(\ln x) + c_2\sin(\ln x),\ x > 0$

7. $y''' - 2y'' - y' + 2y = 6;\quad y = c_1e^x + c_2e^{-x} + c_3e^{2x} + 3$

8. $y^{(4)} - 16y = 0;\quad y = \sin 2x + \cosh 2x$

In Problems 9–16 determine by inspection at least one solution for the given differential equation.

9. $y' = 2x$

10. $\dfrac{dy}{dx} = 5y$

11. $y'' = 1$

12. $y' = y^3 - 8$

13. $y'' = y'$

14. $2y\dfrac{dy}{dx} = 1$

15. $y'' = -y$

16. $y'' = y$

17. Determine an interval on which $y^2 - 2y = x^2 - x - 1$ defines a solution of $2(y - 1)\,dy + (1 - 2x)\,dx = 0$.

18. Explain why the differential equation

$$\left(\dfrac{dy}{dx}\right)^2 = \dfrac{4 - y^2}{4 - x^2}$$

possesses no real solutions for $|x| < 2, |y| > 2$. Are there other regions in the xy-plane for which the equation has no solutions?

19. The conical tank shown in Figure 1.26 loses water out of an orifice at its bottom. If the cross-sectional area of the orifice is $1/4$ ft^2, find the differ-

8 ft

A_w

20 ft

h

$1/4$ ft^2

Figure 1.26

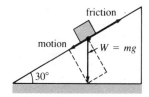

friction

motion

$W = mg$

$30°$

Figure 1.27

ential equation representing the height of the water h at any time. Ignore friction at the orifice.

20. A weight of 96 lb slides down an incline making a 30° angle with the horizontal. If the coefficient of sliding friction is μ, determine the differential equation for the velocity $v(t)$ of the weight at any time. Use the fact that the force of friction opposing the motion is μN, where N is the normal component of the weight. See Figure 1.27.

C H A P T E R 2

FIRST-ORDER DIFFERENTIAL EQUATIONS

2.1 Preliminary Theory

2.2 Separable Variables

2.3 Homogeneous Equations

2.4 Exact Equations

2.5 Linear Equations

[O] **2.6** Equations of Bernoulli,
Ricatti, and Clairaut

[O] **2.7** Substitutions

[O] **2.8** Picard's Method

Chapter 2 Review

Chapter 2 Review Exercises

Important Concepts

Initial-value problem

Initial condition

Existence of a solution

Uniqueness of a solution

Separation of variables

Homogeneous function

Homogeneous equation

Exact differential

Exact equation

Integrating factor

Linear equation

General solution

We are now in a position to solve some differential equations. We begin with first-order differential equations.

If a first-order differential equation can be solved, we shall see that the technique or method for solving it depends on what kind of first-order equation it is. Over the years mathematicians struggled to solve many specialized kinds of equations. Thus there are many methods of solution; what works for one kind of first-order equation does not necessarily apply to another kind of equation. Although we consider solution methods for seven classical types of equations in this chapter, our focus is on four types of equations. Some of these four types are important in applications.

2.1   PRELIMINARY THEORY

Initial-Value Problem

We are often interested in solving a first-order differential equation*

$$\frac{dy}{dx} = f(x, y) \tag{1}$$

subject to a side condition $y(x_0) = y_0$, where x_0 is a number in an interval I and y_0 is an arbitrary real number. The problem

$$\textit{Solve:} \quad \frac{dy}{dx} = f(x, y)$$

$$\textit{Subject to:} \quad y(x_0) = y_0 \tag{2}$$

is called an **initial-value problem**. The side condition is known as an **initial condition**. In geometric terms we are seeking at least one solution of the differential equation defined on some interval I such that the graph of the solution passes through a prescribed point (x_0, y_0). See Figure 2.1.

solutions of the DE

(x_0, y_0)

I

Figure 2.1

EXAMPLE 1

We have seen (page 9) that $y = ce^x$ is a one-parameter family of solutions for $y' = y$ on the interval $(-\infty, \infty)$. If we specify, say, $y(0) = 3$, then substituting $x = 0$, $y = 3$ in the family yields $3 = ce^0 = c$. Thus, as shown in Figure 2.2, the function

$$y = 3e^x$$

is a solution of the initial-value problem

$$y' = y, \qquad y(0) = 3.$$

Had we demanded that a solution of $y' = y$ pass through the point $(1, 3)$ rather than $(0, 3)$, then $y(1) = 3$ would yield $c = 3e^{-1}$ and so $y = 3e^{x-1}$. The graph of this function is also indicated in Figure 2.2. ∎

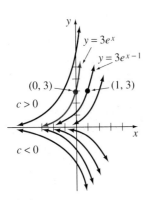

$y = 3e^x$

$y = 3e^{x-1}$

$(0, 3)$

$c > 0$

$(1, 3)$

$c < 0$

Figure 2.2

Two fundamental questions arise in considering an initial-value problem such as (2):

> Does a solution of the problem *exist*?
> If a solution exists, is it *unique*?

---

* In this text we assume that a differential equation $F(x, y, y', \ldots, y^{(n)}) = 0$ can be solved for the highest order derivative: $y^{(n)} = f(x, y, y', \ldots, y^{(n-1)})$. There are exceptions.

In other words, does the differential equation $dy/dx = f(x, y)$ possess a solution whose graph passes through (x_0, y_0) and, if it does, is there precisely *one* such solution?

As the next example shows, the answer to the second question is sometimes no.

Figure 2.3

EXAMPLE 2

You should verify that each of the functions $y = 0$ and $y = x^4/16$ satisfies the differential equation and initial condition in the problem

$$\frac{dy}{dx} = xy^{1/2}, \qquad y(0) = 0.$$

As illustrated in Figure 2.3, the graphs of both functions pass through the same point $(0, 0)$. ∎

It is often desirable to know before tackling an initial-value problem whether a solution exists and, when it does, whether it is the only solution of the problem. The following theorem due to Picard* gives conditions that are sufficient to guarantee that a solution exists and that it is also a unique solution.

> **THEOREM 2.1**   **Existence of a Unique Solution**
> Let R be a rectangular region in the xy-plane defined by $a \leq x \leq b$, $c \leq y \leq d$ that contains the point (x_0, y_0) in its interior. If $f(x, y)$ and $\partial f/\partial y$ are continuous on R, then there exists an interval I centered at x_0 and a unique function $y(x)$ defined on I satisfying the initial-value problem (2).

Figure 2.4

The foregoing is one of the most popular existence and uniqueness theorems for first-order differential equations because the criteria of continuity of $f(x, y)$ and $\partial f/\partial y$ are relatively easy to check. In general, it is not always possible to find a specific interval I on which a solution is defined without actually solving the differential equation (see Problem 16). The geometry of Theorem 2.1 is illustrated in Figure 2.4.

---

*Charles Émile Picard (1856–1941)   Picard was one of the prominent French Mathematicians of the latter nineteenth and early twentieth centuries. He made significant contributions to the fields of differential equations and complex variables. In 1899 Picard lectured at Clark University in Worcester, Massachusetts.

EXAMPLE 3

We saw in Example 2 that the differential equation

$$\frac{dy}{dx} = xy^{1/2}$$

possesses at least two solutions whose graphs pass through $(0, 0)$. The functions

$$f(x, y) = xy^{1/2} \quad \text{and} \quad \frac{\partial f}{\partial y} = \frac{x}{2y^{1/2}}$$

are continuous in the upper half-plane defined by $y > 0$. We conclude from Theorem 2.1 that through any point (x_0, y_0), $y_0 > 0$ (say, for example, $(0, 1)$), there is some interval around x_0 on which the given differential equation has a unique solution. ∎

EXAMPLE 4

Theorem 2.1 guarantees that there exists an interval about $x = 0$ on which $y = 3e^x$ is the only solution of the initial-value problem of Example 1:

$$y' = y, \qquad y(0) = 3.$$

This follows from the fact that $f(x, y) = y$ and $\partial f/\partial y = 1$ are continuous throughout the entire xy-plane. It can be further shown that this interval is $(-\infty, \infty)$. ∎

EXAMPLE 5

For

$$\frac{dy}{dx} = x^2 + y^2$$

we observe that $f(x, y) = x^2 + y^2$ and $\partial f/\partial y = 2y$ are continuous throughout the entire xy-plane. Therefore, through any given point (x_0, y_0) there passes one and only one solution of the differential equation. ∎

Remarks **(i)** You should be aware of the distinction between a solution *existing* and *exhibiting* a solution. Clearly if we find a solution by exhibiting it, we can say that it exists, but on the other hand a solution can exist and we may not be able to display it. From Example 5 we know that a solution of the problem $dy/dx = x^2 + y^2$, $y(0) = 1$, exists on some interval around $x = 0$ and is unique. However, the best we can do in this case is to approximate the solution. See Chapter 9.

(ii) The conditions stated in Theorem 2.1 are *sufficient* but not *necessary*. When $f(x, y)$ and $\partial f/\partial y$ are continuous on a rectangular region R, it must *always* follow that there exists a unique solution of (2) when (x_0, y_0) is a point interior to R. However, if the conditions stated in the hypothesis of the theorem do not hold, then the initial-value problem (2) may still have (a) no solution, (b) more than one solution, or (c) a unique solution. Furthermore, the continuity condition on $\partial f/\partial y$ can be relaxed somewhat without changing the conclusion of the theorem. This results in a stronger theorem but is, unfortunately, not as easy to apply as Theorem 2.1. Indeed, if we are not interested in uniqueness, then a famous theorem due to the Italian mathematician Guiseppe Peano states that the continuity of $f(x, y)$ on R is sufficient to guarantee the existence of at least one solution of $dy/dx = f(x, y)$ through a point (x_0, y_0) interior to R.

2.1 EXERCISES

Answers to odd-numbered problems begin on page A–2.

In Problems 1–10 determine a region of the xy-plane for which the given differential equation would have a unique solution through a point (x_0, y_0) in the region.

1. $\dfrac{dy}{dx} = y^{2/3}$

2. $\dfrac{dy}{dx} = \sqrt{xy}$

3. $x\dfrac{dy}{dx} = y$

4. $\dfrac{dy}{dx} - y = x$

5. $(4 - y^2)y' = x^2$

6. $(1 + y^3)y' = x^2$

7. $(x^2 + y^2)y' = y^2$

8. $(y - x)y' = y + x$

9. $\dfrac{dy}{dx} = x^3 \cos y$

10. $\dfrac{dy}{dx} = (x - 1)e^{y/(x-1)}$

In Problems 11 and 12 determine by inspection at least two solutions of the given initial-value problem.

11. $y' = 3y^{2/3}, \quad y(0) = 0$

12. $x\dfrac{dy}{dx} = 2y, \quad y(0) = 0$

13. By inspection determine a solution of the nonlinear differential equation $y' = y^3$ satisfying $y(0) = 0$. Is the solution unique?

14. By inspection find a solution of the initial-value problem

$$y' = |y - 1|, \qquad y(0) = 1.$$

State why the conditions of Theorem 2.1 do not hold for this differential equation. Although we shall not prove it, the solution to this initial-value problem is unique.

15. Verify that $y = cx$ is a solution of the differential equation $xy' = y$ for every value of the parameter c. Find at least two solutions of the initial-value problem

$$xy' = y, \qquad y(0) = 0.$$

Observe that the piecewise-defined function

$$y = \begin{cases} 0, & x < 0 \\ x, & x \geq 0 \end{cases}$$

satisfies the condition $y(0) = 0$. Is it a solution of the initial-value problem?

16. (a) Consider the differential equation

$$\frac{dy}{dx} = 1 + y^2.$$

Determine a region of the xy-plane for which the equation has a unique solution through a point (x_0, y_0) in the region.

(b) Formally show that $y = \tan x$ satisfies the differential equation and the condition $y(0) = 0$.

(c) Explain why $y = \tan x$ is not a solution of the initial-value problem

$$\frac{dy}{dx} = 1 + y^2, \qquad y(0) = 0,$$

on the interval $(-2, 2)$.

(d) Explain why $y = \tan x$ is a solution of the initial-value problem in part (c) on the interval $(-1, 1)$.

In Problems 17–20 determine whether Theorem 2.1 guarantees that the differential equation $y' = \sqrt{y^2 - 9}$ possesses a unique solution through the given point.

17. $(1, 4)$ **18.** $(5, 3)$

19. $(2, -3)$ **20.** $(-1, 1)$

2.2 SEPARABLE VARIABLES

Note to the student: In solving a differential equation you will often have to utilize, say, integration by parts, partial fractions, or possibly a substitution. It will be worth a few minutes of your time to review some techniques of integration.

We begin our study of the methodology of solving first-order equations with the simplest of all differential equations.

If $g(x)$ is a given continuous function, then the first-order equation

$$\frac{dy}{dx} = g(x) \tag{1}$$

can be solved by integration. The solution of (1) is

$$y = \int g(x)\,dx + c.$$

EXAMPLE I

Solve **(a)** $\dfrac{dy}{dx} = 1 + e^{2x}$ and   **(b)** $\dfrac{dy}{dx} = \sin x$.

Solution   As illustrated above, both equations can be solved by integration.

(a) $y = \displaystyle\int (1 + e^{2x})\,dx = x + \tfrac{1}{2}e^{2x} + c$

(b) $y = \displaystyle\int \sin x\,dx = -\cos x + c$ ∎

Equation (1), as well as its method of solution, is just a special case of the following:

DEFINITION 2.1   Separable Equation
A differential equation of the form

$$\frac{dy}{dx} = \frac{g(x)}{h(y)}$$

is said to be **separable** or to have **separable variables**.

Observe that a separable equation can be written as

$$h(y)\frac{dy}{dx} = g(x). \tag{2}$$

It is seen immediately that (2) reduces to (1) when $h(y) = 1$.

Now if $y = f(x)$ denotes a solution of (2), we must have

$$h(f(x))f'(x) = g(x),$$

and therefore

$$\int h(f(x))f'(x)\,dx = \int g(x)\,dx + c. \tag{3}$$

But $dy = f'(x)\,dx$, so (3) is the same as

$$\int h(y)\,dy = \int g(x)\,dx + c. \tag{4}$$

Method of Solution

Equation (4) indicates the procedure for solving separable differential equations. A one-parameter family of solutions, usually given implicitly, is obtained by integrating both sides of $h(y)\,dy = g(x)\,dx$.

Note   There is no need to use two constants in the integration of a separable equation since

$$\int h(y)\,dy + c_1 = \int g(x)\,dx + c_2$$

$$\int h(y)\,dy = \int g(x)\,dx + c_2 - c_1 = \int g(x)\,dx + c,$$

where c is completely arbitrary. In many instances throughout the following chapters, we shall not hesitate to relabel constants in a manner that may prove convenient for a given equation. For example, multiples of constants or combinations of constants can sometimes be replaced by one constant.

EXAMPLE 2

Solve $(1 + x)\,dy - y\,dx = 0.$

Solution   Dividing by $(1 + x)y$, we can write $dy/y = dx/(1 + x)$, from which it follows that

$$\int \frac{dy}{y} = \int \frac{dx}{1 + x}.$$

$$\ln|y| = \ln|1 + x| + c_1$$

$$y = e^{\ln|1 + x| + c_1}$$

$$= e^{\ln|1 + x|} \cdot e^{c_1}$$

$$= |1 + x|e^{c_1}.$$

$$= \pm e^{c_1}(1 + x) \qquad \boxed{\begin{array}{l} |1 + x| = 1 + x,\ x \geq -1 \\ |1 + x| = -(1 + x),\ x < -1 \end{array}}$$

Relabeling $\pm e^{c_1}$ as c then gives $y = c(1 + x)$.

Alternative Solution   Since each integral results in a logarithm, a judicious choice for the constant of integration is $\ln|c|$ rather than c:

$$\ln|y| = \ln|1 + x| + \ln|c|$$

or

$$\ln|y| = \ln|c(1 + x)|$$

so that

$$y = c(1 + x).$$

Even if not *all* the indefinite integrals are logarithms, it may still be advantageous to use $\ln|c|$. However, no firm rule can be given. ∎

EXAMPLE 3

Solve the initial-value problem

$$\frac{dy}{dx} = -\frac{x}{y}, \qquad y(4) = 3.$$

Solution   From $y\, dy = -x\, dx$ we get

$$\int y\, dy = -\int x\, dx \quad \text{and} \quad \frac{y^2}{2} = -\frac{x^2}{2} + c_1.$$

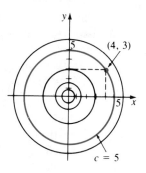

Figure 2.5

This solution can be written as $x^2 + y^2 = c^2$ by replacing the constant $2c_1$ by c^2. The solution represents a family of concentric circles.

Now when $x = 4$, $y = 3$ so that $16 + 9 = 25 = c^2$. Thus the initial-value problem determines $x^2 + y^2 = 25$. In view of Theorem 2.1, we can conclude that it is the only circle of the family passing through the point (4, 3). See Figure 2.5. ∎

EXAMPLE 4

Solve $xe^{-y} \sin x\, dx - y\, dy = 0.$

Solution   After multiplying by e^y, we get

$$x \sin x\, dx = ye^y\, dy.$$

Integration by parts on both sides of the equality gives

$$-x \cos x + \sin x = ye^y - e^y + c.$$ ∎

EXAMPLE 5

Solve $xy^4\, dx + (y^2 + 2)e^{-3x}\, dy = 0.$ **(5)**

Solution   By multiplying the given equation by e^{3x} and dividing by y^4, we obtain

$$xe^{3x}\, dx + \frac{y^2 + 2}{y^4}\, dy = 0 \quad \text{or} \quad xe^{3x}\, dx + (y^{-2} + 2y^{-4})\, dy = 0. \quad \textbf{(6)}$$

Using integration by parts on the first term yields

$$\frac{1}{3} xe^{3x} - \frac{1}{9} e^{3x} - y^{-1} - \frac{2}{3} y^{-3} = c_1.$$

The one-parameter family of solutions can also be written as

$$e^{3x}(3x - 1) = \frac{9}{y} + \frac{6}{y^3} + c, \tag{7}$$

where the constant $9c_1$ is rewritten as c. ∎

Two points are worth mentioning at this time. First, unless it is important or convenient there is no need to try to solve an expression representing a family of solutions for y explicitly in terms of x. Equation (7) shows that this task may present more problems than just the drudgery of symbol pushing. As a consequence it is often the case that the interval over which a solution is valid is not apparent. Second, some care should be exercised when separating variables to make certain that divisors are not zero. A constant solution may sometimes get lost in the shuffle of solving the problem. In Example 5 observe that $y = 0$ is a perfectly good solution of (5) but is not a member of the set of solutions defined by (7).

EXAMPLE 6

Solve the initial-value problem

$$\frac{dy}{dx} = y^2 - 4, \qquad y(0) = -2.$$

Solution   We put the equation into the form

$$\frac{dy}{y^2 - 4} = dx \tag{8}$$

and use partial fractions on the left side. We have

$$\left[\frac{-\frac{1}{4}}{y + 2} + \frac{\frac{1}{4}}{y - 2} \right] dy = dx \tag{9}$$

so that $$-\frac{1}{4} \ln|y + 2| + \frac{1}{4} \ln|y - 2| = x + c_1. \tag{10}$$

Thus $$\ln \left| \frac{y - 2}{y + 2} \right| = 4x + c_2$$

and $$\frac{y - 2}{y + 2} = ce^{4x},$$

where we have replaced $4c_1$ by c_2 and e^{c_2} by c. Finally, solving the last equation for y, we get

$$y = 2\frac{1 + ce^{4x}}{1 - ce^{4x}}. \tag{11}$$

Substituting $x = 0$, $y = -2$ leads to the dilemma

$$-2 = 2\frac{1 + c}{1 - c}$$

$$-1 + c = 1 + c \quad \text{or} \quad -1 = 1.$$

Let us consider the differential equation a little more carefully. The fact is, the equation

$$\frac{dy}{dx} = (y + 2)(y - 2)$$

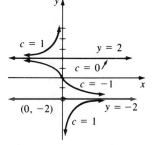

Figure 2.6

is satisfied by two constant functions, namely, $y = -2$ and $y = 2$. Inspection of equations (8), (9), and (10) clearly indicates we must preclude $y = -2$ and $y = 2$ at those steps in our solution. But it is interesting to observe that we can subsequently recover the solution $y = 2$ by setting $c = 0$ in equation (11). However, there is no finite value of c that will ever yield the solution $y = -2$. This latter constant function is the only solution to the original initial-value problem. See Figure 2.6. ∎

If, in Example 6, we had used $\ln|c|$ for the constant of integration, then the form of the one-parameter family of solutions would be

$$y = 2\frac{c + e^{4x}}{c - e^{4x}}. \tag{12}$$

Note that (12) reduces to $y = -2$ when $c = 0$, but now there is no finite value of c that will give the constant solution $y = 2$.

If an initial condition leads to a particular solution by finding a specific value of the parameter c in a family of solutions for a first-order differential equation, it is a natural inclination of most students (and instructors) to relax and be content. In Section 2.1 we saw, however, that a solution of an initial-value problem may not be unique. For example, the problem

$$\frac{dy}{dx} = xy^{1/2}, \qquad y(0) = 0, \tag{13}$$

has at least two solutions, namely, $y = 0$ and $y = x^4/16$. We are now in a position to solve the equation. Separating variables

$$y^{-1/2}\,dy = x\,dx$$

and integrating give

$$2y^{1/2} = \frac{x^2}{2} + c_1 \quad \text{or} \quad y = \left(\frac{x^2}{4} + c\right)^2.$$

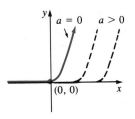

Figure 2.7

When $x = 0$, $y = 0$, so necessarily $c = 0$. Therefore $y = x^4/16$. The solution $y = 0$ was lost by dividing by $y^{1/2}$. In addition, the initial-value problem (13) possesses infinitely more solutions, since for any choice of the parameter $a \geq 0$ the piecewise-defined function

$$y = \begin{cases} 0, & x < a \\ \dfrac{(x^2 - a^2)^2}{16}, & x \geq a \end{cases}$$

satisfies both the differential equation and the initial condition. See Figure 2.7.

Remark   We saw in some of the preceding examples that the constant in the one-parameter family of solutions for a first-order differential equation can be relabeled when convenient. Also, it can easily happen that two individuals solving the same equation arrive at dissimilar answers. For example, by separation of variables we can show that one-parameter families of solutions for $(1 + y^2)\, dx + (1 + x^2)\, dy = 0$ are

$$\arctan x + \arctan y = c \quad \text{or} \quad \arctan x + \arctan y = \arctan c \quad \text{or} \quad \frac{x + y}{1 - xy} = c.$$

As you work your way through the next several sections, bear in mind that families of solutions may be *equivalent* in the sense that one family may be obtained from another by either relabeling the constant or applying algebra and trigonometry.

2.2 EXERCISES

Answers to odd-numbered problems begin on page A–2.

In Problems 1–40 solve the given differential equation by separation of variables.

1. $\dfrac{dy}{dx} = \sin 5x$

2. $\dfrac{dy}{dx} = (x + 1)^2$

3. $dx + e^{3x}\, dy = 0$

4. $dx - x^2\, dy = 0$

5. $(x + 1)\dfrac{dy}{dx} = x + 6$

6. $e^x \dfrac{dy}{dx} = 2x$

7. $xy' = 4y$

8. $\dfrac{dy}{dx} + 2xy = 0$

9. $\dfrac{dy}{dx} = \dfrac{y^3}{x^2}$

10. $\dfrac{dy}{dx} = \dfrac{y + 1}{x}$

11. $\dfrac{dx}{dy} = \dfrac{x^2 y^2}{1 + x}$

12. $\dfrac{dx}{dy} = \dfrac{1 + 2y^2}{y \sin x}$

13. $\dfrac{dy}{dx} = e^{3x + 2y}$

14. $e^x y \dfrac{dy}{dx} = e^{-y} + e^{-2x - y}$

15. $(4y + yx^2)\, dy - (2x + xy^2)\, dx = 0$

16. $(1 + x^2 + y^2 + x^2 y^2)\, dy = y^2\, dx$

17. $2y(x + 1)\, dy = x\, dx$

18. $x^2 y^2\, dy = (y + 1)\, dx$

19. $y \ln x \dfrac{dx}{dy} = \left(\dfrac{y + 1}{x}\right)^2$

20. $\dfrac{dy}{dx} = \left(\dfrac{2y + 3}{4x + 5}\right)^2$

21. $\dfrac{dS}{dr} = kS$

22. $\dfrac{dQ}{dt} = k(Q - 70)$

23. $\dfrac{dP}{dt} = P - P^2$

24. $\dfrac{dN}{dt} + N = Nte^{t+2}$

25. $\sec^2 x\, dy + \csc y\, dx = 0$

26. $\sin 3x\, dx + 2y \cos^3 3x\, dy = 0$

27. $e^y \sin 2x\, dx + \cos x(e^{2y} - y)\, dy = 0$

28. $\sec x\, dy = x \cot y\, dx$

29. $(e^y + 1)^2 e^{-y}\, dx + (e^x + 1)^3 e^{-x}\, dy = 0$

30. $\dfrac{y}{x} \dfrac{dy}{dx} = (1 + x^2)^{-1/2}(1 + y^2)^{1/2}$

31. $(y - yx^2)\dfrac{dy}{dx} = (y + 1)^2$

32. $2\dfrac{dy}{dx} - \dfrac{1}{y} = \dfrac{2x}{y}$

33. $\dfrac{dy}{dx} = \dfrac{xy + 3x - y - 3}{xy - 2x + 4y - 8}$

34. $\dfrac{dy}{dx} = \dfrac{xy + 2y - x - 2}{xy - 3y + x - 3}$

35. $\dfrac{dy}{dx} = \sin x(\cos 2y - \cos^2 y)$

36. $\sec y \dfrac{dy}{dx} + \sin(x - y) = \sin(x + y)$

37. $x\sqrt{1 - y^2}\, dx = dy$

38. $y(4 - x^2)^{1/2}\, dy = (4 + y^2)^{1/2}\, dx$

39. $(e^x + e^{-x})\dfrac{dy}{dx} = y^2$

40. $(x + \sqrt{x})\dfrac{dy}{dx} = y + \sqrt{y}$

In Problems 41–48 solve the given differential equation subject to the indicated initial condition.

41. $(e^{-y} + 1) \sin x\, dx = (1 + \cos x)\, dy, \quad y(0) = 0$

42. $(1 + x^4)\, dy + x(1 + 4y^2)\, dx = 0, \quad y(1) = 0$

43. $y\, dy = 4x(y^2 + 1)^{1/2}\, dx, \quad y(0) = 1$

44. $\dfrac{dy}{dt} + ty = y, \quad y(1) = 3$

45. $\dfrac{dx}{dy} = 4(x^2 + 1), \quad x\left(\dfrac{\pi}{4}\right) = 1$

46. $\dfrac{dy}{dx} = \dfrac{y^2 - 1}{x^2 - 1}, \quad y(2) = 2$

47. $x^2 y' = y - xy, \quad y(-1) = -1$

48. $y' + 2y = 1, \quad y(0) = \dfrac{5}{2}$

In Problems 49 and 50 find a solution of the given differential equation that passes through the indicated points.

49. $\dfrac{dy}{dx} - y^2 = -9$ **(a)** $(0, 0)$ **(b)** $(0, 3)$ **(c)** $\left(\dfrac{1}{3}, 1\right)$

50. $x\dfrac{dy}{dx} = y^2 - y$ **(a)** $(0, 1)$ **(b)** $(0, 0)$ **(c)** $\left(\dfrac{1}{2}, \dfrac{1}{2}\right)$

51. Find a singular solution for the equation in Problem 37.

52. Find a singular solution for the equation in Problem 39.

Often a radical change in the solution of a differential equation corresponds to a very small change in either the initial condition or the equation itself. In Problems 53–56 compare the solutions of the given initial-value problems.

53. $\dfrac{dy}{dx} = (y - 1)^2, \quad y(0) = 1$ **54.** $\dfrac{dy}{dx} = (y - 1)^2, \quad y(0) = 1.01$

55. $\dfrac{dy}{dx} = (y - 1)^2 + 0.01, \quad y(0) = 1$ **56.** $\dfrac{dy}{dx} = (y - 1)^2 - 0.01, \quad y(0) = 1$

A differential equation of the form $dy/dx = f(ax + by + c), b \neq 0$, can always be reduced to an equation with separable variables by means of the substitution $u = ax + by + c$. Use this procedure to solve Problems 57–62.

57. $\dfrac{dy}{dx} = (x + y + 1)^2$ **58.** $\dfrac{dy}{dx} = \dfrac{1 - x - y}{x + y}$

59. $\dfrac{dy}{dx} = \tan^2(x + y)$ **60.** $\dfrac{dy}{dx} = \sin(x + y)$

61. $\dfrac{dy}{dx} = 2 + \sqrt{y - 2x + 3}$ **62.** $\dfrac{dy}{dx} = 1 + e^{y - x + 5}$

2.3 HOMOGENEOUS EQUATIONS

Before considering the concept of a **homogeneous first-order differential** equation and its method of solution, we need to examine closely the nature of a **homogeneous function**. We begin with the definition of this concept.

DEFINITION 2.2   **Homogeneous Function**

If a function f has the property that

$$f(tx, ty) = t^n f(x, y) \tag{1}$$

for some real number n, then f is said to be a **homogeneous function of degree n**.

EXAMPLE I

(a) $f(x, y) = x^2 - 3xy + 5y^2$

$\qquad f(tx, ty) = (tx)^2 - 3(tx)(ty) + 5(ty)^2$

$\qquad\qquad\qquad = t^2x^2 - 3t^2xy + 5t^2y^2$

$\qquad\qquad\qquad = t^2[x^2 - 3xy + 5y^2] = t^2 f(x, y)$

The function is homogeneous of degree two.

(b) $f(x, y) = \sqrt[3]{x^2 + y^2}$

$\qquad f(tx, ty) = \sqrt[3]{t^2x^2 + t^2y^2} = t^{2/3}\sqrt[3]{x^2 + y^2} = t^{2/3}f(x, y)$

The function is homogeneous of degree 2/3.

(c) $f(x, y) = x^3 + y^3 + 1$

$\qquad f(tx, ty) = t^3x^3 + t^3y^3 + 1 \neq t^3f(x, y)$

since $t^3f(x, y) = t^3x^3 + t^3y^3 + t^3$. The function is not homogeneous.

(d) $f(x, y) = \dfrac{x}{2y} + 4$

$\qquad f(tx, ty) = \dfrac{tx}{2ty} + 4 = \dfrac{x}{2y} + 4 = t^0 f(x, y)$

The function is homogeneous of degree zero. ■

As parts (c) and (d) of Example 1 show, a constant added to a function destroys homogeneity, unless the function is homogeneous of degree zero. Also, in many instances a homogeneous function can be recognized by examining the *total degree* of each term.

EXAMPLE 2

(a) $f(x, y) = 6xy^3 - x^2y^2$

degree 1, degree 3 } degree 4

degree 2, degree 2 } degree 4

The function is homogeneous of degree four.

(b) $f(x, y) = x^2 - y$

degree 2, degree 1 } degrees not the same

The function is not homogeneous since the degrees of the two terms are different. ■

If $f(x, y)$ is a homogeneous function of degree n, notice that we can write

$$f(x, y) = x^n f\left(1, \frac{y}{x}\right) \quad \text{and} \quad f(x, y) = y^n f\left(\frac{x}{y}, 1\right), \tag{2}$$

where $f(1, y/x)$ and $f(x/y, 1)$ are both homogeneous of degree zero.

EXAMPLE 3

We see that $f(x, y) = x^2 + 3xy + y^2$ is homogeneous of degree two. Thus

$$f(x, y) = x^2\left[1 + 3\left(\frac{y}{x}\right) + \left(\frac{y}{x}\right)^2\right] = x^2 f\left(1, \frac{y}{x}\right)$$

$$f(x, y) = y^2\left[\left(\frac{x}{y}\right)^2 + 3\left(\frac{x}{y}\right) + 1\right] = y^2 f\left(\frac{x}{y}, 1\right). \qquad \blacksquare$$

A homogeneous first-order differential equation is defined in terms of homogeneous functions.

DEFINITION 2.3 **Homogeneous Equation**

A differential equation of the form

$$M(x, y)\, dx + N(x, y)\, dy = 0 \tag{3}$$

is said to be **homogeneous** if both coefficients M and N are homogeneous functions of the same degree.

In other words, $M(x, y)\, dx + N(x, y)\, dy = 0$ is homogeneous if

$$M(tx, ty) = t^n M(x, y) \quad \text{and} \quad N(tx, ty) = t^n N(x, y).$$

Method of Solution

A homogeneous differential equation $M(x, y)\, dx + N(x, y)\, dy = 0$ can be solved by means of an algebraic substitution. Specifically, *either* substitution $y = ux$ or $x = vy$, where u and v are new dependent variables, *will reduce the equation to a separable first-order differential equation*. To see this, let us substitute $y = ux$ and its differential $dy = u\, dx + x\, du$ into (3):

$$M(x, ux)\, dx + N(x, ux)[u\, dx + x\, du] = 0.$$

Now, by the homogeneity property given in (2), we can write

$$x^n M(1, u)\, dx + x^n N(1, u)[u\, dx + x\, du] = 0$$

or $\qquad\qquad [M(1, u) + uN(1, u)]\, dx + xN(1, u)\, du = 0,$

which gives
$$\frac{dx}{x} + \frac{N(1, u)\, du}{M(1, u) + uN(1, u)} = 0.$$

We hasten to point out that the preceeding formula should not be memorized; rather, the *procedure should be worked through each time*. The proof that the substitution $x = vy$ in (3) also leads to a separable equation is left as an exercise. See Problem 45.

EXAMPLE 4

Solve
$$(x^2 + y^2)\, dx + (x^2 - xy)\, dy = 0.$$

Solution Both $M(x, y)$ and $N(x, y)$ are homogeneous of degree two. If we let $y = ux$, it follows that

$$(x^2 + u^2x^2)\, dx + (x^2 - ux^2)[u\, dx + x\, du] = 0$$

$$x^2(1 + u)\, dx + x^3(1 - u)\, du = 0$$

$$\frac{1 - u}{1 + u}\, du + \frac{dx}{x} = 0$$

$$\left[-1 + \frac{2}{1 + u}\right] du + \frac{dx}{x} = 0 \qquad \Leftarrow \boxed{\text{long division}}$$

After integration the last line gives

$$-u + 2\ln|1 + u| + \ln|x| = \ln|c|$$

$$-\frac{y}{x} + 2\ln\left|1 + \frac{y}{x}\right| + \ln|x| = \ln|c|.$$

Using the properties of logarithms, we can write the preceding solution as

$$\ln\left|\frac{(x + y)^2}{cx}\right| = \frac{y}{x}.$$

The definition of a logarithm then yields

$$(x + y)^2 = cxe^{y/x}. \qquad \blacksquare$$

EXAMPLE 5

Solve
$$(2\sqrt{xy} - y)\, dx - x\, dy = 0.$$

Solution The coefficients $M(x, y)$ and $N(x, y)$ are homogeneous of degree one. If $y = ux$, the differential equation becomes, after simplifying,

$$\frac{du}{2u - 2u^{1/2}} + \frac{dx}{x} = 0.$$

The integral of the first term can be evaluated by the further substitution $t = u^{1/2}$. The result is

$$\frac{dt}{t-1} + \frac{dx}{x} = 0.$$

Integrating yields

$$\ln|t-1| + \ln|x| = \ln|c|$$

$$\ln\left|\sqrt{\frac{y}{x}} - 1\right| + \ln|x| = \ln|c| \quad \Leftarrow \boxed{t = u^{1/2} \quad \text{and} \quad u = y/x}$$

$$x\left(\sqrt{\frac{y}{x}} - 1\right) = c$$

$$\sqrt{xy} - x = c. \qquad \blacksquare$$

By now you may be asking: when should the substitution $x = vy$ be used? Although it can be used for every homogeneous differential equation, in practice we try $x = vy$ whenever the function $M(x, y)$ is simpler than $N(x, y)$. In solving $(x^2 + y^2)\, dx + (x^2 - xy)\, dy = 0$, for example, we know there is no appreciable difference between M and N, so either $y = ux$ or $x = vy$ could be used. Also, it could happen that after using one substitution, we encounter integrals that are difficult or impossible to evaluate in closed form; switching substitutions may result in an easier problem.

EXAMPLE 6

Solve
$$2x^3y\, dx + (x^4 + y^4)\, dy = 0.$$

Solution Each coefficient is a homogeneous function of degree four. Since the coefficient of dx is slightly simpler than the coefficient of dy, we try $x = vy$. After substituting, we simplify the equation

$$2v^3y^4[v\, dy + y\, dv] + (v^4y^4 + y^4)\, dy = 0$$

to
$$\frac{2v^3\, dv}{3v^4 + 1} + \frac{dy}{y} = 0.$$

Integration gives

$$\frac{1}{6}\ln(3v^4 + 1) + \ln|y| = \ln|c_1| \quad \text{or} \quad 3x^4y^2 + y^6 = c,$$

where $c = c_1^6$. Now had the substitution $y = ux$ been used, then

$$\frac{dx}{x} + \frac{u^4 + 1}{u^5 + 3u}\, du = 0.$$

You are urged to reflect on how to evaluate the integral of the second term in the last equation. $\qquad \blacksquare$

A homogeneous differential equation can always be expressed in the alternative form

$$\frac{dy}{dx} = F\left(\frac{y}{x}\right).$$

To see this suppose we write the equation $M(x, y)\, dx + N(x, y)\, dy = 0$ as $dy/dx = f(x, y)$, where

$$f(x, y) = -\frac{M(x, y)}{N(x, y)}.$$

The function $f(x, y)$ must necessarily be homogeneous of degree zero when M and N are homogeneous of degree n. From (2), it follows that

$$f(x, y) = -\frac{x^n M\left(1, \dfrac{y}{x}\right)}{x^n N\left(1, \dfrac{y}{x}\right)} = -\frac{M\left(1, \dfrac{y}{x}\right)}{N\left(1, \dfrac{y}{x}\right)}.$$

The last ratio is recognized as a function of the form $F(y/x)$. We leave it as an exercise to demonstrate that a homogeneous differential equation can also be written as $dy/dx = G(x/y)$. See Problem 47.

EXAMPLE 7

Solve the initial-value problem

$$x\frac{dy}{dx} = y + xe^{y/x}, \qquad y(1) = 1.$$

Solution   By writing the equation as

$$\frac{dy}{dx} = \frac{y}{x} + e^{y/x},$$

we see that the function to the right of the equality is homogeneous of degree zero. From the form of this function we are prompted to use $u = y/x$. After differentiating $y = ux$ by the product rule and substituting, we find

$$u + x\frac{du}{dx} = u + e^u \quad \text{or} \quad e^{-u}\, du = \frac{dx}{x}.$$

Integrating and substituting $u = y/x$ give

$$-e^{-u} + c = \ln|x|$$

$$-e^{-y/x} + c = \ln|x|.$$

Since $y = 1$ when $x = 1$, we get $-e^{-1} + c = 0$ or $c = e^{-1}$. Therefore the solution to the initial-value problem is

$$e^{-1} - e^{-y/x} = \ln|x|. \qquad\blacksquare$$

2.3 **EXERCISES**

Answers to odd-numbered problems begin on page A–3.

In Problems 1–10 determine whether the given function is homogeneous. If so, state the degree of homogeneity.

1. $x^3 + 2xy^2 - y^4/x$

2. $\sqrt{x + y}(4x + 3y)$

3. $\dfrac{x^3 y - x^2 y^2}{(x + 8y)^2}$

4. $\dfrac{x}{y^2 + \sqrt{x^4 + y^4}}$

5. $\cos \dfrac{x^2}{x + y}$

6. $\sin \dfrac{x}{x + y}$

7. $\ln x^2 - 2 \ln y$

8. $\dfrac{\ln x^3}{\ln y^3}$

9. $(x^{-1} + y^{-1})^2$

10. $(x + y + 1)^2$

In Problems 11–30 solve the given differential equation by using an appropriate substitution.

11. $(x - y)\,dx + x\,dy = 0$

12. $(x + y)\,dx + x\,dy = 0$

13. $x\,dx + (y - 2x)\,dy = 0$

14. $y\,dx = 2(x + y)\,dy$

15. $(y^2 + yx)\,dx - x^2\,dy = 0$

16. $(y^2 + yx)\,dx + x^2\,dy = 0$

17. $\dfrac{dy}{dx} = \dfrac{y - x}{y + x}$

18. $\dfrac{dy}{dx} = \dfrac{x + 3y}{3x + y}$

19. $-y\,dx + (x + \sqrt{xy})\,dy = 0$

20. $x\dfrac{dy}{dx} - y = \sqrt{x^2 + y^2}$

21. $2x^2 y\,dx = (3x^3 + y^3)\,dy$

22. $(x^4 + y^4)\,dx - 2x^3 y\,dy = 0$

23. $\dfrac{dy}{dx} = \dfrac{y}{x} + \dfrac{x}{y}$

24. $\dfrac{dy}{dx} = \dfrac{y}{x} + \dfrac{x^2}{y^2} + 1$

25. $y\dfrac{dx}{dy} = x + 4ye^{-2x/y}$

26. $(x^2 e^{-y/x} + y^2)\,dx = xy\,dy$

27. $\left(y + x\cot\dfrac{y}{x}\right)dx - x\,dy = 0.$

28. $\dfrac{dy}{dx} = \dfrac{y}{x}\ln\dfrac{y}{x}$

29. $(x^2 + xy - y^2)\,dx + xy\,dy = 0$

30. $(x^2 + xy + 3y^2)\,dx - (x^2 + 2xy)\,dy = 0$

In Problems 31–44 solve the given differential equation subject to the indicated initial condition.

31. $xy^2 \dfrac{dy}{dx} = y^3 - x^3, \quad y(1) = 2$

32. $(x^2 + 2y^2)\,dx = xy\,dy, \quad y(-1) = 1$

33. $2x^2 \dfrac{dy}{dx} = 3xy + y^2, \quad y(1) = -2$

34. $xy\,dx - x^2\,dy = y\sqrt{x^2 + y^2}\,dy, \quad y(0) = 1$

35. $(x + ye^{y/x})\,dx - xe^{y/x}\,dy = 0, \quad y(1) = 0$

36. $y\,dx + \left(y\cos\dfrac{x}{y} - x\right)dy = 0, \quad y(0) = 2$

37. $(y^2 + 3xy)\,dx = (4x^2 + xy)\,dy, \quad y(1) = 1$

38. $y^3\,dx = 2x^3\,dy - 2x^2y\,dx, \quad y(1) = \sqrt{2}$

39. $(x + \sqrt{xy})\dfrac{dy}{dx} + x - y = x^{-1/2}y^{3/2}, \quad y(1) = 1$

40. $y\,dx + x(\ln x - \ln y - 1)\,dy = 0, \quad y(1) = e$

41. $y^2\,dx + (x^2 + xy + y^2)\,dy = 0, \quad y(0) = 1$

42. $(\sqrt{x} + \sqrt{y})^2\,dx = x\,dy, \quad y(1) = 0$

43. $(x + \sqrt{y^2 - xy})\dfrac{dy}{dx} = y, \quad y(\tfrac{1}{2}) = 1$

44. $\dfrac{dy}{dx} - \dfrac{y}{x} = \cosh\dfrac{y}{x}, \quad y(1) = 0$

45. Suppose $M(x, y)\,dx + N(x, y)\,dy = 0$ is a homogeneous equation. Show that the substitution $x = vy$ reduces the equation to one with separable variables.

46. Suppose $M(x, y)\,dx + N(x, y)\,dy = 0$ is a homogeneous equation. Show that the substitutions $x = r\cos\theta, y = r\sin\theta$ reduce the equation to one with separable variables.

47. Suppose $M(x, y)\,dx + N(x, y)\,dy = 0$ is a homogeneous equation. Show that the equation has the alternative form $dy/dx = G(x/y)$.

48. If $f(x, y)$ is a homogeneous function of degree n, show that

$$x\frac{\partial f}{\partial x} + y\frac{\partial f}{\partial y} = nf.$$

2.4 EXACT EQUATIONS

Although the simple equation

$$y\,dx + x\,dy = 0$$

is both separable and homogeneous, we should recognize that it is also equivalent to the differential of the product of x and y; that is,

$$y\,dx + x\,dy = d(xy) = 0.$$

By integrating we immediately obtain the implicit solution $xy = c$.

From calculus you might remember that if $z = f(x, y)$ is a function having continuous first partial derivatives in a region R of the xy-plane, then its *total differential* is

$$dz = \frac{\partial f}{\partial x}\,dx + \frac{\partial f}{\partial y}\,dy. \tag{1}$$

Now if $f(x, y) = c$, it follows from (1) that

$$\frac{\partial f}{\partial x}\,dx + \frac{\partial f}{\partial y}\,dy = 0. \tag{2}$$

In other words, given a family of curves $f(x, y) = c$, we can generate a first-order differential equation by computing the total differential.

EXAMPLE I

If $x^2 - 5xy + y^3 = c$, then (2) gives

$$(2x - 5y)\,dx + (-5x + 3y^2)\,dy = 0 \quad \text{or} \quad \frac{dy}{dx} = \frac{5y - 2x}{-5x + 3y^2}. \qquad \blacksquare$$

For our purposes, it is more important to turn the problem around; namely, given an equation such as

$$\frac{dy}{dx} = \frac{5y - 2x}{-5x + 3y^2}, \tag{3}$$

can we identify the equation as being equivalent to the statement

$$d(x^2 - 5xy + y^3) = 0?$$

Notice that equation (3) is neither separable nor homogeneous.

DEFINITION 2.4   Exact Equation

A differential expression

$$M(x, y)\,dx + N(x, y)\,dy$$

is an **exact differential** in a region R of the xy-plane if it corresponds to the total differential of some function $f(x, y)$. A differential equation of the form

$$M(x, y)\,dx + N(x, y)\,dy = 0$$

is said to be an **exact equation** if the expression on the left side is an exact differential.

EXAMPLE 2

The equation $x^2y^3 \, dx + x^3y^2 \, dy = 0$ is exact since it is recognized that

$$d(\tfrac{1}{3}x^3y^3) = x^2y^3 \, dx + x^3y^2 \, dy.$$

∎

The following theorem is a test for an exact differential.

THEOREM 2.2   Criterion for an Exact Differential

Let $M(x, y)$ and $N(x, y)$ be continuous and have continuous first partial derivatives in a rectangular region R defined by $a < x < b, c < y < d$. Then a necessary and sufficient condition that

$$M(x, y) \, dx + N(x, y) \, dy$$

be an exact differential is

$$\frac{\partial M}{\partial y} = \frac{\partial N}{\partial x}. \qquad (4)$$

Proof of the Necessity For simplicity let us assume that $M(x, y)$ and $N(x, y)$ have continuous first partial derivatives for all (x, y). Now if the expression $M(x, y) \, dx + N(x, y) \, dy$ is exact, there exists some function f for which

$$M(x, y) \, dx + N(x, y) \, dy = \frac{\partial f}{\partial x} \, dx + \frac{\partial f}{\partial y} \, dy$$

for all (x, y) in R. Therefore

$$M(x, y) = \frac{\partial f}{\partial x}, \qquad N(x, y) = \frac{\partial f}{\partial y},$$

and

$$\frac{\partial M}{\partial y} = \frac{\partial}{\partial y}\left(\frac{\partial f}{\partial x}\right) = \frac{\partial^2 f}{\partial y \, \partial x} = \frac{\partial}{\partial x}\left(\frac{\partial f}{\partial y}\right) = \frac{\partial N}{\partial x}.$$

The equality of the mixed partials is a consequence of the continuity of the first partial derivatives of $M(x, y)$ and $N(x, y)$.   □

The sufficiency part of Theorem 2.2 consists of showing that there exists a function f for which $\partial f/\partial x = M(x, y)$ and $\partial f/\partial y = N(x, y)$ whenever (4) holds. The construction of the function f actually reflects a basic procedure for solving exact equations.

Method of Solution

Given the equation

$$M(x, y)\, dx + N(x, y)\, dy = 0, \tag{5}$$

first show that

$$\frac{\partial M}{\partial y} = \frac{\partial N}{\partial x}.$$

Then assume that

$$\frac{\partial f}{\partial x} = M(x, y)$$

so we can find f by integrating $M(x, y)$ with respect to x, while holding y constant. We write

$$f(x, y) = \int M(x, y)\, dx + g(y), \tag{6}$$

where the arbitrary function $g(y)$ is the "constant" of integration. Now differentiate (6) with respect to y and assume $\partial f / \partial y = N(x, y)$:

$$\frac{\partial f}{\partial y} = \frac{\partial}{\partial y} \int M(x, y)\, dx + g'(y) = N(x, y).$$

This gives

$$g'(y) = N(x, y) - \frac{\partial}{\partial y} \int M(x, y)\, dx. \tag{7}$$

Finally, integrate (7) with respect to y and substitute the result in (6). The solution of the equation is $f(x, y) = c$.

Note Some observations are in order. First, it is important to realize that the expression $N(x, y) - (\partial / \partial y) \int M(x, y)\, dx$ in (7) is independent of x, since

$$\frac{\partial}{\partial x} \left[N(x, y) - \frac{\partial}{\partial y} \int M(x, y)\, dx \right] = \frac{\partial N}{\partial x} - \frac{\partial}{\partial y} \left(\frac{\partial}{\partial x} \int M(x, y)\, dx \right)$$

$$= \frac{\partial N}{\partial x} - \frac{\partial M}{\partial y} = 0.$$

Second, we could just as well start the foregoing procedure with the assumption that $\partial f / \partial y = N(x, y)$. After integrating N with respect to y and then differentiating that result, we find the analogues of (6) and (7) would be, respectively,

$$f(x, y) = \int N(x, y)\, dy + h(x) \quad \text{and} \quad h'(x) = M(x, y) - \frac{\partial}{\partial x} \int N(x, y)\, dy.$$

In either case *none of these formulas should be memorized.* Also, when testing an equation for exactness, make sure it is of form (5). Often a differential equation is written $G(x, y)\, dx = H(x, y)\, dy$. In this case write the equation as $G(x, y)\, dx - H(x, y)\, dy = 0$ and then identify $M(x, y) = G(x, y)$ and $N(x, y) = -H(x, y)$.

EXAMPLE 3

Solve
$$2xy\,dx + (x^2 - 1)\,dy = 0.$$

Solution With $M(x, y) = 2xy$ and $N(x, y) = x^2 - 1$ we have

$$\frac{\partial M}{\partial y} = 2x = \frac{\partial N}{\partial x}.$$

Thus the equation is exact and so, by Theorem 2.2, there exists a function $f(x, y)$ such that

$$\frac{\partial f}{\partial x} = 2xy \quad \text{and} \quad \frac{\partial f}{\partial y} = x^2 - 1.$$

From the first of these equations we obtain, after integrating,

$$f(x, y) = x^2 y + g(y).$$

Taking the partial derivative of the last expression with respect to y and setting the result equal to $N(x, y)$ give

$$\frac{\partial f}{\partial y} = x^2 + g'(y) = x^2 - 1. \qquad \overset{\longleftarrow}{\boxed{N(x, y)}}$$

It follows that $\qquad g'(y) = -1 \quad \text{and} \quad g(y) = -y.$

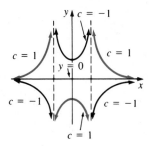

$y \uparrow c = -1$

$c = 1 \qquad c = 1$

$y = 0$

x

$c = -1 \qquad c = -1$

$c = 1$

Figure 2.8

The constant of integration need not be included in the preceding line since the solution is $f(x, y) = c$. Some of the family of curves $x^2 y - y = c$ are given in Figure 2.8.

Note The solution of the equation is *not* $f(x, y) = x^2 y - y$. Rather it is $f(x, y) = c$ or $f(x, y) = 0$ if a constant is used in the integration of $g'(y)$. Observe that the equation could also be solved by separation of variables. ∎

EXAMPLE 4

Solve $\qquad (e^{2y} - y \cos xy)\,dx + (2xe^{2y} - x \cos xy + 2y)\,dy = 0.$

Solution The equation is neither separable nor homogeneous but is exact since

$$\frac{\partial M}{\partial y} = 2e^{2y} + xy \sin xy - \cos xy = \frac{\partial N}{\partial x}.$$

Hence a function $f(x, y)$ exists for which

$$M(x, y) = \frac{\partial f}{\partial x} \quad \text{and} \quad N(x, y) = \frac{\partial f}{\partial y}.$$

Now for variety we shall start with the assumption that $\partial f/\partial y = N(x, y)$;

that is,
$$\frac{\partial f}{\partial y} = 2xe^{2y} - x \cos xy + 2y$$

$$f(x, y) = 2x \int e^{2y}\, dy - x \int \cos xy\, dy + 2 \int y\, dy.$$

Remember, the reason x can come out in front of the symbol $\int$ is that in the integration with respect to y, x is treated as an ordinary constant. It follows that

$$f(x, y) = xe^{2y} - \sin xy + y^2 + h(x)$$

$$\frac{\partial f}{\partial x} = e^{2y} - y \cos xy + h'(x) = e^{2y} - y \cos xy, \qquad \Longleftarrow \boxed{M(x, y)}$$

so that
$$h'(x) = 0 \quad \text{and} \quad h(x) = c.$$

Hence a one-parameter family of solutions is given by

$$xe^{2y} - \sin xy + y^2 + c = 0. \qquad\qquad \blacksquare$$

EXAMPLE 5

Solve the initial-value problem
$$(\cos x \sin x - xy^2)\, dx + y(1 - x^2)\, dy = 0, \qquad y(0) = 2.$$

Solution   The equation is exact since

$$\frac{\partial M}{\partial y} = -2xy = \frac{\partial N}{\partial x}.$$

Now
$$\frac{\partial f}{\partial y} = y(1 - x^2)$$

$$f(x, y) = \frac{y^2}{2}(1 - x^2) + h(x)$$

$$\frac{\partial f}{\partial x} = -xy^2 + h'(x) = \cos x \sin x - xy^2.$$

The last equation implies
$$h'(x) = \cos x \sin x$$

$$h(x) = -\int (\cos x)(-\sin x\, dx) = -\frac{1}{2}\cos^2 x.$$

Thus
$$\frac{y^2}{2}(1 - x^2) - \frac{1}{2}\cos^2 x = c_1$$

or
$$y^2(1 - x^2) - \cos^2 x = c,$$

where $2c_1$ has been replaced by c. The initial condition $y = 2$ when $x = 0$ demands that $4(1) - \cos^2(0) = c$ or that $c = 3$. Thus a solution of the problem is

$$y^2(1 - x^2) - \cos^2 x = 3.$$ ∎

Integrating Factor

It is sometimes possible to convert a nonexact differential equation into an exact equation by multiplying it by a function $\mu(x, y)$ called an **integrating factor**. However, the resulting exact equation

$$\mu M(x, y)\, dx + \mu N(x, y)\, dy = 0$$

may not be equivalent to the original equation in the sense that a solution of one is also a solution of the other. It is possible for a solution to be lost or gained as a result of the multiplication.

EXAMPLE 6

Solve $(x + y)\, dx + x \ln x\, dy = 0$, using $\mu(x, y) = \dfrac{1}{x}$, on $(0, \infty)$.

Solution Let $M(x, y) = x + y$ and $N(x, y) = x \ln x$ so that $\partial M/\partial y = 1$ and $\partial N/\partial x = 1 + \ln x$. The equation is not exact. However, if we multiply the equation by $\mu(x, y) = 1/x$, we obtain

$$\left(1 + \frac{y}{x}\right) dx + \ln x\, dy = 0.$$

From this latter form we make the identifications:

$$M(x, y) = 1 + \frac{y}{x}, \qquad N(x, y) = \ln x, \qquad \frac{\partial M}{\partial y} = \frac{1}{x} = \frac{\partial N}{\partial x}.$$

Therefore the second differential equation is exact. It follows that

$$\frac{\partial f}{\partial x} = 1 + \frac{y}{x} = M(x, y)$$

$$f(x, y) = x + y \ln x + g(y)$$

$$\frac{\partial f}{\partial y} = 0 + \ln x + g'(y) = \ln x$$

and so
$$g'(y) = 0 \quad \text{and} \quad g(y) = c.$$

Hence $f(x, y) = x + y \ln x + c$. It is readily verified that

$$x + y \ln x + c = 0$$

is a solution of both equations on $(0, \infty)$.   ∎

2.4 **EXERCISES**

Answers to odd-numbered problems begin on page A–3.

In Problems 1–24 determine whether the given equation is exact. If exact, solve.

1. $(2x - 1)\, dx + (3y + 7)\, dy = 0$ **2.** $(2x + y)\, dx - (x + 6y)\, dy = 0$

3. $(5x + 4y)\, dx + (4x - 8y^3)\, dy = 0$

4. $(\sin y - y \sin x)\, dx + (\cos x + x \cos y - y)\, dy = 0$

5. $(2y^2 x - 3)\, dx + (2yx^2 + 4)\, dy = 0$

6. $\left(2y - \dfrac{1}{x} + \cos 3x\right)\dfrac{dy}{dx} + \dfrac{y}{x^2} - 4x^3 + 3y \sin 3x = 0$

7. $(x + y)(x - y)\, dx + x(x - 2y)\, dy = 0$

8. $\left(1 + \ln x + \dfrac{y}{x}\right) dx = (1 - \ln x)\, dy$

9. $(y^3 - y^2 \sin x - x)\, dx + (3xy^2 + 2y \cos x)\, dy = 0$

10. $(x^3 + y^3)\, dx + 3xy^2\, dy = 0$

11. $(y \ln y - e^{-xy})\, dx + \left(\dfrac{1}{y} + x \ln y\right) dy = 0$

12. $\dfrac{2x}{y}\, dx - \dfrac{x^2}{y^2}\, dy = 0$

13. $x\dfrac{dy}{dx} = 2xe^x - y + 6x^2$

14. $(3x^2 y + e^y)\, dx + (x^3 + xe^y - 2y)\, dy = 0$

15. $\left(1 - \dfrac{3}{x} + y\right) dx + \left(1 - \dfrac{3}{y} + x\right) dy = 0$

16. $(e^y + 2xy \cosh x)y' + xy^2 \sinh x + y^2 \cosh x = 0$

17. $\left(x^2 y^3 - \dfrac{1}{1 + 9x^2}\right)\dfrac{dx}{dy} + x^3 y^2 = 0$ **18.** $(5y - 2x)y' - 2y = 0$

19. $(\tan x - \sin x \sin y)\, dx + \cos x \cos y\, dy = 0$

20. $(3x \cos 3x + \sin 3x - 3)\, dx + (2y + 5)\, dy = 0$

21. $(1 - 2x^2 - 2y)\dfrac{dy}{dx} = 4x^3 + 4xy$

22. $(2y \sin x \cos x - y + 2y^2 e^{xy^2})\, dx = (x - \sin^2 x - 4xy e^{xy^2})\, dy$

23. $(4x^3 y - 15x^2 - y)\,dx + (x^4 + 3y^2 - x)\,dy = 0$

24. $\left(\dfrac{1}{x} + \dfrac{1}{x^2} - \dfrac{y}{x^2 + y^2}\right) dx + \left(ye^y + \dfrac{x}{x^2 + y^2}\right) dy = 0$

In Problems 25–30 solve the given differential equation subject to the indicated initial condition.

25. $(x + y)^2\,dx + (2xy + x^2 - 1)\,dy = 0, \quad y(1) = 1$

26. $(e^x + y)\,dx + (2 + x + ye^y)\,dy = 0, \quad y(0) = 1$

27. $(4y + 2x - 5)\,dx + (6y + 4x - 1)\,dy = 0, \quad y(-1) = 2$

28. $\left(\dfrac{3y^2 - x^2}{y^5}\right)\dfrac{dy}{dx} + \dfrac{x}{2y^4} = 0, \quad y(1) = 1$

29. $(y^2 \cos x - 3x^2 y - 2x)\,dx + (2y \sin x - x^3 + \ln y)\,dy = 0, \quad y(0) = e$

30. $\left(\dfrac{1}{1 + y^2} + \cos x - 2xy\right)\dfrac{dy}{dx} = y(y + \sin x), \quad y(0) = 1$

In Problems 31–34 find the value of k so that the given differential equation is exact.

31. $(y^3 + kxy^4 - 2x)\,dx + (3xy^2 + 20x^2 y^3)\,dy = 0$

32. $(2x - y \sin xy + ky^4)\,dx - (20xy^3 + x \sin xy)\,dy = 0$

33. $(2xy^2 + ye^x)\,dx + (2x^2 y + ke^x - 1)\,dy = 0$

34. $(6xy^3 + \cos y)\,dx + (kx^2 y^2 - x \sin y)\,dy = 0$

35. Determine a function $M(x, y)$ so that the following differential equation is exact:

$$M(x, y)\,dx + \left(xe^{xy} + 2xy + \dfrac{1}{x}\right)dy = 0.$$

36. Determine a function $N(x, y)$ so that the following differential equation is exact:

$$\left(y^{1/2}x^{-1/2} + \dfrac{x}{x^2 + y}\right)dx + N(x, y)\,dy = 0.$$

In Problems 37–42 solve the given differential equation by verifying that the indicated function $\mu(x, y)$ is an integrating factor.

37. $6xy\,dx + (4y + 9x^2)\,dy = 0, \quad \mu(x, y) = y^2$

38. $-y^2\,dx + (x^2 + xy)\,dy = 0, \quad \mu(x, y) = 1/x^2 y$

39. $(-xy \sin x + 2y \cos x)\,dx + 2x \cos x\,dy = 0, \quad \mu(x, y) = xy$

40. $y(x + y + 1)\,dx + (x + 2y)\,dy = 0, \quad \mu(x, y) = e^x$

41. $(2y^2 + 3x)\,dx + 2xy\,dy = 0, \quad \mu(x, y) = x$

42. $(x^2 + 2xy - y^2)\,dx + (y^2 + 2xy - x^2)\,dy = 0, \quad \mu(x, y) = (x + y)^{-2}$

43. Show that any separable first-order differential equation in the form $h(y)\,dy - g(x)\,dx = 0$ is also exact.

2.5 LINEAR EQUATIONS

In Chapter 1 we defined the general form of a linear differential equation of order n to be

$$a_n(x) \frac{d^n y}{dx^n} + a_{n-1}(x) \frac{d^{n-1} y}{dx^{n-1}} + \cdots + a_1(x) \frac{dy}{dx} + a_0(x)y = g(x).$$

We remind you that linearity means that all coefficients are functions of x only, and that y and all its derivatives are raised to the first power. Now when $n = 1$, we obtain a linear first-order equation.

DEFINITION 2.5   Linear Equation

A differential equation of the form

$$a_1(x) \frac{dy}{dx} + a_0(x)y = g(x)$$

is said to be a **linear equation**.

Dividing by the lead coefficient $a_1(x)$ gives a more useful form of a linear equation:

$$\frac{dy}{dx} + P(x)y = f(x). \tag{1}$$

We seek a solution of (1) on an interval I for which the functions $P(x)$ and $f(x)$ are continuous. In the discussion that follows, we tacitly assume that (1) has a solution.

Integrating Factor

Using differentials, we can rewrite equation (1) as

$$dy + [P(x)y - f(x)]\, dx = 0. \tag{2}$$

Linear equations possess the pleasant property that a function $\mu(x)$ can always be found such that the multiple of (2),

$$\mu(x)\, dy + \mu(x)[P(x)y - f(x)]\, dx = 0, \tag{3}$$

is an exact differential equation. By Theorem 2.2 we know that the left side of equation (3) is an exact differential if

$$\frac{\partial}{\partial x} \mu(x) = \frac{\partial}{\partial y} \mu(x)[P(x)y - f(x)] \tag{4}$$

or
$$\frac{d\mu}{dx} = \mu P(x).$$

This is a separable equation from which we can determine $\mu(x)$. We have

$$\frac{d\mu}{\mu} = P(x)\, dx$$

$$\ln|\mu| = \int P(x)\, dx \tag{5}$$

so that
$$\mu(x) = e^{\int P(x)\, dx}. \tag{6}$$

The function $\mu(x)$ defined in (6) is an **integrating factor** for the linear equation. Note that we need not use a constant of integration in (5) since (3) is unaffected by a constant multiple. Also, $\mu(x) \neq 0$ for every x in I and is continuous and differentiable.

It is interesting to observe that equation (3) is still an exact differential equation even when $f(x) = 0$. In fact, $f(x)$ plays no part in determining $\mu(x)$ since we see from (4) that $(\partial/\partial y)\mu(x)f(x) = 0$. Thus both

$$e^{\int P(x)\, dx}\, dy + e^{\int P(x)\, dx}[P(x)y - f(x)]\, dx$$

and
$$e^{\int P(x)\, dx}\, dy + e^{\int P(x)\, dx}P(x)y\, dx$$

are exact differentials. We now write (3) in the form

$$e^{\int P(x)\, dx}\, dy + e^{\int P(x)\, dx}P(x)y\, dx = e^{\int P(x)\, dx}f(x)\, dx$$

and recognize that we can write this equation as

$$d[e^{\int P(x)\, dx}y] = e^{\int P(x)\, dx}f(x)\, dx.$$

Integrating the last equation gives

$$e^{\int P(x)\, dx}y = \int e^{\int P(x)\, dx}f(x)\, dx + c$$

or
$$y = e^{-\int P(x)\, dx}\int e^{\int P(x)\, dx}f(x)\, dx + ce^{-\int P(x)\, dx}. \tag{7}$$

In other words, if (1) has a solution, it must be of form (7). Conversely, it is a straightforward matter to verify that (7) constitutes a one-parameter family of solutions of equation (1).

Summary of the Method

No attempt should be made to memorize the formula given in (7). The procedure should be followed each time, so for convenience we summarize the results.

Solving a Linear First-Order Equation

(i) To solve a linear first-order equation first put it into the form (1); that is, make the coefficient of dy/dx unity.

(ii) Identify $P(x)$ and find the integrating factor

$$e^{\int P(x)\,dx}.$$

(iii) Multiply the equation obtained in step (i) by the integrating factor:

$$e^{\int P(x)\,dx}\frac{dy}{dx} + P(x)e^{\int P(x)\,dx}y = e^{\int P(x)\,dx}f(x).$$

(iv) The left side of the equation in step (iii) is the derivative of the product of the integrating factor and the dependent variable y; that is,

$$\frac{d}{dx}\left[e^{\int P(x)\,dx}y\right] = e^{\int P(x)\,dx}f(x).$$

(v) Integrate both sides of the equation found in step (iv).

EXAMPLE I

Solve
$$x\frac{dy}{dx} - 4y = x^6 e^x.$$

Solution   Write the equation as

$$\frac{dy}{dx} - \frac{4}{x}y = x^5 e^x \tag{8}$$

by dividing by x. Since $P(x) = -4/x$, the integrating factor is

$$e^{-4\int dx/x} = e^{-4\ln|x|} = e^{\ln x^{-4}} = x^{-4}.$$

Here we have used the basic identity $b^{\log_b N} = N,\ N > 0$. Now we multiply (8) by this term

$$x^{-4}\frac{dy}{dx} - 4x^{-5}y = xe^x \tag{9}$$

and obtain
$$\frac{d}{dx}[x^{-4}y] = xe^x.* \tag{10}$$

---

* You should perform the indicated differentiations a few times in order to be convinced that all equations, such as (8), (9), and (10), are formally equivalent.

It follows from integration by parts that

$$x^{-4}y = xe^x - e^x + c$$

or

$$y = x^5 e^x - x^4 e^x + cx^4.$$ ∎

EXAMPLE 2

Solve

$$\frac{dy}{dx} - 3y = 0.$$

Solution   The equation is already in form (1). Hence the integrating factor is

$$e^{\int (-3)\,dx} = e^{-3x}.$$

Therefore

$$e^{-3x}\frac{dy}{dx} - 3e^{-3x}y = 0$$

$$\frac{d}{dx}\left[e^{-3x}y\right] = 0$$

$$e^{-3x}y = c$$

and so

$$y = ce^{3x}.$$ ∎

General Solution

If it is assumed that $P(x)$ and $f(x)$ are continuous on an interval I and x_0 is any point in the interval, then it follows from Theorem 2.1 that there exists only one solution of the initial-value problem

$$\frac{dy}{dx} + P(x)y = f(x), \qquad y(x_0) = y_0. \tag{11}$$

But we saw earlier that (1) possesses a family of solutions and that every solution of the equation on the interval I is of form (7). Thus, obtaining the solution of (11) is a simple matter of finding an appropriate value of c in (7). Consequently, we are justified in calling (7) the **general solution** of the differential equation. In retrospect, you should recall that in several instances we found singular solutions of nonlinear equations. This cannot happen in the case of a linear equation when proper attention is paid to solving the equation over a common interval on which $P(x)$ and $f(x)$ are continuous.

EXAMPLE 3

Find the general solution of

$$(x^2 + 9)\frac{dy}{dx} + xy = 0.$$

Solution We write $\dfrac{dy}{dx} + \dfrac{x}{x^2+9}\, y = 0.$

The function $P(x) = x/(x^2 + 9)$ is continuous on $(-\infty, \infty)$. Now the integrating factor for the equation is

$$e^{\int x\,dx/(x^2+9)} = e^{\frac{1}{2}\int 2x\,dx/(x^2+9)} = e^{\frac{1}{2}\ln(x^2+9)} = \sqrt{x^2+9}$$

so that

$$\sqrt{x^2+9}\,\frac{dy}{dx} + \frac{x}{\sqrt{x^2+9}}\, y = 0$$

$$\frac{d}{dx}\left[\sqrt{x^2+9}\, y\right] = 0$$

$$\sqrt{x^2+9}\, y = c.$$

Hence the general solution on the interval is

$$y = \frac{c}{\sqrt{x^2+9}}.$$ ∎

EXAMPLE 4

Solve the initial-value problem

$$\frac{dy}{dx} + 2xy = x, \qquad y(0) = -3.$$

Solution The functions $P(x) = 2x$ and $f(x) = x$ are continuous on $(-\infty, \infty)$. The integrating factor is

$$e^{2\int x\,dx} = e^{x^2}$$

so that

$$e^{x^2}\frac{dy}{dx} + 2xe^{x^2}y = xe^{x^2}$$

$$\frac{d}{dx}\left[e^{x^2}y\right] = xe^{x^2}$$

$$e^{x^2}y = \int xe^{x^2}\, dx = \frac{1}{2}e^{x^2} + c.$$

Thus the general solution of the differential equation is

$$y = \frac{1}{2} + ce^{-x^2}.$$

The condition $y(0) = -3$ gives $c = -7/2$, and hence the solution of the initial-value problem on the interval is

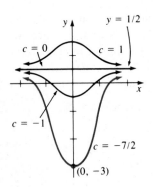

Figure 2.9

$$y = \frac{1}{2} - \frac{7}{2} e^{-x^2}.$$

See Figure 2.9.   ■

EXAMPLE 5

Solve the initial-value problem

$$x \frac{dy}{dx} + y = 2x, \qquad y(1) = 0.$$

Solution   Write the given equation as

$$\frac{dy}{dx} + \frac{1}{x} y = 2$$

and observe that $P(x) = 1/x$ is continuous on any interval not containing the origin. In view of the initial condition, we solve the problem on the interval $(0, \infty)$.

The integrating factor is

$$e^{\int dx/x} = e^{\ln|x|} = x$$

and so

$$\frac{d}{dx} [xy] = 2x$$

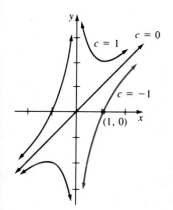

Figure 2.10

gives

$$xy = x^2 + c.$$

The general solution of the equation is

$$y = x + \frac{c}{x}. \tag{12}$$

But $y(1) = 0$ implies $c = -1$. Hence we obtain

$$y = x - \frac{1}{x}, \qquad 0 < x < \infty. \tag{13}$$

Considered as a one-parameter family of curves, the graph of (12) is given in Figure 2.10. The solution (13) of the initial-value problem is indicated by the colored portion of the graph.   ■

EXAMPLE 6

Solve the initial-value problem

$$\frac{dy}{dx} = \frac{1}{x + y^2}, \qquad y(-2) = 0.$$

Solution The given differential equation is not separable, homogeneous, exact, or linear in the variable y. However, if we take the reciprocal, then

$$\frac{dx}{dy} = x + y^2 \quad \text{or} \quad \frac{dx}{dy} - x = y^2.$$

This latter equation is *linear in x*, so the corresponding integrating factor is $e^{-\int dy} = e^{-y}$. Hence,

$$\frac{d}{dy}[e^{-y}x] = y^2 e^{-y}$$

$$e^{-y}x = \int y^2 e^{-y}\, dy.$$

Integrating by parts twice then gives

$$e^{-y}x = -y^2 e^{-y} - 2ye^{-y} - 2e^{-y} + c$$

$$x = -y^2 - 2y - 2 + ce^y.$$

When $x = -2$, $y = 0$, we find $c = 0$ and so

$$x = -y^2 - 2y - 2. \qquad\blacksquare$$

The next example illustrates a way of solving (1) when the function f is discontinuous.

EXAMPLE 7

Find a continuous solution satisfying

$$\frac{dy}{dx} + y = f(x), \qquad \text{where} \quad f(x) = \begin{cases} 1, & 0 \le x \le 1 \\ 0, & x > 1 \end{cases}$$

and the initial condition $y(0) = 0$.

Solution From Figure 2.11 we see that f is discontinuous at $x = 1$. Consequently, we solve the problem in two parts. For $0 \le x \le 1$ we have

$$\frac{dy}{dx} + y = 1$$

$$\frac{d}{dx}[e^x y] = e^x$$

$$y = 1 + c_1 e^{-x}.$$

Since $y(0) = 0$, we must have $c_1 = -1$, and therefore

$$y = 1 - e^{-x}, \qquad 0 \le x \le 1.$$

Figure 2.11

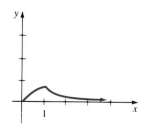

Figure 2.12

For $x > 1$ we then have $\qquad \dfrac{dy}{dx} + y = 0,$

which leads to $\qquad\qquad\qquad y = c_2 e^{-x}.$

Hence we can write $\qquad y = \begin{cases} 1 - e^{-x}, & 0 \le x \le 1 \\ c_2 e^{-x}, & x > 1. \end{cases}$

Now in order that y be a continuous function, we certainly want $\lim_{x \to 1^+} y(x) = y(1)$. This latter requirement is equivalent to $c_2 e^{-1} = 1 - e^{-1}$, or $c_2 = e - 1$. As Figure 2.12 shows, the function

$$y = \begin{cases} 1 - e^{-x}, & 0 \le x \le 1 \\ (e - 1)e^{-x}, & x > 1 \end{cases}$$

is continuous but not differentiable at $x = 1$. ∎

Remark   Formula (7), representing the general solution of (1), actually consists of the sum of two solutions. We define

$$y = y_c + y_p, \tag{14}$$

where

$$y_c = ce^{-\int P(x)\,dx} \quad \text{and} \quad y_p = e^{-\int P(x)\,dx} \int e^{\int P(x)\,dx} f(x)\,dx.$$

The function y_c is readily shown to be the general solution of $y' + P(x)y = 0$, whereas y_p is a particular solution of $y' + P(x)y = f(x)$. As we shall see in Chapter 4, the additivity property of solutions (14) to form a general solution is an intrinsic property of linear equations of any order.

2.5   EXERCISES

Answers to odd-numbered problems begin on page A–4.

In Problems 1–40 find the general solution of the given differential equation. State an interval on which the general solution is defined.

1. $\dfrac{dy}{dx} = 5y$

2. $\dfrac{dy}{dx} + 2y = 0$

3. $3\dfrac{dy}{dx} + 12y = 4$

4. $x\dfrac{dy}{dx} + 2y = 3$

5. $\dfrac{dy}{dx} + y = e^{3x}$

6. $\dfrac{dy}{dx} = y + e^x$

7. $y' + 3x^2 y = x^2$

8. $y' + 2xy = x^3$

9. $x^2 y' + xy = 1$

10. $y' = 2y + x^2 + 5$

11. $(x + 4y^2)\,dy + 2y\,dx = 0$

12. $\dfrac{dx}{dy} = x + y$

13. $x\, dy = (x \sin x - y)\, dx$

14. $(1 + x^2)\, dy + (xy + x^3 + x)\, dx = 0$

15. $(1 + e^x)\dfrac{dy}{dx} + e^x y = 0$

16. $(1 - x^3)\dfrac{dy}{dx} = 3x^2 y$

17. $\cos x\,\dfrac{dy}{dx} + y \sin x = 1$

18. $\dfrac{dy}{dx} + y \cot x = 2 \cos x$

19. $x\dfrac{dy}{dx} + 4y = x^3 - x$

20. $(1 + x)y' - xy = x + x^2$

21. $x^2 y' + x(x + 2)y = e^x$

22. $xy' + (1 + x)y = e^{-x} \sin 2x$

23. $\cos^2 x \sin x\, dy + (y \cos^3 x - 1)\, dx = 0$

24. $(1 - \cos x)\, dy + (2y \sin x - \tan x)\, dx = 0$

25. $y\, dx + (xy + 2x - ye^y)\, dy = 0$

26. $(x^2 + x)\, dy = (x^5 + 3xy + 3y)\, dx$

27. $x\dfrac{dy}{dx} + (3x + 1)y = e^{-3x}$

28. $(x + 1)\dfrac{dy}{dx} + (x + 2)y = 2xe^{-x}$

29. $y\, dx - 4(x + y^6)\, dy = 0$

30. $xy' + 2y = e^x + \ln x$

31. $\dfrac{dy}{dx} + y = \dfrac{1 - e^{-2x}}{e^x + e^{-x}}$

32. $\dfrac{dy}{dx} - y = \sinh x$

33. $y\, dx + (x + 2xy^2 - 2y)\, dy = 0$

34. $y\, dx = (ye^y - 2x)\, dy$

35. $\dfrac{dr}{d\theta} + r \sec \theta = \cos \theta$

36. $\dfrac{dP}{dt} + 2tP = P + 4t - 2$

37. $(x + 2)^2\dfrac{dy}{dx} = 5 - 8y - 4xy$

38. $(x^2 - 1)\dfrac{dy}{dx} + 2y = (x + 1)^2$

39. $y' = (10 - y) \cosh x$

40. $dx = (3e^y - 2x)\, dy$

In Problems 41–54 solve the given differential equation subject to the indicated initial condition.

41. $\dfrac{dy}{dx} + 5y = 20, \quad y(0) = 2$

42. $y' = 2y + x(e^{3x} - e^{2x}), \quad y(0) = 2$

43. $L\dfrac{di}{dt} + Ri = E; \quad L, R, \text{ and } E \text{ constants}, i(0) = i_0$

44. $y\dfrac{dx}{dy} - x = 2y^2, \quad y(1) = 5$

45. $y' + (\tan x)y = \cos^2 x, \quad y(0) = -1$

46. $\dfrac{dQ}{dx} = 5x^4 Q, \quad Q(0) = -7$

47. $\dfrac{dT}{dt} = k(T - 50); \quad k \text{ a constant}, T(0) = 200$

48. $x\,dy + (xy + 2y - 2e^{-x})\,dx = 0, \quad y(1) = 0$

49. $(x + 1)\dfrac{dy}{dx} + y = \ln x, \quad y(1) = 10$

50. $xy' + y = e^x, \quad y(1) = 2$

51. $x(x - 2)y' + 2y = 0, \quad y(3) = 6$

52. $\sin x\,\dfrac{dy}{dx} + (\cos x)y = 0, \quad y\left(-\dfrac{\pi}{2}\right) = 1$

53. $\dfrac{dy}{dx} = \dfrac{y}{y - x}, \quad y(5) = 2$ 　　　　　**54.** $\cos^2 x\,\dfrac{dy}{dx} + y = 1, \quad y(0) = -3$

In Problems 55–58 find a continuous solution satisfying each differential equation and the given initial condition.

55. $\dfrac{dy}{dx} + 2y = f(x), \quad f(x) = \begin{cases} 1, & 0 \le x \le 3 \\ 0, & x > 3 \end{cases}, \quad y(0) = 0$

56. $\dfrac{dy}{dx} + y = f(x), \quad f(x) = \begin{cases} 1, & 0 \le x \le 1 \\ -1, & x > 1 \end{cases}, \quad y(0) = 1$

57. $\dfrac{dy}{dx} + 2xy = f(x), \quad f(x) = \begin{cases} x, & 0 \le x < 1 \\ 0, & x \ge 1 \end{cases}, \quad y(0) = 2$

58. $(1 + x^2)\dfrac{dy}{dx} + 2xy = f(x), \quad f(x) = \begin{cases} x, & 0 \le x < 1 \\ -x, & x \ge 1 \end{cases}, \quad y(0) = 0$

2.6 EQUATIONS OF BERNOULLI, RICATTI, AND CLAIRAUT*

In this section we are not going to study any one particular type of differential equation. Rather, we are going to consider three classical equations that in some instances can be transformed into equations we have already studied.

---

*Jakob Bernoulli (1654–1705)   The Bernoullis were a Swiss family of scholars whose contributions to mathematics, physics, astronomy, and history spanned the sixteenth to the twentieth centuries. Jakob, the elder of the two sons of the patriarch Jacques Bernoulli, made many contributions to the then-new fields of calculus and probability. Originally the second of the two major divisions of calculus was called *calculus summatorius*. In 1696, at Jakob Bernoulli's suggestion, its name was changed to *calculus integralis* or, as we know it today, integral calculus.

Jacobo Francesco Ricatti (1676–1754)   An Italian count, Ricatti was also a mathematician and philosopher.

Alexis Claude Clairaut (1713–1765)   Born in Paris in 1713, Clairaut was a child prodigy who wrote his first book on mathematics at the age of eleven. He was among the first to discover singular solutions of differential equations. Like many mathematicians of his era, Clairaut was also a physicist and astronomer.

Bernoulli's Equation

The differential equation

$$\frac{dy}{dx} + P(x)y = f(x)y^n, \tag{1}$$

where n is any real number, is called **Bernoulli's equation**. For $n = 0$ and $n = 1$, equation (1) is linear in y. Now for $y \neq 0$, (1) can be written as

$$y^{-n}\frac{dy}{dx} + P(x)y^{1-n} = f(x). \tag{2}$$

If we let $w = y^{1-n}$, $n \neq 0$, $n \neq 1$, then

$$\frac{dw}{dx} = (1-n)y^{-n}\frac{dy}{dx}.$$

With these substitutions, (2) can be simplified to the linear equation

$$\frac{dw}{dx} + (1-n)P(x)w = (1-n)f(x). \tag{3}$$

Solving (3) for w and using $y^{1-n} = w$ lead to a solution of (1).

EXAMPLE I

Solve $\dfrac{dy}{dx} + \dfrac{1}{x}y = xy^2.$

Solution   From (1) we identify $P(x) = 1/x$, $f(x) = x$, and $n = 2$. Thus the substitution $w = y^{-1}$ gives

$$\frac{dw}{dx} - \frac{1}{x}w = -x.$$

The integrating factor for this linear equation on, say, $(0, \infty)$ is

$$e^{-\int dx/x} = e^{-\ln|x|} = e^{\ln|x|^{-1}} = x^{-1}.$$

Hence $\dfrac{d}{dx}[x^{-1}w] = -1.$

Integrating this latter form, we get

$$x^{-1}w = -x + c \quad \text{or} \quad w = -x^2 + cx.$$

Since $w = y^{-1}$, we obtain $y = 1/w$ or

$$y = \frac{1}{-x^2 + cx}. \qquad \blacksquare$$

For $n > 0$ note that the trivial solution $y = 0$ is a solution of (1). In Example 1, $y = 0$ is a singular solution of the given equation.

Ricatti's Equation

The nonlinear differential equation

$$\frac{dy}{dx} = P(x) + Q(x)y + R(x)y^2 \qquad (4)$$

is called **Ricattti's equation**. If y_1 is a *known* particular solution of (4), then the substitutions

$$y = y_1 + u \quad \text{and} \quad \frac{dy}{dx} = \frac{dy_1}{dx} + \frac{du}{dx}$$

in (4) lead to the following differential equation for u:

$$\frac{du}{dx} - (Q + 2y_1 R)u = Ru^2. \qquad (5)$$

Since (5) is a Bernoulli equation with $n = 2$, it can in turn be reduced to the linear equation

$$\frac{dw}{dx} + (Q + 2y_1 R)w = -R \qquad (6)$$

by the substitution $w = u^{-1}$. See Problems 25 and 26.

   As Example 2 illustrates, in many cases a solution of a Ricatti equation cannot be expressed in terms of elementary functions.

E X A M P L E **2**

Solve $\qquad\qquad \dfrac{dy}{dx} = 2 - 2xy + y^2.$

Solution   It is easily verified that a particular solution of this equation is $y_1 = 2x$. From (4) we make the identifications $P(x) = 2$, $Q(x) = -2x$, and $R(x) = 1$ and then solve the linear equation (6):

$$\frac{dw}{dx} + (-2x + 4x)w = -1 \quad \text{or} \quad \frac{dw}{dx} + 2xw = -1.$$

The integrating factor for the last equation is e^{x^2}, and so

$$\frac{d}{dx}[e^{x^2}w] = -e^{x^2}.$$

Now the integral $\int_{x_o}^{x} e^{t^2}\, dt$ cannot be expressed in terms of elementary functions.* Thus we write

---

*When an integral $\int f(x)\, dx$ cannot be evaluated in terms of elementary functions, it is customary to write $\int_{x_0}^{x} f(t)\, dt$, where x_0 is a constant. When an initial condition is specified, it is imperative that this form be used.

$$e^{x^2}w = -\int_{x_0}^{x} e^{t^2} \, dt + c \quad \text{or} \quad e^{x^2}\left(\frac{1}{u}\right) = -\int_{x_0}^{x} e^{t^2} \, dt + c,$$

so that

$$u = \frac{e^{x^2}}{c - \int_{x_0}^{x} e^{t^2} \, dt}.$$

A solution of the equation is then $y = 2x + u$. ∎

Clairaut's Equation

It is left as an exercise to show that a solution of **Clairaut's equation**

$$y = xy' + f(y') \tag{7}$$

is the family of straight lines $y = cx + f(c)$, where c is an arbitrary constant. See Problem 30. Furthermore (7) may also possess a solution in parametric form:

$$x = -f'(t), \qquad y = f(t) - tf'(t). \tag{8}$$

This last solution is a singular solution since, if $f''(t) \neq 0$, it cannot be obtained from the family of solutions $y = cx + f(c)$.

Figure 2.13

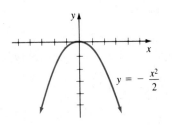

Figure 2.14

EXAMPLE 3

Solve

$$y = xy' + \frac{1}{2}(y')^2.$$

Solution   We first make the identification $f(y') = (1/2)(y')^2$ so that $f(t) = (1/2)t^2$. It follows from the preceding discussion that a family of solutions is

$$y = cx + \frac{1}{2}c^2. \tag{9}$$

The graph of this family is given in Figure 2.13. Since $f'(t) = t$, a singular solution is obtained from (8):

$$x = -t, \qquad y = \frac{1}{2}t^2 - t \cdot t = -\frac{1}{2}t^2.$$

After eliminating the parameter, we find this latter solution is the same as

$$y = -\frac{1}{2}x^2.$$

One can readily see that this function is not part of the family (9). See Figure 2.14. ∎

2.6 EXERCISES

Answers to odd-numbered problems begin on page A–4.

In Problems 1–6 solve the given Bernoulli equation.

1. $x\dfrac{dy}{dx} + y = \dfrac{1}{y^2}$

2. $\dfrac{dy}{dx} - y = e^x y^2$

3. $\dfrac{dy}{dx} = y(xy^3 - 1)$

4. $x\dfrac{dy}{dx} - (1 + x)y = xy^2$

5. $x^2\dfrac{dy}{dx} + y^2 = xy$

6. $3(1 + x^2)\dfrac{dy}{dx} = 2xy(y^3 - 1)$

In Problems 7–10 solve the given differential equation subject to the indicated initial condition.

7. $x^2\dfrac{dy}{dx} - 2xy = 3y^4, \quad y(1) = \dfrac{1}{2}$

8. $y^{1/2}\dfrac{dy}{dx} + y^{3/2} = 1, \quad y(0) = 4$

9. $xy(1 + xy^2)\dfrac{dy}{dx} = 1, \quad y(1) = 0$

10. $2\dfrac{dy}{dx} = \dfrac{y}{x} - \dfrac{x}{y^2}, \quad y(1) = 1$

In Problems 11–16 solve the given Ricatti equation; y_1 is a known solution of the equation.

11. $\dfrac{dy}{dx} = -2 - y + y^2, \quad y_1 = 2$

12. $\dfrac{dy}{dx} = 1 - x - y + xy^2, \quad y_1 = 1$

13. $\dfrac{dy}{dx} = -\dfrac{4}{x^2} - \dfrac{1}{x}y + y^2, \quad y_1 = \dfrac{2}{x}$

14. $\dfrac{dy}{dx} = 2x^2 + \dfrac{1}{x}y - 2y^2, \quad y_1 = x$

15. $\dfrac{dy}{dx} = e^{2x} + (1 + 2e^x)y + y^2, \quad y_1 = -e^x$

16. $\dfrac{dy}{dx} = \sec^2 x - (\tan x)y + y^2, \quad y_1 = \tan x$

17. Solve $\dfrac{dy}{dx} = 6 + 5y + y^2$.

18. Solve $\dfrac{dy}{dx} = 9 + 6y + y^2$.

In Problems 19–24 solve the given Clairaut equation. Obtain a singular solution.

19. $y = xy' + 1 - \ln y'$

20. $y = xy' + (y')^{-2}$

21. $y = x\dfrac{dy}{dx} - \left(\dfrac{dy}{dx}\right)^3$

22. $y = (x + 4)y' + (y')^2$

23. $xy' - y = e^{y'}$

24. $y - xy' = \ln y'$

25. Show that if y_1 is a solution of (4), then the substitution $y = y_1 + u$ in (4) yields (5).

26. Show that (5) can be reduced to (6) by means of the substitution $w = u^{-1}$.

27. When $R(x) = -1$, the Ricatti equation can be written as $y' + y^2 - Q(x)y - P(x) = 0$. Show that the substitution $y = w'/w$ leads to the linear second-order equation $w'' - Q(x)w' - P(x)w = 0$. (When Q and P are also constants, there is little difficulty in solving equations of this type.)

28. An alternative definition of Clairaut's equation is any equation of the form $F(y - xy', y') = 0$.

(a) Show that a family of solutions of the latter equation is

$$F(y - cx, c) = 0.$$

(b) Use the result of part (a) to solve

$$(xy' - y)^3 = (y')^2 + 5.$$

29. Show that $y = cx + f(c)$, where c is an arbitrary constant, is a solution of (7).

30. Show that (8) is a solution of (7). [*Hint:* Differentiate both sides of (7) with respect to x and consider two cases. Use parametric differentiation to show

$$\frac{dy}{dx} = \frac{dy/dt}{dx/dt} = t, \qquad f''(t) \neq 0.$$

Note that since the slope of $y = cx + f(c)$ is constant, the singular solution cannot be obtained from it.]

[O] 2.7 SUBSTITUTIONS

In the preceding sections, we saw that sometimes a differential equation could be transformed by means of a substitution into a form that could then be solved by one of the standard methods. An equation may look different from any of those that we have just studied, but through a judicious change of variables perhaps an apparently difficult problem may be readily solved. Although we can give no firm rules on what, *if any*, substitution to use, a working axiom might be: Try something! It sometimes pays to be clever.

EXAMPLE I

The differential equation

$$y(1 + 2xy)\,dx + x(1 - 2xy)\,dy = 0$$

is not separable, homogeneous, exact, linear, or Bernoulli. However, if we stare at the equation long enough, we might be prompted to try the substitution

$$u = 2xy \quad \text{or} \quad y = \frac{u}{2x}.$$

Since

$$dy = \frac{x \, du - u \, dx}{2x^2},$$

the equation becomes, after we simplify,

$$2u^2 \, dx + (1 - u)x \, du = 0.$$

We recognize the last equation as separable, and so from

$$2\frac{dx}{x} + \frac{1 - u}{u^2} \, du = 0$$

we obtain

$$2 \ln|x| - u^{-1} - \ln|u| = c$$

$$\ln\left|\frac{x}{2y}\right| = c + \frac{1}{2xy}$$

$$\frac{x}{2y} = c_1 e^{1/2xy}$$

$$x = 2c_1 y e^{1/2xy},$$

where e^c was replaced by c_1. We can also replace $2c_1$ by c_2 if desired. ∎

Notice that the differential equation in Example 1 possesses the trivial solution $y = 0$ but that this function is not included in the one-parameter family of solutions.

EXAMPLE 2

Solve

$$2xy \frac{dy}{dx} + 2y^2 = 3x - 6.$$

Solution The presence of the term $2y\dfrac{dy}{dx}$ prompts us to try $u = y^2$ since

$$\frac{du}{dx} = 2y \frac{dy}{dx}.$$

Now

$$x\frac{du}{dx} + 2u = 3x - 6$$

has the linear form

$$\frac{du}{dx} + \frac{2}{x}u = 3 - \frac{6}{x},$$

so that multiplication by the integrating factor $e^{\int (2/x)\,dx} = e^{\ln x^2} = x^2$ gives

$$\frac{d}{dx}[x^2 u] = 3x^2 - 6x$$

$$x^2 u = x^3 - 3x^2 + c \quad \text{or} \quad x^2 y^2 = x^3 - 3x^2 + c. \qquad \blacksquare$$

EXAMPLE 3

Solve
$$x\frac{dy}{dx} - y = \frac{x^3}{y}e^{y/x}.$$

Solution   If we let
$$u = \frac{y}{x},$$

the differential equation can be simplified to

$$ue^{-u}\,du = dx.$$

Integrating by parts and replacing $-c$ by c_1 then yield

$$-ue^{-u} - e^{-u} = x + c$$

$$u + 1 = (c_1 - x)e^u.$$

We then resubstitute $u = y/x$ and simplify:

$$y + x = x(c_1 - x)e^{y/x}. \qquad \blacksquare$$

Some higher-order differential equations can be reduced to first-order equations by a substitution.

EXAMPLE 4

Solve
$$y'' = 2x(y')^2.$$

Solution   If we let $u = y'$ so that $du/dx = y''$, the equation reduces to a separable form. We have

$$\frac{du}{dx} = 2xu^2 \quad \text{or} \quad \frac{du}{u^2} = 2x\,dx$$

$$\int u^{-2}\,du = \int 2x\,dx$$

$$-u^{-1} = x^2 + c_1^2.$$

The constant of integration is written as c_1^2 for convenience. The reason should be obvious in the next few steps. Since $u^{-1} = 1/y'$, it follows that

$$\frac{dy}{dx} = -\frac{1}{x^2 + c_1^2} \quad \text{or} \quad dy = -\frac{dx}{x^2 + c_1^2}$$

$$\int dy = -\int \frac{dx}{x^2 + c_1^2}$$

$$y + c_2 = -\frac{1}{c_1} \tan^{-1} \frac{x}{c_1}. \qquad \blacksquare$$

2.7 EXERCISES

Answers to odd-numbered problems begin on page A–4.

In Problems 1–26 solve the given differential equation by using an appropriate substitution.

1. $xe^{2y} \dfrac{dy}{dx} + e^{2y} = \dfrac{\ln x}{x}$ **2.** $y' + y \ln y = ye^x$

3. $y \, dx + (1 + ye^x) \, dy = 0$

4. $(2 + e^{-x/y}) \, dx + 2\left(1 - \dfrac{x}{y}\right) dy = 0$

5. $\dfrac{dy}{dx} - \dfrac{4}{x} y = 2x^5 e^{y/x^4}$ **6.** $\dfrac{dy}{dx} + x + y + 1 = (x + y)^2 e^{3x}$

7. $2yy' + x^2 + y^2 + x = 0$ **8.** $y' = y + x(y + 1)^2 + 1$

9. $2x \csc 2y \dfrac{dy}{dx} = 2x - \ln(\tan y)$ **10.** $x^2 \dfrac{dy}{dx} + 2xy = x^4 y^2 + 1$

11. $x^4 y^2 y' + x^3 y^3 = 2x^3 - 3$ **12.** $xe^y y' - 2e^y = x^2$

13. $y' + 1 = e^{-(x+y)} \sin x$

14. $\sin y \sinh x \, dx + \cos y \cosh x \, dy = 0$

15. $y \dfrac{dx}{dy} + 2x \ln x = xe^y$ **16.** $x \sin y \dfrac{dy}{dx} + \cos y = -x^2 e^x$

17. $y'' + (y')^2 + 1 = 0$ **18.** $xy'' = y' + x(y')^2$

19. $xy'' = y' + (y')^3$ **20.** $x^2 y'' + (y')^2 = 0$

21. $y' - xy'' - (y'')^3 = 1$ **22.** $y'' = 1 + (y')^2$

23. $xy'' - y' = 0$ **24.** $y'' + (\tan x)y' = 0$

25. $y'' + 2y(y')^3 = 0$ **26.** $y^2 y'' = y'$

$$\left[\text{Hint: Let } u = y' \text{ so that } y'' = \frac{du}{dx} = \frac{du}{dy}\frac{dy}{dx} = \frac{du}{dy}u.\right]$$

27. In calculus the curvature of a curve whose equation is $y = f(x)$ is defined to be the number

$$\kappa = \frac{y''}{[1 + (y')^2]^{3/2}}.$$

Determine a function for which $\kappa = 1$. [*Hint:* For simplicity ignore constants of integration. Also consider a trigonometric substitution.]

[0] 2.8 PICARD'S METHOD

The initial-value problem

$$y' = f(x, y), \qquad y(x_0) = y_0, \tag{1}$$

first considered in Section 2.1 can be written in an alternative manner. Let f be continuous in a region containing the point (x_0, y_0). By integrating both sides of the differential equation with respect to x, we get

$$y(x) = c + \int_{x_0}^{x} f(t, y(t))\, dt.$$

Now

$$y(x_0) = c + \int_{x_0}^{x_0} f(t, y(t))\, dt = c$$

implies $c = y_0$. Thus

$$y(x) = y_0 + \int_{x_0}^{x} f(t, y(t))\, dt. \tag{2}$$

Conversely, if we start with (2), we can obtain (1). In other words, the integral equation (2) and the initial-value problem (1) are equivalent. We now try to solve (2) by a *method of successive approximations.*

Suppose $y_0(x)$ is an arbitrary continuous function that represents a guess or approximation to the solution of (2). Since $f(x, y_0(x))$ is a known function depending solely on x, it can be integrated. With $y(t)$ replaced by $y_0(t)$, the right-hand side of (2) defines another function, which we write as

$$y_1(x) = y_0 + \int_{x_0}^{x} f(t, y_0(t))\, dt.$$

It is hoped that this new function is a better approximation to the solution. When we repeat the procedure, yet another function is given by

$$y_2(x) = y_0 + \int_{x_0}^{x} f(t, y_1(t))\, dt.$$

In this manner we obtain a sequence of functions $y_1(x)$, $y_2(x)$, $y_3(x)$, ... whose nth term is defined by the relation

$$y_n(x) = y_0 + \int_{x_0}^{x} f(t, y_{n-1}(t))\, dt, \qquad n = 1, 2, 3, \ldots. \tag{3}$$

In the application of (3), it is common practice to choose the initial function as $y_0(x) = y_0$. The repetitive use of formula (3) is known as **Picard's method of iteration**.

EXAMPLE 1

Consider the problem

$$y' = y - 1, \qquad y(0) = 2.$$

Use Picard's method to find the approximations y_1, y_2, y_3, y_4.

Solution If we identify $x_0 = 0$, $y_0(x) = 2$, and $f(t, y_{n-1}(t)) = y_{n-1}(t) - 1$, equation (3) becomes

$$y_n(x) = 2 + \int_0^x (y_{n-1}(t) - 1)\, dt, \qquad n = 1, 2, 3, \ldots.$$

Iterating this last expression then gives

$$y_1(x) = 2 + \int_0^x 1 \cdot dt = 2 + x$$

$$y_2(x) = 2 + \int_0^x (1 + t)\, dt = 2 + x + \frac{x^2}{2}$$

$$y_3(x) = 2 + \int_0^x \left(1 + t + \frac{t^2}{2}\right) dt$$

$$= 2 + x + \frac{x^2}{2} + \frac{x^3}{2 \cdot 3}$$

$$y_4(x) = 2 + \int_0^x \left(1 + t + \frac{t^2}{2} + \frac{t^3}{2 \cdot 3}\right) dt$$

$$= 2 + x + \frac{x^2}{2} + \frac{x^3}{2 \cdot 3} + \frac{x^4}{2 \cdot 3 \cdot 4}. \qquad \blacksquare$$

By induction it can be shown in Example 1 that the nth term of the sequence of approximations is

$$y_n(x) = 2 + x + \frac{x^2}{2!} + \frac{x^3}{3!} + \cdots + \frac{x^n}{n!} = 1 + \sum_{k=0}^{n} \frac{x^k}{k!}.$$

From this latter form we recognize that the *limit* of $y_n(x)$ as $n \to \infty$ is $y(x) = 1 + e^x$. It should come as no surprise to note that the function is an exact solution of the given initial-value problem.

You should not be deceived by the relative ease with which the iterates $y_n(x)$ were obtained in the last example. In general, the integration involved in generating each $y_n(x)$ can become complicated very quickly. Nor, for that matter, is it always apparent that the sequence $\{y_n(x)\}$ converges to a nice explicit function. Thus it is fair to ask at this point: Is Picard's method a practical means of solving a first-order equation $y' = f(x, y)$ subject to $y(x_0) = y_0$? In most cases the answer is no. One might ask further, in the spirit of a scientist/engineer: What *is* it good for? The answer is not bound to please: Picard's method of iteration is a theoretical tool used in the consideration of the existence and uniqueness of solutions of differential equations. Under certain conditions on $f(x, y)$ it can be shown that as $n \to \infty$, the sequence $\{y_n(x)\}$ defined by (3) converges to a function $y(x)$ that satisfies the integral equation (2) and hence the initial-value problem (1). Indeed, it is precisely Picard's method of successive approximations that is used in proving Picard's theorem of Section 2.1. However, the proof of Theorem 2.1 uses concepts from advanced calculus and is not presented here. Our purpose in introducing this topic is twofold: to gain an appreciation for the potential of the procedure and to obtain at least a nodding acquaintance with an iterative technique. In Chapter 9 we shall consider other methods for approximating solutions of differential equations that also utilize iteration.

2.8 EXERCISES

Answers to odd-numbered problems begin on page A–5.

In Problems 1–6 use Picard's method to find y_1, y_2, y_3, y_4. Determine the limit of the sequence $\{y_n(x)\}$ as $n \to \infty$.

1. $y' = -y$, $y(0) = 1$ **2.** $y' = x + y$, $y(0) = 1$

3. $y' = 2xy$, $y(0) = 1$ **4.** $y' + 2xy = x$, $y(0) = 0$

5. $y' + y^2 = 0$, $y(0) = 0$ **6.** $y' = 2e^x - y$, $y(0) = 1$

7. (a) Use Picard's method to find y_1, y_2, y_3 for the problem

$$y' = 1 + y^2, \qquad y(0) = 0.$$

(b) Solve the initial-value problem in part (a) by one of the methods of this chapter.

(c) Compare the results of parts (a) and (b).

8. In Picard's method the initial choice $y_0(x) = y_0$ is not necessary. Rework Problem 3 with **(a)** $y_0(x) = k$ a constant and $k \neq 1$, and **(b)** $y_0(x) = x$.

Chapter 2 *REVIEW*

An **initial-value problem** consists of finding a solution of

$$\frac{dy}{dx} = f(x, y), \qquad y(x_0) = y_0,$$

on an interval I containing x_0. If $f(x, y)$ and $\partial f/\partial y$ are continuous in a rectangular region of the xy-plane with (x_0, y_0) in its interior, then we are guaranteed that there exists an interval around x_0 on which the problem has a unique solution.

The method of solution for a first-order differential equation depends on an appropriate classification of the equation. We summarize five cases.

An equation is **separable** if it can be put into the form $h(y) \, dy = g(x) \, dx$. The solution results from integrating both sides of the equation.

If $M(x, y)$ and $N(x, y)$ are **homogeneous functions** of the same degree, then $M(x, y) \, dx + N(x, y) \, dy = 0$ can be reduced to an equation with separable variables by either the substitution $y = ux$ or $x = vy$. The choice of substitution usually depends on which coefficient is simpler.

The differential equation $M(x, y) \, dx + N(x, y) \, dy = 0$ is said to be **exact** if the form $M(x, y) \, dx + N(x, y) \, dy$ is an exact differential. When $M(x, y)$ and $N(x, y)$ are continuous and have continuous first partial derivatives, then $\partial M/\partial y = \partial N/\partial x$ is a necessary and sufficient condition that $M(x, y) \, dx + N(x, y) \, dy$ be exact. This means there exists some function $f(x, y)$ for which $M(x, y) = \partial f/\partial x$ and $N(x, y) = \partial f/\partial y$. The method of solution for an exact equation starts by integrating either of these latter expressions.

If a first-order equation can be put into the form $dy/dx + P(x)y = f(x)$, it is said to be **linear** in the variable y. We solve the equation by first finding the **integrating factor**, $e^{\int P(x) \, dx}$, multiplying both sides of the equation by this factor, and then integrating both sides of

$$\frac{d}{dx} \left[e^{\int P(x) \, dx} y \right] = e^{\int P(x) \, dx} f(x).$$

Bernoulli's equation is $dy/dx + P(x)y = f(x)y^n$, where n is any real number. When $n \neq 0$ and $n \neq 1$, Bernoulli's equation can be reduced to a linear equation by the substitution $w = y^{1-n}$.

In certain circumstances a differential equation can be reduced to one of the familiar forms by an appropriate **substitution**, or **change**, **of variables**. Of course we already know that this is the procedure when solving a homogeneous or a Bernoulli equation. In the general context, no rule on when to use a substitution can be given.

By converting an initial-value problem to an equivalent integral equation, **Picard's method of iteration** is one way of obtaining an approximation to the solution of the problem.

Chapter 2   *REVIEW EXERCISES*

Answers to odd-numbered problems begin on page A–5.

Answer Problems 1–4 without referring back to the text. Fill in the blank or answer true/false.

1. The differential equation $y' = 1/(25 - x^2 - y^2)$ has a unique solution through any point (x_0, y_0) in the region(s) defined by _____.

2. The initial-value problem $xy' = 3y$, $y(0) = 0$, has the solutions $y = x^3$ and _____.

3. The initial-value problem $y' = y^{1/2}$, $y(0) = 0$, has no solution since $\partial f/\partial y$ is discontinuous on the line $y = 0$. _____

4. There exists an interval centered at 2 on which the unique solution of the initial-value problem $y' = (y - 1)^3$, $y(2) = 1$, is $y = 1$. _____

5. Without solving, classify each of the following equations as to: separable, homogeneous, exact, linear, Bernoulli, Ricatti, or Clairaut.

   (a) $\dfrac{dy}{dx} = \dfrac{1}{y - x}$

   (b) $\dfrac{dy}{dx} = \dfrac{x - y}{x}$

   (c) $\left(\dfrac{dy}{dx}\right)^2 + 2y = 2x\dfrac{dy}{dx}$

   (d) $\dfrac{dy}{dx} = \dfrac{1}{x(x - y)}$

   (e) $\dfrac{dy}{dx} = \dfrac{y^2 + y}{x^2 + x}$

   (f) $\dfrac{dy}{dx} = 4 + 5y + y^2$

   (g) $y\,dx = (y - xy^2)\,dy$

   (h) $x\dfrac{dy}{dx} = ye^{x/y} - x$

   (i) $xyy' + y^2 = 2x$

   (j) $2xyy' + y^2 = 2x^2$

   (k) $y\,dx + x\,dy = 0$

   (l) $\left(x^2 + \dfrac{2y}{x}\right)dx = (3 - \ln x^2)\,dy$

   (m) $\dfrac{dy}{dx} = \dfrac{x}{y} + \dfrac{y}{x} + 1$

   (n) $\dfrac{y}{x^2}\dfrac{dy}{dx} + e^{2x^3 + y^2} = 0$

   (o) $y = xy' + (y' - 3)^2$

   (p) $y' + 5y^2 = 3x^4 - 2xy$

6. Solve $(y^2 + 1)\,dx = y\sec^2x\,dy$.

7. Solve $\dfrac{y}{x}\dfrac{dy}{dx} = \dfrac{e^x}{\ln y}$ subject to $y(1) = 1$.

8. Solve $y(\ln x - \ln y)\,dx = (x\ln x - x\ln y - y)\,dy$.

9. Solve $xyy' = 3y^2 + x^2$ subject to $y(-1) = 2$.

10. Solve $(6x + 1)y^2\dfrac{dy}{dx} + 3x^2 + 2y^3 = 0$.

11. Solve $ye^{xy} \dfrac{dx}{dy} + xe^{xy} = 12y^2$ subject to $y(0) = -1$.

12. Solve $x\, dy + (xy + y - x^2 - 2x)\, dx = 0$.

13. Solve $(x^2 + 4) \dfrac{dy}{dx} = 2x - 8xy$ subject to $y(0) = -1$.

14. Solve $(2x + y)y' = 1$.

15. Solve $x \dfrac{dy}{dx} + 4y = x^4 y^2$ subject to $y(1) = 1$.

16. Solve $-xy' + y = (y' + 1)^2$ subject to $y(0) = 0$.

In Problems 17 and 18 solve the given differential equation by means of a substitution.

17. $\dfrac{dy}{dx} + xy^3 \sec \dfrac{1}{y^2} = 0$ **18.** $y'' = x - y'$

19. Use the Picard method to find approximations y_1 and y_2 for
$$y' = x^2 + y^2, \qquad y(0) = 1.$$

20. Solve $y' + 2y = 4$, $y(0) = 3$, by one of the usual methods. Solve the same problem by Picard's method and compare the results.

APPLICATIONS OF FIRST-ORDER DIFFERENTIAL EQUATIONS

3.1 Orthogonal Trajectories
3.2 Applications of Linear Equations

3.3 Applications of Nonlinear Equations

Chapter 3 Review
Chapter 3 Review Exercises
Essay: Population Dynamics

Important Concepts

Orthogonal trajectories
Exponential growth and decay
Half-life
Carbon dating
Response
Transient term
Steady-state term
Logistic equation
Chemical reactions
Escape velocity

In Section 1.2 we saw that a differential equation used to describe the behavior of some real-life system, whether physical, sociological, or even economic, is called a **mathematical model**. Mathematical models for phenomena, such as radioactive decay, population growth, chemical reactions, cooling of bodies, velocity of a falling body, rate of memorization, or current in a series circuit, are often first-order differential equations.

In this chapter we are concerned with solving some of the more commonly occurring linear and nonlinear first-order differential equations that arise in applications.

3.1 ORTHOGONAL TRAJECTORIES

Differential Equation of a Family of Curves

At the end of Section 1.1 we expressed the expectation or, better, the hope that an nth-order ordinary differential will yield an n-parameter family of solutions. On the other hand, suppose we turn the problem around: Starting with an n-parameter family of curves, can we find an associated nth-order differential equation that is entirely free of arbitrary parameters and represents the given family? In most cases the answer is yes.

In the discussion that follows we are interested in finding the differential equation $dy/dx = f(x, y)$ of a one-parameter family of curves.

EXAMPLE 1

Find the differential equation of the family

$$y = c_1 x^3.$$

Solution   Differentiation gives

$$\frac{dy}{dx} = 3c_1 x^2.$$

We can eliminate the parameter c_1 from the equation by using $c_1 = y/x^3$ obtained from the first equation:

$$\frac{dy}{dx} = 3\left(\frac{y}{x^3}\right)x^2 \quad \text{or} \quad \frac{dy}{dx} = 3\,\frac{y}{x}. \qquad \blacksquare$$

Orthogonal Curves

Recall from your study of analytic geometry that two lines L_1 and L_2, which are not parallel to the coordinate axes, are perpendicular if and only if their respective slopes satisfy the relationship $m_1 m_2 = -1$. For this reason, the graphs of $y = (-1/2)x + 1$ and $y = 2x + 4$ are obviously perpendicular. In general, two curves $\mathscr{C}_1$ and $\mathscr{C}_2$ are said to be **orthogonal** at a point if and only if their tangent lines T_1 and T_2 are perpendicular at the point of intersection. See Figure 3.1. Except for the case when T_1 and T_2 are parallel to the coordinate axes, this means the slopes of the tangents are negative reciprocals of each other.

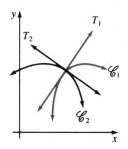

Figure 3.1

EXAMPLE 2

Show that the curves $\mathscr{C}_1$ and $\mathscr{C}_2$ defined by $y = x^3$ and $x^2 + 3y^2 = 4$ are orthogonal at their point(s) of intersection.

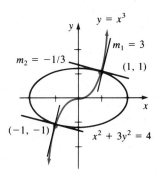

Figure 3.2

Solution In Figure 3.2 it is seen that the points of intersection of the graphs are $(1, 1)$ and $(-1, -1)$. Now the slope of the tangent line to $y = x^3$ at any point is $dy/dx = 3x^2$, so that

$$\frac{dy}{dx}\bigg|_{x=1} = \frac{dy}{dx}\bigg|_{x=-1} = 3.$$

We use implicit differentiation to obtain dy/dx for the second curve:

$$2x + 6y\frac{dy}{dx} = 0 \quad \text{or} \quad \frac{dy}{dx} = -\frac{x}{3y}$$

and therefore

$$\frac{dy}{dx}\bigg|_{(1,1)} = \frac{dy}{dx}\bigg|_{(-1,-1)} = -\frac{1}{3}.$$

Thus, at either $(1, 1)$ or $(-1, -1)$ we have

$$\left(\frac{dy}{dx}\right)_{\mathscr{C}_1} \cdot \left(\frac{dy}{dx}\right)_{\mathscr{C}_2} = -1. \qquad \blacksquare$$

It is easy to show that any curve $\mathscr{C}_1$ in the family $y = c_1 x^3$, $c_1 \neq 0$, is orthogonal to each curve $\mathscr{C}_2$ in the family $x^2 + 3y^2 = c_2$, $c_2 > 0$. From Example 1 we know that the differential equation of the first family is

$$\frac{dy}{dx} = 3\frac{y}{x}.$$

Implicit differentiation of $x^2 + 3y^2 = c_2$ leads to exactly the same differential equation as for $x^2 + 3y^2 = 4$ in Example 2, namely,

$$\frac{dy}{dx} = -\frac{x}{3y}.$$

Hence, at the point (x, y) on each curve,

$$\left(\frac{dy}{dx}\right)_{\mathscr{C}_1} \cdot \left(\frac{dy}{dx}\right)_{\mathscr{C}_2} = \left(\frac{3y}{x}\right)\left(-\frac{x}{3y}\right) = -1.$$

Since the slopes of the tangent lines are negative reciprocals, the curves $\mathscr{C}_1$ and $\mathscr{C}_2$ intersect each other in an orthogonal manner.

This discussion leads to the following definition.

DEFINITION 3.1   Orthogonal Trajectories

When *all* the curves of one family of curves $G(x, y, c_1) = 0$ intersect orthogonally *all* the curves of another family $H(x, y, c_2) = 0$, then the families are said to be **orthogonal trajectories** of each other.

In other words, an orthogonal trajectory is any *one* curve that intersects every curve of another family at right angles.

Figure 3.3

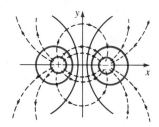

Figure 3.4

EXAMPLE 3

(a) The graph of $y = (-1/2)x + 1$ is an orthogonal trajectory of $y = 2x + c_1$. The families $y = (-1/2)x + c_2$ and $y = 2x + c_1$ are orthogonal trajectories.

(b) The graph of $y = 4x^3$ is an orthogonal trajectory of $x^2 + 3y^2 = c_2$. The families $y = c_1 x^3$ and $x^2 + 3y^2 = c_2$ are orthogonal trajectories.

(c) In Figure 3.3 it is seen that the family of straight lines $y = c_1 x$ through the origin and the family $x^2 + y^2 = c_2$ of concentric circles with center at the origin are orthogonal trajectories. ∎

Orthogonal trajectories occur naturally in the construction of meteorological maps and in the study of electricity and magnetism. For example, in an electric field around two bodies of opposite charge, the lines of force are perpendicular to the equipotential curves (that is, curves along which the potential is constant). The lines of force are indicated in Figure 3.4 by dashed lines.

General Method

To find the orthogonal trajectories of a given family of curves we first find the differential equation

$$\frac{dy}{dx} = f(x, y)$$

that describes the family. The differential equation of the second, and orthogonal, family is then

$$\frac{dy}{dx} = \frac{-1}{f(x, y)}.$$

EXAMPLE 4

Find the orthogonal trajectories of the family of rectangular hyperbolas

$$y = \frac{c_1}{x}.$$

Solution   The derivative of $y = c_1/x$ is

$$\frac{dy}{dx} = \frac{-c_1}{x^2}.$$

Replacing c_1 by $c_1 = xy$ yields the differential equation of the given family:

$$\frac{dy}{dx} = -\frac{y}{x}.$$

The differential equation of the orthogonal family is then

$$\frac{dy}{dx} = \frac{-1}{(-y/x)} = \frac{x}{y}.$$

We solve this last equation by separation of variables:

$$y\,dy = x\,dx$$

$$\int y\,dy = \int x\,dx$$

$$\frac{y^2}{2} = \frac{x^2}{2} + c_2' \quad \text{or} \quad y^2 - x^2 = c_2,$$

where we replaced $2c_2'$ by c_2. We recognize that this solution represents another family of hyperbolas. The graphs of both families, for various values of c_1 and c_2, are given in Figure 3.5. ∎

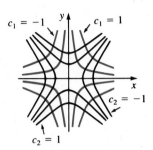

$c_1 = -1$ $c_1 = 1$
$c_2 = -1$
$c_2 = 1$

Figure 3.5

E X A M P L E **5**

Find the orthogonal trajectories of

$$y = \frac{c_1 x}{1 + x}.$$

Solution From the quotient rule we find

$$\frac{dy}{dx} = \frac{c_1}{(1 + x)^2} \quad \text{or} \quad \frac{dy}{dx} = \frac{y}{x(1 + x)},$$

since $c_1 = y(1 + x)/x$. The differential equation of the orthogonal trajectories is then

$$\frac{dy}{dx} = -\frac{x(1 + x)}{y}.$$

Again, by separating variables, we have:

$$y\,dy = -x(1 + x)\,dx$$

$$\int y\,dy = -\int (x + x^2)\,dx$$

$$\frac{y^2}{2} = -\frac{x^2}{2} - \frac{x^3}{3} + c_2' \quad \text{or} \quad 3y^2 + 3x^2 + 2x^3 = c_2. \quad \blacksquare$$

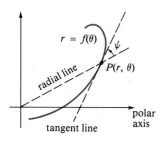

Figure 3.6

Polar Curves

In calculus it is shown that for a graph of a polar equation $r = f(\theta)$,

$$r\frac{d\theta}{dr} = \tan\psi,$$

where ψ is the positive counterclockwise angle between the radial line and the tangent line. See Figure 3.6. It is left as an exercise to show that two polar curves $r = f_1(\theta)$ and $r = f_2(\theta)$ are orthogonal at a point of intersection if and only if

$$(\tan\psi_1)_{\mathscr{C}_1}(\tan\psi_2)_{\mathscr{C}_2} = -1. \tag{1}$$

See Problem 42.

EXAMPLE 6

Find the orthogonal trajectories of

$$r = c_1(1 - \sin\theta).$$

Solution   For the given curve we can write

$$\frac{dr}{d\theta} = -c_1\cos\theta = \frac{-r\cos\theta}{1 - \sin\theta}$$

so that

$$r\frac{d\theta}{dr} = -\frac{1 - \sin\theta}{\cos\theta} = \tan\psi_1.$$

Thus, by (1), the differential equation of the orthogonal trajectories is

$$r\frac{d\theta}{dr} = \frac{\cos\theta}{1 - \sin\theta} = \tan\psi_2.$$

Separating variables then gives

$$\frac{dr}{r} = \frac{1 - \sin\theta}{\cos\theta}\,d\theta = (\sec\theta - \tan\theta)\,d\theta$$

so that

$$\ln|r| = \ln|\sec\theta + \tan\theta| + \ln|\cos\theta| + \ln c_2$$

$$= \ln|c_2(1 + \sin\theta)|.$$

Hence

$$r = c_2(1 + \sin\theta). \qquad\blacksquare$$

3.1 EXERCISES

Answers to odd-numbered problems begin on page A–5.

In Problems 1–26 find the orthogonal trajectories of the given family of curves.

1. $y = c_1 x$

2. $3x + 4y = c_1$

3. $y = c_1 x^2$

4. $y = (x - c_1)^2$

5. $c_1 x^2 + y^2 = 1$

6. $2x^2 + y^2 = c_1^2$

7. $y = c_1 e^{-x}$

8. $y = e^{c_1 x}$

9. $y^2 = c_1 x^3$

10. $y^a = c_1 x^b$, a and b constants

11. $y = \dfrac{x}{1 + c_1 x}$

12. $y = \dfrac{1 + c_1 x}{1 - c_1 x}$

13. $2x^2 + y^2 = 4c_1 x$

14. $x^2 + y^2 = 2c_1 x$

15. $y^3 + 3x^2 y = c_1$

16. $y^2 - x^2 = c_1 x^3$

17. $y = \dfrac{c_1}{1 + x^2}$

18. $y = \dfrac{1}{c_1 + x}$

19. $4y + x^2 + 1 + c_1 e^{2y} = 0$

20. $y = -x - 1 + c_1 e^x$

21. $y = \dfrac{1}{\ln c_1 x}$

22. $y = \ln(\tan x + c_1)$

23. $\sinh y = c_1 x$

24. $y = c_1 \sin x$

25. $x^{1/3} + y^{1/3} = c_1$

26. $x^a + y^a = c_1$, $a \neq 2$

27. Find the member of the orthogonal trajectories for $x + y = c_1 e^y$ that passes through $(0, 5)$.

28. Find the member of the orthogonal trajectories for $3xy^2 = 2 + 3c_1 x$ that passes through $(0, 10)$.

In Problems 29–34 find the orthogonal trajectories of the given polar curves.

29. $r = 2c_1 \cos \theta$

30. $r = c_1(1 + \cos \theta)$

31. $r^2 = c_1 \sin 2\theta$

32. $r = \dfrac{c_1}{1 + \cos \theta}$

33. $r = c_1 \sec \theta$

34. $r = c_1 e^\theta$

35. A family of curves that intersects a given family of curves at a specified constant angle $\alpha \neq \pi/2$ is said to be an isogonal family. The two families are said to be **isogonal trajectories** of each other. If $dy/dx = f(x, y)$ is the differential equation of the given family, show that the differential equation of the isogonal family is

$$\frac{dy}{dx} = \frac{f(x, y) \pm \tan \alpha}{1 \mp f(x, y) \tan \alpha}.$$

In Problems 36–38 use the results of Problem 35 to find the isogonal family that intersects the one-parameter family of straight lines $y = c_1 x$ at the given angle.

36. $\alpha = 45°$

37. $\alpha = 60°$

38. $\alpha = 30°$

A family of curves can be **self-orthogonal** in the sense that a member of the orthogonal trajectories is also a member of the original family. In Problems 39 and 40 show that the given family of curves is self-orthogonal.

39. parabolas $y^2 = c_1(2x + c_1)$

40. confocal conics $\dfrac{x^2}{c_1 + 1} + \dfrac{y^2}{c_1} = 1$

41. Verify that the orthogonal trajectories of the family of curves given by the parametric equations $x = c_1 e^t \cos t$, $y = c_1 e^t \sin t$ are

$$x = c_2 e^{-t} \cos t, \quad y = c_2 e^{-t} \sin t.$$

[*Hint:* $dy/dx = (dy/dt)/(dx/dt)$.]

42. Show that two polar curves $r = f_1(\theta)$ and $r = f_2(\theta)$ are orthogonal at a point of intersection if and only if

$$(\tan \psi_1)_{\mathscr{C}_1}(\tan \psi_2)_{\mathscr{C}_2} = -1.$$

3.2 APPLICATIONS OF LINEAR EQUATIONS

Growth and Decay

The initial-value problem

$$\frac{dx}{dt} = kx, \qquad x(t_0) = x_0, \tag{1}$$

where k is a constant of proportionality, occurs in many physical theories involving either **growth** or **decay**. For example, in biology it is often observed that the rate at which certain bacteria grow is proportional to the number of bacteria present at any time. Over short intervals of time, the population of small animals, such as rodents, can be predicted fairly accurately by the solution of (1). In physics an initial-value problem such as (1) provides a model for approximating the remaining amount of a substance that is disintegrating, or decaying, through radioactivity. The differential equation in (1) could also determine the temperature of a cooling body. In chemistry the amount of a substance remaining during certain reactions is also described by (1).

The constant of proportionality k in (1) is either positive or negative and can be determined from the solution of the problem using a subsequent measurement of x at a time $t_1 > t_0$.

EXAMPLE I

A culture initially has N_0 number of bacteria. At $t = 1$ hour the number of bacteria is measured to be $(3/2)N_0$. If the rate of growth is proportional to the number of bacteria present, determine the time necessary for the number of bacteria to triple.

Solution We first solve the differential equation

$$\frac{dN}{dt} = kN \tag{2}$$

subject to $N(0) = N_0$. Then we use the empirical condition $N(1) = (3/2)N_0$ to determine the constant of proportionality k.

Now (2) is both separable and linear. When it is put into the form

$$\frac{dN}{dt} - kN = 0,$$

we can see by inspection that the integrating factor is e^{-kt}. Multiplying both sides of the equation by this term gives immediately

$$\frac{d}{dt}\left[e^{-kt}N\right] = 0.$$

Integrating both sides of the last equation yields

$$e^{-kt}N = c \quad \text{or} \quad N(t) = ce^{kt}.$$

At $t = 0$ it follows that $N_0 = ce^0 = c$ and so $N(t) = N_0 e^{kt}$. At $t = 1$ we have

$$\frac{3}{2}N_0 = N_0 e^k \quad \text{or} \quad e^k = \frac{3}{2},$$

from which we get to four decimal places

$$k = \ln\left(\frac{3}{2}\right) = 0.4055.$$

Thus

$$N(t) = N_0 e^{0.4055t}.$$

To find the time at which the bacteria have tripled we solve

$$3N_0 = N_0 e^{0.4055t}$$

for t. It follows from this equation that $0.4055t = \ln 3$ and so

$$t = \frac{\ln 3}{0.4055} \approx 2.71 \text{ hours.}$$

See Figure 3.7. ∎

$N(t) = N_0 e^{0.4055t}$

$t = 2.71$

Figure 3.7

Note We can write the function $N(t)$ obtained in Example 1 in an alternative form. From the laws of exponents,

$$N(t) = N_0(e^k)^t = N_0\left(\frac{3}{2}\right)^t$$

since $e^k = 3/2$. This latter solution provides a convenient method for computing $N(t)$ for small positive integral values of t. It also clearly shows the influence of the subsequent experimental observation at $t = 1$ on the solution

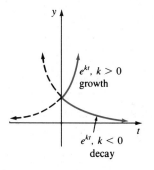

Figure 3.8

for all time. We notice, too, that the actual number of bacteria present at time $t = 0$ is quite irrelevant in finding the time required to triple the number in the culture. The necessary time to triple, say, 100 or 10,000 bacteria is still approximately 2.71 hours.

As shown in Figure 3.8, the exponential function e^{kt} increases as t increases for $k > 0$, and decreases as t increases for $k < 0$. Thus problems describing growth, such as population, bacteria, or even capital, are characterized by a positive value of k, whereas problems involving decay, as in radioactive disintegration, yield a negative k value.

Half-Life

In physics the **half-life** is a measure of the stability of a radioactive substance. The half-life is simply the time it takes for one-half of the atoms in an initial amount A_0 to disintegrate, or transmute, into the atoms of another element. The longer the half-life of a substance, the more stable it is. For example, the half-life of highly radioactive radium, Ra-226, is about 1700 years. In 1700 years one-half of a given quantity of Ra-226 is transmuted into radon, Rn-222. The most commonly occurring uranium isotope, U-238, has a half-life of approximately 4,500,000,000 years. In about 4.5 billion years, one-half of a quantity of U-238 is transmuted into lead, Pb-206.

EXAMPLE **2**

A breeder reactor converts the relatively stable uranium 238 into the isotope plutonium 239. After 15 years it is determined that 0.043% of the initial amount A_0 of the plutonium has disintegrated. Find the half-life of this isotope if the rate of disintegration is proportional to the amount remaining.

Solution   Let $A(t)$ denote the amount of plutonium remaining at any time. As in Example 1, the solution of the initial-value problem

$$\frac{dA}{dt} = kA, \qquad A(0) = A_0,$$

is
$$A(t) = A_0 e^{kt}.$$

If 0.043% of the atoms of A_0 have disintegrated, then 99.957% of the substance remains. To find k we use $0.99957A_0 = A(15)$; that is,

$$0.99957A_0 = A_0 e^{15k}.$$

Solving for k then gives $15k = \ln(0.99957)$, or

$$k = \frac{\ln(0.99957)}{15} = -0.00002867.$$

Hence
$$A(t) = A_0 e^{-0.00002867t}.$$

Now the half-life is the corresponding value of time for which $A(t) = A_0/2$. Solving for t gives

$$\frac{A_0}{2} = A_0 e^{-0.00002867t} \quad \text{or} \quad \frac{1}{2} = e^{-0.00002867t}$$

$$-0.00002867t = \ln\left(\frac{1}{2}\right) = -\ln 2$$

$$t = \frac{\ln 2}{0.00002867} \approx 24{,}180 \text{ years.} \qquad \blacksquare$$

Carbon Dating

About 1950 the chemist Willard Libby devised a method of using radioactive carbon as a means of determining the approximate ages of fossils. The theory of **carbon dating** is based on the fact that the isotope carbon 14 is produced in the atmosphere by the action of cosmic radiation on nitrogen. The ratio of the amount of C-14 to ordinary carbon in the atmosphere appears to be a constant, and as a consequence the proportionate amount of the isotope present in all living organisms is the same as that in the atmosphere. When an organism dies, the absorption of C-14, by either breathing or eating, ceases. Thus, by comparing the proportionate amount of C-14 present, say, in a fossil with the constant ratio found in the atmosphere, it is possible to obtain a reasonable estimation of its age. The method is based on the knowledge that the half-life of the radioactive C-14 is approximately 5600 years. For his work Libby won the Nobel Prize for chemistry in 1960. Libby's method has been used to date wooden furniture in Egyptian tombs and the woven flax wrappings of the Dead Sea scrolls.

EXAMPLE 3

A fossilized bone is found to contain 1/1000 the original amount of C-14. Determine the age of the fossil.

Solution The starting point is again $A(t) = A_0 e^{kt}$. To determine the value of k we use the fact that $A_0/2 = A(5600)$, or $A_0/2 = A_0 e^{5600k}$. Hence we have

$$5600k = \ln\left(\frac{1}{2}\right) = -\ln 2$$

$$k = -\frac{\ln 2}{5600} = -0.00012378.$$

Therefore
$$A(t) = A_0 e^{-0.00012378t}.$$

When $A(t) = A_0/1000$, we have

$$\frac{A_0}{1000} = A_0 e^{-0.00012378t}$$

so that

$$-0.00012378t = \ln\left(\frac{1}{1000}\right) = -\ln 1000$$

$$t = \frac{\ln 1000}{0.00012378} \approx 55{,}800 \text{ years.} \qquad \blacksquare$$

The date found in Example 3 is really at the border of accuracy for this method. The usual carbon 14 technique is limited to about 9 half-lives of the isotope, or about 50,000 years. One reason is that the chemical analysis needed to obtain an accurate measurement of the remaining C-14 becomes somewhat formidable around the point of $A_0/1000$. Also, this analysis demands the destruction of a rather large sample of the specimen. If this measurement is accomplished indirectly, based on the actual radioactivity of the specimen, then it is very difficult to distinguish between the radiation from the fossil and the normal background radiation. But in recent developments, the use of a particle accelerator has enabled scientists to separate the C-14 from the stable C-12 directly. By computing the precise value of the ratio of C-14 to C-12, the accuracy of this method can be extended to 70,000–100,000 years. Other isotopic techniques such as using potassium 40 and argon 40 can give dates of several million years. Nonisotopic methods based on the use of amino acids are also sometimes possible.

Cooling

Newton's law of cooling states that the rate at which the temperature $T(t)$ changes in a cooling body is proportional to the difference between the temperature in the body and the constant temperature T_m of the surrounding medium—that is,

$$\frac{dT}{dt} = k(T - T_m), \qquad (3)$$

where k is a constant of proportionality.

EXAMPLE 4

When a cake is removed from a baking oven, its temperature is measured at 300°F. Three minutes later its temperature is 200°F. How long will it take to cool off to a room temperature of 70°F?

Solution   In (3) we make the identification $T_m = 70$. We must then solve the initial-value problem

(a)

$T(t)$	t (minutes)
75°	20.1
74°	21.3
73°	22.8
72°	24.9
71°	28.6
70.5°	32.3

(b)

Figure 3.9

$$\frac{dT}{dt} = k(T - 70), \qquad T(0) = 300, \tag{4}$$

and determine the value of k so that $T(3) = 200$.

Equation (4) is both linear and separable. Separating variables yields

$$\frac{dT}{T - 70} = k \, dt$$

$$\ln|T - 70| = kt + c_1$$

$$T - 70 = c_2 e^{kt}$$

$$T = 70 + c_2 e^{kt}.$$

When $t = 0$, $T = 300$ so that $300 = 70 + c_2$ gives $c_2 = 230$ and, therefore, $T = 70 + 230e^{kt}$.

From $T(3) = 200$ we find

$$e^{3k} = \frac{13}{23} \quad \text{or} \quad k = \frac{1}{3} \ln \frac{13}{23} = -0.19018.$$

Thus $$T(t) = 70 + 230e^{-0.19018t}. \tag{5}$$

We note that (5) furnishes no finite solution to $T(t) = 70$ since $\lim_{t \to \infty} T(t) = 70$. Yet intuitively we expect the cake to reach the room temperature after a reasonably long period of time. How long is long? Of course, we should not be disturbed by the fact that the model (4) does not quite live up to our physical intuition. Parts (a) and (b) of Figure 3.9 clearly show that the cake will be approximately at room temperature in about one-half hour.

∎

Series Circuits

In a series circuit containing only a resistor and an inductor, Kirchhoff's second law states that the sum of the voltage drop across the inductor ($L(di/dt)$) and the voltage drop across the resistor (iR) is the same as the impressed voltage ($E(t)$) on the circuit. See Figure 3.10.

Thus we obtain the linear differential equation for the current $i(t)$,

$$L\frac{di}{dt} + Ri = E(t), \tag{6}$$

where L and R are constants known as the inductance and the resistance, respectively. The current $i(t)$ is sometimes called the **response** of the system.

The voltage drop across a capacitor with capacitance C is given by $q(t)/C$, where q is the charge on the capacitor. Hence, for the series circuit shown in Figure 3.11, Kirchhoff's second law gives

$$Ri + \frac{1}{C}q = E(t). \tag{7}$$

L-R Series Circuit

Figure 3.10

R-C Series Circuit

Figure 3.11

But current i and charge q are related by $i = dq/dt$, so (7) becomes the linear differential equation

$$R \frac{dq}{dt} + \frac{1}{C} q = E(t). \tag{8}$$

EXAMPLE 5

A 12-volt battery is connected to a series circuit in which the inductance is 1/2 henry and the resistance is 10 ohms. Determine the current i if the initial current is zero.

Solution   From (6) we see that we must solve

$$\frac{1}{2} \frac{di}{dt} + 10i = 12$$

subject to $i(0) = 0$. First, we multiply the differential equation by 2 and read off the integrating factor e^{20t}. We then obtain

$$\frac{d}{dt} [e^{20t}i] = 24e^{20t}$$

$$e^{20t}i = \frac{24}{20} e^{20t} + c$$

$$i = \frac{6}{5} + ce^{-20t}.$$

Now $i(0) = 0$ implies $0 = 6/5 + c$, or $c = -6/5$. Therefore the response is

$$i(t) = \frac{6}{5} - \frac{6}{5} e^{-20t}. \qquad \blacksquare$$

From (7) of Section 2.5 we can write a general solution of (6):

$$i(t) = \frac{e^{-(R/L)t}}{L} \int e^{(R/L)t} E(t) \, dt + ce^{-(R/L)t}. \tag{9}$$

In particular, when $E(t) = E_0$ is a constant, (9) becomes

$$i(t) = \frac{E_0}{R} + ce^{-(R/L)t}. \tag{10}$$

Note that as $t \to \infty$, the second term in equation (10) approaches zero. Such a term is usually called a **transient term**; any remaining terms are called the **steady-state** part of the solution. In this case E_0/R is also called the **steady-state current**; for large values of time it then appears that the current in the circuit is simply governed by Ohm's law ($E = iR$).

Mixture Problem

The mixing of two fluids sometimes gives rise to a linear first-order differential equation. In the next example we consider the mixture of two salt solutions with different concentrations.

EXAMPLE 6

Initially 50 pounds of salt is dissolved in a large tank holding 300 gallons of water. A brine solution is pumped into the tank at a rate of 3 gallons per minute, and the well-stirred solution is then pumped out at the same rate. See Figure 3.12. If the concentration of the solution entering is 2 pounds per gallon, determine the amount of salt in the tank at any time. How much salt is present after 50 minutes? after a long time?

constant 300 gal.

Figure 3.12

Solution   Let $A(t)$ be the amount of salt (in pounds) in the tank at any time. For problems of this sort, the net rate at which $A(t)$ changes is given by

$$\frac{dA}{dt} = \left(\begin{array}{c} \text{rate of} \\ \text{substance entering} \end{array}\right) - \left(\begin{array}{c} \text{rate of} \\ \text{substance leaving} \end{array}\right) = R_1 - R_2. \quad \textbf{(11)}$$

Now the rate at which the salt enters the tank is, in pounds per minute,

$$R_1 = (3 \text{ gal/min}) \cdot (2 \text{ lb/gal}) = 6 \text{ lb/min},$$

whereas the rate at which salt is leaving is

$$R_2 = (3 \text{ gal/min}) \cdot \left(\frac{A}{300} \text{ lb/gal}\right) = \frac{A}{100} \text{ lb/min}.$$

Thus equation (11) becomes

$$\frac{dA}{dt} = 6 - \frac{A}{100}, \quad \textbf{(12)}$$

A = 600

500   t

(a)

which we solve subject to the initial condition $A(0) = 50$.
Since the integrating factor is $e^{t/100}$, we can write (12) as

$$\frac{d}{dt}\left[e^{t/100}A\right] = 6e^{t/100}$$

and therefore

$$e^{t/100}A = 600e^{t/100} + c$$

$$A = 600 + ce^{-t/100}. \quad \textbf{(13)}$$

When $t = 0$, $A = 50$, so we find that $c = -550$. Finally, we obtain

$$A(t) = 600 - 550e^{-t/100}. \quad \textbf{(14)}$$

t(minutes)	A(lbs)
50	266.41
100	397.67
150	477.27
200	525.57
300	572.62
400	589.93

(b)

Figure 3.13

At $t = 50$ we find $A(50) = 266.41$ pounds. Also, as $t \to \infty$ it is seen from (14) and Figure 3.13 that $A \to 600$. Of course this is what we would expect; over a long period of time the number of pounds of salt in the solution must be

$$(300 \text{ gal})(2 \text{ lb/gal}) = 600 \text{ lb.} \qquad \blacksquare$$

In Example 6 we assumed that the rate at which the solution was pumped in was the same as the rate at which the solution was pumped out. However, this need not be the case; the mixed brine solution could be pumped out at a rate faster or slower than the rate at which the other solution is pumped in. The resulting differential equation in this latter situation is linear with a variable coefficient.

EXAMPLE 7

If the well-stirred solution in Example 6 is pumped out at a slower rate of 2 gallons per minute, then the solution is *accumulating* at a rate of

$$(3 - 2) \text{ gal/min} = 1 \text{ gal/min}.$$

After t minutes there are $300 + t$ gallons of brine in the tank. The rate at which the salt is leaving is then

$$R_2 = (2 \text{ gal/min}) \cdot \left(\frac{A}{300 + t} \text{ lb/gal} \right) = \frac{2A}{300 + t} \text{ lb/min}.$$

Hence equation (11) becomes

$$\frac{dA}{dt} = 6 - \frac{2A}{300 + t} \quad \text{or} \quad \frac{dA}{dt} + \frac{2A}{300 + t} = 6.$$

Finding the integrating factor and solving the last equation, we get

$$A(t) = 2(300 + t) + c(300 + t)^{-2}.$$

The initial condition $A(0) = 50$ yields $c = -4.95 \times 10^7$ and so

$$A(t) = 2(300 + t) - (4.95 \times 10^7)(300 + t)^{-2}. \qquad \blacksquare$$

Remark Consider the differential equation in Example 1 that describes the growth of bacteria. The solution $N(t) = N_0 e^{0.4055t}$ of the initial-value problem $dN/dt = kN$, $N(t_0) = N_0$, is of course a continuous function. But in the example we are talking about a population of bacteria and so common sense dictates that N take on only positive integer values. Moreover, the population does not necessarily grow continuously, that is, every second, every microsecond, and so on, as predicted by the function $N(t) = N_0 e^{0.4055t}$; there may be time intervals $[t_1, t_2]$ over which there is no growth at all. Perhaps, then, the graph shown in Figure 3.14(a) is a more realistic description of N than that given by the graph of an exponential function. The point is, in many instances a mathematical model describes a system in only approximate terms. It is often more convenient than accurate to use a continuous function to describe a discrete phenomenon. However, for some purposes we may be satisfied if

Figure 3.14

our model describes the system fairly accurately when viewed macroscopically in time as in Figures 3.14(b) and (c) rather than microscopically.

3.2   EXERCISES

Answers to odd-numbered problems begin on page A–6.

1. The population of a certain community is known to increase at a rate proportional to the number of people present at any time. If the population has doubled in 5 years, how long will it take to triple? to quadruple?

2. Suppose it is known that the population of the community in Problem 1 is 10,000 after 3 years. What was the initial population? What will be the population in 10 years?

3. The population of a town grows at a rate proportional to the population at any time. Its initial population of 500 increases by 15% in 10 years. What will be the population in 30 years?

4. The population of bacteria in a culture grows at a rate proportional to the number of bacteria present at any time. After 3 hours it is observed that there are 400 bacteria present. After 10 hours there are 2000 bacteria present. What is the initial number of bacteria?

5. The radioactive isotope of lead, Pb-209, decays at a rate proportional to the amount present at any time and has a half-life of 3.3 hours. If 1 gram of lead is present initially, how long will it take for 90% of the lead to decay?

6. Initially there were 100 milligrams of a radioactive substance present. After 6 hours the mass decreased by 3%. If the rate of decay is proportional to the amount of the substance present at any time, find the amount remaining after 24 hours.

7. Determine the half-life of the radioactive substance described in Problem 6.

8. Show that the half-life of a radioactive substance is, in general,

$$t = \frac{(t_2 - t_1)\ln 2}{\ln(A_1/A_2)},$$

where $A_1 = A(t_1)$ and $A_2 = A(t_2)$, $t_1 < t_2$.

9. When a vertical beam of light passes through a transparent substance, the rate at which its intensity I decreases is proportional to $I(t)$, where t represents the thickness of the medium (in feet). In clear seawater the intensity 3 feet below the surface is 25% of the initial intensity I_0 of the incident beam. What is the intensity of the beam 15 feet below the surface?

10. When interest is compounded continuously, the amount of money S increases at a rate proportional to the amount present at any time: $dS/dt = rS$, where r is the annual rate of interest (see (26) of Section 1.2).

    (a) Find the amount of money accrued at the end of 5 years when $5000 is deposited in a savings account drawing $5\frac{3}{4}\%$ annual interest compounded continuously.

    (b) In how many years will the initial sum deposited be doubled?

    (c) Use a hand calculator to compare the number obtained in part (a) with the value

$$S = 5000\left(1 + \frac{0.0575}{4}\right)^{5(4)}.$$

    This value represents the amount accrued when interest is compounded quarterly.

11. In a piece of burned wood, or charcoal, it was found that 85.5% of the C-14 had decayed. Use the information in Example 3 to determine the approximate age of the wood. (It is precisely these data that archaeologists used to date prehistoric paintings in a cave in Lascaux, France.)

12. A thermometer is taken from an inside room to the outside where the air temperature is 5°F. After 1 minute the thermometer reads 55°F, and after 5 minutes the reading is 30°F. What is the initial temperature of the room?

13. A thermometer is removed from a room where the air temperature is 70°F to the outside where the temperature is 10°F. After $\frac{1}{2}$ minute the thermometer reads 50°F. What is the reading at $t = 1$ minute? How long will it take for the thermometer to reach 15°F?

14. Formula (3) also holds when an object absorbs heat from the surrounding medium. If a small metal bar whose initial temperature is 20°C is dropped into a container of boiling water, how long will it take for the bar to reach 90°C if it is known that its temperature increased 2° in 1 second? How long will it take the bar to reach 98°C?

15. A 30-volt electromotive force is applied to an L-R series circuit in which

the inductance is 0.1 henry and the resistance is 50 ohms. Find the current $i(t)$ if $i(0) = 0$. Determine the current as $t \to \infty$.

16. Solve equation (6) under the assumption that $E(t) = E_0 \sin \omega t$ and $i(0) = i_0$.

17. A 100-volt electromotive force is applied to an R-C series circuit in which the resistance is 200 ohms and the capacitance is 10^{-4} farad. Find the charge $q(t)$ on the capacitor if $q(0) = 0$. Find the current $i(t)$.

18. A 200-volt electromotive force is applied to an R-C series circuit in which the resistance is 1000 ohms and the capacitance is 5×10^{-6} farad. Find the charge $q(t)$ on the capacitor if $i(0) = 0.4$. Determine the charge and current at $t = 0.005$ second. Determine the charge as $t \to \infty$.

19. An electromotive force

$$E(t) = \begin{cases} 120, & 0 \le t \le 20 \\ 0, & t > 20 \end{cases}$$

is applied to an L-R series circuit in which the inductance is 20 henry and the resistance is 2 ohms. Find the current $i(t)$ if $i(0) = 0$.

20. Suppose an R-C series circuit has a variable resistor. If the resistance at any time t is given by $R = k_1 + k_2 t$, where $k_1 > 0$ and $k_2 > 0$ are known constants, then (8) becomes

$$(k_1 + k_2 t) \frac{dq}{dt} + \frac{1}{C} q = E(t).$$

Show that if $E(t) = E_0$ and $q(0) = q_0$, then

$$q(t) = E_0 C + (q_0 - E_0 C) \left(\frac{k_1}{k_1 + k_2 t} \right)^{1/Ck_2}.$$

21. A tank contains 200 liters of fluid in which 30 g of salt is dissolved. Brine containing 1 g of salt per liter is then pumped into the tank at a rate of 4 liters per minute; the well-mixed solution is pumped out at the same rate. Find the number of grams of salt $A(t)$ in the tank at any time.

22. Solve Problem 21 assuming pure water is pumped into the tank.

23. A large tank is filled with 500 gallons of pure water. Brine containing 2 lb of salt per gallon is pumped into the tank at a rate of 5 gallons per minute. The well-mixed solution is pumped out at the same rate. Find the number of pounds of salt $A(t)$ in the tank at any time.

24. Solve Problem 23 under the assumption that the solution is pumped out at a faster rate of 10 gallons per minute. When is the tank empty?

25. A large tank is partially filled with 100 gallons of fluid in which 10 lb of salt is dissolved. Brine containing $\frac{1}{2}$ lb of salt per gallon is pumped into the tank at a rate of 6 gallons per minute. The well-mixed solution is then pumped out at a slower rate of 4 gallons per minute. Find the number of pounds of salt in the tank after 30 minutes.

26. Beer containing 6% alcohol per gallon is pumped into a vat that initially contains 400 gallons of beer at 3% alcohol. The rate at which the beer is pumped in is 3 gallons per minute, whereas the mixed liquid is pumped out at a rate of 4 gallons per minute. Find the number of gallons of alcohol $A(t)$ in the tank at any time. What is the percentage of alcohol in the tank after 60 minutes? When is the tank empty?

Miscellaneous Applications

27. The differential equation governing the velocity v of a falling mass m subjected to air resistance proportional to the instantaneous velocity is

$$m\frac{dv}{dt} = mg - kv,$$

where k is a positive constant of proportionality.

(a) Solve the equation subject to the initial condition $v(0) = v_0$.

(b) Determine the limiting, or terminal, velocity of the weight.

(c) If distance s is related to velocity $ds/dt = v$, find an explicit expression for s if it is further known that $s(0) = s_0$.

28. The rate at which a drug disseminates into the bloodstream is governed by the differential equation

$$\frac{dX}{dt} = A - BX,$$

where A and B are positive constants. The function $X(t)$ describes the concentration of the drug in the bloodstream at any time t. Find the limiting value of X as $t \to \infty$. At what time is the concentration one-half this limiting value? Assume that $X(0) = 0$.

29. A heart pacemaker, as shown in Figure 3.15, consists of a battery, a capacitor, and the heart as a resistor. When the switch S is at P the capacitor charges; when S is at Q the capacitor discharges, sending an electrical stimulus to the heart. During this time the voltage E applied to the heart is given by

$$\frac{dE}{dt} = -\frac{1}{RC}E, \qquad t_1 < t < t_2,$$

where R and C are constants. Determine $E(t)$ if $E(t_1) = E_0$. (Of course, the opening and closing of the switch are periodic in time, to simulate the natural heartbeat.)

Figure 3.15

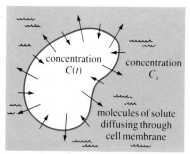

Figure 3.16

30. Suppose a cell is suspended in a solution containing a solute of constant concentration C_s. Suppose further that the cell has constant volume V and that the area of its permeable membrane is the constant A. By Fick's law the rate of change of its mass m is directly proportional to the area A and the difference $C_s - C(t)$, where $C(t)$ is the concentration of the solute inside the cell at any time t. Find $C(t)$ if $m = VC(t)$ and $C(0) = C_0$. See Figure 3.16.

31. In one model of the changing population $P(t)$ of a community, it is assumed that

$$\frac{dP}{dt} = \frac{dB}{dt} - \frac{dD}{dt},$$

where dB/dt and dD/dt are the birth and death rates, respectively.

(a) Solve for $P(t)$ if

$$\frac{dB}{dt} = k_1 P \quad \text{and} \quad \frac{dD}{dt} = k_2 P.$$

(b) Analyze the cases $k_1 > k_2$, $k_1 = k_2$, and $k_1 < k_2$.

32. The differential equation

$$\frac{dP}{dt} = (k \cos t)P,$$

where k is a positive constant, is often used as a model of a population that undergoes yearly seasonal fluctuations. Solve for $P(t)$ and graph the solution. Assume $P(0) = P_0$.

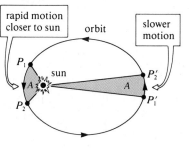

Figure 3.17

33. In polar coordinates the angular momentum of a moving body of mass m is defined to be $L = mr^2(d\theta/dt)$. Assume that the coordinates of the body are (r_1, θ_1) and (r_2, θ_2) at times $t = a$ and $t = b$, $a < b$, respectively. If L is constant, show that the area A swept out by r is $A = L(b - a)/2m$. When the sun is taken to be at the origin, this proves **Kepler's second law of planetary motion**: The radius vector joining the sun sweeps out equal areas in equal intervals of time. See Figure 3.17.

34. When forgetfulness is taken into account, the rate of memorization of a subject is given by

$$\frac{dA}{dt} = k_1(M - A) - k_2 A,$$

where $k_1 > 0$, $k_2 > 0$, $A(t)$ is the amount of material memorized in time t, M is the total amount to be memorized, and $M - A$ is the amount remaining to be memorized. Solve for $A(t)$ and graph the solution. Assume $A(0) = 0$. Find the limiting value of A as $t \to \infty$ and interpret the result.

3.3 APPLICATIONS OF NONLINEAR EQUATIONS

We have seen that if a population P is described by

$$\frac{dP}{dt} = kP, \qquad k > 0, \tag{1}$$

then $P(t)$ exhibits unbounded exponential growth. In many instances this differential equation provides an unrealistic model of the growth of a population; that is, what is actually observed differs substantially from what is predicted.

Around 1840 the Belgian mathematician-biologist P. F. Verhulst was concerned with mathematical formulations for predicting the human populations of various countries. One of the equations he studied was

$$\frac{dP}{dt} = P(a - bP), \tag{2}$$

where a and b are positive constants. Equation (2) came to be known as the **logistic equation** and its solution is called the **logistic function** (the graph of which is naturally called a logistic curve).

Equation (1) does not provide a very accurate model for population growth when the population itself is very large. Overcrowded conditions with the resulting detrimental effects on the environment, such as pollution and excessive and competitive demands for food and fuel, can have an inhibitive effect on the population growth. If a, $a > 0$, is a constant average birth rate, let us assume that the average death rate is proportional to the population $P(t)$ at any time. Thus, if $(1/P)(dP/dt)$ is the rate of growth per individual in a population, then

$$\frac{1}{P}\frac{dP}{dt} = \left(\begin{array}{c} average \\ birth\ rate \end{array}\right) - \left(\begin{array}{c} average \\ death\ rate \end{array}\right) = a - bP, \tag{3}$$

where b is a positive constant of proportionality. Cross multiplying (3) by P immediately gives (2).

As we shall now see, the solution of (2) is bounded as $t \to \infty$. If we rewrite (2) as $dP/dt = aP - bP^2$, the term $-bP^2$, $b > 0$, can be interpreted as an "inhibition" or "competition" term. Also, in most applications, the positive constant a is much larger than the constant b.

Logistic curves have proved to be quite accurate in predicting the growth patterns, in a limited space, of certain types of bacteria, protozoa, water fleas (*Daphnia*), and fruit flies (*Drosophila*). We have already seen equation (2) in the form $dx/dt = kx(n + 1 - x)$, $k > 0$. This differential equation provides a reasonable model for describing the spread of an epidemic brought about by initially introducing an infected individual into a static population. The solution $x(t)$ represents the number of individuals infected with the disease at any time (see Example 11, Section 1.2). Sociologists and even business analysts have borrowed this latter model to study the spread of information and the impact of advertising in certain centers of population.

Solution

One method for solving equation (2) is separation of variables.* By using partial fractions, we can write

$$\frac{dP}{P(a - bP)} = dt$$

$$\left[\frac{1/a}{P} + \frac{b/a}{a - bP} \right] dP = dt$$

$$\frac{1}{a} \ln|P| - \frac{1}{a} \ln|a - bP| = t + c$$

$$\ln \left| \frac{P}{a - bP} \right| = at + ac$$

$$\frac{P}{a - bP} = c_1 e^{at}. \tag{4}$$

It follows from the last equation that

$$P(t) = \frac{ac_1 e^{at}}{1 + bc_1 e^{at}} = \frac{ac_1}{bc_1 + e^{-at}}. \tag{5}$$

Now if we are given the initial condition $P(0) = P_0$, $P_0 \neq a/b$,[†] equation (4) implies $c_1 = P_0/(a - bP_0)$. Substituting this value in (5) and simplifying then give

$$P(t) = \frac{aP_0}{bP_0 + (a - bP_0)e^{-at}}. \tag{6}$$

Graphs of $P(t)$

The basic shape of the graph of the logistic function $P(t)$ can be obtained without too much effort. Although the variable t usually represents time and we are seldom concerned with applications in which $t < 0$, it is nonetheless

---

* In the form $(dP/dt) - aP = -bP^2$ you might recognize the logistic equation as a special case of Bernoulli's equation (see Section 2.6).

† Notice that $P = a/b$ is a singular solution of equation (2).

Figure 3.18

Figure 3.19

of some interest to include this interval when displaying the various graphs of P. From (6) we see that,

$$P(t) \to \frac{aP_0}{bP_0} = \frac{a}{b} \text{ as } t \to \infty \quad \text{and} \quad P(t) \to 0 \text{ as } t \to -\infty.$$

Now differentiating (2) by the product rule gives

$$\frac{d^2P}{dt^2} = P\left(-b\frac{dP}{dt}\right) + (a - bP)\frac{dP}{dt}$$

$$= \frac{dP}{dt}(a - 2bP)$$

$$= P(a - bP)(a - 2bP)$$

$$= 2b^2P\left(P - \frac{a}{b}\right)\left(P - \frac{a}{2b}\right). \tag{7}$$

From calculus recall that the points where $d^2P/dt^2 = 0$ are possible points of inflection, but $P = 0$ and $P = a/b$ can obviously be ruled out. Hence $P = a/2b$ is the only possible ordinate value at which the concavity of the graph can change. For $0 < P < a/2b$ it follows from (7) that $P'' > 0$, and $a/2b < P < a/b$ implies $P'' < 0$. Thus, as we read from left to right, the graph changes from concave up to concave down at the point corresponding to $P = a/2b$. When the initial value satisfies $0 < P_0 < a/2b$, the graph of $P(t)$ assumes the shape of an S, as we see in Figure 3.18(a). For $a/2b < P_0 < a/b$ the graph is still S-shaped but the point of inflection occurs at a negative value of t, as shown in Figure 3.18(b).

If $P_0 > a/b$, equation (7) shows that $P'' > 0$ for all t in the domain of $P(t)$ for which $P > 0$. When $P < 0$, equation (7) implies $P'' < 0$. However, $P = 0$ is not a point of inflection since, whenever $a - bP_0 < 0$, an inspection of (6) reveals a vertical asymptote at

$$t = -\frac{1}{a}\ln\left(\frac{bP_0}{bP_0 - a}\right).$$

The graph of $P(t)$ in this case is given in Figure 3.19.

EXAMPLE I

Suppose a student carrying a flu virus returns to an isolated college campus of 1000 students. If it is assumed that the rate at which the virus spreads is proportional not only to the number x of infected students but also to the number of students not infected, determine the number of infected students after 6 days if it is further observed that after 4 days $x(4) = 50$.

Solution Assuming that no one leaves the campus throughout the duration of the disease, we must solve the initial-value problem

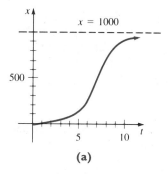

(a)

t (days)	x (number infected)
4	50 (observed)
5	124
6	276
7	507
8	735
9	882
10	953

(b)

Figure 3.20

$$\frac{dx}{dt} = kx(1000 - x), \qquad x(0) = 1.$$

By making the identifications $a = 1000k$ and $b = k$, we have immediately from (6) that

$$x(t) = \frac{1000k}{k + 999ke^{-1000kt}} = \frac{1000}{1 + 999e^{-1000kt}}. \tag{8}$$

Now, using the information $x(4) = 50$, we determine k from

$$50 = \frac{1000}{1 + 999e^{-4000k}}.$$

We find

$$k = \frac{-1}{4000} \ln \frac{19}{999} = 0.0009906.$$

Thus (8) becomes $\quad x(t) = \dfrac{1000}{1 + 999e^{-0.9906t}}.$

Finally

$$x(6) = \frac{1000}{1 + 999e^{-5.9436}} = 276 \text{ students.}$$

Additional calculated values of $x(t)$ are given in the table in Figure 3.20. ∎

Gompertz Curves

A modification of the logistic equation is

$$\frac{dP}{dt} = P(a - b \ln P), \tag{9}$$

where a and b are constants. It is readily shown by separation of variables (see Problem 5) that a solution of (9) is

$$P(t) = e^{a/b}e^{-ce^{-bt}}, \tag{10}$$

where c is an arbitrary constant. We note that when $b > 0$, $P \to e^{a/b}$ as $t \to \infty$, whereas for $b < 0$, $c > 0$, $P \to 0$ as $t \to \infty$. The graph of the function (10), called a **Gompertz curve**, is quite similar to the graph of the logistic function. Figure 3.21 shows two possibilities for the graph of $P(t)$.

Functions such as (10) are encountered, for example, in studies of the growth or decline of certain populations, in the growth of solid tumors, in actuarial predictions, and in the study of growth of revenue in the sale of a commercial product.

Figure 3.21

Chemical Reactions

The disintegration of a radioactive substance, governed by equation (1) of the preceding section, is said to be a **first-order reaction**. In chemistry a few reactions follow the same empirical law: If the molecules of a substance A

decompose into smaller molecules, it is a natural assumption that the rate at which this decomposition takes place is proportional to the amount of the first substance that has not undergone conversion; that is, if $X(t)$ is the amount of substance A remaining at any time, then

$$\frac{dX}{dt} = kX,$$

where k is negative, since X is decreasing. An example of a first-order chemical reaction is the conversion of t-butyl chloride into t-butyl alcohol:

$$(CH_3)_3CCl + NaOH \rightarrow (CH_3)_3COH + NaCl.$$

Only the concentration of the t-butyl chloride controls the rate of reaction. Now in the reaction

$$CH_3Cl + NaOH \rightarrow CH_3OH + NaCl,$$

for every molecule of methyl chloride one molecule of sodium hydroxide is consumed, thus forming one molecule of methyl alcohol and one molecule of sodium chloride. In this case the rate at which the reaction proceeds is proportional to the product of the remaining concentrations of CH_3Cl and of NaOH. If X denotes the amount of CH_3OH formed and α and β are the given amounts of the first two chemicals A and B, then the instantaneous amounts not converted to chemical C are $\alpha - X$ and $\beta - X$, respectively. Hence the rate of formation of C is given by

$$\frac{dX}{dt} = k(\alpha - X)(\beta - X), \tag{11}$$

where k is a constant of proportionality. A reaction described by equation (11) is said to be of **second-order**.

EXAMPLE 2

A compound C is formed when two chemicals A and B are combined. The resulting reaction between the two chemicals is such that for each gram of A, 4 grams of B are used. It is observed that 30 grams of the compound C are formed in 10 minutes. Determine the amount of C at any time if the rate of the reaction is proportional to the amounts of A and B remaining and if initially there are 50 grams of A and 32 grams of B. How much of the compound C is present at 15 minutes? Interpret the solution as $t \rightarrow \infty$.

Solution Let $X(t)$ denote the number of grams of the compound C present at any time. Clearly $X(0) = 0$ and $X(10) = 30$.

Now for example, if there are 2 grams of compound C, we must have used, say, a grams of A and b grams of B so that

$$a + b = 2 \quad \text{and} \quad b = 4a.$$

Thus we must use $a = 2/5 = 2(1/5)$ grams of chemical A and $b = 8/5 = 2(4/5)$ grams of B. In general, for X grams of C we must use

$$\frac{X}{5} \text{ grams of } A \quad \text{and} \quad \frac{4}{5} X \text{ grams of } B.$$

The amounts of A and B remaining at any time are then

$$50 - \frac{X}{5} \quad \text{and} \quad 32 - \frac{4}{5} X,$$

respectively.

Now we know that the rate at which chemical C is formed satisfies

$$\frac{dX}{dt} \propto \left(50 - \frac{X}{5} \right)\left(32 - \frac{4}{5} X \right).$$

To simplify the subsequent algebra, we factor $1/5$ from the first term and $4/5$ from the second, and then introduce the constant of proportionality:

$$\frac{dX}{dt} = k(250 - X)(40 - X).$$

By separation of variables and partial fractions, we can write

$$\frac{dX}{(250 - X)(40 - X)} = k \, dt$$

$$-\frac{1/210}{250 - X} \, dX + \frac{1/210}{40 - X} \, dX = k \, dt$$

$$\ln \left| \frac{250 - X}{40 - X} \right| = 210kt + c_1$$

$$\frac{250 - X}{40 - X} = c_2 e^{210kt}. \tag{12}$$

When $t = 0$, $X = 0$, so it follows at this point that $c_2 = 25/4$. Using $X = 30$ at $t = 10$, we find

$$210k = \frac{1}{10} \ln \frac{88}{25} = 0.1258.$$

With this information we solve (12) for X:

$$X(t) = 1000 \frac{1 - e^{-0.1258t}}{25 - 4e^{-0.1258t}}. \tag{13}$$

The behavior of X as a function of time is displayed in Figure 3.22. It is clear from the accompanying table and equation (13) that $X \to 40$ as $t \to \infty$. This

(a)

t (minutes)	X (grams)
10	30 (measured)
15	34.78
20	37.25
25	38.54
30	39.22
35	39.59

(b)

Figure 3.22

means there are 40 grams of compound C formed, leaving

$$50 - \tfrac{1}{5}(40) = 42 \text{ grams of chemical } A$$

and

$$32 - \tfrac{4}{5}(40) = 0 \text{ grams of chemical } B. \qquad \blacksquare$$

Law of Mass Action

The preceding example can be generalized in the following manner. Suppose that a grams of substance A are combined with b grams of substance B. If there are M parts of A and N parts of B formed in the compound, then the amounts of substances A and B remaining at any time are, respectively,

$$a - \frac{M}{M + N} X \quad \text{and} \quad b - \frac{N}{M + N} X.$$

Thus

$$\frac{dX}{dt} \propto \left[a - \frac{M}{M + N} X \right]\left[b - \frac{N}{M + N} X \right]. \qquad \textbf{(14)}$$

Proceeding as before, if we factor out $M/(M + N)$ from the first term and $N/(M + N)$ from the second term, the resulting differential equation is the same as (11):

$$\frac{dX}{dt} = k(\alpha - X)(\beta - X), \qquad \textbf{(15)}$$

where

$$\alpha = \frac{a(M + N)}{M} \quad \text{and} \quad \beta = \frac{b(M + N)}{N}.$$

Chemists refer to reactions described by equation (15) as the **law of mass action**.

When $\alpha \neq \beta$, it is readily shown (see Problem 9) that a solution of (15) is

$$\frac{1}{\alpha - \beta} \ln\left| \frac{\alpha - X}{\beta - X} \right| = kt + c. \qquad \textbf{(16)}$$

When we assume the natural initial condition $X(0) = 0$, equation (16) yields the explicit solution

$$X(t) = \frac{\alpha\beta[1 - e^{(\alpha - \beta)kt}]}{\beta - \alpha e^{(\alpha - \beta)kt}}. \qquad \textbf{(17)}$$

Without loss of generality we assume in (17) that $\beta > \alpha$ or $\alpha - \beta < 0$. Since $X(t)$ is an increasing function, we expect $k > 0$, and so it follows immediately from (17) that $X \to \alpha$ as $t \to \infty$.

Escape Velocity

In Example 1 of Section 1.2 we saw that the differential equation of a free-falling object of mass m near the surface of the earth is given by

$$m\frac{d^2s}{dt^2} = -mg \quad \text{or simply} \quad \frac{d^2s}{dt^2} = -g,$$

where s represents the distance from the surface of the earth to the object and the positive direction is considered to be upward. In other words, the underlying assumption here is that the distance s to the object is small when compared with the radius R of the earth; put yet another way, the distance y from the center of the earth to the object is approximately the same as R. If, on the other hand, the distance y to an object, such as a rocket or a space probe, is large compared to R, then we combine Newton's second law of motion and his universal law of gravitation to derive a differential equation in the variable y. The solution of this differential equation can be used to determine the minimum velocity, the so-called **escape velocity**, needed by a rocket to break free of the earth's gravitational attraction.

EXAMPLE 3

A rocket is shot vertically upward from the ground as shown in Figure 3.23. If the positive direction is upward and air resistance is ignored, then the differential equation of motion after fuel burnout is

$$m\frac{d^2y}{dt^2} = -k\frac{mM}{y^2} \quad \text{or} \quad \frac{d^2y}{dt^2} = -k\frac{M}{y^2}, \tag{18}$$

where k is a constant of proportionality, y is the distance from the center of the earth to the rocket, M is the mass of the earth, and m is the mass of the rocket. To determine the constant k, we use the fact that when $y = R$,

$$k\frac{mM}{R^2} = mg \quad \text{or} \quad k = \frac{gR^2}{M}.$$

Thus the last equation in (18) becomes

$$\frac{d^2y}{dt^2} = -g\frac{R^2}{y^2}. \tag{19}$$

Although this is not a first-order equation, if we write the acceleration as

$$\frac{d^2y}{dt^2} = \frac{dv}{dt} = \frac{dv}{dy}\frac{dy}{dt} = v\frac{dv}{dy},$$

then (19) becomes first-order in v; that is,

$$v\frac{dv}{dy} = -g\frac{R^2}{y^2}.$$

This last equation can be solved by separation of variables. From

$$\int v\, dv = -gR^2 \int y^{-2}\, dy \quad \text{we get} \quad \frac{v^2}{2} = g\frac{R^2}{y} + c. \tag{20}$$

Figure 3.23

If we assume that the velocity is $v = v_0$ at burnout and that $y \approx R$ at that instant, we can obtain the (approximate) value of c. From (20) we find $c = -gR + v_0^2/2$. Substituting this value in (20) and multiplying the resulting equation by 2 yield

$$v^2 = 2g\frac{R^2}{y} - 2gR + v_0^2. \tag{21}$$

■

You might object, correctly, that in Example 3 we have not really solved the original equation for y. Actually the solution (21) gives quite a bit of information. Now that we have done the hard part, we leave the actual determination of the escape velocity of the earth as an exercise. See Problem 11.

3.3 **EXERCISES**   *Answers to odd-numbered problems begin on page A–6.*

1. The number of supermarkets $C(t)$ throughout the country that are using a computerized checkout system is described by the initial-value problem

$$\frac{dC}{dt} = C(1 - 0.0005C), \qquad C(0) = 1,$$

where $t > 0$. How many supermarkets are using the computerized method when $t = 10$? How many companies are estimated to adopt the new procedure over a long period of time?

2. The number of people $N(t)$ in a community who are exposed to a particular advertisement is governed by the logistic equation. Initially $N(0) = 500$, and it is observed that $N(1) = 1000$. If it is predicted that the limiting number of people in the community who will see the advertisement is 50,000, determine $N(t)$ at any time.

3. The population $P(t)$ at any time in a suburb of a large city is governed by the initial-value problem

$$\frac{dP}{dt} = P(10^{-1} - 10^{-7} P), \qquad P(0) = 5000,$$

where t is measured in months. What is the limiting value of the population? At what time will the population be equal to one-half of this limiting value?

4. Find a solution of the **modified logistic equation**

$$\frac{dP}{dt} = P(a - bP)(1 - cP^{-1}), \qquad a, b, c > 0.$$

5. **(a)** Solve equation (9):

$$\frac{dP}{dt} = P(a - b \ln P).$$

(b) Determine the value of c in equation (10) if $P(0) = P_0$.

6. Assuming $0 < P_0 < e^{a/b}$ and $a > 0$, use equation (9) to find the ordinate of the point of inflection for a Gompertz curve.

7. Two chemicals A and B are combined to form a chemical C. The rate or velocity of the reaction is proportional to the product of the instantaneous amounts of A and B not converted to chemical C. Initially there are 40 grams of A and 50 grams of B, and for each gram of B, 2 grams of A are used. It is observed that 10 grams of C are formed in 5 minutes. How much is formed in 20 minutes? What is the limiting amount of C after a long time? How much of chemicals A and B remains after a long time?

8. Solve Problem 7 if 100 grams of chemical A are present initially. At what time is chemical C half-formed?

9. Obtain a solution of the equation

$$\frac{dX}{dt} = k(\alpha - X)(\beta - X)$$

governing second-order reactions in the two cases $\alpha \neq \beta$ and $\alpha = \beta$.

10. In a third-order chemical reaction the number of grams X of a compound obtained by combining three chemicals is governed by

$$\frac{dX}{dt} = k(\alpha - X)(\beta - X)(\gamma - X).$$

Solve the equation under the assumption $\alpha \neq \beta \neq \gamma$.

11. **(a)** Use equation (21) to show that the escape velocity of the rocket is given by $v_0 = \sqrt{2gR}$. [*Hint:* Take $y \to \infty$ in (21) and assume $v > 0$ for all time t.]

(b) The result in part (a) holds for any body in the solar system. Use the values $g = 32$ ft/s and $R = 4000$ miles to show that the escape velocity from the earth is (approximately) $v_0 = 25{,}000$ mi/h.

(c) Find the escape velocity from the moon if the acceleration of gravity is 0.165 g and $R = 1080$ miles.

Miscellaneous Applications

12. In Example 7 of Section 1.2 we saw that the differential equation describing the shape of a wire of constant linear density w hanging under its own weight is

$$\frac{d^2 y}{dx^2} = \frac{w}{T_1} \sqrt{1 + \left(\frac{dy}{dx}\right)^2},$$

where T_1 is the horizontal tension in the wire at its lowest point. Using the substitution $p = dy/dx$, solve this equation subject to the initial conditions $y(0) = 1$, $y'(0) = 0$.

13. An equation similar to that given in Problem 12 is

$$x \frac{d^2y}{dx^2} = \frac{v_1}{v_2} \sqrt{1 + \left(\frac{dy}{dx}\right)^2}.$$

In this case the equation arises in the study of the shape of the path that a pursuer, traveling at a speed v_2, must take in order to intercept a prey traveling at speed v_1. Use the same substitution as in Problem 12 and the initial conditions $y(1) = 0$, $y'(1) = 0$ to solve the equation. Consider the two cases $v_1 = v_2$ and $v_1 \neq v_2$.

14. According to **Stefan's law** of radiation, the rate of change of temperature from a body at absolute temperature T is

$$\frac{dT}{dt} = k(T^4 - T_m^4),$$

where T_m is the absolute temperature of the surrounding medium. Find a solution of this differential equation. It can be shown that when $T - T_m$ is small compared to T_m, this particular equation is closely approximated by Newton's law of cooling (Equation (3), Section 3.2).

15. The height h of water that is flowing through an orifice at the bottom of a cylindrical tank is given by

$$\frac{dh}{dt} = -\frac{A_o}{A_w} \sqrt{2gh}, \qquad g = 32 \text{ ft/s}^2,$$

where A_w and A_o are the cross-sectional areas of the water and orifice, respectively (see Example 8, Section 1.2). Solve the equation if the initial height of the water is 20 ft and $A_w = 50 \text{ ft}^2$ and $A_o = 1/4 \text{ ft}^2$. At what time is the tank empty?

16. The nonlinear differential equation

$$\left(\frac{dr}{dt}\right)^2 = \frac{2\mu}{r} + 2h,$$

where μ and h are nonnegative constants, arises in the study of the two-body problem of celestial mechanics. Here the variable r represents the distance between the two masses. Solve the equation in the two cases $h = 0$ and $h > 0$.

17. Solve the differential equation of the **tractrix**

$$\frac{dy}{dx} = -\frac{y}{\sqrt{s^2 - y^2}}$$

(see Problem 19, Exercises 1.2). Assume that the initial point on the y-axis is $(0, 10)$ and the length of rope is $s = 10$ ft.

18. A body of mass m falling through a viscous medium encounters a resisting force proportional to the square of its instantaneous velocity. In this situation the differential equation for the velocity $v(t)$ at any time is

$$m\frac{dv}{dt} = mg - kv^2,$$

where k is a positive constant of proportionality. Solve the equation subject to $v(0) = v_0$. What is the limiting velocity of the falling body?

19. The differential equation

$$x\left(\frac{dx}{dy}\right)^2 + 2y\frac{dx}{dy} = x,$$

where $x = x(y)$, occurs in the study of optics. The equation describes the type of plane curve that will reflect all incoming light rays to the same point (see Problem 15, Exercises 1.2). Show that the curve must be a parabola. [*Hint:* Use the substitution $w = x^2$ and then re-examine Section 2.6.]

20. Solve the equation in Problem 19 with the aid of the quadratic formula.

21. The equations of Lotka and Volterra*

$$\frac{dy}{dt} = y(\alpha - \beta x)$$

$$\frac{dx}{dt} = x(-\gamma + \delta y),$$

where α, β, γ, and δ are positive constants, occur in the analysis of the biological balance of two species of animals such as predators and prey (for example, foxes and rabbits). Here $x(t)$ and $y(t)$ denote the populations of the two species at any time. Although no explicit solutions of the system exist, solutions can be found relating the two populations at any time. Divide the first equation by the second and solve the resulting nonlinear first-order differential equation.

---

*A.J. Lotka (1880–1949)   Lotka, born in Austria, was an American biomathematician.

Vito Volterra (1860–1940)   Born in Ancona, Italy, Vito Volterra showed an early aptitude for mathematics. He studied calculus on his own initiative and investigated problems in gravitation at the age of twelve. Although his education was a constant financial struggle, Volterra quickly attained prominence as a scientist and mathematician. He was also an active politician and was appointed Senator of the Kingdom of Italy in 1905. Volterra became interested in the applications of mathematics to ecology in the mid-1920s and formulated this system of differential equations in an attempt to explain the variations in the fish population in the Mediterranean as a result of predator–prey interactions. (Lotka, working independently, arrived at the same system of equations and published the result in 1925 in his text *Elements of Physical Biology*.) Through his research into mathematical models of population, Volterra established the groundwork for a field of mathematics known as integral equations. A man of principle, Volterra refused to sign a loyalty oath to the fascist regime of Benito Mussolini and eventually resigned his chair of mathematics at the University of Rome and all his memberships in Italian scientific societies.

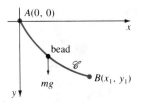

Figure 3.24

22. A classical problem in the calculus of variations is to find the shape of a curve $\mathscr{C}$ such that a bead, under the influence of gravity, will slide from $A(0, 0)$ to $B(x_1, y_1)$ in the least time. See Figure 3.24. It can be shown that the differential equation for the shape of the path is $y[1 + (y')^2] = k$, where k is a constant. First solve for dx in terms of y and dy, and then use the substitution $y = k \sin^2 \theta$ to obtain the parametric form of the solution. The curve $\mathscr{C}$ turns out to be a cycloid.

23. The initial-value problem describing the motion of a simple pendulum released from rest from an angle $\theta_0 > 0$ is

$$\frac{d^2\theta}{dt^2} + \frac{g}{l} \sin \theta = 0$$

$$\theta(0) = \theta_0, \qquad \left. \frac{d\theta}{dt} \right|_{t=0} = 0.$$

(a) Obtain the first-order equation

$$\left(\frac{d\theta}{dt} \right)^2 = \frac{2g}{l} (\cos \theta - \cos \theta_0).$$

[*Hint:* Multiply the given equation by $2 \, d\theta/dt$.]

(b) Use the equation in part (a) to show that the period of motion is

$$T = 2 \sqrt{\frac{2l}{g}} \int_0^{\theta_0} \frac{d\theta}{\sqrt{\cos \theta - \cos \theta_0}}.$$

Chapter 3 REVIEW

If every curve in a one-parameter family of curves $G(x, y, c_1) = 0$ is orthogonal to every curve in a second one-parameter family $H(x, y, c_2) = 0$, we say that the two families are **orthogonal trajectories**. Two curves are orthogonal if their tangent lines are perpendicular at a point of intersection. When given a family, we find its differential equation $dy/dx = f(x, y)$ by differentiating the equation $G(x, y, c_1) = 0$ and eliminating the parameter c_1. The differential equation of the second and orthogonal family is then $dy/dx = -1/f(x, y)$. We solve this latter equation by the methods of Chapter 2.

In the mathematical analysis of population growth, radioactive decay, or chemical mixtures, we often encounter **linear** differential equations such as

$$\frac{dx}{dt} = kx \quad \text{and} \quad \frac{dx}{dt} = a + bx$$

or **nonlinear** differential equations such as

$$\frac{dx}{dt} = x(a - bx) \quad \text{and} \quad \frac{dx}{dt} = k(\alpha - x)(\beta - x).$$

You should be able to solve these particular equations without hesitation. It is never a good idea simply to memorize solutions of differential equations.

Chapter 3   REVIEW EXERCISES

Answers to odd-numbered problems begin on page A–8.

1. Find the orthogonal trajectories of the family of curves $y(x^3 + c_1) = 3$.

2. Find the orthogonal trajectory to the family $y = 4x + 1 + c_1 e^{4x}$ passing through the point $(0, 0)$.

3. Find the orthogonal trajectories of the family of parabolas opening in the y direction with vertex at $(1, 2)$.

4. Show that if a population expands at a rate proportional to the number of people present at any time, then the doubling time of the population is $T = (\ln 2)/k$, where k is the positive growth rate. This is known as the **Law of Malthus**.

5. In March 1976 the world population reached 4 billion. A popular news magazine has predicted that with an average yearly growth rate of 1.8%, the world population will be 8 billion in 45 years. How does this value compare with that predicted by the model that says the rate of increase is proportional to the population at any time?

6. Air containing 0.06% carbon dioxide is pumped into a room whose volume is 8000 ft^3. The rate at which the air is pumped in is 2000 ft^3/min, and the circulated air is then pumped out at the same rate. If there is an initial concentration of 0.2% carbon dioxide, determine the subsequent amount in the room at any time. What is the concentration at 10 minutes? What is the steady-state or equilibrium concentration of carbon dioxide?

7. The populations of two species of animals are described by the nonlinear system of first-order differential equations

$$\frac{dx}{dt} = k_1 x(\alpha - x), \qquad \frac{dy}{dt} = k_2 xy.$$

Solve for x and y in terms of t.

8. A projectile is shot vertically into the air with an initial velocity of v_0 ft/s. Assuming that air resistance is proportional to the square of the instantaneous velocity, the motion is described by the pair of differential equations:

$$m\frac{dv}{dt} = -mg - kv^2, \qquad k > 0,$$

positive y-axis up, origin at ground level so that $v = v_0$ at $y = 0$, and

$$m\frac{dv}{dt} = mg - kv^2, \qquad k > 0,$$

positive y-axis down, origin at the maximum height so that $v = 0$ at $y = h$. The first and second equations describe the motion of the projectile when rising and falling, respectively.

(a) Determine the limiting, or terminal, velocity of the falling projectile. Compare this terminal velocity with that obtained in Problem 27 in Exercises 3.2.

(b) Prove that the impact velocity v_i of the projectile is less than the initial velocity v_0. It can also be shown that the time t_1 needed to attain its maximum height h is less than the time t_2 that it takes to fall from this height. See Figure 3.25.

Figure 3.25

9. Consider Newton's law of cooling $dT/dt = k(T - T_m)$, $k < 0$, where the temperature of the surrounding medium T_m changes with time. Suppose the initial temperature of a body is T_1 and the initial temperature of the surrounding medium is T_2 and $T_m = T_2 + B(T_1 - T)$, where $B > 0$ is a constant.

(a) Find the temperature of the body at any time t.

(b) What is the limiting value of the temperature as $t \to \infty$?

(c) What is the limiting value of T_m as $t \to \infty$?

10. An L-R series circuit has a variable inductor with the inductance defined by

$$L = \begin{cases} 1 - \dfrac{t}{10}, & 0 \le t < 10 \\ 0, & t \ge 10 \end{cases}$$

Find the current $i(t)$ if the resistance is 0.2 ohm, the impressed voltage is $E(t) = 4$, and $i(0) = 0$. Graph $i(t)$.

POPULATION DYNAMICS

Michael Olinick
Department of Mathematics and Computer Science
Middlebury College

The fact that ecology is essentially a mathematical subject is becoming ever more widely accepted," writes Evelyn C. Pielou [2]. "Ecologists everywhere are attempting to formulate and solve their problems by mathematical reasoning." Historically, the first and perhaps most important branch of mathematical ecology is the investigation of population dynamics: how populations grow and decline. First-order differential equations have been a critically important tool in these studies.

Many attempts to model population growth begin with the assumption that the rate of population growth is dependent on the size of the population. If P represents the population at time t, then these models all have the form

$$\frac{dP}{dt} = f(P),$$

where f is some function of the population level P. How should f be selected?

The central figure in the history of population is the Reverend Thomas Robert Malthus (1766–1834). Malthus was an honors graduate in mathematics at Cambridge University, an ordained minister in the Church of England, and a professor of history and political economy. In a seminal work, *An Essay on the Principle of Population* [3], Malthus argued that the appropriate form of $f(P)$, at least when the population is small, should be a constant multiple of P; that is,

$$\frac{dP}{dt} = rP,$$

where r is a constant.

As we have seen, this model yields **exponential growth**, since the solution of the differential equation is

$$P(t) = P_0 e^{rt}.$$

One characteristic of exponential growth is **constant doubling time**: it takes exactly the same amount of time, $(\ln 2)/r$, for the population to double in size from P_0 to $2P_0$ regardless of the size of P_0. Another way to examine exponential growth is to examine the population at successive time units:

$$P(0), P(1), P(2), P(3), \ldots, P(k), P(k + 1), \ldots.$$

Since $P(k) = P_0 e^{rk} = P_0(e^r)^k$, these populations form a **geometric sequence**

$$a, ac, ac^2, ac^3, ac^4, ac^5, \ldots$$

with initial term $a = P_0$ and constant ratio $c = e^r$. Malthus, in fact, begins his famous essay with the observation, "In taking a view of animated nature, we cannot fail to be struck with a prodigious power of increase in plants and animals . . . whether they increase slowly or rapidly, their natural tendency must be to increase in a geometrical ratio, that is, by multiplication; and at whatever rate they are increasing during any one period, if no further obstacles be opposed to them, they must proceed in a geometrical progression." After carefully examining figures collected during the first censuses of the United States and looking at data from other countries, Malthus concluded that "the natural progress of population" was exponential in nature with a doubling time of about 25 years for humans.

Since this model asserts that there is no limit to the number of individuals in this population, it is clear that the exponential model is not a completely realistic picture. The exponential model *may* be a realistic one for the growth of some populations over relatively short time intervals. The population of the United States during the period from 1790 to 1860 grew at such an exponential pace, with an annual growth rate of about 3%.

It's instructive to examine the actual U.S. census figures alongside those given by an exponential growth model. In Table 1 we show the actual and predicted populations in millions. The "error" column displays the difference

TABLE I

Year	Actual Population	Predicted Population	Error	% Error
1790	3.929	3.929	0.000	0.00
1800	5.308	5.285	0.023	0.43
1810	7.24	7.110	0.130	1.80
1820	9.638	9.564	0.074	0.76
1830	12.866	12.866	0.000	0.00
1840	17.069	17.307	−0.238	−1.40
1850	23.192	23.282	−0.090	−0.39
1860	31.433	31.319	0.114	0.36
1870	38.558	42.131	−3.573	−9.27
1880	50.156	56.675	−6.519	−13.00
1890	62.948	76.240	−13.292	−21.12

between the actual and predicted numbers. The final column, "% error," is derived from the ratio of the error to the actual population. The "actual population" is that reported by the U.S. Census Bureau. The "predicted population" is generated from the equation

$$P(t) = 3.929e^{0.029655t}.$$

From the table we see that the exponential growth model fits the actual census data quite closely for the 70-year period beginning in 1790; the largest error is less than 2%. "Reality" and the model's predictions begin to diverge by 1870; the model had no way of predicting the Civil War, which raged from 1861 to 1865 and during which more than half a million American young men lost their lives. We have cut off the table in 1890. The disparity between the actual and predicted populations becomes enormously large in the twentieth century, reaching a staggering 495% error by 1990! One of the consequences we can draw from this model is that if the U.S. population had continued to grow at the same rate it had grown in the first half of the nineteenth century, we would have today a nation six times more populous than the one we do have. That's worth contemplating next time you're stuck in a traffic jam or waiting to check out at your local supermarket!

There are many generalizations of this model. In the **logistic growth model**, first developed by the Belgian mathematician Pierre-Francois Verhulst (1804–1849), we assume that r is not a constant, but a variable that *decreases* in a simple linear fashion as the population increases. Thus we can represent r as $a - bP$, where a and b are positive constants. This yields the model $dP/dt = (a - bP)P$, which has a solution of the form

$$P(t) = \frac{K}{1 + e^{d-at}}.$$

Although Verhulst attempted to test his model on actual population data, he was frustrated by the inaccurate census information available in the early 1840s when he carried out his studies. Because the existing data on population were too inadequate to form any effective test of the logistic model at the time, Verhulst's work lay forgotten for nearly 80 years. It was rediscovered independently by two American scientists working at Johns Hopkins University, Raymond Pearl and Lowell J. Reed.

In 1920 Pearl and Reed examined how closely the U.S. population growth curve followed a logistic curve. Using data from the censuses of 1790, 1850, and 1910 to find values for K, d, and a, they found that the logistic equation

$$P(t) = \frac{197.274}{1 + e^{3.896 - 0.031t}}$$

matched well the actual population figures for the 120-year period beginning in 1790. In fact, the logistic model gives an excellent portrayal of the changes in the U.S. population from 1790 through 1950. Table 2 shows the comparison between the predictions of this logistic model and U.S. census data.

TABLE 2

Year	Predicted Population	Census Population	Error	% Error
1790	3.929	3.929	0.000	0.00
1800	5.336	5.308	0.028	0.53
1810	7.228	7.240	−0.012	−0.17
1820	9.757	9.638	0.119	1.23
1830	13.109	12.866	0.243	1.89
1840	17.506	17.069	0.437	2.56
1850	23.192	23.192	0.000	0.00
1860	30.412	31.433	−1.021	−3.25
1870	39.372	38.558	0.814	2.11
1880	50.177	50.156	0.021	0.04
1890	62.769	62.948	−0.179	−0.28
1900	76.870	75.996	0.874	1.15
1910	91.972	91.972	0.000	0.00
1920	107.395	105.711	1.684	1.59
1930	122.398	122.775	−0.377	−0.31
1940	136.318	131.669	4.649	3.53
1950	148.678	150.697	−2.019	−1.34
1960	159.230	179.323	−20.093	−11.20
1970	167.944	203.185	−35.241	−17.34
1980	174.941	226.546	−51.605	−22.78
1990	180.437	248.710	−68.273	−27.45

We have excellent agreement between the model's predictions and the observed population between 1790 and 1950. The largest error is about 3.5%. While the model predicts a leveling off of U.S. inhabitants that should continue after mid-century, the actual data show the "baby boom" of the 1950s, which increased the population by almost 30 million in 10 years.

It is remarkable that a relatively simple model such as the logistic one can give such accurate results for a period of 160 years. The particular logistic curve calculated by Pearl and Reed could have been derived as early as 1911, when the results of the 1910 census were published. Their equation could have been used for 40 years to give accurate population projections that would have been useful for government planning.

How has the population of the United States grown in the last half of the twentieth century? Can a simple model explain the observed changes and give us a reasonable prediction for the near future? As populations become large, we often observe a slowing down of the rate of increase. Various reasons for this have been given, the principal one being competition for limited resources.

Let's look at one of the simplest mathematical assumptions we can make: the growth rate is **inversely** proportional to the population. Our mathematical model looks like

$$\frac{dP}{dt} = \frac{b}{P} \quad \text{with} \quad P(0) = P_0,$$

where b is a positive constant.

The differential equation is easily solved by separating the variables and integrating. We obtain

$$P(t) = \sqrt{2bt + P_0^2}.$$

If b is positive, this model predicts that the population will continue to increase without bound.

If the population at a later time t_1 is observed to be P_1, then we can determine the value of the parameter b as

$$b = \frac{P_1^2 - P_0^2}{2t_1}.$$

As a test of this simple model, we will look at the U.S. census figures. Taking $t = 0$ to correspond with the 1950 figure of 150.697 million and $t = 1$ for the 1960 census of 179.323 million, we obtain a value of 4723.58 for b. Thus our fitted model has the form

$$P(t) = \sqrt{9447.16t + (150.697)^2}$$

where each unit increase in t represents 10 years.

With these values, we can compare the predictions of the model with the best estimates of the U.S. Census Bureau for the years 1970, 1980, and 1990:

Year	Actual Population	Predicted Population	% Error
1970	203.302	203.970	0.33
1980	226.505	225.945	0.24
1990	248.710	245.964	1.10

(Here both actual and predicted populations are given in millions; the percentage error is obtained by comparing the absolute value of the difference between actual and predicted values to the actual population level.)

We observe that this simple model yields surprisingly accurate results for a least one "real world" set of data. Our value of $t = 0$ corresponds to the April 1, 1950, census date. The U.S. Census Bureau makes population projections for the first day of July. A recent projection for July 1, 2000, is 268.266 million. Using the corresponding value for t of 5.025, our model predicts a population of 264.918 million. The difference is about 1.25%.

More accurate models of the growth of the U.S. population might be obtained by refining the logistic models in several different ways. The function

$f(P)$—chosen to be quadratic in the logistic model—might be taken to be a polynomial of higher degree so that higher-order effects of the size of the population on the growth rate could be included. Additional factors might be attached to the differential equation to incorporate the concept that the rate of change of population is a function not only of population but of time as well; that is, you could examine models of the form

$$\frac{dP}{dt} = f(P, t).$$

We should note that the **Gompertz model** can also be viewed as a generalization of the exponential model $dP/dt = rP$, with the constant r replaced by a variable $r(t)$ which itself decreases at a constant percentage rate; that is,

$$\frac{dr}{dt} = \alpha r,$$

where α is a negative constant. This gives r as $r_0 e^{\alpha t}$, so the Gompertz model takes the form

$$\frac{dP}{dt} = r_0 e^{\alpha t} P.$$

Generalizations of the Gompertz model include approaches where $r(t)$ is taken to be a polynomial in t of a fixed degree.

Demographers are using increasingly complex and sophisticated mathematical models of both deterministic and probabilistic character to study changes in population growth in the past and to make projections about the future. Although there are many different approaches to the creation of such models, most of them are extensions or generalizations of a first-order nonlinear differential equation.

REFERENCES

1. Pearl, Raymond, and Lowell Reed. "On the Rate of Growth of the United States Population Since 1790 and Its Mathematical Representation." *Proceedings of the National Academy of Sciences* **6** (1920): 275–288.

2. Pielou, E. C. *An Introduction to Mathematical Ecology*, second edition. New York: Wiley, 1977.

3. Malthus, Thomas R. *An Essay on the Principle of Population and a Summary View of the Principle of Population*. Baltimore: Penguin, 1970.

C H A P T E R 4

LINEAR DIFFERENTIAL EQUATIONS OF HIGHER ORDER

4.1 Preliminary Theory

4.2 Constructing a Second Solution from a Known Solution

4.3 Homogeneous Linear Equations with Constant Coefficients

4.4 Undetermined Coefficients— Superposition Approach

4.5 Differential Operators

4.6 Undetermined Coefficients— Annihilator Approach

4.7 Variation of Parameters

Chapter 4 Review

Chapter 4 Review Exercises

Essay: Chaos

Important Concepts

Initial-value problem

Initial conditions

Boundary-value problem

Boundary conditions

Linear dependence

Linear independence

Wronskian

Homogeneous equation

Nonhomogeneous equation

Superposition principle

Fundamental set of solutions

General solution

Complete solution

Particular solution

Complementary function

Reduction of order

Auxiliary equation

Characteristic equation

Euler's formula

Undetermined coefficients

Differential operator

Annihilator operator

Variation of parameters

We turn now to the solution of differential equations of order two or higher. Although we can solve some *nonlinear* first-order equations by the techniques considered in Chapter 2, nonlinear equations of higher order generally defy solution. This does not mean a nonlinear equation has no solution, but rather there are no rules or methods whereby its solution can be exhibited in terms of elementary or other kinds of functions. As a consequence, in attempting to solve higher-order equations, we shall confine our attention to *linear* equations.

We begin this chapter by first examining the underlying theory of linear equations. As we did in Section 2.5, we set conditions on the differential equation under which we can obtain its *general solution*. Recall, a general solution contains all solutions of the equation on some interval. For the remainder of the chapter we then develop methods for obtaining a general solution for a linear equation with *constant*

Continues

Continued

coefficients. We shall see that our ability to solve an *n*th-order linear differential equation with constant coefficients hinges on our ability to solve an *n*th-degree polynomial equation. The method of solution of linear equations with *nonconstant coefficients* will be taken up in Chapter 6.

4.1   PRELIMINARY THEORY

We begin the discussion of higher-order differential equations, as we did with first-order equations, with the notion of an initial-value problem. However, we confine our attention to linear differential equations.

4.1.1   Initial-Value and Boundary-Value Problems

Initial-Value Problem

For a linear *n*th-order differential equation, the problem

$$\textit{Solve:} \quad a_n(x)\frac{d^n y}{dx^n} + a_{n-1}(x)\frac{d^{n-1}y}{dx^{n-1}} + \cdots + a_1(x)\frac{dy}{dx} + a_0(x)y = g(x)$$

(1)

$$\textit{Subject to:} \quad y(x_0) = y_0, \quad y'(x_0) = y_0', \ldots, y^{(n-1)}(x_0) = y_0^{(n-1)},$$

where $y_0, y_0', \ldots, y_0^{(n-1)}$ are arbitrary constants, is called an **initial-value problem**. The specified values $y(x_0) = y_0, y'(x_0) = y_0', \ldots, y^{(n-1)}(x_0) = y_0^{(n-1)}$ are called **initial conditions**. We seek a solution on some interval I containing x_0.

In the case of a linear second-order equation, a solution of the initial-value problem

$$a_2(x)\frac{d^2 y}{dx^2} + a_1(x)\frac{dy}{dx} + a_0(x)y = g(x), \qquad y(x_0) = y_0, \quad y'(x_0) = y_0',$$

is a function satisfying the differential equation on I whose graph passes through (x_0, y_0) such that the slope of the curve at the point is the number y_0'. See Figure 4.1.

The next theorem gives sufficient conditions for the existence of a unique solution to (1).

solutions of the DE

$m = y_0'$

(x_0, y_0)

I

Figure 4.1

> **THEOREM 4.1**   **Existence of a Unique Solution**
>
> Let $a_n(x)$, $a_{n-1}(x)$, ..., $a_1(x)$, $a_0(x)$, and $g(x)$ be continuous on an interval I and let $a_n(x) \neq 0$ for every x in this interval. If $x = x_0$ is any point in this interval, then a solution $y(x)$ of the initial-value problem (1) exists on the interval and is unique.

EXAMPLE 1

You should verify that the function $y = 3e^{2x} + e^{-2x} - 3x$ is a solution of the initial-value problem

$$y'' - 4y = 12x, \qquad y(0) = 4, \quad y'(0) = 1.$$

Now the differential equation is linear, the coefficients as well as $g(x) = 12x$ are continuous, and $a_2(x) = 1 \neq 0$ on any interval containing $x = 0$. We conclude from Theorem 4.1 that the given function is the unique solution. ■

EXAMPLE 2

The initial-value problem

$$3y''' + 5y'' - y' + 7y = 0, \qquad y(1) = 0, \quad y'(1) = 0, \quad y''(1) = 0,$$

possesses the trivial solution $y = 0$. Since the third-order equation is linear with constant coefficients, it follows that all the conditions of Theorem 4.1 are fulfilled. Hence $y = 0$ is the *only* solution on any interval containing $x = 1$. ■

EXAMPLE 3

The function $y = \frac{1}{4} \sin 4x$ is a solution of the initial-value problem

$$y'' + 16y = 0, \qquad y(0) = 0, \quad y'(0) = 1.$$

It follows from Theorem 4.1 that on any interval containing $x = 0$ the solution is unique. ■

The requirements in Theorem 4.1 that $a_i(x)$, $i = 0, 1, 2, \ldots, n$, be continuous and $a_n(x) \neq 0$ for every x in I are both important. Specifically, if $a_n(x) = 0$ for some x in the interval, then the solution of a linear initial-value problem may not be unique or even exist.

EXAMPLE 4

Verify that the function $y = cx^2 + x + 3$ is a solution of the initial-value problem

$$x^2y'' - 2xy' + 2y = 6, \qquad y(0) = 3, \quad y'(0) = 1,$$

on the interval $(-\infty, \infty)$ for any choice of the parameter c.

Solution   Since $y' = 2cx + 1$ and $y'' = 2c$, it follows that

$$x^2y'' - 2xy' + 2y = x^2(2c) - 2x(2cx + 1) + 2(cx^2 + x + 3)$$
$$= 2cx^2 - 4cx^2 - 2x + 2cx^2 + 2x + 6 = 6.$$

Also $y(0) = c(0)^2 + 0 + 3 = 3$

and $y'(0) = 2c(0) + 1 = 1.$ ∎

Although the differential equation in Example 4 is linear and the coefficients and $g(x) = 6$ are continuous everywhere, the obvious difficulties are that $a_2(x) = x^2$ is zero at $x = 0$ and that the initial conditions are also imposed at $x = 0$.

Boundary-Value Problem

Another type of problem consists of solving a differential equation of order two or greater in which the dependent variable y or its derivatives are specified at *different points*. A problem such as

$$\textit{Solve:} \quad a_2(x)\frac{d^2y}{dx^2} + a_1(x)\frac{dy}{dx} + a_0(x)y = g(x)$$

$$\textit{Subject to:} \quad y(a) = y_0, \quad y(b) = y_1$$

is called a **two-point boundary-value problem**, or simply a **boundary-value problem**. The specified values $y(a) = y_0$ and $y(b) = y_1$ are called **boundary conditions**. A solution of the foregoing problem is a function satisfying the differential equation on some interval I, containing a and b, whose graph passes through the two points (a, y_0) and (b, y_1). See Figure 4.2.

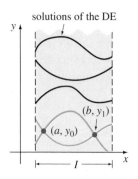

solutions of the DE

Figure 4.2

EXAMPLE 5

You should verify that on the interval $(0, \infty)$ the function $y = 3x^2 - 6x + 3$ satisfies both the differential equation and the boundary conditions of the two-point boundary-value problem

$$x^2y'' - 2xy' + 2y = 6, \qquad y(1) = 0, \quad y(2) = 3.$$ ∎

For a second-order differential equation, other pairs of boundary conditions could be

$$y'(a) = y_0', \qquad y(b) = y_1;$$

$$y(a) = y_0, \qquad y'(b) = y_1';$$

or

$$y'(a) = y_0', \qquad y'(b) = y_1',$$

where y_0, y_0', y_1, and y_1' denote arbitrary constants. These three pairs of conditions are just special cases of the general boundary conditions

$$\alpha_1 y(a) + \beta_1 y'(a) = \gamma_1$$

$$\alpha_2 y(b) + \beta_2 y'(b) = \gamma_2.$$

The next examples show that even when the conditions of Theorem 4.1 are fulfilled, a boundary-value problem may have

  (i) several solutions (as suggested in Figure 4.2),

  (ii) a unique solution, or

  (iii) no solution at all.

EXAMPLE 6

In Example 6 of Section 1.1, we saw that a two-parameter family of solutions for the differential equation $y'' + 16y = 0$ is

$$y = c_1 \cos 4x + c_2 \sin 4x.$$

Suppose we now wish to determine that solution of the equation that further satisfies the boundary conditions

$$y(0) = 0, \qquad y\left(\frac{\pi}{2}\right) = 0.$$

Observe that the first condition

$$0 = c_1 \cos 0 + c_2 \sin 0$$

implies $c_1 = 0$, so that $y = c_2 \sin 4x$. But when $x = \pi/2$, we have

$$0 = c_2 \sin 2\pi.$$

Since $\sin 2\pi = 0$, this latter condition is satisfied for any choice of c_2, so it follows that a solution of the problem

$$y'' + 16y = 0, \qquad y(0) = 0, \quad y\left(\frac{\pi}{2}\right) = 0,$$

is the one-parameter family

$$y = c_2 \sin 4x.$$

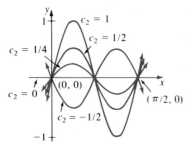

Figure 4.3

As Figure 4.3 shows, there are an infinite number of functions satisfying the differential equation whose graphs pass through the two points $(0, 0)$ and $(\pi/2, 0)$.

If the boundary conditions were $y(0) = 0$ and $y(\pi/8) = 0$, then necessarily c_1 and c_2 would both equal zero. Thus $y = 0$ would be a solution of this new boundary-value problem. In fact, as we shall see later on in this section, it is the only solution. ∎

EXAMPLE 7

The boundary-value problem

$$y'' + 16y = 0, \qquad y(0) = 0, \quad y\left(\frac{\pi}{2}\right) = 1,$$

has no solution in the family $y = c_1 \cos 4x + c_2 \sin 4x$. As in Example 6, the condition $y(0) = 0$ still implies that $c_1 = 0$. Thus $y = c_2 \sin 4x$ so that when $x = \pi/2$, we obtain the contradiction $1 = c_2 \cdot 0 = 0$. ∎

Boundary-value problems are often encountered in the applications of partial differential equations.

4.1.2   Linear Dependence and Linear Independence

The next two concepts are basic to the study of linear differential equations.

DEFINITION 4.1   Linear Dependence

A set of functions $f_1(x), f_2(x), \ldots, f_n(x)$ is said to be **linearly dependent** on an interval I if there exist constants $c_1, c_2, \ldots, c_n$, not all zero, such that

$$c_1 f_1(x) + c_2 f_2(x) + \cdots + c_n f_n(x) = 0$$

for every x in the interval.

DEFINITION 4.2   Linear Independence

A set of functions $f_1(x), f_2(x), \ldots, f_n(x)$ is said to be **linearly independent** on an interval I if it is not linearly dependent on the interval.

In other words, a set of functions is linearly independent on an interval if the only constants for which

$$c_1 f_1(x) + c_2 f_2(x) + \cdots + c_n f_n(x) = 0,$$

for every x in the interval, are $c_1 = c_2 = \cdots = c_n = 0$.

It is easy to understand these definitions in the case of two functions $f_1(x)$ and $f_2(x)$. If the functions are linearly dependent on an interval, then there exist constants c_1 and c_2 that are not both zero such that for every x in the interval,

$$c_1 f_1(x) + c_2 f_2(x) = 0.$$

Therefore, if we assume that $c_1 \neq 0$, it follows that

$$f_1(x) = -\frac{c_2}{c_1} f_2(x);$$

that is, *if two functions are linearly dependent, then one is simply a constant multiple of the other.* Conversely, if $f_1(x) = c_2 f_2(x)$ for some constant c_2, then

$$(-1) \cdot f_1(x) + c_2 f_2(x) = 0$$

for every x on some interval. Hence the functions are linearly dependent since at least one of the constants (namely, $c_1 = -1$) is not zero. We conclude that *two functions are linearly independent when neither is a constant multiple of the other* on an interval.

EXAMPLE 8

The functions $f_1(x) = \sin 2x$ and $f_2(x) = \sin x \cos x$ are linearly dependent on the interval $(-\infty, \infty)$ since

$$c_1 \sin 2x + c_2 \sin x \cos x = 0$$

is satisfied for every real x if we choose $c_1 = 1/2$ and $c_2 = -1$. (Recall the trigonometric identity $\sin 2x = 2 \sin x \cos x$.) ∎

EXAMPLE 9

The functions $f_1(x) = x$ and $f_2(x) = |x|$ are linearly independent on the interval $(-\infty, \infty)$. Inspection of Figure 4.4 should convince you that neither function is a constant multiple of the other. Thus in order to have $c_1 f_1(x) + c_2 f_2(x) = 0$ for every real x, we must choose $c_1 = 0$ and $c_2 = 0$. ∎

In the consideration of linear dependence or linear independence, the interval on which the functions are defined is important. The functions $f_1(x) = x$ and $f_2(x) = |x|$ in Example 9 are linearly dependent on the interval $(0, \infty)$ since

$$c_1 x + c_2 |x| = c_1 x + c_2 x = 0$$

is satisfied for any nonzero choice of c_1 and c_2 for which $c_1 = -c_2$.

(a)

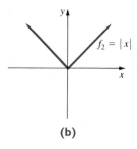

(b)

Figure 4.4

EXAMPLE 10

The functions $f_1(x) = \cos^2 x$, $f_2(x) = \sin^2 x$, $f_3(x) = \sec^2 x$, $f_4(x) = \tan^2 x$ are linearly dependent on the interval $(-\pi/2, \pi/2)$ since

$$c_1 \cos^2 x + c_2 \sin^2 x + c_3 \sec^2 x + c_4 \tan^2 x = 0,$$

when $c_1 = c_2 = 1$, $c_3 = -1$, $c_4 = 1$. We note that $\cos^2 x + \sin^2 x = 1$ and $1 + \tan^2 x = \sec^2 x$. ∎

A set of functions $f_1(x), f_2(x), \ldots, f_n(x)$ is linearly dependent on an interval if at least one function can be expressed as a linear combination of the remaining functions.

EXAMPLE 11

The functions $f_1(x) = \sqrt{x} + 5$, $f_2(x) = \sqrt{x} + 5x$, $f_3(x) = x - 1$, $f_4(x) = x^2$ are linearly dependent on the interval $(0, \infty)$ since f_2 can be written as a linear combination of f_1, f_3, and f_4. Observe that

$$f_2(x) = 1 \cdot f_1(x) + 5 \cdot f_3(x) + 0 \cdot f_4(x)$$

for every x in the interval $(0, \infty)$. ∎

Wronskian

The following theorem provides a sufficient condition for the linear independence of n functions on an interval. Each function is assumed to be differentiable at least $n - 1$ times.

THEOREM 4.2 **Criterion for Linearly Independent Functions**

Suppose $f_1(x), f_2(x), \ldots, f_n(x)$ possess at least $n - 1$ derivatives. If the determinant

$$\begin{vmatrix} f_1 & f_2 & \cdots & f_n \\ f_1' & f_2' & \cdots & f_n' \\ \vdots & \vdots & & \vdots \\ f_1^{(n-1)} & f_2^{(n-1)} & \cdots & f_n^{(n-1)} \end{vmatrix}$$

is not zero for at least one point in the interval I, then the functions $f_1(x), f_2(x), \ldots, f_n(x)$ are linearly independent on the interval.

The determinant in the preceding theorem is denoted by

$$W(f_1(x), f_2(x), \ldots, f_n(x))$$

and is called the **Wronskian*** of the functions.

Proof   We prove Theorem 4.2 by contradiction for the case when $n = 2$. Assume that $W(f_1(x_0), f_2(x_0)) \neq 0$ for a fixed x_0 in the interval I and that $f_1(x)$ and $f_2(x)$ are linearly dependent on the interval. The fact that the functions are linearly dependent means there exist constants c_1 and c_2, not both zero, for which

$$c_1 f_1(x) + c_2 f_2(x) = 0$$

for every x in I. Differentiating this combination then gives

$$c_1 f'_1(x) + c_2 f'_2(x) = 0.$$

Thus we obtain the system of linear equations

$$c_1 f_1(x) + c_2 f_2(x) = 0$$

$$c_1 f'_1(x) + c_2 f'_2(x) = 0. \tag{2}$$

But the linear dependence of f_1 and f_2 implies that (2) possesses a nontrivial solution for each x in the interval. Hence

$$W(f_1(x), f_2(x)) = \begin{vmatrix} f_1(x) & f_2(x) \\ f'_1(x) & f'_2(x) \end{vmatrix} = 0$$

for every x in I.[†] This contradicts the assumption that $W(f_1(x_0), f_2(x_0)) \neq 0$. We conclude that f_1 and f_2 are linearly independent.   □

COROLLARY

If $f_1(x), f_2(x), \ldots, f_n(x)$ possess at least $n - 1$ derivatives and are linearly dependent on I, then

$$W(f_1(x), f_2(x), \ldots, f_n(x)) = 0$$

for every x in the interval.

---

*Josef Maria Hoëne Wronski (1778–1853)   Born in Poland and educated in Germany, Wronski lived most of his life in France. More a philosopher than a mathematician, he believed that absolute truth could be attained through mathematics. Wronski's only noteworthy contribution to mathematics was the above determinant. Always an eccentric, he eventually went insane.

---

† See Appendix III for a review of determinants.

EXAMPLE 12

The functions $f_1(x) = \sin^2 x$ and $f_2(x) = 1 - \cos 2x$ are linearly dependent on $(-\infty, \infty)$. (Why?) By the preceding corollary, $W(\sin^2 x, 1 - \cos 2x) = 0$ for every real number. To see this we observe

$$W(\sin^2 x, 1 - \cos 2x) = \begin{vmatrix} \sin^2 x & 1 - \cos 2x \\ 2 \sin x \cos x & 2 \sin 2x \end{vmatrix}$$

$$= 2 \sin^2 x \sin 2x - 2 \sin x \cos x$$

$$+ 2 \sin x \cos x \cos 2x$$

$$= \sin 2x [2 \sin^2 x - 1 + \cos 2x]$$

$$= \sin 2x [2 \sin^2 x - 1 + \cos^2 x - \sin^2 x]$$

$$= \sin 2x [\sin^2 x + \cos^2 x - 1] = 0$$

Here we have used the trigonometric identities $\sin 2x = 2 \sin x \cos x$, $\cos 2x = \cos^2 x - \sin^2 x$, and $\sin^2 x + \cos^2 x = 1$. ∎

EXAMPLE 13

For $f_1(x) = e^{m_1 x}$, $f_2(x) = e^{m_2 x}$, $m_1 \neq m_2$,

$$W(e^{m_1 x}, e^{m_2 x}) = \begin{vmatrix} e^{m_1 x} & e^{m_2 x} \\ m_1 e^{m_1 x} & m_2 e^{m_2 x} \end{vmatrix} = (m_2 - m_1) e^{(m_1 + m_2)x} \neq 0$$

for every real value of x. Thus f_1 and f_2 are linearly independent on any interval of the x-axis. ∎

EXAMPLE 14

If α and β are real numbers, $\beta \neq 0$, then $y_1 = e^{\alpha x} \cos \beta x$ and $y_2 = e^{\alpha x} \sin \beta x$ are linearly independent on any interval of the x-axis since

$$W(e^{\alpha x} \cos \beta x, e^{\alpha x} \sin \beta x) = \begin{vmatrix} e^{\alpha x} \cos \beta x & e^{\alpha x} \sin \beta x \\ -\beta e^{\alpha x} \sin \beta x + \alpha e^{\alpha x} \cos \beta x & \beta e^{\alpha x} \cos \beta x + \alpha e^{\alpha x} \sin \beta x \end{vmatrix}$$

$$= \beta e^{2\alpha x} (\cos^2 \beta x + \sin^2 \beta x) = \beta e^{2\alpha x} \neq 0.$$

Notice when $\alpha = 0$ we see that $\cos \beta x$ and $\sin \beta x$, $\beta \neq 0$, are also linearly independent on any interval of the x-axis. ∎

EXAMPLE 15

The functions $f_1(x) = e^x$, $f_2(x) = xe^x$, and $f_3(x) = x^2 e^x$ are linearly independent on any interval of the x-axis since

$$W(e^x, xe^x, x^2e^x) = \begin{vmatrix} e^x & xe^x & x^2e^x \\ e^x & xe^x + e^x & x^2e^x + 2xe^x \\ e^x & xe^x + 2e^x & x^2e^x + 4xe^x + 2e^x \end{vmatrix} = 2e^{3x}$$

is not zero for any real value of x. ∎

EXAMPLE 16

In Example 9 we saw that $f_1(x) = x$ and $f_2(x) = |x|$ are linearly independent on $(-\infty, \infty)$; however, we cannot compute the Wronskian since f_2 is not differentiable at $x = 0$. ∎

We leave it as an exercise to show that a set of functions $f_1(x)$, $f_2(x), \ldots, f_n(x)$ could be linearly independent on some interval and yet have a vanishing Wronskian. See Problem 30. In other words, if $W(f_1(x), f_2(x), \ldots, f_n(x)) = 0$ for every x in an interval, it does not necessarily mean that the functions are linearly dependent.

4.1.3 Solutions of Linear Equations

Homogeneous Equations

A linear nth-order differential equation of the form

$$a_n(x)\frac{d^ny}{dx^n} + a_{n-1}(x)\frac{d^{n-1}y}{dx^{n-1}} + \cdots + a_1(x)\frac{dy}{dx} + a_0(x)y = 0 \qquad (3)$$

is said to be **homogeneous**, whereas

$$a_n(x)\frac{d^ny}{dx^n} + a_{n-1}(x)\frac{d^{n-1}y}{dx^{n-1}} + \cdots + a_1(x)\frac{dy}{dx} + a_0(x)y = g(x), \qquad (4)$$

$g(x)$ not identically zero, is said to be **nonhomogeneous**.

The word *homogeneous* in this context does not refer to coefficients that are homogeneous functions. See Section 2.3.

EXAMPLE 17

(a) The equation $\qquad\qquad 2y'' + 3y' - 5y = 0$

is a homogeneous linear second-order ordinary differential equation.

(b) The equation $\qquad\quad x^3y''' - 2xy'' + 5y' + 6y = e^x$

is a nonhomogeneous linear third-order ordinary differential equation. ∎

We shall see in the latter part of this section, as well as in the subsequent sections of this chapter, that in order to solve a nonhomogeneous equation (4), we must first solve the associated homogeneous equation (3).

Note   To avoid needless repetition throughout the remainder of this text we shall, as a matter of course, make the following important assumptions when giving definitions and proving theorems about the linear equations (3) and (4). On some common interval I,

- the coefficients $a_i(x)$, $i = 0, 1, \ldots, n$, are continuous;
- the right-hand member $g(x)$ is continuous; and
- $a_n(x) \neq 0$ for every x in the interval.

Superposition Principle

In the next theorem we see that the sum, or **superposition**, of two or more solutions of a homogeneous linear differential equation is also a solution.

> **THEOREM 4.3   Superposition Principle—Homogeneous Equations**
>
> Let $y_1, y_2, \ldots, y_k$ be solutions of the homogeneous linear nth-order differential equation (3) on an interval I. Then the linear combination
>
> $$y = c_1 y_1(x) + c_2 y_2(x) + \cdots + c_k y_k(x), \tag{5}$$
>
> where the c_i, $i = 1, 2, \ldots, k$, are arbitrary constants, is also a solution on the interval.

Proof   We prove the case when $n = k = 2$. Let $y_1(x)$ and $y_2(x)$ be solutions of

$$a_2(x)y'' + a_1(x)y' + a_0(x)y = 0.$$

If we define $y = c_1 y_1(x) + c_2 y_2(x)$, then

$$a_2(x)[c_1 y_1'' + c_2 y_2''] + a_1(x)[c_1 y_1' + c_2 y_2'] + a_0(x)[c_1 y_1 + c_2 y_2]$$

$$= c_1 \underbrace{[a_2(x)y_1'' + a_1(x)y_1' + a_0(x)y_1]}_{\text{zero}} + c_2 \underbrace{[a_2(x)y_2'' + a_1(x)y_2' + a_0(x)y_2]}_{\text{zero}}$$

$$= c_1 \cdot 0 + c_2 \cdot 0 = 0. \qquad \square$$

> **COROLLARIES**
>
> **(A)**   A constant multiple $y = c_1 y_1(x)$ of a solution $y_1(x)$ of a homogeneous linear differential equation is also a solution.
>
> **(B)**   A homogeneous linear differential equation always possesses the trivial solution $y = 0$.

The superposition principle defined by (5) and its special case given in Corollary (A) are properties that nonlinear differential equations, in general, do not possess. See Problems 31 and 32.

EXAMPLE 18

The functions $y_1 = x^2$ and $y_2 = x^2 \ln x$

are both solutions of the homogeneous third-order equation

$$x^3 y''' - 2xy' + 4y = 0$$

on the interval $(0, \infty)$. By the superposition principle the linear combination

$$y = c_1 x^2 + c_2 x^2 \ln x$$

is also a solution of the equation on the interval.   ■

EXAMPLE 19

The functions $y_1 = e^x$, $y_2 = e^{2x}$, and $y_3 = e^{3x}$ all satisfy the homogeneous equation

$$\frac{d^3 y}{dx^3} - 6\frac{d^2 y}{dx^2} + 11\frac{dy}{dx} - 6y = 0$$

on $(-\infty, \infty)$. By Theorem 4.3 another solution is

$$y = c_1 e^x + c_2 e^{2x} + c_3 e^{3x}.$$ ■

EXAMPLE 20

The function $y = x^2$ is a solution of the homogeneous linear equation

$$x^2 y'' - 3xy' + 4y = 0$$

on $(0, \infty)$. Hence $y = cx^2$ is also a solution. For various values of c we see that $y = 3x^2$, $y = ex^2$, $y = 0, \ldots$ are all solutions of the equation on the interval.

■

Linearly Independent Solutions

We are interested in determining when n solutions, $y_1, y_2, \ldots, y_n$, of the **homogeneous** differential equation (3) are linearly independent. Surprisingly the nonvanishing of the Wronskian of a set of n such solutions on an interval I is both necessary and sufficient for linear independence.

THEOREM 4.4 **Criterion for Linearly Independent Solutions**

Let $y_1, y_2, \ldots, y_n$ be n solutions of the homogeneous linear nth-order differential equation (3) on an interval I. Then the set of solutions is **linearly independent** on I if and only if

$$W(y_1, y_2, \ldots, y_n) \neq 0$$

for every x in the interval.

Proof We prove Theorem 4.4 for the case when $n = 2$. First, if $W(y_1, y_2) \neq 0$ for every x in I, it follows immediately from Theorem 4.2 that y_1 and y_2 are linearly independent. Next, we must show that if y_1 and y_2 are linearly independent solutions of a homogeneous linear second-order differential equation, then $W(y_1, y_2) \neq 0$ for every x in I. To show this, let us suppose y_1 and y_2 are linearly independent and there is some fixed x_0 in I for which $W(y_1(x_0), y_2(x_0)) = 0$. Hence there must exist c_1 and c_2, not both zero, such that

$$c_1 y_1(x_0) + c_2 y_2(x_0) = 0$$

$$c_1 y_1'(x_0) + c_2 y_2'(x_0) = 0. \tag{6}$$

If we define $\qquad\qquad y(x) = c_1 y_1(x) + c_2 y_2(x),$

then in view of (6), $y(x)$ must also satisfy

$$y(x_0) = 0, \qquad y'(x_0) = 0. \tag{7}$$

But the identically zero function satisfies both the differential equation and the initial conditions (7), and thus by Theorem 4.1 it is the only solution. In other words, $y = 0$ or

$$c_1 y_1(x) + c_2 y_2(x) = 0$$

for every x in I. This contradicts the assumption that y_1 and y_2 are linearly independent on the interval. $\qquad\qquad\qquad\qquad\qquad\qquad\qquad\qquad \square$

From the foregoing discussion we conclude that when $y_1, y_2, \ldots, y_n$ are n solutions of (3) on an interval I, the Wronskian either is identically zero or is never zero on the interval.

DEFINITION 4.3 **Fundamental Set of Solutions**

Any set $y_1, y_2, \ldots, y_n$ of n linearly independent solutions of the homogeneous linear nth-order differential equation (3) on an interval I is said to be a **fundamental set of solutions** on the interval.

THEOREM 4.5

Let $y_1, y_2, \ldots, y_n$ be a fundamental set of solutions of the homogeneous linear nth-order differential equation (3) on an interval I. Then for any solution $Y(x)$ of (3) on I, constants $C_1, C_2, \ldots, C_n$ can be found so that

$$Y = C_1 y_1(x) + C_2 y_2(x) + \cdots + C_n y_n(x).$$

Proof   We prove the case when $n = 2$. Let Y be a solution, and let y_1 and y_2 be linearly independent solutions of

$$a_2(x)y'' + a_1(x)y' + a_0(x)y = 0$$

on an interval I. Suppose $x = t$ is a point in this interval for which $W(y_1(t), y_2(t)) \neq 0$. Suppose also that the values of $Y(t)$ and $Y'(t)$ are given by

$$Y(t) = k_1, \qquad Y'(t) = k_2.$$

If we now examine the system of equations

$$C_1 y_1(t) + C_2 y_2(t) = k_1$$

$$C_1 y_1'(t) + C_2 y_2'(t) = k_2,$$

it follows that we can determine C_1 and C_2 uniquely provided the determinant of the coefficients satisfies

$$\begin{vmatrix} y_1(t) & y_2(t) \\ y_1'(t) & y_2'(t) \end{vmatrix} \neq 0.$$

But this latter determinant is simply the Wronskian evaluated at $x = t$, and by assumption $W \neq 0$. If we now define the function

$$G(x) = C_1 y_1(x) + C_2 y_2(x),$$

we observe:

(i) $G(x)$ satisfies the differential equation since it is the superposition of two known solutions y_1 and y_2.

(ii) $G(x)$ satisfies the initial conditions

$$G(t) = C_1 y_1(t) + C_2 y_2(t) = k_1$$

$$G'(t) = C_1 y_1'(t) + C_2 y_2'(t) = k_2.$$

(iii) $Y(x)$ satisfies the *same* linear equation and the *same* initial conditions.

Since the solution of this linear initial-value problem is unique (Theorem 4.1), we have $Y(x) = G(x)$, or

$$Y(x) = C_1 y_1(x) + C_2 y_2(x). \qquad \square$$

The basic question of whether a fundamental set of solutions exists for a linear equation is answered in the next theorem.

THEOREM 4.6   Existence of a Fundamental Set

There exists a fundamental set of solutions for the homogeneous linear nth-order differential equation (3) on an interval I.

The proof of this result follows from Theorem 4.1. The justification of Theorem 4.6 in the special case of second-order equations is left as an exercise.

Since we have shown that any solution of (3) is obtained from a linear combination of functions in a fundamental set of solutions, we are able to make the following definition.

DEFINITION 4.4   General Solution—Homogeneous Equations

Let $y_1, y_2, \ldots, y_n$ be a fundamental set of solutions of the homogeneous linear nth-order differential equation (3) on an interval I. The **general solution** of the equation on the interval is defined to be

$$y = c_1 y_1(x) + c_2 y_2(x) + \cdots + c_n y_n(x),$$

where the c_i, $i = 1, 2, \ldots, n$, are arbitrary constants.

Recall, the general solution as defined in Section 1.1 is also called the **complete solution** of the differential equation.

EXAMPLE 21

The second-order equation $y'' - 9y = 0$ possesses two solutions

$$y_1 = e^{3x} \quad \text{and} \quad y_2 = e^{-3x}.$$

Since

$$W(e^{3x}, e^{-3x}) = \begin{vmatrix} e^{3x} & e^{-3x} \\ 3e^{3x} & -3e^{-3x} \end{vmatrix} = -6 \neq 0$$

for every value of x, y_1 and y_2 form a fundamental set of solutions on $(-\infty, \infty)$. The general solution of the differential equation on the interval is

$$y = c_1 e^{3x} + c_2 e^{-3x}.$$ ∎

EXAMPLE 22

You should verify that the function $y = 4 \sinh 3x - 5e^{-3x}$ also satisfies the differential equation in Example 21. By choosing $c_1 = 2$, $c_2 = -7$ in the general solution $y = c_1 e^{3x} + c_2 e^{-3x}$, we obtain

$$y = 2e^{3x} - 7e^{-3x}$$

$$= 2e^{3x} - 2e^{-3x} - 5e^{-3x}$$

$$= 4\left(\frac{e^{3x} - e^{-3x}}{2}\right) - 5e^{-3x}$$

$$= 4 \sinh 3x - 5e^{-3x}.$$ ■

EXAMPLE 23

The functions $y_1 = e^x$, $y_2 = e^{2x}$, and $y_3 = e^{3x}$ satisfy the third-order equation

$$\frac{d^3 y}{dx^3} - 6 \frac{d^2 y}{dx^2} + 11 \frac{dy}{dx} - 6y = 0.$$

Since

$$W(e^x, e^{2x}, e^{3x}) = \begin{vmatrix} e^x & e^{2x} & e^{3x} \\ e^x & 2e^{2x} & 3e^{3x} \\ e^x & 4e^{2x} & 9e^{3x} \end{vmatrix} = 2e^{6x} \neq 0$$

for every real value of x, y_1, y_2, and y_3 form a fundamental set of solutions on $(-\infty, \infty)$. We conclude that

$$y = c_1 e^x + c_2 e^{2x} + c_3 e^{3x}$$

is the general solution of the differential equation on the interval.   ■

Nonhomogeneous Equations

We now turn our attention to defining the general solution of a **nonhomogeneous** linear equation. Any function y_p, free of arbitrary parameters, that satisfies (4) is said to be a **particular solution** of the equation (sometimes also called a **particular integral**).

EXAMPLE 24

(a) A particular solution of

$$y'' + 9y = 27$$

is $y_p = 3$ since $y_p'' = 0$ and $0 + 9y_p = 9(3) = 27$.

(b) $y_p = x^3 - x$ is a particular solution of

$$x^2 y'' + 2xy' - 8y = 4x^3 + 6x$$

since $y'_p = 3x^2 - 1$, $y''_p = 6x$, and

$$x^2 y''_p + 2xy'_p - 8y_p = x^2(6x) + 2x(3x^2 - 1) - 8(x^3 - x) = 4x^3 + 6x.$$

∎

THEOREM 4.7

Let $y_1, y_2, \ldots, y_k$ be solutions of the homogeneous linear nth-order differential equation (3) on an interval I and let y_p be any solution of the nonhomogeneous equation (4) on the same interval. Then

$$y = c_1 y_1(x) + c_2 y_2(x) + \cdots + c_k y_k(x) + y_p(x)$$

is also a solution of the nonhomogeneous equation on the interval for any constants $c_1, c_2, \ldots, c_k$.

We can now prove the following analogue of Theorem 4.5 for nonhomogeneous differential equations.

THEOREM 4.8

Let y_p be a given solution of the nonhomogeneous linear nth-order differential equation (4) on an interval I and let $y_1, y_2, \ldots, y_n$ be a fundamental set of solutions of the associated homogeneous equation (3) on the interval. Then for any solution $Y(x)$ of (4) on I, constants $C_1, C_2, \ldots, C_n$ can be found so that

$$Y = C_1 y_1(x) + C_2 y_2(x) + \cdots + C_n y_n(x) + y_p(x).$$

Proof We prove the case when $n = 2$. Suppose Y and y_p are both solutions of

$$a_2(x)y'' + a_1(x)y' + a_0(x)y = g(x).$$

If we define a function u by $u(x) = Y(x) - y_p(x)$, then

$$a_2(x)u'' + a_1(x)u' + a_0(x)u$$
$$= a_2(x)[Y'' - y''_p] + a_1(x)[Y' - y'_p] + a_0(x)[Y - y_p]$$
$$= a_2(x)Y'' + a_1(x)Y' + a_0(x)Y - [a_2(x)y''_p + a_1(x)y'_p + a_0(x)y_p]$$
$$= g(x) - g(x) = 0.$$

Therefore, in view of Definition 4.4 and Theorem 4.5, we can write

$$u(x) = C_1 y_1(x) + C_2 y_2(x)$$

$$Y(x) - y_p(x) = C_1 y_1(x) + C_2 y_2(x)$$

or
$$Y(x) = C_1 y_1(x) + C_2 y_2(x) + y_p(x).$$ □

Thus we come to the last definition of this section.

DEFINITION 4.5   **General Solution—Nonhomogeneous Equations**

Let y_p be a given solution of the nonhomogeneous linear nth-order differential equation (4) on an interval I and let

$$y_c = c_1 y_1(x) + c_2 y_2(x) + \cdots + c_n y_n(x)$$

denote the general solution of the associated homogeneous equation (3) on the interval. The **general solution** of the nonhomogeneous equation on the interval is defined to be

$$y = c_1 y_1(x) + c_2 y_2(x) + \cdots + c_n y_n(x) + y_p(x) = y_c(x) + y_p(x).$$

Complementary Function

In Definition 4.5 the linear combination

$$y_c(x) = c_1 y_1(x) + c_2 y_2(x) + \cdots + c_n y_n(x),$$

which is the general solution of (3), is called the **complementary function** for equation (4). In other words, the general solution of a nonhomogeneous linear differential equation is

$$y = complementary\ function + any\ particular\ solution.$$

E X A M P L E   **25**

By substitution, the function $y_p = -\dfrac{11}{12} - \dfrac{1}{2} x$ is readily shown to be a particular solution of the nonhomogeneous equation

$$\frac{d^3 y}{dx^3} - 6 \frac{d^2 y}{dx^2} + 11 \frac{dy}{dx} - 6y = 3x. \tag{8}$$

In order to write the general solution of (8), we must also be able to solve the associated homogeneous equation

$$\frac{d^3 y}{dx^3} - 6 \frac{d^2 y}{dx^2} + 11 \frac{dy}{dx} - 6y = 0.$$

But in Example 23 we saw that the general solution of this latter equation on the interval $(-\infty, \infty)$ was

$$y_c = c_1 e^x + c_2 e^{2x} + c_3 e^{3x}.$$

Hence the general solution of (8) on the interval is

$$y = y_c + y_p = c_1 e^x + c_2 e^{2x} + c_3 e^{3x} - \frac{11}{12} - \frac{1}{2} x. \qquad \blacksquare$$

Another Superposition Principle

The last theorem of this discussion will be useful in Section 4.4 when we consider a method for finding particular solutions of nonhomogeneous equations.

THEOREM 4.9   Superposition Principle—Nonhomogeneous Equations

Let $y_{p_1}, y_{p_2}, \ldots, y_{p_k}$ be k particular solutions of the linear nth-order differential equation (4) on an interval I corresponding, in turn, to k distinct functions $g_1, g_2, \ldots, g_k$. That is, suppose y_{p_i} denotes a particular solution of the corresponding differential equation

$$a_n(x)y^{(n)} + a_{n-1}(x)y^{(n-1)} + \cdots + a_1(x)y' + a_0(x)y = g_i(x),$$

where $i = 1, 2, \ldots, k$. Then

$$y_p = y_{p_1}(x) + y_{p_2}(x) + \cdots + y_{p_k}(x)$$

is a particular solution of

$$a_n(x)y^{(n)} + a_{n-1}(x)y^{(n-1)} + \cdots + a_1(x)y' + a_0(x)y$$
$$= g_1(x) + g_2(x) + \cdots + g_k(x).$$

We leave the proof of this result when $k = 2$ as an exercise. See Problem 50.

EXAMPLE 26

You should verify that

$y_{p_1} = -4x^2$ is a particular solution of $y'' - 3y' + 4y = -16x^2 + 24x - 8$

$y_{p_2} = e^{2x}$ is a particular solution of $y'' - 3y' + 4y = 2e^{2x}$

$y_{p_3} = xe^x$ is a particular solution of $y'' - 3y' + 4y = 2xe^x - e^x.$

It follows from Theorem 4.9 that the superposition of y_{p_1}, y_{p_2}, and y_{p_3},

$$y = y_{p_1} + y_{p_2} + y_{p_3} = -4x^2 + e^{2x} + xe^x,$$

is a solution of

$$y'' - 3y' + 4y = \underbrace{-16x^2 + 24x - 8}_{g_1(x)} + \underbrace{2e^{2x}}_{g_2(x)} + \underbrace{2xe^x - e^x}_{g_3(x)}. \qquad \blacksquare$$

Before we actually start solving homogeneous and nonhomogeneous linear differential equations, we need the one additional bit of theory presented in the next section.

Remark   A physical system that changes with time and whose mathematical model is a linear differential equation

$$a_n(t)y^{(n)} + a_{n-1}(t)y^{(n-1)} + \cdots + a_1(t)y' + a_0(t)y = g(t)$$

is said to be a **linear system**. The values of the variables $y(t)$, $y'(t)$, ..., $y^{(n-1)}(t)$ at a specific time t_0 describe the **state** of the system. The function g is variously called the **input function, forcing function**, or **excitation function**. A solution $y(t)$ of the differential equation is said to be the **output** or **response** of the system. The dependence of the output on the input is illustrated in Figure 4.5.

Figure 4.5

In order that a physical system be a linear system it is necessary that the superposition principle (Theorem 4.9) hold in the system; that is, the response of the system to a superposition of inputs is a superposition of outputs.

4.1 **EXERCISES**   *Answers to odd-numbered problems begin on page A–8.*

[4.1.1]   **1.** Given that

$$y = c_1 e^x + c_2 e^{-x}$$

is a two-parameter family of solutions of $y'' - y = 0$ on the interval $(-\infty, \infty)$, find a member of the family satisfying the initial conditions $y(0) = 0$, $y'(0) = 1$.

2. Find a solution of the differential equation in Problem 1 satisfying the boundary conditions $y(0) = 0$, $y(1) = 1$.

3. Given that
$$y = c_1 e^{4x} + c_2 e^{-x}$$
is a two-parameter family of solutions of $y'' - 3y' - 4y = 0$ on the interval $(-\infty, \infty)$, find a member of the family satisfying the initial conditions $y(0) = 1$, $y'(0) = 2$.

4. Given that
$$y = c_1 + c_2 \cos x + c_3 \sin x$$
is a three-parameter family of solutions of $y''' + y' = 0$ on the interval $(-\infty, \infty)$, find a member of the family satisfying the initial conditions $y(\pi) = 0$, $y'(\pi) = 2$, $y''(\pi) = -1$.

5. Given that
$$y = c_1 x + c_2 x \ln x$$
is a two-parameter family of solutions of $x^2 y'' - xy' + y = 0$ on the interval $(-\infty, \infty)$, find a member of the family satisfying the initial conditions $y(1) = 3$, $y'(1) = -1$.

6. Given that
$$y = c_1 + c_2 x^2$$
is a two-parameter family of solutions of $xy'' - y' = 0$ on the interval $(-\infty, \infty)$, show that constants c_1 and c_2 cannot be found so that a member of the family satisfies the initial conditions $y(0) = 0$, $y'(0) = 1$. Explain why this does not violate Theorem 4.1.

7. Find two members of the family of solutions of $xy'' - y' = 0$ given in Problem 6 satisfying the initial conditions $y(0) = 0$, $y'(0) = 0$.

8. Find a member of the family of solutions of $xy'' - y' = 0$ given in Problem 6 satisfying the boundary conditions $y(0) = 1$, $y'(1) = 6$. Does Theorem 4.1 guarantee that this solution is unique?

9. Given that
$$y = c_1 e^x \cos x + c_2 e^x \sin x$$
is a two-parameter family of solutions of $y'' - 2y' + 2y = 0$ on the interval $(-\infty, \infty)$, determine whether a member of the family can be found that satisfies the conditions

(a) $y(0) = 1$, $y'(0) = 0$
(b) $y(0) = 1$, $y(\pi) = -1$
(c) $y(0) = 1$, $y(\pi/2) = 1$
(d) $y(0) = 0$, $y(\pi) = 0$.

10. Given that
$$y = c_1 x^2 + c_2 x^4 + 3$$
is a two-parameter family of solutions of $x^2 y'' - 5xy' + 8y = 24$ on the interval $(-\infty, \infty)$, determine whether a member of the family can be found that satisfies the conditions

(a) $y(-1) = 0$, $y(1) = 4$
(b) $y(0) = 1$, $y(1) = 2$
(c) $y(0) = 3$, $y(1) = 0$
(d) $y(1) = 3$, $y(2) = 15$.

In Problems 11 and 12 find an interval around $x = 0$ for which the given initial-value problem has a unique solution.

11. $(x - 2)y'' + 3y = x$; $y(0) = 0$, $y'(0) = 1$

12. $y'' + (\tan x)y = e^x$; $y(0) = 1$, $y'(0) = 0$

13. Given that $y = c_1 \cos \lambda x + c_2 \sin \lambda x$ is a family of solutions of the differential equation $y'' + \lambda^2 y = 0$, determine the values of the parameter λ for which the boundary-value problem

$$y'' + \lambda^2 y = 0, \qquad y(0) = 0, \quad y(\pi) = 0,$$

has nontrivial solutions.

14. Determine the values of the parameter λ for which the boundary-value problem

$$y'' + \lambda^2 y = 0, \qquad y(0) = 0, \quad y(5) = 0,$$

has nontrivial solutions. See Problem 13.

[4.1.2] In Problems 15–22 determine whether the given functions are linearly independent or dependent on $(-\infty, \infty)$.

15. $f_1(x) = x$, $\quad f_2(x) = x^2$, $\quad f_3(x) = 4x - 3x^2$

16. $f_1(x) = 0$, $\quad f_2(x) = x$, $\quad f_3(x) = e^x$

17. $f_1(x) = 5$, $\quad f_2(x) = \cos^2 x$, $\quad f_3(x) = \sin^2 x$

18. $f_1(x) = \cos 2x$, $\quad f_2(x) = 1$, $\quad f_3(x) = \cos^2 x$

19. $f_1(x) = x$, $\quad f_2(x) = x - 1$, $\quad f_3(x) = x + 3$

20. $f_1(x) = 2 + x$, $\quad f_2(x) = 2 + |x|$

21. $f_1(x) = 1 + x$, $\quad f_2(x) = x$, $\quad f_3(x) = x^2$

22. $f_1(x) = e^x$, $\quad f_2(x) = e^{-x}$, $\quad f_3(x) = \sinh x$

In Problems 23–28 show by computing the Wronskian that the given functions are linearly independent on the indicated interval.

23. $x^{1/2}, x^2$; $\quad (0, \infty)$

24. $1 + x, x^3$; $\quad (-\infty, \infty)$

25. $\sin x, \csc x$; $\quad (0, \pi)$

26. $\tan x, \cot x$; $\quad (0, \pi/2)$

27. e^x, e^{-x}, e^{4x}; $\quad (-\infty, \infty)$

28. $x, x \ln x, x^2 \ln x$; $\quad (0, \infty)$

29. Observe that for the functions $f_1(x) = 2$ and $f_2(x) = e^x$,

$$1 \cdot f_1(0) - 2 \cdot f_2(0) = 0.$$

Does this imply that f_1 and f_2 are linearly dependent on any interval containing $x = 0$?

30. (a) Show graphically that $f_1(x) = x^2$ and $f_2(x) = x|x|$ are linearly independent on $(-\infty, \infty)$.

(b) Show that $W(f_1(x), f_2(x)) = 0$ for every real number.

[4.1.3] **31. (a)** Verify that $y = 1/x$ is a solution of the nonlinear differential equation $y'' = 2y^3$ on the interval $(0, \infty)$.

(b) Show that the constant multiple $y = c/x$ is not a solution of the equation when $c \neq 0, \pm 1$.

32. (a) Verify that $y_1 = 1$ and $y_2 = \ln x$ are solutions of the nonlinear differential equation $y'' + (y')^2 = 0$ on the interval $(0, \infty)$.

(b) Is $y_1 + y_2$ a solution of the equation? Is $c_1 y_1 + c_2 y_2$, c_1 and c_2 arbitrary, a solution of the equation?

In Problems 33–40 verify that the given functions form a fundamental set of solutions of the differential equation on the indicated interval. Form the general solution.

33. $y'' - y' - 12y = 0;$ $e^{-3x}, e^{4x}, (-\infty, \infty)$

34. $y'' - 4y = 0;$ $\cosh 2x, \sinh 2x, (-\infty, \infty)$

35. $y'' - 2y' + 5y = 0;$ $e^x \cos 2x, e^x \sin 2x, (-\infty, \infty)$

36. $4y'' - 4y' + y = 0;$ $e^{x/2}, xe^{x/2}, (-\infty, \infty)$

37. $x^2 y'' - 6xy' + 12y = 0;$ $x^3, x^4, (0, \infty)$

38. $x^2 y'' + xy' + y = 0;$ $\cos(\ln x), \sin(\ln x), (0, \infty)$

39. $x^3 y''' + 6x^2 y'' + 4xy' - 4y = 0;$ $x, x^{-2}, x^{-2} \ln x, (0, \infty)$

40. $y^{(4)} + y'' = 0;$ $1, x, \cos x, \sin x, (-\infty, \infty)$

In Problems 41–44 verify that the given two-parameter family of functions is the general solution of the nonhomogeneous differential equation on the indicated interval.

41. $y'' - 7y' + 10y = 24e^x$
$y = c_1 e^{2x} + c_2 e^{5x} + 6e^x, (-\infty, \infty)$

42. $y'' + y = \sec x$
$y = c_1 \cos x + c_2 \sin x + x \sin x + (\cos x) \ln(\cos x), (-\pi/2, \pi/2)$

43. $y'' - 4y' + 4y = 2e^{2x} + 4x - 12$
$y = c_1 e^{2x} + c_2 xe^{2x} + x^2 e^{2x} + x - 2, (-\infty, \infty)$

44. $2x^2 y'' + 5xy' + y = x^2 - x$
$y = c_1 x^{-1/2} + c_2 x^{-1} + \dfrac{1}{15} x^2 - \dfrac{1}{6} x, (0, \infty)$

45. (a) Verify that $y_1 = x^3$ and $y_2 = |x|^3$ are linearly independent solutions of the differential equation $x^2 y'' - 4xy' + 6y = 0$ on $(-\infty, \infty)$.

(b) Show that $W(y_1, y_2) = 0$ for every real number.

(c) Does the result of part (b) violate Theorem 4.4?

(d) Verify that $Y_1 = x^3$ and $Y_2 = x^2$ are also linearly independent solutions of the differential equation on the interval $(-\infty, \infty)$.

(e) Find a solution of the equation satisfying $y(0) = 0, y'(0) = 0$.

(f) By the superposition principle both linear combinations

$$y = c_1 y_1 + c_2 y_2 \quad \text{and} \quad y = c_1 Y_1 + c_2 Y_2$$

are solutions of the differential equation. Is one, both, or neither the general solution of the differential equation on $(-\infty, \infty)$?

46. Consider the second-order differential equation

$$a_2(x)y'' + a_1(x)y' + a_0(x)y = 0, \tag{9}$$

where $a_2(x)$, $a_1(x)$, and $a_0(x)$ are continuous on an interval I and $a_2(x) \neq 0$ for every x in the interval. From Theorem 4.1 there exists only one solution y_1 of the equation satisfying $y(x_0) = 1$ and $y'(x_0) = 0$, where x_0 is a point in I. Similarly there exists a unique solution y_2 of the equation satisfying $y(x_0) = 0$ and $y'(x_0) = 1$. Show that y_1 and y_2 form a fundamental set of solutions of the differential equation on the interval I.

47. Let y_1 and y_2 be two solutions of (9).

(a) If $W(y_1, y_2)$ is the Wronskian of y_1 and y_2, show that

$$a_2(x)\frac{dW}{dx} + a_1(x)W = 0.$$

(b) Derive **Abel's formula***

$$W = ce^{-\int [a_1(x)/a_2(x)]\, dx},$$

where c is a constant.

(c) Using an alternative form of Abel's formula,

$$W = ce^{-\int_{x_0}^{x} [a_1(t)/a_2(t)]\, dt},$$

for x_0 in I, show that

$$W(y_1, y_2) = W(x_0)e^{-\int_{x_0}^{x} [a_1(t)/a_2(t)]\, dt}.$$

(d) Show that if $W(x_0) = 0$, then $W = 0$ for every x in I, whereas if $W(x_0) \neq 0$, then $W \neq 0$ for every x in the interval.

*****Niels Henrik Abel** (1802–1829)   Abel was a brilliant Norwegian mathematician whose tragic death at age 26 due to tuberculosis was an inestimable loss for mathematics. His greatest achievement was the solution of a problem that baffled mathematicians for centuries: he showed that a general fifth-degree polynomial equation cannot be solved algebraically—that is, in terms of radicals. Abel's contemporary, the Frenchman Evariste Galois, then proved that it was impossible to solve *any* general polynomial equation of degree greater than four in an algebraic manner. Galois is another tragic figure in the history of mathematics; a political activist, he was killed in a duel at the age of 22.

In Problems 48 and 49 use the results of Problem 47.

48. If y_1 and y_2 are two solutions of

$$(1 - x^2)y'' - 2xy' + n(n + 1)y = 0$$

on $(-1, 1)$, show that $W(y_1, y_2) = c/(1 - x^2)$, where c is a constant.

49. In Chapter 6 we shall see that the solutions y_1 and y_2 of $xy'' + y' + xy = 0$, $0 < x < \infty$, are infinite series. Suppose we consider initial conditions

$$y_1(x_0) = k_1, \qquad y_1'(x_0) = k_2$$

and

$$y_2(x_0) = k_3, \qquad y_2'(x_0) = k_4$$

for $x_0 > 0$. Show that

$$W(y_1, y_2) = \frac{(k_1 k_4 - k_2 k_3)x_0}{x}.$$

50. Suppose the mathematical model of a linear system is given by

$$a_2(t)\frac{d^2y}{dt^2} + a_1(t)\frac{dy}{dt} + a_0(t) = E(t)$$

If y_1 is a response of the system to an input $E_1(t)$ and y_2 is a response of the same system to an input $E_2(t)$, show that $y_1 + y_2$ is a response of the system to the input $E_1(t) + E_2(t)$.

4.2 CONSTRUCTING A SECOND SOLUTION FROM A KNOWN SOLUTION

Reduction of Order

It is one of the more interesting as well as important facts in the study of linear *second-order* differential equations that we can construct a second solution from a *known* solution. Suppose $y_1(x)$ is a nonzero solution of the equation

$$a_2(x)y'' + a_1(x)y' + a_0(x)y = 0. \tag{1}$$

We assume, as we did in the preceding section, that the coefficients in (1) are continuous and $a_2(x) \neq 0$ for every x in some interval I. The process we use to find a second solution $y_2(x)$ consists of **reducing the order** of equation (1) to a first-order equation. For example, it is easily verified that $y_1 = e^x$ satisfies the differential equation $y'' - y = 0$. If we try to determine a solution of the form $y = u(x)e^x$, then

$$y' = ue^x + e^x u'$$

$$y'' = ue^x + 2e^x u' + e^x u''$$

and so

$$y'' - y = e^x(u'' + 2u') = 0.$$

Since $e^x \neq 0$, this last equation requires that $u'' + 2u' = 0$.

If we let $w = u'$, then the latter equation is recognized as a linear first-order equation $w' + 2w = 0$. Using the integrating factor e^{2x}, we can write

$$\frac{d}{dx}[e^{2x}w] = 0$$

$$w = c_1 e^{-2x} \quad \text{or} \quad u' = c_1 e^{-2x}.$$

Thus

$$u = -\frac{c_1}{2} e^{-2x} + c_2$$

and so

$$y = u(x)e^x = -\frac{c_1}{2} e^{-x} + c_2 e^x.$$

By picking $c_2 = 0$ and $c_1 = -2$, we obtain the second solution $y_2 = e^{-x}$. Since $W(e^x, e^{-x}) \neq 0$ for every x, the solutions are linearly independent on $(-\infty, \infty)$ and thus the expression for y is actually the general solution of the given equation.

EXAMPLE I

Given that $y_1 = x^3$ is a solution of $x^2 y'' - 6y = 0$, use reduction of order to find a second solution on the interval $(0, \infty)$.

Solution   Define

$$y = u(x)x^3$$

so that

$$y' = 3x^2 u + x^3 u'$$

$$y'' = x^3 u'' + 6x^2 u' + 6xu$$

$$x^2 y'' - 6y = x^2(x^3 u'' + 6x^2 u' + 6xu) - 6ux^3$$

$$= x^5 u'' + 6x^4 u' = 0$$

provided $u(x)$ is a solution of

$$x^5 u'' + 6x^4 u' = 0 \quad \text{or} \quad u'' + \frac{6}{x} u' = 0.$$

If $w = u'$, we obtain the linear first-order equation

$$w' + \frac{6}{x} w = 0,$$

which possesses the integrating factor $e^{6 \int dx/x} = e^{6 \ln x} = x^6$. Now

$$\frac{d}{dx}[x^6 w] = 0 \quad \text{gives} \quad x^6 w = c_1.$$

Therefore
$$w = u' = \frac{c_1}{x^6}$$

$$u = -\frac{c_1}{5x^5} + c_2$$

$$y = u(x)x^3 = -\frac{c_1}{5x^2} + c_2 x^3.$$

Choosing $c_2 = 0$ and $c_1 = -5$ yields the second solution $y_2 = 1/x^2$.   ∎

General Case

Suppose we divide by $a_2(x)$ in order to put equation (1) in the standard form

$$y'' + P(x)y' + Q(x)y = 0, \tag{2}$$

where $P(x)$ and $Q(x)$ are continuous on some interval I. Let us suppose further that $y_1(x)$ is a known solution of (2) on I and that $y_1(x) \neq 0$ for every x in the interval. If we define $y = u(x)y_1(x)$, it follows that

$$y' = uy_1' + y_1 u'$$

$$y'' = uy_1'' + 2y_1' u' + y_1 u''$$

$$y'' + Py' + Qy = u\underbrace{[y_1'' + Py_1' + Qy_1]}_{\text{zero}} + y_1 u'' + (2y_1' + Py_1)u' = 0.$$

This implies that we must have

$$y_1 u'' + (2y_1' + Py_1)u' = 0$$

or
$$y_1 w' + (2y_1' + Py_1)w = 0, \tag{3}$$

where we have let $w = u'$. Observe that equation (3) is both linear and separable. Applying the latter technique, we obtain

$$\frac{dw}{w} + 2\frac{y_1'}{y_1}\,dx + P\,dx = 0$$

$$\ln|w| + 2\ln|y_1| = -\int P\,dx + c$$

$$\ln|wy_1^2| = -\int P\,dx + c$$

$$wy_1^2 = c_1 e^{-\int P\,dx}$$

$$w = u' = c_1 \frac{e^{-\int P\,dx}}{y_1^2}.$$

Integrating again gives $\quad u = c_1 \displaystyle\int \frac{e^{-\int P\,dx}}{y_1^2}\,dx + c_2\quad$ and therefore

$$y = u(x)y_1(x) = c_1 y_1(x) \int \frac{e^{-\int P(x)\,dx}}{y_1^2(x)}\,dx + c_2 y_1(x).$$

By choosing $c_2 = 0$ and $c_1 = 1$, we find that a second solution of equation (2) is

$$y_2 = y_1(x) \int \frac{e^{-\int P(x)\,dx}}{y_1^2(x)}\,dx. \tag{4}$$

It makes a good review exercise in differentiation to start with formula (4) and actually verify that equation (2) is satisfied.

Now $y_1(x)$ and $y_2(x)$ are linearly independent since

$$W(y_1(x),\, y_2(x)) = \begin{vmatrix} y_1 & y_1 \displaystyle\int \frac{e^{-\int P\,dx}}{y_1^2}\,dx \\[2ex] y_1' & \dfrac{e^{-\int P\,dx}}{y_1} + y_1' \displaystyle\int \frac{e^{-\int P\,dx}}{y_1^2}\,dx \end{vmatrix} = e^{-\int P\,dx}$$

is not zero on any interval on which $y_1(x)$ is not zero.*

EXAMPLE 2

The function $y_1 = x^2$ is a solution of $x^2 y'' - 3xy' + 4y = 0$. Find the general solution on the interval $(0, \infty)$.

Solution   Since the equation has the alternative form

$$y'' - \frac{3}{x}\,y' + \frac{4}{x^2}\,y = 0,$$

we find from (4) $\quad y_2 = x^2 \displaystyle\int \frac{e^{3\int dx/x}}{x^4}\,dx \qquad \Longleftarrow \boxed{e^{3\int dx/x} = e^{\ln x^3} = x^3}$

$$= x^2 \int \frac{dx}{x} = x^2 \ln x.$$

The general solution on $(0, \infty)$ is given by $y = c_1 y_1 + c_2 y_2$; that is,

$$y = c_1 x^2 + c_2 x^2 \ln x. \qquad\blacksquare$$

---

* Alternatively, if $y_2 = u(x)y_1$, then $W(y_1, y_2) = u'(y_1)^2 \neq 0$, since $y_1 \neq 0$ for every x on some interval. If $u' = 0$, then $u = $ constant.

EXAMPLE 3

It can be verified that $y_1 = \dfrac{\sin x}{\sqrt{x}}$ is a solution of $x^2 y'' + xy' + (x^2 - \frac{1}{4})y = 0$ on $(0, \pi)$. Find a second solution.

Solution   First put the equation into the form

$$y'' + \frac{1}{x}y' + \left(1 - \frac{1}{4x^2}\right)y = 0.$$

Then from (4) we have

$$y_2 = \frac{\sin x}{\sqrt{x}} \int \frac{e^{-\int dx/x}}{\left(\dfrac{\sin x}{\sqrt{x}}\right)^2} \, dx \qquad \boxed{e^{-\int dx/x} = e^{\ln x^{-1}} = x^{-1}}$$

$$= \frac{\sin x}{\sqrt{x}} \int \csc^2 x \, dx$$

$$= \frac{\sin x}{\sqrt{x}}(-\cot x) = -\frac{\cos x}{\sqrt{x}}.$$

Since the differential equation is homogeneous, we can disregard the negative sign and take the second solution to be $y_2 = (\cos x)/\sqrt{x}$.   ∎

Observe that $y_1(x)$ and $y_2(x)$ of Example 3 are linearly independent solutions of the given differential equation on the larger interval $(0, \infty)$.

Remark   We have derived and illustrated how to use (4) because you will see this formula again in the next section and in Section 6.1. We use (4) simply to save time in obtaining a desired result. Your instructor will tell you whether you should memorize (4) or whether you should know the first principles of reduction of order.

4.2 EXERCISES

Answers to odd-numbered problems begin on page A–9.

In Problems 1–30 find a second solution of each differential equation. Use reduction of order or formula (4) as instructed. Assume an appropriate interval of validity.

1. $y'' + 5y' = 0$; $y_1 = 1$
2. $y'' - y' = 0$; $y_1 = 1$
3. $y'' - 4y' + 4y = 0$; $y_1 = e^{2x}$
4. $y'' + 2y' + y = 0$; $y_1 = xe^{-x}$
5. $y'' + 16y = 0$; $y_1 = \cos 4x$
6. $y'' + 9y = 0$; $y_1 = \sin 3x$
7. $y'' - y = 0$; $y_1 = \cosh x$
8. $y'' - 25y = 0$; $y_1 = e^{5x}$
9. $9y'' - 12y' + 4y = 0$; $y_1 = e^{2x/3}$
10. $6y'' + y' - y = 0$; $y_1 = e^{x/3}$

11. $x^2y'' - 7xy' + 16y = 0;$ $y_1 = x^4$

12. $x^2y'' + 2xy' - 6y = 0;$ $y_1 = x^2$

13. $xy'' + y' = 0;$ $y_1 = \ln x$

14. $4x^2y'' + y = 0;$ $y_1 = x^{1/2}\ln x$

15. $(1 - 2x - x^2)y'' + 2(1 + x)y' - 2y = 0;$ $y_1 = x + 1$

16. $(1 - x^2)y'' - 2xy' = 0;$ $y_1 = 1$

17. $x^2y'' - xy' + 2y = 0;$ $y_1 = x\sin(\ln x)$

18. $x^2y'' - 3xy' + 5y = 0;$ $y_1 = x^2\cos(\ln x)$

19. $(1 + 2x)y'' + 4xy' - 4y = 0;$ $y_1 = e^{-2x}$

20. $(1 + x)y'' + xy' - y = 0;$ $y_1 = x$

21. $x^2y'' - xy' + y = 0;$ $y_1 = x$ **22.** $x^2y'' - 20y = 0;$ $y_1 = x^{-4}$

23. $x^2y'' - 5xy' + 9y = 0;$ $y_1 = x^3\ln x$

24. $x^2y'' + xy' + y = 0;$ $y_1 = \cos(\ln x)$

25. $x^2y'' - 4xy' + 6y = 0;$ $y_1 = x^2 + x^3$

26. $x^2y'' - 7xy' - 20y = 0;$ $y_1 = x^{10}$

27. $(3x + 1)y'' - (9x + 6)y' + 9y = 0;$ $y_1 = e^{3x}$

28. $xy'' - (x + 1)y' + y = 0;$ $y_1 = e^x$

29. $y'' - 3(\tan x)y' = 0;$ $y_1 = 1$ **30.** $xy'' - (2 + x)y' = 0;$ $y_1 = 1$

In Problems 31–34 use the method of reduction of order to find a solution of the given nonhomogeneous equation. The indicated function $y_1(x)$ is a solution of the associated homogeneous equation. Determine a second solution of the homogeneous equation and a particular solution of the nonhomogeneous equation.

31. $y'' - 4y = 2;$ $y_1 = e^{-2x}$ **32.** $y'' + y' = 1;$ $y_1 = 1$

33. $y'' - 3y' + 2y = 5e^{3x};$ $y_1 = e^x$

34. $y'' - 4y' + 3y = x;$ $y_1 = e^x$

35. Verify by direct substitution that formula (4) satisfies equation (2).

4.3 HOMOGENEOUS LINEAR EQUATIONS WITH CONSTANT COEFFICIENTS

We have seen that the linear first-order equation $dy/dx + ay = 0$, where a is a constant, has the exponential solution $y = c_1e^{-ax}$ on $(-\infty, \infty)$. Therefore, it is natural to seek to determine whether exponential solutions exist on $(-\infty, \infty)$ for higher-order equations such as

$$a_ny^{(n)} + a_{n-1}y^{(n-1)} + \cdots + a_2y'' + a_1y' + a_0y = 0, \tag{1}$$

where the a_i, $i = 0, 1, \ldots, n$, are constants. The surprising fact is that *all* solutions of (1) are exponential functions or constructed out of exponential functions. We begin by considering the special case of the second-order equation

$$ay'' + by' + cy = 0. \tag{2}$$

Auxiliary Equation

If we try a solution of the form $y = e^{mx}$, then $y' = me^{mx}$ and $y'' = m^2 e^{mx}$ so that equation (2) becomes

$$am^2 e^{mx} + bme^{mx} + ce^{mx} = 0 \quad \text{or} \quad e^{mx}(am^2 + bm + c) = 0.$$

Because e^{mx} is never zero for real values of x, it is apparent that the only way that this exponential function can satisfy the differential equation is to choose m so that it is a root of the quadratic equation

$$am^2 + bm + c = 0. \tag{3}$$

This latter equation is called the **auxiliary equation**, or **characteristic equation**, of the differential equation (2). We consider three cases, namely, the solutions of the auxiliary equation corresponding to distinct real roots, real but equal roots, and a conjugate pair of complex roots.

    CASE I   DISTINCT REAL ROOTS   Under the assumption that the auxiliary equation (3) has two unequal real roots m_1 and m_2, we find two solutions

$$y_1 = e^{m_1 x} \quad \text{and} \quad y_2 = e^{m_2 x}.$$

We have seen that these functions are linearly independent on $(-\infty, \infty)$ (see Example 13, Section 4.1) and hence form a fundamental set. It follows that the general solution of (2) on this interval is

$$y = c_1 e^{m_1 x} + c_2 e^{m_2 x}. \tag{4}$$

    CASE II   REPEATED REAL ROOTS   When $m_1 = m_2$ we necessarily obtain only one exponential solution $y_1 = e^{m_1 x}$. However, it follows immediately from the discussion of Section 4.2 that a second solution is

$$y_2 = e^{m_1 x} \int \frac{e^{-(b/a)x}}{e^{2m_1 x}}\, dx. \tag{5}$$

But from the quadratic formula, we have $m_1 = -b/2a$ since the only way to have $m_1 = m_2$ is to have $b^2 - 4ac = 0$. In view of the fact that $2m_1 = -b/a$, (5) becomes

$$y_2 = e^{m_1 x} \int \frac{e^{2m_1 x}}{e^{2m_1 x}}\, dx = e^{m_1 x} \int dx = x e^{m_1 x}.$$

The general solution of (2) is then

$$y = c_1 e^{m_1 x} + c_2 x e^{m_1 x}. \tag{6}$$

CASE III CONJUGATE COMPLEX ROOTS If m_1 and m_2 are complex, then we can write

$$m_1 = \alpha + i\beta \quad \text{and} \quad m_2 = \alpha - i\beta,$$

where α and $\beta > 0$ are real and $i^2 = -1$. Formally there is no difference between this case and Case I, and hence

$$y = C_1 e^{(\alpha + i\beta)x} + C_2 e^{(\alpha - i\beta)x}.$$

However, in practice we prefer to work with real functions instead of complex exponentials. To this end we use Euler's formula:*

$$e^{i\theta} = \cos\theta + i\sin\theta,$$

where θ is any real number. It follows from this formula that

$$e^{i\beta x} = \cos\beta x + i\sin\beta x \quad \text{and} \quad e^{-i\beta x} = \cos\beta x - i\sin\beta x, \tag{7}$$

where we have used $\cos(-\beta x) = \cos\beta x$ and $\sin(-\beta x) = -\sin\beta x$. Note that by first adding and then subtracting the two equations in (7), we obtain, respectively,

$$e^{i\beta x} + e^{-i\beta x} = 2\cos\beta x \quad \text{and} \quad e^{i\beta x} - e^{-i\beta x} = 2i\sin\beta x.$$

Since $y = C_1 e^{(\alpha + i\beta)x} + C_2 e^{(\alpha - i\beta)x}$ is a solution of (2) for any choice of the constants C_1 and C_2, the choices $C_1 = C_2 = 1$ and $C_1 = 1, C_2 = -1$ give, in turn, two solutions:

$$y_1 = e^{(\alpha + i\beta)x} + e^{(\alpha - i\beta)x} \quad \text{and} \quad y_2 = e^{(\alpha + i\beta)x} - e^{(\alpha - i\beta)x}.$$

---

*__Leonhard Euler__ (1707–1783) A man with a prodigious memory and phenomenal powers of concentration, Euler had almost universal interests; he was a theologian, physicist, astronomer, linguist, physiologist, classical scholar, and, primarily, mathematician. Euler is considered to be a true genius of the era. In mathematics, he made lasting contributions to algebra, trigonometry, analytic geometry, calculus, calculus of variations, differential equations, complex variables, number theory, and topology. The volume of his mathematical output did not seem to be affected by the distractions of thirteen children or the fact that he was totally blind for the last seventeen years of his life. Euler wrote over 700 papers and 32 books on mathematics and was responsible for introducing many of the symbols (such as e, π, and $i = \sqrt{-1}$) and notations that are still used (such as $f(x)$, Σ, $\sin x$, and $\cos x$). Euler was born in Basel, Switzerland, on April 15, 1707, and died of a stroke in St. Petersburg on September 18, 1783, while serving in the court of the Russian empress Catherine the Great.

See Appendix IV for a review of complex numbers and a derivation of Euler's formula.

But $\qquad$ $y_1 = e^{\alpha x}(e^{i\beta x} + e^{-i\beta x}) = 2e^{\alpha x}\cos\beta x$

and $\qquad$ $y_2 = e^{\alpha x}(e^{i\beta x} - e^{-i\beta x}) = 2ie^{\alpha x}\sin\beta x.$

Hence, from Corollary (A) of Theorem 4.3 the last two results show that the *real* functions $e^{\alpha x}\cos\beta x$ and $e^{\alpha x}\sin\beta x$ are solutions of (2). Moreover, from Example 14 of Section 4.1 we have $W(e^{\alpha x}\cos\beta x, e^{\alpha x}\sin\beta x) = \beta e^{2\alpha x} \neq 0$, $\beta > 0$, and so we can conclude that $e^{\alpha x}\cos\beta x$ and $e^{\alpha x}\sin\beta x$ themselves form a fundamental set of solutions of the differential equation on $(-\infty, \infty)$. By the superposition principle, the general solution is

$$
\begin{aligned}
y &= c_1 e^{\alpha x}\cos\beta x + c_2 e^{\alpha x}\sin\beta x \\
&= e^{\alpha x}(c_1 \cos\beta x + c_2 \sin\beta x).
\end{aligned}
\tag{8}
$$

EXAMPLE 1

Solve the following differential equations:

(a) $2y'' - 5y' - 3y = 0$ $\qquad$ **(b)** $y'' - 10y' + 25y = 0$ $\qquad$ **(c)** $y'' + y' + y = 0$

Solution **(a)** $2m^2 - 5m - 3 = (2m + 1)(m - 3) = 0$

$$m_1 = -\tfrac{1}{2}, \qquad m_2 = 3$$

$$y = c_1 e^{-x/2} + c_2 e^{3x}$$

(b) $m^2 - 10m + 25 = (m - 5)^2 = 0$

$$m_1 = m_2 = 5$$

$$y = c_1 e^{5x} + c_2 x e^{5x}$$

(c) $m^2 + m + 1 = 0$

$$m_1 = -\frac{1}{2} + \frac{\sqrt{3}}{2}i, \qquad m_2 = -\frac{1}{2} - \frac{\sqrt{3}}{2}i$$

$$y = e^{-x/2}\left(c_1 \cos\frac{\sqrt{3}}{2}x + c_2 \sin\frac{\sqrt{3}}{2}x\right) \qquad\blacksquare$$

EXAMPLE 2

Solve the initial-value problem

$$y'' - 4y' + 13y = 0, \qquad y(0) = -1, \quad y'(0) = 2.$$

Solution The roots of the auxiliary equation $m^2 - 4m + 13 = 0$ are $m_1 = 2 + 3i$ and $m_2 = 2 - 3i$ so that

$$y = e^{2x}(c_1 \cos 3x + c_2 \sin 3x).$$

The condition $y(0) = -1$ implies

$$-1 = e^0(c_1 \cos 0 + c_2 \sin 0) = c_1,$$

from which we can write

$$y = e^{2x}(-\cos 3x + c_2 \sin 3x).$$

Differentiating this latter expression and using $y'(0) = 2$ give

$$y' = e^{2x}(3 \sin 3x + 3c_2 \cos 3x) + 2e^{2x}(-\cos 3x + c_2 \sin 3x)$$

$$2 = 3c_2 - 2$$

and so $c_2 = 4/3$. Hence

$$y = e^{2x}\left(-\cos 3x + \frac{4}{3} \sin 3x\right).$$ ∎

EXAMPLE 3

The two equations

$$y'' + k^2 y = 0 \tag{9}$$

$$y'' - k^2 y = 0 \tag{10}$$

are frequently encountered in the study of applied mathematics. For the former differential equation, the auxiliary equation $m^2 + k^2 = 0$ has the roots $m_1 = ki$ and $m_2 = -ki$. It follows from (8) that the general solution of (9) is

$$y = c_1 \cos kx + c_2 \sin kx. \tag{11}$$

The differential equation (10) has the auxiliary equation $m^2 - k^2 = 0$, with real roots $m_1 = k$ and $m_2 = -k$, so that its general solution is

$$y = c_1 e^{kx} + c_2 e^{-kx}. \tag{12}$$

Notice that if we choose $c_1 = c_2 = 1/2$ in (12), then

$$y = \frac{e^{kx} + e^{-kx}}{2} = \cosh kx$$

is also a solution of (10). Furthermore, if $c_1 = 1/2$, $c_2 = -1/2$, then (12) becomes

$$y = \frac{e^{kx} - e^{-kx}}{2} = \sinh kx.$$

Since $\cosh kx$ and $\sinh kx$ are linearly independent on any interval of the x-axis, they form a fundamental set. Thus an alternative form for the general solution of (10) is

$$y = c_1 \cosh kx + c_2 \sinh kx. \tag{13}$$

∎

Higher-Order Equations

In general, to solve an nth-order differential equation

$$a_n y^{(n)} + a_{n-1} y^{(n-1)} + \cdots + a_2 y'' + a_1 y' + a_0 y = 0, \tag{14}$$

where the a_i, $i = 0, 1, \ldots, n$, are real constants, we must solve an nth-degree polynominal equation

$$a_n m^n + a_{n-1} m^{n-1} + \cdots + a_2 m^2 + a_1 m + a_0 = 0 \tag{15}$$

If all the roots of (15) are real and distinct, then the general solution of (14) is

$$y = c_1 e^{m_1 x} + c_2 e^{m_2 x} + \cdots + c_n e^{m_n x}. \tag{16}$$

It is somewhat harder to summarize the analogues of Cases II and III because the roots of an auxiliary equation of degree greater than two can occur in many combinations. For example, a fifth-degree equation could have five distinct real roots, or three distinct real and two complex roots, or one real and four complex roots, or five real but equal roots, or five real roots but two of them equal, and so on. When m_1 is a root of multiplicity k of an nth-degree auxiliary equation (that is, k roots are equal to m_1), it can be shown that the linearly independent solutions are

$$e^{m_1 x}, xe^{m_1 x}, x^2 e^{m_1 x}, \ldots, x^{k-1} e^{m_1 x}$$

and the general solution must contain the linear combination

$$c_1 e^{m_1 x} + c_2 xe^{m_1 x} + c_3 x^2 e^{m_1 x} + \cdots + c_k x^{k-1} e^{m_1 x}.$$

Lastly, it should be remembered that when the coefficients are real, complex roots of an auxiliary equation always appear in conjugate pairs. Thus, for example, a cubic polynomial equation can have at most two complex roots.

E X A M P L E   **4**

Solve $\qquad\qquad y''' + 3y'' - 4y = 0.$

Solution   It should be apparent from inspection of

$$m^3 + 3m^2 - 4 = 0$$

that one root is $m_1 = 1$. Now if we divide $m^3 + 3m^2 - 4$ by $m - 1$, we find

$$m^3 + 3m^2 - 4 = (m - 1)(m^2 + 4m + 4) = (m - 1)(m + 2)^2,$$

and so the other roots are $m_2 = m_3 = -2$. Thus the general solution is

$$y = c_1 e^x + c_2 e^{-2x} + c_3 x e^{-2x}.$$ ∎

Of course, the most difficult aspect of solving constant-coefficient equations is finding the roots of auxiliary equations of degree greater than two. As illustrated in Example 4, one way to solve an equation is to guess a root m_1. If we have found one root m_1, we then know from the **factor theorem** that $m - m_1$ is a factor of the polynomial. By dividing the polynomial by $m - m_1$, we obtain the factorization $(m - m_1)Q(m)$. We then try to find the roots of the quotient $Q(m)$. The algebraic technique of **synthetic division** is also very helpful in finding **rational** roots of polynomial equations. Specifically, if $m_1 = p/q$ is a **rational real root** (p and q integers, p/q in lowest terms) of an auxiliary equation

$$a_n m^n + \cdots + a_1 m + a_0 = 0$$

with integer coefficients, then p is a factor of a_0 and q is a factor of a_n. Thus to determine whether a polynomial equation has rational roots we need only examine all ratios of each factor of a_0 to each factor of a_n. In this manner we construct a list of all the possible rational roots of the equation. We test each of these numbers by synthetic division. If the remainder is zero, the number m_1 being tested is a root of the equation and thus $m - m_1$ is a factor of the polynomial.

The next example illustrates this method.

EXAMPLE 5

Solve $\qquad\qquad 3y''' + 5y'' + 10y' - 4y = 0.$

Solution   The auxiliary equation is

$$3m^3 + 5m^2 + 10m - 4 = 0.$$

All the factors of $a_0 = -4$ and $a_n = 3$ are

$$p: \pm 1, \pm 2, \pm 4 \quad \text{and} \quad q: \pm 1, \pm 3,$$

respectively. Therefore the possible rational roots of the auxiliary equation are

$$\frac{p}{q}: -1, 1, -2, 2, -4, 4, -\frac{1}{3}, \frac{1}{3}, -\frac{2}{3}, \frac{2}{3}, -\frac{4}{3}, \frac{4}{3}.$$

Testing each of these numbers in turn by synthetic division, we eventually find

coefficients of auxiliary equation

$$\begin{array}{r|rrrr} \frac{1}{3} & 3 & 5 & 10 & -4 \\ & & 1 & 2 & 4 \\ \hline & 3 & 6 & 12 & 0 \end{array} \text{ remainder}$$

Consequently $m_1 = 1/3$ is a root. Furthermore, you should verify that it is the *only* rational root. The numbers highlighted in the preceding division are the coefficients of the quotient. Thus the auxiliary equation can be written as

$$\left(m - \frac{1}{3}\right)(3m^2 + 6m + 12) = 0 \quad \text{or} \quad (3m - 1)(m^2 + 2m + 4) = 0.$$

Solving $m^2 + 2m + 4 = 0$ by the quadratic formula leads to the complex roots $m_2 = -1 + \sqrt{3}\,i$ and $m_3 = -1 - \sqrt{3}\,i$. Thus the general solution of the differential equation is

$$y = c_1 e^{x/3} + e^{-x}(c_2 \cos \sqrt{3}\,x + c_3 \sin \sqrt{3}\,x). \quad \blacksquare$$

EXAMPLE 6

Solve
$$\frac{d^4y}{dx^4} + 2\frac{d^2y}{dx^2} + y = 0.$$

Solution   The auxiliary equation

$$m^4 + 2m^2 + 1 = (m^2 + 1)^2 = 0$$

has roots $m_1 = m_3 = i$ and $m_2 = m_4 = -i$. Thus from Case II the solution is

$$y = C_1 e^{ix} + C_2 e^{-ix} + C_3 x e^{ix} + C_4 x e^{-ix}.$$

By Euler's formula the grouping $C_1 e^{ix} + C_2 e^{-ix}$ can be rewritten as

$$c_1 \cos x + c_2 \sin x$$

after a relabeling of constants. Similarly, $x(C_3 e^{ix} + C_4 e^{-ix})$ can be expressed as $x(c_3 \cos x + c_4 \sin x)$. Hence the general solution is

$$y = c_1 \cos x + c_2 \sin x + c_3 x \cos x + c_4 x \sin x. \quad \blacksquare$$

Example 6 illustrates a special case when the auxiliary equation has repeated complex roots. In general, if $m_1 = \alpha + i\beta$ is a complex root of multiplicity k of an auxiliary equation with real coefficients, then its conjugate $m_2 = \alpha - i\beta$ is also a root of multiplicity k. From the $2k$ complex-valued solutions

$$e^{(\alpha + i\beta)x}, \, xe^{(\alpha + i\beta)x}, \, x^2 e^{(\alpha + i\beta)x}, \ldots, x^{k-1} e^{(\alpha + i\beta)x}$$

$$e^{(\alpha - i\beta)x}, \, xe^{(\alpha - i\beta)x}, \, x^2 e^{(\alpha - i\beta)x}, \ldots, x^{k-1} e^{(\alpha - i\beta)x}$$

we conclude, with the aid of Euler's formula, that the general solution of the corresponding differential equation must then contain a linear combination of the $2k$ real linearly independent solutions

$$e^{\alpha x} \cos \beta x, \, xe^{\alpha x} \cos \beta x, \, x^2 e^{\alpha x} \cos \beta x, \ldots, x^{k-1} e^{\alpha x} \cos \beta x$$

$$e^{\alpha x} \sin \beta x, \, xe^{\alpha x} \sin \beta x, \, x^2 e^{\alpha x} \sin \beta x, \ldots, x^{k-1} e^{\alpha x} \sin \beta x.$$

In Example 6 we identify $k = 2$, $\alpha = 0$, and $\beta = 1$.

4.3 EXERCISES

Answers to odd-numbered problems begin on page A–9.

In Problems 1–36 find the general solution of the given differential equation.

1. $4y'' + y' = 0$

2. $2y'' - 5y' = 0$

3. $y'' - 36y = 0$

4. $y'' - 8y = 0$

5. $y'' + 9y = 0$

6. $3y'' + y = 0$

7. $y'' - y' - 6y = 0$

8. $y'' - 3y' + 2y = 0$

9. $\dfrac{d^2y}{dx^2} + 8\dfrac{dy}{dx} + 16y = 0$

10. $\dfrac{d^2y}{dx^2} - 10\dfrac{dy}{dx} + 25y = 0$

11. $y'' + 3y' - 5y = 0$

12. $y'' + 4y' - y = 0$

13. $12y'' - 5y' - 2y = 0$

14. $8y'' + 2y' - y = 0$

15. $y'' - 4y' + 5y = 0$

16. $2y'' - 3y' + 4y = 0$

17. $3y'' + 2y' + y = 0$

18. $2y'' + 2y' + y = 0$

19. $y''' - 4y'' - 5y' = 0$

20. $4y''' + 4y'' + y' = 0$

21. $y''' - y = 0$

22. $y''' + 5y'' = 0$

23. $y''' - 5y'' + 3y' + 9y = 0$

24. $y''' + 3y'' - 4y' - 12y = 0$

25. $y''' + y'' - 2y = 0$

26. $y''' - y'' - 4y = 0$

27. $y''' + 3y'' + 3y' + y = 0$

28. $y''' - 6y'' + 12y' - 8y = 0$

29. $\dfrac{d^4y}{dx^4} + \dfrac{d^3y}{dx^3} + \dfrac{d^2y}{dx^2} = 0$

30. $\dfrac{d^4y}{dx^4} - 2\dfrac{d^2y}{dx^2} + y = 0$

31. $16\dfrac{d^4y}{dx^4} + 24\dfrac{d^2y}{dx^2} + 9y = 0$

32. $\dfrac{d^4y}{dx^4} - 7\dfrac{d^2y}{dx^2} - 18y = 0$

33. $\dfrac{d^5y}{dx^5} - 16\dfrac{dy}{dx} = 0$

34. $\dfrac{d^5y}{dx^5} - 2\dfrac{d^4y}{dx^4} + 17\dfrac{d^3y}{dx^3} = 0$

35. $\dfrac{d^5y}{dx^5} + 5\dfrac{d^4y}{dx^4} - 2\dfrac{d^3y}{dx^3} - 10\dfrac{d^2y}{dx^2} + \dfrac{dy}{dx} + 5y = 0$

36. $2\dfrac{d^5y}{dx^5} - 7\dfrac{d^4y}{dx^4} + 12\dfrac{d^3y}{dx^3} + 8\dfrac{d^2y}{dx^2} = 0$

In Problems 37–52 solve the given differential equation subject to the indicated initial conditions.

37. $y'' + 16y = 0$, $y(0) = 2, y'(0) = -2$

38. $y'' - y = 0$, $y(0) = y'(0) = 1$

39. $y'' + 6y' + 5y = 0$, $y(0) = 0, y'(0) = 3$

40. $y'' - 8y' + 17y = 0$, $y(0) = 4, y'(0) = -1$

41. $2y'' - 2y' + y = 0$, $y(0) = -1, y'(0) = 0$

42. $y'' - 2y' + y = 0$, $y(0) = 5, y'(0) = 10$

43. $y'' + y' + 2y = 0$, $y(0) = y'(0) = 0$

44. $4y'' - 4y' - 3y = 0,$ $y(0) = 1, y'(0) = 5$

45. $y'' - 3y' + 2y = 0,$ $y(1) = 0, y'(1) = 1$

46. $y'' + y = 0,$ $y(\pi/3) = 0, y'(\pi/3) = 2$

47. $y''' + 12y'' + 36y' = 0,$ $y(0) = 0, y'(0) = 1, y''(0) = -7$

48. $y''' + 2y'' - 5y' - 6y = 0,$ $y(0) = y'(0) = 0, y''(0) = 1$

49. $y''' - 8y = 0,$ $y(0) = 0, y'(0) = -1, y''(0) = 0$

50. $\dfrac{d^4y}{dx^4} = 0,$ $y(0) = 2, y'(0) = 3, y''(0) = 4, y'''(0) = 5$

51. $\dfrac{d^4y}{dx^4} - 3\dfrac{d^3y}{dx^3} + 3\dfrac{d^2y}{dx^2} - \dfrac{dy}{dx} = 0,$ $y(0) = y'(0) = 0, y''(0) = y'''(0) = 1$

52. $\dfrac{d^4y}{dx^4} - y = 0,$ $y(0) = y'(0) = y''(0) = 0, y'''(0) = 1$

In Problems 53–56 solve the given differential equation subject to the indicated boundary conditions.

53. $y'' - 10y' + 25y = 0,$ $y(0) = 1, y(1) = 0$

54. $y'' + 4y = 0,$ $y(0) = 0, y(\pi) = 0$

55. $y'' + y = 0,$ $y'(0) = 0, y'\left(\dfrac{\pi}{2}\right) = 2$

56. $y'' - y = 0,$ $y(0) = 1, y'(1) = 0$

57. The roots of an auxiliary equation are $m_1 = 4, m_2 = m_3 = -5.$ What is the corresponding differential equation?

58. The roots of an auxiliary equation are $m_1 = -\frac{1}{2}, m_2 = 3 + i, m_3 = 3 - i.$ What is the corresponding differential equation?

In Problems 59 and 60 find the general solution of the given equation if it is known that y_1 is a solution.

59. $y''' - 9y'' + 25y' - 17y = 0;$ $y_1 = e^x$

60. $y''' + 6y'' + y' - 34y = 0;$ $y_1 = e^{-4x} \cos x$

In Problems 61–64 determine a homogeneous linear differential equation with constant coefficients having the given solutions.

61. $4e^{6x}, 3e^{-3x}$

62. $10 \cos 4x, -5 \sin 4x$

63. $3, 2x, -e^{7x}$

64. $8 \sinh 3x, 12 \cosh 3x$

65. Use the facts

$$i = \left(\dfrac{\sqrt{2}}{2} + \dfrac{\sqrt{2}}{2}i\right)^2 \quad \text{and} \quad -i = \left(\dfrac{\sqrt{2}}{2} - \dfrac{\sqrt{2}}{2}i\right)^2$$

to solve the differential equation

$$\frac{d^4y}{dx^4} + y = 0.$$

[*Hint:* Write the auxiliary equation $m^4 + 1 = 0$ as $(m^2 + 1)^2 - 2m^2 = 0$. See what happens when you factor.]

4.4 UNDETERMINED COEFFICIENTS—SUPERPOSITION APPROACH

> ***To the instructor*** In this section the method of undetermined coefficients is developed from the viewpoint of the superposition principle for nonhomogeneous differential equations (Theorem 4.9). In Section 4.6 an entirely different approach to this method will be presented, one utilizing the concept of differential annihilator operators. Take your pick.

To obtain the general solution of a nonhomogeneous linear differential equation we must do two things:

(i) Find the complementary function y_c.

(ii) Find *any* particular solution y_p of the nonhomogeneous equation.

Recall from the discussion of Section 4.1 that a particular solution is any function, free of arbitrary constants, that satisfies the differential equation identically. The general solution of a nonhomogeneous equation on an interval is then $y = y_c + y_p$.

As in the Section 4.3 we begin with second-order equations, but in this case nonhomogeneous equations of the form

$$ay'' + by' + cy = g(x), \tag{1}$$

where a, b, and c are constants. Although **the method of undetermined coefficients** presented in this section is *not* limited to second-order equations, it *is* limited to nonhomogeneous linear equations

- that have constant coefficients, and
- where $g(x)$ is a constant k, a polynomial function, an exponential function $e^{\alpha x}$, $\sin \beta x$, $\cos \beta x$, or finite sums and products of these functions.

Note Strictly speaking $g(x) = k$ (a constant) is a polynomial function. Since a constant function is probably not the first thing that comes to mind when you think of polynomial functions, for emphasis we continue to use the redundancy "constant functions, polynomials, ..."

The following are some examples of the types of input functions $g(x)$ that are appropriate for this discussion:

$$g(x) = 10$$

$$g(x) = x^2 - 5x$$

$$g(x) = 15x - 6 + 8e^{-4x}$$

$$g(x) = \sin 3x - 5x \cos 2x$$

$$g(x) = e^x \cos x - (3x^2 - 1)e^{-x}$$

and so on. That is, $g(x)$ is a linear combination of functions of the type

$$k \text{ (constant)}, \quad x^n, \quad x^n e^{\alpha x}, \quad x^n e^{\alpha x} \cos \beta x, \quad \text{and} \quad x^n e^{\alpha x} \sin \beta x,$$

where n is a nonnegative integer and α and β are real numbers. The method of undetermined coefficients is not applicable to equations of form (1) when

$$g(x) = \ln x, \quad g(x) = \frac{1}{x}, \quad g(x) = \tan x, \quad g(x) = \sin^{-1}x,$$

and so on. Differential equations with this latter kind of input function will be considered in Section 4.7.

The set of functions that consists of constants, polynomials, exponentials $e^{\alpha x}$, sines, and cosines has the remarkable property that derivatives of their sums and products are again sums and products of constants, polynomials, exponentials $e^{\alpha x}$, sines, and cosines. Since the linear combination of derivatives $ay_p'' + by_p' + cy_p$ must be identically equal to $g(x)$, it seems reasonable to assume then that y_p *has the same form as* $g(x)$. This assumption could be better characterized as an educated conjecture or guess. The next two examples illustrate the basic method.

E X A M P L E I

Solve $$y'' + 4y' - 2y = 2x^2 - 3x + 6. \tag{2}$$

Solution   Step 1.   We first solve the associated homogeneous equation $y'' + 4y' - 2y = 0$. From the quadratic formula we find that the roots of the auxiliary equation $m^2 + 4m - 2 = 0$ are $m_1 = -2 - \sqrt{6}$ and $m_2 = -2 + \sqrt{6}$. Hence the complementary function is

$$y_c = c_1 e^{-(2+\sqrt{6})x} + c_2 e^{(-2+\sqrt{6})x}.$$

Step 2.   Now since the input function $g(x)$ is a quadratic polynomial, let us assume a particular solution that is also in the form of a quadratic polynomial:

$$y_p = Ax^2 + Bx + C.$$

We seek to determine *specific* coefficients A, B, and C for which y_p is a solution of (2). Substituting y_p and the derivatives

$$y_p' = 2Ax + B \quad \text{and} \quad y_p'' = 2A$$

into the given differential equation (2), we get

$$y_p'' + 4y_p' - 2y_p = 2A + 8Ax + 4B - 2Ax^2 - 2Bx - 2C = 2x^2 - 3x + 6.$$

Since the last equation is supposed to be an identity, the coefficients of like powers of x must be equal:

That is,
$$-2A = 2$$

$$8A - 2B = -3$$

$$2A + 4B - 2C = 6.$$

Solving this system of equations leads to the values $A = -1$, $B = -5/2$, and $C = -9$. Thus a particular solution is

$$y_p = -x^2 - \frac{5}{2}x - 9.$$

Step 3.   The general solution of the given equation is

$$y = y_c + y_p = c_1 e^{-(2+\sqrt{6})x} + c_2 e^{(-2+\sqrt{6})x} - x^2 - \frac{5}{2}x - 9. \qquad \blacksquare$$

EXAMPLE 2

Find a particular solution of $\qquad y'' - y' + y = 2 \sin 3x$.

Solution   A natural first guess for a particular solution would be $A \sin 3x$. But since successive differentiations of $\sin 3x$ produce $\sin 3x$ *and* $\cos 3x$, we are prompted instead to assume a particular solution that includes both of these terms:

$$y_p = A \cos 3x + B \sin 3x.$$

Differentiating y_p and substituting the results into the differential equation give, after regrouping,

$$y_p'' - y_p' + y_p = (-8A - 3B) \cos 3x + (3A - 8B) \sin 3x = 2 \sin 3x$$

or

From the resulting system of equations

$$-8A - 3B = 0$$

$$3A - 8B = 2$$

we get $A = 6/73$ and $B = -16/73$. A particular solution of the equation is

$$y_p = \frac{6}{73} \cos 3x - \frac{16}{73} \sin 3x.$$

As we mentioned, the form that we assume for the particular solution y_p is an educated guess; it is not a blind guess. This educated guess must take into consideration not only the types of functions that make up $g(x)$, but also, as we shall see in Example 4, the functions that make up the complementary function y_c.

EXAMPLE 3

Solve $y'' - 2y' - 3y = 4x - 5 + 6xe^{2x}.$ (3)

Solution Step 1.   First, the solution of the associated homogeneous equation $y'' - 2y' - 3y = 0$ is found to be $y_c = c_1 e^{-x} + c_2 e^{3x}$.

Step 2.   Next, the presence of $4x - 5$ in $g(x)$ suggests that the particular solution includes a linear polynomial. Furthermore, since the derivative of the product xe^{2x} produces $2xe^{2x}$ and e^{2x}, we also assume that the particular solution includes both xe^{2x} and e^{2x}. In other words, g is the sum of two basic kinds of functions:

$$g(x) = g_1(x) + g_2(x) = polynomial + exponentials.$$

Correspondingly, the superposition principle for nonhomogeneous equations (Theorem 4.9) suggests that we seek a particular solution

$$y_p = y_{p_1} + y_{p_2},$$

where $y_{p_1} = Ax + B$ and $y_{p_2} = Cxe^{2x} + De^{2x}$. Substituting

$$y_p = Ax + B + Cxe^{2x} + De^{2x}$$

into the given equation (3) and grouping like terms give

$$y_p'' - 2y_p' - 3y_p = -3Ax - 2A - 3B - 3Cxe^{2x} + (2C - 3D)e^{2x} = 4x - 5 + 6xe^{2x}. \quad (4)$$

From this identity we obtain the system of four equations in four unknowns:

$$-3A = 4$$

$$-2A - 3B = -5$$

$$-3C = 6$$

$$2C - 3D = 0.$$

The last equation in this system results from the interpretation that the coefficient of e^{2x} in the right member of (4) is zero. Solving, we find $A = -4/3$, $B = 23/9$, $C = -2$, and $D = -4/3$. Consequently,

$$y_p = -\frac{4}{3}x + \frac{23}{9} - 2xe^{2x} - \frac{4}{3}e^{2x}.$$

Step 3. The general solution of the equation is

$$y = c_1 e^{-x} + c_2 e^{3x} - \frac{4}{3}x + \frac{23}{9} - \left(2x + \frac{4}{3}\right)e^{2x}. \qquad \blacksquare$$

In light of the superposition principle (Theorem 4.9) we can also approach Example 3 from the viewpoint of solving two simpler problems. You should verify that substituting

$$y_{p_1} = Ax + B \qquad \text{into} \quad y'' - 2y' - 3y = 4x - 5$$

and $\qquad y_{p_2} = Cxe^{2x} + De^{2x} \quad$ into $\quad y'' - 2y' - 3y = 6xe^{2x}$

yields in turn $y_{p_1} = -(4/3)x + 23/9$ and $y_{p_2} = -(2x + 4/3)e^{2x}$. A particular solution of (3) is then $y_p = y_{p_1} + y_{p_2}$.

The next example illustrates that sometimes the "obvious" assumption for the form of y_p is not a correct assumption.

EXAMPLE 4

Find a particular solution of $\qquad y'' - 5y' + 4y = 8e^x$.

Solution Differentiation of e^x produces no new functions. Thus, proceeding as we did in the earlier examples, we can reasonably assume a particular solution of the form

$$y_p = Ae^x.$$

But in this case substitution of this expression into the differential equation yields the contradictory statement

$$0 = 8e^x,$$

and so we have clearly made the wrong guess for y_p.

The difficulty here is apparent upon examining the complementary function $y_c = c_1e^x + c_2e^{4x}$. Observe that our assumption Ae^x is already present in y_c. This means that e^x is a solution of the associated homogeneous differential equation, and a constant multiple Ae^x when substituted into the differential equation necessarily produces zero.

What then should be the form of y_p? Inspired by Case II of Section 4.3, let's see whether we can find a particular solution of the form

$$y_p = Axe^x.$$

Using $y'_p = Axe^x + Ae^x$ and $y''_p = Axe^x + 2Ae^x$, we obtain

$$y''_p - 5y'_p + 4y_p = Axe^x + 2Ae^x - 5Axe^x - 5Ae^x + 4Axe^x = 8e^x$$

or $$-3Ae^x = 8e^x.$$

From this last equation we see that the value of A is now determined as $A = -8/3$. Therefore

$$y_p = -\frac{8}{3}xe^x$$

must be a particular solution of the given equation.   ∎

The difference in the procedures used in Examples 1–3 and in Example 4 suggests that we consider two cases. The first case reflects the situation in Examples 1–3.

CASE I   *No function in the assumed particular solution is a solution of the associated homogeneous differential equation.*

In the following table we illustrate some specific examples of $g(x)$ in (1) along with the corresponding form of the particular solution. We are, of course, taking for granted that no function in the assumed particular solution y_p is duplicated by a function in the complementary function y_c.

Trial Particular Solutions

$g(x)$	Form of y_p
1. 1 (any constant)	A
2. $5x + 7$	$Ax + B$
3. $3x^2 - 2$	$Ax^2 + Bx + C$
4. $x^3 - x + 1$	$Ax^3 + Bx^2 + Cx + D$
5. $\sin 4x$	$A\cos 4x + B\sin 4x$
6. $\cos 4x$	$A\cos 4x + B\sin 4x$
7. e^{5x}	Ae^{5x}
8. $(9x - 2)e^{5x}$	$(Ax + B)e^{5x}$
9. x^2e^{5x}	$(Ax^2 + Bx + C)e^{5x}$
10. $e^{3x}\sin 4x$	$Ae^{3x}\cos 4x + Be^{3x}\sin 4x$
11. $5x^2\sin 4x$	$(Ax^2 + Bx + C)\cos 4x + (Dx^2 + Ex + F)\sin 4x$
12. $xe^{3x}\cos 4x$	$(Ax + B)e^{3x}\cos 4x + (Cx + D)e^{3x}\sin 4x$

EXAMPLE 5

Determine the form of a particular solution of

> **(a)** $y'' - 8y' + 25y = 5x^3 e^{-x} - 7e^{-x}$ and   **(b)** $y'' + 4y = x \cos x$.

Solution   **(a)** We can write

$$g(x) = (5x^3 - 7)e^{-x}.$$

Using entry 9 in the table as a model, we assume a particular solution of the form

$$y_p = (Ax^3 + Bx^2 + Cx + D)e^{-x}.$$

Note that there is no duplication between the terms in y_p and the terms in the complementary function $y_c = e^{4x}(c_1 \cos 3x + c_2 \sin 3x)$.

(b) The function $g(x) = x \cos x$ is similar to entry 11 in the table, except of course we use a linear rather than a quadratic polynomial and $\cos x$ and $\sin x$ instead of $\cos 4x$ and $\sin 4x$ in the form of y_p:

$$y_p = (Ax + B) \cos x + (Cx + D) \sin x.$$

Again observe that there is no duplication of terms between y_p and $y_c = c_1 \cos 2x + c_2 \sin 2x$. ∎

If $g(x)$ consists of a sum of, say, m terms of the kind listed in the table, then (as in Example 3) the assumption for a particular solution y_p consists of the sum of the trial forms $y_{p_1}, y_{p_2}, \ldots, y_{p_m}$ corresponding to these terms:

$$y_p = y_{p_1} + y_{p_2} + \cdots + y_{p_m}.$$

Put another way:

> *The form of y_p is a linear combination of all linearly independent functions that are generated by repeated differentiations of $g(x)$.*

EXAMPLE 6

Determine the form of a particular solution of

$$y'' - 9y' + 14y = 3x^2 - 5 \sin 2x + 7xe^{6x}.$$

Solution

Corresponding to $3x^2$ we assume: $y_{p_1} = Ax^2 + Bx + C$

Corresponding to $-5 \sin 2x$ we assume: $y_{p_2} = D \cos 2x + E \sin 2x$

Corresponding to $7xe^{6x}$ we assume: $y_{p_3} = (Fx + G)e^{6x}$

The assumption for the particular solution is then

$$y_p = y_{p_1} + y_{p_2} + y_{p_3} = Ax^2 + Bx + C + D \cos 2x + E \sin 2x + (Fx + G)e^{6x}.$$

No term in this assumption duplicates a term in $y_c = c_1 e^{2x} + c_2 e^{7x}$. ∎

CASE II *A function in the assumed particular solution is also a solution of the associated homogeneous differential equation.*

The next example is similar to Example 4.

EXAMPLE 7

Find a particular solution of $y'' - 2y' + y = e^x$.

Solution The complementary function is $y_c = c_1 e^x + c_2 x e^x$. As in Example 4, the assumption $y_p = A e^x$ will fail since it is apparent from y_c that e^x is a solution of the associated homogeneous equation $y'' - 2y' + y = 0$. Moreover, we will not be able to find a particular solution of the form $y_p = A x e^x$ since the term $x e^x$ is also duplicated in y_c. We next try

$$y_p = A x^2 e^x.$$

Substituting into the given differential equation yields

$$2A e^x = e^x \quad \text{and so} \quad A = \frac{1}{2}.$$

Thus a particular solution is

$$y_p = \frac{1}{2} x^2 e^x.$$

∎

Suppose again that $g(x)$ consists of m terms of the kind given in the table and suppose further that the usual assumption for a particular solution is

$$y_p = y_{p_1} + y_{p_2} + \cdots + y_{p_m},$$

where the y_{p_i}, $i = 1, 2, \ldots, m$, are the trial particular solution forms corresponding to these terms. Under the circumstance described in Case II, we can make up the following **general rule**:

> If any y_{p_i} contains terms that duplicate terms in y_c, then that y_{p_i} must be multiplied by x^n, where n is the smallest positive integer that eliminates that duplication.

EXAMPLE 8

Solve the initial-value problem

$$y'' + y = 4x + 10 \sin x, \qquad y(\pi) = 0, \quad y'(\pi) = 2.$$

Solution The solution of the associated homogeneous equation $y'' + y = 0$ is

$$y_c = c_1 \cos x + c_2 \sin x.$$

Now since $g(x)$ is the sum of a linear polynomial and a sine function, our normal assumption for y_p from entries 2 and 5 of the trial solutions table would be the sum of $y_{p_1} = Ax + B$ and $y_{p_2} = C \cos x + D \sin x$:

$$y_p = Ax + B + C \cos x + D \sin x. \tag{5}$$

But there is an obvious duplication of the terms $\cos x$ and $\sin x$ in this assumed form and two terms in the complementary function. This duplication can be eliminated by simply multiplying y_{p_2} by x. Instead of (5) we now use

$$y_p = Ax + B + Cx \cos x + Dx \sin x. \tag{6}$$

Differentiating this expression and substituting the results into the differential equation give

$$y_p'' + y_p = Ax + B - 2C \sin x + 2D \cos x = 4x + 10 \sin x$$

and so

$$A = 4$$

$$B = 0$$

$$-2C = 10$$

$$2D = 0.$$

The solutions of the system are immediate: $A = 4, B = 0, C = -5$, and $D = 0$. Therefore from (6) we obtain

$$y_p = 4x - 5x \cos x.$$

The general solution of the given equation is

$$y = y_c + y_p = c_1 \cos x + c_2 \sin x + 4x - 5x \cos x.$$

We now apply the prescribed initial conditions to the general solution of the equation. First, $y(\pi) = c_1 \cos \pi + c_2 \sin \pi + 4\pi - 5\pi \cos \pi = 0$ yields $c_1 = 9\pi$ since $\cos \pi = -1$ and $\sin \pi = 0$. Next, from the derivative

$$y' = -9\pi \sin x + c_2 \cos x + 4 + 5x \sin x - 5 \cos x$$

and $\quad y'(\pi) = -9\pi \sin \pi + c_2 \cos \pi + 4 + 5\pi \sin \pi - 5 \cos \pi = 2$

we find $c_2 = 7$. The solution of the initial value is then

$$y = 9\pi \cos x + 7 \sin x + 4x - 5x \cos x. \qquad \blacksquare$$

EXAMPLE 9

Solve $\qquad y'' - 6y' + 9y = 6x^2 + 2 - 12e^{3x}.$

Solution The complementary function is

$$y_c = c_1 e^{3x} + c_2 x e^{3x}$$

and based on entries 3 and 7 of the table, the usual assumption for a particular solution would be

$$y_p = \underbrace{Ax^2 + Bx + C}_{y_{p_1}} + \underbrace{De^{3x}}_{y_{p_2}}.$$

Inspection of these functions shows that the one term in y_{p_2} is duplicated in y_c. If we multiply y_{p_2} by x, we note that the term xe^{3x} is still part of y_c. But multiplying y_{p_2} by x^2 eliminates all duplications. Thus the operative form of a particular solution is

$$y_p = Ax^2 + Bx + C + Dx^2e^{3x}.$$

Differentiating this last form, substituting into the differential equation, and collecting like terms give

$$y_p'' - 6y_p' + 9y_p = 9Ax^2 + (-12A + 9B)x + 2A - 6B + 9C + 2De^{3x} = 6x^2 + 2 - 12e^{3x}.$$

It follows from this identity that $A = 2/3$, $B = 8/9$, $C = 2/3$, and $D = -6$. Hence the general solution $y = y_c + y_p$ is

$$y = c_1e^{3x} + c_2xe^{3x} + \frac{2}{3}x^2 + \frac{8}{9}x + \frac{2}{3} - 6x^2e^{3x}. \qquad \blacksquare$$

Higher-Order Equations

The method of undetermined coefficients given here is not restricted to second-order equations but can be used as well with higher-order equations

$$a_ny^{(n)} + a_{n-1}y^{(n-1)} + \cdots + a_1y' + a_0y = g(x)$$

with constant coefficients. It is only necessary that $g(x)$ consist of the proper kinds of functions discussed above.

EXAMPLE 10

Solve
$$y''' + y'' = e^x \cos x.$$

Solution   From the characteristic equation $m^3 + m^2 = 0$ we find $m_1 = m_2 = 0$ and $m_3 = -1$. Hence the complementary solution of the equation is $y_c = c_1 + c_2x + c_3e^{-x}$. With $g(x) = e^x \cos x$ we see from entry 10 of the table of trial particular solutions that we should assume

$$y_p = Ae^x \cos x + Be^x \sin x.$$

Since there are no functions in y_p that duplicate functions in the complementary solution, we proceed in the usual manner. From

$$y_p''' + y_p'' = (-2A + 4B)e^x \cos x + (-4A - 2B)e^x \sin x = e^x \cos x$$

we get
$$-2A + 4B = 1$$
$$-4A - 2B = 0.$$

This system gives $A = -1/10$ and $B = 1/5$, so that a particular solution is
$$y_p = -\frac{1}{10} e^x \cos x + \frac{1}{5} e^x \sin x.$$

The general solution of the equation is
$$y = y_c + y_p = c_1 + c_2 x + c_3 e^{-x} - \frac{1}{10} e^x \cos x + \frac{1}{5} e^x \sin x. \qquad \blacksquare$$

EXAMPLE II

Determine the form of a particular solution of
$$y^{(4)} + y''' = 1 - e^{-x}.$$

Solution Comparing the complementary function
$$y_c = c_1 + c_2 x + c_3 x^2 + c_4 e^{-x}$$

with our normal assumption for a particular solution
$$y_p = \underset{\boxed{y_{p_1}}}{\underbrace{A}} + \underset{\boxed{y_{p_2}}}{\underbrace{Be^{-x}}},$$

we see that the duplications between y_c and y_p are eliminated when y_{p_1} is multiplied by x^3 and y_{p_2} is multiplied by x. Thus the correct assumption for a particular solution is
$$y_p = Ax^3 + Bxe^{-x}. \qquad \blacksquare$$

4.4 EXERCISES

Answers to odd-numbered problems begin on page A–10.

In Problems 1–26 solve the given differential equation by undetermined coefficients.

1. $y'' + 3y' + 2y = 6$
2. $4y'' + 9y = 15$
3. $y'' - 10y' + 25y = 30x + 3$
4. $y'' + y' - 6y = 2x$
5. $\frac{1}{4} y'' + y' + y = x^2 - 2x$
6. $y'' - 8y' + 20y = 100x^2 - 26xe^x$
7. $y'' + 3y = -48x^2 e^{3x}$
8. $4y'' - 4y' - 3y = \cos 2x$
9. $y'' - y' = -3$
10. $y'' + 2y' = 2x + 5 - e^{-2x}$

11. $y'' - y' + \frac{1}{4}y = 3 + e^{x/2}$

12. $y'' - 16y = 2e^{4x}$

13. $y'' + 4y = 3 \sin 2x$

14. $y'' + 4y = (x^2 - 3) \sin 2x$

15. $y'' + y = 2x \sin x$

16. $y'' - 5y' = 2x^3 - 4x^2 - x + 6$

17. $y'' - 2y' + 5y = e^x \cos 2x$

18. $y'' - 2y' + 2y = e^{2x}(\cos x - 3 \sin x)$

19. $y'' + 2y' + y = \sin x + 3 \cos 2x$

20. $y'' + 2y' - 24y = 16 - (x + 2)e^{4x}$

21. $y''' - 6y'' = 3 - \cos x$

22. $y''' - 2y'' - 4y' + 8y = 6xe^{2x}$

23. $y''' - 3y'' + 3y' - y = x - 4e^x$

24. $y''' - y'' - 4y' + 4y = 5 - e^x + e^{2x}$

25. $y^{(4)} + 2y'' + y = (x - 1)^2$

26. $y^{(4)} - y'' = 4x + 2xe^{-x}$

In Problems 27 and 28 use a trigonometric identity as an aid in finding a particular solution of the given differential equation.

27. $y'' + 4y = 8 \sin^2 x$

28. $y'' + y = \sin x \cos 2x$

In Problems 29–40 solve the given differential equation subject to the indicated initial conditions.

29. $y'' + y = -2, \quad y\left(\frac{\pi}{8}\right) = \frac{1}{2}, y'\left(\frac{\pi}{8}\right) = 2$

30. $2y'' + 3y' - 2y = 14x^2 - 4x - 11, \quad y(0) = 0, y'(0) = 0$

31. $5y'' + y' = -6x, \quad y(0) = 0, y'(0) = -10$

32. $y'' + 4y' + 4y = (3 + x)e^{-2x}, \quad y(0) = 2, y'(0) = 5$

33. $y'' + 4y' + 5y = 35e^{-4x}, \quad y(0) = -3, y'(0) = 1$

34. $y'' - y = \cosh x, \quad y(0) = 2, y'(0) = 12$

35. $\dfrac{d^2x}{dt^2} + \omega^2 x = F_0 \sin \omega t, \quad x(0) = 0, x'(0) = 0$

36. $\dfrac{d^2x}{dt^2} + \omega^2 x = F_0 \cos \gamma t, \quad x(0) = 0, x'(0) = 0$

37. $y'' + y = \cos x - \sin 2x, \quad y\left(\frac{\pi}{2}\right) = 0, y'\left(\frac{\pi}{2}\right) = 0$

38. $y'' - 2y' - 3y = 2 \cos^2 x, \quad y(0) = -\frac{1}{3}, y'(0) = 0$

39. $y''' - 2y'' + y' = 2 - 24e^x + 40e^{5x}, \quad y(0) = \frac{1}{2}, y'(0) = \frac{5}{2}, y''(0) = -\frac{9}{2}$

40. $y''' + 8y = 2x - 5 + 8e^{-2x}, \quad y(0) = -5, y'(0) = 3, y''(0) = -4$

In Problems 41 and 42 solve the given differential equation subject to the indicated boundary conditions.

41. $y'' + y = x^2 + 1, \quad y(0) = 5, y(1) = 0$

42. $y'' - 2y' + 2y = 2x - 2, \quad y(0) = 0, y(\pi) = \pi$

43. In applications the input function $g(x)$ is often discontinuous. Solve the initial-value problem

$$y'' + 4y = g(x), \qquad y(0) = 1, \quad y'(0) = 2,$$

where
$$g(x) = \begin{cases} \sin x, & 0 \le x \le \pi/2 \\ 0, & x > \pi/2 \end{cases}.$$

[*Hint:* Solve the problem on the two intervals and then find a solution so that y and y' are continuous at $x = \pi/2$.]

4.5 DIFFERENTIAL OPERATORS

In calculus, differentiation is often denoted by the capital letter D; that is,

$$\frac{dy}{dx} = Dy.$$

The symbol D is called a **differential operator** and it transforms a differentiable function into another function; for example,

$$D(e^{4x}) = 4e^{4x}, \qquad D(5x^3 - 6x^2) = 15x^2 - 12x, \qquad D(\cos 2x) = -2\sin 2x.$$

The differential operator D also possesses a linearity property; D operating on a linear combination of two differentiable functions is the same as the linear combination of D operating on the individual functions. In symbols, this means

$$D\{af(x) + bg(x)\} = a\,Df(x) + b\,Dg(x), \tag{1}$$

where a and b are constants. Because of (1) we say that D is a **linear differential operator**.

Higher-Order Derivatives

Higher-order derivatives can be expressed in terms of D in a natural manner:

$$\frac{d}{dx}\left(\frac{dy}{dx}\right) = \frac{d^2y}{dx^2} = D(Dy) = D^2y \quad \text{and in general} \quad \frac{d^ny}{dx^n} = D^ny,$$

where y represents a sufficiently differentiable function. Polynomial expressions involving D such as

$$D + 3, \quad D^2 + 3D - 4, \quad \text{and} \quad 5D^3 - 6D^2 + 4D + 9$$

are also linear differential operators.

Differential Equations

Any linear differential equation can be expressed in terms of the D notation. For example, a second-order differential equation with constant coefficients $ay'' + by' + cy = g(x)$ can be written as

$$aD^2y + bDy + cy = g(x) \quad \text{or} \quad (aD^2 + bD + c)y = g(x).$$

If we further define $L = aD^2 + bD + c$, then the last equation can be written compactly as

$$L(y) = g(x).$$

The operator $L = aD^2 + bD + c$ is said to be a **second-order linear differential operator with constant coefficients**.

EXAMPLE 1

The differential equation $y'' + y' + 2y = 5x - 3$ can be written as

$$(D^2 + D + 2)y = 5x - 3. \qquad \blacksquare$$

An nth-order linear differential operator

$$L = a_nD^n + a_{n-1}D^{n-1} + \cdots + a_1D + a_0$$

with constant coefficients can be factored whenever the characteristic polynomial $a_nm^n + a_{n-1}m^{n-1} + \cdots + a_1m + a_0$ factors.* For example, if we treat D as an algebraic quantity, then $D^2 + 5D + 6$ can be factored as $(D + 2)(D + 3)$ or as $(D + 3)(D + 2)$. In other words, for a function $y = f(x)$ possessing a second derivative

$$(D^2 + 5D + 6)y = (D + 2)(D + 3)y = (D + 3)(D + 2)y.$$

To see why this is so, let $w = (D + 3)y = y' + 3y$, then

$$(D + 2)w = Dw + 2w = \underbrace{(y'' + 3y')}_{Dw} + \underbrace{(2y' + 6y)}_{2w} = y'' + 5y' + 6y$$

Similarly, if we let $w = (D + 2)y = y' + 2y$, then

$$(D + 3)w = Dw + 3w = \underbrace{(y'' + 2y')}_{Dw} + \underbrace{(3y' + 6y)}_{3w} = y'' + 5y' + 6y.$$

---

* If one is willing to use complex numbers, then a differential operator with constant coefficients can *always* be factored. We are, however, primarily concerned with writing differential equations in operator form with real coefficients.

This illustrates a general property:

Factors of a linear differential operator with constant coefficients commute.

EXAMPLE 2

(a) The operator $D^2 - 1$ can be written as

$$(D + 1)(D - 1) \quad \text{or} \quad (D - 1)(D + 1).$$

(b) The operator $D^2 + D + 2$ in Example 1 does not factor with real numbers. ∎

EXAMPLE 3

The differential equation $y'' + 4y' + 4y = 0$ is written as

$$(D^2 + 4D + 4)y = 0 \quad \text{or} \quad (D + 2)(D + 2)y = 0 \quad \text{or} \quad (D + 2)^2 y = 0. \quad ∎$$

Annihilator Operator

If L is a linear differential operator with constant coefficients and $y = f(x)$ is a sufficiently differentiable function such that

$$L(y) = 0,$$

then L is said to be an **annihilator** of the function. For example, if $y = k$ (a constant), then $Dk = 0$. Also, $D^2 x = 0$, $D^3 x^2 = 0$, and so on.

> The differential operator D^n annihilates each of the functions
> $$1, x, x^2, \ldots, x^{n-1}.$$
> (2)

As an immediate consequence of (2) and the fact that differentiation can be done term by term, a polynomial

$$c_0 + c_1 x + \cdots + c_{n-1} x^{n-1}$$

can be annihilated by finding an operator that annihilates the highest power of x.

EXAMPLE 4

Find a differential operator that annihilates $1 - 5x^2 + 8x^3$.

Solution   From (2) we know that $D^4x^3 = 0$, and so it follows that

$$D^4(1 - 5x^2 + 8x^3) = 0. \qquad\blacksquare$$

Note   The functions that are annihilated by an nth-order linear differential operator L are simply those functions that can be obtained from the general solution of the homogeneous differential equation $L(y) = 0$.

> The differential operator $(D - \alpha)^n$ annihilates each of the functions   (3)
>
> $$e^{\alpha x}, xe^{\alpha x}, x^2e^{\alpha x}, \ldots, x^{n-1}e^{\alpha x}.$$

To see this, note that the auxiliary equation of the homogeneous equation $(D - \alpha)^n y = 0$ is $(m - \alpha)^n = 0$. Since α is a root of multiplicity n, the general solution is

$$y = c_1 e^{\alpha x} + c_2 xe^{\alpha x} + \cdots + c_n x^{n-1}e^{\alpha x}. \qquad (4)$$

EXAMPLE 5

Find an annihilator operator for **(a)** e^{5x} and **(b)** $4e^{2x} - 6xe^{2x}$.

Solution   **(a)** From (3) with $\alpha = 5$ and $n = 1$, we see

$$(D - 5)e^{5x} = 0.$$

(b) From (3) and (4) with $\alpha = 2$ and $n = 2$, we have

$$(D - 2)^2(4e^{2x} - 6xe^{2x}) = 0. \qquad\blacksquare$$

When α and β are real numbers, the quadratic formula reveals that $[m^2 - 2\alpha m + (\alpha^2 + \beta^2)]^n = 0$ has complex roots $\alpha + i\beta$, $\alpha - i\beta$, both of multiplicity n. From the discussion at the end of Section 4.3, we have the next result.

> The differential operator $[D^2 - 2\alpha D + (\alpha^2 + \beta^2)]^n$ annihilates each of the functions
>
> $$e^{\alpha x}\cos\beta x, xe^{\alpha x}\cos\beta x, x^2e^{\alpha x}\cos\beta x, \ldots, x^{n-1}e^{\alpha x}\cos\beta x,$$ (5)
>
> $$e^{\alpha x}\sin\beta x, xe^{\alpha x}\sin\beta x, x^2e^{\alpha x}\sin\beta x, \ldots, x^{n-1}e^{\alpha x}\sin\beta x.$$

EXAMPLE 6

From (5), with $\alpha = -1$, $\beta = 2$, and $n = 1$, we see that

$$(D^2 + 2D + 5)e^{-x}\cos 2x = 0 \quad \text{and} \quad (D^2 + 2D + 5)e^{-x}\sin 2x = 0. \qquad\blacksquare$$

Since $y_1(x) = e^{-x} \cos 2x$ and $y_2(x) = e^{-x} \sin 2x$ are the two linearly independent functions in the general solution of $(D^2 + 2D + 5)y = 0$, the linear operator $D^2 + 2D + 5$ also annihilates any linear combination of these functions, such as $5e^{-x} \cos 2x - 9e^{-x} \sin 2x$.

EXAMPLE 7

From (5), with $\alpha = 0$, $\beta = 1$, and $n = 2$, it is seen that the differential operator $(D^2 + 1)^2$ or $D^4 + 2D^2 + 1$ annihilates $\cos x$, $x \cos x$, $\sin x$, and $x \sin x$. Moreover, $(D^2 + 1)^2$ annihilates any linear combination of these functions. ∎

When $\alpha = 0$ and $n = 1$, a special case of (5) is

$$(D^2 + \beta^2)\begin{cases} \cos \beta x \\ \sin \beta x \end{cases} = 0. \tag{6}$$

We are often interested in annihilating the sum of two or more functions. As seen in Examples 4–7, if L is a linear differential operator such that $L(y_1) = 0$ and $L(y_2) = 0$, then L annihilates the linear combination $c_1 y_1(x) + c_2 y_2(x)$. This is a direct consequence of Theorem 4.3. Let us now suppose that L_1 and L_2 are linear differential operators with constant coefficients such that L_1 annihilates $y_1(x)$ and L_2 annihilates $y_2(x)$ but $L_1(y_2) \neq 0$ and $L_2(y_1) \neq 0$. Then the *product* of differential operators $L_1 L_2$ annihilates the sum $c_1 y_1(x) + c_2 y_2(x)$. We can easily demonstrate this using linearity and the fact that $L_1 L_2 = L_2 L_1$:

$$L_1 L_2(y_1 + y_2) = L_1 L_2(y_1) + L_1 L_2(y_2)$$
$$= L_2 L_1(y_1) + L_1 L_2(y_2) \tag{7}$$
$$= L_2[\underbrace{L_1(y_1)}_{\text{zero}}] + L_1[\underbrace{L_2(y_2)}_{\text{zero}}] = 0.$$

EXAMPLE 8

Find a differential operator that annihilates $7 - x + 6 \sin 3x$.

Solution From (2) and (6) we have, respectively,

$$D^2(7 - x) = 0 \quad \text{and} \quad (D^2 + 9) \sin 3x = 0.$$

It follows from (7) that the operator $D^2(D^2 + 9)$ annihilates the given linear combination. ∎

EXAMPLE 9

Find a differential operator that annihilates $e^{-3x} + xe^x$.

Solution   From (3),

$$(D + 3)e^{-3x} = 0 \quad \text{and} \quad (D - 1)^2 xe^x = 0.$$

Hence the product of the two operations $(D + 3)(D - 1)^2$ annihilates the given linear combination.   ∎

Remark   The differential operator that annihilates a function is not unique. For example, we know that $D - 5$ annihilates e^{5x}, but so do differential operators of higher order such as $(D - 5)(D + 1)$ and $(D - 5)D^2$. (Verify this.) When we seek a differential annihilator for a function $y = f(x)$, we want the operator of lowest possible order that does the job.

4.5 EXERCISES

Answers to odd-numbered problems begin on page A–10.

In Problems 1–6 write the given differential equation in the form $L(y) = g(x)$, where L is a differential operator with constant coefficients

1. $\dfrac{dy}{dx} + 5y = 9 \sin x$

2. $4\dfrac{dy}{dx} + 8y = x + 3$

3. $3y'' - 5y' + y = e^x$

4. $y''' - 2y'' + 7y' - 6y = 1 - \sin x$

5. $y''' - 4y'' + 5y' = 4x$

6. $y^{(4)} - 2y'' + y = e^{-3x} + e^{2x}$

In Problems 7–16, if possible, factor the given differentiable operator.

7. $9D^2 - 4$

8. $D^2 - 5$

9. $D^2 - 4D - 12$

10. $2D^2 - 3D - 2$

11. $D^3 + 10D^2 + 25D$

12. $D^3 + 4D$

13. $D^3 + 2D^2 - 13D + 10$

14. $D^3 + 4D^2 + 3D$

15. $D^4 + 8D$

16. $D^4 - 8D^2 + 16$

In Problems 17–20 verify that the given differential operator annihilates the indicated function.

17. D^4; $y = 10x^3 - 2x$

18. $2D - 1$; $y = 4e^{x/2}$

19. $(D - 2)(D + 5)$; $y = 4e^{2x}$

20. $D^2 + 64$; $y = 2 \cos 8x - 5 \sin 8x$

In Problems 21–32 find a differential operator that annihilates the given function.

21. $1 + 6x - 2x^3$

22. $x^3(1 - 5x)$

23. $1 + 7e^{2x}$

24. $x + 3xe^{6x}$

25. $\cos 2x$

26. $1 + \sin x$

27. $13x + 9x^2 - \sin 4x$

28. $8x - \sin x + 10 \cos 5x$

29. $e^{-x} + 2xe^x - x^2 e^x$

30. $(2 - e^x)^2$

31. $3 + e^x \cos 2x$ **32.** $e^{-x} \sin x - e^{2x} \cos x$

In Problems 33–40 find linearly independent functions that are annihilated by the given differential operator.

33. D^5 **34.** $D^2 + 4D$

35. $(D - 6)(2D + 3)$ **36.** $D^2 - 9D - 36$

37. $D^2 + 5$ **38.** $D^2 - 6D + 10$

39. $D^3 - 10D^2 + 25D$ **40.** $D^2(D - 5)(D - 7)$

4.6 UNDETERMINED COEFFICIENTS—ANNIHILATOR APPROACH

To obtain the general solution of a nonhomogeneous linear differential equation we must do two things:

   (i) Find the complementary function y_c.

   (ii) Find *any* particular solution y_p of the nonhomogeneous equation.

Recall from the discussion of Section 4.1 that a particular solution is any function, free of arbitrary constants, that satisfies the differential equation identically. The general solution of a nonhomogeneous equation on an interval is then $y = y_c + y_p$.

   If L denotes a linear differential operator of the form $a_n D^n + a_{n-1} D^{n-1} + \cdots + a_1 D + a_0$, then a nonhomogeneous linear differential equation can be written simply as

$$L(y) = g(x). \tag{1}$$

The **method of undetermined coefficients** presented in this section is limited to nonhomogeneous linear equations

- that have constant coefficients, and
- where $g(x)$ is a constant k, a polynomial function, an exponential function $e^{\alpha x}$, $\sin \beta x$, $\cos \beta x$, or finite sums and products of these functions.

Note Strictly speaking, $g(x) = k$ (a constant) is a polynomial function. Since a constant function is probably not the first thing that comes to mind when you think of polynomial functions, for emphasis we continue to use the redundancy "constant functions, polynomials, ..."

   The following are some examples of the types of input functions $g(x)$ that are appropriate for this discussion:

$$g(x) = 10$$

$$g(x) = x^2 - 5x$$

$$g(x) = 15x - 6 + 8e^{4x}$$

$$g(x) = \sin 3x - 5x \cos 2x$$

$$g(x) = e^x \cos x - (3x^2 - 1)e^{-x}$$

and so on. In other words, $g(x)$ is a linear combination of functions of the form

$$k \text{ (constant)}, \quad x^m, \quad x^m e^{\alpha x}, \quad x^m e^{\alpha x} \cos \beta x, \quad \text{and} \quad x^m e^{\alpha x} \sin \beta x,$$

where m is a nonnegative integer and α and β are real numbers. The method of undetermined coefficients is not applicable to equations of the form (1) when

$$g(x) = \ln x, \quad g(x) = \frac{1}{x}, \quad g(x) = \tan x, \quad g(x) = \sin^{-1} x,$$

and so on. Differential equations with this latter kind of input function will be considered in Section 4.7.

As we saw in Section 4.5, a linear combination of functions of the type $k, x^m, x^m e^{\alpha x}, x^m e^{\alpha x} \cos \beta x$, and $x^m e^{\alpha x} \sin \beta x$ is precisely that kind of function that can be annihilated by an operator L_1 (of lowest order) consisting of a product of operators such as $D^n, (D - \alpha)^n$, and $(D^2 - 2\alpha D + \alpha^2 + \beta^2)^n$. Applying L_1 to both members of (1) yields

$$L_1 L(y) = L_1(g(x)) = 0. \tag{2}$$

By solving the *homogeneous higher-order* equation $L_1 L(y) = 0$, we can discover the form of a particular solution y_p for the original nonhomogeneous equation $L(y) = g(x)$.

The next several examples illustrate the method. The general solution of each equation is defined on the interval $(-\infty, \infty)$.

EXAMPLE 1

Solve
$$\frac{d^2 y}{dx^2} + 3\frac{dy}{dx} + 2y = 4x^2. \tag{3}$$

Solution   **Step 1.**   We first solve the homogeneous equation

$$\frac{d^2 y}{dx^2} + 3\frac{dy}{dx} + 2y = 0.$$

From the auxiliary equation $m^2 + 3m + 2 = (m + 1)(m + 2) = 0$, we find $m_1 = -1, m_2 = -2$, and so the complementary function is

$$y_c = c_1 e^{-x} + c_2 e^{-2x}.$$

Step 2.   In view of (2) of Section 4.5, (3) can be rendered homogeneous by taking three derivatives of each side of the equation. In other words,

$$D^3(D^2 + 3D + 2)y = 4D^3 x^2 = 0, \tag{4}$$

since $D^3x^2 = 0$. The auxiliary equation of (4),

$$m^3(m^2 + 3m + 2) = 0 \quad \text{or} \quad m^3(m + 1)(m + 2) = 0,$$

has roots $0, 0, 0, -1$, and -2. Thus its general solution must be

$$y = c_1 + c_2x + c_3x^2 + \boxed{c_4e^{-x} + c_5e^{-2x}}. \tag{5}$$

The terms in the box in (5) constitute the complementary function of the original equation (3). We can then argue that a particular solution y_p of (3) should also satisfy equation (4). This means that the terms remaining in (5) must be the basic structure of y_p:

$$y_p = A + Bx + Cx^2, \tag{6}$$

where, for convenience, we have replaced c_1, c_2, and c_3 by A, B, and C, respectively. For (6) to be a particular solution of (3), it is necessary to find *specific* coefficients A, B, and C. Differentiating (6), we have

$$y_p' = B + 2Cx, \qquad y_p'' = 2C,$$

and substitution into (3) then gives

$$y_p'' + 3y_p' + 2y_p = 2C + 3B + 6Cx + 2A + 2Bx + 2Cx^2 = 4x^2.$$

Since the last equation is supposed to be an identity, the coefficients of like powers of x must be equal:

That is,

$$2C = 4$$

$$2B + 6C = 0 \tag{7}$$

$$2A + 3B + 2C = 0.$$

Solving (7) gives $A = 7$, $B = -6$, and $C = 2$. Thus $y_p = 7 - 6x + 2x^2$.

Step 3. The general solution of (3) is $y = y_c + y_p$, or

$$y = c_1e^{-x} + c_2e^{-2x} + 7 - 6x + 2x^2. \qquad \blacksquare$$

EXAMPLE 2

Solve $\qquad\qquad y'' - 3y' = 8e^{3x} + 4\sin x.$ (8)

Solution **Step 1.** The auxiliary equation for the homogeneous equation

$y'' - 3y' = 0$ is $m(m - 3) = 0$ and so

$$y_c = c_1 + c_2 e^{3x}.$$

Step 2. Now since $(D - 3)e^{3x} = 0$ and $(D^2 + 1) \sin x = 0$, we apply the differential operator $(D - 3)(D^2 + 1)$ to both sides of (8):

$$(D - 3)(D^2 + 1)(D^2 - 3D)y = 0. \tag{9}$$

The auxiliary equation of (9) is

$$(m - 3)(m^2 + 1)(m^2 - 3m) = 0 \quad \text{or} \quad m(m - 3)^2(m^2 + 1) = 0.$$

Thus $y = \boxed{c_1 + c_2 e^{3x}} + c_3 x e^{3x} + c_4 \cos x + c_5 \sin x.$ \tag{10}

After excluding the linear combination of terms in the box that corresponds to y_c, we arrive at the form of y_p:

$$y_p = Axe^{3x} + B \cos x + C \sin x.$$

Substituting y_p in (8) and simplifying yield

$$y_p'' - 3y_p' = 3Ae^{3x} + (-B - 3C) \cos x + (3B - C) \sin x$$
$$= 8e^{3x} + 4 \sin x.$$

Equating coefficients gives

$$3A = 8$$
$$-B - 3C = 0$$
$$3B - C = 4.$$

We find $A = 8/3$, $B = 6/5$, $C = -2/5$, and consequently,

$$y_p = \frac{8}{3} xe^{3x} + \frac{6}{5} \cos x - \frac{2}{5} \sin x.$$

Step 3. The general solution of (8) is then

$$y = c_1 + c_2 e^{3x} + \frac{8}{3} xe^{3x} + \frac{6}{5} \cos x - \frac{2}{5} \sin x, \qquad ■$$

EXAMPLE 3

Solve $y'' + 8y = 5x + 2e^{-x}.$ \tag{11}

Solution From (2) and (3) of Section 4.5 we know that $D^2 x = 0$ and $(D + 1)e^{-x} = 0$, respectively. Hence we apply $D^2(D + 1)$ to (11):

$$D^2(D + 1)(D^2 + 8)y = 0.$$

It is seen that

$$y = \boxed{c_1 \cos 2\sqrt{2}x + c_2 \sin 2\sqrt{2}x} + c_3 + c_4 x + c_5 e^{-x}$$

and so

$$y_p = A + Bx + Ce^{-x}.$$

Substituting y_p into (11) yields

$$y_p'' + 8y_p = 8A + 8Bx + 9Ce^{-x} = 5x + 2e^{-x}.$$

This implies $A = 0$, $B = 5/8$, and $C = 2/9$, so that the general solution of (1) is

$$y = c_1 \cos 2\sqrt{2}x + c_2 \sin 2\sqrt{2}x + \frac{5}{8}x + \frac{2}{9}e^{-x}. \qquad \blacksquare$$

EXAMPLE 4

Solve

$$y'' + y = x \cos x - \cos x. \qquad (12)$$

Solution   In Example 7 of Section 4.5 we saw that $x \cos x$ and $\cos x$ are annihilated by the operator $(D^2 + 1)^2$. Thus

$$(D^2 + 1)^2(D^2 + 1)y = 0 \quad \text{or} \quad (D^2 + 1)^3 y = 0.$$

Since i and $-i$ are both complex roots of multiplicity 3 of the auxiliary equation of the last differential equation, we conclude

$$y = \boxed{c_1 \cos x + c_2 \sin x} + c_3 x \cos x + c_4 x \sin x + c_5 x^2 \cos x + c_6 x^2 \sin x.$$

We substitute

$$y_p = Ax \cos x + Bx \sin x + Cx^2 \cos x + Ex^2 \sin x$$

into (12) and simplify:

$$y_p'' + y_p = 4Ex \cos x - 4Cx \sin x + (2B + 2C) \cos x + (-2A + 2E) \sin x$$

$$= x \cos x - \cos x.$$

Equating coefficients gives the equations

$$4E = 1$$

$$-4C = 0$$

$$2B + 2C = -1$$

$$-2A + 2E = 0,$$

from which we find $E = 1/4$, $C = 0$, $B = -1/2$, and $A = 1/4$. Hence the general solution of (12) is

$$y = c_1 \cos x + c_2 \sin x + \frac{1}{4}x \cos x - \frac{1}{2}x \sin x + \frac{1}{4}x^2 \sin x. \qquad \blacksquare$$

EXAMPLE 5

Determine the form of a particular solution for

$$y'' - 2y' + y = 10e^{-2x} \cos x. \tag{13}$$

Solution   The complementary function for the given equation is $y_c = c_1 e^x + c_2 x e^x$.

Now from (5) of Section 4.5, with $\alpha = -2$, $\beta = 1$, and $n = 1$, we know that

$$(D^2 + 4D + 5)e^{-2x} \cos x = 0.$$

Applying the operator $D^2 + 4D + 5$ to (13) gives

$$(D^2 + 4D + 5)(D^2 - 2D + 1)y = 0. \tag{14}$$

Since the roots of the auxiliary equation of (14) are $-2 - i$, $-2 + i$, 1, and 1,

$$y = \boxed{c_1 e^x + c_2 x e^x} + c_3 e^{-2x} \cos x + c_4 e^{-2x} \sin x.$$

Hence a particular solution of (13) can be found with the form

$$y_p = Ae^{-2x} \cos x + Be^{-2x} \sin x. \qquad \blacksquare$$

EXAMPLE 6

Determine the form of a particular solution for

$$y''' - 4y'' + 4y' = 5x^2 - 6x + 4x^2 e^{2x} + 3e^{5x}. \tag{15}$$

Solution   Observe that

$$D^3(5x^2 - 6x) = 0, \quad (D - 2)^3 x^2 e^{2x} = 0, \quad \text{and} \quad (D - 5)e^{5x} = 0.$$

Therefore $D^3(D - 2)^3(D - 5)$ applied to (15) gives

$$D^3(D - 2)^3(D - 5)(D^3 - 4D^2 + 4D)y = 0$$

or

$$D^4(D - 2)^5(D - 5)y = 0.$$

The roots of the auxiliary equation for the last differential equation are easily seen to be $0, 0, 0, 0, 2, 2, 2, 2, 2$, and 5. Hence

$$y = \boxed{c_1} + c_2 x + c_3 x^2 + c_4 x^3 + \boxed{c_5 e^{2x} + c_6 x e^{2x}} + c_7 x^2 e^{2x} + c_8 x^3 e^{2x} + c_9 x^4 e^{2x} + c_{10} e^{5x}. \tag{16}$$

Since the linear combination $c_1 + c_5 e^{2x} + c_6 x e^{2x}$ can be taken as the complementary function of (15), the remaining terms in (16) give the form of a particular solution of the differential equation:

$$y_p = Ax + Bx^2 + Cx^3 + Ex^2 e^{2x} + Fx^3 e^{2x} + Gx^4 e^{2x} + He^{5x}. \qquad \blacksquare$$

Summary of the Method

For your convenience the method of undetermined coefficients is summarized here.

Undetermined Coefficients—Annihilator Approach

The differential equation $L(y) = g(x)$ has constant coefficients and the function $g(x)$ consists of finite sums and products of constants, polynomials, exponential functions $e^{\alpha x}$, sines, and cosines.

(i) Find the complementary solution y_c for the homogeneous equation $L(y) = 0$.

(ii) Operate on both sides of the nonhomogeneous equation $L(y) = g(x)$ with a differential operator L_1 that annihilates the function $g(x)$.

(iii) Find the general solution of the higher-order homogeneous differential equation $L_1 L(y) = 0$.

(iv) Delete all those terms from the solution in step (iii) that are duplicated in the complementary solution y_c found in step (i). Form a linear combination y_p of the terms that remain. This is the form of a particular solution of $L(y) = g(x)$.

(v) Substitute y_p found in step (iv) into $L(y) = g(x)$. Match coefficients of the various functions on each side of the equality and solve the resulting system of equations for the unknown coefficients in y_p.

(vi) With the particular solution found in step (v), form the general solution $y = y_c + y_p$ of the given differential equation.

4.6 EXERCISES

Answers to odd-numbered problems begin on page A–10.

In Problems 1–32 solve the given differential equation by undetermined coefficients.

1. $y'' - 9y = 54$

2. $2y'' - 7y' + 5y = -29$

3. $y'' + y' = 3$

4. $y''' + 2y'' + y' = 10$

5. $y'' + 4y' + 4y = 2x + 6$

6. $y'' + 3y' = 4x - 5$

7. $y''' + y'' = 8x^2$

8. $y'' - 2y' + y = x^3 + 4x$

9. $y'' - y' - 12y = e^{4x}$

10. $y'' + 2y' + 2y = 5e^{6x}$

11. $y'' - 2y' - 3y = 4e^x - 9$

12. $y'' + 6y' + 8y = 3e^{-2x} + 2x$

13. $y'' + 25y = 6 \sin x$

14. $y'' + 4y = 4 \cos x + 3 \sin x - 8$

15. $y'' + 6y' + 9y = -xe^{4x}$

16. $y'' + 3y' - 10y = x(e^x + 1)$

17. $y'' - y = x^2 e^x + 5$

18. $y'' + 2y' + y = x^2 e^{-x}$

19. $y'' - 2y' + 5y = e^x \sin x$

20. $y'' + y' + \frac{1}{4}y = e^x(\sin 3x - \cos 3x)$

21. $y'' + 25y = 20 \sin 5x$

22. $y'' + y = 4 \cos x - \sin x$

23. $y'' + y' + y = x \sin x$

24. $y'' + 4y = \cos^2 x$

25. $y''' + 8y'' = -6x^2 + 9x + 2$

26. $y''' - y'' + y' - y = xe^x - e^{-x} + 7$

27. $y''' - 3y'' + 3y' - y = e^x - x + 16$

28. $2y''' - 3y'' - 3y' + 2y = (e^x + e^{-x})^2$

29. $y^{(4)} - 2y''' + y'' = e^x + 1$

30. $y^{(4)} - 4y'' = 5x^2 - e^{2x}$

31. $16y^{(4)} - y = e^{x/2}$

32. $y^{(4)} - 5y'' + 4y = 2 \cosh x - 6$

In Problems 33–40 solve the given differential equation subject to the indicated initial conditions.

33. $y'' - 64y = 16, \quad y(0) = 1, y'(0) = 0$

34. $y'' + y' = x, \quad y(0) = 1, y'(0) = 0$

35. $y'' - 5y' = x - 2, \quad y(0) = 0, y'(0) = 2$

36. $y'' + 5y' - 6y = 10e^{2x}, \quad y(0) = 1, y'(0) = 1$

37. $y'' + y = 8 \cos 2x - 4 \sin x, \quad y(\pi/2) = -1, y'(\pi/2) = 0$

38. $y''' - 2y'' + y' = xe^x + 5, \quad y(0) = 2, y'(0) = 2, y''(0) = -1$

39. $y'' - 4y' + 8y = x^3, \quad y(0) = 2, y'(0) = 4$

40. $y^{(4)} - y''' = x + e^x, \quad y(0) = 0, y'(0) = 0, y''(0) = 0, y'''(0) = 0$

In Problems 41 and 42 determine the form of a particular solution for the given differential equation.

41. $y'' - y = e^x(2 + 3x \cos 2x)$

42. $y'' + y' = 9 - e^{-x} + x^2 \sin x$

43. Show that the operator $(xD - 1)(D + 4)$ is not the same as the operator $(D + 4)(xD - 1)$.

44. Prove that the differential equation

$$a_n y^{(n)} + a_{n-1} y^{(n-1)} + \cdots + a_1 y' + a_0 y = k,$$

k a constant, $a_0 \neq 0$, has the particular solution $y_p = k/a_0$.

4.7 VARIATION OF PARAMETERS

Linear First-Order Equation Revisited

In Chapter 2 we saw that the general solution of the linear first-order differential equation

$$\frac{dy}{dx} + P(x)y = f(x),$$ (1)

where $P(x)$ and $f(x)$ are continuous on an interval I, is

$$y = e^{-\int P(x)\,dx} \int e^{\int P(x)\,dx} f(x)\,dx + c_1 e^{-\int P(x)\,dx}. \tag{2}$$

Now (2) has the form $y = y_c + y_p$, where $y_c = c_1 e^{-\int P(x)\,dx}$ is a solution of

$$\frac{dy}{dx} + P(x)y = 0 \tag{3}$$

and

$$y_p = e^{-\int P(x)\,dx} \int e^{\int P(x)\,dx} f(x)\,dx \tag{4}$$

is a particular solution of (1). As a means of motivating an additional method for solving nonhomogeneous linear equations of higher order, we rederive (4) by a method known as **variation of parameters**. The basic procedure is essentially that used in Section 4.2.

Suppose y_1 is a known solution of (3); that is,

$$\frac{dy_1}{dx} + P(x)y_1 = 0.$$

We proved in Section 2.5 that $y_1 = e^{-\int P(x)\,dx}$ is a solution, and since the differential equation is linear, its general solution is $y = c_1 y_1(x)$. Variation of parameters consists of finding a function u_1 such that

$$y_p = u_1(x)y_1(x)$$

is a particular solution of (1). In other words, we replace the parameter c_1 by a variable u_1.

Substituting $y_p = u_1 y_1$ into (1) gives

$$\frac{d}{dx}[u_1 y_1] + P(x)u_1 y_1 = f(x)$$

$$u_1\frac{dy_1}{dx} + y_1\frac{du_1}{dx} + P(x)u_1 y_1 = f(x)$$

$$u_1\underbrace{\left[\frac{dy_1}{dx} + P(x)y_1\right]}_{\text{zero}} + y_1\frac{du_1}{dx} = f(x)$$

so that

$$y_1\frac{du_1}{dx} = f(x).$$

By separating variables, we find

$$du_1 = \frac{f(x)}{y_1(x)}\,dx \quad\text{and}\quad u_1 = \int \frac{f(x)}{y_1(x)}\,dx,$$

from which it follows that

$$y = u_1 y_1 = y_1 \int \frac{f(x)}{y_1(x)}\, dx.$$

From the definition of y_1, we see that the last result is identical to (4).

Second-Order Equations

To adapt the foregoing procedure to a linear second-order differential equation

$$a_2(x)y'' + a_1(x)y' + a_0(x)y = g(x), \tag{5}$$

we put (5) in the standard form

$$y'' + P(x)y' + Q(x)y = f(x) \tag{6}$$

by dividing through by $a_2(x)$. Here we assume that $P(x)$, $Q(x)$, and $f(x)$ are continuous on some interval I. Equation (6) is the analogue of (1). As we know, when $P(x)$ and $Q(x)$ are constants, we have absolutely no difficulty in writing y_c.

Suppose y_1 and y_2 form a fundamental set of solutions on I of the associated homogeneous form of (6); that is,

$$y_1'' + P(x)y_1' + Q(x)y_1 = 0 \quad \text{and} \quad y_2'' + P(x)y_2' + Q(x)y_2 = 0.$$

Now we ask: Can two functions u_1 and u_2 be found so that

$$y_p = u_1(x)y_1(x) + u_2(x)y_2(x)$$

is a particular solution of (6)? Notice that our assumption for y_p is the same as $y_c = c_1 y_1 + c_2 y_2$, but we have replaced c_1 and c_2 by the "variable parameters" u_1 and u_2. Because we seek to determine two unknown functions, reason dictates that we need two equations. As in the introductory discussion leading to the discovery of (4), one of these equations results from substituting $y_p = u_1 y_1 + u_2 y_2$ into the given differential equation (6). The other equation that we impose is

$$y_1 u_1' + y_2 u_2' = 0. \tag{7}$$

This equation is an assumption that is made to simplify the first derivative and, thereby, the second derivative of y_p. Using the product rule to differentiate y_p, we get

$$y_p' = u_1 y_1' + y_1 u_1' + u_2 y_2' + y_2 u_2' = u_1 y_1' + u_2 y_2' + \overbrace{y_1 u_1' + y_2 u_2'}^{\text{zero from (7)}} \tag{8}$$

and so
$$y_p' = u_1 y_1' + u_2 y_2'.$$

Continuing, we find
$$y_p'' = u_1 y_1'' + y_1' u_1' + u_2 y_2'' + y_2' u_2'.$$

Substitution of these results in (6) yields

$$y_p'' + Py_p' + Qy_p = u_1y_1'' + y_1'u_1' + u_2y_2'' + y_2'u_2' + Pu_1y_1' + Pu_2y_2' + Qu_1y_1 + Qu_2y_2$$

$$= u_1\underbrace{[y_1'' + Py_1' + Qy_1]}_{\boxed{\text{zero}}} + u_2\underbrace{[y_2'' + Py_2' + Qy_2]}_{\boxed{\text{zero}}} + y_1'u_1' + y_2'u_2' = f(x).$$

In other words, u_1 and u_2 must be functions that also satisfy the condition

$$y_1'u_1' + y_2'u_2' = f(x). \tag{9}$$

Equations (7) and (9) constitute a linear system of equations for determining the derivatives u_1' and u_2'. By Cramer's rule,* the solution of

$$y_1u_1' + y_2u_2' = 0$$

$$y_1'u_1' + y_2'u_2' = f(x)$$

can be expressed in terms of determinants:

$$u_1' = \frac{W_1}{W} \quad \text{and} \quad u_2' = \frac{W_2}{W}, \tag{10}$$

where

$$W = \begin{vmatrix} y_1 & y_2 \\ y_1' & y_2' \end{vmatrix}, \qquad W_1 = \begin{vmatrix} 0 & y_2 \\ f(x) & y_2' \end{vmatrix}, \qquad \text{and} \quad W_2 = \begin{vmatrix} y_1 & 0 \\ y_1' & f(x) \end{vmatrix} \tag{11}$$

The determinant W is recognized as the Wronskian of y_1 and y_2. By linear independence of y_1 and y_2 on I, we know that $W(y_1(x), y_2(x)) \neq 0$ for every x in the interval.

Summary of the Method

Usually it is not a good idea to memorize formulas in lieu of understanding a procedure. However, the foregoing procedure is too long and complicated to use each time we wish to solve a differential equation. In this case it is more efficient to simply use the formulas in (10). Thus, to solve $a_2y'' + a_1y' + a_0y = g(x)$, first find the complementary function $y_c = c_1y_1 + c_2y_2$ and then compute the Wronskian

$$W = \begin{vmatrix} y_1 & y_2 \\ y_1' & y_2' \end{vmatrix}.$$

By dividing by a_2, we put the equation into the form $y'' + Py' + Qy = f(x)$ to determine $f(x)$. We find u_1 and u_2 by integrating $u_1' = W_1/W$ and $u_2' = W_2/W$, where W_1 and W_2 are defined in (11). A particular solution is $y_p = u_1y_1 + u_2y_2$. The general solution of the equation is then $y = y_c + y_p$.

---

* See Appendix III for a review of Cramer's rule.

EXAMPLE 1

Solve $$y'' - 4y' + 4y = (x + 1)e^{2x}.$$

Solution Since the auxiliary equation is $m^2 - 4m + 4 = (m - 2)^2 = 0$, we have

$$y_c = c_1 e^{2x} + c_2 x e^{2x}.$$

Identifying $y_1 = e^{2x}$ and $y_2 = x e^{2x}$, we next compute the Wronskian

$$W(e^{2x}, xe^{2x}) = \begin{vmatrix} e^{2x} & xe^{2x} \\ 2e^{2x} & 2xe^{2x} + e^{2x} \end{vmatrix} = e^{4x}.$$

Since the given differential equation is already in form (6) (that is, the coefficient of y'' is 1), we identify $f(x) = (x + 1)e^{2x}$. From (11) we obtain

$$W_1 = \begin{vmatrix} 0 & xe^{2x} \\ (x + 1)e^{2x} & 2xe^{2x} + e^{2x} \end{vmatrix} = -(x + 1)xe^{4x}$$

$$W_2 = \begin{vmatrix} e^{2x} & 0 \\ 2e^{2x} & (x + 1)e^{2x} \end{vmatrix} = (x + 1)e^{4x}$$

and so from (10),

$$u_1' = -\frac{(x + 1)xe^{4x}}{e^{4x}} = -x^2 - x, \qquad u_2' = \frac{(x + 1)e^{4x}}{e^{4x}} = x + 1.$$

It follows that

$$u_1 = -\frac{x^3}{3} - \frac{x^2}{2} \quad \text{and} \quad u_2 = \frac{x^2}{2} + x.$$

Therefore $$y_p = \left(-\frac{x^3}{3} - \frac{x^2}{2} \right) e^{2x} + \left(\frac{x^2}{2} + x \right) xe^{2x}$$

$$= \left(\frac{x^3}{6} + \frac{x^2}{2} \right) e^{2x}.$$

Hence $$y = y_c + y_p = c_1 e^{2x} + c_2 x e^{2x} + \left(\frac{x^3}{6} + \frac{x^2}{2} \right) e^{2x}. \qquad ■$$

EXAMPLE 2

Solve $$4y'' + 36y = \csc 3x.$$

Solution We first put the equation in the standard form (6) by dividing by 4:

$$y'' + 9y = \frac{1}{4} \csc 3x.$$

Since the roots of the auxiliary equation $m^2 + 9 = 0$ are $m_1 = 3i$ and $m_2 = -3i$, the complementary function is

$$y_c = c_1 \cos 3x + c_2 \sin 3x.$$

Using $y_1 = \cos 3x$, $y_2 = \sin 3x$, and $f(x) = (1/4) \csc 3x$, we find

$$W(\cos 3x, \sin 3x) = \begin{vmatrix} \cos 3x & \sin 3x \\ -3 \sin 3x & 3 \cos 3x \end{vmatrix} = 3$$

$$W_1 = \begin{vmatrix} 0 & \sin 3x \\ \frac{1}{4} \csc 3x & 3 \cos 3x \end{vmatrix} = -\frac{1}{4} \qquad \boxed{\csc 3x = \dfrac{1}{\sin 3x}}$$

$$W_2 = \begin{vmatrix} \cos 3x & 0 \\ -3 \sin 3x & \frac{1}{4} \csc 3x \end{vmatrix} = \frac{1}{4} \frac{\cos 3x}{\sin 3x}.$$

Integrating

$$u_1' = \frac{W_1}{W} = -\frac{1}{12} \quad \text{and} \quad u_2' = \frac{W_2}{W} = \frac{1}{12} \frac{\cos 3x}{\sin 3x}$$

gives

$$u_1 = -\frac{1}{12} x \quad \text{and} \quad u_2 = \frac{1}{36} \ln|\sin 3x|.$$

Thus a particular solution is

$$y_p = -\frac{1}{12} x \cos 3x + \frac{1}{36} (\sin 3x) \ln|\sin 3x|.$$

The general solution of the equation is

$$y = y_c + y_p = c_1 \cos 3x + c_2 \sin 3x - \frac{1}{12} x \cos 3x + \frac{1}{36} (\sin 3x) \ln|\sin 3x|. \quad \textbf{(12)}$$

■

Equation (12) represents the general solution of the differential equation on, say, the interval $(0, \pi/6)$.

Constants of Integration

When computing the indefinite integrals of u_1' and u_2', we need not introduce any constants. This is because

$$y = y_c + y_p = c_1 y_1 + c_2 y_2 + (u_1 + a_1) y_1 + (u_2 + b_1) y_2$$

$$= (c_1 + a_1) y_1 + (c_2 + b_1) y_2 + u_1 y_1 + u_2 y_2$$

$$= C_1 y_1 + C_2 y_2 + u_1 y_1 + u_2 y_2.$$

EXAMPLE 3

Solve
$$y'' - y = \frac{1}{x}.$$

Solution   The auxiliary equation $m^2 - 1 = 0$ yields $m_1 = -1$ and $m_2 = 1$. Therefore

$$y_c = c_1 e^x + c_2 e^{-x}$$

$$W(e^x, e^{-x}) = \begin{vmatrix} e^x & e^{-x} \\ e^x & -e^{-x} \end{vmatrix} = -2$$

$$u_1' = -\frac{e^{-x}(1/x)}{-2}, \qquad u_1 = \frac{1}{2} \int_{x_0}^x \frac{e^{-t}}{t} \, dt$$

$$u_2' = \frac{e^x(1/x)}{-2}, \qquad u_2 = -\frac{1}{2} \int_{x_0}^x \frac{e^t}{t} \, dt.$$

It is well known that the integrals defining u_1 and u_2 cannot be expressed in terms of elementary functions. Hence we write

$$y_p = \frac{1}{2} e^x \int_{x_0}^x \frac{e^{-t}}{t} \, dt - \frac{1}{2} e^{-x} \int_{x_0}^x \frac{e^t}{t} \, dt,$$

and so

$$y = y_c + y_p = c_1 e^x + c_2 e^{-x} + \frac{1}{2} e^x \int_{x_0}^x \frac{e^{-t}}{t} \, dt - \frac{1}{2} e^{-x} \int_{x_0}^x \frac{e^t}{t} \, dt. \qquad \blacksquare$$

In Example 3 we can integrate on any interval $x_0 \leq t \leq x$ not containing the origin.

Higher-Order Equations

The method we have just examined for nonhomogeneous second-order differential equations can be generalized to nth-order linear equations that have been put into the form

$$y^{(n)} + P_{n-1}(x)y^{(n-1)} + \cdots + P_1(x)y' + P_0(x)y = f(x). \tag{13}$$

If $y = c_1 y_1 + c_2 y_2 + \cdots + c_n y_n$ is the complementary function for (13), then a particular solution is

$$y_p = u_1(x)y_1(x) + u_2(x)y_2(x) + \cdots + u_n(x)y_n(x),$$

where the u_k', $k = 1, 2, \ldots, n$, are determined by the n equations

$$y_1 u_1' + \qquad y_2 u_2' + \cdots + \qquad y_n u_n' = 0$$
$$y_1' u_1' + \qquad y_2' u_2' + \cdots + \qquad y_n' u_n' = 0$$
$$\vdots$$
$$y_1^{(n-1)} u_1' + y_2^{(n-1)} u_2' + \cdots + y_n^{(n-1)} u_n' = f(x).$$

The first $n - 1$ equations in this system, like (7), are assumptions made to simplify the first $n - 1$ derivatives of y_p. The last equation of the system results from substituting the nth derivative of y_p and the simplified lower derivatives into (13). In this case, Cramer's rule gives

$$u'_k = \frac{W_k}{W}, \qquad k = 1, 2, \ldots, n,$$

where W is the Wronskian of $y_1, y_2, \ldots, y_n$, and W_k is the determinant obtained by replacing the kth column of the Wronskian by the column

$$\begin{matrix} 0 \\ 0 \\ \vdots \\ 0 \\ f(x). \end{matrix}$$

When $n = 2$ we get (10) and (11).

Remarks **(i)** Variation of parameters has a distinct advantage over the method of undetermined coefficients in that it will *always* yield a particular solution y_p, provided the related homogeneous equation can be solved. The present method is not limited to a function $f(x)$, which is a combination of the four types of functions listed on page 169. Also, variation of parameters, unlike undetermined coefficients, is applicable to differential equations with variable coefficients.

In the problems that follow do not hesitate to simplify the form of y_p. Depending on how the antiderivatives of u'_1 and u'_2 are found, you may not obtain the same y_p as given in the answer section. For example, in Problem 3, both $y_p = \frac{1}{2} \sin x - \frac{1}{2}x \cos x$ and $y_p = \frac{1}{4} \sin x - \frac{1}{2}x \cos x$ are valid answers. In either case, the general solution $y = y_c + y_p$ simplifies to $y = c_1 \cos x + c_2 \sin x - \frac{1}{2}x \cos x$. Why?

(ii) In Problems 25–28 you are asked to solve initial-value problems. Be sure to apply the initial conditions to the general solution $y = y_c + y_p$. Students often make the mistake of applying the initial conditions to only the complementary function y_c since it is that part of the solution that contains the constants. Review Example 8 in Section 4.4 for the correct procedure.

4.7 EXERCISES

Answers to odd-numbered problems begin on page A–11.

In Problems 1–24 solve each differential equation by variation of parameters. State an interval on which the general solution is defined.

1. $y'' + y = \sec x$ **2.** $y'' + y = \tan x$

3. $y'' + y = \sin x$ **4.** $y'' + y = \sec x \tan x$

5. $y'' + y = \cos^2 x$ **6.** $y'' + y = \sec^2 x$

7. $y'' - y = \cosh x$

8. $y'' - y = \sinh 2x$

9. $y'' - 4y = e^{2x}/x$

10. $y'' - 9y = 9x/e^{3x}$

11. $y'' + 3y' + 2y = 1/(1 + e^x)$

12. $y'' - 3y' + 2y = e^{3x}/(1 + e^x)$

13. $y'' + 3y' + 2y = \sin e^x$

14. $y'' - 2y' + y = e^x \arctan x$

15. $y'' - 2y' + y = e^x/(1 + x^2)$

16. $y'' - 2y' + 2y = e^x \sec x$

17. $y'' + 2y' + y = e^{-x} \ln x$

18. $y'' + 10y' + 25y = e^{-10x}/x^2$

19. $3y'' - 6y' + 30y = e^x \tan 3x$

20. $4y'' - 4y' + y = e^{x/2}\sqrt{1 - x^2}$

21. $y''' + y' = \tan x$

22. $y''' + 4y' = \sec 2x$

23. $y''' - 2y'' - y' + 2y = e^{3x}$

24. $2y''' - 6y'' = x^2$

In Problems 25–28 solve each differential equation by variations of parameters subject to the initial conditions $y(0) = 1$, $y'(0) = 0$.

25. $4y'' - y = xe^{x/2}$

26. $2y'' + y' - y = x + 1$

27. $y'' + 2y' - 8y = 2e^{-2x} - e^{-x}$

28. $y'' - 4y' + 4y = (12x^2 - 6x)e^{2x}$

29. Given that $y_1 = x$ and $y_2 = x \ln x$, form a fundamental set of solutions of $x^2 y'' - xy' + y = 0$ on $(0, \infty)$. Find the general solution of

$$x^2 y'' - xy' + y = 4x \ln x.$$

30. Given that $y_1 = x^2$ and $y_2 = x^3$, form a fundamental set of solutions of $x^2 y'' - 4xy' + 6y = 0$ on $(0, \infty)$. Find the general solution of

$$x^2 y'' - 4xy' + 6y = \frac{1}{x}.$$

31. Given that $y_1 = x^{-1/2} \cos x$ and $y_2 = x^{-1/2} \sin x$, form a fundamental set of solutions of $x^2 y'' + xy' + (x^2 - \frac{1}{4})y = 0$ on $(0, \infty)$. Find the general solution of

$$x^2 y'' + xy' + \left(x^2 - \frac{1}{4}\right)y = x^{3/2}.$$

32. Given that $y_1 = \cos(\ln x)$ and $y_2 = \sin(\ln x)$ are known linearly independent solutions of $x^2 y'' + xy' + y = 0$ on $(0, \infty)$:

(a) Find a particular solution of

$$x^2 y'' + xy' + y = \sec(\ln x).$$

(b) Give the general solution of the equation and state an interval of validity. [*Hint:* It is *not* $(0, \infty)$. Why?]

33. (a) Use undetermined coefficients to find a particular solution of

$$y'' + 2y' + y = 4x^2 - 3.$$

(b) Use variation of parameters to find a particular solution of

$$y'' + 2y' + y = \frac{e^{-x}}{x}.$$

(c) Use the superposition principle (Theorem 4.9) to find a particular solution of

$$y'' + 2y' + y = 4x^2 - 3 + \frac{e^{-x}}{x}.$$

34. Use the method outlined in Problem 33 to find a particular solution of

$$y'' + y = 2x - e^{3x} + \cot x.$$

Chapter 4 *REVIEW*

We summarize the important results of this chapter for **linear second-order** differential equations.

The equation

$$a_2(x)y'' + a_1(x)y' + a_0(x)y = 0 \tag{1}$$

is said to be **homogeneous**, whereas

$$a_2(x)y'' + a_1(x)y' + a_0(x)y = g(x), \tag{2}$$

$g(x)$ not identically zero, is **nonhomogeneous**. In the consideration of the linear equations (1) and (2), we assume that $a_2(x), a_1(x), a_0(x)$, and $g(x)$ are continuous on an interval I and that $a_2(x) \neq 0$ for every x in the interval. Under these assumptions there exists a unique solution of (2) satisfying the **initial conditions** $y(x_0) = y_0, y'(x_0) = y'_0$, where x_0 is a point in I.

The **Wronskian** of two differentiable functions $f_1(x)$ and $f_2(x)$ is the determinant

$$W(f_1(x), f_2(x)) = \begin{vmatrix} f_1(x) & f_2(x) \\ f'_1(x) & f'_2(x) \end{vmatrix}.$$

When $W \neq 0$ for at least one point in an interval, the functions are **linearly independent** on the interval. If the functions are **linearly dependent** on the interval, then $W = 0$ for every x in the interval.

In solving the homogeneous equation (1), we want linearly independent solutions. A necessary and sufficient condition that two solutions y_1 and y_2 are linearly independent on I is that $W(y_1, y_2) \neq 0$ for every x in I. We say y_1 and y_2 form a **fundamental set** on I when they are linearly independent solutions of (1) on the interval. For *any* two solutions y_1 and y_2, the **superposition principle** states that the linear combination $c_1 y_1 + c_2 y_2$ is also a solution of (1). When y_1 and y_2 form a fundamental set, the function $y = c_1 y_1 + c_2 y_2$ is called the **general solution** of (1). The **general solution** of (2) is $y = y_c + y_p$, where y_c is the **complementary function**, or general solution, of (1) and y_p is any **particular solution** of (2).

To solve (1) in the case $ay'' + by' + cy = 0$, a, b, and c constants, we first solve the **auxiliary equation** $am^2 + bm + c = 0$. There are three forms of the general solution depending on the three possible ways in which the roots of the auxiliary equation can occur.

Roots	General Solution
1. m_1 and m_2: real and distinct	$y = c_1 e^{m_1 x} + c_2 e^{m_2 x}$
2. m_1 and m_2: real but $m_1 = m_2$	$y = c_1 e^{m_1 x} + c_2 x e^{m_1 x}$
3. m_1 and m_2: complex	
$\quad m_1 = \alpha + i\beta$, $\quad m_2 = \alpha - i\beta$	$y = e^{\alpha x}(c_1 \cos \beta x + c_2 \sin \beta x)$

To solve a nonhomogeneous differential equation we use either the method of **undetermined coefficients** or **variation of parameters** to find a particular solution y_p. The former procedure is limited to differential equations $ay'' + by' + cy = g(x)$, where a, b, and c are constants and $g(x)$ is a constant, a polynomial, $e^{\alpha x}$, $\cos \beta x$, $\sin \beta x$, or finite sums and products of these functions.

Chapter 4 *REVIEW EXERCISES*

Answers to odd-numbered problems begin on page A–11.

Answer Problems 1–10 without referring back to the text. Fill in the blank or answer true/false. In some cases there can be more than one correct answer.

1. The only solution of $y'' + x^2 y = 0$, $y(0) = 0$, $y'(0) = 0$, is _____.

2. If two differentiable functions $f_1(x)$ and $f_2(x)$ are linearly independent on an interval, then $W(f_1(x), f_2(x)) \neq 0$ for at least one point in the interval. _____

3. Two functions $f_1(x)$ and $f_2(x)$ are linearly independent on an interval if one is not a constant multiple of the other. _____

4. The functions $f_1(x) = x^2$, $f_2(x) = 1 - x^2$, and $f_3(x) = 2 + x^2$ are linearly _____ on the interval $(-\infty, \infty)$.

5. The functions $f_1(x) = x^2$ and $f_2(x) = x|x|$ are linearly independent on the interval _____, whereas they are linearly dependent on the interval _____.

6. Two solutions y_1 and y_2 of $y'' + y' + y = 0$ are linearly dependent if $W(y_1, y_2) = 0$ for every real value of x. _____

7. A constant multiple of a solution of a differential equation is also a solution. _____

8. A fundamental set of two solutions of $(x-2)y'' + y = 0$ exists on any interval not containing the point _____.

9. For the method of undetermined coefficients, the assumed form of the particular solution y_p for $y'' - y = 1 + e^x$ is _____.

10. A differential operator that annihilates $e^{2x}(x + \sin x)$ is _____.

In Problems 11 and 12 find a second solution for the differential equation given that $y_1(x)$ is a known solution.

11. $y'' + 4y = 0, \quad y_1 = \cos 2x$

12. $xy'' - 2(x+1)y' + (x+2)y = 0, \quad y_1 = e^x$

In Problems 13–18 find the general solution of each differential equation.

13. $y'' - 2y' - 2y = 0$

14. $2y'' + 2y' + 3y = 0$

15. $y''' + 10y'' + 25y' = 0$

16. $2y''' + 9y'' + 12y' + 5y = 0$

17. $3y''' + 10y'' + 15y' + 4y = 0$

18. $2\dfrac{d^4y}{dx^4} + 3\dfrac{d^3y}{dx^3} + 2\dfrac{d^2y}{dx^2} + 6\dfrac{dy}{dx} - 4y = 0$

In Problems 19–22 solve each differential equation by the method of undetermined coefficients.

19. $y'' - 3y' + 5y = 4x^3 - 2x$

20. $y'' - 2y' + y = x^2 e^x$

21. $y''' - 5y'' + 6y' = 2\sin x + 8$

22. $y''' - y'' = 6$

In Problems 23 and 24 solve the given differential equation subject to the indicated conditions.

23. $y'' - 2y' + 2y = 0, \quad y(\pi/2) = 0, \, y(\pi) = -1$

24. $y'' - y = x + \sin x, \quad y(0) = 2, \, y'(0) = 3$

In Problems 25 and 26 solve each differential equation by the method of variation of parameters.

25. $y'' - 2y' + 2y = e^x \tan x$

26. $y'' - y = 2e^x/(e^x + e^{-x})$

In Problems 27 and 28 solve the given differential equation subject to the indicated conditions.

27. $(2D^3 - 13D^2 + 24D - 9)y = 36, \quad y(0) = -4, \, y'(0) = 0, \, y''(0) = \frac{5}{2}$

28. $y'' + y = \sec^3 x, \quad y(0) = 1, \, y'(0) = \frac{1}{2}$

CHAOS

John H. Hubbard
Department of Mathematics
Cornell University

A famous science fiction story tells of a politician who, soon after winning an election, takes a trip in a time travel machine back to the days of the dinosaurs. While there, and though much admonished not to disturb anything, he bends a blade of grass. When he returns to his real present, he finds that in this modified world he has lost the election.

This is what mathematicians mean when they say that a system exhibits chaos: tiny changes in the initial state of a system can decisively affect the outcome. One speaks of the butterfly effect. Can the fluttering of the wings of a butterfly in Japan have a decisive effect on the weather in the United States a month later?

Most people would probably dismiss the suggestion as preposterous without a second thought. But I think it is true, at least if the time span of one month is increased to six months, and I propose here to give some quantitative reasons for my conclusions.

It is not obvious how you would go about justifying the reality of the butterfly effect. We do not have a time machine available; we cannot go back six weeks, catch a butterfly (while disturbing nothing else, whatever that means), and then come back and observe the consequences. We need to take a more devious tack.

To help you follow the argument, I will describe a "toy model" that clearly exhibits "butterfly behavior," and within which the notions that I will introduce can be understood.

Consider the (purely mathematical) system in which, at each tick of the clock, an angle is doubled. A state of the system is an angle, and it evolves by doubling again and again. In symbols, you could describe the system as a sequence of angles

$$\theta_0, \theta_1, \ldots,$$

where θ_0 is the initial state of the system and $\theta_{n+1} = 2\theta_n$.

This system exhibits butterfly behavior. If θ_0 is perturbed by one-billionth of a turn, then the state after 30 ticks of the clock is completely unknown. Indeed, the uncertainty in our knowledge of the state of the system *doubles* at each tick; after 30 ticks, our uncertainty has grown to

$$\frac{2^{30}}{1,000,000,000},$$

which is slightly greater than 1. Our uncertainty is now more than a turn; we don't know anything anymore.

The example above brings in a key notion in all descriptions of chaos: entropy. This is essentially the rate at which information dissipates. There are many ways of describing this rate with precision, and they go by various names (for example, Lyapunov exponents), but for the purposes of this article, I will stick with the doubling time: **the amount of time it takes for a small uncertainty to double**.

How would we estimate this time for the system formed by the weather? Of course, we cannot "choose two initial states for the weather an epsilon apart and measure how fast they diverge," but we can do something like it. We can go back over history and find times when the weather was very similar. Then we can see how long it took for the weather pattern to diverge. This has been done and leads to a doubling time of about $2\frac{1}{2}$ days.

You can also go to the meteorology department of a big research institution (in practice, the Institut de Météorologie de l'université de Paris-VI) and ask what the doubling time is for their best computer models. You get much the same figure.

The next question that must be faced is: What corresponds to the number one-billionth above? What proportion of the size of the system (the atmosphere) is our disturbance (one butterfly)? One (perhaps contestable) way of estimating this is simply to measure the ratio of the masses. We might guess that a butterfly weighs 1 gram (rather a heavy butterfly), and it turns out that the atmosphere has a mass of about 5×10^{21} grams.

The atmospheric pressure is very nearly 1 kg/cm^2, meaning that there is 1 kg of air above every square centimeter of the earth. The area of a sphere of radius r is $4\pi r^2$, and the radius of the earth is about 6000 km. Thus the mass of the atmosphere is about

$$1000 \times 4 \times \pi \times (6 \times 10^8)^2 \text{ grams.}$$

So a butterfly isn't one-billionth the size of the system; it is more nearly a thousand-billion-billionth.

Since 5×10^{21} is approximately 2^{72}, it should take about 72 doubling periods for the effects of an individual butterfly to induce perturbations on a global scale.

One consequence of this analysis is that long-term weather prediction should be completely impossible. It is inconceivable that anyone could ever know the state of the atmosphere to within the effect of one butterfly, or even on a scale a thousand billion times larger. Perturbations on that scale decisively affect the global weather within a month.

Physicists, chemists, astronomers, and mathematicians have now shown that a vast collection of systems appear to display "chaos," in the sense that they are expanding and have a doubling time for errors.

One example is due to the meteorologist E. Lorenz, whose discovery may be credited with starting this entire line of thought. In 1961 he was running a simulation of the weather. The computers of the time were primitive, so the data had to be drastically simplified for any computation to be possible. He observed various behaviors from his model, apparently quite satisfactory, until one day he decided to examine something he had already computed over a longer time. He typed in what he thought were the original initial conditions, went off to get a cup of coffee, and when he came back he found that his new "weather" was unrelated to the previous run of his model.

He eventually realized that the difference was due to the fact that he had entered the initial conditions with fewer decimals than in the initial run. After further simplification of his model, Lorenz found that the following set of three coupled differential equations exhibits the same sort of chaotic behavior:

$$x' = 10(y - x)$$

$$y' = 28x - y - xz$$

$$z' = \frac{8}{3}z + xy.$$

Much further work on these equations and others in R^3 have shown that chaotic behavior and fractal attractors are common.

The presence of chaos plays havoc with predictions, but it is sometimes useful; sometimes chaos can be translated into control.

NASA is not able to build rockets with enough fuel to give the speeds required to go huge distances. So they delicately bounced the spacecraft off Venus, stealing a bit of Venus's potential energy to give the spacecraft the fantastic speeds required. Only because the tiny variations in the trajectory due to the guidance rockets can be amplified to enormous variations in speed and trajectory is such a scheme feasible. But imagine how difficult this makes the long-term predictions about the orbits of comets.

The presence of chaos also has philosophical consequences, for instance, about the conflict of determinism and free will. How can humans have free will if the universe is completely governed by determinstic laws?

Well, if the equations exhibit chaos, then it follows that you cannot know that they are deterministic, however long you observe the system. If you were to observe a sequence of "doubled angles," each with 15 decimal places, you could never know whether you are observing a sequence of exact doubles of some angle or the same system perturbed at a scale smaller than 10^{-15} (such as round-off error).

Similarly, if the brain were not following exactly the laws of physics but were perturbed (by spiritual forces, free will, god) at a scale unmeasurable without drastically affecting the system (I doubt anyone reacts the same way with electrodes in his brain), then you could never know it, but over a time scale of perhaps 4 seconds, you could decisively alter all decisions. Making precise sense of the above would require knowing something about the doubling time of the brain. Introspection tells me this should be perhaps 0.1 second, the time of an elementary realization. Perhaps some day neurology will give a more credible estimate. If this is roughly accurate, then 4 seconds is 40 doubling times later, and $2^{40} \approx 10^{12}$ is approximately the number of neurons in the brain.

Personally, I do not think that there is a god tinkering with the laws of physics in my brain, but this is not knowable. Chaos prevents us from knowing such things.

More generally, although I am as firm a scientist as you could hope to find, I think that the world is essentially incomprehensible, with all sorts of tiny events having enormous consequences that no one can hope to foresee or understand.

If you find preposterous the idea that your smallest actions probably influence the entire world after a short time, consider the following. The time span used is much longer, about 1200 years, but I think the implications are much the same. Who were your ancestors in the year 800 A.D.? You have two parents, four grandparents, and so on. The number of ancestors should be 2^n in the nth generation. The year 800 is 50–60 generations ago, leading to $2^{50} \approx 10^{15}$ ancestors at that time. Of course this is a million times larger than the population of the world today, and a much larger multiple of the population at that time. This shows among other things that there must have been a lot of inbreeding.

Now think about it from the point of view of a woman in the year 800. In order for the population to be sustained, and even grow, she had to have at least an average of 2 children, 4 grandchildren, and so on. Of course, her line might become extinct, but if that were to happen, then it must happen soon, after fewer than three generations. The probability of having no descendants can be evaluated over the short term and might be 1/3 or 1/2. And once the descendance has become large, it tends to become the entire population quite rapidly (although the influence of inbreeding slows this down considerably).

If you put these two points of view together, you will see that essentially everyone of European ancestry is descended from every person alive in Europe in the year 800 who has any descendants at all. For instance, we are all descended from Charlemagne, but also from about half of the peasants alive at the time. If just one of those couples had had one child fewer, *everyone* would be genetically different.

Of course, this doesn't make sense; long ago the paths of history would have diverged so that no detailed comparisons are possible. But certainly the smallest variation in the sexual behavior of anyone at that time would certainly have affected the world in incalculable ways.

CHAPTER 5

APPLICATIONS OF SECOND-ORDER DIFFERENTIAL EQUATIONS: VIBRATIONAL MODELS

5.1 Simple Harmonic Motion
5.2 Damped Motion
5.3 Forced Motion

5.4 Electric Circuits and Other Analogous Systems

Chapter 5 Review
Chapter 5 Review Exercises
Essay: Tacoma Narrows
      Suspension Bridge Collapse

Important Concepts

Free motion
Simple harmonic motion
Free undamped motion
Period
Frequency
Equation of motion
Amplitude
Phase angle
Free damped motion
Overdamping
Critical damping
Underdamping
Forced motion
Transient term
Transient solution
Steady-state solution
Pure resonance
Resonance
Resonance curve
Electrical vibrations
Overdamped circuit
Critically damped circuit
Underdamped circuit
Steady-state current
Reactance
Impedance
Extreme displacement

A single differential equation can serve as a mathematical model for many different phenomena. Forms of the linear second-order differential equation $ay'' + by' + cy = f(t)$ appear in the analysis of problems in physics, engineering, chemistry, and biology.

In this chapter our primary focus is on one application: the motion of a mass attached to a spring. We shall see what the individual terms ay'', by', cy, and $f(t)$ of the differential equation $ay'' + by' + cy = f(t)$ mean in the context of this vibrational system. We shall also see that, except for terminology and physical interpretation of the terms ay'', by', cy, and $f(t)$, the mathematics of a series circuit is identical to that of a vibrating spring-mass system. Our goal, of course, is not to study all possible applications but to acquaint you with the mathematical procedures that are common to these problems.

5.1 SIMPLE HARMONIC MOTION

Hooke's Law

Suppose, as in Figure 5.1(b), a mass m_1 is attached to a flexible spring suspended from a rigid support. When m_1 is replaced with a different mass m_2, the amount of stretch, or elongation, of the spring will of course be different.

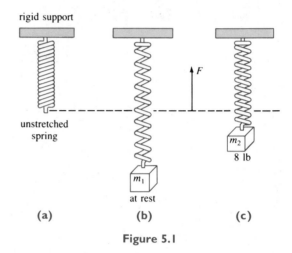

Figure 5.1

By Hooke's law,* the spring itself exerts a restoring force F opposite to the direction of elongation and proportional to the amount of elongation s. Simply stated, $F = ks$, where k is a constant of proportionality. Although masses with different weights stretch a spring by different amounts, the spring is essentially characterized by the number k. For example, if a mass weighing 10 lb stretches a spring by 1/2 ft, then

$$10 = k\left(\frac{1}{2}\right) \quad \text{implies} \quad k = 20 \text{ lb/ft.}$$

Necessarily then a mass weighing 8 lb stretches the same spring 2/5 ft.

---

*Robert Hooke (1635–1703)   An English physicist and inventor, Hooke published this law in 1658. The idea of attaching a spring to a balance wheel, causing oscillatory motion that enabled a clock to mark units of time, is usually attributed to Hooke. The concept of the balance spring led to the invention of the pocket watch by Christian Huygens in 1674. Hooke accused Huygens of stealing his invention. Irascible and contentious, Hooke charged many of his colleagues, notably Isaac Newton, with plagiarism.

Newton's Second Law

After a mass m is attached to a spring, it stretches the spring by an amount s and attains a position of equilibrium at which its weight W is balanced by the restoring force ks. Recall from Section 1.2 that weight is defined by

$$W = mg,$$

where the mass is measured in slugs, kilograms, or grams and $g = 32 \text{ ft/s}^2$, 9.8 m/s^2, or 980 cm/s^2, respectively. As indicated in Figure 5.2(b), the condition of equilibrium is $mg = ks$, or $mg - ks = 0$. If the mass is now displaced by an amount x from its equilibrium position and released, the net force F in this dynamic case is given by **Newton's second law of motion** $F = ma$, where a is the acceleration d^2x/dt^2. Assuming that there are no retarding forces acting on the system and assuming that the mass vibrates free of other external influencing forces—**free motion**—we can equate F to the resultant force of the weight and the restoring force:

$$m\frac{d^2x}{dt^2} = -k(s + x) + mg$$

$$= -kx + \underbrace{mg - ks}_{\text{zero}} = -kx. \tag{1}$$

The negative sign in (1) indicates that the restoring force of the spring acts opposite to the direction of motion. Furthermore, we shall adopt the convention that displacements measured *below* the equilibrium position are positive. See Figure 5.3.

(a)   **(b)**   **(c)**

Figure 5.2   **Figure 5.3**

Differential Equation of Free Undamped Motion

By dividing (1) by the mass m, we obtain the second-order differential equation

$$\frac{d^2x}{dt^2} + \frac{k}{m}x = 0 \tag{2}$$

or

$$\frac{d^2x}{dt^2} + \omega^2 x = 0, \tag{3}$$

where $\omega^2 = k/m$. Equation (3) is said to describe **simple harmonic motion**, or **free undamped motion**. There are two obvious initial conditions associated with (3):

$$x(0) = \alpha, \qquad x'(0) = \beta, \tag{4}$$

representing the amount of initial displacement and the initial velocity, respectively. For example, if $\alpha > 0$, $\beta < 0$, the mass starts from a point *below* the equilibrium position with an imparted *upward* velocity. If $\alpha < 0$, $\beta = 0$, the mass is released from *rest* from a point $|\alpha|$ units *above* the equilibrium position, and so on.

Solution and Equation of Motion

To solve equation (3) we note that the solutions of the auxiliary equation $m^2 + \omega^2 = 0$ are the complex numbers

$$m_1 = \omega i, \qquad m_2 = -\omega i.$$

Thus from (8) of Section 4.3, we find the general solution of (3) to be

$$x(t) = c_1 \cos \omega t + c_2 \sin \omega t. \tag{5}$$

The **period** of free vibrations described by (5) is $T = 2\pi/\omega$ and the **frequency** is $f = 1/T = \omega/2\pi$.* For example, for $x(t) = 2\cos 3t - 4\sin 3t$, the period is $2\pi/3$ and the frequency is $3/2\pi$. The former number means that the graph of $x(t)$ repeats every $2\pi/3$ units; the latter number means that there are 3 cycles of the graph every 2π units or, equivalently, the mass undergoes $3/2\pi$ complete vibrations per unit time. In addition, it can be shown that the period $2\pi/\omega$ is the time interval between two successive maxima of $x(t)$. Keep in mind that a maximum of $x(t)$ is a positive displacement corresponding to the mass attaining a maximum distance *below* the equilibrium position, whereas a mini-

---

* Sometimes the number ω is called the *circular frequency* of vibrations. For free undamped motion the numbers $2\pi/\omega$ and $\omega/2\pi$ are also referred to as the *natural period* and *natural frequency*, respectively.

mum of $x(t)$ is a negative displacement corresponding to the mass attaining a maximum height *above* the equilibrium position. We shall refer to either case as an **extreme displacement** of the mass. Finally, when the initial conditions (4) are used to determine the constants c_1 and c_2 in (5), we say that the resulting particular solution is the **equation of motion**.

EXAMPLE I

Solve and interpret the initial-value problem

$$\frac{d^2x}{dt^2} + 16x = 0, \qquad x(0) = 10, \quad x'(0) = 0.$$

Solution   The problem is equivalent to pulling a mass on a spring down 10 units below the equilibrium position, holding it until $t = 0$, and then releasing it from rest. Applying the initial conditions to the solution

$$x(t) = c_1 \cos 4t + c_2 \sin 4t$$

gives

$$x(0) = 10 = c_1 \cdot 1 + c_2 \cdot 0$$

so that $c_1 = 10$, and hence

$$x(t) = 10 \cos 4t + c_2 \sin 4t$$

$$\frac{dx}{dt} = -40 \sin 4t + 4c_2 \cos 4t$$

$$x'(0) = 0 = 4c_2 \cdot 1.$$

The latter equation implies that $c_2 = 0$; therefore, the equation of motion is $x(t) = 10 \cos 4t$.

   The solution clearly shows that once the system is set in motion, it stays in motion with the mass bouncing back and forth 10 units on either side of the equilibrium position $x = 0$. As shown in Figure 5.4(b), the period of oscillation is $2\pi/4 = \pi/2$ seconds.

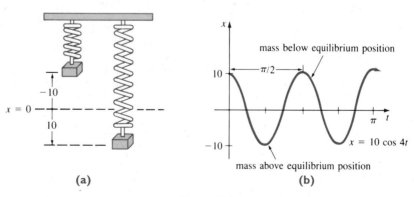

(a)   (b)

Figure 5.4   ■

EXAMPLE 2

A mass weighing 2 lb stretches a spring 6 inches. At $t = 0$ the mass is released from a point 8 inches below the equilibrium position with an upward velocity of 4/3 ft/s. Determine the function $x(t)$ that describes the subsequent free motion.

Solution   Since we are using the engineering system of units, the measurements given in terms of inches must be converted to feet: 6 inches = 1/2 foot; 8 inches = 2/3 foot. In addition we must convert the units of weight given in pounds to units of mass. From $m = W/g$ we have

$$m = \frac{2}{32} = \frac{1}{16} \text{ slug.}$$

Also, from Hooke's law,

$$2 = k\left(\frac{1}{2}\right) \quad \text{implies} \quad k = 4 \text{ lb/ft.}$$

Hence the analogues of (1) and (2) are, respectively,

$$\frac{1}{16} \frac{d^2x}{dt^2} = -4x \quad \text{and} \quad \frac{d^2x}{dt^2} + 64x = 0.$$

The initial displacement and initial velocity are given by

$$x(0) = \frac{2}{3}, \qquad x'(0) = -\frac{4}{3},$$

where the negative sign in the last condition is a consequence of the fact that the mass is given an initial velocity in the negative or upward direction.

Now $\omega^2 = 64$ or $\omega = 8$ so that the general solution of the differential equation is

$$x(t) = c_1 \cos 8t + c_2 \sin 8t. \tag{6}$$

Applying the initial conditions to (6), we obtain

$$x(0) = \frac{2}{3} = c_1 \cdot 1 + c_2 \cdot 0, \qquad \left(c_1 = \frac{2}{3}\right)$$

$$x(t) = \frac{2}{3} \cos 8t + c_2 \sin 8t$$

$$x'(t) = -\frac{16}{3} \sin 8t + 8c_2 \cos 8t$$

$$x'(0) = -\frac{4}{3} = -\frac{16}{3} \cdot 0 + 8c_2 \cdot 1, \qquad \left(c_2 = -\frac{1}{6}\right).$$

Thus the equation of motion is

$$x(t) = \frac{2}{3} \cos 8t - \frac{1}{6} \sin 8t. \tag{7}$$

∎

Note The distinction between weight and mass is often blurred. Thus one often speaks of both the motion of a mass on a spring and the motion of a weight on a spring.

Alternative Form of x(t)

When $c_1 \neq 0$ and $c_2 \neq 0$, the actual **amplitude** A of free vibrations is not obvious from inspection of equation (5). For example, although the mass in Example 2 is initially displaced 2/3 ft beyond the equilibrium position, the amplitude of vibrations is a number larger than 2/3. Hence it is often convenient to convert a solution of form (5) to the simpler form

$$x(t) = A \sin(\omega t + \phi), \tag{8}$$

where

$$A = \sqrt{c_1^2 + c_2^2}$$

and ϕ is a **phase angle** defined by

$$\left. \begin{array}{l} \sin \phi = \dfrac{c_1}{A} \\[2mm] \cos \phi = \dfrac{c_2}{A} \end{array} \right\} \quad \tan \phi = \dfrac{c_1}{c_2}. \tag{9}$$

To verify this we expand (8) by the addition formula for the sine function:

$$A \sin \omega t \cos \phi + A \cos \omega t \sin \phi = (A \sin \phi) \cos \omega t + (A \cos \phi) \sin \omega t. \tag{10}$$

It follows from Figure 5.5 that if ϕ is defined by

$$\sin \phi = \frac{c_1}{\sqrt{c_1^2 + c_2^2}} = \frac{c_1}{A}, \qquad \cos \phi = \frac{c_2}{\sqrt{c_1^2 + c_2^2}} = \frac{c_2}{A},$$

then (10) becomes

$$A \frac{c_1}{A} \cos \omega t + A \frac{c_2}{A} \sin \omega t = c_1 \cos \omega t + c_2 \sin \omega t = x(t).$$

Figure 5.5

E X A M P L E 3

In view of the foregoing discussion, we can write the solution (7) in Example 2

$$x(t) = \frac{2}{3} \cos 8t - \frac{1}{6} \sin 8t \quad \text{alternatively as} \quad x(t) = A \sin(8t + \phi).$$

The amplitude is given by

$$A = \sqrt{\left(\frac{2}{3}\right)^2 + \left(-\frac{1}{6}\right)^2} = \frac{\sqrt{17}}{6} \approx 0.69 \text{ ft.}$$

One should exercise some care when finding the phase angle ϕ defined by (9). In this case

$$\tan \phi = \frac{2/3}{-1/6} = -4$$

and a scientific hand calculator would give

$$\tan^{-1}(-4) = -1.326 \text{ radians.*}$$

But this angle is located in the fourth quadrant and therefore contradicts the fact that $\sin \phi > 0$ and $\cos \phi < 0$ (recall, $c_1 > 0$ and $c_2 < 0$). Hence we must take ϕ to be the second-quadrant angle

$$\phi = \pi + (-1.326) = 1.816 \text{ radians.}$$

Thus we have
$$x(t) = \frac{\sqrt{17}}{6} \sin(8t + 1.816). \tag{11}$$

∎

Form (8) is very useful since it is easy to find the values of time for which the graph of $x(t)$ crosses the positive t-axis (the line $x = 0$). We observe that $\sin(\omega t + \phi) = 0$ when

$$\omega t + \phi = n\pi,$$

where n is a nonnegative integer.

$$x = \frac{\sqrt{17}}{6} \sin(8t + 1.816)$$

Figure 5.6

EXAMPLE 4

For motion described by $x(t) = (\sqrt{17}/6) \sin(8t + 1.816)$, find the first value of time for which the mass passes through the equilibrium position heading downward.

Solution The values $t_1, t_2, t_3, \ldots$ for which $\sin(8t + 1.816) = 0$ are determined from

$$8t_1 + 1.816 = \pi, \qquad 8t_2 + 1.816 = 2\pi, \qquad 8t_3 + 1.816 = 3\pi, \ldots.$$

We find $t_1 = 0.166$, $t_2 = 0.558$, $t_3 = 0.951, \ldots$, respectively.

Figure 5.6 shows that the mass passes through $x = 0$ heading downward (namely, toward $x > 0$) the first time at $t_2 = 0.558$ second. ∎

---

* The range of the inverse tangent is $-\pi/2 < \tan^{-1}x < \pi/2$.

5.1 EXERCISES

Answers to odd-numbered problems begin on page A–11.

In Problems 1 and 2 state in words a possible physical interpretation of the given initial-value problem.

1. $\dfrac{4}{32}x'' + 3x = 0$

$x(0) = -3,\ x'(0) = -2$

2. $\dfrac{1}{16}x'' + 4x = 0$

$x(0) = 0.7,\ x'(0) = 0$

In Problems 3–8 write the solution of the given initial-value problem in form (8).

3. $x'' + 25x = 0$

$x(0) = -2,\ x'(0) = 10$

4. $\dfrac{1}{2}x'' + 8x = 0$

$x(0) = 1,\ x'(0) = -2$

5. $x'' + 2x = 0$

$x(0) = -1,\ x'(0) = -2\sqrt{2}$

6. $\dfrac{1}{4}x'' + 16x = 0$

$x(0) = 4,\ x'(0) = 16$

7. $0.1x'' + 10x = 0$

$x(0) = 1,\ x'(0) = 1$

8. $x'' + x = 0$

$x(0) = -4,\ x'(0) = 3$

9. The period of free undamped oscillations of a mass on a spring is $\pi/4$ second. If the spring constant is 16 lb/ft, what is the numerical value of the weight?

10. A spring is suspended from a ceiling. When a mass weighing 60 lb is attached, the spring is stretched $1/2$ ft. The mass is removed and a person grabs the end of the spring and proceeds to bounce up and down with a period of 1 second. How much does the person weigh?

11. A 4-lb weight is attached to a spring whose spring constant is 16 lb/ft. What is the period of simple harmonic motion?

12. A 20-kg mass is attached to a spring. If the frequency of simple harmonic motion is $2/\pi$ vibrations/second, what is the spring constant k? What is the frequency of simple harmonic motion if the original mass is replaced with an 80-kg mass?

13. A 24-lb weight, attached to the end of a spring, stretches it 4 in. Find the equation of motion if the weight is released from rest from a point 3 in. above the equilibrium position.

14. Determine the equation of motion if the weight in Problem 13 is released from the equilibrium position with an initial downward velocity of 2 ft/s.

15. A 20-lb weight stretches a spring 6 in. The weight is released from rest 6 in. below the equilibrium position.

   (a) Find the position of the weight at $t = \pi/12,\ \pi/8,\ \pi/6,\ \pi/4,\ 9\pi/32$ seconds.

(b) What is the velocity of the weight when $t = 3\pi/16$ second? In which direction is the weight heading at this instant?

(c) At what times does the weight pass through the equilibrium position?

16. A force of 400 newtons stretches a spring 2 m. A mass of 50 kg is attached to the end of the spring and released from the equilibrium position with an upward velocity of 10 m/s. Find the equation of motion.

17. Another spring whose constant is 20 N/m is suspended from the same rigid support but parallel to the spring-mass system in Problem 16. A mass of 20 kg is attached to the second spring and both masses are released from the equilibrium position with an upward velocity of 10 m/s.

   (a) Which mass exhibits the greater amplitude of motion?

   (b) Which mass is moving faster at $t = \pi/4$ second? at $\pi/2$ seconds?

   (c) At what times are the two masses in the same position? Where are the masses at these times? In which directions are they moving?

18. A 32-lb weight stretches a spring 2 ft. Determine the amplitude and period of motion if the weight is released 1 ft above the equilibrium position with an initial upward velocity of 2 ft/s. How many complete vibrations will the weight have completed at the end of 4π seconds?

19. An 8-lb weight attached to a spring exhibits simple harmonic motion. Determine the equation of motion if the spring constant is 1 lb/ft and if the weight is released 6 in. below the equilibrium position with a downward velocity of $3/2$ ft/s. Express the solution in form (8).

20. A mass weighing 10 lb stretches a spring $1/4$ ft. This mass is removed and replaced with a mass of 1.6 slugs, which is released $1/3$ ft above the equilibrium position with a downward velocity of $5/4$ ft/s. Express the solution in form (8). At what times does the mass attain a displacement below the equilibrium position numerically equal to one-half the amplitude?

21. A 64-lb weight attached to the end of a spring stretches it 0.32 ft. From a position 8 in. above the equilibrium position the weight is given a downward velocity of 5 ft/s.

   (a) Find the equation of motion.

   (b) What are the amplitude and period of motion?

   (c) How many complete vibrations will the weight have completed at the end of 3π seconds?

   (d) At what time does the weight pass through the equilibrium position heading downward for the second time?

   (e) At what time does the weight attain its extreme displacement on either side of the equilibrium position?

   (f) What is the position of the weight at $t = 3$ seconds?

   (g) What is the instantaneous velocity at $t = 3$ seconds?

(h) What is the acceleration at $t = 3$ seconds?

(i) What is the instantaneous velocity at the times when the weight passes through the equilibrium position?

(j) At what times is the weight 5 in. below the equilibrium position?

(k) At what times is the weight 5 in. below the equilibrium position heading in the upward direction?

22. A mass of 1 slug is suspended from a spring whose characteristic spring constant is 9 lb/ft. Initially the mass starts from a point 1 ft above the equilibrium position with an upward velocity of $\sqrt{3}$ ft/s. Find the times for which the mass is heading downward at a velocity of 3 ft/s.

23. Under some circumstances when two parallel springs, with constants k_1 and k_2, support a single weight W, the **effective spring constant** of the system is given by $k = 4k_1 k_2/(k_1 + k_2)$.* A 20-lb weight stretches one spring 6 in. and another spring 2 in. The springs are attached to a common rigid support and then to a metal plate. As shown in Figure 5.7, the 20-lb weight is attached to the center of the plate in the double spring arrangement. Determine the effective spring constant of this system. Find the equation of motion if the weight is released from the equilibrium position with a downward velocity of 2 ft/s.

k_1 k_2

20 lb

Figure 5.7

24. A certain weight stretches one spring 1/3 ft and another spring 1/2 ft. The two springs are attached to a common rigid support in a manner indicated in Problem 23 and Figure 5.7. The first weight is set aside, and an 8-lb weight is attached to the double spring arrangement and the system is set in motion. If the period of motion is $\pi/15$ second, determine the numerical value of the first weight.

25. If x_0 and v_0 are the initial position and velocity, respectively, of a weight exhibiting simple harmonic motion, show that the amplitude of vibrations is

$$A = \sqrt{x_0^2 + \left(\frac{v_0}{\omega}\right)^2}.$$

26. Show that any linear combination $x(t) = c_1 \cos \omega t + c_2 \sin \omega t$ can also be written in the form

$$x(t) = A \cos(\omega t + \phi), \qquad \text{where} \quad A = \sqrt{c_1^2 + c_2^2}$$

and

$$\sin \phi = -\frac{c_2}{A}, \qquad \cos \phi = \frac{c_1}{A}.$$

---

* If the two springs have the same natural length and if the plate is guided either by rails or by an external force so that the extensions of both springs are always the same, then the effective spring constant is simply $k = k_1 + k_2$.

27. Express the solution of Problem 3 in the form of the cosine function given in Problem 26.

28. Show that when a weight attached to a spring exhibits simple harmonic motion, the maximum value of the speed (that is, $|v(t)|$) occurs when the weight is passing through the equilibrium position.

29. A weight attached to a spring exhibits simple harmonic motion. Show that the maximum acceleration of the weight occurs at an extreme displacement and has the magnitude $4\pi^2 A/T^2$, where A is the amplitude and T is the period of free vibrations.

30. Use (8) to prove that the time interval between two successive maxima of $x(t)$ is $2\pi/\omega$.

5.2 DAMPED MOTION

(a)

(b)

Figure 5.8

The discussion of free harmonic motion is somewhat unrealistic since the motion described by equation (2) of Section 5.1 assumes that no retarding forces are acting on the moving mass. Unless the mass is suspended in a perfect vacuum, there will be at least a resisting force due to the surrounding medium. For example, as Figure 5.8 shows, the mass m could be suspended in a viscous medium or connected to a dashpot damping device.

Differential Equation of Motion with Damping

In the study of mechanics, damping forces acting on a body are considered to be proportional to a power of the instantaneous velocity. In particular, we shall assume throughout the subsequent discussion that this force is given by a constant multiple of dx/dt.* When no other external forces are impressed on the system, it follows from Newton's second law that

$$m\frac{d^2x}{dt^2} = -kx - \beta\frac{dx}{dt}, \tag{1}$$

where β is a positive *damping constant* and the negative sign is a consequence of the fact that the damping force acts in a direction opposite to the motion.

Dividing (1) by the mass m, we find the differential equation of **free damped motion** is

$$\frac{d^2x}{dt^2} + \frac{\beta}{m}\frac{dx}{dt} + \frac{k}{m}x = 0 \tag{2}$$

---

* In many instances, such as problems in hydrodynamics, the damping force is proportional to $(dx/dt)^2$.

or
$$\frac{d^2x}{dt^2} + 2\lambda \frac{dx}{dt} + \omega^2 x = 0. \tag{3}$$

In equation (3) we make the identifications

$$2\lambda = \frac{\beta}{m}, \qquad \omega^2 = \frac{k}{m}. \tag{4}$$

The symbol 2λ is used only for algebraic convenience since the auxiliary equation is $m^2 + 2\lambda m + \omega^2 = 0$ and the corresponding roots are then

$$m_1 = -\lambda + \sqrt{\lambda^2 - \omega^2}, \qquad m_2 = -\lambda - \sqrt{\lambda^2 - \omega^2}.$$

We can now distinguish three possible cases depending on the algebraic sign of $\lambda^2 - \omega^2$. Since each solution contains the *damping factor* $e^{-\lambda t}$, $\lambda > 0$, the displacements of the mass become negligible for large time.

CASE I $\lambda^2 - \omega^2 > 0$. In this situation the system is said to be **overdamped**, since the damping coefficient β is large when compared to the spring constant k. The corresponding solution of (3) is

$$x(t) = c_1 e^{m_1 t} + c_2 e^{m_2 t}$$

or
$$x(t) = e^{-\lambda t}(c_1 e^{\sqrt{\lambda^2 - \omega^2}\,t} + c_2 e^{-\sqrt{\lambda^2 - \omega^2}\,t}). \tag{5}$$

This equation represents a smooth and nonoscillatory motion. Figure 5.9 shows two possible graphs of $x(t)$.

CASE II $\lambda^2 - \omega^2 = 0$. The system is said to be **critically damped**, since any slight decrease in the damping force would result in oscillatory motion. The general solution of (3) is

$$x(t) = c_1 e^{m_1 t} + c_2 t e^{m_1 t}$$

or
$$x(t) = e^{-\lambda t}(c_1 + c_2 t). \tag{6}$$

Some graphs of typical motion are given in Figure 5.10. Notice that the motion is quite similar to that of an overdamped system. It is also apparent from (6) that the mass can pass through the equilibrium position at most one time.*

(a)

(b)

Figure 5.9

(a)

(b)

Figure 5.10

---

* An examination of the derivatives of (5) and (6) would show that these functions can have at most one relative maximum or one relative minimum for $t > 0$.

CASE III $\lambda^2 - \omega^2 < 0$. In this case the system is said to be **underdamped**, since the damping coefficient is small compared to the spring constant. The roots m_1 and m_2 are now complex:

$$m_1 = -\lambda + \sqrt{\omega^2 - \lambda^2}\,i, \qquad m_2 = -\lambda - \sqrt{\omega^2 - \lambda^2}\,i,$$

and so the general solution of equation (3) is

$$x(t) = e^{-\lambda t}(c_1 \cos \sqrt{\omega^2 - \lambda^2}\,t + c_2 \sin \sqrt{\omega^2 - \lambda^2}\,t). \tag{7}$$

As indicated in Figure 5.11, the motion described by (7) is oscillatory; but because of the coefficient $e^{-\lambda t}$, the amplitudes of vibration $\to 0$ as $t \to \infty$.

Figure 5.11

EXAMPLE 1

It is readily verified that the solution of the initial-value problem

$$\frac{d^2x}{dt^2} + 5\frac{dx}{dt} + 4x = 0, \qquad x(0) = 1, \quad x'(0) = 1,$$

is

$$x(t) = \frac{5}{3}e^{-t} - \frac{2}{3}e^{-4t}. \tag{8}$$

The problem can be interpreted as representing the overdamped motion of a mass on a spring. The mass starts from a position 1 unit *below* the equilibrium position with a *downward* velocity of 1 ft/s.

To graph $x(t)$ we find the value of t for which the function has an extremum, that is, the value of time for which the first derivative (velocity) is zero. Differentiating (8) gives

$$x'(t) = -\frac{5}{3}e^{-t} + \frac{8}{3}e^{-4t}$$

so that $x'(t) = 0$ implies

$$e^{3t} = \frac{8}{5} \quad \text{or} \quad t = \frac{1}{3}\ln\frac{8}{5} = 0.157.$$

It follows from the first derivative test, as well as our physical intuition, that $x(0.157) = 1.069$ ft is actually a maximum. In other words, the mass attains an extreme displacement of 1.069 ft below the equilibrium position.

We should also check to see whether the graph crosses the t-axis, that is, whether the mass passes through the equilibrium position. This cannot happen in this instance since the equation $x(t) = 0$, or $e^{3t} = 2/5$, has the physically irrelevant solution $t = (1/3)\ln(2/5) = -0.305$.

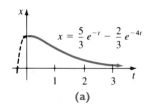

$$x = \frac{5}{3}e^{-t} - \frac{2}{3}e^{-4t}$$

(a)

t	$x(t)$
1	0.601
1.5	0.370
2	0.225
2.5	0.137
3	0.083

(b)

Figure 5.12

The graph of $x(t)$, along with some other pertinent data, is given in Figure 5.12. ■

EXAMPLE 2

An 8-lb weight stretches a spring 2 ft. Assuming a damping force numerically equal to two times the instantaneous velocity acts on the system, determine the equation of motion if the weight is released from the equilibrium position with an upward velocity of 3 ft/s.

Solution From Hooke's law we have

$$8 = k(2), \qquad k = 4 \text{ lb/ft}$$

and from $m = W/g$,

$$m = \frac{8}{32} = \frac{1}{4} \text{ slug}.$$

Thus the differential equation of motion is

$$\frac{1}{4}\frac{d^2x}{dt^2} = -4x - 2\frac{dx}{dt} \quad \text{or} \quad \frac{d^2x}{dt^2} + 8\frac{dx}{dt} + 16x = 0. \tag{9}$$

The initial conditions are

$$x(0) = 0, \qquad x'(0) = -3.$$

Now the auxiliary equation for (9) is

$$m^2 + 8m + 16 = (m + 4)^2 = 0$$

so that $m_1 = m_2 = -4$. Hence the system is critically damped and

$$x(t) = c_1 e^{-4t} + c_2 t e^{-4t}. \tag{10}$$

The initial condition $x(0) = 0$ immediately demands that $c_1 = 0$, whereas using $x'(0) = -3$ gives $c_2 = -3$. Thus the equation of motion is

$$x(t) = -3t e^{-4t}. \tag{11}$$

To graph $x(t)$ we proceed as in Example 1:

$$x'(t) = -3(-4t e^{-4t} + e^{-4t})$$
$$= -3e^{-4t}(1 - 4t).$$

Clearly $x'(t) = 0$ when $t = 1/4$. The corresponding extreme displacement is

$$x\left(\frac{1}{4}\right) = -3\left(\frac{1}{4}\right)e^{-1} = -0.276 \text{ ft}.$$

As shown in Figure 5.13, we interpret this value to mean that the weight reaches a maximum height of 0.276 ft above the equilibrium position. ■

$t = 1/4$

-0.276

maximum height
above equilibrium position

Figure 5.13

EXAMPLE 3

A 16-lb weight is attached to a 5-ft-long spring. At equilibrium the spring measures 8.2 ft. If the weight is pushed up and released from rest at a point 2 ft above the equilibrium position, find the displacements $x(t)$ if it is further known that the surrounding medium offers a resistance numerically equal to the instantaneous velocity.

Solution   The elongation of the spring after the weight is attached is $8.2 - 5 = 3.2$ ft, so it follows from Hooke's law that

$$16 = k(3.2), \qquad k = 5 \text{ lb/ft.}$$

In addition,

$$m = \frac{16}{32} = \frac{1}{2} \text{ slug,}$$

so that the differential equation is given by

$$\frac{1}{2}\frac{d^2 x}{dt^2} = -5x - \frac{dx}{dt} \quad \text{or} \quad \frac{d^2 x}{dt^2} + 2\frac{dx}{dt} + 10x = 0. \tag{12}$$

This latter equation is solved subject to the conditions

$$x(0) = -2, \qquad x'(0) = 0.$$

Proceeding, we find that the roots of $m^2 + 2m + 10 = 0$ are $m_1 = -1 + 3i$ and $m_2 = -1 - 3i$, which then implies the system is underdamped and

$$x(t) = e^{-t}(c_1 \cos 3t + c_2 \sin 3t). \tag{13}$$

Now $x(0) = -2 = c_1$

$$x(t) = e^{-t}(-2 \cos 3t + c_2 \sin 3t)$$

$$x'(t) = e^{-t}(6 \sin 3t + 3c_2 \cos 3t) - e^{-t}(-2 \cos 3t + c_2 \sin 3t)$$

$$x'(0) = 0 = 3c_2 + 2,$$

which gives $c_2 = -2/3$. Thus we finally obtain

$$x(t) = e^{-t}\left(-2 \cos 3t - \frac{2}{3} \sin 3t\right). \tag{14}$$

∎

Alternative Form of the Solution

In a manner identical to the procedure used in Section 5.1, we can write any solution

$$x(t) = e^{-\lambda t}(c_1 \cos \sqrt{\omega^2 - \lambda^2}\, t + c_2 \sin \sqrt{\omega^2 - \lambda^2}\, t)$$

in the alternative form

$$x(t) = Ae^{-\lambda t}\sin(\sqrt{\omega^2 - \lambda^2}\,t + \phi), \tag{15}$$

where $A = \sqrt{c_1^2 + c_2^2}$ and the phase angle ϕ is determined from the equations

$$\sin\phi = \frac{c_1}{A}, \qquad \cos\phi = \frac{c_2}{A}, \qquad \tan\phi = \frac{c_1}{c_2}.$$

The coefficient $Ae^{-\lambda t}$ is sometimes called the **damped amplitude** of vibrations. Because (15) is not a periodic function, the number $2\pi/\sqrt{\omega^2 - \lambda^2}$ is called the **quasi period** and $\sqrt{\omega^2 - \lambda^2}/2\pi$ is the **quasi frequency**. The quasi period is the time interval between two successive maxima of $x(t)$.

To graph an equation such as (15), we first find the intercepts $t_1, t_2, \ldots, t_k, \ldots$; that is, for some integer n we must solve

$$\sqrt{\omega^2 - \lambda^2}\,t + \phi = n\pi$$

for t. It follows that

$$t = \frac{n\pi - \phi}{\sqrt{\omega^2 - \lambda^2}}. \tag{16}$$

In addition we note that $|x(t)| \leq Ae^{-\lambda t}$ since

$$|\sin(\sqrt{\omega^2 - \lambda^2}\,t + \phi)| \leq 1.$$

Indeed, the graph of (15) touches the graphs of $\pm Ae^{-\lambda t}$ at the values $t_1^*, t_2^*, \ldots, t_k^*, \ldots$ for which

$$\sin(\sqrt{\omega^2 + \lambda^2}\,t + \phi) = \pm 1.$$

This means $\sqrt{\omega^2 - \lambda^2}\,t + \phi$ must be an odd multiple of $\pi/2$,

$$\sqrt{\omega^2 - \lambda^2}\,t + \phi = (2n + 1)\frac{\pi}{2}$$

$$t = \frac{(2n + 1)\pi/2 - \phi}{\sqrt{\omega^2 - \lambda^2}}. \tag{17}$$

For example, if we are asked to graph $x(t) = e^{-0.5t}\sin(2t - \pi/3)$, we find the intercepts on the *positive* t-axis by solving

$$2t_1 - \frac{\pi}{3} = 0, \qquad 2t_2 - \frac{\pi}{3} = \pi, \qquad 2t_3 - \frac{\pi}{3} = 2\pi, \ldots,$$

which gives, respectively,

$$t_1 = \frac{\pi}{6}, \qquad t_2 = \frac{4\pi}{6}, \qquad t_3 = \frac{7\pi}{6}, \ldots.$$

Notice that even though $x(t)$ is not periodic, the difference between the successive roots is $t_k - t_{k-1} = \pi/2$ units or one-half the quasi period of $2\pi/2 = \pi$ seconds. Also $\sin(2t - \pi/3) = \pm 1$ at the solutions of

$$2t_1^* - \frac{\pi}{3} = \frac{\pi}{2}, \qquad 2t_2^* - \frac{\pi}{3} = \frac{3\pi}{2}, \qquad 2t_3^* - \frac{\pi}{3} = \frac{5\pi}{2}, \ldots$$

or

$$t_1^* = \frac{5\pi}{12}, \qquad t_2^* = \frac{11\pi}{12}, \qquad t_3^* = \frac{17\pi}{12}, \ldots.$$

It is readily shown that the difference between the successive t_k^* values is also $\pi/2$.* The graph of $x(t)$ is given in Figure 5.14.

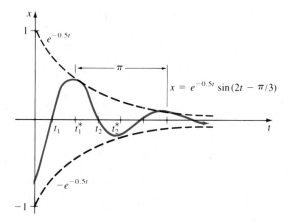

Figure 5.14

EXAMPLE 4

Using (15), we can write the solution of the initial-value problem

$$\frac{d^2x}{dt^2} + 2\frac{dx}{dt} + 10x = 0, \qquad x(0) = -2, \quad x'(0) = 0,$$

in Example 3 in the form $x(t) = Ae^{-t}\sin(3t + \phi)$.
From (14) we have $c_1 = -2, c_2 = -2/3$ so that

$$A = \sqrt{4 + \frac{4}{9}} = \frac{2}{3}\sqrt{10}$$

$$\tan \phi = \frac{-2}{-2/3} = 3 \quad \text{and} \quad \tan^{-1}(3) = 1.249 \text{ radians.}$$

---

* We note that the values of t for which the graph of $x(t)$ touches the exponential graphs are *not* the values for which the function attains its relative extrema.

But since $\sin \phi < 0$ and $\cos \phi < 0$, we take ϕ to be the third-quadrant angle $\phi = \pi + 1.249 = 4.391$ radians. Hence

$$x(t) = \frac{2}{3}\sqrt{10}\,e^{-t}\sin(3t + 4.391).$$

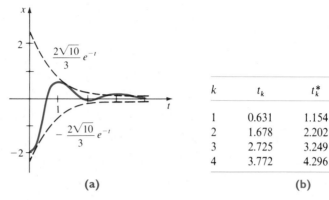

k	t_k	t_k^*	$x(t_k^*)$
1	0.631	1.154	0.665
2	1.678	2.202	−0.233
3	2.725	3.249	0.082
4	3.772	4.296	−0.029

(a)   (b)

Figure 5.15

The graph of this function is given in Figure 5.15. The values of t_k and t_k^* given in the accompanying table are the intercepts and the points at which the graph of $x(t)$ touches the graphs of $\pm(2/3)\sqrt{10}\,e^{-t}$, respectively. In this example the quasi period is $2\pi/3$ seconds, and so the difference between the successive t_k (and the successive t_k^*) is $\pi/3$ units. ∎

5.2   EXERCISES

Answers to odd-numbered problems begin on page A–12.

In Problems 1 and 2 give a possible physical interpretation of the given initial-value problem.

1. $\dfrac{1}{16}x'' + 2x' + x = 0$

$x(0) = 0,\ x'(0) = -1.5$

2. $\dfrac{16}{32}x'' + x' + 2x = 0$

$x(0) = -2,\ x'(0) = 1$

In Problems 3–6 the given figure represents the graph of an equation of motion for a mass on a spring. The spring-mass system is damped. Use the graph to determine:

(a) whether the initial displacement of the mass is above or below the equilibrium position; and

(b) whether the mass is initially released from rest heading downward or heading upward.

3.

Figure 5.16

4.

Figure 5.17

5.

Figure 5.18

6.

Figure 5.19

7. A 4-lb weight is attached to a spring whose constant is 2 lb/ft. The medium offers a resistance to the motion of the weight numerically equal to the instantaneous velocity. If the weight is released from a point 1 ft above the equilibrium position with a downward velocity of 8 ft/s, determine the time that the weight passes through the equilibrium position. Find the time for which the weight attains its extreme displacement from the equilibrium position. What is the position of the weight at this instant?

8. A 4-ft spring measures 8 ft long after an 8-lb weight is attached to it. The medium through which the weight moves offers a resistance numerically equal to $\sqrt{2}$ times the instantaneous velocity. Find the equation of motion if the weight is released from the equilibrium position with a downward velocity of 5 ft/s. Find the time for which the weight attains its extreme displacement from the equilibrium position. What is the position of the weight at this instant?

9. A 1-kg mass is attached to a spring whose constant is 16 N/m and the entire system is then submerged in a liquid that imparts a damping force numerically equal to 10 times the instantaneous velocity. Determine the equations of motion if

(a) the weight is released from rest 1 m below the equilibrium position; and

(b) the weight is released 1 m below the equilibrium position with an upward velocity of 12 m/s.

10. In parts (a) and (b) of Problem 9 determine whether the weight passes through the equilibrium position. In each case find the time at which the

weight attains its extreme displacement from the equilibrium position. What is the position of the weight at this instant?

11. A force of 2 lb stretches a spring 1 ft. A 3.2-lb weight is attached to the spring and the system is then immersed in a medium that imparts a damping force numerically equal to 0.4 times the instantaneous velocity.

(a) Find the equation of motion if the weight is released from rest 1 ft above the equilibrium position.

(b) Express the equation of motion in the form given in (15).

(c) Find the first time for which the weight passes through the equilibrium position heading upward.

12. After a 10-lb weight is attached to a 5-ft spring, the spring measures 7 ft long. The 10-lb weight is removed and replaced with an 8-lb weight and the entire system is placed in a medium offering a resistance numerically equal to the instantaneous velocity.

(a) Find the equation of motion if the weight is released 1/2 ft below the equilibrium position with a downward velocity of 1 ft/s.

(b) Express the equation of motion in the form given in (15).

(c) Find the times for which the weight passes through the equilibrium position heading downward.

(d) Graph the equation of motion.

13. A 10-lb weight attached to a spring stretches it 2 ft. The weight is attached to a dashpot damping device that offers a resistance numerically equal to β ($\beta > 0$) times the instantaneous velocity. Determine the values of the damping constant β so that the subsequent motion is (a) overdamped, (b) critically damped, and (c) underdamped.

14. A 24-lb weight stretches a spring 4 ft. The subsequent motion takes place in a medium offering a resistance numerically equal to β ($\beta > 0$) times the instantaneous velocity. If the weight starts from the equilibrium position with an upward velocity of 2 ft/s, show that if $\beta > 3\sqrt{2}$, the equation of motion is

$$x(t) = \frac{-3}{\sqrt{\beta^2 - 18}} e^{-2\beta t/3} \sinh \frac{2}{3} \sqrt{\beta^2 - 18}\, t.$$

15. A mass of 40 g stretches a spring 10 cm. A damping device imparts a resistance to motion numerically equal to 560 (measured in dynes/(cm/s)) times the instantaneous velocity. Find the equation of motion if the mass is released from the equilibrium position with a downward velocity of 2 cm/s.

16. Find the equation of motion for the mass in Problem 15 if the damping constant is doubled.

17. A mass of 1 slug is attached to a spring whose constant is 9 lb/ft. The medium offers a resistance to the motion numerically equal to 6 times the instantaneous velocity. The mass is released from a point 8 in. above the equilibrium position with a downward velocity of v_0 ft/s. Determine the values of v_0 such that the mass will subsequently pass through the equilibrium position.

18. The quasi period of an underdamped, vibrating 1-slug mass on a spring is $\pi/2$ seconds. If the spring constant is 25 lb/ft, find the damping constant β.

19. In the case of underdamped motion show that the difference in times between two successive positive maxima of the equation of motion is $2\pi/\sqrt{\omega^2 - \lambda^2}$.

20. Use (16) to show that the time interval between successive intercepts of (15) is one-half the quasi period.

21. Use (17) to show that the time interval between successive values of t for which the graph of (15) touches the graphs of $\pm Ae^{-\lambda t}$ is one-half the quasi period.

22. Use equation (17) to show that the intercepts of the graph of $x(t) = Ae^{-\lambda t} \sin(\sqrt{\omega^2 - \lambda^2}\, t + \phi)$ are halfway between the values of t for which the graph of $x(t)$ touches the graphs of $\pm Ae^{-\lambda t}$. The values of t for which $x(t)$ is a maximum or minimum are not located halfway between the intercepts of the graph of $x(t)$. Verify this last statement by considering the function $x(t) = e^{-t} \sin(t + \pi/4)$.

23. In the case of underdamped motion show that the ratio between two consecutive maximum (or minimum) displacements x_n and x_{n+2} is the constant

$$\frac{x_n}{x_{n+2}} = e^{2\pi\lambda/\sqrt{\omega^2 - \lambda^2}}.$$

The number $\delta = \ln(x_n/x_{n+2}) = 2\pi\lambda/\sqrt{\omega^2 - \lambda^2}$ is called the **logarithmic decrement**.

24. The logarithmic decrement defined in Problem 23 is an indicator of the rate at which the motion is damped out.

    (a) Describe the motion of an underdamped system if δ is a very small positive number.

    (b) Compute the logarithmic decrement for the motion described in Problem 12.

5.3 FORCED MOTION

With Damping

Suppose we now take into consideration an external force $f(t)$ acting on a vibrating mass on a spring. For example, $f(t)$ could represent a driving force causing an oscillatory vertical motion of the support of the spring. See Fig-

Figure 5.20

ure 5.20. The inclusion of $f(t)$ in the formulation of Newton's second law gives the differential equation of **forced motion**

$$m\frac{d^2x}{dt^2} = -kx - \beta\frac{dx}{dt} + f(t), \tag{1}$$

$$\frac{d^2x}{dt^2} + \frac{\beta}{m}\frac{dx}{dt} + \frac{k}{m}x = \frac{f(t)}{m}, \tag{2}$$

or

$$\frac{d^2x}{dt^2} + 2\lambda\frac{dx}{dt} + \omega^2x = F(t), \tag{3}$$

where $F(t) = f(t)/m$ and, as in the preceding section, $2\lambda = \beta/m$, $\omega^2 = k/m$. To solve the latter nonhomogeneous equation we can use either the method of undetermined coefficients or variation of parameters.

EXAMPLE I

Interpret and solve the initial-value problem

$$\frac{1}{5}\frac{d^2x}{dt^2} + 1.2\frac{dx}{dt} + 2x = 5\cos 4t, \qquad x(0) = \frac{1}{2}, \quad x'(0) = 0. \tag{4}$$

Solution   We can interpret the problem to represent a vibrational system consisting of a mass ($m = 1/5$ slug or kilogram) attached to a spring ($k = 2$ lb/ft or N/m). The mass is released from rest 1/2 unit (foot or meter) below the equilibrium position. The motion is damped ($\beta = 1.2$) and is being driven by an external periodic ($T = \pi/2$ seconds) force beginning at $t = 0$. Intuitively we would expect that even with damping the system will remain in motion until such time as the forcing function is "turned off," in which case the amplitudes diminish. However, as the problem is given, $f(t) = 5\cos 4t$ will remain "on" forever.

We first multiply (4) by 5 and solve the homogeneous equation

$$\frac{dx^2}{dt^2} + 6\frac{dx}{dt} + 10x = 0$$

by the usual methods. Since $m_1 = -3 + i$, $m_2 = -3 - i$, it follows that

$$x_c(t) = e^{-3t}(c_1\cos t + c_2\sin t).$$

Using the method of undetermined coefficients, we assume a particular solution of the form $x_p(t) = A\cos 4t + B\sin 4t$. Now

$$x'_p = -4A\sin 4t + 4B\cos 4t$$

$$x''_p = -16A\cos 4t - 16B\sin 4t$$

so that

$$x_p'' + 6x_p' + 10x_p = -16A \cos 4t - 16B \sin 4t - 24A \sin 4t$$

$$+ 24B \cos 4t + 10A \cos 4t + 10B \sin 4t$$

$$= (-6A + 24B) \cos 4t + (-24A - 6B) \sin 4t$$

$$= 25 \cos 4t.$$

The resulting system of equations

$$-6A + 24B = 25$$

$$-24A - 6B = 0$$

yields $A = -25/102$ and $B = 50/51$. It follows that

$$x(t) = e^{-3t}(c_1 \cos t + c_2 \sin t) - \frac{25}{102} \cos 4t + \frac{50}{51} \sin 4t. \tag{5}$$

When we set $t = 0$ in the above equation, we obtain $c_1 = 38/51$. By differentiating the expression and then setting $t = 0$, we also find that $c_2 = -86/51$. Therefore the equation of motion is

$$x(t) = e^{-3t}\left(\frac{38}{51} \cos t - \frac{86}{51} \sin t\right) - \frac{25}{102} \cos 4t + \frac{50}{51} \sin 4t. \tag{6}$$

■

Transient and Steady-State Terms

Notice that the complementary function

$$x_c(t) = e^{-3t}\left(\frac{38}{51} \cos t - \frac{86}{51} \sin t\right)$$

in Example 1 possesses the property that

$$\lim_{t \to \infty} x_c(t) = 0.$$

Since $x_c(t)$ becomes negligible (namely, $\to 0$) as $t \to \infty$, it is said to be a **transient term**, or **transient solution**. Thus for large time the displacements of the weight in the preceding problem are closely approximated by the particular solution $x_p(t)$. This latter function is also called the **steady-state solution**. When F is a periodic function, such as $F(t) = F_0 \sin \gamma t$ or $F(t) = F_0 \cos \gamma t$, the general solution of (3) consists of

$$x(t) = transient + steady\text{-}state.$$

EXAMPLE 2

The solution to the initial-value problem

$$\frac{d^2x}{dt^2} + 2\frac{dx}{dt} + 2x = 4 \cos t + 2 \sin t, \qquad x(0) = 0, \quad x'(0) = 3,$$

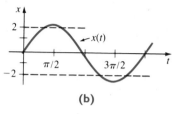

Figure 5.21

is readily shown to be

$$x = x_c + x_p = \underbrace{e^{-t} \sin t}_{\text{transient}} + \underbrace{2 \sin t}_{\text{steady-state}}.$$

Inspection of Figure 5.21 shows that the effect of the transient term on the solution is, in this case, negligible for about $t > 2\pi$. ∎

Without Damping

With a periodic impressed force and no damping force, there is no transient term in the solution of a problem. Also, we shall see that a periodic impressed force with a frequency near, or the same as, the frequency of free undamped vibrations can cause a severe problem in any oscillatory mechanical system.

EXAMPLE 3

Solve the initial-value problem

$$\frac{d^2x}{dt^2} + \omega^2 x = F_0 \sin \gamma t, \qquad x(0) = 0, \quad x'(0) = 0, \tag{7}$$

where F_0 is a constant.

Solution   The complementary function is $x_c(t) = c_1 \cos \omega t + c_2 \sin \omega t$. To obtain a particular solution we assume $x_p(t) = A \cos \gamma t + B \sin \gamma t$, so that

$$x_p' = -A\gamma \sin \gamma t + B\gamma \cos \gamma t$$

$$x_p'' = -A\gamma^2 \cos \gamma t - B\gamma^2 \sin \gamma t$$

$$x_p'' + \omega^2 x_p = A(\omega^2 - \gamma^2) \cos \gamma t + B(\omega^2 - \gamma^2) \sin \gamma t$$

$$= F_0 \sin \gamma t.$$

It follows that $A(\omega^2 - \gamma^2) = 0$, $B(\omega^2 - \gamma^2) = F_0$, and so

$$A = 0, \qquad B = \frac{F_0}{\omega^2 - \gamma^2} \quad (\gamma \neq \omega).$$

Therefore

$$x_p(t) = \frac{F_0}{\omega^2 - \gamma^2} \sin \gamma t.$$

Applying the given initial conditions to the general solution

$$x(t) = c_1 \cos \omega t + c_2 \sin \omega t + \frac{F_0}{\omega^2 - \gamma^2} \sin \gamma t$$

yields $c_1 = 0$ and $c_2 = -\gamma F_0/\omega(\omega^2 - \gamma^2)$. Thus the solution is

$$x(t) = \frac{F_0}{\omega(\omega^2 - \gamma^2)} (-\gamma \sin \omega t + \omega \sin \gamma t), \qquad \gamma \neq \omega. \tag{8}$$

∎

Pure Resonance

Although equation (8) is not defined for $\gamma = \omega$, it is interesting to observe that its limiting value as $\gamma \to \omega$ can be obtained by applying L'Hôpital's rule. This limiting process is analogous to "tuning in" the frequency of the driving force $(\gamma/2\pi)$ to the frequency of free vibrations $(\omega/2\pi)$. Intuitively we expect that over a length of time we should be able to substantially increase the amplitudes of vibration.* For $\gamma = \omega$ we define the solution to be

$$x(t) = \lim_{\gamma \to \omega} F_0 \frac{-\gamma \sin \omega t + \omega \sin \gamma t}{\omega(\omega^2 - \gamma^2)}$$

$$= F_0 \lim_{\gamma \to \omega} \frac{\dfrac{d}{d\gamma}(-\gamma \sin \omega t + \omega \sin \gamma t)}{\dfrac{d}{d\gamma}(\omega^3 - \omega\gamma^2)}$$

$$= F_0 \lim_{\gamma \to \omega} \frac{-\sin \omega t + \omega t \cos \gamma t}{-2\omega\gamma}$$

$$= F_0 \frac{-\sin \omega t + \omega t \cos \omega t}{-2\omega^2}$$

$$= \frac{F_0}{2\omega^2} \sin \omega t - \frac{F_0}{2\omega} t \cos \omega t. \tag{9}$$

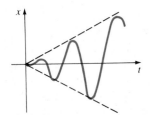

Figure 5.22

As suspected, when $t \to \infty$, the displacements become large; in fact, $|x(t_n)| \to \infty$ when $t_n = n\pi/\omega$, $n = 1, 2, \ldots$. The phenomenon we have just described is known as **pure resonance**. The graph given in Figure 5.22 shows typical motion in this case.

In conclusion it should be noted that there is no actual need to use a limiting process on (8) to obtain the solution for $\gamma = \omega$. Alternatively, equation (9) follows by solving the initial-value problem

$$\frac{d^2x}{dt^2} + \omega^2 x = F_0 \sin \omega t, \qquad x(0) = 0, \quad x'(0) = 0,$$

directly by conventional methods.

Resonance Curve

In the case of underdamped vibrations, the general solution of the differential equation

$$\frac{d^2x}{dt^2} + 2\lambda \frac{dx}{dt} + \omega^2 x = F_0 \sin \gamma t \tag{10}$$

---

* Forgetting about the damping effects of shock absorbers, the situation is roughly equivalent to a number of passengers jumping up and down in the back of a bus in time with the natural vertical motion caused by equally spaced faults (such as cracks) in the road. Theoretically these passengers could upset the bus—assuming they are not kicked off first!

can be shown to be

$$x(t) = Ae^{-\lambda t} \sin(\sqrt{\omega^2 - \lambda^2}\, t + \phi) + \frac{F_0}{\sqrt{(\omega^2 - \gamma^2)^2 + 4\lambda^2\gamma^2}} \sin(\gamma t + \theta), \quad \textbf{(11)}$$

where $A = \sqrt{c_1^2 + c_2^2}$ and the phase angles ϕ and θ are, respectively, defined by

$$\sin\phi = \frac{c_1}{A}, \qquad \cos\phi = \frac{c_2}{A}$$

$$\sin\theta = \frac{-2\lambda\gamma}{\sqrt{(\omega^2 - \gamma^2)^2 + 4\lambda^2\gamma^2}}, \qquad \cos\theta = \frac{\omega^2 - \gamma^2}{\sqrt{(\omega^2 - \gamma^2)^2 + 4\lambda^2\gamma^2}}.$$

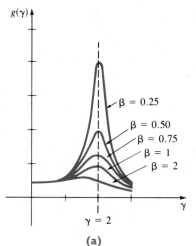

$\gamma = 2$

(a)

β	γ_1	$g(\gamma_1)$
2	1.41	0.58
1	1.87	1.03
0.75	1.93	1.36
0.50	1.97	2.02
0.25	1.99	4.01

(b)

Figure 5.23

EXAMPLE 4

Inspection of (11) shows that $x_c(t)$ is transient when damping is present, and hence for large values of time the solution is closely approximated by the steady-state solution

$$x_p(t) = g(\gamma)\sin(\gamma t + \theta),$$

where we define

$$g(\gamma) = \frac{F_0}{\sqrt{(\omega^2 - \gamma^2)^2 + 4\lambda^2\gamma^2}}. \qquad \textbf{(12)}$$

Although the amplitude of x_p is bounded as $t \to \infty$, it is easily shown that the maximum oscillations will occur at the value $\gamma_1 = \sqrt{\omega^2 - 2\lambda^2}$ (see Problem 11). Thus when the frequency of the external force is $\sqrt{\omega^2 - 2\lambda^2}/2\pi$, the system is said to be in **resonance**.

In the specific case $k = 4$, $m = 1$, $F_0 = 2$, $g(\gamma)$ becomes

$$g(\gamma) = \frac{2}{\sqrt{(4 - \gamma^2)^2 + \beta^2\gamma^2}}. \qquad \textbf{(13)}$$

Figure 5.23(a) shows the graph of (13) for various values of the damping coefficients β. This family of graphs is called the **resonance curve** of the system. Observe the behavior of the amplitudes $g(\gamma)$ as $\beta \to 0$, that is, as the system approaches pure resonance. ∎

Remark If a mechanical system were actually described by a function such as (9) of this section, it would necessarily fail. Large oscillations of a weight on a spring would eventually force the spring beyond its elastic limit. One might argue too that the resonating model presented in Figure 5.22 is completely unrealistic since it ignores the retarding effects of ever-present damping forces. Although it is true that pure resonance cannot occur when the smallest amount of damping is taken into consideration, large and equally destructive amplitudes of vibration (although bounded as $t \to \infty$) can occur.

If you have ever looked out an airplane window while in flight, you have probably observed that the wings on an airplane are not perfectly rigid. A reasonable amount of flex or flutter is not only tolerated but necessary to prevent the wing from snapping like a piece of peppermint stick candy. In late 1959 and early 1960 two commercial plane crashes involving a relatively new model of propjet occurred, illustrating the destructive effects of large mechanical oscillations.

The unusual aspect of these crashes was that they both happened while the planes were in mid-flight. Barring midair collisions, the safest period during any flight is when the plane has attained its cruising altitude. It is well known that a plane is most vulnerable to an accident when it is least maneuverable, namely, during either take-off or landing. So, having two planes simply fall out of the sky was not only a tragedy but an embarrassment to the aircraft industry and a thoroughly puzzling problem to aerodynamic engineers. In crashes of this sort, a structural failure of some kind is immediately suspected. After a massive technical investigation, the problem was eventually traced in each case to an outboard engine and engine housing. Roughly, it was determined that when each plane surpassed a critical speed of approximately 400 mph, a propeller and engine began to wobble, causing a gyroscopic force that could not be quelled or damped by the engine housing. This external vibrational force was then transferred to the already oscillating wing. This, in itself, need not have been destructively dangerous since aircraft wings are designed to withstand the stress of unusual and excessive forces. (In fact the particular wing in question was so incredibly strong that test engineers and pilots who were deliberately trying to snap a wing under every conceivable flight condition failed to do so.) But, unfortunately, after a short period of time during which the engine wobbled rapidly, the frequency of the impressed force actually slowed to a point at which it approached and finally coincided with the maximum frequency of wing flutter (around 3 cycles per second). The resulting resonance finally accomplished what the test engineers could not do; namely, the amplitudes of wing flutter became large enough to snap the wing. See Figure 5.24.

The problem was solved in two steps. All models of this particular plane were required to fly at speeds substantially below 400 mph until each plane could be modified by considerably strengthening (or stiffening) the engine housings. A strengthened engine housing was shown to be able to impart a damping effect capable of preventing the critical resonance phenomenon even in the unlikely event of a subsequent engine wobble.*

You may be aware that soldiers usually do not march in step across bridges. The reason for breaking stride is simply to avoid any possibility of

normal flutter

wing snaps

large flutter

Figure 5.24

---

* For a fascinating nontechnical account of the investigation, see Robert J. Serling, *Loud and Clear* (New York: Dell, 1970), Chapter 5.

Figure 5.25 © 1988 Memtek Products

resonance occurring between the natural vibrations inherent in the bridge's structure and the frequency of the external force of a multitude of feet stomping in unison on the bridge.

Acoustic vibrations can be equally as destructive as large mechanical vibrations. In television commercials, jazz singers have inflicted destruction on the lowly wine glass. See Figure 5.25. Sounds from organs and piccolos have been known to crack windows.

> As the horns blew, the people began to shout. When they heard the signal horn, they raised a tremendous shout. The wall collapsed. . . . *Joshua 6:20*

Did the power of acoustic resonance cause the walls of Jericho to tumble down? This is the conjecture of some contemporary scholars.

The phenomenon of resonance is not always destructive, however. For example, it is resonance of an electrical circuit that enables a radio to be tuned to a specific station.

5.3 EXERCISES

Answers to odd-numbered problems begin on page A–13.

1. A 16-lb weight stretches a spring 8/3 ft. Initially the weight starts from rest 2 ft below the equilibrium position and the subsequent motion takes place in a medium that offers a damping force numerically equal to 1/2 the instantaneous velocity. Find the equation of motion if the weight is driven by an external force equal to $f(t) = 10 \cos 3t$.

2. A mass of 1 slug is attached to a spring whose constant is 5 lb/ft. Initially the mass is released 1 ft below the equilibrium position with a downward velocity of 5 ft/s, and the subsequent motion takes place in a medium that offers a damping force numerically equal to 2 times the instantaneous velocity.

    (a) Find the equation of motion if the mass is driven by an external force equal to $f(t) = 12 \cos 2t + 3 \sin 2t$.

    (b) Graph the transient and steady-state solutions on the same coordinate axes.

    (c) Graph the equation of motion.

3. A mass of 1 slug, when attached to a spring, stretches it 2 ft and then comes to rest in the equilibrium position. Starting at $t = 0$, an external force equal to $f(t) = 8 \sin 4t$ is applied to the system. Find the equation of motion if the surrounding medium offers a damping force numerically equal to 8 times the instantaneous velocity.

4. In Problem 3 determine the equation of motion if the external force is $f(t) = e^{-t} \sin 4t$. Analyze the displacements for $t \to \infty$.

5. When a mass of 2 kilograms is attached to a spring whose constant is 32 N/m, it comes to rest in the equilibrium position. Starting at $t = 0$,

Figure 5.26

a force equal to $f(t) = 68e^{-2t}\cos 4t$ is applied to the system. Find the equation of motion in the absence of damping.

6. In Problem 5 write the equation of motion in the form $x(t) = A\sin(\omega t + \phi) + Be^{-2t}\sin(4t + \theta)$. What is the amplitude of vibrations after a very long time?

7. A mass m is attached to the end of a spring whose constant is k. After the mass reaches equilibrium, its support begins to oscillate vertically about a horizontal line L according to a formula $h(t)$. The value of h represents the distance in feet measured from L. See Figure 5.26. Determine the differential equation of motion if the entire system moves through a medium offering a damping force numerically equal to $\beta(dx/dt)$.

8. Solve the differential equation of the preceding problem if the spring is stretched 4 ft by a weight of 16 lb, and $\beta = 2$, $h(t) = 5\cos t$, $x(0) = x'(0) = 0$.

9. A mass of 100 g is attached to a spring whose constant is 1600 dynes/cm. After the mass reaches equilibrium, its support oscillates according to the formula $h(t) = \sin 8t$, where h represents displacement from its original position. See Problem 7 and Figure 5.26.

   (a) In the absence of damping, determine the equation of motion if the mass starts from rest from the equilibrium position.

   (b) At what times does the mass pass through the equilibrium position?

   (c) At what times does the mass attain its extreme displacements?

   (d) What are the maximum and minimum displacements?

   (e) Graph the equation of motion.

10. Show that the general solution of equation (10) is given by (11).

11. (a) Prove that $g(\gamma)$ given in (13) of Example 4 has a maximum value at $\gamma_1 = \sqrt{\omega^2 - 2\lambda^2}$. [*Hint:* Differentiate with respect to γ.]

   (b) What is the maximum value of $g(\gamma)$ at resonance?

12. (a) If $k = 3$ lb/ft and $m = 1$ slug, use the information in Example 4 to show that the system is underdamped when the damping coefficient β satisfies $0 < \beta < 2\sqrt{3}$ but that resonance can occur only if $0 < \beta < \sqrt{6}$.

   (b) Construct the resonance curve of the system when $F_0 = 3$.

13. A mass of 1/2 slug is suspended on a spring whose constant is 6 lb/ft. The system is set in motion in a medium offering a damping force numerically equal to twice the instantaneous velocity. Find the steady-state solution if an external force $f(t) = 40\sin 2t$ is applied to the system starting at $t = 0$. Write this solution in the form of a constant multiple of $\sin(2t + \theta)$.

14. Verify that the mechanical system described in Problem 13 is in resonance. Show that the amplitude of the steady-state solution is the maximum value of $g(\gamma)$ described in Problem 11.

15. **(a)** Show that the solution of the initial-value problem

$$\frac{d^2x}{dt^2} + \omega^2 x = F_0 \cos \gamma t, \qquad x(0) = 0, \quad x'(0) = 0,$$

is

$$x(t) = \frac{F_0}{\omega^2 - \gamma^2} (\cos \gamma t - \cos \omega t).$$

(b) Evaluate

$$\lim_{\gamma \to \omega} \frac{F_0}{\omega^2 - \gamma^2} (\cos \gamma t - \cos \omega t).$$

16. Compare the result obtained in part (b) of Problem 15 with the solution obtained using variation of parameters when the external force is $F_0 \cos \omega t$.

In Problems 17 and 18 solve the given initial-value problem.

17. $\dfrac{d^2x}{dt^2} + 4x = -5 \sin 2t + 3 \cos 2t, \quad x(0) = -1, x'(0) = 1$

18. $\dfrac{d^2x}{dt^2} + 9x = 5 \sin 3t, \quad x(0) = 2, x'(0) = 0$

19. **(a)** Show that $x(t)$ given in part (a) of Problem 15 can be written in the form

$$x(t) = \frac{-2F_0}{\omega^2 - \gamma^2} \sin \frac{1}{2} (\gamma - \omega)t \sin \frac{1}{2} (\gamma + \omega)t.$$

(b) If we define $\varepsilon = \frac{1}{2}(\gamma - \omega)$, show that when ε is small an *approximate* solution is

$$x(t) = \frac{F_0}{2\varepsilon\gamma} \sin \varepsilon t \sin \gamma t.$$

When ε is small, the frequency $\gamma/2\pi$ of the impressed force is close to the frequency $\omega/2\pi$ of free vibrations. When this occurs, the motion is as indicated in Figure 5.27. Oscillations of this kind are called *beats* and are due to the fact that the frequency of $\sin \varepsilon t$ is quite small in comparison to the frequency of $\sin \gamma t$. The dashed curves, or *envelope* of the graph of $x(t)$, are obtained from the graphs of $\pm (F_0/2\varepsilon\gamma) \sin \varepsilon t$. Use a computer with various values of F_0, ε, and γ to verify the graph in Figure 5.27.

Figure 5.27

(c) Evaluate
$$\lim_{\varepsilon \to 0} \frac{F_0}{2\varepsilon\gamma} \sin \varepsilon t \sin \gamma t.$$

20. Show that the solution of the initial-value problem

$$\frac{d^2x}{dt^2} + 25x = 10 \cos 7t, \qquad x(0) = 0, \quad x'(0) = 0,$$

is $x(t) = \frac{5}{6} \sin t \sin 6t$. Use a graphics calculator or computer to obtain the graph of $x(t)$.

5.4 ELECTRIC CIRCUITS AND OTHER ANALOGOUS SYSTEMS

L-R-C Series Circuits

As mentioned in the introduction to this chapter, many different physical systems can be described by a linear second-order differential equation similar to the differential equation of forced motion with damping:

$$m\frac{d^2x}{dt^2} + \beta\frac{dx}{dt} + kx = f(t). \tag{1}$$

If $i(t)$ denotes current in an ***L-R-C* series electrical circuit** shown in Figure 5.28(a), then the voltage drops across the inductor, resistor, and capacitor are as shown in Figure 5.28(b). By Kirchhoff's second law, the sum of these voltages equals the voltage $E(t)$ impressed on the circuit; that is,

$$L\frac{di}{dt} + Ri + \frac{1}{C}q = E(t). \tag{2}$$

But the charge $q(t)$ on the capacitor is related to the current $i(t)$ by $i = dq/dt$ and so (2) becomes the linear second-order differential equation

$$L\frac{d^2q}{dt^2} + R\frac{dq}{dt} + \frac{1}{C}q = E(t). \tag{3}$$

(a)

| Inductor inductance L: henrys (h) voltage drop across: $L\dfrac{di}{dt}$ | Resistor resistance R: ohms (Ω) voltage drop across: iR | Capacitor capacitance C: farads (f) voltage drop across: $\dfrac{1}{C}q$ |

(b)

Figure 5.28

The nomenclature used in the analysis of circuits is similar to that used to describe spring-mass systems.

If $E(t) = 0$, the **electrical vibrations** of the circuit are said to be *free*. Since the auxiliary equation for (3) is $Lm^2 + Rm + 1/C = 0$, there will be three forms of the solution when $R \neq 0$, depending on the value of the discriminant $R^2 - 4L/C$. We say that the circuit is

overdamped if	$R^2 - 4L/C > 0$,
critically damped if	$R^2 - 4L/C = 0$,
and **underdamped** if	$R^2 - 4L/C < 0$.

In each of these three cases, the general solution of (3) contains the factor $e^{-Rt/2L}$ and so $q(t) \to 0$ as $t \to \infty$. In the underdamped case when $q(0) = q_0$, the charge on the capacitor oscillates as it decays; in other words, the capacitor is charging and discharging as $t \to \infty$. When $E(t) = 0$ and $R = 0$, the circuit is said to be undamped and the electrical vibrations do not approach zero as t increases without bound; the response of the circuit is **simple harmonic**.

EXAMPLE I

Consider an *L-C* series circuit in which $E(t) = 0$. Determine the charge $q(t)$ on the capacitor for $t > 0$ if its initial charge is q_0 and if initially there is no current flowing in the circuit.

Solution In an *L-C* circuit there is no resistor, so from (3) we obtain

$$L \frac{d^2q}{dt^2} + \frac{1}{C} q = 0.$$

The initial conditions are $q(0) = q_0$ and $i(0) = 0$. Since $q'(t) = i(t)$, the latter condition is the same as $q'(0) = 0$. The general solution of the differential equation is

$$q(t) = c_1 \cos \frac{1}{\sqrt{LC}} t + c_2 \sin \frac{1}{\sqrt{LC}} t.$$

Now the initial conditions imply $c_1 = q_0$ and $c_2 = 0$, so that

$$q(t) = q_0 \cos \frac{1}{\sqrt{LC}} t. \qquad \blacksquare$$

In Example 1 if we want to find the current in the circuit, we use $i(t) = q'(t)$:

$$i(t) = -\frac{q_0}{\sqrt{LC}} \sin \frac{1}{\sqrt{LC}} t.$$

EXAMPLE 2

Find the charge $q(t)$ on the capacitor in an L-R-C series circuit when $L = 0.25$ henry, $R = 10$ ohms, $C = 0.001$ farad, $E(t) = 0$, $q(0) = q_0$ coulombs, and $i(0) = 0$.

Solution Since $1/C = 1000$, equation (3) becomes

$$\frac{1}{4}q'' + 10q' + 1000q = 0 \quad \text{or} \quad q'' + 40q' + 4000q = 0.$$

Solving this homogeneous equation in the usual manner, we find that the circuit is underdamped and

$$q(t) = e^{-20t}(c_1 \cos 60t + c_2 \sin 60t).$$

Applying the initial conditions, we find $c_1 = q_0$ and $c_2 = q_0/3$. Thus the solution is given by

$$q(t) = q_0 e^{-20t}\left(\cos 60t + \frac{1}{3} \sin 60t\right). \qquad \blacksquare$$

The solution in Example 2 can be written as a single sine function using the method discussed in Section 5.2. From (15) of that section, we find

$$q(t) = \frac{q_0\sqrt{10}}{3} e^{-20t} \sin(60t + 1.249).$$

When there is an impressed voltage $E(t)$ on the circuit, the electrical vibrations are said to be *forced*. Note in Example 1 that the free electrical vibrations are simple harmonic with period $2\pi/(1/\sqrt{LC}) = 2\pi\sqrt{LC}$ and frequency $1/(2\pi\sqrt{LC})$. If a periodic voltage $E(t)$ with the same frequency were impressed on the circuit, the system would be in **resonance**. In the case when $R \neq 0$, the complementary function $q_c(t)$ of (3) is called a **transient solution**. If $E(t)$ is periodic or a constant, then the particular solution $q_p(t)$ of (3) is a **steady-state solution**.

EXAMPLE 3

Find the steady-state solution $q_p(t)$ and the **steady-state current** in an L-R-C series circuit when the impressed voltage is $E(t) = E_0 \sin \gamma t$.

Solution The steady-state solution $q_p(t)$ is a particular solution of the differential equation

$$L\frac{d^2q}{dt^2} + R\frac{dq}{dt} + \frac{1}{C}q = E_0 \sin \gamma t.$$

Using the method of undetermined coefficients, we assume a particular solution of the form

$$q_p(t) = A \sin \gamma t + B \cos \gamma t. \tag{4}$$

Substituting (4) into the differential equation, simplifying, and equating coefficients give

$$A = \frac{E_0(L\gamma - 1/C\gamma)}{-\gamma \left[L^2\gamma^2 - \dfrac{2L}{C} + \dfrac{1}{C^2\gamma^2} + R^2 \right]}$$

and

$$B = \frac{E_0 R}{-\gamma \left[L^2\gamma^2 - \dfrac{2L}{C} + \dfrac{1}{C^2\gamma^2} + R^2 \right]}.$$

It is convenient to express A and B in terms of some new symbols. If

$$X = L\gamma - \frac{1}{C\gamma}, \quad \text{then} \quad X^2 = L^2\gamma^2 - \frac{2L}{C} + \frac{1}{C^2\gamma^2}.$$

If $\quad Z = \sqrt{X^2 + R^2}, \quad$ then $\quad Z^2 = L^2\gamma^2 - \dfrac{2L}{C} + \dfrac{1}{C^2\gamma^2} + R^2.$

Therefore $A = E_0 X/(-\gamma Z^2)$ and $B = E_0 R/(-\gamma Z^2)$ and so the steady-state charge is

$$q_p(t) = -\frac{E_0 X}{\gamma Z^2} \sin \gamma t - \frac{E_0 R}{\gamma Z^2} \cos \gamma t.$$

Now the steady-state current is given by $i_p(t) = q'_p(t)$:

$$i_p(t) = \frac{E_0}{Z} \left(\frac{R}{Z} \sin \gamma t - \frac{X}{Z} \cos \gamma t \right). \tag{5}$$

■

The quantities $X = L\gamma - 1/C\gamma$ and $Z = \sqrt{X^2 + R^2}$ defined in Example 3 are called, respectively, the **reactance** and **impedance** of the circuit. Both the reactance and the impedance are measured in ohms.

Twisted Shaft

The differential equation governing the torsional motion of a weight suspended from the end of an elastic shaft is

$$I \frac{d^2\theta}{dt^2} + c \frac{d\theta}{dt} + k\theta = T(t). \tag{6}$$

As shown in Figure 5.29, the function $\theta(t)$ represents the amount of twist of the weight at any time.

Figure 5.29

By comparing equations (3) and (6) with (1), we see that, with the exception of terminology, there is absolutely no difference between the mathematics of vibrating springs, simple series circuits, and torsional vibrations. The following table gives a comparison of the analogous parts of these three kinds of systems.

Mechanical	Series Electrical	Torsional
m (mass)	L (inductance)	I (moment of inertia)
β (damping)	R (resistance)	c (damping)
k (spring constant)	$\dfrac{1}{C}$ (reciprocal of capacitance— called elastance)	k (elastic shaft constant)
$f(t)$ (applied force)	$E(t)$ (impressed voltage)	$T(t)$ (applied torque)

Simple Pendulum

In Example 3 of Section 1.2 it was seen that the angular displacements θ of a simple pendulum are described by the nonlinear second-order equation

$$\frac{d^2\theta}{dt^2} + \frac{g}{l}\sin\theta = 0,$$

where l is the length of the pendulum rod. For small displacements $\sin\theta$ is replaced by θ, and the resulting differential equation

$$\frac{d^2\theta}{dt^2} + \frac{g}{l}\theta = 0 \tag{7}$$

indicates that the pendulum exhibits simple harmonic motion. Inspection of the solution (7) reveals that the period of small oscillations is given by the familiar formula from physics $T = 2\pi\sqrt{l/g}$.

5.4 EXERCISES

Answers to odd-numbered problems begin on page A–13.

In Problems 1 and 2 find the charge on the capacitor and the current in the given L-C series circuit. Assume $q(0) = 0$ and $i(0) = 0$.

1. $L = 1$ henry, $C = \dfrac{1}{16}$ farad, $E(t) = 60$ volts

2. $L = 5$ henrys, $C = 0.01$ farad, $E(t) = 20t$ volts

In Problems 3 and 4 without solving (3) determine whether the given L-R-C series circuit is overdamped, critically damped, or underdamped.

3. $L = 3$ henrys, $R = 10$ ohms, $C = 0.1$ farad

4. $L = 1$ henry, $R = 20$ ohms, $C = 0.01$ farad

5. Find the charge on the capacitor in an L-R-C series circuit at $t = 0.01$ second when $L = 0.05$ henry, $R = 2$ ohms, $C = 0.01$ farad, $E(t) = 0$ volts, $q(0) = 5$ coulombs, and $i(0) = 0$ amperes. Determine the first time at which the charge on the capacitor is equal to zero.

6. Find the charge on the capacitor in an L-R-C series circuit when $L = 1/4$ henry, $R = 20$ ohms, $C = 1/300$ farad, $E(t) = 0$ volts, $q(0) = 4$ coulombs, and $i(0) = 0$ amperes. Is the charge on the capacitor ever equal to zero?

In Problems 7 and 8 find the charge on the capacitor and the current in the given L-R-C series circuit. Find the maximum charge on the capacitor.

7. $L = 5/3$ henrys, $R = 10$ ohms, $C = 1/30$ farad, $E(t) = 300$ volts, $q(0) = 0$ coulombs, $i(0) = 0$ amperes.

8. $L = 1$ henry, $R = 100$ ohms, $C = 0.0004$ farad, $E(t) = 30$ volts, $q(0) = 0$ coulombs, $i(0) = 2$ amperes.

9. Find the steady-state charge and the steady-state current in an L-R-C series circuit when $L = 1$ henry, $R = 2$ ohms, $C = 0.25$ farad, and $E(t) = 50 \cos t$ volts.

10. Show that the amplitude of the steady-state current in the L-R-C series circuit in Example 3 is given by E_0/Z, where Z is the impedance of the circuit.

11. Show that the steady-state current in an L-R-C series circuit when $L = 1/2$ henry, $R = 20$ ohms, $C = 0.001$ farad, and $E(t) = 100 \sin 60t$ volts is given by $i_p(t) = (4.160) \sin (60t - 0.588)$. [*Hint:* Use Problem 10.]

12. Find the steady-state current in an L-R-C series circuit when $L = 1/2$ henry, $R = 20$ ohms, $C = 0.001$ farad, and $E(t) = 100 \sin 60t + 200 \cos 40t$ volts.

13. Find the charge on the capacitor in an L-R-C series circuit when $L = 1/2$ henry, $R = 10$ ohms, $C = 0.01$ farad, $E(t) = 150$ volts, $q(0) = 1$ coulomb, and $i(0) = 0$ amperes. What is the charge on the capacitor after a long time?

14. Show that if L, R, C, and E_0 are constant, then the amplitude of the steady-state current in Example 3 is a maximum when $\gamma = 1/\sqrt{LC}$. What is the maximum amplitude?

15. Show that if L, R, E_0, and γ are constant, then the amplitude of the steady-state current in Example 3 is a maximum when the capacitance is $C = 1/L\gamma^2$.

16. Find the charge on the capacitor and the current in an L-C circuit when $L = 0.1$ henry, $C = 0.1$ farad, $E(t) = 100 \sin \gamma t$ volts, $q(0) = 0$ coulombs, and $i(0) = 0$ amperes.

17. Find the charge on the capacitor and the current in an L-C circuit when $E(t) = E_0 \cos \gamma t$ volts, $q(0) = q_0$ coulombs, and $i(0) = i_0$ amperes.

18. In Problem 17 find the current when the circuit is in resonance.

19. Find the equation of motion describing the small displacements $\theta(t)$ of a simple pendulum of length 2 ft released at $t = 0$ with a displacement of $1/2$ radian to the right of the vertical and angular velocity of $2\sqrt{3}$ ft/s to the right. What are the amplitude, period, and frequency of motion?

20. In Problem 19 at what times does the pendulum pass through its equilibrium position? At what times does the pendulum attain its extreme angular displacements on either side of its equilibrium position?

Chapter 5   *REVIEW*

When a mass is attached to a spring, it stretches to a position where the restoring force ks of the spring is balanced by the weight mg. Any subsequent motion is then measured x units (feet in the engineering system) above or below this **equilibrium position**. When the mass is above the equilibrium position, we adopt the convention that $x < 0$, whereas when the mass is below the equilibrium position, we take $x > 0$.

The differential equation of motion is obtained by equating Newton's second law $F = ma = m(d^2x/dt^2)$ with the net force acting on the mass at any time. We distinguish three cases.

CASE I   The equation

$$m\frac{d^2x}{dt^2} = -kx \quad \text{or} \quad \frac{d^2x}{dt^2} + \omega^2 x = 0, \tag{1}$$

where $\omega^2 = k/m$, describes the motion under the assumptions that no damping force and no external impressed forces are acting on the system. The solution of (1) is $x(t) = c_1 \cos \omega t + c_2 \sin \omega t$ and the mass is said to exhibit **simple harmonic motion**. The constants c_1 and c_2 are determined by the initial position $x(0)$ and the initial velocity $x'(0)$ of the mass.

CASE II   When a damping force is present, the differential equation becomes

$$m\frac{d^2x}{dt^2} = -kx - \beta\frac{dx}{dt} \quad \text{or} \quad \frac{d^2x}{dt^2} + 2\lambda\frac{dx}{dt} + \omega^2 x = 0, \tag{2}$$

where $\beta > 0$, $2\lambda = \beta/m$, and $\omega^2 = k/m$. The resulting motion is said to be **overdamped**, **critically damped**, or **underdamped** accordingly as $\lambda^2 - \omega^2 > 0$, $\lambda^2 - \omega^2 = 0$, or $\lambda^2 - \omega^2 < 0$. The respective solutions of (2) are then

$$x(t) = c_1 e^{m_1 t} + c_2 e^{m_2 t},$$

where $m_1 = -\lambda + \sqrt{\lambda^2 - \omega^2}$, $m_2 = -\lambda - \sqrt{\lambda^2 - \omega^2}$;

$$x(t) = c_1 e^{m_1 t} + c_2 t e^{m_1 t},$$

where $m_1 = -\lambda$; and

$$x(t) = e^{-\lambda t}(c_1 \cos \sqrt{\omega^2 - \lambda^2}\, t + c_2 \sin \sqrt{\omega^2 - \lambda^2}\, t).$$

In each case the damping force is responsible for the displacements becoming negligible for large time; that is, $x \to 0$ as $t \to \infty$.

The motion described in Cases I and II is said to be **free motion**.

CASE III When an external force is impressed on the system for $t > 0$, the differential equation becomes

$$m\frac{d^2x}{dt^2} = -kx - \beta\frac{dx}{dt} + f(t) \quad \text{or} \quad \frac{d^2x}{dt^2} + 2\lambda\frac{dx}{dt} + \omega^2 x = F(t), \quad \textbf{(3)}$$

where λ and ω^2 are defined in Case II. The solution of the nonhomogeneous equation (3) is $x(t) = x_c + x_p$.

Since the complementary function x_c always contains the factor $e^{-\lambda t}$, it will be **transient**; that is, $x_c \to 0$ as $t \to \infty$. If $F(t)$ is periodic, then x_p will be a **steady-state solution**.

In the absence of a damping force, an impressed periodic force can cause the amplitudes of vibration to become very large. If the frequency of the external force is the same as the frequency $\omega/2\pi$ of free vibrations, we say that the system is in a state of **pure resonance**. In this case the amplitudes of vibrations become unbounded as $t \to \infty$. In the presence of a damping force, amplitudes of oscillatory motion are always bounded. However, large and potentially destructive amplitudes can occur.

When a series circuit containing an inductor, resistor, and capacitor is driven by an electromotive force $E(t)$, the resulting differential equations for the charge $q(t)$ or the current $i(t)$ are quite similar to equation (3). Hence the analysis of such circuits is the same as outlined above.

Chapter 5 *REVIEW EXERCISES*

Answers to odd-numbered problems begin on page A–14.

Answer Problems 1–9 without referring back to the text. Fill in the blank or answer true/false.

1. If a 10-lb weight stretches a spring 2.5 ft, a 32-lb weight will stretch it _____ ft.

2. The period of simple harmonic motion of an 8-lb weight attached to a spring whose constant is 6.25 lb/ft is _____ seconds.

3. The differential equation of a weight on a spring is $x'' + 16x = 0$. If the weight is released at $t = 0$ from 1 m above the equilibrium position with a downward velocity of 3 m/s, the amplitude of vibrations is _____ m.

4. Pure resonance cannot take place in the presence of a damping force. _____

5. In the presence of damping, the displacements of a weight on a spring will always approach zero as $t \to \infty$. _____

6. A weight on a spring whose motion is critically damped can possibly pass through the equilibrium position twice. _____

7. At critical damping any increase in damping will result in an _____ system.

8. When simple harmonic motion is described by $x = (\sqrt{2}/2) \sin(2t + \phi)$, the phase angle ϕ is _____ when $x(0) = -1/2$ and $x'(0) = 1$.

9. A 16-lb weight attached to a spring exhibits simple harmonic motion. If the frequency of oscillations is $3/2\pi$ vibrations/second, the spring constant is _____.

10. A 12-lb weight stretches a spring 2 ft. The weight is released from a point 1 ft below the equilibrium position with an upward velocity of 4 ft/s.

    (a) Find the equation describing the resulting simple harmonic motion.

    (b) What are the amplitude, period, and frequency of motion?

    (c) At what times does the weight return to the point 1 ft below the equilibrium position?

    (d) At what times does the weight pass through the equilibrium position moving upward? moving downward?

    (e) What is the velocity of the weight at $t = 3\pi/16$ second?

    (f) At what times is the velocity zero?

11. A force of 2 lb stretches a spring 1 ft. With one end held fixed, an 8-lb weight is attached to the other end and the system lies on a table that imparts a frictional force numerically equal to $3/2$ times the instantaneous velocity. Initially the weight is displaced 4 in. above the equilibrium position and released from rest. Find the equation of motion if the motion takes place along a horizontal straight line that is taken as the x-axis.

12. A 32-lb weight stretches a spring 6 in. The weight moves through a medium offering a damping force numerically equal to β times the instantaneous velocity. Determine the values of β for which the system will exhibit oscillatory motion.

13. A spring with constant $k = 2$ is suspended in a liquid that offers a damping force numerically equal to 4 times the instantaneous velocity. If a mass m is suspended from the spring, determine the values of m for which the subsequent free motion is nonoscillatory.

14. The vertical motion of a weight attached to a spring is described by the initial-value problem

$$\frac{1}{4}\frac{d^2x}{dt^2} + \frac{dx}{dt} + x = 0, \qquad x(0) = 4, \quad x'(0) = 2.$$

Determine the maximum vertical displacement.

15. A 4-lb weight stretches a spring 18 in. A periodic force equal to $f(t) = \cos \gamma t + \sin \gamma t$ is impressed on the system starting at $t = 0$. In the absence of a damping force, for what value of γ will the system be in a state of pure resonance?

16. Find a particular solution for $\dfrac{d^2x}{dt^2} + 2\lambda \dfrac{dx}{dt} + \omega^2 x = A$, where A is a constant force.

17. A 4-lb weight is suspended from a spring whose constant is 3 lb/ft. The entire system is immersed in a fluid offering a damping force numerically equal to the instantaneous velocity. Beginning at $t = 0$, an external force equal to $f(t) = e^{-t}$ is impressed on the system. Determine the equation of motion if the weight is released from rest at a point 2 ft below the equilibrium position.

18. (a) Two springs are attached in series as shown in Figure 5.30. If the mass of each spring is ignored, show that the effective spring constant k is given by $1/k = 1/k_1 + 1/k_2$.

(b) A weight of W lb stretches one spring $1/2$ ft and stretches a different spring $1/4$ ft. The two springs are attached as in the figure and the weight W is then attached to the double spring. Assume that the motion is free and that there is no damping force present. Determine the equation of motion if the weight is released at a point 1 ft below the equilibrium position with a downward velocity of $2/3$ ft/s.

(c) Show that the maximum speed of the weight is $(2/3)\sqrt{3g + 1}$.

19. A series circuit contains an inductance of $L = 1$ henry, a capacitance of $C = 10^{-4}$ farad, and an electromotive force of $E(t) = 100 \sin 50t$ volts. Initially the charge q and current i are zero.

(a) Find the equation for the charge at any time.

(b) Find the equation for the current at any time.

(c) Find the times for which the charge on the capacitor is zero.

20. Show that the current $i(t)$ in an L-R-C series circuit satisfies the differential equation

$$L\frac{d^2i}{dt^2} + R\frac{di}{dt} + \frac{1}{C}i = E'(t),$$

where $E'(t)$ denotes the derivative of $E(t)$.

Figure 5.30

TACOMA NARROWS SUSPENSION BRIDGE COLLAPSE

Gilbert N. Lewis
Department of Mathematical and Computer Sciences
Michigan Technological University

As can be seen from equation (9) and Figure 5.22 of Section 5.3, the oscillations in the case of resonance become very large as time increases. In a physical system this would be catastrophic, since the ever-increasing amplitudes of oscillation would tear the system apart. Two historical examples of this have already been given (airplane wings and soldiers marching across bridges). In those examples, periodic forces of the same frequency as the natural frequency of the structures were set up in the direction of the vibrations, leading to the resonant destruction of the structures.

Another example that has, until very recently, been used to illustrate the phenomenon of resonance is the collapse of the Tacoma Narrows Bridge in the state of Washington. The bridge was completed and open to traffic in the summer of 1940. It was soon noticed that large vibrations of the roadbed were induced when wind blew across it. "Galloping Gertie," as the bridge was called, became a tourist attraction; people liked to watch it vibrate and even take an exciting roller coaster ride across it. Finally, on November 7, 1940, the entire span was shaken apart by the large vibrations and collapsed. See Figure 5.31.

Figure 5.31

roadway

wind

vertical forces

Figure 5.32

For fifty years, the popular and easily explained reason for the collapse was resonance. It was thought that, as the wind blew horizontally across the roadway, vortices (wind swirls) were shed from the downwind side of the bridge alternately from above and below, thus setting up a periodic vertical force acting in the same direction as the vibration of the bridge. See Figure 5.32. It was further hypothesized that the frequency of this periodic force exactly matched the natural frequency of the bridge, thus setting up large amplitude vibrations (see equation (9)) and causing the bridge to fail. This explanation was (perhaps erroneously) attributed to the noted engineer von Karman. In his autobiography, he explained that the bridge collapse was indeed due to the von Karman vortices [4]. However, in a technical report to the Federal Works Agency, he and his coauthors concluded "it is very improbable that resonance with alternating vortices plays an important role in the oscillations of suspension bridges" [1]. Unfortunately, the resonance explanation has remained firmly entrenched in the popular and mathematical literature.

Resonance is a linear phenomenon (notice the linear differential equation (7) of Section 5.3). In addition, it is entirely dependent on an exact match between the natural frequency of the bridge (or portions thereof) and the frequency of any externally applied periodic forces. Furthermore, it requires that absolutely no damping is present in the system ($\lambda = 0$ in equation (3) of Section 5.3). It should not be surprising, therefore, that resonance was *not* the dominant factor in the collapse of the Tacoma Narrows Bridge.

If not resonance, then what? Recent research has provided an alternative explanation for the collapse of the bridge. Lazer and McKenna (see [3] for a good review article) contend that nonlinear effects, and not linear resonance, were the main factors leading to the large oscillations of the bridge [2]. There is no doubt that the wind blowing across the roadbed provided the external driving force that caused the motion. This force could even have been partly due to the shedding of vortices, as von Karman suggested. However, nonlinear interactions within the bridge and between the bridge and the external forces are a more probable culprit for the cause of the collapse.

In the linear theory, the supporting cables act as rigid elastic rods (springs). As such, the mathematical model gives rise to the linear differential equation (3), or equation (7) of Section 5.3, if no damping is present. This latter case leading to resonance is the only possible situation in the linear theory in which small external forces could lead to large amplitude vibrations. As has been mentioned, this scenario is highly unlikely.

On the other hand, nonlinear effects could yield large amplitude vibrations from small amplitude forces. They could also explain the transition from one-dimensional axial (lengthwise) oscillations to transverse (across the roadway) oscillations, which were ultimately responsible for the collapse of the bridge.

The basic idea in the Lazer–McKenna model is the following. When the vertical supporting cables are under tension (the weight of the roadbed is pulling down on them), the cables act as rigid elastic linear rods, in which case the differential equation is linear. However, once oscillations are set up in the bridge system by external forces (wind and possibly earthquakes), the cables will not always be under tension, and there will be only gravity acting downward on the roadbed. In other words, the Hooke's law term, $(k/m)x$ in equation (2) of Section 5.3, will be missing. This transition from one type of linear differential equation to another is one source of the nonlinearity. Other sources of nonlinearity in suspension bridges might include the nonlinear and nonsymmetric design of the bridge or interactions of the supporting cables with the main cables or supporting towers. This nonlinearity is compounded by the fact that different cables in the bridge system may be under tension at different times. The net result of this nonlinearity, contend Lazer and McKenna, may be large amplitude oscillations under moderate external forces.

One other aspect of nonlinear equations is their unpredictability. This might explain, for example, why the bridge was observed to undergo large amplitude oscillations under light wind conditions yet seemed to be perfectly stable under higher wind conditions.

As Lazer and McKenna point out, the nonlinear theory of suspension bridges has not been thoroughly developed yet. However, numerical simulations of their suspension bridge model agree with actual observations. It seems likely that this approach will yield a more accurate explanation of the failure of the Tacoma Narrows Suspension Bridge.

REFERENCES

1. Amann, O. H., T. von Karman, and G. B. Woodruff. *The Failure of the Tacoma Narrows Bridge.* Federal Works Agency, 1941.

2. Lazer, A. C., and P. J. McKenna. "Large-Amplitude Periodic Oscillations in Suspension Bridges: Some New Connections with Nonlinear Analysis." *SIAM Review* 32 (Dec. 1990): 537–578.

3. Peterson, I. "Rock and Roll Bridge." *Science News* 137 (1991): 344–346.

4. von Karman, T. *The Wind and Beyond, Theodore von Karman, Pioneer in Aviation and Pathfinder in Space.* Boston: Little, Brown, 1967.

C H A P T E R 6

DIFFERENTIAL EQUATIONS WITH VARIABLE COEFFICIENTS

6.1 Cauchy-Euler Equation
6.2 Review of Power Series; Power Series Solutions
6.3 Solutions About Ordinary Points

6.4 Solutions About Singular Points
6.5 Two Special Equations

Chapter 6 Review
Chapter 6 Review Exercises

Important Concepts

Cauchy-Euler equation
Auxiliary equation
Power series
Power series solutions
Recurrence relation
Ordinary point
Singular point
Regular singular point
Irregular singular point
Method of Frobenius
Indicial equation
Indicial roots
Bessel's equation
Parametric Bessel equation
Bessel functions
Legendre's equation
Legendre polynomials

Up to now we have solved linear differential equations of order two or higher only in the case when the equation has constant coefficients. In applications, higher-order linear equations with nonconstant coefficients are just as important, if not more so, as equations with constant coefficients. For example, if we wish to find the temperature or potential u in the region bounded between two concentric spheres as shown in the accompanying figure, then under some circumstances we have to solve the differential equation

$$r \frac{d^2u}{dr^2} + 2 \frac{du}{dr} = 0,$$

where the variable $r > 0$ represents the radial distance measured outward from the center of the spheres.

Continues

Continued

Differential equations with variable coefficients such as

$$x^2 y'' + xy' + (x^2 - v^2)y = 0,$$

$$(1 - x^2)y'' - 2xy' + n(n + 1)y = 0,$$

and $\qquad\qquad y'' - 2xy' + 2ny = 0$

occur in applications ranging from potential problems, temperature distributions, and vibrational phenomena, to quantum mechanics.

The same ease with which we solved constant-coefficient differential equations does not carry over to equations with variable coefficients. In fact we cannot expect to be able to express the solution of even a simple linear equation such as $y'' - 2xy = 0$ in terms of the familiar elementary functions constructed from powers of x, sines, cosines, logarithms, and exponentials. The best we can do for the equation $y'' - 2xy = 0$ is to find a solution in the form of an infinite series. However, there is one type of variable-coefficient differential equation, called the Cauchy-Euler equation, whose general solution can always be written in terms of elementary functions. We begin the chapter with this equation.

6.1 CAUCHY-EULER EQUATION

Any linear differential equation of the form

$$a_n x^n \frac{d^n y}{dx^n} + a_{n-1} x^{n-1} \frac{d^{n-1} y}{dx^{n-1}} + \cdots + a_1 x \frac{dy}{dx} + a_0 y = g(x),$$

where $a_n, a_{n-1}, \ldots, a_0$ are constants, is said to be a **Cauchy-Euler equation,*** or **equidimensional equation.** The distinguishing characteristic of this type of differential equation is that the *degree* of each monomial coefficient matches the *order* of differentiation:

$$a_n \overset{\text{same}}{x^n} \frac{d^n y}{dx^n} + a_{n-1} \overset{\text{same}}{x^{n-1}} \frac{d^{n-1} y}{dx^{n-1}} + \cdots.$$

For the sake of discussion, we confine our attention to solving the homogeneous second-order equation

$$ax^2 \frac{d^2y}{dx^2} + bx \frac{dy}{dx} + cy = 0.$$

The solution of higher-order equations follows analogously. Also, we can solve the nonhomogeneous equation

$$ax^2 \frac{d^2y}{dx^2} + bx \frac{dy}{dx} + cy = g(x)$$

by variation of parameters once we have determined the complementary function $y_c(x)$.

Note The coefficient of d^2y/dx^2 is zero at $x = 0$. Hence in order to guarantee that the fundamental results of Theorem 4.1 are applicable to the Cauchy-Euler equation, we shall confine our attention to finding the general solution on the interval $(0, \infty)$. Solutions on the interval $(-\infty, 0)$ can be obtained by substituting $t = -x$ in the differential equation.

Method of Solution

We try a solution of the form $y = x^m$, where m is to be determined. The first and second derivatives are, respectively,

$$\frac{dy}{dx} = mx^{m-1} \quad \text{and} \quad \frac{d^2y}{dx^2} = m(m-1)x^{m-2}.$$

Consequently the differential equation becomes

$$ax^2 \frac{d^2y}{dx^2} + bx \frac{dy}{dx} + cy = ax^2 \cdot m(m-1)x^{m-2} + bx \cdot mx^{m-1} + cx^m$$

$$= am(m-1)x^m + bmx^m + cx^m$$

$$= x^m(am(m-1) + bm + c).$$

Thus $y = x^m$ is a solution of the differential equation whenever m is a solution of the **auxiliary equation**

$$am(m-1) + bm + c = 0 \quad \text{or} \quad am^2 + (b-a)m + c = 0. \tag{1}$$

---

*Augustin-Louis Cauchy (1789–1857)   Born during a period of upheaval in French history, Augustin-Louis Cauchy was destined to initiate a revolution of his own—in mathematics. For many original contributions, but especially for his efforts in clarifying mathematical obscurities and his incessant demand for satisfactory definitions and rigorous proofs of theorems, Cauchy is often called "the father of modern analysis." A prolific writer whose output was surpassed by only a few, Cauchy produced nearly 800 papers in astronomy, physics, and mathematics. It was he who developed the concept of convergence of an infinite series and the theory of functions of a complex variable. The same mind that was always open and inquiring in science and mathematics was narrow and unquestioning in many other areas. Outspoken and arrogant, Cauchy was passionate on political and religious issues. His stands on these issues often alienated him from his colleagues.

There are three different cases to be considered, depending on whether the roots of this quadratic equation are real and distinct, real and equal, or complex. In the last case, the roots appear as a conjugate pair.

CASE I DISTINCT REAL ROOTS   Let m_1 and m_2 denote the real roots of (1) such that $m_1 \neq m_2$. Then

$$y_1 = x^{m_1} \quad \text{and} \quad y_2 = x^{m_2}$$

form a fundamental set of solutions. Hence the general solution is

$$y = c_1 x^{m_1} + c_2 x^{m_2}. \tag{2}$$

EXAMPLE I

Solve
$$x^2 \frac{d^2 y}{dx^2} - 2x \frac{dy}{dx} - 4y = 0.$$

Solution Rather than just memorizing equation (1), it is preferable to assume $y = x^m$ as the solution a few times in order to understand the origin and the difference between this new form of the auxiliary equation and that obtained in Chapter 4. Differentiate twice

$$\frac{dy}{dx} = mx^{m-1}, \qquad \frac{d^2 y}{dx^2} = m(m-1)x^{m-2}$$

and substitute back into the differential equation:

$$x^2 \frac{d^2 y}{dx^2} - 2x \frac{dy}{dx} - 4y = x^2 \cdot m(m-1)x^{m-2} - 2x \cdot mx^{m-1} - 4x^m$$

$$= x^m(m(m-1) - 2m - 4)$$

$$= x^m(m^2 - 3m - 4) = 0$$

if $m^2 - 3m - 4 = 0$. Now $(m+1)(m-4) = 0$ implies $m_1 = -1$, $m_2 = 4$

so that
$$y = c_1 x^{-1} + c_2 x^4. \qquad \blacksquare$$

CASE II REPEATED REAL ROOTS   If the roots of (1) are repeated, that is, $m_1 = m_2$, then we obtain only one solution, namely, $y = x^{m_1}$. When the roots of the quadratic equation $am^2 + (b-a)m + c = 0$ are equal, the discriminant of the coefficients is necessarily zero. It follows from the quadratic formula that the root must be $m_1 = -(b-a)/2a$.

Now we can construct a second solution y_2, using (4) of Section 4.2. We first write the Cauchy-Euler equation in the form

$$\frac{d^2 y}{dx^2} + \frac{b}{ax} \frac{dy}{dx} + \frac{c}{ax^2} y = 0$$

and make the identification $P(x) = b/ax$. Thus

$$y_2 = x^{m_1} \int \frac{e^{-\int (b/ax)\,dx}}{(x^{m_1})^2}\,dx$$

$$= x^{m_1} \int \frac{e^{-(b/a)\ln x}}{x^{2m_1}}\,dx$$

$$= x^{m_1} \int x^{-b/a} \cdot x^{-2m_1}\,dx \qquad \Longleftarrow \boxed{e^{-(b/a)\ln x} = e^{\ln x^{-b/a}} = x^{-b/a}}$$

$$= x^{m_1} \int x^{-b/a} \cdot x^{(b-a)/a}\,dx \qquad \Longleftarrow \boxed{-2m_1 = (b-a)/a}$$

$$= x^{m_1} \int \frac{dx}{x} = x^{m_1} \ln x.$$

The general solution is then

$$y = c_1 x^{m_1} + c_2 x^{m_1} \ln x. \tag{3}$$

EXAMPLE 2

Solve
$$4x^2 \frac{d^2 y}{dx^2} + 8x \frac{dy}{dx} + y = 0.$$

Solution The substitution $y = x^m$ yields

$$4x^2 \frac{d^2 y}{dx^2} + 8x \frac{dy}{dx} + y = x^m(4m(m-1) + 8m + 1) = x^m(4m^2 + 4m + 1) = 0$$

when $4m^2 + 4m + 1 = 0$ or $(2m + 1)^2 = 0$. Since $m_1 = -1/2$, the general solution is

$$y = c_1 x^{-1/2} + c_2 x^{-1/2} \ln x. \qquad \blacksquare$$

For higher-order equations, if m_1 is a root of multiplicity k, then it can be shown that

$$x^{m_1}, \quad x^{m_1} \ln x, \quad x^{m_1}(\ln x)^2, \quad \ldots, \quad x^{m_1}(\ln x)^{k-1}$$

are k linearly independent solutions. Correspondingly, the general solution of the differential equation must then contain a linear combination of these k solutions.

CASE III CONJUGATE COMPLEX ROOTS   If the roots of (1) are the conjugate pair

$$m_1 = \alpha + i\beta, \qquad m_2 = \alpha - i\beta,$$

where α and $\beta > 0$ are real, then a solution is

$$y = C_1 x^{\alpha + i\beta} + C_2 x^{\alpha - i\beta}.$$

But, as in the case of equations with constant coefficients, when the roots of the auxiliary equation are complex, we wish to write the solution in terms of real functions only. We note the identity

$$x^{i\beta} = (e^{\ln x})^{i\beta} = e^{i\beta \ln x},$$

which, by Euler's formula, is the same as

$$x^{i\beta} = \cos(\beta \ln x) + i \sin(\beta \ln x).$$

Similarly we have

$$x^{-i\beta} = \cos(\beta \ln x) - i \sin(\beta \ln x).$$

Adding and subtracting the last two results yield, respectively,

$$x^{i\beta} + x^{-i\beta} = 2 \cos(\beta \ln x)$$

and
$$x^{i\beta} - x^{-i\beta} = 2i \sin(\beta \ln x).$$

From the fact that $y = C_1 x^{\alpha + i\beta} + C_2 x^{\alpha - i\beta}$ is a solution of $ax^2 y'' + bxy' + cy = 0$ for any values of the constants C_1 and C_2 we see that

$$y_1 = x^{\alpha}(x^{i\beta} + x^{-i\beta}), \qquad (C_1 = C_2 = 1)$$

$$y_2 = x^{\alpha}(x^{i\beta} - x^{-i\beta}), \qquad (C_1 = 1, C_2 = -1)$$

or
$$y_1 = 2x^{\alpha}(\cos(\beta \ln x))$$

$$y_2 = 2ix^{\alpha}(\sin(\beta \ln x))$$

are also solutions. Since $W(x^{\alpha} \cos(\beta \ln x), x^{\alpha} \sin(\beta \ln x)) = \beta x^{2\alpha - 1} \neq 0$, $\beta > 0$, on the interval $(0, \infty)$, we conclude that

$$y_1 = x^{\alpha} \cos(\beta \ln x) \quad \text{and} \quad y_2 = x^{\alpha} \sin(\beta \ln x)$$

constitute a fundamental set of real solutions of the differential equation. Hence, the general solution is

$$y = x^{\alpha}[c_1 \cos(\beta \ln x) + c_2 \sin(\beta \ln x)]. \tag{4}$$

EXAMPLE 3

Solve the initial-value problem

$$x^2 \frac{d^2 y}{dx^2} + 3x \frac{dy}{dx} + 3y = 0, \qquad y(1) = 1, \quad y'(1) = -5.$$

Solution We have

$$x^2 \frac{d^2y}{dx^2} + 3x \frac{dy}{dx} + 3y = x^m(m(m-1) + 3m + 3) = x^m(m^2 + 2m + 3) = 0$$

when $m^2 + 2m + 3 = 0$. From the quadratic formula we find $m_1 = -1 + \sqrt{2}\,i$ and $m_2 = -1 - \sqrt{2}\,i$. If we make the identifications $\alpha = -1$ and $\beta = \sqrt{2}$, we see from (4) that the general solution of the differential equation is

$$y = x^{-1}[c_1 \cos(\sqrt{2}\ \ln x) + c_2 \sin(\sqrt{2}\ \ln x)].$$

By applying the conditions $y(1) = 1$, $y'(1) = -5$ to the foregoing solution, we find, in turn, $c_1 = 1$ and $c_2 = -2\sqrt{2}$. Thus the solution to the initial-value problem is

$$y = x^{-1}[\cos(\sqrt{2}\ \ln x) - 2\sqrt{2}\ \sin(\sqrt{2}\ \ln x)].$$

The graph of this solution, obtained with the aid of computer software, is given in Figure 6.1. ∎

Figure 6.1

EXAMPLE 4

Solve the third-order Cauchy-Euler equation

$$x^3 \frac{d^3y}{dx^3} + 5x^2 \frac{d^2y}{dx^2} + 7x \frac{dy}{dx} + 8y = 0.$$

Solution The first three derivatives of $y = x^m$ are

$$\frac{dy}{dx} = mx^{m-1}, \qquad \frac{d^2y}{dx^2} = m(m-1)x^{m-2}, \qquad \frac{d^3y}{dx^3} = m(m-1)(m-2)x^{m-3},$$

so that the given differential equation becomes

$$x^3 \frac{d^3y}{dx^3} + 5x^2 \frac{d^2y}{dx^2} + 7x \frac{dy}{dx} + 8y = x^3m(m-1)(m-2)x^{m-3} + 5x^2m(m-1)x^{m-2} + 7xmx^{m-1} + 8x^m$$

$$= x^m(m(m-1)(m-2) + 5m(m-1) + 7m + 8)$$

$$= x^m(m^3 + 2m^2 + 4m + 8).$$

In this case we see that $y = x^m$ will be a solution of the differential equation, provided m is a root of the cubic equation

$$m^3 + 2m^2 + 4m + 8 = 0 \quad \text{or} \quad (m+2)(m^2 + 4) = 0.$$

The roots are: $m_1 = -2$, $m_2 = 2i$, $m_3 = -2i$. Hence the general solution is

$$y = c_1 x^{-2} + c_2 \cos(2 \ln x) + c_3 \sin(2 \ln x).$$ ∎

EXAMPLE 5

Solve the nonhomogeneous equation

$$x^2 y'' - 3xy' + 3y = 2x^4 e^x.$$

Solution The substitution $y = x^m$ leads to the auxiliary equation

$$m(m - 1) - 3m + 3 = 0 \quad \text{or} \quad (m - 1)(m - 3) = 0.$$

Thus

$$y_c = c_1 x + c_2 x^3.$$

Before using variation of parameters to find a particular solution $y_p = u_1 y_1 + u_2 y_2$, recall that the formulas $u_1' = W_1/W$ and $u_2' = W_2/W$, where

$$W_1 = \begin{vmatrix} 0 & y_2 \\ f(x) & y_2' \end{vmatrix}, \quad W_2 = \begin{vmatrix} y_1 & 0 \\ y_1' & f(x) \end{vmatrix}, \quad \text{and } W \text{ is the Wronskian of } y_1 \text{ and } y_2,$$

were derived under the assumption that the differential equation has been put into the special form $y'' + P(x)y' + Q(x)y = f(x)$. Therefore we divide the given equation by x^2, and from

$$y'' - \frac{3}{x} y' + \frac{3}{x^2} y = 2x^2 e^x$$

we make the identification $f(x) = 2x^2 e^x$. Now with $y_1 = x$, $y_2 = x^3$, and

$$W = \begin{vmatrix} x & x^3 \\ 1 & 3x^2 \end{vmatrix} = 2x^3, \quad W_1 = \begin{vmatrix} 0 & x^3 \\ 2x^2 e^x & 3x^2 \end{vmatrix} = -2x^5 e^x, \quad W_2 = \begin{vmatrix} x & 0 \\ 1 & 2x^2 e^x \end{vmatrix} = 2x^3 e^x$$

we find

$$u_1' = -\frac{2x^5 e^x}{2x^3} = -x^2 e^x \quad \text{and} \quad u_2' = \frac{2x^3 e^x}{2x^3} = e^x.$$

The integral of the latter function is immediate, but in the case of u_1' we integrate by parts twice. The results are

$$u_1 = -x^2 e^x + 2x e^x - 2e^x \quad \text{and} \quad u_2 = e^x.$$

Hence

$$y_p = u_1 y_1 + u_2 y_2$$
$$= (-x^2 e^x + 2x e^x - 2e^x)x + e^x x^3 = 2x^2 e^x - 2x e^x.$$

Finally we have

$$y = y_c + y_p = c_1 x + c_2 x^3 + 2x^2 e^x - 2x e^x.$$

Alternative Method of Solution

Any Cauchy-Euler differential equation can be reduced to an equation with constant coefficients by means of the substitution $x = e^t$. The next example illustrates this method.

EXAMPLE 6

Solve
$$x^2 \frac{d^2y}{dx^2} - x \frac{dy}{dx} + y = \ln x.$$

Solution With the substitution $x = e^t$ or $t = \ln x$, it follows that

$$\frac{dy}{dx} = \frac{dy}{dt}\frac{dt}{dx} = \frac{1}{x}\frac{dy}{dt} \qquad \Longleftarrow \boxed{\text{chain rule}}$$

$$\frac{d^2y}{dx^2} = \frac{1}{x}\frac{d}{dx}\left(\frac{dy}{dt}\right) + \frac{dy}{dt}\left(-\frac{1}{x^2}\right) \qquad \Longleftarrow \boxed{\text{product and chain rule}}$$

$$= \frac{1}{x}\left(\frac{d^2y}{dt^2}\frac{1}{x}\right) + \frac{dy}{dt}\left(-\frac{1}{x^2}\right) = \frac{1}{x^2}\left(\frac{d^2y}{dt^2} - \frac{dy}{dt}\right).$$

Substituting in the given differential equation and simplifying yield

$$\frac{d^2y}{dt^2} - 2\frac{dy}{dt} + y = t.$$

Since this last equation has constant coefficients, its auxiliary equation is $m^2 - 2m + 1 = 0$, or $(m-1)^2 = 0$. Thus we obtain $y_c = c_1 e^t + c_2 t e^t$.

By undetermined coefficients we try a particular solution of the form $y_p = A + Bt$. This assumption leads to $-2B + A + Bt = t$ so that $A = 2$ and $B = 1$. Using $y = y_c + y_p$, we get

$$y = c_1 e^t + c_2 t e^t + 2 + t;$$

and so the general solution of the original differential equation on the interval $(0, \infty)$ is

$$y = c_1 x + c_2 x \ln x + 2 + \ln x. \qquad \blacksquare$$

6.1 EXERCISES

Answers to odd-numbered problems begin on page A–14.

In Problems 1–22 solve the given differential equation.

1. $x^2 y'' - 2y = 0$ **2.** $4x^2 y'' + y = 0$

3. $xy'' + y' = 0$ **4.** $xy'' - y' = 0$

5. $x^2 y'' + xy' + 4y = 0$ **6.** $x^2 y'' + 5xy' + 3y = 0$

7. $x^2 y'' - 3xy' - 2y = 0$ **8.** $x^2 y'' + 3xy' - 4y = 0$

9. $25x^2 y'' + 25xy' + y = 0$ **10.** $4x^2 y'' + 4xy' - y = 0$

11. $x^2 y'' + 5xy' + 4y = 0$ **12.** $x^2 y'' + 8xy' + 6y = 0$

13. $x^2 y'' - xy' + 2y = 0$ **14.** $x^2 y'' - 7xy' + 41y = 0$

15. $3x^2y'' + 6xy' + y = 0$

16. $2x^2y'' + xy' + y = 0$

17. $x^3y''' - 6y = 0$

18. $x^3y''' + xy' - y = 0$

19. $x^3\dfrac{d^3y}{dx^3} - 2x^2\dfrac{d^2y}{dx^2} - 2x\dfrac{dy}{dx} + 8y = 0$

20. $x^3\dfrac{d^3y}{dx^3} - 2x^2\dfrac{d^2y}{dx^2} + 4x\dfrac{dy}{dx} - 4y = 0$

21. $x\dfrac{d^4y}{dx^4} + 6\dfrac{d^3y}{dx^3} = 0$

22. $x^4\dfrac{d^4y}{dx^4} + 6x^3\dfrac{d^3y}{dx^3} + 9x^2\dfrac{d^2y}{dx^2} + 3x\dfrac{dy}{dx} + y = 0$

In Problems 23–26 solve the given differential equation subject to the indicated initial conditions.

23. $x^2y'' + 3xy' = 0$, $y(1) = 0, y'(1) = 4$

24. $x^2y'' - 5xy' + 8y = 0$, $y(2) = 32, y'(2) = 0$

25. $x^2y'' + xy' + y = 0$, $y(1) = 1, y'(1) = 2$

26. $x^2y'' - 3xy' + 4y = 0$, $y(1) = 5, y'(1) = 3$

In Problems 27 and 28 solve the given differential equation subject to the indicated initial conditions. [*Hint:* Let $t = -x$.]

27. $4x^2y'' + y = 0$, $y(-1) = 2, y'(-1) = 4$

28. $x^2y'' - 4xy' + 6y = 0$, $y(-2) = 8, y'(-2) = 0$

Solve Problems 29–34 by variation of parameters.

29. $xy'' + y' = x$

30. $xy'' - 4y' = x^4$

31. $2x^2y'' + 5xy' + y = x^2 - x$

32. $x^2y'' - 2xy' + 2y = x^4e^x$

33. $x^2y'' - xy' + y = 2x$

34. $x^2y'' - 2xy' + 2y = x^3\ln x$

In Problems 35–40 solve the given differential equation by means of the substitution $x = e^t$.

35. $x^2\dfrac{d^2y}{dx^2} + 10x\dfrac{dy}{dx} + 8y = x^2$

36. $x^2y'' - 4xy' + 6y = \ln x^2$

37. $x^2y'' - 3xy' + 13y = 4 + 3x$

38. $2x^2y'' - 3xy' - 3y = 1 + 2x + x^2$

39. $x^2y'' + 9xy' - 20y = 5/x^3$

40. $x^3\dfrac{d^3y}{dx^3} - 3x^2\dfrac{d^2y}{dx^2} + 6x\dfrac{dy}{dx} - 6y = 3 + \ln x^3$

Figure 6.2

41. Consider two concentric spheres of radius $r = a$ and $r = b$, $a < b$, as shown in Figure 6.2. The temperature $u(r)$ in the region between the spheres is determined from the boundary-value problem

$$r\frac{d^2u}{dr^2} + 2\frac{du}{dr} = 0, \qquad u(a) = u_0, \quad u(b) = u_1,$$

where u_0 and u_1 are constant. Solve for $u(r)$.

42. The temperature $u(r)$ in the circular ring shown in Figure 6.3 is determined from the boundary-value problem

$$r\frac{d^2u}{dr^2} + \frac{du}{dr} = 0, \qquad u(a) = u_0, \quad u(b) = u_1,$$

where u_0 and u_1 are constant. Show that

$$u(r) = \frac{u_0 \ln(r/b) - u_1 \ln(r/a)}{\ln(a/b)}.$$

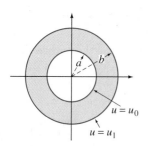

Figure 6.3

In Problems 43–45 solve the given differential equation.

43. $(x-1)^2\dfrac{d^2y}{dx^2} - 2(x-1)\dfrac{dy}{dx} - 4y = 0$ [*Hint:* Let $t = x - 1$.]

44. $(3x+4)^2y'' + 10(3x+4)y' + 9y = 0$

45. $(x+2)^2y'' + (x+2)y' + y = 0$

6.2 REVIEW OF POWER SERIES; POWER SERIES SOLUTIONS

Review of Power Series

The preceding section notwithstanding, most linear differential equations with variable coefficients cannot be solved in terms of elementary functions. A standard technique for solving higher-order linear differential equations with variable coefficients is to try to find a solution in the form of an infinite series. Often the solution can be found in the form of a power series. Because of this, it is appropriate to list some of the more important facts about power series. However, for an in-depth review of the infinite series concept you should consult a calculus text.

- **Definition of a Power Series**   A **power series** in $x - a$ is an infinite series of the form

$$\sum_{n=0}^{\infty} c_n(x-a)^n.$$

A series such as this is also said to be a **power series centered at a**. For example, $\displaystyle\sum_{n=1}^{\infty} \frac{(-1)^{n+1}}{n^2}x^n$ is a power series in x; the series is centered at zero.

- **Convergence** For a specified value of x a power series is a series of constants. If the series equals a finite real constant for the given x, then the series is said to **converge** at x. If the series does not converge at x, it is said to **diverge** at x.

- **Interval of Convergence** Every power series has an **interval of convergence**. The interval of convergence is the set of all numbers for which the series converges.

- **Radius of Convergence** Every interval of convergence has a **radius of convergence** R. For a power series $\sum_{n=0}^{\infty} c_n(x - a)^n$ we have just three possibilities:

  (i) The series converges only at its center a. In this case $R = 0$.

  (ii) The series converges for all x satisfying $|x - a| < R$, where $R > 0$. The series diverges for $|x - a| > R$.

  (iii) The series converges for all x. In this case we write $R = \infty$.

- **Convergence at an Endpoint** Recall, the absolute-value inequality $|x - a| < R$ is equivalent to $-R < x - a < R$, or $a - R < x < a + R$. If a power series converges for $|x - a| < R$, where $R > 0$, it may or may not converge at the endpoints of the interval $a - R < x < a + R$. Figure 6.4 shows four possible intervals of convergence.

(a) $[a - R, a + R]$
Series converges
at both endpoints.

(b) $(a - R, a + R)$
Series diverges
at both endpoints.

(c) $[a - R, a + R)$
Series converges at $a - R$,
diverges at $a + R$.

(d) $(a - R, a + R]$
Series diverges at $a - R$,
converges at $a + R$.

Figure 6.4

- **Absolute Convergence** Within its interval of convergence a power series converges absolutely. In other words, for x in the interval of convergence the series of absolute values $\sum_{n=0}^{\infty} |c_n||(x - a)^n|$ converges.

- **Finding the Interval of Convergence**   Convergence of a power series can often be determined by the **ratio test**:

$$\lim_{n \to \infty} \left| \frac{c_{n+1}}{c_n} \right| |x - a| = L.$$

The series will converge absolutely for those values of x for which $L < 1$. From this test we see that the radius of convergence is given by

$$R = \lim_{n \to \infty} \left| \frac{c_n}{c_{n+1}} \right| \tag{1}$$

provided the limit exists.

- **A Power Series Represents a Function**   A power series represents a function

$$f(x) = \sum_{n=0}^{\infty} c_n(x - a)^n = c_0 + c_1(x - a) + c_2(x - a)^2 + c_3(x - a)^3 + \cdots$$

whose domain is the interval of convergence of the series. If the series has a radius of convergence $R > 0$, then f is continuous, differentiable, and integrable on the interval $(a - R, a + R)$. Moreover, $f'(x)$ and $\int f(x)\, dx$ can be found from term-by-term differentiation and integration:

$$f'(x) = c_1 + 2c_2(x - a) + 3c_3(x - a)^2 + \cdots = \sum_{n=1}^{\infty} nc_n(x - a)^{n-1}$$

$$\int f(x)\, dx = C + c_0(x - a) + c_1 \frac{(x - a)^2}{2} + c_2 \frac{(x - a)^3}{3} + \cdots = C + \sum_{n=0}^{\infty} c_n \frac{(x - a)^{n+1}}{n + 1}.$$

Although the radius of convergence for both these series is R, the interval of convergence may differ from the original series in that convergence at an endpoint may be either lost by differentiation or gained through integration.

- **Series That Are Identically Zero**   If $\sum_{n=0}^{\infty} c_n(x - a)^n = 0$, $R > 0$, for all real numbers x in the interval of convergence, then $c_n = 0$ for all n.
- **Analytic at a Point**   In calculus it is seen that functions such as e^x, $\cos x$, and $\ln(x - 1)$ can be represented by power series by expansions in either Maclaurin or Taylor series. We say that a function f is **analytic at point a** if it can be represented by a power series in $x - a$ with a positive radius of convergence. The notion of analyticity at a point will be important in Sections 6.3 and 6.4.
- **Arithmetic of Power Series**   Power series can be combined through the operations of addition, multiplication, and division. The procedures for power series are similar to the way in which two polynomials are added, multiplied, and divided; that is, we add coefficients of like powers of x, use the distributive law and collect like terms, and perform long division. For example, if the power series $f(x) = \sum_{n=0}^{\infty} c_n x^n$

and $g(x) = \sum_{n=0}^{\infty} b_n x^n$ both converge for $|x| < R$, then

$$f(x) + g(x) = \sum_{n=0}^{\infty} (c_n + b_n)x^n$$

and

$$f(x)g(x) = c_0 b_0 + (c_0 b_1 + c_1 b_0)x + (c_0 b_2 + c_1 b_1 + c_2 b_0)x^2 + \cdots.$$

EXAMPLE I

Find the interval of convergence of the power series $\sum_{n=1}^{\infty} \dfrac{(x - 3)^n}{2^n n}$.

Solution The power series is centered at 3. From (1), the radius of convergence is

$$R = \lim_{n \to \infty} \frac{2^{n+1}(n + 1)}{2^n n} = 2.$$

The series converges absolutely for $|x - 3| < 2$, or $1 < x < 5$. At the left endpoint $x = 1$ we find that the series of constants $\sum_{n=1}^{\infty}((-1)^n/n)$ is convergent by the alternating series test. At the right endpoint $x = 5$ we find that the series is the divergent harmonic series $\sum_{n=1}^{\infty}(1/n)$. Thus the interval of convergence is $[1, 5)$. ∎

EXAMPLE 2

Find the first four terms of a power series in x for $e^x \cos x$.

Solution From calculus the Maclaurin series for e^x and $\cos x$ are, respectively,

$$e^x = 1 + x + \frac{x^2}{2} + \frac{x^3}{6} + \frac{x^4}{24} + \cdots \quad \text{and} \quad \cos x = 1 - \frac{x^2}{2} + \frac{x^4}{24} + \cdots.$$

Multiplying out and collecting like terms yield

$$e^x \cos x = \left(1 + x + \frac{x^2}{2} + \frac{x^3}{6} + \frac{x^4}{24} + \cdots\right)\left(1 - \frac{x^2}{2} + \frac{x^4}{24} - \cdots\right)$$

$$= 1 + (1)x + \left(-\frac{1}{2} + \frac{1}{2}\right)x^2 + \left(-\frac{1}{2} + \frac{1}{6}\right)x^3 + \left(\frac{1}{24} - \frac{1}{4} + \frac{1}{24}\right)x^4 + \cdots$$

$$= 1 + x - \frac{x^3}{3} - \frac{x^4}{6} + \cdots. \quad ∎$$

In Example 2 the interval of convergence for the Maclaurin series for both e^x and $\cos x$ is $(-\infty, \infty)$. Consequently the interval of convergence for the power series for $e^x \cos x$ is also $(-\infty, \infty)$.

EXAMPLE 3

Find the first four terms of a power series in x for $\sec x$.

Solution   We will use the Maclaurin series for $\cos x$ given in Example 2 and then use long division. Since $\sec x = 1/\cos x$, we have

$$
\cos x = 1 - \frac{x^2}{2} + \frac{x^4}{24} - \frac{x^6}{720} + \cdots \overline{\smash{\big)}\; 1}
$$

$$
\begin{array}{r}
1 + \dfrac{x^2}{2} + \dfrac{5x^4}{24} + \dfrac{61x^6}{720} + \cdots \\[2ex]
\hline
1 - \dfrac{x^2}{2} + \dfrac{x^4}{24} - \dfrac{x^6}{720} + \cdots \\[2ex]
\hline
\dfrac{x^2}{2} - \dfrac{x^4}{24} + \dfrac{x^6}{720} - \cdots \\[2ex]
\dfrac{x^2}{2} - \dfrac{x^4}{4} + \dfrac{x^6}{48} - \cdots \\[2ex]
\hline
\dfrac{5x^4}{24} - \dfrac{7x^6}{360} + \cdots \\[2ex]
\dfrac{5x^4}{24} - \dfrac{5x^6}{48} + \cdots \\[2ex]
\hline
\dfrac{61x^6}{720} - \cdots
\end{array}
$$

Thus
$$
\sec x = 1 + \frac{x^2}{2} + \frac{5x^4}{24} + \frac{61x^6}{720} + \cdots. \tag{2}
$$

The interval of convergence of this series is $(-\pi/2, \pi/2)$. (Why?)   ∎

The procedures illustrated in Examples 2 and 3 are obviously tedious to do by hand. Problems of this sort can be done using a computer algebra system (CAS) such as Mathematica. When you type the command: Series[Sec[x], {x, 0, 8}] and enter, Mathematica immediately gives the result obtained in (2).

For the remainder of this section, as well as this chapter, it is important that you become adept at simplifying the sum of two or more power series, each series expressed in summation (sigma) notation, to an expression with a single $\sum$. This often requires a shift of the summation indices.

EXAMPLE 4

Write $\displaystyle\sum_{n=1}^{\infty} 2nc_n x^{n-1} + \sum_{n=0}^{\infty} 6c_n x^{n+1}$ as one series.

Solution   In order to add the series, we require that both summation indices start with the same number and that the powers of x in each series be "in phase"; that is, if one series starts with a multiple of, say, x to the first power, then we want the other series to start with the same power. By writing

| series starts with x for $n = 2$ | series starts with x for $n = 0$ |

$$\sum_{n=1}^{\infty} 2nc_n x^{n-1} + \sum_{n=0}^{\infty} 6c_n x^{n+1} = 2 \cdot 1 \cdot c_1 x^0 + \sum_{n=2}^{\infty} 2nc_n x^{n-1} + \sum_{n=0}^{\infty} 6c_n x^{n+1}, \tag{3}$$

We have both series on the right side start with x^1. To get the same summation index we are inspired by the exponents of x; we let $k = n - 1$ in the first series and at the same time let $k = n + 1$ in the second series. Thus the right side of (3) becomes

$$2c_1 + \sum_{k=1}^{\infty} 2(k+1)c_{k+1} x^k + \sum_{k=1}^{\infty} 6c_{k-1} x^k. \tag{4}$$

Recall that the summation index is a "dummy" variable. The fact that $k = n - 1$ in one case and $k = n + 1$ in the other should cause no confusion if you keep in mind that it is the *value* of the summation index that is important. In both cases k takes on the same successive values $1, 2, 3, \ldots$ for $n = 2, 3, 4, \ldots$ (for $k = n - 1$) and $n = 0, 1, 2, \ldots$ (for $k = n + 1$).

We are now in a position to add the series in (4) term by term:

$$\sum_{n=1}^{\infty} 2nc_n x^{n-1} + \sum_{n=0}^{\infty} 6c_n x^{n+1} = 2c_1 + \sum_{k=1}^{\infty} [2(k+1)c_{k+1} + 6c_{k-1}]x^k. \tag{5}$$

If you are not convinced, then write out a few terms on both sides of (5). ■

Powers Series Solution of a Differential Equation

We saw in Section 1.1 that the function $y = e^{x^2}$ is an explicit solution of the linear first-order differential equation

$$\frac{dy}{dx} - 2xy = 0. \tag{6}$$

By replacing x by x^2 in the Maclaurin series for e^x, we can write the solution of (6) as

$$y = \sum_{n=0}^{\infty} \frac{x^{2n}}{n!}.$$

This last series converges for all real values of x. In other words, knowing the solution in advance, we were able to find an infinite series solution of the differential equation.

We now propose to obtain a **power series solution** of equation (6) directly; the method of attack is similar to the technique of undetermined coefficients.

ILLUSTRATED APPLICATIONS

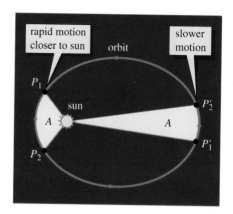

After painstaking observations, Johannes Kepler formulated and then published in 1609 three laws of planetary motion. The first law, probably his most famous, states that a planet revolves around the sun in an elliptical orbit with the sun at one focus. His second law of planetary motion states that the radius vector joining a planet with the sun sweeps out equal areas in equal intervals of time. See Problem 33, Exercises 3.2.

Around 1840, the Belgian mathematician P. F. Verhulst studied the nonlinear first-order differential equation $dP/dt = P(a - bP)$, called the *logistic equation*, in an attempt to model the population growth of various countries. This equation also provides a reasonable model for other growth phenomena such as the spread of flu throughout a static population, as found on the college campus illustrated here. See page 24, Section 3.3, and the essay at the end of Chapter 3.

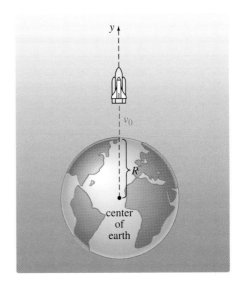

A ball or rocket projected straight up into the sky will fall back to earth under the influence of gravity unless it is given a sufficiently great velocity. The second-order differential equation $m\,d^2y/dt^2 = -kMm/y^2$ can be used to determine the so-called "escape velocity" an object needs to break free from the gravitational attraction of any celestial body such as the earth. The escape velocity is independent of the mass of the object, so on earth the escape velocity for a ball is the same as that for a rocket. See Section 3.3.

The motion of a mass attached to a spring is described by a linear second-order differential equation of the form $x'' + 2\lambda x' + \omega^2 x = f(t)$. An example of such a spring/mass system is the spring suspension system of some models of cars and trucks. See Chapter 5.

Normal flutter

Wing snaps

Large flutter

In late 1959 and early 1960, two commercial airplane crashes occurred involving the popular Lockheed Electra, a new model four-engine propjet. Braniff Flight 542 crashed near Buffalo, Texas, in September, 1959, and Northwest Flight 710 went down near Tell City, Indiana, in March, 1960. In both cases a wing had separated from the plane while at cruising altitude. These crashes illustrate the destructive effects of large mechanical oscillations or resonance. See Section 5.3.

A single differential equation can serve as the mathematical model for different phenomena. The differential equation governing the charge $q(t)$ on a capacitor in an *L-R-C* series circuit has the same form as the differential equation governing the displacement $x(t)$ of a vibrating mass on a spring. See Sections 5.3 and 5.4.

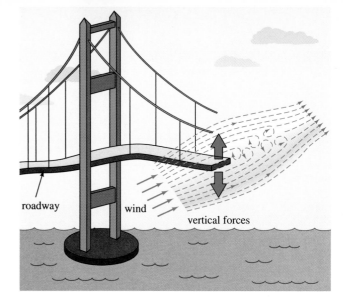

For years the 1940 collapse of the Tacoma Narrows Bridge was attributed to a resonance condition induced by wind blowing across its roadway and causing periodic vertical forces to act in the same direction as the vibration of the bridge. Recent studies suggest a different cause for the collapse. See the essay following Chapter 5. Compare the girder stiffened roadway in this photograph, showing the reconstructed bridge, with the delicate 1940 design evident in the photographs on page 253.

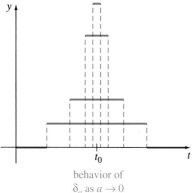

behavior of
δ_a as $a \to 0$

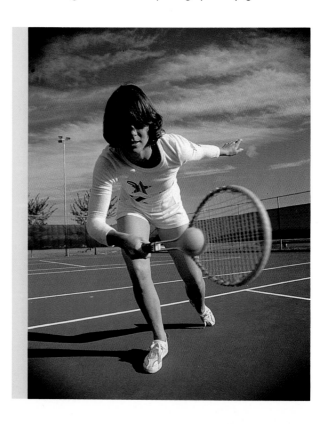

Mechanical systems are often acted upon by a sharp external force known as an impulsive force. This kind of force, which occurs in the collision of two objects, is a force of very large and approximately constant magnitude that acts on the system for a short time. A racket smashing into a tennis ball imparts a tremendous force to the ball but is in physical contact with it for only a fraction of a second. Analogous situations include a club striking a golf ball, lightning striking an airplane wing, and a proton striking the nucleus of an atom. See Section 7.6.

EXAMPLE 5

Find a solution of $dy/dx - 2xy = 0$ in the form of a power series in x.

Solution   If we assume that a solution of the given equation exists in the form

$$y = \sum_{n=0}^{\infty} c_n x^n, \tag{7}$$

we pose the question: Can we determine coefficients c_n for which the power series converges to a function satisfying (6)? Formal* term-by-term differentiation of (7) gives

$$\frac{dy}{dx} = \sum_{n=0}^{\infty} n c_n x^{n-1} = \sum_{n=1}^{\infty} n c_n x^{n-1}.$$

Note that since the first term in the first series (corresponding to $n = 0$) is zero, we begin the summation with $n = 1$. Using the last result and assumption (7), we find

$$\frac{dy}{dx} - 2xy = \sum_{n=1}^{\infty} n c_n x^{n-1} - \sum_{n=0}^{\infty} 2 c_n x^{n+1}. \tag{8}$$

We would like to add the two series in (9). To this end we write

$$\frac{dy}{dx} - 2xy = 1 \cdot c_1 x^0 + \sum_{n=2}^{\infty} n c_n x^{n-1} - \sum_{n=0}^{\infty} 2 c_n x^{n+1} \tag{9}$$

and then proceed as in Example 4 by letting $k = n - 1$ in the first series and $k = n + 1$ in the second. The right side of (8) becomes

$$c_1 + \sum_{k=1}^{\infty} (k + 1) c_{k+1} x^k - \sum_{k=1}^{\infty} 2 c_{k-1} x^k.$$

After we add the series termwise, it follows that

$$\frac{dy}{dx} - 2xy = c_1 + \sum_{k=1}^{\infty} [(k + 1) c_{k+1} - 2 c_{k-1}] x^k = 0. \tag{10}$$

Hence in order to have (10) identically zero it is necessary that the coefficients satisfy

$$c_1 = 0 \quad \text{and} \quad (k + 1) c_{k+1} - 2 c_{k-1} = 0, \qquad k = 1, 2, 3, \ldots. \tag{11}$$

Equation (11) provides a **recurrence relation** that determines the c_k. Since $k + 1 \neq 0$ for all the indicated values of k, we can write (11) as

$$c_{k+1} = \frac{2 c_{k-1}}{k + 1}. \tag{12}$$

---

* At this point we do not know the interval of convergence.

Iteration of this last formula then gives

$$k = 1, \qquad c_2 = \frac{2}{2} c_0 = c_0$$

$$k = 2, \qquad c_3 = \frac{2}{3} c_1 = 0$$

$$k = 3, \qquad c_4 = \frac{2}{4} c_2 = \frac{1}{2} c_0 = \frac{1}{2!} c_0$$

$$k = 4, \qquad c_5 = \frac{2}{5} c_3 = 0$$

$$k = 5, \qquad c_6 = \frac{2}{6} c_4 = \frac{1}{3 \cdot 2!} c_0 = \frac{1}{3!} c_0$$

$$k = 6, \qquad c_7 = \frac{2}{7} c_5 = 0$$

$$k = 7, \qquad c_8 = \frac{2}{8} c_6 = \frac{1}{4 \cdot 3!} c_0 = \frac{1}{4!} c_0$$

and so on. Thus from the original assumption (7), we find

$$y = \sum_{n=0}^{\infty} c_n x^n = c_0 + c_1 x + c_2 x^2 + c_3 x^3 + c_4 x^4 + c_5 x^5 + \cdots$$

$$= c_0 + 0 + c_0 x^2 + 0 + \frac{1}{2!} c_0 x^4 + 0 + \frac{1}{3!} c_0 x^6 + 0 + \cdots$$

$$= c_0 \left[1 + x^2 + \frac{1}{2!} x^4 + \frac{1}{3!} x^6 + \cdots \right] = c_0 \sum_{n=0}^{\infty} \frac{x^{2n}}{n!}. \qquad \textbf{(13)}$$

Since the iteration of (12) leaves c_0 completely undetermined, we have in fact found the general solution of (6).   ∎

The differential equation in Example 5, like the differential equation in the following example, can be easily solved by prior methods. The point of these two examples is to prepare you for the techniques considered in Sections 6.3 and 6.4.

EXAMPLE 6

Find solutions of $4y'' + y = 0$ in the form of powers series in x.

Solution   If $y = \sum_{n=0}^{\infty} c_n x^n$, then

$$y' = \sum_{n=0}^{\infty} nc_n x^{n-1} = \sum_{n=1}^{\infty} nc_n x^{n-1}$$

and

$$y'' = \sum_{n=1}^{\infty} n(n-1)c_n x^{n-2} = \sum_{n=2}^{\infty} n(n-1)c_n x^{n-2}.$$

Substituting the expressions for y'' and y back into the differential equation gives

$$4y'' + y = \underbrace{\sum_{n=2}^{\infty} 4n(n-1)c_n x^{n-2} + \sum_{n=0}^{\infty} c_n x^n}_{\boxed{\text{both series start with } x^0}}.$$

If we substitute $k = n - 2$ in the first series and $k = n$ in the second, we get (after using, in turn, $n = k + 2$ and $n = k$)

$$4y'' + y = \sum_{k=0}^{\infty} 4(k+2)(k+1)c_{k+2} x^k + \sum_{k=0}^{\infty} c_k x^k$$

$$= \sum_{k=0}^{\infty} [4(k+2)(k+1)c_{k+2} + c_k]x^k = 0.$$

From this last identity we conclude that

$$4(k+2)(k+1)c_{k+2} + c_k = 0$$

or

$$c_{k+2} = \frac{-c_k}{4(k+2)(k+1)}, \qquad k = 0, 1, 2, \dots.$$

From iteration of this recurrence relation it follows that

$$c_2 = \frac{-c_0}{4 \cdot 2 \cdot 1} = -\frac{c_0}{2^2 \cdot 2!}$$

$$c_3 = \frac{-c_1}{4 \cdot 3 \cdot 2} = -\frac{c_1}{2^2 \cdot 3!}$$

$$c_4 = \frac{-c_2}{4 \cdot 4 \cdot 3} = \frac{c_0}{2^4 \cdot 4!}$$

$$c_5 = \frac{-c_3}{4 \cdot 5 \cdot 4} = \frac{c_1}{2^4 \cdot 5!}$$

$$c_6 = \frac{-c_4}{4 \cdot 6 \cdot 5} = -\frac{c_0}{2^6 \cdot 6!}$$

$$c_7 = \frac{-c_5}{4 \cdot 7 \cdot 6} = -\frac{c_1}{2^6 \cdot 7!}$$

and so forth. This iteration leaves both c_0 and c_1 arbitrary. From the original assumption we have

$$y = c_0 + c_1 x + c_2 x^2 + c_3 x^3 + c_4 x^4 + c_5 x^5 + c_6 x^6 + c_7 x^7 + \cdots$$

$$= c_0 + c_1 x - \frac{c_0}{2^2 \cdot 2!} x^2 - \frac{c_1}{2^2 \cdot 3!} x^3 + \frac{c_0}{2^4 \cdot 4!} x^4 + \frac{c_1}{2^4 \cdot 5!} x^5 - \frac{c_0}{2^6 \cdot 6!} x^6 - \frac{c_1}{2^6 \cdot 7!} x^7 + \cdots$$

or

$$y = c_0 \left[1 - \frac{1}{2^2 \cdot 2!} x^2 + \frac{1}{2^4 \cdot 4!} x^4 - \frac{1}{2^6 \cdot 6!} x^6 + \cdots \right]$$

$$+ c_1 \left[x - \frac{1}{2^2 \cdot 3!} x^3 + \frac{1}{2^4 \cdot 5!} x^5 - \frac{1}{2^6 \cdot 7!} x^7 + \cdots \right]$$

is a general solution. When the series are written in summation notation,

$$y_1(x) = c_0 \sum_{k=0}^{\infty} \frac{(-1)^k}{(2k)!} \left(\frac{x}{2} \right)^{2k} \quad \text{and} \quad y_2(x) = 2c_1 \sum_{k=0}^{\infty} \frac{(-1)^k}{(2k+1)!} \left(\frac{x}{2} \right)^{2k+1},$$

the ratio test can be applied to show that both series converge for all x. You might also recognize the Maclaurin series as $y_1(x) = c_0 \cos(x/2)$ and $y_2(x) = 2c_1 \sin(x/2)$. ∎

6.2 EXERCISES

Answers to odd-numbered problems begin on page A–14.

In Problems 1–10 find the interval of convergence of the given power series.

1. $\displaystyle\sum_{n=1}^{\infty} \frac{(-1)^n}{n} x^n$

2. $\displaystyle\sum_{n=1}^{\infty} \frac{x^n}{n^2}$

3. $\displaystyle\sum_{k=1}^{\infty} \frac{2^k}{k} x^k$

4. $\displaystyle\sum_{k=0}^{\infty} \frac{5^k}{k!} x^k$

5. $\displaystyle\sum_{n=1}^{\infty} \frac{(x-3)^n}{n^3}$

6. $\displaystyle\sum_{n=1}^{\infty} \frac{(x+7)^n}{\sqrt{n}}$

7. $\displaystyle\sum_{k=1}^{\infty} \frac{(-1)^k}{10^k} (x-5)^k$

8. $\displaystyle\sum_{k=1}^{\infty} \frac{k}{(k+2)^2} (x-4)^k$

9. $\displaystyle\sum_{k=0}^{\infty} k! 2^k x^k$

10. $\displaystyle\sum_{k=0}^{\infty} \frac{k-1}{k^{2k}} x^k$

In Problems 11–20 find the first four terms of a power series in x for the given function. Calculate the series by hand or use a CAS as instructed.

11. $e^x \sin x$

12. $e^{-x} \cos x$

13. $\sin x \cos x$

14. $e^x \ln(1-x)$

15. $\left(x - \dfrac{x^3}{3} + \dfrac{x^5}{5} - \dfrac{x^7}{7} + \cdots \right)^2$

16. $\left(1 - \dfrac{x^2}{2} + \dfrac{x^4}{3} - \dfrac{x^6}{4} + \cdots \right)^2$

17. $\tan x$

18. $\dfrac{1}{e^x + e^{-x}}$

19. $\dfrac{1}{1 - \dfrac{x^2}{2} + \dfrac{x^4}{3} - \dfrac{x^6}{4} + \cdots}$

20. $\dfrac{1}{\left(1 - \dfrac{x^2}{2} + \dfrac{x^4}{3} - \dfrac{x^6}{4} + \cdots\right)^2}$

In Problems 21–30 solve each differential equation in the manner of the previous chapters and then compare the results with the solutions obtained by assuming a power series solution $y = \sum_{n=0}^{\infty} c_n x^n$.

21. $y' + y = 0$

22. $y' = 2y$

23. $y' - x^2 y = 0$

24. $y' + x^3 y = 0$

25. $(1 - x)y' - y = 0$

26. $(1 + x)y' - 2y = 0$

27. $y'' + y = 0$

28. $y'' - y = 0$

29. $y'' = y'$

30. $2y'' + y' = 0$

6.3 SOLUTIONS ABOUT ORDINARY POINTS

Suppose the linear second-order differential equation

$$a_2(x)y'' + a_1(x)y' + a_0(x)y = 0 \qquad \textbf{(1)}$$

is put into the form

$$y'' + P(x)y' + Q(x)y = 0 \qquad \textbf{(2)}$$

by dividing by the leading coefficient $a_2(x)$. We make the following definition.

DEFINITION 6.1   Ordinary and Singular Points

A point x_0 is said to be an **ordinary point** of the differential equation (1) if both $P(x)$ and $Q(x)$ are analytic* at x_0. A point that is not an ordinary point is said to be a **singular point** of the equation.

EXAMPLE I

Every finite value of x is an ordinary point of

$$y'' + (e^x)y' + (\sin x)y = 0.$$

---

* See page 269.

In particular we see that $x = 0$ is an ordinary point since

$$e^x = 1 + \frac{x}{1!} + \frac{x^2}{2!} + \cdots \quad \text{and} \quad \sin x = x - \frac{x^3}{3!} + \frac{x^5}{5!} - \cdots$$

converge for all finite values of x.   ∎

EXAMPLE 2

The differential equation $xy'' + (\sin x)y = 0$ has an ordinary point at $x = 0$ since it can be shown that $Q(x) = (\sin x)/x$ possesses the power series expansion

$$Q(x) = 1 - \frac{x^2}{3!} + \frac{x^4}{5!} - \frac{x^6}{7!} + \cdots$$

that converges for all finite values of x.   ∎

EXAMPLE 3

The differential equation $y'' + (\ln x)y = 0$ has a singular point at $x = 0$ because $Q(x) = \ln x$ possesses no power series in x.   ∎

Polynomial Coefficients

Primarily we shall be concerned with the case when (1) has *polynomial* coefficients. As a consequence of Definition 6.1, we note that when $a_2(x)$, $a_1(x)$, and $a_0(x)$ are polynomials with *no common factors*, a point $x = x_0$ is

(i) an ordinary point if $a_2(x_0) \neq 0$, or

(ii) a singular point if $a_2(x_0) = 0$.

EXAMPLE 4

(a) The singular points of the equation $(x^2 - 1)y'' + 2xy' + 6y = 0$ are the solutions of $x^2 - 1 = 0$ or $x = \pm 1$. All other finite values of x are ordinary points.

(b) Singular points need not be real numbers. The equation $(x^2 + 1)y'' + xy' - y = 0$ has singular points at the solutions of $x^2 + 1 = 0$, namely, $x = \pm i$. All other finite values of x, real or complex, are ordinary points.   ∎

EXAMPLE 5

The Cauchy-Euler equation $ax^2y'' + bxy' + cy = 0$, where a, b, and c are constants, has a singular point at $x = 0$. All other finite values of x, real or complex, are ordinary points. ∎

For our purposes ordinary points and singular points will always be finite points. It is possible for a differential equation to have, say, a singular point at infinity. (See Problem 40, Exercises 6.4.)

We state the following theorem about the existence of power series solutions without proof.

THEOREM 6.1   **Existence of Power Series Solutions**

If $x = x_0$ is an ordinary point of the differential equation (2), we can always find two linearly independent solutions in the form of power series centered at x_0:

$$y = \sum_{n=0}^{\infty} c_n(x - x_0)^n. \tag{3}$$

A series solution converges at least for $|x - x_0| < R$, where R is the distance from x_0 to the closest singular point (real or complex).

A solution of a differential equation of the form given in (3) is said to be a solution *about* the ordinary point x_0. The distance R given in Theorem 6.1 is the minimum value for the radius of convergence. A differential equation could have a finite singular point and yet a solution could be valid for all x; for example, the differential equation may possess a polynomial solution.

To solve a linear second-order equation such as (2) we find two sets of coefficients c_n so that we have two distinct power series $y_1(x)$ and $y_2(x)$, both expanded about the same ordinary point x_0. The procedure used to solve a second-order equation is the same as that used in Example 6 of Section 6.2; that is, we assume a solution $y = \sum_{n=0}^{\infty} c_n(x - x_0)^n$ and then determine the c_n. The general solution of the differential equation is $y = C_1y_1(x) + C_2y_2(x)$; in fact, it can be shown that $C_1 = c_0$ and $C_2 = c_1$, where c_0 and c_1 are arbitrary.

Note For the sake of simplicity, we assume an ordinary point is always located at $x = 0$, since, if not, the substitution $t = x - x_0$ translates the value $x = x_0$ to $t = 0$.

EXAMPLE 6

Solve $\qquad\qquad\qquad\qquad y'' - 2xy = 0.$

Solution We see that $x = 0$ is an ordinary point of the equation. Since there are no finite singular points, Theorem 6.1 guarantees two solutions of the form $y = \sum_{n=0}^{\infty} c_n x^n$ convergent for $|x| < \infty$. Proceeding, we write

$$y' = \sum_{n=0}^{\infty} nc_n x^{n-1} = \sum_{n=1}^{\infty} nc_n x^{n-1}$$

$$y'' = \sum_{n=1}^{\infty} n(n-1)c_n x^{n-2} = \sum_{n=2}^{\infty} n(n-1)c_n x^{n-2},$$

where we have used the fact that the first term in each series, corresponding to $n = 0$ and $n = 1$, respectively, is zero. Therefore

$$y'' - 2xy = \sum_{n=2}^{\infty} n(n-1)c_n x^{n-2} - \sum_{n=0}^{\infty} 2c_n x^{n+1}$$

$$= 2 \cdot 1 c_2 x^0 + \underbrace{\sum_{n=3}^{\infty} n(n-1)c_n x^{n-2} - \sum_{n=0}^{\infty} 2c_n x^{n+1}}_{\boxed{\text{both series start with } x}}.$$

Letting $k = n - 2$ in the first series and $k = n + 1$ in the second, we have

$$y'' - 2xy = 2c_2 + \sum_{k=1}^{\infty} (k+2)(k+1)c_{k+2} x^k - \sum_{k=1}^{\infty} 2c_{k-1} x^k$$

$$= 2c_2 + \sum_{k=1}^{\infty} [(k+2)(k+1)c_{k+2} - 2c_{k-1}]x^k = 0.$$

We must then have

$$2c_2 = 0 \quad \text{and} \quad (k+2)(k+1)c_{k+2} - 2c_{k-1} = 0.$$

The last expression is the same as

$$c_{k+2} = \frac{2c_{k-1}}{(k+2)(k+1)}, \qquad k = 1, 2, 3, \ldots.$$

Iteration gives
$$c_3 = \frac{2c_0}{3 \cdot 2}$$

$$c_4 = \frac{2c_1}{4 \cdot 3}$$

$$c_5 = \frac{2c_2}{5 \cdot 4} = 0$$

$$c_6 = \frac{2c_3}{6 \cdot 5} = \frac{2^2}{6 \cdot 5 \cdot 3 \cdot 2} c_0$$

$$c_7 = \frac{2c_4}{7 \cdot 6} = \frac{2^2}{7 \cdot 6 \cdot 4 \cdot 3} c_1$$

$$c_8 = \frac{2c_5}{8 \cdot 7} = 0$$

$$c_9 = \frac{2c_6}{9 \cdot 8} = \frac{2^3}{9 \cdot 8 \cdot 6 \cdot 5 \cdot 3 \cdot 2} c_0$$

$$c_{10} = \frac{2c_7}{10 \cdot 9} = \frac{2^3}{10 \cdot 9 \cdot 7 \cdot 6 \cdot 4 \cdot 3} c_1$$

$$c_{11} = \frac{2c_8}{11 \cdot 10} = 0$$

and so on. It should be apparent that both c_0 and c_1 are arbitrary. Now

$$y = c_0 + c_1 x + c_2 x^2 + c_3 x^3 + c_4 x^4 + c_5 x^5 + c_6 x^6 + c_7 x^7 + c_8 x^8$$
$$+ c_9 x^9 + c_{10} x^{10} + c_{11} x^{11} + \cdots$$

$$= c_0 + c_1 x + 0 + \frac{2}{3 \cdot 2} c_0 x^3 + \frac{2}{4 \cdot 3} c_1 x^4 + 0 + \frac{2^2}{6 \cdot 5 \cdot 3 \cdot 2} c_0 x^6$$

$$+ \frac{2^2}{7 \cdot 6 \cdot 4 \cdot 3} c_1 x^7 + 0 + \frac{2^3}{9 \cdot 8 \cdot 6 \cdot 5 \cdot 3 \cdot 2} c_0 x^9$$

$$+ \frac{2^3}{10 \cdot 9 \cdot 7 \cdot 6 \cdot 4 \cdot 3} c_1 x^{10} + 0 + \cdots$$

$$= c_0 \left[1 + \frac{2}{3 \cdot 2} x^3 + \frac{2^2}{6 \cdot 5 \cdot 3 \cdot 2} x^6 + \frac{2^3}{9 \cdot 8 \cdot 6 \cdot 5 \cdot 3 \cdot 2} x^9 + \cdots \right]$$

$$+ c_1 \left[x + \frac{2}{4 \cdot 3} x^4 + \frac{2^2}{7 \cdot 6 \cdot 4 \cdot 3} x^7 + \frac{2^3}{10 \cdot 9 \cdot 7 \cdot 6 \cdot 4 \cdot 3} x^{10} + \cdots \right].$$

∎

Although the pattern of the coefficients in Example 6 should be clear, it is sometimes useful to write the solutions in terms of summation notation. By using the properties of the factorial, we can write

$$y_1(x) = c_0 \left[1 + \sum_{k=1}^{\infty} \frac{2^k [1 \cdot 4 \cdot 7 \cdots (3k - 2)]}{(3k)!} x^{3k} \right]$$

and
$$y_2(x) = c_1 \left[x + \sum_{k=1}^{\infty} \frac{2^k [2 \cdot 5 \cdot 8 \cdots (3k - 1)]}{(3k + 1)!} x^{3k+1} \right].$$

In this form the ratio test can be used to show that each series converges for $|x| < \infty$.

EXAMPLE 7

Solve $$(x^2 + 1)y'' + xy' - y = 0.$$

Solution   Since the singular points are $x = \pm i$, a power series solution will converge at least for $|x| < 1$.* The assumption $y = \sum_{n=0}^{\infty} c_n x^n$ leads to

$$(x^2 + 1) \sum_{n=2}^{\infty} n(n - 1)c_n x^{n-2} + x \sum_{n=1}^{\infty} nc_n x^{n-1} - \sum_{n=0}^{\infty} c_n x^n$$

$$= \sum_{n=2}^{\infty} n(n - 1)c_n x^n + \sum_{n=2}^{\infty} n(n - 1)c_n x^{n-2}$$

$$+ \sum_{n=1}^{\infty} nc_n x^n - \sum_{n=0}^{\infty} c_n x^n$$

$$= 2c_2 x^0 - c_0 x^0 + 6c_3 x + c_1 x - c_1 x + \underbrace{\sum_{n=2}^{\infty} n(n - 1)c_n x^n}_{k = n}$$

$$+ \underbrace{\sum_{n=4}^{\infty} n(n - 1)c_n x^{n-2}}_{k = n - 2} + \underbrace{\sum_{n=2}^{\infty} nc_n x^n}_{k = n} - \underbrace{\sum_{n=2}^{\infty} c_n x^n}_{k = n}$$

$$= 2c_2 - c_0 + 6c_3 x$$

$$+ \sum_{k=2}^{\infty} [k(k - 1)c_k + (k + 2)(k + 1)c_{k+2} + kc_k - c_k]x^k$$

$$= 2c_2 - c_0 + 6c_3 x$$

$$+ \sum_{k=2}^{\infty} [(k + 1)(k - 1)c_k + (k + 2)(k + 1)c_{k+2}]x^k = 0.$$

Thus $$2c_2 - c_0 = 0$$

$$c_3 = 0$$

$$(k + 1)(k - 1)c_k + (k + 2)(k + 1)c_{k+2} = 0$$

or, after we divide by $(k + 2)(k + 1)$,

$$c_2 = \frac{1}{2} c_0$$

$$c_3 = 0$$

$$c_{k+2} = \frac{1 - k}{k + 2} c_k, \qquad k = 2, 3, 4, \ldots.$$

---

* The **modulus** or magnitude of the complex number $x = i$ is $|x| = 1$. If $x = a + bi$ is a singular point, then $|x| = \sqrt{a^2 + b^2}$. See Appendix IV.

Iteration of the last formula gives

$$c_4 = -\frac{1}{4}c_2 = -\frac{1}{2 \cdot 4}c_0 = -\frac{1}{2^2 2!}c_0$$

$$c_5 = -\frac{2}{5}c_3 = 0$$

$$c_6 = -\frac{3}{6}c_4 = \frac{3}{2 \cdot 4 \cdot 6}c_0 = \frac{1 \cdot 3}{2^3 3!}c_0$$

$$c_7 = -\frac{4}{7}c_5 = 0$$

$$c_8 = -\frac{5}{8}c_6 = -\frac{3 \cdot 5}{2 \cdot 4 \cdot 6 \cdot 8}c_0 = -\frac{1 \cdot 3 \cdot 5}{2^4 4!}c_0$$

$$c_9 = -\frac{6}{9}c_7 = 0$$

$$c_{10} = -\frac{7}{10}c_8 = \frac{3 \cdot 5 \cdot 7}{2 \cdot 4 \cdot 6 \cdot 8 \cdot 10}c_0 = \frac{1 \cdot 3 \cdot 5 \cdot 7}{2^5 5!}c_0$$

and so on. Therefore

$$y = c_0 + c_1 x + c_2 x^2 + c_3 x^3 + c_4 x^4 + c_5 x^5 + c_6 x^6 + c_7 x^7 + c_8 x^8 + \cdots$$

$$= c_1 x + c_0\left[1 + \frac{1}{2}x^2 - \frac{1}{2^2 2!}x^4 + \frac{1 \cdot 3}{2^3 3!}x^6 - \frac{1 \cdot 3 \cdot 5}{2^4 4!}x^8 + \frac{1 \cdot 3 \cdot 5 \cdot 7}{2^5 5!}x^{10} - \cdots\right].$$

The solutions are

$$y_1(x) = c_0\left[1 + \frac{1}{2}x^2 + \sum_{n=2}^{\infty}(-1)^{n-1}\frac{1 \cdot 3 \cdot 5 \cdots (2n-3)}{2^n n!}x^{2n}\right], \qquad |x| < 1$$

$$y_2(x) = c_1 x. \qquad\qquad\qquad\qquad\qquad\qquad\qquad\qquad\qquad\quad \blacksquare$$

EXAMPLE 8

If we seek a solution $y = \sum_{n=0}^{\infty} c_n x^n$ for the equation

$$y'' - (1 + x)y = 0,$$

we obtain $c_2 = c_0/2$ and the three-term recurrence relation

$$c_{k+2} = \frac{c_k + c_{k-1}}{(k+1)(k+2)}, \qquad k = 1, 2, 3, \ldots.$$

To simplify the iteration we can first choose $c_0 \neq 0$, $c_1 = 0$; this yields one solution. The other solution follows from next choosing $c_0 = 0$, $c_1 \neq 0$. With the first assumption we find

$$c_2 = \frac{1}{2}c_0$$

$$c_3 = \frac{c_1 + c_0}{2 \cdot 3} = \frac{c_0}{2 \cdot 3} = \frac{1}{6}c_0$$

$$c_4 = \frac{c_2 + c_1}{3 \cdot 4} = \frac{c_0}{2 \cdot 3 \cdot 4} = \frac{1}{24}c_0$$

$$c_5 = \frac{c_3 + c_2}{4 \cdot 5} = \frac{c_0}{4 \cdot 5}\left[\frac{1}{2 \cdot 3} + \frac{1}{2}\right] = \frac{1}{30}c_0$$

and so on. Thus one solution is

$$y_1(x) = c_0\left[1 + \frac{1}{2}x^2 + \frac{1}{6}x^3 + \frac{1}{24}x^4 + \frac{1}{30}x^5 + \cdots\right].$$

Similarly if we choose $c_0 = 0$, then

$$c_2 = 0$$

$$c_3 = \frac{c_1 + c_0}{2 \cdot 3} = \frac{c_1}{2 \cdot 3} = \frac{1}{6}c_1$$

$$c_4 = \frac{c_2 + c_1}{3 \cdot 4} = \frac{c_1}{3 \cdot 4} = \frac{1}{12}c_1$$

$$c_5 = \frac{c_3 + c_2}{4 \cdot 5} = \frac{c_1}{2 \cdot 3 \cdot 4 \cdot 5} = \frac{1}{120}c_1$$

and so on. Hence another solution is

$$y_2(x) = c_1\left[x + \frac{1}{6}x^3 + \frac{1}{12}x^4 + \frac{1}{120}x^5 + \cdots\right].$$

Each series converges for all finite values of x. ∎

Nonpolynomial Coefficients

The next example illustrates how to find a power series solution about an ordinary point of a differential equation when its coefficients are not polynomials. In this example we see an application of multiplication of two power series that was discussed in Section 6.2.

EXAMPLE 9

Solve
$$y'' + (\cos x)y = 0.$$

Solution Since $\cos x = 1 - \dfrac{x^2}{2!} + \dfrac{x^4}{4!} - \dfrac{x^6}{6!} + \cdots$, it is seen that $x = 0$ is an ordinary point. Thus the assumption $y = \sum_{n=0}^{\infty} c_n x^n$ leads to

$$y'' + (\cos x)y = \sum_{n=2}^{\infty} n(n-1)c_n x^{n-2} + \left(1 - \frac{x^2}{2!} + \frac{x^4}{4!} - \cdots\right)\sum_{n=0}^{\infty} c_n x^n$$

$$= (2c_2 + 6c_3 x + 12c_4 x^2 + 20c_5 x^3 + \cdots)$$

$$+ \left(1 - \frac{x^2}{2} + \frac{x^4}{24} - \cdots\right)(c_0 + c_1 x + c_2 x^2 + c_3 x^3 + \cdots)$$

$$= 2c_2 + c_0 + (6c_3 + c_1)x + \left(12c_4 + c_2 - \frac{1}{2}c_0\right)x^2 + \left(20c_5 + c_3 - \frac{1}{2}c_1\right)x^3 + \cdots.$$

Since the last line is to be identically zero, we must have

$$2c_2 + c_0 = 0$$

$$6c_3 + c_1 = 0$$

$$12c_4 + c_2 - \frac{1}{2}c_0 = 0$$

$$20c_5 + c_3 - \frac{1}{2}c_1 = 0$$

and so on. Since c_0 and c_1 are arbitrary, we find

$$y_1(x) = c_0\left[1 - \frac{1}{2}x^2 + \frac{1}{12}x^4 - \cdots\right] \quad \text{and} \quad y_2(x) = c_1\left[x - \frac{1}{6}x^3 + \frac{1}{30}x^5 - \cdots\right].$$

Since the differential equation has no singular points, both series converge for all finite values of x. ∎

6.3 EXERCISES

Answers to odd-numbered problems begin on page A–15.

In Problems 1–14 for each differential equation find two linearly independent power series solutions about the ordinary point $x = 0$.

1. $y'' = xy$

2. $y'' + x^2 y = 0$

3. $y'' - 2xy' + y = 0$

4. $y'' - xy' + 2y = 0$

5. $y'' + x^2 y' + xy = 0$

6. $y'' + 2xy' + 2y = 0$

7. $(x - 1)y'' + y' = 0$

8. $(x + 2)y'' + xy' - y = 0$

9. $(x^2 - 1)y'' + 4xy' + 2y = 0$

10. $(x^2 + 1)y'' - 6y = 0$

11. $(x^2 + 2)y'' + 3xy' - y = 0$

12. $(x^2 - 1)y'' + xy' - y = 0$

13. $y'' - (x + 1)y' - y = 0$

14. $y'' - xy' - (x + 2)y = 0$

In Problems 15–18 use the power series method to solve the given differential equation subject to the indicated initial conditions.

15. $(x - 1)y'' - xy' + y = 0$, $y(0) = -2$, $y'(0) = 6$

16. $(x + 1)y'' - (2 - x)y' + y = 0$, $y(0) = 2$, $y'(0) = -1$

17. $y'' - 2xy' + 8y = 0$, $y(0) = 3$, $y'(0) = 0$

18. $(x^2 + 1)y'' + 2xy' = 0$, $y(0) = 0$, $y'(0) = 1$

In Problems 19–22 use the procedure illustrated in Example 9 to find two power series solutions of the given differential equation about the ordinary point $x = 0$.

19. $y'' + (\sin x)y = 0$

20. $xy'' + (\sin x)y = 0$ [*Hint:* See Example 2.]

21. $y'' + e^{-x}y = 0$ **22.** $y'' + e^{x}y' - y = 0$

In Problems 23 and 24 use the power series method to solve the nonhomogeneous equation.

23. $y'' - xy = 1$ **24.** $y'' - 4xy' - 4y = e^{x}$

25. The differential equation $y'' - 2xy' + 2ny = 0$ is known as **Hermite's equation.*** When $n \geq 0$ is an integer, Hermite's equation has a polynomial solution. Hermite polynomials have some importance in the study of quantum mechanics. Obtain the polynomial solutions corresponding to $n = 1$ and $n = 2$.

26. In the analysis of a uniform thin column of length L that is buckling under its own weight, the following boundary-value problem is encountered:

$$\theta'' + \frac{\delta g}{EI}(L - x)\theta = 0, \qquad \theta(0) = 0, \quad \theta'(L) = 0.$$

Here E is Young's modulus, I is the cross-sectional moment of inertia, δ is the constant linear density, x is distance measured along the column, and $\theta(x)$ is the angular deflection of the column from the vertical at a point $P(x)$. See Figure 6.5. Obtain a power series solution of the differential equation that satisfies the condition $\theta'(L) = 0$. For convenience define $\lambda^2 = \delta g L / EI$ and change the variable by letting $t = L - x$.

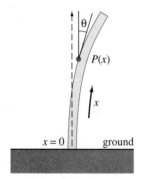

Figure 6.5

6.4 SOLUTIONS ABOUT SINGULAR POINTS

6.4.1   Regular Singular Points; Method of Frobenius—Case I

We saw in the preceding section that there is no basic problem in finding a power series solution of

$$a_2(x)y'' + a_1(x)y' + a_0(x)y = 0 \tag{1}$$

---

* Named after the French mathematician **Charles Hermite** (1822–1901).

about an ordinary point $x = x_0$. However, when $x = x_0$ is a singular point, it is not always possible to find a solution of the form $y = \sum_{n=0}^{\infty} c_n(x - x_0)^n$; it turns out that we *may* be able to find a solution of the form $y = \sum_{n=0}^{\infty} c_n(x - x_0)^{n+r}$, where r is a constant that must be determined.

Regular and Irregular Singular Points

Singular points are further classified as either regular or irregular. To define these concepts we again put (1) into the standard form

$$y'' + P(x)y' + Q(x)y = 0. \tag{2}$$

DEFINITION 6.2 **Regular and Irregular Singular Points**

A singular point $x = x_0$ of equation (1) is said to be a **regular singular point** if both $(x - x_0)P(x)$ and $(x - x_0)^2 Q(x)$ are analytic at x_0. A singular point that is not regular is said to be an **irregular singular point** of the equation.

Polynomial Coefficients

In the case in which the coefficients in (1) are polynomials with no common factors, Definition 6.2 is equivalent to the following.

Let $a_2(x_0) = 0$. Form $P(x)$ and $Q(x)$ by reducing $a_1(x)/a_2(x)$ and $a_0(x)/a_2(x)$ to lowest terms, respectively. If the factor $(x - x_0)$ appears *at most* to the first power in the denominator of $P(x)$ and *at most* to the second power in the denominator of $Q(x)$, then $x = x_0$ is a regular singular point.

EXAMPLE 1

It should be clear that $x = -2$ and $x = 2$ are singular points of the equation

$$(x^2 - 4)^2 y'' + (x - 2)y' + y = 0.$$

Dividing the equation by $(x^2 - 4)^2 = (x - 2)^2(x + 2)^2$, we find that

$$P(x) = \frac{1}{(x - 2)(x + 2)^2} \quad \text{and} \quad Q(x) = \frac{1}{(x - 2)^2(x + 2)^2}.$$

We now test $P(x)$ and $Q(x)$ at each singular point.

In order that $x = -2$ be a regular singular point, the factor $x + 2$ can appear at most to the first power in the denominator of $P(x)$, and can appear at most to the second power in the denominator of $Q(x)$. Inspection of $P(x)$ and $Q(x)$ shows that the first condition does not obtain, and so we conclude that $x = -2$ is an irregular singular point.

In order that $x = 2$ be a regular singular point, the factor $x - 2$ can appear at most to the first power in the denominator of $P(x)$ and can appear at most to the second power in the denominator of $Q(x)$. Further inspection of $P(x)$ and $Q(x)$ shows that both these conditions are satisfied, so $x = 2$ is a regular singular point.   ∎

E X A M P L E **2**

Both $x = 0$ and $x = -1$ are singular points of the differential equation

$$x^2(x + 1)^2 y'' + (x^2 - 1)y' + 2y = 0.$$

Inspection of

$$P(x) = \frac{x - 1}{x^2(x + 1)} \quad \text{and} \quad Q(x) = \frac{2}{x^2(x + 1)^2}$$

shows that $x = 0$ is an irregular singular point since $(x - 0)$ appears to the second power in the denominator of $P(x)$. Note, however, that $x = -1$ is a regular singular point.   ∎

E X A M P L E **3**

(a) $x = 1$ and $x = -1$ are regular singular points of

$$(1 - x^2)y'' - 2xy' + 30y = 0.$$

(b) $x = 0$ is an irregular singular point of

$$x^3 y'' - 2xy' + 5y = 0$$

since

$$Q(x) = \frac{5}{x^3}.$$

(c) $x = 0$ is a regular singular point of

$$xy'' - 2xy' + 5y = 0$$

since

$$P(x) = -2 \quad \text{and} \quad Q(x) = \frac{5}{x}.$$ ∎

In part (c) of Example 3 notice that $(x - 0)$ and $(x - 0)^2$ do not even appear in the denominators of $P(x)$ and $Q(x)$, respectively. Remember, these factors can appear at most in this fashion. For a singular point $x = x_0$, any nonnegative power of $(x - x_0)$ less than one (namely, zero) and nonnegative power less than two (namely, zero and one) in the denominators of $P(x)$ and $Q(x)$, respectively, imply x_0 is a regular singular point.

Also, recall that singular points can be complex numbers. It should be apparent that both $x = 3i$ and $x = -3i$ are regular singular points of the equation $(x^2 + 9)y'' - 3xy' + (1 - x)y = 0$ since

$$P(x) = \frac{-3x}{(x - 3i)(x + 3i)} \quad \text{and} \quad Q(x) = \frac{1 - x}{(x - 3i)(x + 3i)}.$$

EXAMPLE 4

From our discussion of the Cauchy-Euler equation in Section 6.1, we can show that $y_1 = x^2$ and $y_2 = x^2 \ln x$ are solutions of the equation $x^2 y'' - 3xy' + 4y = 0$ on the interval $(0, \infty)$. If the procedure of Theorem 6.1 is attempted at the regular singular point $x = 0$ (that is, an assumed solution of the form $y = \sum_{n=0}^{\infty} c_n x^n$), we would succeed in obtaining only the solution $y_1 = x^2$. The fact that we would not obtain the second solution is not really surprising since $\ln x$ does not possess a Taylor series expansion about $x = 0$. It follows that $y_2 = x^2 \ln x$ does not have a power series in x. ∎

EXAMPLE 5

The differential equation $\quad 6x^2 y'' + 5xy' + (x^2 - 1)y = 0 \quad$ has a regular singular point at $x = 0$ but does not possess *any* solution of the form $y = \sum_{n=0}^{\infty} c_n x^n$. By the procedure that we shall now consider it can be shown, however, that there exist two series solutions of the form

$$y = \sum_{n=0}^{\infty} c_n x^{n + 1/2} \quad \text{and} \quad y = \sum_{n=0}^{\infty} c_n x^{n - 1/3}. \qquad ∎$$

Method of Frobenius

To solve a differential equation such as (1) about a regular singular point we employ the following theorem due to Frobenius.*

---

*Ferdinand Georg Frobenius (1848–1917)** Although the basic idea of this series method can be traced back to Euler, the German mathematician Ferdinand Frobenius was the first to prove the result, which he published in 1878. Frobenius made many contributions to the field of analysis, but his name appears more in texts on abstract algebra than in texts on differential equations. His most significant contributions to mathematics were in the field of group theory.

> **THEOREM 6.2   Frobenius' Theorem**
>
> If $x = x_0$ is a regular singular point of equation (1), then there exists at least one series solution of the form
>
> $$y = (x - x_0)^r \sum_{n=0}^{\infty} c_n(x - x_0)^n = \sum_{n=0}^{\infty} c_n(x - x_0)^{n+r}, \qquad (3)$$
>
> where the number r is a constant that must be determined. The series will converge at least on some interval $0 < x - x_0 < R$.

Note, the words *at least* in the second line of Theorem 6.2. This means that, in contrast to Theorem 6.1, we are *not* guaranteed two solutions of the indicated form. The **method of Frobenius** consists of identifying a regular singular point x_0, substituting $y = \sum_{n=0}^{\infty} c_n(x - x_0)^{n+r}$ into the differential equation, and determining the unknown exponent r and the coefficients c_n.

As in the preceding section, for the sake of simplicity we shall always assume $x_0 = 0$.

EXAMPLE 6

Since $x = 0$ is a regular singular point of the differential equation

$$3xy'' + y' - y = 0, \qquad (4)$$

we try a solution of the form

$$y = \sum_{n=0}^{\infty} c_n x^{n+r}.$$

Now

$$y' = \sum_{n=0}^{\infty} (n + r)c_n x^{n+r-1}$$

$$y'' = \sum_{n=0}^{\infty} (n + r)(n + r - 1)c_n x^{n+r-2}$$

so that

$$3xy'' + y' - y = 3 \sum_{n=0}^{\infty} (n + r)(n + r - 1)c_n x^{n+r-1}$$

$$+ \sum_{n=0}^{\infty} (n + r)c_n x^{n+r-1} - \sum_{n=0}^{\infty} c_n x^{n+r}$$

$$= \sum_{n=0}^{\infty} (n + r)(3n + 3r - 2)c_n x^{n+r-1} - \sum_{n=0}^{\infty} c_n x^{n+r}$$

$$= x^r \left[r(3r - 2)c_0 x^{-1} + \underbrace{\sum_{n=1}^{\infty} (n + r)(3n + 3r - 2)c_n x^{n-1}}_{k = n - 1} - \underbrace{\sum_{n=0}^{\infty} c_n x^n}_{k = n} \right]$$

$$= x^r \left[r(3r - 2)c_0 x^{-1} + \sum_{k=0}^{\infty} [(k + r + 1)(3k + 3r + 1)c_{k+1} - c_k] x^k \right] = 0$$

which implies
$$r(3r - 2)c_0 = 0$$

$$(k + r + 1)(3k + 3r + 1)c_{k+1} - c_k = 0, \qquad k = 0, 1, 2, \ldots. \qquad (5)$$

Since nothing is gained by taking $c_0 = 0$, we must then have

$$r(3r - 2) = 0 \qquad (6)$$

and
$$c_{k+1} = \frac{c_k}{(k + r + 1)(3k + 3r + 1)}, \qquad k = 0, 1, 2, \ldots. \qquad (7)$$

The two values of r that satisfy (6), $r_1 = 2/3$ and $r_2 = 0$, when substituted in (7), give two different recurrence relations:

$$r_1 = \frac{2}{3}, \qquad c_{k+1} = \frac{c_k}{(3k + 5)(k + 1)}, \qquad k = 0, 1, 2, \ldots, \qquad (8)$$

and
$$r_2 = 0, \qquad c_{k+1} = \frac{c_k}{(k + 1)(3k + 1)}, \qquad k = 0, 1, 2, \ldots. \qquad (9)$$

Iteration of (8) gives
$$c_1 = \frac{c_0}{5 \cdot 1}$$

$$c_2 = \frac{c_1}{8 \cdot 2} = \frac{c_0}{2! 5 \cdot 8}$$

$$c_3 = \frac{c_2}{11 \cdot 3} = \frac{c_0}{3! 5 \cdot 8 \cdot 11}$$

$$c_4 = \frac{c_3}{14 \cdot 4} = \frac{c_0}{4! 5 \cdot 8 \cdot 11 \cdot 14}$$

$$\vdots$$

$$c_n = \frac{c_0}{n! 5 \cdot 8 \cdot 11 \cdots (3n + 2)}, \qquad n = 1, 2, 3, \ldots,$$

whereas iteration of (9) yields

$$c_1 = \frac{c_0}{1 \cdot 1}$$

$$c_2 = \frac{c_1}{2 \cdot 4} = \frac{c_0}{2! 1 \cdot 4}$$

$$c_3 = \frac{c_2}{3 \cdot 7} = \frac{c_0}{3! 1 \cdot 4 \cdot 7}$$

$$c_4 = \frac{c_3}{4 \cdot 10} = \frac{c_0}{4! 1 \cdot 4 \cdot 7 \cdot 10}$$

$$\vdots$$

$$c_n = \frac{c_0}{n! 1 \cdot 4 \cdot 7 \cdots (3n - 2)}, \qquad n = 1, 2, 3, \ldots.$$

Thus we obtain two series solutions

$$y_1 = c_0 x^{2/3} \left[1 + \sum_{n=1}^{\infty} \frac{1}{n! 5 \cdot 8 \cdot 11 \cdots (3n + 2)} x^n \right] \tag{10}$$

and $\qquad y_2 = c_0 x^0 \left[1 + \sum_{n=1}^{\infty} \frac{1}{n! 1 \cdot 4 \cdot 7 \cdots (3n - 2)} x^n \right].$ $\qquad$ **(11)**

By the ratio test it can be demonstrated that both (10) and (11) converge for all finite values of x. Also it should be clear from the form of (10) and (11) that neither series is a constant multiple of the other and, therefore, $y_1(x)$ and $y_2(x)$ are linearly independent solutions on the x-axis. Hence by the superposition principle

$$y = C_1 y_1(x) + C_2 y_2(x) = C_1 \left[x^{2/3} + \sum_{n=1}^{\infty} \frac{1}{n! 5 \cdot 8 \cdot 11 \cdots (3n + 2)} x^{n + 2/3} \right]$$

$$+ C_2 \left[1 + \sum_{n=1}^{\infty} \frac{1}{n! 1 \cdot 4 \cdot 7 \cdots (3n - 2)} x^n \right], \qquad |x| < \infty.$$

is another solution of (4). On any interval not containing the origin, this combination represents the general solution of the differential equation. ∎

Although Example 6 illustrates the general procedure for using the method of Frobenius, we hasten to point out that we may not always be able to find two solutions so readily or for that matter find two solutions that are infinite series consisting entirely of powers of x.

Indicial Equation

Equation (6) is called the **indicial equation** of the problem, and the values $r_1 = 2/3$ and $r_2 = 0$ are called the **indicial roots**, or **exponents**, of the singularity. In general, if $x = 0$ is a regular singular point of (1), then the functions $xP(x)$ and $x^2 Q(x)$ obtained from (2) are analytic at zero; that is, the expansions

$$xP(x) = p_0 + p_1 x + p_2 x^2 + \cdots$$

$$x^2 Q(x) = q_0 + q_1 x + q_2 x^2 + \cdots \tag{12}$$

are valid on intervals that have a positive radius of convergence. After substituting $y = \sum_{n=0}^{\infty} c_n x^{n+r}$ in (1) or (2) and simplifying, the indicial equation is a quadratic equation in r that results from equating the *total coefficient of the lowest power of x to zero*. It is left as an exercise to show that the general indicial equation is

$$r(r-1) + p_0 r + q_0 = 0. \tag{13}$$

See Problem 38. We then solve the latter equation for the two values of the exponents and substitute these values into a recurrence relation such as (7). Theorem 6.2 guarantees that at least one solution of the assumed series form can be found.

EXAMPLE 7

The differential equation

$$xy'' + 3y' - y = 0 \tag{14}$$

has a regular singular point at $x = 0$. The method of Frobenius yields

$$xy'' + 3y' - y = x^r \left[r(r+2)c_0 x^{-1} + \sum_{k=0}^{\infty} [(k+r+1)(k+r+3)c_{k+1} - c_k]x^k \right] = 0$$

so that the indicial equation and exponents are $r(r+2) = 0$ and $r_1 = 0$, $r_2 = -2$, respectively.

Since

$$(k+r+1)(k+r+3)c_{k+1} - c_k = 0, \qquad k = 0, 1, 2, \ldots, \tag{15}$$

it follows that when $r_1 = 0$,

$$c_{k+1} = \frac{c_k}{(k+1)(k+3)}$$

$$c_1 = \frac{c_0}{1 \cdot 3}$$

$$c_2 = \frac{c_1}{2 \cdot 4} = \frac{2c_0}{2!4!}$$

$$c_3 = \frac{c_2}{3 \cdot 5} = \frac{2c_0}{3!5!}$$

$$c_4 = \frac{c_3}{4 \cdot 6} = \frac{2c_0}{4!6!}$$

$$\vdots$$

$$c_n = \frac{2c_0}{n!(n+2)!}, \qquad n = 1, 2, 3, \ldots.$$

Thus one series solution is

$$y_1 = c_0 x^0 \left[1 + \sum_{n=1}^{\infty} \frac{2}{n!(n+2)!} x^n \right]$$

$$= c_0 \sum_{n=0}^{\infty} \frac{2}{n!(n+2)!} x^n, \qquad |x| < \infty. \tag{16}$$

Now when $r_2 = -2$, (15) becomes

$$(k-1)(k+1)c_{k+1} - c_k = 0, \tag{17}$$

but note here that we *do not divide* by $(k-1)(k+1)$ immediately since this term is zero for $k = 1$. However, we use the recurrence relation (17) for the cases $k = 0$ and $k = 1$:

$$-1 \cdot 1 c_1 - c_0 = 0 \quad \text{and} \quad 0 \cdot 2 c_2 - c_1 = 0.$$

The latter equation implies that $c_1 = 0$ and so the former equation implies that $c_0 = 0$. Continuing, we find

$$c_{k+1} = \frac{c_k}{(k-1)(k+1)}, \qquad k = 2, 3, 4, \dots,$$

and so

$$c_3 = \frac{c_2}{1 \cdot 3}$$

$$c_4 = \frac{c_3}{2 \cdot 4} = \frac{2c_2}{2!4!}$$

$$c_5 = \frac{c_4}{3 \cdot 5} = \frac{2c_2}{3!5!}$$

$$\vdots$$

$$c_n = \frac{2c_2}{(n-2)!n!}, \qquad n = 2, 3, 4, \dots.$$

Thus

$$y_2 = c_2 x^{-2} \sum_{n=2}^{\infty} \frac{2}{(n-2)!n!} x^n. \tag{18}$$

However, close inspection of (18) reveals that y_2 is simply a constant multiple of (16). To see this, let $k = n - 2$ in (18). We conclude that the method of Frobenius gives only one series solution of (14). ∎

Cases of Indicial Roots

When using the method of Frobenius, we usually distinguish three cases corresponding to the nature of the indicial roots. For the sake of discussion let us suppose that r_1 and r_2 are the *real* solutions of the indicial equation and that, when appropriate, r_1 *denotes the largest root*.

CASE I ROOTS NOT DIFFERING BY AN INTEGER If r_1 and r_2 are distinct and do not differ by an integer, then there exist two linearly independent solutions of equation (1) of the form

$$y_1 = \sum_{n=0}^{\infty} c_n x^{n+r_1}, \qquad c_0 \neq 0, \tag{19a}$$

$$y_2 = \sum_{n=0}^{\infty} b_n x^{n+r_2}, \qquad b_0 \neq 0. \tag{19b}$$

E X A M P L E **8**

Solve $\qquad\qquad 2xy'' + (1 + x)y' + y = 0.$ \hfill (20)

Solution If $y = \sum_{n=0}^{\infty} c_n x^{n+r}$, then

$$2xy'' + (1 + x)y' + y = 2\sum_{n=0}^{\infty} (n + r)(n + r - 1)c_n x^{n+r-1} + \sum_{n=0}^{\infty} (n + r)c_n x^{n+r-1}$$

$$+ \sum_{n=0}^{\infty} (n + r)c_n x^{n+r} + \sum_{n=0}^{\infty} c_n x^{n+r}$$

$$= \sum_{n=0}^{\infty} (n + r)(2n + 2r - 1)c_n x^{n+r-1} + \sum_{n=0}^{\infty} (n + r + 1)c_n x^{n+r}$$

$$= x^r \left[r(2r - 1)c_0 x^{-1} + \underbrace{\sum_{n=1}^{\infty} (n + r)(2n + 2r - 1)c_n x^{n-1}}_{k = n - 1} + \underbrace{\sum_{n=0}^{\infty} (n + r + 1)c_n x^{n}}_{k = n} \right]$$

$$= x^r \left[r(2r - 1)c_0 x^{-1} + \sum_{k=0}^{\infty} [(k + r + 1)(2k + 2r + 1)c_{k+1} + (k + r + 1)c_k]x^k \right] = 0,$$

which implies $\qquad\qquad r(2r - 1) = 0$ \hfill (21)

$$(k + r + 1)(2k + 2r + 1)c_{k+1} + (k + r + 1)c_k = 0, \qquad k = 0, 1, 2, \ldots . \tag{22}$$

For $r_1 = 1/2$, we can divide by $k + 3/2$ in (22) to obtain

$$c_{k+1} = \frac{-c_k}{2(k + 1)}$$

$$c_1 = \frac{-c_0}{2 \cdot 1}$$

$$c_2 = \frac{-c_1}{2 \cdot 2} = \frac{c_0}{2^2 \cdot 2!}$$

$$c_3 = \frac{-c_2}{2 \cdot 3} = \frac{-c_0}{2^3 \cdot 3!}$$

$$\vdots$$

$$c_n = \frac{(-1)^n c_0}{2^n n!}, \qquad n = 1, 2, 3, \ldots.$$

Thus we have
$$y_1 = c_0 x^{1/2} \left[1 + \sum_{n=1}^{\infty} \frac{(-1)^n}{2^n n!} x^n \right]$$

$$= c_0 \sum_{n=0}^{\infty} \frac{(-1)^n}{2^n n!} x^{n + 1/2}, \tag{23}$$

which converges for $x \geq 0$. As given, the series is not meaningful for $x < 0$ because of the presence of $x^{1/2}$.

Now for $r_2 = 0$, (22) becomes

$$c_{k+1} = \frac{-c_k}{2k + 1}$$

$$c_1 = \frac{-c_0}{1}$$

$$c_2 = \frac{-c_1}{3} = \frac{c_0}{1 \cdot 3}$$

$$c_3 = \frac{-c_2}{5} = \frac{-c_0}{1 \cdot 3 \cdot 5}$$

$$c_4 = \frac{-c_3}{7} = \frac{c_0}{1 \cdot 3 \cdot 5 \cdot 7}$$

$$\vdots$$

$$c_n = \frac{(-1)^n c_0}{1 \cdot 3 \cdot 5 \cdot 7 \cdots (2n - 1)}, \qquad n = 1, 2, 3, \ldots.$$

We conclude that a second solution to (20) is

$$y_2 = c_0 \left[1 + \sum_{n=1}^{\infty} \frac{(-1)^n}{1 \cdot 3 \cdot 5 \cdot 7 \cdots (2n - 1)} x^n \right], \qquad |x| < \infty. \tag{24}$$

On the interval $(0, \infty)$, the general solution is $y = C_1 y_1(x) + C_2 y_2(x)$.   ■

6.4.2   *Method of Frobenius—Cases II and III*

When the roots of the indicial equation differ by a positive integer, we may or may not be able to find two solutions of (1) having form (3). If not, then one solution corresponding to the smaller root contains a logarithmic term.

When the exponents are equal, a second solution *always* contains a logarithm. This latter situation is analogous to the solutions of the Cauchy-Euler differential equation when the roots of the auxiliary equation are equal. We have the next two cases.

CASE II   ROOTS DIFFERING BY A POSITIVE INTEGER   If $r_1 - r_2 = N$, where N is a positive integer, then there exist two linearly independent solutions of equation (1) of the form

$$y_1 = \sum_{n=0}^{\infty} c_n x^{n+r_1}, \qquad c_0 \neq 0 \tag{25a}$$

$$y_2 = Cy_1(x) \ln x + \sum_{n=0}^{\infty} b_n x^{n+r_2}, \qquad b_0 \neq 0, \tag{25b}$$

where C is a constant that could be zero.

CASE III   EQUAL INDICIAL ROOTS   If $r_1 = r_2$, there always exist two linearly independent solutions of equation (1) of the form

$$y_1 = \sum_{n=0}^{\infty} c_n x^{n+r_1}, \qquad c_0 \neq 0 \tag{26a}$$

$$y_2 = y_1(x) \ln x + \sum_{n=1}^{\infty} b_n x^{n+r_1}. \tag{26b}$$

EXAMPLE 9

Solve
$$xy'' + (x - 6)y' - 3y = 0. \tag{27}$$

Solution   The assumption $y = \sum_{n=0}^{\infty} c_n x^{n+r}$ leads to

$$xy'' + (x - 6)y' - 3y$$

$$= \sum_{n=0}^{\infty} (n + r)(n + r - 1)c_n x^{n+r-1} - 6 \sum_{n=0}^{\infty} (n + r)c_n x^{n+r-1} + \sum_{n=0}^{\infty} (n + r)c_n x^{n+r} - 3 \sum_{n=0}^{\infty} c_n x^{n+r}$$

$$= x^r \left[r(r - 7)c_0 x^{-1} + \underbrace{\sum_{n=1}^{\infty} (n + r)(n + r - 7)c_n x^{n-1}}_{k = n-1} + \underbrace{\sum_{n=0}^{\infty} (n + r - 3)c_n x^n}_{k = n} \right]$$

$$= x^r \left[r(r - 7)c_0 x^{-1} + \sum_{k=0}^{\infty} [(k + r + 1)(k + r - 6)c_{k+1} + (k + r - 3)c_k]x^k \right] = 0.$$

Thus $r(r - 7) = 0$ so that $r_1 = 7, r_2 = 0, r_1 - r_2 = 7$, and

$$(k + r + 1)(k + r - 6)c_{k+1} + (k + r - 3)c_k = 0, \qquad k = 0, 1, 2, \ldots. \tag{28}$$

For the smaller root $r_2 = 0$, (28) becomes

$$(k + 1)(k - 6)c_{k+1} + (k - 3)c_k = 0. \tag{29}$$

Since $k - 6 = 0$ when $k = 6$, we do not divide by this term until $k > 6$. We find

$$1 \cdot (-6)c_1 + (-3)c_0 = 0$$

$$2 \cdot (-5)c_2 + (-2)c_1 = 0$$

$$3 \cdot (-4)c_3 + (-1)c_2 = 0$$

$$4 \cdot (-3)c_4 + 0 \cdot c_3 = 0$$

$$5 \cdot (-2)c_5 + 1 \cdot c_4 = 0$$

$$6 \cdot (-1)c_6 + 2 \cdot c_5 = 0$$

$$7 \cdot 0c_7 + 3 \cdot c_6 = 0$$

implies $c_4 = c_5 = c_6 = 0$ but c_0 and c_7 can be chosen arbitrarily

Hence

$$c_1 = -\frac{1}{2}c_0$$

$$c_2 = -\frac{1}{5}c_1 = \frac{1}{10}c_0 \qquad \text{(30)}$$

$$c_3 = -\frac{1}{12}c_2 = -\frac{1}{120}c_0$$

and for $k \geq 7$,

$$c_{k+1} = \frac{-(k-3)c_k}{(k+1)(k-6)}$$

$$c_8 = \frac{-4}{8 \cdot 1}c_7$$

$$c_9 = \frac{-5}{9 \cdot 2}c_8 = \frac{4 \cdot 5}{2! 8 \cdot 9}c_7$$

$$c_{10} = \frac{-6}{10 \cdot 3}c_9 = \frac{-4 \cdot 5 \cdot 6}{3! 8 \cdot 9 \cdot 10}c_7$$

$$\vdots$$

$$c_n = \frac{(-1)^{n+1} 4 \cdot 5 \cdot 6 \cdots (n-4)}{(n-7)! 8 \cdot 9 \cdot 10 \cdots n}c_7, \qquad n = 8, 9, 10, \ldots. \qquad \text{(31)}$$

If we choose $c_7 = 0$ and $c_0 \neq 0$, we obtain the polynomial solution

$$y_1 = c_0\left[1 - \frac{1}{2}x + \frac{1}{10}x^2 - \frac{1}{120}x^3\right], \qquad \text{(32)}$$

but when $c_7 \neq 0$ and $c_0 = 0$, it follows that a second, though infinite series, solution is

$$y_2 = c_7 \left[x^7 + \sum_{n=8}^{\infty} \frac{(-1)^{n+1} 4 \cdot 5 \cdot 6 \cdots (n-4)}{(n-7)! \, 8 \cdot 9 \cdot 10 \cdots n} x^n \right]$$

$$= c_7 \left[x^7 + \sum_{k=1}^{\infty} \frac{(-1)^k 4 \cdot 5 \cdot 6 \cdots (k+3)}{k! \, 8 \cdot 9 \cdot 10 \cdots (k+7)} x^{k+7} \right], \qquad |x| < \infty. \quad \textbf{(33)}$$

Finally, the general solution of (27) on the interval $(0, \infty)$ is

$$y = C_1 y_1(x) + C_2 y_2(x)$$

$$= C_1 \left[1 - \frac{1}{2} x + \frac{1}{10} x^2 - \frac{1}{120} x^3 \right] + C_2 \left[x^7 + \sum_{k=1}^{\infty} \frac{(-1)^k 4 \cdot 5 \cdot 6 \cdots (k+3)}{k! \, 8 \cdot 9 \cdot 10 \cdots (k+7)} x^{k+7} \right].$$

∎

It is interesting to observe that in Example 9 the larger root $r_1 = 7$ was not used. Had we done so, we would have obtained a series solution of the form*

$$y = \sum_{n=0}^{\infty} c_n x^{n+7}, \quad \textbf{(34)}$$

where the c_n are defined by (28) with $r_1 = 7$:

$$c_{k+1} = \frac{-(k+4)}{(k+8)(k+1)} c_k, \qquad k = 0, 1, 2, \dots.$$

Iteration of this latter recurrence relation then would yield only *one* solution, namely, the solution given by (33) (with c_0 playing the part of c_7).

When the roots of the indicial equation differ by a positive integer, the second solution *may* contain a logarithm. In practice this is something we do not know in advance but is determined after we have found the indicial roots and have carefully examined the recurrence relation that defines the coefficients c_n. As the foregoing example shows, we just may be lucky enough to find two solutions that involve only powers of x. On the other hand, if we fail to find a second series-type solution, we can always use the fact that

$$y_2 = y_1(x) \int \frac{e^{-\int P(x) \, dx}}{y_1^2(x)} \, dx \quad \textbf{(35)}$$

is also a solution of the equation $y'' + P(x)y' + Q(x)y = 0$, whenever y_1 is a known solution (see Section 4.2).

---

*Observe that both (33) and (34) start with the power x^7. In Case II it is always a good idea to work with the smaller root first.

EXAMPLE 10

Find the general solution of $xy'' + 3y' - y = 0$.

Solution Recall from Example 7 that the method of Frobenius provides only one solution to this equation, namely,

$$y_1 = \sum_{n=0}^{\infty} \frac{2}{n!(n+2)!} x^n = 1 + \frac{1}{3}x + \frac{1}{24}x^2 + \frac{1}{360}x^3 + \cdots. \tag{36}$$

From (35) we obtain a second solution:

$$y_2 = y_1(x) \int \frac{e^{-\int (3/x)\,dx}}{y_1^2(x)}\,dx = y_1(x) \int \frac{dx}{x^3\left[1 + \dfrac{1}{3}x + \dfrac{1}{24}x^2 + \dfrac{1}{360}x^3 + \cdots\right]^2}$$

$$= y_1(x) \int \frac{dx}{x^3\left[1 + \dfrac{2}{3}x + \dfrac{7}{36}x^2 + \dfrac{1}{30}x^3 + \cdots\right]} \qquad \Longleftarrow \boxed{\text{squaring}}$$

$$= y_1(x) \int \frac{1}{x^3}\left[1 - \dfrac{2}{3}x + \dfrac{1}{4}x^2 - \dfrac{19}{270}x^3 + \cdots\right] dx \qquad \Longleftarrow \boxed{\text{long division}}$$

$$= y_1(x) \int \left[\dfrac{1}{x^3} - \dfrac{2}{3x^2} + \dfrac{1}{4x} - \dfrac{19}{270} + \cdots\right] dx$$

$$= y_1(x) \left[-\dfrac{1}{2x^2} + \dfrac{2}{3x} + \dfrac{1}{4}\ln x - \dfrac{19}{270}x + \cdots\right]$$

or

$$y_2 = \frac{1}{4}y_1(x)\ln x + y_1(x)\left[-\dfrac{1}{2x^2} + \dfrac{2}{3x} - \dfrac{19}{270}x + \cdots\right]. \tag{37}$$

Hence on the interval $(0, \infty)$, the general solution is

$$y = C_1 y_1(x) + C_2\left[\frac{1}{4}y_1(x)\ln x + y_1(x)\left(-\dfrac{1}{2x^2} + \dfrac{2}{3x} - \dfrac{19}{270}x + \cdots\right)\right], \tag{38}$$

where $y_1(x)$ is defined by (36).   ∎

Alternative Procedure

There are several alternative procedures to formula (35) when the method of Frobenius fails to provide a second series solution. Although the next method is somewhat tedious, it is nonetheless straightforward. The basic idea is to assume a solution either of the form (25b) or (26b) and determine coefficients b_n in terms of the coefficients c_n that define the known solution $y_1(x)$.

EXAMPLE 11

The smaller of the two indicial roots for the equation $xy'' + 3y' - y = 0$ is $r_2 = -2$. From (25b) we now assume a second solution

$$y_2 = y_1 \ln x + \sum_{n=0}^{\infty} b_n x^{n-2}, \qquad (39)$$

where

$$y_1 = \sum_{n=0}^{\infty} \frac{2}{n!(n+2)!} x^n. \qquad (40)$$

Differentiation of (39) gives

$$y_2' = \frac{y_1}{x} + y_1' \ln x + \sum_{n=0}^{\infty} (n-2)b_n x^{n-3}$$

$$y_2'' = -\frac{y_1}{x^2} + \frac{2y_1'}{x} + y_1'' \ln x + \sum_{n=0}^{\infty} (n-2)(n-3)b_n x^{n-4}$$

so that

$$xy_2'' + 3y_2' - y_2 = \ln x \underbrace{[xy_1'' + 3y_1' - y_1]}_{\text{zero}} + 2y_1' + \frac{2y_1}{x} + \sum_{n=0}^{\infty} (n-2)(n-3)b_n x^{n-3}$$

$$+ 3 \sum_{n=0}^{\infty} (n-2)b_n x^{n-3} - \sum_{n=0}^{\infty} b_n x^{n-2}$$

$$= 2y_1' + \frac{2y_1}{x} + \sum_{n=0}^{\infty} (n-2)nb_n x^{n-3} - \sum_{n=0}^{\infty} b_n x^{n-2}, \qquad (41)$$

where we have combined the first two summations and used the fact that $xy_1'' + 3y_1' - y_1 = 0$,

By differentiating (40), we can write (41) as

$$\sum_{n=0}^{\infty} \frac{4n}{n!(n+2)!} x^{n-1} + \sum_{n=0}^{\infty} \frac{4}{n!(n+2)!} x^{n-1} + \sum_{n=0}^{\infty} (n-2)nb_n x^{n-3} - \sum_{n=0}^{\infty} b_n x^{n-2}$$

$$= 0(-2)b_0 x^{-3} + (-b_0 - b_1)x^{-2} + \underbrace{\sum_{n=0}^{\infty} \frac{4(n+1)}{n!(n+2)!} x^{n-1}}_{k=n} + \underbrace{\sum_{n=2}^{\infty} (n-2)nb_n x^{n-3}}_{k=n-2} - \underbrace{\sum_{n=1}^{\infty} b_n x^{n-2}}_{k=n-1}$$

$$= -(b_0 + b_1)x^{-2} + \sum_{k=0}^{\infty} \left[\frac{4(k+1)}{k!(k+2)!} + k(k+2)b_{k+2} - b_{k+1} \right] x^{k-1}. \qquad (42)$$

Setting (42) equal to zero then gives $b_1 = -b_0$ and

$$\frac{4(k+1)}{k!(k+2)!} + k(k+2)b_{k+2} - b_{k+1} = 0, \qquad \text{for } k = 0, 1, 2, \ldots. \qquad (43)$$

When $k = 0$ in (43), we have $2 + 0 \cdot 2b_2 - b_1 = 0$ so that $b_1 = 2, b_0 = -2$, but b_2 is arbitrary.

Rewriting (43) as

$$b_{k+2} = \frac{b_{k+1}}{k(k+2)} - \frac{4(k+1)}{k!(k+2)!k(k+2)} \tag{44}$$

and evaluating for $k = 1, 2, \ldots$ give

$$b_3 = \frac{b_2}{3} - \frac{4}{9}$$

$$b_4 = \frac{1}{8}b_3 - \frac{1}{32} = \frac{1}{24}b_2 - \frac{25}{288}$$

and so on. Thus we can finally write

$$
\begin{aligned}
y_2 &= y_1 \ln x + b_0 x^{-2} + b_1 x^{-1} + b_2 + b_3 x + \cdots \\
&= y_1 \ln x - 2x^{-2} + 2x^{-1} + b_2 + \left(\frac{b_2}{3} - \frac{4}{9}\right)x + \cdots,
\end{aligned} \tag{45}
$$

where b_2 is arbitrary.   ∎

Equivalent Solutions

At this point you may be wondering whether (37) and (45) are really equivalent. If we choose $C_2 = 4$ in (38), then

$$
\begin{aligned}
y_2 &= y_1 \ln x + y_1\left(-\frac{2}{x^2} + \frac{8}{3x} - \frac{38}{135}x + \cdots\right) \\
&= y_1 \ln x + \left(1 + \frac{1}{3}x + \frac{1}{24}x^2 + \frac{1}{360}x^3 + \cdots\right)\left(-\frac{2}{x^2} + \frac{8}{3x} - \frac{38}{135}x + \cdots\right) \\
&= y_1 \ln x - 2x^{-2} + 2x^{-1} + \frac{29}{36} - \frac{19}{108}x + \cdots,
\end{aligned} \tag{46}
$$

which is precisely what we obtain from (45) if b_2 is chosen as $29/36$.

The next example illustrates the case when the indicial roots are equal.

EXAMPLE 12

Find the general solution of

$$xy'' + y' - 4y = 0. \tag{47}$$

Solution The assumption $y = \sum_{n=0}^{\infty} c_n x^{n+r}$ leads to

$$xy'' + y' - 4y = \sum_{n=0}^{\infty} (n+r)(n+r-1)c_n x^{n+r-1} + \sum_{n=0}^{\infty} (n+r)c_n x^{n+r-1} - 4\sum_{n=0}^{\infty} c_n x^{n+r}$$

$$= \sum_{n=0}^{\infty} (n+r)^2 c_n x^{n+r-1} - 4\sum_{n=0}^{\infty} c_n x^{n+r}$$

$$= x^r \left[r^2 c_0 x^{-1} + \underbrace{\sum_{n=1}^{\infty} (n+r)^2 c_n x^{n-1}}_{k=n-1} - 4\underbrace{\sum_{n=0}^{\infty} c_n x^n}_{k=n} \right]$$

$$= x^r \left[r^2 c_0 x^{-1} + \sum_{k=0}^{\infty} [(k+r+1)^2 c_{k+1} - 4c_k] x^k \right] = 0.$$

Therefore $r^2 = 0$, and so the indicial roots are equal: $r_1 = r_2 = 0$. Moreover, we have

$$(k+r+1)^2 c_{k+1} - 4c_k = 0, \qquad k = 0, 1, 2, \ldots. \qquad \textbf{(48)}$$

Clearly the root $r_1 = 0$ will only yield one solution corresponding to the coefficients defined by the iteration of

$$c_{k+1} = \frac{4c_k}{(k+1)^2}, \qquad k = 0, 1, 2, \ldots.$$

The result is
$$y_1 = c_0 \sum_{n=0}^{\infty} \frac{4^n}{(n!)^2} x^n, \qquad |x| < \infty. \qquad \textbf{(49)}$$

To obtain the second linearly independent solution we set $c_0 = 1$ in (49) and then use (35):

$$y_2 = y_1(x) \int \frac{e^{-\int (1/x)\,dx}}{y_1^2(x)}\,dx = y_1(x) \int \frac{dx}{x\left[1 + 4x + 4x^2 + \dfrac{16}{9}x^3 + \cdots\right]^2}$$

$$= y_1(x) \int \frac{dx}{x\left[1 + 8x + 24x^2 + \dfrac{16}{9}x^3 + \cdots\right]}$$

$$= y_1(x) \int \frac{1}{x}\left[1 - 8x + 40x^2 - \frac{1472}{9}x^3 + \cdots\right] dx$$

$$= y_1(x) \int \left[\frac{1}{x} - 8 + 40x - \frac{1472}{9}x^2 + \cdots\right] dx$$

$$= y_1(x)\left[\ln x - 8x + 20x^2 - \frac{1472}{27}x^3 + \cdots\right]. \qquad \textbf{(50)}$$

Thus on the interval $(0, \infty)$ the general solution of (47) is

$$y = C_1 y_1(x) + C_2 \left[y_1(x) \ln x + y_1(x)\left(-8x + 20x^2 - \frac{1472}{27}x^3 + \cdots \right) \right], \quad (51)$$

where $y_1(x)$ is defined by (49). ■

As in Case II we can also determine $y_2(x)$ of Example 11 directly from assumption (26b).

Remarks **(i)** We purposely have not considered two further complications when solving a differential equation such as (1) about a point x_0 for which $a_2(x_0) = 0$. When using (3), it is quite possible that the roots of the indicial equation could turn out to be complex numbers. When the exponents r_1 and r_2 are complex, the statement $r_1 > r_2$ is meaningless and must be replaced with $\mathrm{Re}(r_1) > \mathrm{Re}(r_2)$ (for example, if $r = \alpha + i\beta$, then $\mathrm{Re}(r) = \alpha$). In particular, when the indicial equation has real coefficients, the complex roots will be a conjugate pair $r_1 = \alpha + i\beta, r_2 = \alpha - i\beta$, and $r_1 - r_2 = 2i\beta \neq$ integer. Thus for $x_0 = 0$ there will always exist two solutions

$$y_1 = \sum_{n=0}^{\infty} c_n x^{n+r_1} \quad \text{and} \quad y_2 = \sum_{n=0}^{\infty} b_n x^{n+r_2}.$$

Unfortunately both solutions give complex values of y for each real choice of x. This latter difficulty can be surmounted by the superposition principle. Since a combination of solutions is also a solution to the differential equation, we could form appropriate combinations of $y_1(x)$ and $y_2(x)$ to yield real solutions (see Case III of the solution of the Cauchy-Euler equation).

(ii) If $x = 0$ is an irregular singular point, it should be noted that we may not be able to find *any* solution of the form $y = \sum_{n=0}^{\infty} c_n x^{n+r}$.

(iii) In the advanced study of differential equations it is sometimes important to examine the nature of a singular point at ∞. A differential equation is said to have a singular point at ∞ if, after the substitution $z = 1/x$, the resulting equation has a singular point at $z = 0$. For example, the differential equation $y'' + xy = 0$ has no finite singular points. However, by the chain rule the substitution $z = 1/x$ transforms the equation into

$$z^5 \frac{d^2 y}{dz^2} + 2z^4 \frac{dy}{dz} + y = 0.$$

(Verify this.) Inspection of $P(z) = 2/z$ and $Q(z) = 1/z^5$ shows that $z = 0$ is an irregular singular point of the equation. Hence ∞ is an irregular singular point. See Problem 40.

6.4 **EXERCISES** *Answers to odd-numbered problems begin on page A–15.*

[6.4.1] In Problems 1–10 determine the singular points of each differential equation. Classify each singular point as regular or irregular.

1. $x^3 y'' + 4x^2 y' + 3y = 0$

2. $xy'' - (x + 3)^{-2}y = 0$

3. $(x^2 - 9)^2 y'' + (x + 3)y' + 2y = 0$

4. $y'' - \dfrac{1}{x} y' + \dfrac{1}{(x-1)^3} y = 0$

5. $(x^3 + 4x)y'' - 2xy' + 6y = 0$

6. $x^2(x - 5)^2 y'' + 4xy' + (x^2 - 25)y = 0$

7. $(x^2 + x - 6)y'' + (x + 3)y' + (x - 2)y = 0$

8. $x(x^2 + 1)^2 y'' + y = 0$

9. $x^3(x^2 - 25)(x - 2)^2 y'' + 3x(x - 2)y' + 7(x + 5)y = 0$

10. $(x^3 - 2x^2 - 3x)^2 y'' + x(x - 3)^2 y' - (x + 1)y = 0$

In Problems 11–22 show that the indicial roots do not differ by an integer. Use the method of Frobenius to obtain two linearly independent series solutions about the regular singular point $x_0 = 0$. Form the general solution on $(0, \infty)$.

11. $2xy'' - y' + 2y = 0$

12. $2xy'' + 5y' + xy = 0$

13. $4xy'' + \dfrac{1}{2} y' + y = 0$

14. $2x^2 y'' - xy' + (x^2 + 1)y = 0$

15. $3xy'' + (2 - x)y' - y = 0$

16. $x^2 y'' - \left(x - \dfrac{2}{9}\right)y = 0$

17. $2xy'' - (3 + 2x)y' + y = 0$

18. $x^2 y'' + xy' + \left(x^2 - \dfrac{4}{9}\right)y = 0$

19. $9x^2 y'' + 9x^2 y' + 2y = 0$

20. $2x^2 y'' + 3xy' + (2x - 1)y = 0$

21. $2x^2 y'' - x(x - 1)y' - y = 0$

22. $x(x - 2)y'' + y' - 2y = 0$

[6.4.2] In Problems 23–34 show that the indicial roots differ by an integer. Use the method of Frobenius to obtain two linearly independent series solutions about the regular singular point $x_0 = 0$. Form the general solution on $(0, \infty)$.

23. $xy'' + 2y' - xy = 0$

24. $x^2 y'' + xy' + \left(x^2 - \dfrac{1}{4}\right)y = 0$

25. $x(x - 1)y'' + 3y' - 2y = 0$

26. $y'' + \dfrac{3}{x} y' - 2y = 0$

27. $xy'' + (1 - x)y' - y = 0$

28. $xy'' + y = 0$

29. $xy'' + y' + y = 0$

30. $xy'' - xy' + y = 0$

31. $x^2 y'' + x(x - 1)y' + y = 0$

32. $xy'' + y' - 4xy = 0$

33. $xy'' + (x - 1)y' - 2y = 0$

34. $xy'' - y' + x^3 y = 0$

In Problems 35 and 36 note that $x_0 = 0$ is an irregular singular point of each equation. In each case determine whether the method of Frobenius yields a solution.

35. $x^3 y'' + y = 0$ **36.** $x^2 y'' - y' + y = 0$

37. Solve the Cauchy-Euler equation

$$x^2 y'' + 3xy' - 8y = 0$$

by the method of Frobenius.

38. If $x = 0$ is a regular singular point, use (12) in (2) to show that (13) is the indicial equation obtained from the method of Frobenius.

39. Use (13) to find the indicial equation and exponents of

$$x^2 y'' + \left(\frac{5}{3} x + x^2\right) y' - \frac{1}{3} y = 0.$$

40. (a) Show that the differential equation $x^2 y'' - 4y = 0$ has a singular point at ∞. [*Hint:* See page 304.]

    (b) Classify the singular point at ∞ as either regular or irregular.

6.5 TWO SPECIAL EQUATIONS

The two equations

$$x^2 y'' + xy' + (x^2 - v^2)y = 0 \tag{1}$$

$$(1 - x^2)y'' - 2xy' + n(n + 1)y = 0 \tag{2}$$

occur frequently in advanced studies in applied mathematics, physics, and engineering. They are called **Bessel's equation** and **Legendre's equation**, respectively.* In solving (1) we shall assume $v \geq 0$; whereas in (2) we shall consider only the case when n is a nonnegative integer. Since we seek series solutions

---

*Friedrich Wilhelm Bessel (1784–1846)   Bessel was a German astronomer who in 1838 was the first to measure the distance to a star (61 Cygni). In 1840 he predicted the existence of a planetary mass beyond the orbit of Uranus. The planet Neptune was discovered six years later. Bessel was also the first person to calculate the orbit of Halley's comet. Although Bessel certainly studied equation (1) in his work on planetary motion, the differential equation and its solution were probably discovered by Daniel Bernoulli in his research on determining the displacements of an oscillating chain.

Adrien Marie Legendre (1752–1833)   A French mathematician, Legendre is best remembered for spending almost forty years of his life studying and calculating elliptic integrals. However, the particular polynomial solutions of the equation that bears his name were encountered in his studies of gravitation.

of each equation about $x = 0$, we observe that the origin is a regular singular point of Bessel's equation, but it is an ordinary point of Legendre's equation.

6.5.1 Solution of Bessel's Equation

If we assume $y = \sum_{n=0}^{\infty} c_n x^{n+r}$, then

$$x^2 y'' + xy' + (x^2 - v^2)y = \sum_{n=0}^{\infty} c_n(n+r)(n+r-1)x^{n+r} + \sum_{n=0}^{\infty} c_n(n+r)x^{n+r} + \sum_{n=0}^{\infty} c_n x^{n+r+2}$$

$$- v^2 \sum_{n=0}^{\infty} c_n x^{n+r}$$

$$= c_0(r^2 - r + r - v^2)x^r + x^r \sum_{n=1}^{\infty} c_n[(n+r)(n+r-1)$$

$$+ (n+r) - v^2]x^n + x^r \sum_{n=0}^{\infty} c_n x^{n+2}$$

$$= c_0(r^2 - v^2)x^r + x^r \sum_{n=1}^{\infty} c_n[(n+r)^2 - v^2]x^n + x^r \sum_{n=0}^{\infty} c_n x^{n+2}. \quad \textbf{(3)}$$

From (3) we see that the indicial equation is $r^2 - v^2 = 0$, so that the indicial roots are $r_1 = v$ and $r_2 = -v$. When $r_1 = v$, (3) becomes

$$x^v \sum_{n=1}^{\infty} c_n n(n+2v)x^n + x^v \sum_{n=0}^{\infty} c_n x^{n+2}$$

$$= x^v \left[(1+2v)c_1 x + \underbrace{\sum_{n=2}^{\infty} c_n n(n+2v)x^n}_{k=n-2} + \underbrace{\sum_{n=0}^{\infty} c_n x^{n+2}}_{k=n} \right]$$

$$= x^v \left[(1+2v)c_1 x + \sum_{k=0}^{\infty} [(k+2)(k+2+2v)c_{k+2} + c_k]x^{k+2} \right] = 0.$$

Therefore by the usual argument, we can write

$$(1+2v)c_1 = 0$$

$$(k+2)(k+2+2v)c_{k+2} + c_k = 0$$

or $\qquad c_{k+2} = \dfrac{-c_k}{(k+2)(k+2+2v)}, \qquad k = 0, 1, 2, \ldots. \qquad \textbf{(4)}$

The choice $c_1 = 0$ in (4) implies $c_3 = c_5 = c_7 = \cdots = 0$, so for $k = 0, 2, 4, \ldots$, we find, after letting $k + 2 = 2n$, $n = 1, 2, 3, \ldots$, that

$$c_{2n} = -\frac{c_{2n-2}}{2^2 n(n+v)}. \qquad \textbf{(5)}$$

Thus $c_2 = -\dfrac{c_0}{2^2 \cdot 1 \cdot (1 + v)}$

$$c_4 = -\frac{c_2}{2^2 \cdot 2(2 + v)} = \frac{c_0}{2^4 \cdot 1 \cdot 2(1 + v)(2 + v)}$$

$$c_6 = -\frac{c_4}{2^2 \cdot 3(3 + v)} = -\frac{c_0}{2^6 \cdot 1 \cdot 2 \cdot 3(1 + v)(2 + v)(3 + v)}$$

$$\vdots$$

$$c_{2n} = \frac{(-1)^n c_0}{2^{2n} n!(1 + v)(2 + v) \cdots (n + v)}, \qquad n = 1, 2, 3, \ldots . \tag{6}$$

It is standard practice to choose c_0 to be a specific value, namely,

$$c_0 = \frac{1}{2^v \Gamma(1 + v)},$$

where $\Gamma(1 + v)$ is the Gamma function. See Appendix I. Since this latter function possesses the convenient property $\Gamma(1 + \alpha) = \alpha \Gamma(\alpha)$, we can reduce the indicated product in the denominator of (6) to one term. For example,

$$\Gamma(1 + v + 1) = (1 + v)\Gamma(1 + v)$$

$$\Gamma(1 + v + 2) = (2 + v)\Gamma(2 + v)$$

$$= (2 + v)(1 + v)\Gamma(1 + v).$$

Hence we can write (6) as

$$c_{2n} = \frac{(-1)^n}{2^{2n+v} n!(1 + v)(2 + v) \cdots (n + v)\Gamma(1 + v)}$$

$$= \frac{(-1)^n}{2^{2n+v} n!\Gamma(1 + v + n)}, \qquad n = 0, 1, 2, \ldots .$$

It follows that one solution is

$$y = \sum_{n=0}^{\infty} c_{2n} x^{2n+v} = \sum_{n=0}^{\infty} \frac{(-1)^n}{n!\Gamma(1 + v + n)} \left(\frac{x}{2}\right)^{2n+v}.$$

If $v \geq 0$, the series converges at least on the interval $[0, \infty)$.

Bessel Functions of the First Kind

The foregoing series solution is usually denoted by $J_v(x)$:

$$J_v(x) = \sum_{n=0}^{\infty} \frac{(-1)^n}{n!\Gamma(1 + v + n)} \left(\frac{x}{2}\right)^{2n+v}. \tag{7}$$

Also, for the second exponent $r_2 = -v$, we obtain, in exactly the same manner,

$$J_{-v}(x) = \sum_{n=0}^{\infty} \frac{(-1)^n}{n!\Gamma(1 - v + n)} \left(\frac{x}{2}\right)^{2n-v}. \tag{8}$$

The functions $J_v(x)$ and $J_{-v}(x)$ are called **Bessel functions of the first kind** of order v and $-v$, respectively. Depending on the value of v, (8) may contain negative powers of x and hence converges on $(0, \infty)$.*

Now some care must be taken in writing the general solution of (1). When $v = 0$, it is apparent that (7) and (8) are the same. If $v > 0$ and $r_1 - r_2 = v - (-v) = 2v$ is not a positive integer, it follows from Case I of Section 6.4 that $J_v(x)$ and $J_{-v}(x)$ are linearly independent solutions of (1) on $(0, \infty)$, and so the general solution of the interval would be $y = c_1 J_v(x) + c_2 J_{-v}(x)$. But we also know from Case II of Section 6.4 that when $r_1 - r_2 = 2v$ is a positive integer, a second series solution of (1) *may* exist. In this second case we distinguish two possibilities. When $v = m =$ positive integer, $J_{-m}(x)$ defined by (8) and $J_m(x)$ are not linearly independent solutions. It can be shown that J_{-m} is a constant multiple of J_m (see Property (i) on page 310). In addition, $r_1 - r_2 = 2v$ can be a positive integer when v is half an odd positive integer. It can be shown in this latter event that $J_v(x)$ and $J_{-v}(x)$ are linearly independent. In other words, the general solution of (1) on $(0, \infty)$ is

$$y = c_1 J_v(x) + c_2 J_{-v}(x), \qquad v \neq \text{integer.} \tag{9}$$

Figure 6.6

The graphs of $y = J_0(x)$ and $y = J_1(x)$ are given in Figure 6.6. Observe that the graphs of J_0 and J_1 resemble damped cosine and sine graphs, respectively.†

EXAMPLE I

Find the general solution of the equation

$$x^2 y'' + xy' + \left(x^2 - \frac{1}{4}\right) y = 0$$

on $(0, \infty)$.

Solution   We identify $v^2 = 1/4$ and so $v = 1/2$. From (9) we see that the general solution of the differential equation is

$$y = c_1 J_{1/2}(x) + c_2 J_{-1/2}(x). \qquad \blacksquare$$

Bessel Functions of the Second Kind

If $v \neq$ integer, the function defined by the linear combination

$$Y_v(x) = \frac{\cos v\pi J_v(x) - J_{-v}(x)}{\sin v\pi} \tag{10}$$

---

* By replacing x by $|x|$, the series given in (7) and (8) converge for $0 < |x| < \infty$.

† Bessel functions belong to a class of functions that are called "almost periodic."

and the function $J_v(x)$ are linearly independent solutions of (1). Thus another form of the general solution of (1) is $y = c_1 J_v(x) + c_2 Y_v(x)$, provided $v \neq$ integer. As $v \to m$, m an integer, (10) has the indeterminate form 0/0. However, it can be shown by L'Hôpital's rule that $\lim_{v \to m} Y_v(x)$ exists. Moreover, the function

$$Y_m(x) = \lim_{v \to m} Y_v(x)$$

and $J_m(x)$ are linearly independent solutions of $x^2 y'' + xy' + (x^2 - m^2)y = 0$. Hence for *any* value of v the general solution of (1) on $(0, \infty)$ can be written as

$$y = c_1 J_v(x) + c_2 Y_v(x). \tag{11}$$

$Y_v(x)$ is sometimes called **Neumann's function;*** more commonly, $Y_v(x)$ is called the **Bessel function of the second kind** of order v. Figure 6.7 shows the graphs of $Y_0(x)$ and $Y_1(x)$.

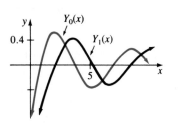

Figure 6.7

EXAMPLE 2

Find the general solution of the equation

$$x^2 y'' + xy' + (x^2 - 9)y = 0$$

on $(0, \infty)$.

Solution Since $v^2 = 9$, we identify $v = 3$. It follows from (11) that the general solution of the differential equation is

$$y = c_1 J_3(x) + c_2 Y_3(x). \qquad \blacksquare$$

Parametric Bessel Equation

By replacing x by λx in (1) and using the chain rule, we obtain an alternative form of Bessel's equation known as the **parametric Bessel equation**:

$$x^2 y'' + xy' + (\lambda^2 x^2 - v^2)y = 0. \tag{12}$$

The general solution of (12) is

$$y = c_1 J_v(\lambda x) + c_2 Y_v(\lambda x). \tag{13}$$

Properties

We list below a few of the more useful properties of Bessel functions of order m, $m = 0, 1, 2, \ldots$:

(i) $J_{-m}(x) = (-1)^m J_m(x)$

(ii) $J_m(-x) = (-1)^m J_m(x)$

(iii) $J_m(0) = 0, \quad m > 0$

---

* The function in (10) is also denoted by $N_v(x)$ in honor of the German mathematician C. Neumann (1832–1925), who investigated its properties.

(iv) $J_0(0) = 1$

(v) $\lim_{x \to 0^+} Y_m(x) = -\infty$

Note that Property (ii) indicates $J_m(x)$ is an even function if m is an even integer and an odd function if m is an odd integer. The graphs of $Y_0(x)$ and $Y_1(x)$ in Figure 6.7 illustrate Property (v): $Y_m(x)$ is unbounded at the origin. This last fact is not obvious from (10). It can be shown either from (10) or by the methods of Section 6.4 that for $x > 0$,

$$Y_0(x) = \frac{2}{\pi} J_0(x)\left[\gamma + \ln\frac{x}{2}\right] - \frac{2}{\pi}\sum_{k=1}^{\infty} \frac{(-1)^k}{(k!)^2}\left(1 + \frac{1}{2} + \cdots + \frac{1}{k}\right)\left(\frac{x}{2}\right)^{2k},$$

where $\gamma = 0.57721566\ldots$ is called the **Euler-Mascheroni constant**. Observe that because of the presence of the logarithmic term, $Y_0(x)$ is discontinuous at $x = 0$.

Numerical Values of $J_0(x)$ and $J_1(x)$

Functional values of $J_0(x)$ and $J_1(x)$ for some selected values of x are given in Table 6.1. The zeros of these two functions, namely, the values of x for

TABLE 6.1   Numerical Values of J_0 and J_1

x	$J_0(x)$	$J_1(x)$
0	1.0000	0.0000
1	0.7652	0.4401
2	0.2239	0.5767
2.405	0.0000	0.5191
3	−0.2601	0.3391
3.832	−0.4028	0.0000
4	−0.3971	−0.0660
5	−0.1776	−0.3276
5.520	0.0000	−0.3403
6	0.1506	−0.2767
7	0.3001	−0.0047
7.016	0.3001	0.0000
8	0.1717	0.2346
8.654	0.0000	0.2715
9	−0.0903	0.2453
10	−0.2459	0.0435
10.173	−0.2497	0.0000
11	−0.1712	−0.1768
11.792	0.0000	−0.2325
12	0.0477	−0.2234
13	0.2069	−0.0703
13.323	0.2184	0.0000
14	0.1711	0.1334
14.931	0.0000	0.2065

which $J_0(x) = 0$ and $J_1(x) = 0$, are important in some applications of partial differential equations. Numerical values of $\int_0^x J_0(r)\, dr$ are also important.*

Recurrence Relation

Recurrence formulas that relate Bessel functions of different orders are important in theory and in applications. In the next example we derive a **differential recurrence relation**.

EXAMPLE 3

Derive the formula

$$xJ_v'(x) = vJ_v(x) - xJ_{v+1}(x).$$

Solution It follows from (7) that

$$J_v(x) = \sum_{n=0}^{\infty} \frac{(-1)^n}{n!\,\Gamma(1+v+n)}\left(\frac{x}{2}\right)^{2n+v}$$

$$= v\sum_{n=0}^{\infty} \frac{(-1)^n}{n!\,\Gamma(1+v+n)}\left(\frac{x}{2}\right)^{2n+v} + 2\sum_{n=0}^{\infty} \frac{(-1)^n n}{n!\,\Gamma(1+v+n)}\left(\frac{x}{2}\right)^{2n+v}$$

$$= vJ_v(x) + x\underbrace{\sum_{n=1}^{\infty} \frac{(-1)^n}{(n-1)!\,\Gamma(1+v+n)}\left(\frac{x}{2}\right)^{2n+v-1}}_{\boxed{k=n-1}}$$

$$= vJ_v(x) - x\sum_{k=0}^{\infty} \frac{(-1)^k}{k!\,\Gamma(2+v+k)}\left(\frac{x}{2}\right)^{2k+v+1}$$

$$= vJ_v(x) - xJ_{v+1}(x).\qquad\blacksquare$$

The result in Example 3 can be written in an alternative form. Dividing $xJ_v'(x) - vJ_v(x) = -xJ_{v+1}(x)$ by x gives

$$J_v'(x) - \frac{v}{x}J_v(x) = -J_{v+1}(x).$$

This last expression is recognized as a linear first-order differential equation in $J_v(x)$. Multiplying both sides of the equality by the integrating factor x^{-v} then yields

$$\frac{d}{dx}\left[x^{-v}J_v(x)\right] = -x^{-v}J_{v+1}(x). \tag{14}$$

---

* See *Handbook of Mathematical Functions*, edited by Milton Abramowitz and Irene A. Stegun (New York: Dover Publications, 1972).

We leave it as an exercise to derive a similar formula:

$$\frac{d}{dx}\left[x^{\nu}J_{\nu}(x)\right] = x^{\nu}J_{\nu-1}(x). \tag{15}$$

(See Problem 20.)

When $v = $ half an odd integer, $J_{\nu}(x)$ can be expressed in terms of $\sin x$, $\cos x$, and powers of x. Such Bessel functions are called **spherical Bessel functions**.

EXAMPLE 4

Find an alternative expression for $J_{1/2}(x)$. Use the fact that $\Gamma(\tfrac{1}{2}) = \sqrt{\pi}$.

Solution   With $v = 1/2$, we have from (7)

$$J_{1/2}(x) = \sum_{n=0}^{\infty} \frac{(-1)^n}{n!\,\Gamma(1+\frac{1}{2}+n)}\left(\frac{x}{2}\right)^{2n+1/2}.$$

Now in view of the property $\Gamma(1+\alpha) = \alpha\Gamma(\alpha)$ we obtain

$$n = 0, \qquad \Gamma\left(1+\frac{1}{2}\right) = \frac{1}{2}\Gamma\left(\frac{1}{2}\right) = \frac{1}{2}\sqrt{\pi}$$

$$n = 1, \qquad \Gamma\left(1+\frac{3}{2}\right) = \frac{3}{2}\Gamma\left(\frac{3}{2}\right) = \frac{3}{2^2}\sqrt{\pi}$$

$$n = 2, \qquad \Gamma\left(1+\frac{5}{2}\right) = \frac{5}{2}\Gamma\left(\frac{5}{2}\right) = \frac{5\cdot 3}{2^3}\sqrt{\pi} = \frac{5\cdot 4\cdot 3\cdot 2\cdot 1}{2^3 4\cdot 2}\sqrt{\pi} = \frac{5!}{2^5 2!}\sqrt{\pi}$$

$$n = 3, \qquad \Gamma\left(1+\frac{7}{2}\right) = \frac{7}{2}\Gamma\left(\frac{7}{2}\right) = \frac{7\cdot 5!}{2^6 2!}\sqrt{\pi} = \frac{7\cdot 6\cdot 5!}{2^6\cdot 6\cdot 2!}\sqrt{\pi} = \frac{7!}{2^7 3!}\sqrt{\pi}.$$

In general,
$$\Gamma\left(1+\frac{1}{2}+n\right) = \frac{(2n+1)!}{2^{2n+1}n!}\sqrt{\pi}.$$

Hence
$$J_{1/2}(x) = \sum_{n=0}^{\infty} \frac{(-1)^n}{n!\,\dfrac{(2n+1)!\sqrt{\pi}}{2^{2n+1}n!}}\left(\frac{x}{2}\right)^{2n+\frac{1}{2}}$$

$$= \sqrt{\frac{2}{\pi x}}\sum_{n=0}^{\infty} \frac{(-1)^n}{(2n+1)!}x^{2n+1}.$$

Since the series in the last line is the Maclaurin series for $\sin x$, we have shown that

$$J_{1/2}(x) = \sqrt{\frac{2}{\pi x}}\sin x. \qquad \blacksquare$$

6.5.2   Solution of Legendre's Equation

Since $x = 0$ is an ordinary point of equation (2), we assume a solution of the form $y = \sum_{k=0}^{\infty} c_k x^k$. Therefore

$$(1 - x^2)y'' - 2xy' + n(n + 1)y = (1 - x^2)\sum_{k=0}^{\infty} c_k k(k - 1)x^{k-2} - 2\sum_{k=0}^{\infty} c_k k x^k + n(n + 1)\sum_{k=0}^{\infty} c_k x^k$$

$$= \sum_{k=2}^{\infty} c_k k(k - 1)x^{k-2} - \sum_{k=2}^{\infty} c_k k(k - 1)x^k - 2\sum_{k=1}^{\infty} c_k k x^k + n(n + 1)\sum_{k=0}^{\infty} c_k x^k$$

$$= [n(n + 1)c_0 + 2c_2]x^0 + [n(n + 1)c_1 - 2c_1 + 6c_3]x$$

$$+ \underbrace{\sum_{k=4}^{\infty} c_k k(k - 1)x^{k-2}}_{\boxed{j = k - 2}} - \underbrace{\sum_{k=2}^{\infty} c_k k(k - 1)x^k}_{\boxed{j = k}}$$

$$- \underbrace{2\sum_{k=2}^{\infty} c_k k x^k}_{\boxed{j = k}} + \underbrace{n(n + 1)\sum_{k=2}^{\infty} c_k x^k}_{\boxed{j = k}}$$

$$= [n(n + 1)c_0 + 2c_2] + [(n - 1)(n + 2)c_1 + 6c_3]x$$

$$+ \sum_{j=2}^{\infty} [(j + 2)(j + 1)c_{j+2} + (n - j)(n + j + 1)c_j]x^j = 0$$

implies that

$$n(n + 1)c_0 + 2c_2 = 0$$

$$(n - 1)(n + 2)c_1 + 6c_3 = 0$$

$$(j + 2)(j + 1)c_{j+2} + (n - j)(n + j + 1)c_j = 0$$

or

$$c_2 = -\frac{n(n + 1)}{2!}c_0$$

$$c_3 = -\frac{(n - 1)(n + 2)}{3!}c_1$$

$$c_{j+2} = -\frac{(n - j)(n + j + 1)}{(j + 2)(j + 1)}c_j, \qquad j = 2, 3, 4, \ldots. \tag{16}$$

Iterating (16) gives

$$c_4 = -\frac{(n - 2)(n + 3)}{4 \cdot 3}c_2 = \frac{(n - 2)n(n + 1)(n + 3)}{4!}c_0$$

$$c_5 = -\frac{(n-3)(n+4)}{5\cdot 4}\, c_3 = \frac{(n-3)(n-1)(n+2)(n+4)}{5!}\, c_1$$

$$c_6 = -\frac{(n-4)(n+5)}{6\cdot 5}\, c_4 = -\frac{(n-4)(n-2)n(n+1)(n+3)(n+5)}{6!}\, c_0$$

$$c_7 = -\frac{(n-5)(n+6)}{7\cdot 6}\, c_5$$

$$= -\frac{(n-5)(n-3)(n-1)(n+2)(n+4)(n+6)}{7!}\, c_1$$

and so on. Thus for at least $|x| < 1$, we obtain two linearly independent power series solutions

$$y_1(x) = c_0\left[1 - \frac{n(n+1)}{2!}x^2 + \frac{(n-2)n(n+1)(n+3)}{4!}x^4\right.$$

$$\left. - \frac{(n-4)(n-2)n(n+1)(n+3)(n+5)}{6!}x^6 + \cdots\right]$$

$$\tag{17}$$

$$y_2(x) = c_1\left[x - \frac{(n-1)(n+2)}{3!}x^3 + \frac{(n-3)(n-1)(n+2)(n+4)}{5!}x^5\right.$$

$$\left. - \frac{(n-5)(n-3)(n-1)(n+2)(n+4)(n+6)}{7!}x^7 + \cdots\right].$$

Notice that if n is an even integer, the first series terminates, whereas $y_2(x)$ is an infinite series. For example, if $n = 4$, then

$$y_1(x) = c_0\left[1 - \frac{4\cdot 5}{2!}x^2 + \frac{2\cdot 4\cdot 5\cdot 7}{4!}x^4\right] = c_0\left[1 - 10x^2 + \frac{35}{3}x^4\right].$$

Similarly, when n is an odd integer, the series for $y_2(x)$ terminates with x^n; that is, *when n is a nonnegative integer we obtain an nth-degree polynomial solution* of Legendre's equation.

Since we know that a constant multiple of a solution of Legendre's equation is also a solution, it is traditional to choose specific values for c_0 and c_1, depending on whether n is an even or odd positive integer, respectively. For $n = 0$ we choose $c_0 = 1$ and for $n = 2, 4, 6, \ldots,$

$$c_0 = (-1)^{n/2}\frac{1\cdot 3\cdots(n-1)}{2\cdot 4\cdots n},$$

whereas for $n = 1$ we choose $c_1 = 1$ and for $n = 3, 5, 7, \ldots,$

$$c_1 = (-1)^{(n-1)/2}\frac{1\cdot 3\cdots n}{2\cdot 4\cdots(n-1)}.$$

For example, when $n = 4$, we have

$$y_1(x) = (-1)^{4/2} \frac{1 \cdot 3}{2 \cdot 4} \left[1 - 10x^2 + \frac{35}{3} x^4 \right]$$

$$= \frac{3}{8} - \frac{30}{8} x^2 + \frac{35}{8} x^4$$

$$= \frac{1}{8} (35x^4 - 30x^2 + 3).$$

Legendre Polynomials

These specific nth-degree polynomial solutions are called **Legendre polynomials** and are denoted by $P_n(x)$. From the series for $y_1(x)$ and $y_2(x)$ and from the above choices of c_0 and c_1, we find that the first several Legendre polynomials are

$$P_0(x) = 1, \qquad\qquad P_1(x) = x,$$

$$P_2(x) = \frac{1}{2} (3x^2 - 1), \qquad\qquad P_3(x) = \frac{1}{2} (5x^3 - 3x), \qquad\qquad \text{(18)}$$

$$P_4(x) = \frac{1}{8} (35x^4 - 30x^2 + 3), \qquad P_5(x) = \frac{1}{8} (63x^5 - 70x^3 + 15x).$$

Remember, $P_0(x)$, $P_1(x)$, $P_2(x)$, $P_3(x)$, ... are, in turn, particular solutions of the differential equations

$$n = 0, \qquad (1 - x^2)y'' - 2xy' = 0$$

$$n = 1, \qquad (1 - x^2)y'' - 2xy' + 2y = 0$$

$$n = 2, \qquad (1 - x^2)y'' - 2xy' + 6y = 0 \qquad\qquad \text{(19)}$$

$$n = 3, \qquad (1 - x^2)y'' - 2xy' + 12y = 0$$

$$\vdots$$

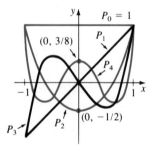

Figure 6.8

The graphs of the first four Legendre polynomials on the interval $-1 \le x \le 1$ are given in Figure 6.8.

Properties

The following properties of the Legendre polynomials are apparent in (18) and Figure 6.8:

(i) $P_n(-x) = (-1)^n P_n(x)$

(ii) $P_n(1) = 1$

(iii) $P_n(-1) = (-1)^n$

(iv) $P_n(0) = 0, \quad n = 1, 3, 5, \ldots$

(v) $P'_n(0) = 0, \quad n = 0, 2, 4, \ldots$

Property (i) indicates that $P_n(x)$ is an even or odd function according to whether n is even or odd.

Recurrence Relation

Recurrence relations that relate Legendre polynomials of different degrees are also very important in some aspects of their application. We shall derive one such relation using the formula

$$(1 - 2xt + t^2)^{-1/2} = \sum_{n=0}^{\infty} P_n(x)t^n. \tag{20}$$

The function on the left is called a **generating function** for the Legendre polynomials. Its derivation follows from the binomial theorem and is left as an exercise. (See Problem 43.)

Differentiating both sides of (20) with respect to t gives

$$(1 - 2xt + t^2)^{-3/2}(x - t) = \sum_{n=0}^{\infty} nP_n(x)t^{n-1} = \sum_{n=1}^{\infty} nP_n(x)t^{n-1}$$

so that after multiplying by $1 - 2xt + t^2$, we have

$$(x - t)(1 - 2xt + t^2)^{-1/2} = (1 - 2xt + t^2) \sum_{n=1}^{\infty} nP_n(x)t^{n-1}$$

or
$$(x - t) \sum_{n=0}^{\infty} P_n(x)t^n = (1 - 2xt + t^2) \sum_{n=1}^{\infty} nP_n(x)t^{n-1}. \tag{21}$$

We multiply out and rewrite (21) as

$$\sum_{n=0}^{\infty} xP_n(x)t^n - \sum_{n=0}^{\infty} P_n(x)t^{n+1} - \sum_{n=1}^{\infty} nP_n(x)t^{n-1} + 2x \sum_{n=1}^{\infty} nP_n(x)t^n - \sum_{n=1}^{\infty} nP_n(x)t^{n+1} = 0$$

or
$$x + x^2 t + \sum_{n=2}^{\infty} xP_n(x)t^n - t - \sum_{n=1}^{\infty} P_n(x)t^{n+1} - x - 2\left(\frac{3x^2 - 1}{2}\right)t$$

$$- \sum_{n=3}^{\infty} nP_n(x)t^{n-1} + 2x^2 t + 2x \sum_{n=2}^{\infty} nP_n(x)t^n - \sum_{n=1}^{\infty} nP_n(x)t^{n+1} = 0.$$

Observing the appropriate cancellations, simplifying, and changing the summation indices give

$$\sum_{k=2}^{\infty} [-(k + 1)P_{k+1}(x) + (2k + 1)xP_k(x) - kP_{k-1}(x)]t^k = 0.$$

Equating the total coefficient of t^k to zero gives the three-term recurrence relation

$$(k + 1)P_{k+1}(x) - (2k + 1)xP_k(x) + kP_{k-1}(x) = 0, \quad k = 2, 3, 4, \ldots. \tag{22}$$

This formula is also valid when $k = 1$.

In (18) we listed the first six Legendre polynomials. If, say, we wish to find $P_6(x)$ we could use (22) with $k = 5$. This relation then expresses $P_6(x)$ in terms of the known quantities $P_4(x)$ and $P_5(x)$. (See Problem 45.)

6.5 EXERCISES

Answers to odd-numbered problems begin on page A–16.

[6.5.1] In Problems 1–8 find the general solution of the given differential equation on $(0, \infty)$.

1. $x^2 y'' + xy' + \left(x^2 - \dfrac{1}{9} \right) y = 0$ 2. $x^2 y'' + xy' + (x^2 - 1)y = 0$

3. $4x^2 y'' + 4xy' + (4x^2 - 25)y = 0$

4. $16x^2 y'' + 16xy' + (16x^2 - 1)y = 0$

5. $xy'' + y' + xy = 0$ 6. $\dfrac{d}{dx}[xy'] + \left(x - \dfrac{4}{x} \right)y = 0$

7. $x^2 y'' + xy' + (9x^2 - 4)y = 0$ 8. $x^2 y'' + xy' + \left(36x^2 - \dfrac{1}{4} \right)y = 0$

9. Use the change of variables $y = x^{-1/2}v(x)$ to find the general solution of the equation
$$x^2 y'' + 2xy' + \lambda^2 x^2 y = 0, \qquad x > 0.$$

10. Verify that the differential equation
$$xy'' + (1 - 2n)y' + xy = 0, \qquad x > 0,$$
possesses the particular solution $y = x^n J_n(x)$.

11. Verify that the differential equation
$$xy'' + (1 + 2n)y' + xy = 0, \qquad x > 0,$$
possesses the particular solution $y = x^{-n} J_n(x)$.

12. Verify that the differential equation
$$x^2 y'' + \left(\lambda^2 x^2 - v^2 + \dfrac{1}{4} \right)y = 0, \qquad x > 0,$$
possesses the particular solution $y = \sqrt{x} J_v(\lambda x)$, where $\lambda > 0$.

In Problems 13–18 use the results of Problems 10, 11, and 12 to find a particular solution of the given differential equation on $(0, \infty)$.

13. $y'' + y = 0$ 14. $xy'' - y' + xy = 0$

15. $xy'' + 3y' + xy = 0$ 16. $4x^2 y'' + (16x^2 + 1)y = 0$

17. $x^2 y'' + (x^2 - 2)y = 0$ 18. $xy'' - 5y' + xy = 0$

In Problems 19–22 derive the given recurrence relation.

19. $xJ'_v(x) = -vJ_v(x) + xJ_{v-1}(x)$ [*Hint:* $2n + v = 2(n + v) - v$.]

20. $\dfrac{d}{dx}[x^v J_v(x)] = x^v J_{v-1}(x)$

21. $2vJ_v(x) = xJ_{v+1}(x) + xJ_{v-1}(x)$ **22.** $2J'_v(x) = J_{v-1}(x) - J_{v+1}(x)$

In Problems 23–26 use (14) or (15) to obtain the given result.

23. $\displaystyle\int_0^x rJ_0(r)\,dr = xJ_1(x)$ **24.** $J'_0(x) = J_{-1}(x) = -J_1(x)$

25. $\displaystyle\int x^n J_0(x)\,dx = x^n J_1(x) + (n-1)x^{n-1}J_0(x) - (n-1)^2\int x^{n-2}J_0(x)\,dx$

26. $\displaystyle\int x^3 J_0(x)\,dx = x^3 J_1(x) + 2x^2 J_0(x) - 4xJ_1(x) + c$

27. Proceed as in Example 4 and express $J_{-1/2}(x)$ in terms of $\cos x$ and a power of x.

In Problems 28–33 use the recurrence relation given in Problem 21 and the results obtained in Problem 27 and Example 4 to express the given Bessel function in terms of $\sin x$, $\cos x$, and powers of x.

28. $J_{3/2}(x)$

29. $J_{-3/2}(x)$ **30.** $J_{5/2}(x)$

31. $J_{-5/2}(x)$ **32.** $J_{7/2}(x)$

33. $J_{-7/2}(x)$

34. Show that $i^{-v}J_v(ix)$, $i^2 = -1$, is a real function. The function defined by $I_v(x) = i^{-v}J_v(ix)$ is called a **modified Bessel function of the first kind** of order v.

35. Find the general solution of the differential equation
$$x^2 y'' + xy' - (x^2 + v^2)y = 0, \qquad x > 0, \qquad v \neq \text{integer}.$$
[*Hint:* $i^2 x^2 = -x^2$.]

36. If $y_1 = J_0(x)$ is one solution of the zero-order Bessel equation, verify that another solution is
$$y_2 = J_0(x)\ln x + \frac{x^2}{4} - \frac{3x^4}{128} + \frac{11x^6}{13,824} - \cdots.$$

37. Use (8) with $v = m$, where m is a positive integer, and the fact that $1/\Gamma(N) = 0$, where N is a negative integer, to show that
$$J_{-m}(x) = (-1)^m J_m(x).$$

38. Use (7) with $v = m$, where m is a nonnegative integer, to show that
$$J_m(-x) = (-1)^m J_m(x).$$

[6.5.2] **39.** **(a)** Use the explicit solutions $y_1(x)$ and $y_2(x)$ of Legendre's equation and the appropriate choices of c_0 and c_1 to find the Legendre polynomials $P_6(x)$ and $P_7(x)$.

(b) Write the explicit differential equations for which $P_6(x)$ and $P_7(x)$ are particular solutions.

40. Show that Legendre's equation has an alternative form

$$\frac{d}{dx}\left[(1 - x^2)\frac{dy}{dx}\right] + n(n + 1)y = 0.$$

41. Show that the equation

$$\sin\theta \frac{d^2 y}{d\theta^2} + \cos\theta \frac{dy}{d\theta} + n(n + 1)(\sin\theta)y = 0$$

can be transformed in Legendre's equation by means of the substitution $x = \cos\theta$.

42. The general Legendre polynomial can be written as

$$P_n(x) = \sum_{k=0}^{[n/2]} \frac{(-1)^k (2n - 2k)!}{2^n k!(n - k)!(n - 2k)!} x^{n-2k},$$

where $[n/2]$ is the greatest integer not greater than $n/2$. Verify the results for $n = 0, 1, 2, 3, 4, 5$.

43. Use the binomial series to formally show that

$$(1 - 2xt + t^2)^{-1/2} = \sum_{n=0}^{\infty} P_n(x)t^n.$$

44. Use Problem 43 to show that $P_n(1) = 1$ and $P_n(-1) = (-1)^n$.

45. Use the recurrence relation (22) and $P_0(x) = 1$, $P_1(x) = x$ to generate the next five Legendre polynomials.

46. The Legendre polynomials are also generated by **Rodrigues' formula***

$$P_n(x) = \frac{1}{2^n n!} \frac{d^n}{dx^n} (x^2 - 1)^n.$$

Verify the results for $n = 0, 1, 2, 3$.

47. Use the explicit Legendre polynomials $P_0(x)$, $P_1(x)$, $P_2(x)$, and $P_3(x)$ to evaluate $\int_{-1}^{1} P_n^2(x)\, dx$ for $n = 0, 1, 2, 3$. Generalize the results.

48. Use the explicit Legendre polynomials $P_0(x)$, $P_1(x)$, $P_2(x)$, and $P_3(x)$ to evaluate $\int_{-1}^{1} P_n(x)P_m(x)\, dx$ for $n \neq m$. Generalize the results.

---

*Olinde Rodrigues (1794–1851) Rodrigues was a French banker and amateur mathematician. In mathematics he is remembered solely for the discovery of this one formula in 1816. In politics he is remembered as the financial backer and disciple of Count de Saint-Simon, the founder of French socialism.

49. We know that $y_1 = x$ is a solution of Legendre's equation when $n = 1$, $(1 - x^2)y'' - 2xy' + 2y = 0$. Show that a second linearly independent solution on the interval $-1 < x < 1$ is

$$y_2 = \frac{x}{2} \ln\left(\frac{1 + x}{1 - x}\right) - 1.$$

Chapter 6 *REVIEW*

The remarkable characteristic of a Cauchy-Euler equation is that even though it is a differential equation with variable coefficients, it can be solved in terms of elementary functions. A **second-order Cauchy-Euler** equation is any differential equation of the form $ax^2y'' + bxy' + cy = g(x)$, where a, b, and c are constants. To solve the homogeneous equation we try a solution of the form $y = x^m$ and this in turn leads to an algebraic **auxiliary equation** $am(m - 1) + bm + c = 0$. Accordingly, when the roots are real and distinct, real and equal, or complex conjugates, the general solutions on the interval $(0, \infty)$ are, respectively,

$$y = c_1 x^{m_1} + c_2 x^{m_2}$$

$$y = c_1 x^{m_1} + c_2 x^{m_1} \ln x$$

and
$$y = x^{\alpha}[c_1 \cos(\beta \ln x) + c_2 \sin(\beta \ln x)].$$

We say that $x = 0$ is an **ordinary point** of the linear second-order differential equation $a_2(x)y'' + a_1(x)y' + a_0(x)y = 0$ when $a_2(0) \neq 0$ and $a_2(x)$, $a_1(x)$, $a_0(x)$ are polynomials having no common factors. Every solution has the form of a power series in x, $y = \sum_{n=0}^{\infty} c_n x^n$. To find the coefficients c_n we substitute the basic assumption into the differential equation, and after appropriate algebraic manipulations we determine a **recurrence relation** by equating to zero the combined total coefficient of x^k. Iteration of the recurrence relation yields two distinct sets of coefficients, one set containing the arbitrary coefficient c_0 and the other containing c_1. Using each set of coefficients, we form two linearly independent solutions $y_1(x)$ and $y_2(x)$. A solution is valid at least on an interval defined by $|x| < R$, where R is the distance from $x = 0$ to the closest singular point of the equation. If $a_2(0) = 0$, then $x = 0$ is a **singular point**. Singular points are classified as either **regular** or **irregular**. To determine whether $x = 0$ is a regular singular point, we examine the denominators of the rational functions P and Q that result when the equation is put into the form $y'' + P(x)y' + Q(x)y = 0$. It is understood that $a_1(x)/a_2(x)$ and $a_0(x)/a_2(x)$ are reduced to lowest terms. If x appears *at most* to the first power in the denominator of $P(x)$ and *at most* to the second power in the denominator of $Q(x)$, then $x = 0$ is a regular singular point. Around the regular singular point $x = 0$, the **method of Frobenius** guarantees that there exists *at least one* solution of the form $y = \sum_{n=0}^{\infty} c_n x^{n+r}$. The exponent r is a root of a quadratic **indicial equation**. When the indicial roots r_1 and r_2 $(r_1 > r_2)$ satisfy $r_1 - r_2 \neq$ an integer, then we can always find *two* linearly independent solutions of the assumed form. When $r_1 - r_2 =$ a positive integer, then we could *possibly* find two solutions, but when $r_1 - r_2 = 0$, or $r_1 = r_2$, we can find only one solution of the form $y = \sum_{n=0}^{\infty} c_n x^{n+r}$.

Bessel's equation $x^2 y'' + xy' + (x^2 - v^2)y = 0$ has a regular singular point at $x = 0$, whereas $x = 0$ is an ordinary point of **Legendre's equation** $(1 - x^2)y'' - 2xy' + n(n + 1)y = 0$. The latter equation possesses a polynomial solution when n is a nonnegative integer.

Chapter 6 *REVIEW EXERCISES*

Answers to odd-numbered problems begin on page A–18.

In Problems 1–4 solve the given Cauchy-Euler equation.

1. $6x^2 y'' + 5xy' - y = 0$

2. $2x^3 y''' + 19x^2 y'' + 39xy' + 9y = 0$

3. $x^2 y'' - 4xy' + 6y = 2x^4 + x^2$ **4.** $x^2 y'' - xy' + y = x^3$

5. Specify the ordinary points of $(x^3 - 8)y'' - 2xy' + y = 0$.

6. Specify the singular points of $(x^4 - 16)y'' + 2y = 0$.

In Problems 7–10 specify the regular and irregular singular points of the given differential equation.

7. $(x^3 - 10x^2 + 25x)y'' + y' = 0$ **8.** $(x^3 - 10x^2 + 25x)y'' + y = 0$

9. $x^2(x^2 - 9)^2 y'' - (x^2 - 9)y' + xy = 0$

10. $x(x^2 + 1)^3 y'' + y' - 8xy = 0$

In Problems 11 and 12 specify an interval around $x = 0$ for which a power series solution of the given differential equation will converge.

11. $y'' - xy' + 6y = 0$ **12.** $(x^2 - 4)y'' - 2xy' + 9y = 0$

In Problems 13–16 for each differential equation find two power series solutions about the ordinary point $x = 0$.

13. $y'' + xy = 0$ **14.** $y'' - 4y = 0$

15. $(x - 1)y'' + 3y = 0$ **16.** $y'' - x^2 y' + xy = 0$

In Problems 17–22 find two linearly independent solutions of each equation.

17. $2x^2 y'' + xy' - (x + 1)y = 0$ **18.** $2xy'' + y' + y = 0$

19. $x(1 - x)y'' - 2y' + y = 0$ **20.** $x^2 y'' - xy' + (x^2 + 1)y = 0$

21. $xy'' - (2x - 1)y' + (x - 1)y = 0$ **22.** $x^2 y'' - x^2 y' + (x^2 - 2)y = 0$

23. Without referring to Section 6.5 use the method of Frobenius to obtain a solution of the Bessel equation for $v = 0$: $xy'' + y' + xy = 0$.

LAPLACE TRANSFORM

7.1 Laplace Transform

7.2 Inverse Transform

7.3 Translation Theorems and Derivatives of a Transform

7.4 Transforms of Derivatives, Integrals, and Periodic Functions

7.5 Applications

[O] **7.6** Dirac Delta Function

Chapter 7 Review

Chapter 7 Review Exercises

Important Concepts

Linear operation

Laplace transform

Linear transform

Piecewise continuous

Exponential order

Inverse Laplace transform

First translation theorem

Unit step function

Second translation theorem

Convolution

Convolution theorem

Volterra integral equation

Integrodifferential equation

Unit impulse

Dirac delta function

In this chapter we examine the definition and properties of an integral known as the **Laplace transform**.

We shall see in Section 7.5 that when the Laplace transform is applied to a linear nth-order differential equation with constant coefficients

$$a_n \frac{d^n y}{dt^n} + a_{n-1} \frac{d^{n-1} y}{dt^{n-1}} + \cdots + a_1 \frac{dy}{dt} + a_0 y = g(t),$$

the differential equation is transformed into an *algebraic equation* that involves the conditions $y(0)$, $y'(0)$, $y''(0), \ldots, y^{(n-1)}(0)$. As a consequence of this property, the Laplace transform is well suited to the solution of certain kinds of initial-value problems.

Recall that in physical systems such as a spring-mass system or a series electrical circuit, the right-hand member in the differential equations

$$m \frac{d^2 x}{dt^2} + \beta \frac{dx}{dt} + kx = f(t)$$

$$L \frac{d^2 q}{dt^2} + R \frac{dq}{dt} + \frac{1}{C} q = E(t)$$

Continues

Continued

is a driving function and represents either an external force $f(t)$ or an impressed voltage $E(t)$. In Chapter 5 we solved problems in which the functions f and E were continuous. However, piecewise continuous driving functions are not uncommon. For example, the impressed voltage on a circuit could be

or

In this case, solving the differential equation of the circuit is difficult but not impossible. The Laplace transform is an invaluable aid in solving problems such as these.

7.1 LAPLACE TRANSFORM

In elementary calculus you learned that differentiation and integration transform a function into another function. For example, the function $f(x) = x^2$ is transformed, in turn, into a linear function, a family of cubic polynomial functions, and a constant by the operations of differentiation, indefinite integration, and definite integration:

$$\frac{d}{dx} x^2 = 2x, \qquad \int x^2 \, dx = \frac{x^3}{3} + c, \qquad \int_0^3 x^2 \, dx = 9.$$

Moreover, these three operations possess the **linearity property**. This means that for any constants α and β,

$$\frac{d}{dx}\left[\alpha f(x) + \beta g(x)\right] = \alpha\,\frac{d}{dx}\,f(x) + \beta\,\frac{d}{dx}\,g(x)$$

$$\int\left[\alpha f(x) + \beta g(x)\right]dx = \alpha\int f(x)\,dx + \beta\int g(x)\,dx \qquad (1)$$

$$\int_a^b\left[\alpha f(x) + \beta g(x)\right]dx = \alpha\int_a^b f(x)\,dx + \beta\int_a^b g(x)\,dx$$

provided each derivative and integral exists.

If $f(x, y)$ is a function of two variables, then a definite integral of f with respect to one of the variables leads to a function of the other variable. For example, by holding y constant, we see that $\int_1^2 2xy^2\,dx = 3y^2$. Similarly, a definite integral such as $\int_a^b K(s, t)f(t)\,dt$ transforms a function $f(t)$ into a function of the variable s. We are particularly interested in **integral transforms** of this last kind, where the interval of integration is the unbounded interval $[0, \infty)$.

Basic Definition

If $f(t)$ is defined for $t \geq 0$, then the improper integral

$$\int_0^\infty K(s, t)f(t)\,dt$$

is defined by means of a limit:

$$\int_0^\infty K(s, t)f(t)\,dt = \lim_{b\to\infty}\int_0^b K(s, t)f(t)\,dt.$$

If the limit exists, the integral exists or is convergent; if the limit does not exist, the integral does not exist and is said to be divergent. The foregoing limit will, in general, exist for only certain values of the variable s. The choice $K(s, t) = e^{-st}$ gives us an especially important integral transform.

DEFINITION 7.1   Laplace Transform

Let f be a function defined for $t \geq 0$. Then the integral

$$\mathscr{L}\{f(t)\} = \int_0^\infty e^{-st}f(t)\,dt \qquad (2)$$

is said to be the **Laplace transform*** of f provided the integral converges.

When the defining integral (2) converges, the result is a function of s. In general discussion, we shall use a lowercase letter to denote the function

---

* Please see note on page 326.

being transformed and the corresponding capital letter to denote its Laplace transform; for example,

$$\mathscr{L}\{f(t)\} = F(s), \qquad \mathscr{L}\{g(t)\} = G(s), \qquad \mathscr{L}\{y(t)\} = Y(s).$$

EXAMPLE 1

Evaluate $\mathscr{L}\{1\}$.

Solution
$$\mathscr{L}\{1\} = \int_0^\infty e^{-st}(1)\, dt = \lim_{b \to \infty} \int_0^b e^{-st}\, dt$$

$$= \lim_{b \to \infty} \frac{-e^{-st}}{s}\bigg|_0^b = \lim_{b \to \infty} \frac{-e^{-sb} + 1}{s}$$

$$= \frac{1}{s}$$

provided $s > 0$. In other words, when $s > 0$, the exponent $-sb$ is negative and $e^{-sb} \to 0$ as $b \to \infty$. When $s < 0$, the integral is divergent. ∎

The use of the limit sign becomes somewhat tedious, so we shall adopt the notation $|_0^\infty$ as a shorthand to writing $\lim_{b \to \infty}(\ )|_0^b$. For example,

$$\mathscr{L}\{1\} = \int_0^\infty e^{-st}\, dt = \frac{-e^{-st}}{s}\bigg|_0^\infty = \frac{1}{s}, \qquad s > 0,$$

where it is understood that at the upper limit we mean $e^{-st} \to 0$ as $t \to \infty$ for $s > 0$.

$\mathscr{L}$ a Linear Transform

For a sum of functions we can write

$$\int_0^\infty e^{-st}[\alpha f(t) + \beta g(t)]\, dt = \alpha \int_0^\infty e^{-st} f(t)\, dt + \beta \int_0^\infty e^{-st} g(t)\, dt,$$

---

*Pierre Simon Marquis de Laplace (1749–1827) A noted mathematician, physicist, and astronomer, Laplace was called by some of his enthusiastic contemporaries the "Newton of France." Although Laplace made use of the integral transform (2) in his work in probability theory, it is likely that the integral was first discovered by Euler. Laplace's noted treatises were *Mécanique Céleste* and *Théorie Analytique des Probabilités*. Born into a poor farming family, Laplace became a friend of Napoleon but was elevated to the nobility by Louis XVIII after the Restoration.

whenever both integrals converge. Hence it follows that

$$\mathscr{L}\{\alpha f(t) + \beta g(t)\} = \alpha\mathscr{L}\{f(t)\} + \beta\mathscr{L}\{g(t)\}$$
$$= \alpha F(s) + \beta G(s). \tag{3}$$

Because of the property given in (3), $\mathscr{L}$ is said to be a **linear transform**, or **linear operator**.

Sufficient Conditions for Existence of $\mathscr{L}\{f(t)\}$

The integral that defines the Laplace transform does not have to converge. For example, neither $\mathscr{L}\{1/t\}$ nor $\mathscr{L}\{e^{t^2}\}$ exists. Sufficient conditions that guarantee the existence of $\mathscr{L}\{f(t)\}$ are that f be piecewise continuous on $[0, \infty)$ and that f be of exponential order for $t > T$. Recall that a function f is **piecewise continuous** on $[0, \infty)$ if, in any interval $0 \leq a \leq t \leq b$, there are at most a finite number of points t_k, $k = 1, 2, \ldots, n$ $(t_{k-1} < t_k)$, at which f has finite discontinuities and is continuous on each open interval $t_{k-1} < t < t_k$. See Figure 7.1. The concept of **exponential order** is defined in the following manner.

Figure 7.1

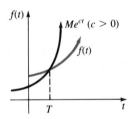

Figure 7.2

DEFINITION 7.2 **Exponential Order**

A function f is said to be of **exponential order** if there exist numbers c, $M > 0$, and $T > 0$ such that $|f(t)| \leq Me^{ct}$ for $t > T$.

If, for example, f is an *increasing* function, then the condition $|f(t)| \leq Me^{ct}$, $t > T$, simply states that the graph of f on the interval (T, ∞) does not grow faster than the graph of the exponential function Me^{ct}, where c is a positive constant. See Figure 7.2. The functions $f(t) = t$, $f(t) = e^{-t}$, and $f(t) = 2\cos t$ are all of exponential order for $t > 0$ since we have, respectively,

$$|t| \leq e^t, \qquad |e^{-t}| \leq e^t, \qquad |2\cos t| \leq 2e^t.$$

A comparison of the graphs on the interval $(0, \infty)$ is given in Figure 7.3.

(a) (b) (c)

Figure 7.3

Figure 7.4

A function such as $f(t) = e^{t^2}$ is not of exponential order since, as shown in Figure 7.4, its graph grows faster than any positive linear power of e for $t > c > 0$.

A positive integral power of t is always of exponential order since for $c > 0$,

$$|t^n| \leq Me^{ct} \quad \text{or} \quad \left|\frac{t^n}{e^{ct}}\right| \leq M \qquad \text{for } t > T$$

is equivalent to showing that $\lim_{t \to \infty} t^n/e^{ct}$ is finite for $n = 1, 2, 3, \ldots$. The result follows by n applications of L'Hôpital's rule.

THEOREM 7.1 **Sufficient Conditions for Existence**

Let $f(t)$ be piecewise continuous on the interval $[0, \infty)$ and of exponential order for $t > T$; then $\mathscr{L}\{f(t)\}$ exists for $s > c$.

Proof

$$\mathscr{L}\{f(t)\} = \int_0^\infty e^{-st}f(t)\, dt$$

$$= \int_0^T e^{-st}f(t)\, dt + \int_T^\infty e^{-st}f(t)\, dt = I_1 + I_2.$$

The integral I_1 exists because it can be written as a sum of integrals over intervals for which $e^{-st}f(t)$ is continuous. Now

$$|I_2| \leq \int_T^\infty |e^{-st}f(t)|\, dt \leq M \int_T^\infty e^{-st}e^{ct}\, dt$$

$$= M \int_T^\infty e^{-(s-c)t}\, dt = -M \left.\frac{e^{-(s-c)t}}{s-c}\right|_T^\infty$$

$$= M \frac{e^{-(s-c)T}}{s-c} \qquad \text{for } s > c.$$

Since $\int_T^\infty Me^{-(s-c)t}\, dt$ converges, the integral $\int_T^\infty |e^{-st}f(t)|\, dt$ converges by the comparison test for improper integrals. This, in turn, implies that I_2 exists for $s > c$. The existence of I_1 and I_2 implies that $\mathscr{L}\{f(t)\} = \int_0^\infty e^{-st}f(t)\, dt$ exists for $s > c$. $\quad\square$

Throughout this entire chapter we shall be concerned only with functions that are both piecewise continuous and of exponential order. We note, however, that these conditions are sufficient but not necessary for the existence of a Laplace transform. The function $f(t) = t^{-1/2}$ is not piecewise continuous on the interval $[0, \infty)$, but its Laplace transform exists. See Problem 44.

EXAMPLE 2

Evaluate $\mathscr{L}\{t\}$.

Solution From Definition 7.1 we have

$$\mathscr{L}\{t\} = \int_0^\infty e^{-st} t \, dt.$$

Integrating by parts and using $\lim_{t \to \infty} te^{-st} = 0$, $s > 0$, along with the result of Example 1, we obtain

$$\mathscr{L}\{t\} = \frac{-te^{-st}}{s}\Big|_0^\infty + \frac{1}{s}\int_0^\infty e^{-st}\,dt = \frac{1}{s}\mathscr{L}\{1\} = \frac{1}{s}\left(\frac{1}{s}\right) = \frac{1}{s^2}. \qquad \blacksquare$$

EXAMPLE 3

Evaluate $\mathscr{L}\{e^{-3t}\}$.

Solution From Definition 7.1 we have

$$\begin{aligned}
\mathscr{L}\{e^{-3t}\} &= \int_0^\infty e^{-st} e^{-3t}\,dt \\
&= \int_0^\infty e^{-(s+3)t}\,dt \\
&= \frac{-e^{-(s+3)t}}{s+3}\Big|_0^\infty \\
&= \frac{1}{s+3}, \qquad s > -3.
\end{aligned}$$

The result follows from the fact that $\lim_{t \to \infty} e^{-(s+3)t} = 0$ for $s + 3 > 0$ or $s > -3$. $\qquad \blacksquare$

EXAMPLE 4

Evaluate $\mathscr{L}\{\sin 2t\}$.

Solution From Definition 7.1 and integration by parts we have

$$\begin{aligned}
\mathscr{L}\{\sin 2t\} &= \int_0^\infty e^{-st} \sin 2t \, dt \\
&= \frac{-e^{-st}\sin 2t}{s}\Big|_0^\infty + \frac{2}{s}\int_0^\infty e^{-st}\cos 2t\,dt
\end{aligned}$$

$$= \frac{2}{s} \int_0^\infty e^{-st} \cos 2t \, dt, \qquad s > 0 \qquad \boxed{\text{Laplace transform of } \sin 2t}$$

$$= \frac{2}{s} \left[\frac{-e^{-st} \cos 2t}{s} \Big|_0^\infty - \frac{2}{s} \int_0^\infty e^{-st} \sin 2t \, dt \right]$$

$$= \frac{2}{s^2} - \frac{4}{s^2} \mathscr{L}\{\sin 2t\}. \qquad \boxed{\lim_{t \to \infty} e^{-st} \cos 2t = 0, s > 0}$$

Now we solve for $\mathscr{L}\{\sin 2t\}$:

$$\left[1 + \frac{4}{s^2} \right] \mathscr{L}\{\sin 2t\} = \frac{2}{s^2}$$

$$\mathscr{L}\{\sin 2t\} = \frac{2}{s^2 + 4}, \qquad s > 0. \qquad ∎$$

EXAMPLE 5

Evaluate $\mathscr{L}\{3t - 5 \sin 2t\}$.

Solution From Examples 2 and 4 and the linearity property of the Laplace transform we can write

$$\mathscr{L}\{3t - 5 \sin 2t\} = 3\mathscr{L}\{t\} - 5\mathscr{L}\{\sin 2t\}$$

$$= 3 \cdot \frac{1}{s^2} - 5 \cdot \frac{2}{s^2 + 4}$$

$$= \frac{-7s^2 + 12}{s^2(s^2 + 4)}, \qquad s > 0. \qquad ∎$$

EXAMPLE 6

Evaluate **(a)** $\mathscr{L}\{te^{-2t}\}$ and **(b)** $\mathscr{L}\{t^2 e^{-2t}\}$.

Solution **(a)** From Definition 7.1 and integration by parts we have

$$\mathscr{L}\{te^{-2t}\} = \int_0^\infty e^{-st}(te^{-2t}) \, dt$$

$$= \int_0^\infty te^{-(s+2)t} \, dt$$

$$= \frac{-te^{-(s+2)t}}{s+2} \Big|_0^\infty + \frac{1}{s+2} \int_0^\infty e^{-(s+2)t} \, dt$$

$$= \frac{-e^{-(s+2)t}}{(s+2)^2}\Big|_0^\infty, \qquad s > -2$$

$$= \frac{1}{(s+2)^2}, \qquad s > -2.$$

(b) Again, integration by parts gives

$$\mathscr{L}\{t^2 e^{-2t}\} = \frac{-t^2 e^{-(s+2)t}}{s+2}\Big|_0^\infty + \frac{2}{s+2}\int_0^\infty t e^{-(s+2)t}\, dt$$

$$= \frac{2}{s+2}\int_0^\infty e^{-st}(te^{-2t})\, dt, \qquad s > -2$$

$$= \frac{2}{s+2}\mathscr{L}\{te^{-2t}\} = \frac{2}{s+2}\left[\frac{1}{(s+2)^2}\right] \qquad \Longleftarrow \boxed{\text{from part (a)}}$$

$$= \frac{2}{(s+2)^3}, \qquad s > -2. \qquad \blacksquare$$

EXAMPLE 7

Evaluate $\mathscr{L}\{f(t)\}$ for $\qquad f(t) = \begin{cases} 0, & 0 \le t < 3 \\ 2, & t \ge 3. \end{cases}$

Solution This piecewise continuous function is shown in Figure 7.5. Since f is defined in two pieces, $\mathscr{L}\{f(t)\}$ is expressed as the sum of two integrals:

$$\mathscr{L}\{f(t)\} = \int_0^\infty e^{-st} f(t)\, dt$$

$$= \int_0^3 e^{-st} f(t)\, dt + \int_3^\infty e^{-st} f(t)\, dt$$

$$= \int_0^3 e^{-st}(0)\, dt + \int_3^\infty e^{-st}(2)\, dt$$

$$= -\frac{2e^{-st}}{s}\Big|_3^\infty$$

$$= \frac{2e^{-3s}}{s}, \qquad s > 0. \qquad \blacksquare$$

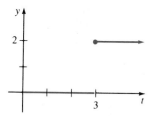

Figure 7.5

We state the generalization of some of the preceding examples by means of the next theorem. From this point on we shall also refrain from stating any restrictions on s; it is understood that s is sufficiently restricted to guarantee the convergence of the appropriate Laplace transform.

THEOREM 7.2   **Transforms of Some Basic Functions**

(a) $\mathscr{L}\{1\} = \dfrac{1}{s}$

(b) $\mathscr{L}\{t^n\} = \dfrac{n!}{s^{n+1}}, \qquad n = 1, 2, 3, \ldots$

(c) $\mathscr{L}\{e^{at}\} = \dfrac{1}{s-a}$

(d) $\mathscr{L}\{\sin kt\} = \dfrac{k}{s^2 + k^2}$

(e) $\mathscr{L}\{\cos kt\} = \dfrac{s}{s^2 + k^2}$

(f) $\mathscr{L}\{\sinh kt\} = \dfrac{k}{s^2 - k^2}$

(g) $\mathscr{L}\{\cosh kt\} = \dfrac{s}{s^2 - k^2}$

Part (b) of Theorem 7.2 can be justified in the following manner: Integration by parts yields

$$\mathscr{L}\{t^n\} = \int_0^\infty e^{-st} t^n \, dt$$

$$= -\frac{1}{s} e^{-st} t^n \Big|_0^\infty + \frac{n}{s} \int_0^\infty e^{-st} t^{n-1} \, dt$$

$$= \frac{n}{s} \int_0^\infty e^{-st} t^{n-1} \, dt$$

or
$$\mathscr{L}\{t^n\} = \frac{n}{s}\, \mathscr{L}\{t^{n-1}\}, \qquad n = 1, 2, 3, \ldots.$$

Now $\mathscr{L}\{1\} = 1/s$, so it follows by iteration that

$$\mathscr{L}\{t\} = \frac{1}{s}\, \mathscr{L}\{1\} = \frac{2}{s^2}$$

$$\mathscr{L}\{t^2\} = \frac{2}{s}\, \mathscr{L}\{t\} = \frac{2}{s}\left(\frac{1}{s^2}\right) = \frac{2}{s^3}$$

$$\mathscr{L}\{t^3\} = \frac{3}{s}\, \mathscr{L}\{t^2\} = \frac{3}{s}\left(\frac{2}{s^3}\right) = \frac{3!}{s^4}.$$

Although a rigorous proof requires mathematical induction, it seems reasonable to conclude from the foregoing results that in general

$$\mathscr{L}\{t^n\} = \frac{n}{s}\,\mathscr{L}\{t^{n-1}\} = \frac{n}{s}\left[\frac{(n-1)!}{s^n}\right] = \frac{n!}{s^{n+1}}.$$

The justifications of parts (f) and (g) of Theorem 7.2 are left to you. See Problems 33 and 34.

EXAMPLE 8

Evaluate $\mathscr{L}\{\sin^2 t\}$.

Solution With the aid of a trigonometric identity, linearity, and parts (a) and (e) of Theorem 7.2, we obtain

$$\mathscr{L}\{\sin^2 t\} = \mathscr{L}\left\{\frac{1 - \cos 2t}{2}\right\} = \frac{1}{2}\,\mathscr{L}\{1\} - \frac{1}{2}\,\mathscr{L}\{\cos 2t\}$$

$$= \frac{1}{2}\cdot\frac{1}{s} - \frac{1}{2}\cdot\frac{s}{s^2 + 4}$$

$$= \frac{2}{s(s^2 + 4)}. \qquad\blacksquare$$

7.1 EXERCISES

Answers to odd-numbered problems begin on page A–18.

In Problems 1–18 use Definition 7.1 to find $\mathscr{L}\{f(t)\}$.

1. $f(t) = \begin{cases} -1, & 0 \le t < 1 \\ 1, & t \ge 1 \end{cases}$

2. $f(t) = \begin{cases} 4, & 0 \le t < 2 \\ 0, & t \ge 2 \end{cases}$

3. $f(t) = \begin{cases} t, & 0 \le t < 1 \\ 1, & t \ge 1 \end{cases}$

4. $f(t) = \begin{cases} 2t + 1, & 0 \le t < 1 \\ 0, & t \ge 1 \end{cases}$

5. $f(t) = \begin{cases} \sin t, & 0 \le t < \pi \\ 0, & t \ge \pi \end{cases}$

6. $f(t) = \begin{cases} 0, & 0 \le t < \pi/2 \\ \cos t, & t \ge \pi/2 \end{cases}$

7.

Figure 7.6

8.

Figure 7.7

9.

Figure 7.8

10.

Figure 7.9

11. $f(t) = e^{t+7}$

12. $f(t) = e^{-2t-5}$

13. $f(t) = te^{4t}$

14. $f(t) = t^2 e^{3t}$

15. $f(t) = e^{-t} \sin t$

16. $f(t) = e^t \cos t$

17. $f(t) = t \cos t$

18. $f(t) = t \sin t$

In Problems 19–42 use Theorem 7.2 to find $\mathscr{L}\{f(t)\}$.

19. $f(t) = 2t^4$

20. $f(t) = t^5$

21. $f(t) = 4t - 10$

22. $f(t) = 7t + 3$

23. $f(t) = t^2 + 6t - 3$

24. $f(t) = -4t^2 + 16t + 9$

25. $f(t) = (t + 1)^3$

26. $f(t) = (2t - 1)^3$

27. $f(t) = 1 + e^{4t}$

28. $f(t) = t^2 - e^{-9t} + 5$

29. $f(t) = (1 + e^{2t})^2$

30. $f(t) = (e^t - e^{-t})^2$

31. $f(t) = 4t^2 - 5 \sin 3t$

32. $f(t) = \cos 5t + \sin 2t$

33. $f(t) = \sinh kt$

34. $f(t) = \cosh kt$

35. $f(t) = e^t \sinh t$

36. $f(t) = e^{-t} \cosh t$

37. $f(t) = \sin 2t \cos 2t$

38. $f(t) = \cos^2 t$

39. $f(t) = \cos t \cos 2t$ [*Hint:* Examine $\cos(t_1 \pm t_2)$.]

40. $f(t) = \sin t \sin 2t$

41. $f(t) = \sin t \cos 2t$ [*Hint:* Examine $\sin(t_1 \pm t_2)$.]

42. $f(t) = \sin^3 t$ [*Hint:* $\sin^3 t = \sin t \sin^2 t$.]

43. The **gamma function** is defined by the integral

$$\Gamma(\alpha) = \int_0^{\infty} t^{\alpha-1} e^{-t} \, dt, \qquad \alpha > 0.$$

See Appendix I. Show that $\mathscr{L}\{t^{\alpha}\} = \dfrac{\Gamma(\alpha + 1)}{s^{\alpha+1}}, \alpha > -1$.

In Problems 44–46 use the result of Problem 43 to find $\mathscr{L}\{f(t)\}$.

44. $f(t) = t^{-1/2}$ **45.** $f(t) = t^{1/2}$ **46.** $f(t) = t^{3/2}$

47. Show that the function $f(t) = 1/t^2$ does not possess a Laplace transform.
$\left[\text{\textit{Hint:}} \ \mathscr{L}\{f(t)\} = \displaystyle\int_0^1 e^{-st} f(t) \, dt + \int_1^{\infty} e^{-st} f(t) \, dt. \ \text{Use the definition of} \right.$
an improper integral to show that $\displaystyle\int_0^1 e^{-st} f(t) \, dt$ does not exist. $\Bigg]$

48. Show that if the functions f and g are of exponential order for $t > T$, then the product fg is of exponential order for $t > T$.

7.2 INVERSE TRANSFORM

In the preceding section we were concerned with the problem of transforming a function $f(t)$ into another function $F(s)$ by means of the integral $\int_0^\infty e^{-st} f(t)\, dt$. We denoted this symbolically by $\mathscr{L}\{f(t)\} = F(s)$. We now turn the problem around; namely, given $F(s)$, find the function $f(t)$ corresponding to this transform. We say $f(t)$ is the **inverse Laplace transform** of $F(s)$ and write

$$f(t) = \mathscr{L}^{-1}\{F(s)\}.$$

The analogue of Theorem 7.2 for the inverse transform is the following:

THEOREM 7.3   **Some Inverse Transforms**

(a) $1 = \mathscr{L}^{-1}\left\{\dfrac{1}{s}\right\}$,

(b) $t^n = \mathscr{L}^{-1}\left\{\dfrac{n!}{s^{n+1}}\right\}$, $n = 1, 2, 3, \ldots$

(c) $e^{at} = \mathscr{L}^{-1}\left\{\dfrac{1}{s-a}\right\}$

(d) $\sin kt = \mathscr{L}^{-1}\left\{\dfrac{k}{s^2 + k^2}\right\}$

(e) $\cos kt = \mathscr{L}^{-1}\left\{\dfrac{s}{s^2 + k^2}\right\}$

(f) $\sinh kt = \mathscr{L}^{-1}\left\{\dfrac{k}{s^2 - k^2}\right\}$

(g) $\cosh kt = \mathscr{L}^{-1}\left\{\dfrac{s}{s^2 - k^2}\right\}$

$\mathscr{L}^{-1}$ a Linear Transform

We shall assume that the inverse Laplace transform is itself a linear transform;* that is, for constants α and β,

$$\mathscr{L}^{-1}\{\alpha F(s) + \beta G(s)\} = \alpha \mathscr{L}^{-1}\{F(s)\} + \beta \mathscr{L}^{-1}\{G(s)\},$$

where F and G are the transforms of some functions f and g.

---

* The inverse Laplace transform is actually another integral. However, evaluation of this integral demands the use of complex variables, which is beyond the scope of this text.

The inverse Laplace transform of a function $F(s)$ *may* not be unique. It is possible that $\mathscr{L}\{f_1(t)\} = \mathscr{L}\{f_2(t)\}$ and yet $f_1 \neq f_2$. See Problems 35 and 36. For our purposes this is not as bad as it appears. If f_1 and f_2 are piecewise continuous on $[0, \infty)$ and of exponential order for $t > 0$ and if $\mathscr{L}\{f_1(t)\} = \mathscr{L}\{f_2(t)\}$, then it can be proved that the functions f_1 and f_2 are *essentially* the same; that is, they can differ only at points of discontinuity. However, if f_1 and f_2 are continuous on $[0, \infty)$ and $\mathscr{L}\{f_1(t)\} = \mathscr{L}\{f_2(t)\}$, then $f_1 = f_2$ on the interval.

EXAMPLE I

Evaluate $\mathscr{L}^{-1}\left\{\dfrac{1}{s^5}\right\}$.

Solution To match the form given in part (b) of Theorem 7.3, we identify $n = 4$ and then multiply and divide by 4!. It follows that

$$\mathscr{L}^{-1}\left\{\frac{1}{s^5}\right\} = \frac{1}{4!}\,\mathscr{L}^{-1}\left\{\frac{4!}{s^5}\right\} = \frac{1}{24}\,t^4. \qquad \blacksquare$$

EXAMPLE 2

Evaluate $\mathscr{L}^{-1}\left\{\dfrac{1}{s^2 + 64}\right\}$.

Solution Identifying $k^2 = 64$, we multiply and divide by 8 and use part (d) of Theorem 7.3:

$$\mathscr{L}^{-1}\left\{\frac{1}{s^2 + 64}\right\} = \frac{1}{8}\,\mathscr{L}^{-1}\left\{\frac{8}{s^2 + 64}\right\} = \frac{1}{8}\sin 8t. \qquad \blacksquare$$

EXAMPLE 3

Evaluate $\mathscr{L}^{-1}\left\{\dfrac{3s + 5}{s^2 + 7}\right\}$.

Solution The given function of s can be written as two expressions by means of termwise division:

$$\frac{3s + 5}{s^2 + 7} = \frac{3s}{s^2 + 7} + \frac{5}{s^2 + 7}.$$

From the linearity property of the inverse transform and parts (e) and (d) of Theorem 7.3, we then have

$$\mathscr{L}^{-1}\left\{\frac{3s+5}{s^2+7}\right\} = 3\mathscr{L}^{-1}\left\{\frac{s}{s^2+7}\right\} + \frac{5}{\sqrt{7}}\,\mathscr{L}^{-1}\left\{\frac{\sqrt{7}}{s^2+7}\right\}$$

$$= 3\cos\sqrt{7}\,t + \frac{5}{\sqrt{7}}\sin\sqrt{7}\,t. \qquad \blacksquare$$

Partial Fractions

The use of **partial fractions** is very important in finding inverse Laplace transforms. Here we review three basic cases of that theory. For example, the denominators of

(i) $F(s) = \dfrac{1}{(s-1)(s+2)(s+4)}$ **(ii)** $F(s) = \dfrac{s+1}{s^2(s+2)^3}$ **(iii)** $F(s) = \dfrac{3s-2}{s^3(s^2+4)}$

contain, respectively, only distinct linear factors, repeated linear factors, and a quadratic factor that is irreducible.*

E X A M P L E **4**

Evaluate $\mathscr{L}^{-1}\left\{\dfrac{1}{(s-1)(s+2)(s+4)}\right\}$.

Solution   There exist unique constants A, B, and C so that

$$\frac{1}{(s-1)(s+2)(s+4)} = \frac{A}{s-1} + \frac{B}{s+2} + \frac{C}{s+4}$$

$$= \frac{A(s+2)(s+4) + B(s-1)(s+4) + C(s-1)(s+2)}{(s-1)(s+2)(s+4)}.$$

Since the denominators are identical, the numerators are identical:

$$1 = A(s+2)(s+4) + B(s-1)(s+4) + C(s-1)(s+2).$$

By comparing coefficients of powers of s on both sides of the equality, we know that the last equation is equivalent to a system of three equations in the three unknowns A, B, and C. However, you might recall the following shortcut for determining these unknowns. If we set $s=1$, $s=-2$, and $s=-4$, the zeros of the common denominator $(s-1)(s+2)(s+4)$, we obtain, in turn,

$$1 = A(3)(5), \qquad A = 1/15,$$

$$1 = B(-3)(2), \qquad B = -1/6,$$

$$1 = C(-5)(-2), \qquad C = 1/10.$$

---

* *Irreducible* means that the quadratic factor has no real zeros.

Hence we can write

$$\frac{1}{(s-1)(s+2)(s+4)} = \frac{1/15}{s-1} - \frac{1/6}{s+2} + \frac{1/10}{s+4}$$

and thus, from part (c) of Theorem 7.3,

$$\mathscr{L}^{-1}\left\{\frac{1}{(s-1)(s+2)(s+4)}\right\} = \frac{1}{15}\mathscr{L}^{-1}\left\{\frac{1}{s-1}\right\} - \frac{1}{6}\mathscr{L}^{-1}\left\{\frac{1}{s+2}\right\} + \frac{1}{10}\mathscr{L}^{-1}\left\{\frac{1}{s+4}\right\}$$

$$= \frac{1}{15}e^t - \frac{1}{6}e^{-2t} + \frac{1}{10}e^{-4t}. \qquad \blacksquare$$

EXAMPLE 5

Evaluate $\mathscr{L}^{-1}\left\{\dfrac{s+1}{s^2(s+2)^3}\right\}$.

Solution   Assume

$$\frac{s+1}{s^2(s+2)^3} = \frac{A}{s} + \frac{B}{s^2} + \frac{C}{s+2} + \frac{D}{(s+2)^2} + \frac{E}{(s+2)^3}$$

so that

$$s+1 = As(s+2)^3 + B(s+2)^3 + Cs^2(s+2)^2 + Ds^2(s+2) + Es^2.$$

Setting $s=0$ and $s=-2$ gives $B = 1/8$ and $E = -1/4$, respectively. By equating the coefficients of s^4, s^3, and s, we obtain

$$0 = A + C$$

$$0 = 6A + B + 4C + D$$

$$1 = 8A + 12B,$$

from which it follows that $A = -1/16$, $C = 1/16$, and $D = 0$. Hence from parts (a), (b), and (c) of Theorem 7.3,

$$\mathscr{L}^{-1}\left\{\frac{s+1}{s^2(s+2)^3}\right\} = \mathscr{L}^{-1}\left\{-\frac{1/16}{s} + \frac{1/8}{s^2} + \frac{1/16}{s+2} - \frac{1/4}{(s+2)^3}\right\}$$

$$= -\frac{1}{16}\mathscr{L}^{-1}\left\{\frac{1}{s}\right\} + \frac{1}{8}\mathscr{L}^{-1}\left\{\frac{1}{s^2}\right\} + \frac{1}{16}\mathscr{L}^{-1}\left\{\frac{1}{s+2}\right\} - \frac{1}{8}\mathscr{L}^{-1}\left\{\frac{2}{(s+2)^3}\right\}$$

$$= -\frac{1}{16} + \frac{1}{8}t + \frac{1}{16}e^{-2t} - \frac{1}{8}t^2e^{-2t}.$$

Here we have also used $\mathscr{L}^{-1}\{2/(s+2)^3\} = t^2e^{-2t}$ from Example 6 of Section 7.1. $\qquad \blacksquare$

EXAMPLE 6

Evaluate $\mathscr{L}^{-1}\left\{\dfrac{3s-2}{s^3(s^2+4)}\right\}$.

Solution Assume

$$\frac{3s-2}{s^3(s^2+4)} = \frac{A}{s} + \frac{B}{s^2} + \frac{C}{s^3} + \frac{Ds+E}{s^2+4}$$

so that

$$3s-2 = As^2(s^2+4) + Bs(s^2+4) + C(s^2+4) + (Ds+E)s^3.$$

Setting $s = 0$ gives immediately $C = -1/2$. Now the coefficients of s^4, s^3, s^2, and s are, respectively,

$$0 = A + D, \qquad 0 = B + E, \qquad 0 = 4A + C, \qquad 3 = 4B,$$

from which we obtain $B = 3/4, E = -3/4, A = 1/8$, and $D = -1/8$. Therefore from parts (a), (b), (e), and (d) of Theorem 7.3, we have

$$\mathscr{L}^{-1}\left\{\frac{3s-2}{s^3(s^2+4)}\right\} = \mathscr{L}^{-1}\left\{\frac{1/8}{s} + \frac{3/4}{s^2} - \frac{1/2}{s^3} + \frac{-s/8-3/4}{s^2+4}\right\}$$

$$= \frac{1}{8}\mathscr{L}^{-1}\left\{\frac{1}{s}\right\} + \frac{3}{4}\mathscr{L}^{-1}\left\{\frac{1}{s^2}\right\} - \frac{1}{4}\mathscr{L}^{-1}\left\{\frac{2}{s^3}\right\}$$

$$- \frac{1}{8}\mathscr{L}^{-1}\left\{\frac{s}{s^2+4}\right\} - \frac{3}{8}\mathscr{L}^{-1}\left\{\frac{2}{s^2+4}\right\}$$

$$= \frac{1}{8} + \frac{3}{4}t - \frac{1}{4}t^2 - \frac{1}{8}\cos 2t - \frac{3}{8}\sin 2t. \qquad \blacksquare$$

Not every arbitrary function of s is a Laplace transform of a piecewise function of exponential order.

THEOREM 7.4 **Behavior of F(s) as $s \to \infty$**

Let $f(t)$ be piecewise continuous on $[0, \infty)$ and of exponential order for $t > T$; then

$$\lim_{s \to \infty} \mathscr{L}\{f(t)\} = 0.$$

Proof Since $f(t)$ is piecewise continuous on $0 \le t \le T$, it is necessarily bounded on this interval:

$$|f(t)| \le M_1 = M_1 e^{0t}.$$

Also $$|f(t)| \le M_2 e^{\gamma t}$$

for $t > T$. If M denotes the maximum of $\{M_1, M_2\}$ and c denotes the maximum of $\{0, \gamma\}$, then

$$|\mathscr{L}\{f(t)\}| \le \int_0^\infty e^{-st}|f(t)|\, dt$$

$$\le M \int_0^\infty e^{-st} \cdot e^{ct}\, dt$$

$$= -M \frac{e^{-(s-c)t}}{s-c}\Big|_0^\infty = \frac{M}{s-c}$$

for $s > c$. As $s \to \infty$, we have $|\mathscr{L}\{f(t)\}| \to 0$ and so $\mathscr{L}\{f(t)\} \to 0$. ☐

EXAMPLE 7

The functions $F_1(s) = s^2$ and $F_2(s) = s/(s+1)$ are not the Laplace transforms of piecewise continuous functions of exponential order since

$$F_1(s) \nrightarrow 0 \quad \text{and} \quad F_2(s) \nrightarrow 0$$

as $s \to \infty$. We say that $\mathscr{L}^{-1}\{F_1(s)\}$ and $\mathscr{L}^{-1}\{F_2(s)\}$ do not exist. ∎

Remark There is another way of determining the coefficients in a partial fraction decomposition in the special case when $\mathscr{L}\{f(t)\} = F(s)$ is a quotient of polynomials $P(s)/Q(s)$ and $Q(s)$ is a product of *distinct* linear factors:

$$F(s) = \frac{P(s)}{(s-r_1)(s-r_2)\cdots(s-r_n)}.$$

Let's illustrate by means of a specific example. From the theory of partial fractions we know there exist unique constants A, B, and C such that

$$\frac{s^2 + 4s - 1}{(s-1)(s-2)(s+3)} = \frac{A}{s-1} + \frac{B}{s-2} + \frac{C}{s+3}. \tag{1}$$

Suppose we then multiply both sides of this last expression by, say, $s - 1$, simplify, and then set $s = 1$. Since the coefficients of B and C are zero we get

$$\frac{s^2 + 4s - 1}{(s-2)(s+3)}\bigg|_{s=1} = A \quad \text{or} \quad A = -1.$$

Written another way,

$$\frac{s^2 + 4s - 1}{\boxed{(s-1)}(s-2)(s+3)}\bigg|_{s=1} = A$$

where we have shaded or *covered up* the factor that canceled when the left side of (1) was multiplied by $s - 1$. We *do not evaluate this covered-up factor* at

$s = 1$. Now to obtain B and C we simply evaluate the left member of (1) while covering, in turn, $s - 2$ and $s + 3$:

$$\frac{s^2 + 4s - 1}{(s - 1)\boxed{(s - 2)}(s + 3)}\bigg|_{s=2} = B \quad \text{or} \quad B = \frac{11}{5}$$

$$\frac{s^2 + 4s - 1}{(s - 1)(s - 2)\boxed{(s + 3)}}\bigg|_{s=-3} = C \quad \text{or} \quad C = -\frac{1}{5}.$$

Note carefully that in the calculation of C we evaluated at $s = -3$. By filling in the details in arriving at this last expression you will see why this is so. You should also verify by other means that

$$\frac{s^2 + 4s - 1}{(s - 1)(s - 2)(s + 3)} = \frac{-1}{s - 1} + \frac{11/5}{s - 2} + \frac{-1/5}{s + 3}.$$

This **cover-up method** is a simplified version of a result known as **Heaviside's expansion theorem***.

7.2 EXERCISES

Answers to odd-numbered problems begin on page A–19.

In Problems 1–34 use Theorem 7.3 to find the given inverse transform.

1. $\mathscr{L}^{-1}\left\{\dfrac{1}{s^3}\right\}$

2. $\mathscr{L}^{-1}\left\{\dfrac{1}{s^4}\right\}$

3. $\mathscr{L}^{-1}\left\{\dfrac{1}{s^2} - \dfrac{48}{s^5}\right\}$

4. $\mathscr{L}^{-1}\left\{\left(\dfrac{2}{s} - \dfrac{1}{s^3}\right)^2\right\}$

---

*Oliver Heaviside (1850–1925) Many of the results that we present in this chapter were devised by the English electrical engineer Oliver Heaviside and are set forth in his 1899 treatise *Electromagnetic Theory*. Heaviside originally utilized the Laplace transform as a means of solving linear differential equations with constant coefficients that arose in his investigation of transmission lines. Since many of his results lacked formal proof, the Heaviside operational calculus, as his procedures came to be called, initially met with scorn from mathematicians. A curmudgeon, Heaviside, in turn, called these "establishment" mathematicians "woodenheaded." When Heaviside, using his symbolic methods, was able to obtain answers to problems that mathematicians could not solve, their scorn turned to censure and his papers were no longer published in mathematics journals. Heaviside was also the discoverer of a layer of maximum electron density, called the Heaviside layer, in the upper atmosphere that reflects radio waves back to earth. He lived the last years of his life as a recluse in poverty, forgotten by the scientific community. He died in an unheated house in 1925.

True to their nature, mathematicians eventually seized his ideas, put them on a sound mathematical foundation, and then proceeded to generalize them into an abstract theory.

5. $\mathcal{L}^{-1}\left\{\dfrac{(s+1)^3}{s^4}\right\}$

6. $\mathcal{L}^{-1}\left\{\dfrac{(s+2)^2}{s^3}\right\}$

7. $\mathcal{L}^{-1}\left\{\dfrac{1}{s^2}-\dfrac{1}{s}+\dfrac{1}{s-2}\right\}$

8. $\mathcal{L}^{-1}\left\{\dfrac{4}{s}+\dfrac{6}{s^5}-\dfrac{1}{s+8}\right\}$

9. $\mathcal{L}^{-1}\left\{\dfrac{1}{4s+1}\right\}$

10. $\mathcal{L}^{-1}\left\{\dfrac{1}{5s-2}\right\}$

11. $\mathcal{L}^{-1}\left\{\dfrac{5}{s^2+49}\right\}$

12. $\mathcal{L}^{-1}\left\{\dfrac{10s}{s^2+16}\right\}$

13. $\mathcal{L}^{-1}\left\{\dfrac{4s}{4s^2+1}\right\}$

14. $\mathcal{L}^{-1}\left\{\dfrac{1}{4s^2+1}\right\}$

15. $\mathcal{L}^{-1}\left\{\dfrac{1}{s^2-16}\right\}$

16. $\mathcal{L}^{-1}\left\{\dfrac{10s}{s^2-25}\right\}$

17. $\mathcal{L}^{-1}\left\{\dfrac{2s-6}{s^2+9}\right\}$

18. $\mathcal{L}^{-1}\left\{\dfrac{s-1}{s^2+2}\right\}$

19. $\mathcal{L}^{-1}\left\{\dfrac{1}{s^2+3s}\right\}$

20. $\mathcal{L}^{-1}\left\{\dfrac{s+1}{s^2-4s}\right\}$

21. $\mathcal{L}^{-1}\left\{\dfrac{s}{s^2+2s-3}\right\}$

22. $\mathcal{L}^{-1}\left\{\dfrac{1}{s^2+s-20}\right\}$

23. $\mathcal{L}^{-1}\left\{\dfrac{0.9s}{(s-0.1)(s+0.2)}\right\}$

24. $\mathcal{L}^{-1}\left\{\dfrac{s-3}{(s-\sqrt{3})(s+\sqrt{3})}\right\}$

25. $\mathcal{L}^{-1}\left\{\dfrac{s}{(s-2)(s-3)(s-6)}\right\}$

26. $\mathcal{L}^{-1}\left\{\dfrac{s^2+1}{s(s-1)(s+1)(s-2)}\right\}$

27. $\mathcal{L}^{-1}\left\{\dfrac{2s+4}{(s-2)(s^2+4s+3)}\right\}$

28. $\mathcal{L}^{-1}\left\{\dfrac{s+1}{(s^2-4s)(s+5)}\right\}$

29. $\mathcal{L}^{-1}\left\{\dfrac{1}{s^2(s^2+4)}\right\}$

30. $\mathcal{L}^{-1}\left\{\dfrac{s-1}{s^2(s^2+1)}\right\}$

31. $\mathcal{L}^{-1}\left\{\dfrac{s}{(s^2+4)(s+2)}\right\}$

32. $\mathcal{L}^{-1}\left\{\dfrac{1}{s^4-9}\right\}$

33. $\mathcal{L}^{-1}\left\{\dfrac{1}{(s^2+1)(s^2+4)}\right\}$

34. $\mathcal{L}^{-1}\left\{\dfrac{6s+3}{(s^2+1)(s^2+4)}\right\}$

The inverse Laplace transform may not be unique. In Problems 35 and 36 evaluate $\mathcal{L}\{f(t)\}$.

35. $f(t)=\begin{cases}1, & t\ge 0,\, t\ne 1,\, t\ne 2\\ 3, & t=1\\ 4, & t=2\end{cases}$

36. $f(t)=\begin{cases}e^{3t}, & t\ge 0,\, t\ne 5\\ 1, & t=5\end{cases}$

7.3 TRANSLATION THEOREMS AND DERIVATIVES OF A TRANSFORM

It is not convenient to use Definition 7.1 each time we wish to find the Laplace transform of a function $f(t)$. For example, the integration by parts involved in evaluating, say, $\mathscr{L}\{e^t t^2 \sin 3t\}$ is formidable to say the least. In the discussion that follows, we present several labor-saving theorems; these, in turn, enable us to build up a more extensive list of transforms without the necessity of using the definition of the Laplace transform. Indeed, we shall see that evaluating transforms such as $\mathscr{L}\{e^{4t} \cos 6t\}$, $\mathscr{L}\{t^3 \sin 2t\}$, and $\mathscr{L}\{t^{10}e^{-t}\}$ is fairly straightforward, provided we know $\mathscr{L}\{\cos 6t\}$, $\mathscr{L}\{\sin 2t\}$, and $\mathscr{L}\{t^{10}\}$, respectively. Though extensive tables can be constructed, and we have included a table in Appendix II, it is nonetheless a good idea to know the Laplace transforms of basic functions such as t^n, e^{at}, $\sin kt$, $\cos kt$, $\sinh kt$, and $\cosh kt$.

If we know $\mathscr{L}\{f(t)\} = F(s)$, we can compute the Laplace transform $\mathscr{L}\{e^{at}f(t)\}$ with no additional effort other than *translating*, or *shifting*, $F(s)$ to $F(s - a)$. This result is known as the **first translation theorem** or **first shifting theorem**.

THEOREM 7.5   First Translation Theorem

If a is any real number, then

$$\mathscr{L}\{e^{at}f(t)\} = F(s - a),$$

where $F(s) = \mathscr{L}\{f(t)\}$.

Proof   The proof is immediate, since by Definition 7.1

$$\mathscr{L}\{e^{at}f(t)\} = \int_0^\infty e^{-st}e^{at}f(t)\,dt = \int_0^\infty e^{-(s-a)t}f(t)\,dt = F(s - a). \qquad \square$$

If we consider s a real variable, then the graph of $F(s - a)$ is the graph of $F(s)$ shifted on the s-axis by the amount $|a|$ units. If $a > 0$, the graph of $F(s)$ is shifted a units to the right, whereas if $a < 0$, the graph is shifted $|a|$ units to the left. See Figure 7.10.

For emphasis it is sometimes useful to use the symbolism

$$\mathscr{L}\{e^{at}f(t)\} = \mathscr{L}\{f(t)\}_{s \to s-a}$$

where $s \to s - a$ means that we replace s in $F(s)$ by $s - a$.

$s = a, a > 0$

shift on s-axis

Figure 7.10

EXAMPLE 1

Evaluate **(a)** $\mathscr{L}\{e^{5t}t^3\}$ and **(b)** $\mathscr{L}\{e^{-2t}\cos 4t\}$.

Solution The results follow from Theorem 7.5.

(a) $\mathscr{L}\{e^{5t}t^3\} = \mathscr{L}\{t^3\}_{s \to s-5}$

$$= \frac{3!}{s^4}\Big|_{s \to s-5} = \frac{6}{(s-5)^4}.$$

(b) $\mathscr{L}\{e^{-2t}\cos 4t\} = \mathscr{L}\{\cos 4t\}_{s \to s+2}$ ⟵ $a = -2$ so $s - a = s - (-2) = s + 2$

$$= \frac{s}{s^2 + 16}\Big|_{s \to s+2} = \frac{s+2}{(s+2)^2 + 16}.$$ ∎

Inverse Form of the First Translation Theorem

The inverse form of Theorem 7.5 can be written

$$e^{at}f(t) = \mathscr{L}^{-1}\{F(s-a)\} = \mathscr{L}^{-1}\{F(s)|_{s \to s-a}\}, \tag{1}$$

where $f(t) = \mathscr{L}^{-1}\{F(s)\}$.

EXAMPLE 2

Evaluate $\mathscr{L}^{-1}\left\{\dfrac{s}{s^2 + 6s + 11}\right\}$.

Solution

$$\mathscr{L}^{-1}\left\{\frac{s}{s^2 + 6s + 11}\right\} = \mathscr{L}^{-1}\left\{\frac{s}{(s+3)^2 + 2}\right\}$$ ⟵ completion of square

$$= \mathscr{L}^{-1}\left\{\frac{s+3-3}{(s+3)^2 + 2}\right\}$$ ⟵ adding zero in the numerator

$$= \mathscr{L}^{-1}\left\{\frac{s+3}{(s+3)^2 + 2} - \frac{3}{(s+3)^2 + 2}\right\}$$ ⟵ termwise division

$$= \mathscr{L}^{-1}\left\{\frac{s+3}{(s+3)^2 + 2}\right\} - 3\mathscr{L}^{-1}\left\{\frac{1}{(s+3)^2 + 2}\right\}$$

$$= \mathscr{L}^{-1}\left\{\frac{s}{s^2 + 2}\Big|_{s \to s+3}\right\} - \frac{3}{\sqrt{2}}\mathscr{L}^{-1}\left\{\frac{\sqrt{2}}{s^2 + 2}\Big|_{s \to s+3}\right\}$$

$$= e^{-3t}\cos\sqrt{2}\,t - \frac{3}{\sqrt{2}}e^{-3t}\sin\sqrt{2}\,t.$$ ⟵ from (1) and Theorem 7.3 ∎

EXAMPLE 3

Evaluate $\mathcal{L}^{-1}\left\{\dfrac{1}{(s-1)^3} + \dfrac{1}{s^2+2s-8}\right\}$.

Solution Completing the square in the second denominator and using linearity yield

$$\mathcal{L}^{-1}\left\{\frac{1}{(s-1)^3} + \frac{1}{s^2+2s-8}\right\} = \mathcal{L}^{-1}\left\{\frac{1}{(s-1)^3} + \frac{1}{(s+1)^2-9}\right\}$$

$$= \frac{1}{2!}\,\mathcal{L}^{-1}\left\{\frac{2!}{(s-1)^3}\right\} + \frac{1}{3}\,\mathcal{L}^{-1}\left\{\frac{3}{(s+1)^2-9}\right\}$$

$$= \frac{1}{2!}\,\mathcal{L}^{-1}\left\{\frac{2!}{s^3}\Big|_{s\to s-1}\right\} + \frac{1}{3}\,\mathcal{L}^{-1}\left\{\frac{3}{s^2-9}\Big|_{s\to s+1}\right\}$$

$$= \frac{1}{2}\,e^t t^2 + \frac{1}{3}\,e^{-t}\sinh 3t. \qquad\blacksquare$$

Unit Step Function

In engineering one frequently encounters functions that can be either "on" or "off." For example, an external force acting on a mechanical system or a voltage impressed on a circuit can be turned off after a period of time. It is thus convenient to define a special function called the **unit step function**.

DEFINITION 7.3 **Unit Step Function**

The function $\mathcal{U}(t-a)$ is defined to be

$$\mathcal{U}(t-a) = \begin{cases} 0, & 0 \le t < a \\ 1, & t \ge a. \end{cases}$$

Notice that we define $\mathcal{U}(t-a)$ only on the nonnegative t-axis since this is all that we are concerned with in the study of the Laplace transform. In a broader sense, $\mathcal{U}(t-a) = 0$ for $t < a$.

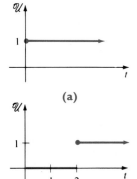

(a)

(b)

Figure 7.11

EXAMPLE 4

Graph **(a)** $\mathcal{U}(t)$ and **(b)** $\mathcal{U}(t-2)$.

Solution **(a)** $\mathcal{U}(t) = 1$, $\quad t \ge 0$ **(b)** $\mathcal{U}(t-2) = \begin{cases} 0, & 0 \le t < 2 \\ 1, & t \ge 2. \end{cases}$

The respective graphs are given in Figure 7.11. $\qquad\blacksquare$

When multiplied by another function defined for $t \geq 0$, the unit step function "turns off" a portion of the graph of the function. For example, Figure 7.12 illustrates the graph of $\sin t, t \geq 0$, when multiplied by $\mathscr{U}(t - 2\pi)$:

$$f(t) = \sin t \; \mathscr{U}(t - 2\pi) = \begin{cases} 0, & 0 \leq t < 2\pi \\ \sin t, & t \geq 2\pi. \end{cases}$$

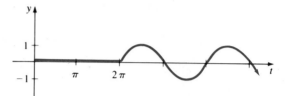

Figure 7.12

The unit step function can also be used to write piecewise-defined functions in a compact form. For instance, the piecewise-defined function

$$f(t) = \begin{cases} g(t), & 0 \leq t < a \\ h(t), & t \geq a \end{cases} \tag{2}$$

can be written as

$$f(t) = g(t) - g(t)\mathscr{U}(t - a) + h(t)\mathscr{U}(t - a). \tag{3}$$

To verify this, we use the definition of $\mathscr{U}(t - a)$:

$$f(t) = \begin{cases} g(t) - g(t) \cdot 0 + h(t) \cdot 0, & 0 \leq t < a \\ g(t) - g(t) \cdot 1 + h(t) \cdot 1, & t \geq a. \end{cases}$$

Similarly, a function of the type

$$f(t) = \begin{cases} 0, & 0 \leq t < a \\ g(t), & a \leq t < b \\ 0, & t \geq b \end{cases} \tag{4}$$

can be written

$$f(t) = g(t)[\mathscr{U}(t - a) - \mathscr{U}(t - b)]. \tag{5}$$

EXAMPLE 5

The voltage in a circuit is given by

$$E(t) = \begin{cases} 20t, & 0 \leq t < 5 \\ 0, & t \geq 5. \end{cases}$$

Graph $E(t)$. Express $E(t)$ in terms of unit step functions.

Solution The graph of this piecewise-defined function is given in Figure 7.13. Now from (2) and (3) with $g(t) = 20t$ and $h(t) = 0$, we get

$$E(t) = 20t - 20t\,\mathscr{U}(t - 5).$$

Figure 7.13

EXAMPLE 6

Consider the function $y = f(t)$ defined by $f(t) = t^3$. Compare the graphs of

(a) $f(t) = t^3$,

(b) $f(t) = t^3$, $t \geq 0$,

(c) $f(t - 2)$, $t \geq 0$

(d) $f(t - 2)\mathscr{U}(t - 2)$, $t \geq 0$.

Solution The respective graphs are given in Figure 7.14.

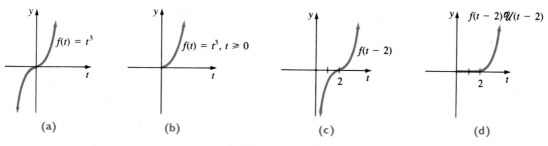

(a) (b) (c) (d)

Figure 7.14 ∎

In general, if $a > 0$, then the graph of $y = f(t - a)$ is the graph of $y = f(t)$, $t \geq 0$, shifted a units to the right on the t-axis. However, when $y = f(t - a)$ is multiplied by the unit step function $\mathscr{U}(t - a)$ in the manner illustrated in part (d) of Example 6, then the graph of the function

$$y = f(t - a)\mathscr{U}(t - a) \tag{6}$$

coincides with the graph of $y = f(t - a)$ for $t \geq a$ but is identically zero for $0 \leq t < a$. See Figure 7.15.

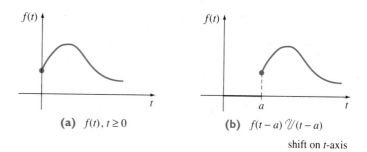

(a) $f(t), t \geq 0$

(b) $f(t - a)\,\mathscr{U}(t - a)$
shift on t-axis

Figure 7.15

We saw in Theorem 7.5 that an exponential multiple of $f(t)$ results in a translation or shift of the transform $F(s)$ on the s-axis. In the next theorem we see that whenever $F(s)$ is multiplied by an appropriate exponential function, the inverse transform of this product is the shifted function given in (6). This result is called the **second translation theorem**, or **second shifting theorem**.

> **THEOREM 7.6** **Second Translation Theorem**
>
> If a is a positive constant, then
>
> $$\mathscr{L}\{f(t-a)\mathscr{U}(t-a)\} = e^{-as}F(s),$$
>
> where $F(s) = \mathscr{L}\{f(t)\}$.

Proof From Definition 7.1 we have

$$\mathscr{L}\{f(t-a)\mathscr{U}(t-a)\} = \int_0^\infty e^{-st}f(t-a)\mathscr{U}(t-a)\,dt$$

$$= \int_0^a e^{-st}f(t-a)\underbrace{\mathscr{U}(t-a)}_{\substack{\text{zero for}\\ 0 \le t < a}}\,dt + \int_a^\infty e^{-st}f(t-a)\underbrace{\mathscr{U}(t-a)}_{\substack{\text{one for}\\ t \ge a}}\,dt$$

$$= \int_a^\infty e^{-st}f(t-a)\,dt$$

Now let $v = t - a$, $dv = dt$; then

$$\mathscr{L}\{f(t-a)\mathscr{U}(t-a)\} = \int_0^\infty e^{-s(v+a)}f(v)\,dv$$

$$= e^{-as}\int_0^\infty e^{-sv}f(v)\,dv = e^{-as}\mathscr{L}\{f(t)\}. \qquad \square$$

EXAMPLE 7

Evaluate $\mathscr{L}\{(t-2)^3\mathscr{U}(t-2)\}$.

Solution With the identification $a = 2$, it follows from Theorem 7.6 that

$$\mathscr{L}\{(t-2)^3\mathscr{U}(t-2)\} = e^{-2s}\mathscr{L}\{t^3\}$$

$$= e^{-2s}\frac{3!}{s^4}$$

$$= \frac{6}{s^4}e^{-2s}. \qquad \blacksquare$$

We often wish to find the Laplace transform of just the unit step function. This can be found from either Definition 7.1 or Theorem 7.6. If we identify $f(t) = 1$ in Theorem 7.6, then $f(t-a) = 1$, $F(s) = \mathscr{L}\{1\} = 1/s$, and so

$$\mathscr{L}\{\mathscr{U}(t-a)\} = \frac{e^{-as}}{s}. \qquad (7)$$

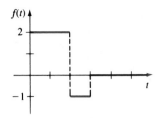

Figure 7.16

EXAMPLE 8

Find the Laplace transform of the function shown in Figure 7.16.

Solution With the aid of the unit step function, we can write

$$f(t) = 2 - 3\mathcal{U}(t - 2) + \mathcal{U}(t - 3).$$

Using linearity and the result in (7) it follows that

$$\mathcal{L}\{f(t)\} = \mathcal{L}\{2\} - 3\mathcal{L}\{\mathcal{U}(t-2)\} + \mathcal{L}\{\mathcal{U}(t-3)\}$$

$$= \frac{2}{s} - 3\frac{e^{-2s}}{s} + \frac{e^{-3s}}{s}. \qquad \blacksquare$$

EXAMPLE 9

Evaluate $\mathcal{L}\{\sin t\,\mathcal{U}(t - 2\pi)\}$.

Solution With $a = 2\pi$ we have from Theorem 7.6

$$\mathcal{L}\{\sin t\,\mathcal{U}(t - 2\pi)\} = \mathcal{L}\{\sin(t - 2\pi)\mathcal{U}(t - 2\pi)\} \qquad \Leftarrow \boxed{\sin t \text{ has period } 2\pi}$$

$$= e^{-2\pi s}\mathcal{L}\{\sin t\}$$

$$= \frac{e^{-2\pi s}}{s^2 + 1}. \qquad \blacksquare$$

EXAMPLE 10

Find the Laplace transform of the function shown in Figure 7.17.

Solution An equation of the straight line is found to be $y = 2t - 3$. To "turn off" this graph on the interval $0 \le t < 1$ we form $(2t - 3)\mathcal{U}(t - 1)$. Now Theorem 7.6 is not immediately applicable in the evaluation of $\mathcal{L}\{(2t-3)\mathcal{U}(t-1)\}$, since the function being transformed lacks the precise form $f(t - a)\mathcal{U}(t - a)$. However, by rewriting $2t - 3$ in terms of $t - 1$,

$$2t - 3 = 2(t - 1) - 1,$$

we can identify $f(t - 1) = 2(t - 1) - 1$ and consequently $f(t) = 2t - 1$. We are now in a position to apply the second translation theorem:

$$\mathcal{L}\{(2t - 3)\mathcal{U}(t - 1)\} = \mathcal{L}\{(2(t - 1) - 1)\mathcal{U}(t - 1)\}$$

$$= e^{-s}\mathcal{L}\{2t - 1\}$$

$$= e^{-s}\left(\frac{2}{s^2} - \frac{1}{s}\right). \qquad \blacksquare$$

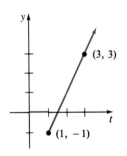

Figure 7.17

Inverse Form of the Second Translation Theorem

The inverse form of Theorem 7.6 is

$$f(t - a)\mathcal{U}(t - a) = \mathcal{L}^{-1}\{e^{-as}F(s)\}, \tag{8}$$

where $a > 0$ and $f(t) = \mathcal{L}^{-1}\{F(s)\}$.

EXAMPLE 11

Evaluate $\mathcal{L}^{-1}\left\{\dfrac{e^{-\pi s/2}}{s^2 + 9}\right\}$.

Solution   We identify $a = \dfrac{\pi}{2}$ and $f(t) = \mathcal{L}^{-1}\left\{\dfrac{1}{s^2 + 9}\right\} = \dfrac{1}{3}\sin 3t$. Thus from (8),

$$\mathcal{L}^{-1}\left\{\frac{e^{-\pi s/2}}{s^2 + 9}\right\} = \frac{1}{3}\mathcal{L}^{-1}\left\{\frac{3}{s^2 + 9}\right\}_{t \to t - \pi/2}\mathcal{U}\left(t - \frac{\pi}{2}\right)$$

$$= \frac{1}{3}\sin 3\left(t - \frac{\pi}{2}\right)\mathcal{U}\left(t - \frac{\pi}{2}\right)$$

$$= \frac{1}{3}\cos 3t\,\mathcal{U}\left(t - \frac{\pi}{2}\right). \qquad \blacksquare$$

If $F(s) = \mathcal{L}\{f(t)\}$ and if we assume that interchanging of differentiation and integration is possible, then

$$\frac{d}{ds} F(s) = \frac{d}{ds}\int_0^\infty e^{-st}f(t)\,dt$$

$$= \int_0^\infty \frac{\partial}{\partial s}\left[e^{-st}f(t)\right]dt$$

$$= -\int_0^\infty e^{-st}tf(t)\,dt$$

$$= -\mathcal{L}\{tf(t)\};$$

that is,

$$\mathcal{L}\{tf(t)\} = -\frac{d}{ds}\mathcal{L}\{f(t)\}.$$

Similarly, $\mathcal{L}\{t^2 f(t)\} = \mathcal{L}\{t \cdot tf(t)\} = -\dfrac{d}{ds}\mathcal{L}\{tf(t)\}$

$$= -\frac{d}{ds}\left(-\frac{d}{ds}\mathcal{L}\{f(t)\}\right) = \frac{d^2}{ds^2}\mathcal{L}\{f(t)\}.$$

The preceding two cases suggest the general result for $\mathscr{L}\{t^n f(t)\}$.

THEOREM 7.7   **Derivatives of Transforms**

For $n = 1, 2, 3, \ldots,$

$$\mathscr{L}\{t^n f(t)\} = (-1)^n \frac{d^n}{ds^n} F(s),$$

where $F(s) = \mathscr{L}\{f(t)\}$.

EXAMPLE 12

Evaluate **(a)** $\mathscr{L}\{te^{3t}\}$, **(b)** $\mathscr{L}\{t \sin kt\}$, **(c)** $\mathscr{L}\{t^2 \sin kt\}$, and **(d)** $\mathscr{L}\{te^{-t} \cos t\}$

Solution   We make use of results (c), (d), and (e) of Theorem 7.2.

(a) Note in this first example we could also use the first translation theorem. To apply Theorem 7.7 we identify $n = 1$ and $f(t) = e^{3t}$:

$$\mathscr{L}\{te^{3t}\} = -\frac{d}{ds} \mathscr{L}\{e^{3t}\} = -\frac{d}{ds}\left(\frac{1}{s-3}\right) = \frac{1}{(s-3)^2}.$$

(b)
$$\mathscr{L}\{t \sin kt\} = -\frac{d}{ds} \mathscr{L}\{\sin kt\}$$

$$= -\frac{d}{ds}\left(\frac{k}{s^2 + k^2}\right)$$

$$= \frac{2ks}{(s^2 + k^2)^2}$$

(c) With $n = 2$ in Theorem 7.7 this transform can be written

$$\mathscr{L}\{t^2 \sin kt\} = \frac{d^2}{ds^2} \mathscr{L}\{\sin kt\}$$

and so by carrying out the two derivatives we obtain the result. Alternatively, we can make use of the result already obtained in part (b). Since $t^2 \sin kt = t(t \sin kt)$, we have

$$\mathscr{L}\{t^2 \sin kt\} = -\frac{d}{ds} \mathscr{L}\{t \sin kt\}$$

$$= -\frac{d}{ds}\left(\frac{2ks}{(s^2 + k^2)^2}\right). \quad \Longleftarrow \boxed{\text{from part (b)}}$$

Differentiating and simplifying then give

$$\mathscr{L}\{t^2 \sin kt\} = \frac{6ks^2 - 2k^3}{(s^2 + k^2)^3}.$$

(d)
$$\mathscr{L}\{te^{-t}\cos t\} = -\frac{d}{ds}\mathscr{L}\{e^{-t}\cos t\}$$

$$= -\frac{d}{ds}\mathscr{L}\{\cos t\}_{s\to s+1} \quad \Longleftarrow \boxed{\begin{array}{l}\text{first translation}\\\text{theorem}\end{array}}$$

$$= -\frac{d}{ds}\left(\frac{s+1}{(s+1)^2+1}\right)$$

$$= \frac{(s+1)^2-1}{[(s+1)^2+1]^2} \qquad \blacksquare$$

7.3 EXERCISES

Answers to odd-numbered problems begin on page A–19.

In Problems 1–44 find either $F(s)$ or $f(t)$ as indicated.

1. $\mathscr{L}\{te^{10t}\}$ **2.** $\mathscr{L}\{te^{-6t}\}$

3. $\mathscr{L}\{t^3e^{-2t}\}$ **4.** $\mathscr{L}\{t^{10}e^{-7t}\}$

5. $\mathscr{L}\{e^t\sin 3t\}$ **6.** $\mathscr{L}\{e^{-2t}\cos 4t\}$

7. $\mathscr{L}\{e^{5t}\sinh 3t\}$ **8.** $\mathscr{L}\left\{\dfrac{\cosh t}{e^t}\right\}$

9. $\mathscr{L}\{t(e^t+e^{2t})^2\}$ **10.** $\mathscr{L}\{e^{2t}(t-1)^2\}$

11. $\mathscr{L}\{e^{-t}\sin^2 t\}$ **12.** $\mathscr{L}\{e^t\cos^2 3t\}$

13. $\mathscr{L}^{-1}\left\{\dfrac{1}{(s+2)^3}\right\}$ **14.** $\mathscr{L}^{-1}\left\{\dfrac{1}{(s-1)^4}\right\}$

15. $\mathscr{L}^{-1}\left\{\dfrac{1}{s^2-6s+10}\right\}$ **16.** $\mathscr{L}^{-1}\left\{\dfrac{1}{s^2+2s+5}\right\}$

17. $\mathscr{L}^{-1}\left\{\dfrac{s}{s^2+4s+5}\right\}$ **18.** $\mathscr{L}^{-1}\left\{\dfrac{2s+5}{s^2+6s+34}\right\}$

19. $\mathscr{L}^{-1}\left\{\dfrac{s}{(s+1)^2}\right\}$ **20.** $\mathscr{L}^{-1}\left\{\dfrac{5s}{(s-2)^2}\right\}$

21. $\mathscr{L}^{-1}\left\{\dfrac{2s-1}{s^2(s+1)^3}\right\}$ **22.** $\mathscr{L}^{-1}\left\{\dfrac{(s+1)^2}{(s+2)^4}\right\}$

23. $\mathscr{L}\{(t-1)\mathscr{U}(t-1)\}$ **24.** $\mathscr{L}\{e^{2-t}\mathscr{U}(t-2)\}$

25. $\mathscr{L}\{t\mathscr{U}(t-2)\}$ **26.** $\mathscr{L}\{(3t+1)\mathscr{U}(t-3)\}$

27. $\mathscr{L}\{\cos 2t\,\mathscr{U}(t-\pi)\}$ **28.** $\mathscr{L}\left\{\sin t\,\mathscr{U}\left(t-\dfrac{\pi}{2}\right)\right\}$

29. $\mathscr{L}\{(t-1)^3e^{t-1}\mathscr{U}(t-1)\}$ **30.** $\mathscr{L}\{te^{t-5}\mathscr{U}(t-5)\}$

31. $\mathscr{L}^{-1}\left\{\dfrac{e^{-2s}}{s^3}\right\}$ **32.** $\mathscr{L}^{-1}\left\{\dfrac{(1+e^{-2s})^2}{s+2}\right\}$

33. $\mathcal{L}^{-1}\left\{\dfrac{e^{-\pi s}}{s^2 + 1}\right\}$

34. $\mathcal{L}^{-1}\left\{\dfrac{se^{-\pi s/2}}{s^2 + 4}\right\}$

35. $\mathcal{L}^{-1}\left\{\dfrac{e^{-s}}{s(s + 1)}\right\}$

36. $\mathcal{L}^{-1}\left\{\dfrac{e^{-2s}}{s^2(s - 1)}\right\}$

37. $\mathcal{L}\{t \cos 2t\}$

38. $\mathcal{L}\{t \sinh 3t\}$

39. $\mathcal{L}\{t^2 \sinh t\}$

40. $\mathcal{L}\{t^2 \cos t\}$

41. $\mathcal{L}\{te^{2t} \sin 6t\}$

42. $\mathcal{L}\{te^{-3t} \cos 3t\}$

43. $\mathcal{L}^{-1}\left\{\dfrac{s}{(s^2 + 1)^2}\right\}$

44. $\mathcal{L}^{-1}\left\{\dfrac{s + 1}{(s^2 + 2s + 2)^2}\right\}$

In Problems 45–50 match the given graph with one of the functions in
(a)–**(f)**. The graph of $f(t)$ is given in Figure 7.18.

(a) $f(t) - f(t)\mathcal{U}(t - a)$

(b) $f(t - b)\mathcal{U}(t - b)$

(c) $f(t)\mathcal{U}(t - a)$

(d) $f(t) - f(t)\mathcal{U}(t - b)$

(e) $f(t)\mathcal{U}(t - a) - f(t)\mathcal{U}(t - b)$

(f) $f(t - a)\mathcal{U}(t - a) - f(t - a)\mathcal{U}(t - b)$

Figure 7.18

45.

Figure 7.19

46.

Figure 7.20

47. $f(t)$

Figure 7.21

48. $f(t)$

Figure 7.22

49. $f(t)$

Figure 7.23

50. $f(t)$

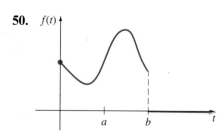

Figure 7.24

In Problems 51–58 write each function in terms of unit step functions. Find the Laplace transform of the given function.

51. $f(t) = \begin{cases} 2, & 0 \le t < 3 \\ -2, & t \ge 3 \end{cases}$

52. $f(t) = \begin{cases} 1, & 0 \le t < 4 \\ 0, & 4 \le t < 5 \\ 1, & t \ge 5 \end{cases}$

53. $f(t) = \begin{cases} 0, & 0 \le t < 1 \\ t^2, & t \ge 1 \end{cases}$

54. $f(t) = \begin{cases} 0, & 0 \le t < \dfrac{3\pi}{2} \\ \sin t, & t \ge \dfrac{3\pi}{2} \end{cases}$

55. $f(t) = \begin{cases} t, & 0 \le t < 2 \\ 0, & t \ge 2 \end{cases}$

56. $f(t) = \begin{cases} \sin t, & 0 \le t < 2\pi \\ 0, & t \ge 2\pi \end{cases}$

57. $f(t)$

rectangular pulse

Figure 7.25

58. $f(t)$

staircase function

Figure 7.26

In Problems 59 and 60 sketch the graph of the given function.

59. $f(t) = \mathscr{L}^{-1}\left\{\dfrac{1}{s^2} - \dfrac{e^{-s}}{s^2}\right\}$ **60.** $f(t) = \mathscr{L}^{-1}\left\{\dfrac{2}{s} - \dfrac{3e^{-s}}{s^2} + \dfrac{5e^{-2s}}{s^2}\right\}$

In Problems 61–64 use Theorem 7.7 in the form ($n = 1$)

$$f(t) = -\frac{1}{t}\mathscr{L}^{-1}\left\{\frac{d}{ds}F(s)\right\}$$

to evaluate the given inverse Laplace transform.

61. $\mathscr{L}^{-1}\left\{\ln\dfrac{s-3}{s+1}\right\}$ **62.** $\mathscr{L}^{-1}\left\{\ln\dfrac{s^2+1}{s^2+4}\right\}$

63. $\mathscr{L}^{-1}\left\{\dfrac{\pi}{2} - \tan^{-1}\dfrac{s}{2}\right\}$ **64.** $\mathscr{L}^{-1}\left\{\dfrac{1}{s} - \cot^{-1}\dfrac{4}{s}\right\}$

7.4 TRANSFORMS OF DERIVATIVES, INTEGRALS, AND PERIODIC FUNCTIONS

Our goal is to use the Laplace transform to solve certain kinds of differential equations. To that end we need to evaluate quantities such as $\mathscr{L}\{dy/dt\}$ and $\mathscr{L}\{d^2y/dt^2\}$. For example, if f' is continuous for $t \geq 0$, then integration by parts gives

$$\mathscr{L}\{f'(t)\} = \int_0^\infty e^{-st}f'(t)\,dt$$

$$= e^{-st}f(t)\Big|_0^\infty + s\int_0^\infty e^{-st}f(t)\,dt$$

$$= -f(0) + s\mathscr{L}\{f(t)\}$$

or $\mathscr{L}\{f'(t)\} = sF(s) - f(0).$ (1)

Here we have assumed that $e^{-st}f(t) \to 0$ as $t \to \infty$. Similarly,

$$\mathscr{L}\{f''(t)\} = \int_0^\infty e^{-st}f''(t)\,dt$$

$$= e^{-st}f'(t)\Big|_0^\infty + s\int_0^\infty e^{-st}f'(t)\,dt$$

$$= -f'(0) + s\mathscr{L}\{f'(t)\}$$

$$= s[sF(s) - f(0)] - f'(0)$$

or $\mathscr{L}\{f''(t)\} = s^2F(s) - sf(0) - f'(0).$ (2)

The results in (1) and (2) are special cases of the next theorem, which gives the Laplace transform of the nth derivative of f. The proof is omitted.

> **THEOREM 7.8   Transform of a Derivative**
>
> If $f(t), f'(t), \ldots, f^{(n-1)}(t)$ are continuous on $[0, \infty)$ and are of exponential order and if $f^{(n)}(t)$ is piecewise continuous on $[0, \infty)$, then
>
> $$\mathcal{L}\{f^{(n)}(t)\} = s^n F(s) - s^{n-1} f(0) - s^{n-2} f'(0) - \cdots - f^{(n-1)}(0),$$
>
> where $F(s) = \mathcal{L}\{f(t)\}$.

EXAMPLE I

Observe that the sum $kt \cos kt + \sin kt$ is the derivative of $t \sin kt$. Hence

$$\mathcal{L}\{kt \cos kt + \sin kt\} = \mathcal{L}\left\{\frac{d}{dt}(t \sin kt)\right\}$$

$$= s\mathcal{L}\{t \sin kt\} \qquad \Leftarrow \boxed{\text{by (1)}}$$

$$= s\left(-\frac{d}{ds}\mathcal{L}\{\sin kt\}\right) \qquad \Leftarrow \boxed{\text{from Theorem 7.7}}$$

$$= s\left(\frac{2ks}{(s^2 + k^2)^2}\right) = \frac{2ks^2}{(s^2 + k^2)^2}. \qquad \blacksquare$$

Convolution

If functions f and g are piecewise continuous on $[0, \infty)$, then the **convolution** of f and g, denoted by $f * g$, is given by the integral

$$f * g = \int_0^t f(\tau)g(t - \tau)\, d\tau.$$

It is left as an exercise to show that $\int_0^t f(\tau)g(t - \tau)\, d\tau = \int_0^t f(t - \tau)g(\tau)d\tau$; that is,

$$f * g = g * f.$$

See Problem 29. This means that the convolution of two functions is commutative.

EXAMPLE 2

The convolution of $f(t) = e^t$ and $g(t) = \sin t$ is

$$e^t * \sin t = \int_0^t e^\tau \sin(t - \tau)\, d\tau \tag{3}$$

$$= \frac{1}{2}(-\sin t - \cos t + e^t). \tag{4}$$

$\blacksquare$

It is possible to find the Laplace transform of the convolution of two functions, such as (3), without actually evaluating the integral as we did in (4). The result that follows is known as the **convolution theorem**.

THEOREM 7.9 **Convolution Theorem**

Let $f(t)$ and $g(t)$ be piecewise continuous on $[0, \infty)$ and of exponential order; then

$$\mathscr{L}\{f * g\} = \mathscr{L}\{f(t)\}\mathscr{L}\{g(t)\} = F(s)G(s).$$

Proof Let $\quad F(s) = \mathscr{L}\{f(t)\} = \displaystyle\int_0^\infty e^{-s\tau}f(\tau)\,d\tau$

and $\quad\quad\quad\quad G(s) = \mathscr{L}\{g(t)\} = \displaystyle\int_0^\infty e^{-s\beta}g(\beta)\,d\beta.$

Proceeding formally, we have

$$F(s)G(s) = \left(\int_0^\infty e^{-s\tau}f(\tau)\,d\tau\right)\left(\int_0^\infty e^{-s\beta}g(\beta)\,d\beta\right)$$

$$= \int_0^\infty \int_0^\infty e^{-s(\tau+\beta)}f(\tau)g(\beta)\,d\tau\,d\beta$$

$$= \int_0^\infty f(\tau)\,d\tau \int_0^\infty e^{-s(\tau+\beta)}g(\beta)\,d\beta.$$

Holding τ fixed, we let $t = \tau + \beta$, $dt = d\beta$ so that

$$F(s)G(s) = \int_0^\infty f(\tau)\,d\tau \int_\tau^\infty e^{-st}g(t - \tau)\,dt.$$

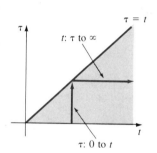

Figure 7.27

In the $t\tau$-plane we are integrating over the shaded region in Figure 7.27. Since f and g are piecewise continuous on $[0, \infty)$ and of exponential order, it is possible to interchange the order of integration:

$$F(s)G(s) = \int_0^\infty e^{-st}\,dt \int_0^t f(\tau)g(t - \tau)\,d\tau$$

$$= \int_0^\infty e^{-st}\left\{\int_0^t f(\tau)g(t - \tau)\,d\tau\right\}dt = \mathscr{L}\{f * g\}. \quad \square$$

When $g(t) = 1$ and $G(s) = 1/s$, the convolution theorem implies that the Laplace transform of the integral of a function f is

$$\mathscr{L}\left\{\int_0^t f(\tau)\,d\tau\right\} = \frac{F(s)}{s}. \tag{5}$$

EXAMPLE 3

Evaluate $\mathscr{L}\left\{\int_0^t e^{\tau}\sin(t-\tau)\,d\tau\right\}$.

Solution With $f(t)=e^t$ and $g(t)=\sin t$, the convolution theorem states that the Laplace transform of the convolution of f and g is the product of their Laplace transforms:

$$\mathscr{L}\left\{\int_0^t e^{\tau}\sin(t-\tau)\,d\tau\right\}=\mathscr{L}\{e^t\}\cdot\mathscr{L}\{\sin t\}$$

$$=\frac{1}{s-1}\cdot\frac{1}{s^2+1}=\frac{1}{(s-1)(s^2+1)}. \qquad \blacksquare$$

Inverse Form of the Convolution Theorem

The convolution theorem is sometimes useful in finding the inverse Laplace transform of a product of two Laplace transforms. From Theorem 7.9 we have

$$\mathscr{L}^{-1}\{F(s)G(s)\}=f*g. \qquad (6)$$

EXAMPLE 4

Evaluate $\mathscr{L}^{-1}\left\{\dfrac{1}{(s-1)(s+4)}\right\}$.

Solution Partial fractions could be used, but if we identify

$$F(s)=\frac{1}{s-1} \quad \text{and} \quad G(s)=\frac{1}{s+4},$$

then $\mathscr{L}^{-1}\{F(s)\}=f(t)=e^t$ and $\mathscr{L}^{-1}\{G(s)\}=g(t)=e^{-4t}$.

Hence from (6) we obtain

$$\mathscr{L}^{-1}\left\{\frac{1}{(s-1)(s+4)}\right\}=\int_0^t f(\tau)g(t-\tau)\,d\tau=\int_0^t e^{\tau}e^{-4(t-\tau)}\,d\tau$$

$$=e^{-4t}\int_0^t e^{5\tau}\,d\tau$$

$$=e^{-4t}\frac{1}{5}e^{5\tau}\Big|_0^t$$

$$=\frac{1}{5}e^t-\frac{1}{5}e^{-4t}. \qquad \blacksquare$$

EXAMPLE 5

Evaluate $\mathscr{L}^{-1}\left\{\dfrac{1}{(s^2 + k^2)^2}\right\}$.

Solution   Let $\qquad\qquad F(s) = G(s) = \dfrac{1}{s^2 + k^2}$

so that $\qquad\qquad f(t) = g(t) = \dfrac{1}{k}\mathscr{L}^{-1}\left\{\dfrac{k}{s^2 + k^2}\right\} = \dfrac{1}{k}\sin kt$.

In this case (6) gives

$$\mathscr{L}^{-1}\left\{\dfrac{1}{(s^2 + k^2)^2}\right\} = \dfrac{1}{k^2}\int_0^t \sin k\tau \sin k(t - \tau)\, d\tau. \tag{7}$$

Now recall from trigonometry that

$$\cos(A + B) = \cos A \cos B - \sin A \sin B$$

and $\qquad\qquad \cos(A - B) = \cos A \cos B + \sin A \sin B.$

Subtracting the first from the second gives the identity

$$\sin A \sin B = \dfrac{1}{2}[\cos(A - B) - \cos(A + B)].$$

If we set $A = k\tau$ and $B = k(t - \tau)$, we can carry out the integration in (7):

$$\mathscr{L}^{-1}\left\{\dfrac{1}{(s^2 + k^2)^2}\right\} = \dfrac{1}{2k^2}\int_0^t [\cos k(2\tau - t) - \cos kt]\, d\tau$$

$$= \dfrac{1}{2k^2}\left[\dfrac{1}{2k}\sin k(2\tau - t) - \tau \cos kt\right]_0^t$$

$$= \dfrac{\sin kt - kt \cos kt}{2k^3}. \qquad\blacksquare$$

Transform of a Periodic Function

If a periodic function has period T, $T > 0$, then $f(t + T) = f(t)$. The Laplace transform of a **periodic function** can be obtained by an integration over one period.

THEOREM 7.10   Transform of a Periodic Function

Let $f(t)$ be piecewise continuous on $[0, \infty)$ and of exponential order. If $f(t)$ is periodic with period T, then

$$\mathscr{L}\{f(t)\} = \dfrac{1}{1 - e^{-sT}}\int_0^T e^{-st}f(t)\, dt. \tag{8}$$

Proof   Write the Laplace transform as two integrals:

$$\mathscr{L}\{f(t)\} = \int_0^T e^{-st}f(t)\, dt + \int_T^\infty e^{-st}f(t)\, dt. \qquad (9)$$

When we let $t = u + T$, the last integral in (9) becomes

$$\int_T^\infty e^{-st}f(t)\, dt = \int_0^\infty e^{-s(u+T)}f(u+T)\, du = e^{-sT}\int_0^\infty e^{-su}f(u)\, du = e^{-sT}\mathscr{L}\{f(t)\}.$$

Hence (9) is

$$\mathscr{L}\{f(t)\} = \int_0^T e^{-st}f(t)\, dt + e^{-sT}\mathscr{L}\{f(t)\}.$$

Solving for $\mathscr{L}\{f(t)\}$ yields the desired result:

$$\mathscr{L}\{f(t)\} = \frac{1}{1 - e^{-sT}}\int_0^T e^{-st}f(t)\, dt. \qquad \square$$

EXAMPLE 6

Find the Laplace transform of the periodic function shown in Figure 7.28.

Solution   On the interval $0 \le t < 2$ the function can be defined by

$$f(t) = \begin{cases} t, & 0 \le t < 1 \\ 0, & 1 \le t < 2 \end{cases}$$

and outside the interval by $f(t + 2) = f(t)$. Identifying $T = 2$, we use (8) and integration by parts to obtain

$$\mathscr{L}\{f(t)\} = \frac{1}{1 - e^{-2s}}\int_0^2 e^{-st}f(t)\, dt \qquad (10)$$

$$= \frac{1}{1 - e^{-2s}}\left[\int_0^1 e^{-st}t\, dt + \int_1^2 e^{-st}0\, dt\right]$$

$$= \frac{1}{1 - e^{-2s}}\left[-\frac{e^{-s}}{s} + \frac{1 - e^{-s}}{s^2}\right] \qquad (11)$$

$$= \frac{1 - (s + 1)e^{-s}}{s^2(1 - e^{-2s})}. \qquad \blacksquare$$

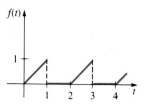

Figure 7.28

The result in (11) of Example 6 can be obtained without actually integrating by making use of the second translation theorem. If we define

$$g(t) = \begin{cases} t, & 0 \le t < 1 \\ 0, & t \ge 1, \end{cases}$$

then $f(t) = g(t)$ on the interval $[0, T]$, where $T = 2$. But we can express g in terms of unit step functions as

$$g(t) = t - t\,\mathcal{U}(t - 1) = t - (t - 1)\mathcal{U}(t - 1) - \mathcal{U}(t - 1).$$

Thus (10) can be written as

$$\mathcal{L}\{f(t)\} = \frac{1}{1 - e^{-2s}}\, \mathcal{L}\{g(t)\}$$

$$= \frac{1}{1 - e^{-2s}}\, \mathcal{L}\{t - (t - 1)\mathcal{U}(t - 1) - \mathcal{U}(t - 1)\}$$

$$= \frac{1}{1 - e^{-2s}}\left[\frac{1}{s^2} - \frac{1}{s^2}e^{-s} - \frac{1}{s}e^{-s}\right].$$

Inspection of the expression inside the brackets reveals that it is identical to (11).

7.4 EXERCISES

Answers to odd-numbered problems begin on page A–19.

1. Use the result $(d/dt)e^t = e^t$ and (1) of this section to evaluate $\mathcal{L}\{e^t\}$.

2. Use the result $(d/dt)\cos^2 t = -\sin 2t$ and (1) of this section to evaluate $\mathcal{L}\{\cos^2 t\}$.

In Problems 3 and 4 suppose a function $y(t)$ has the properties that $y(0) = 1$ and $y'(0) = -1$. Find the Laplace transform of the given expression.

3. $y'' + 3y'$ 　　　　　　　　　　　　**4.** $y'' - 4y' + 5y$

In Problems 5 and 6 suppose a function $y(t)$ has the properties that $y(0) = 2$ and $y'(0) = 3$. Solve for the Laplace transform $\mathcal{L}\{y(t)\} = Y(s)$.

5. $y'' - 2y' + y = 0$ 　　　　　　　　**6.** $y'' + y = 1$

In Problems 7–20 evaluate the given Laplace transform without evaluating the integral.

7. $\mathcal{L}\left\{\displaystyle\int_0^t e^\tau\, d\tau\right\}$ 　　　　　　　　**8.** $\mathcal{L}\left\{\displaystyle\int_0^t \cos \tau\, d\tau\right\}$

9. $\mathcal{L}\left\{\displaystyle\int_0^t e^{-\tau} \cos \tau\, d\tau\right\}$ 　　　　**10.** $\mathcal{L}\left\{\displaystyle\int_0^t \tau \sin \tau\, d\tau\right\}$

11. $\mathcal{L}\left\{\displaystyle\int_0^t \tau e^{t-\tau}\, d\tau\right\}$ 　　　　　**12.** $\mathcal{L}\left\{\displaystyle\int_0^t \sin \tau \cos(t - \tau)\, d\tau\right\}$

13. $\mathcal{L}\left\{t \displaystyle\int_0^t \sin \tau\, d\tau\right\}$ 　　　　　**14.** $\mathcal{L}\left\{t \displaystyle\int_0^t \tau e^{-\tau}\, d\tau\right\}$

15. $\mathscr{L}\{1 * t^3\}$ **16.** $\mathscr{L}\{1 * e^{-2t}\}$

17. $\mathscr{L}\{t^2 * t^4\}$ **18.** $\mathscr{L}\{t^2 * te^t\}$

19. $\mathscr{L}\{e^{-t} * e^t \cos t\}$ **20.** $\mathscr{L}\{e^{2t} * \sin t\}$

In Problems 21 and 22 suppose $\mathscr{L}^{-1}\{F(s)\} = f(t)$. Find the inverse Laplace transform of the given function.

21. $\dfrac{1}{s + 5} F(s)$ **22.** $\dfrac{s}{s^2 + 4} F(s)$

In Problems 23–28 use (6) to find $f(t)$.

23. $\mathscr{L}^{-1}\left\{\dfrac{1}{s(s + 1)}\right\}$ **24.** $\mathscr{L}^{-1}\left\{\dfrac{1}{s(s^2 + 1)}\right\}$

25. $\mathscr{L}^{-1}\left\{\dfrac{1}{(s + 1)(s - 2)}\right\}$ **26.** $\mathscr{L}^{-1}\left\{\dfrac{1}{(s + 1)^2}\right\}$

27. $\mathscr{L}^{-1}\left\{\dfrac{s}{(s^2 + 4)^2}\right\}$ **28.** $\mathscr{L}^{-1}\left\{\dfrac{1}{(s^2 + 4s + 5)^2}\right\}$

29. Prove the commutative property of the convolution integral $f * g = g * f$.

30. Prove the distributive property of the convolution integral $f * (g + h) = f * g + f * h$.

In Problems 31–38 use Theorem 7.10 to find the Laplace transform of the given periodic function.

31.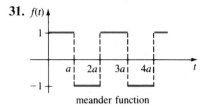

meander function

Figure 7.29

32.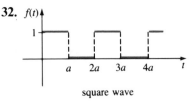

square wave

Figure 7.30

33.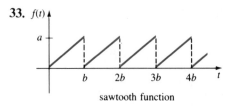

sawtooth function

Figure 7.31

34.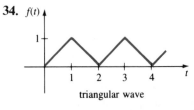

triangular wave

Figure 7.32

35. $f(t)$

36. $f(t)$

full-wave rectification of $\sin t$

half-wave rectification of $\sin t$

Figure 7.33

Figure 7.34

37. $f(t) = \sin t$
$f(t + 2\pi) = f(t)$

38. $f(t) = \cos t$
$f(t + 2\pi) = f(t)$

7.5 APPLICATIONS

Since $\mathscr{L}\{y^{(n)}(t)\}$, $n > 1$, depends on $y(t)$ and its $n - 1$ derivatives evaluated at $t = 0$, the Laplace transform is ideally suited to initial-value problems for linear differential equations with constant coefficients. This kind of differential equation can be reduced to an *algebraic equation* in the transformed function $Y(s)$. To see this, consider the initial-value problem

$$a_n \frac{d^n y}{dt^n} + a_{n-1} \frac{d^{n-1} y}{dt^{n-1}} + \cdots + a_1 \frac{dy}{dt} + a_0 y = g(t)$$

$$y(0) = y_0, \quad y'(0) = y_0', \quad \ldots, \quad y^{(n-1)}(0) = y_0^{(n-1)},$$

where a_i, $i = 0, 1, \ldots, n$, and $y_0, y_0', \ldots, y_0^{(n-1)}$ are constants. By the linearity property of the Laplace transform, we can write

$$a_n \mathscr{L}\left\{\frac{d^n y}{dt^n}\right\} + a_{n-1} \mathscr{L}\left\{\frac{d^{n-1} y}{dt^{n-1}}\right\} + \cdots + a_0 \mathscr{L}\{y\} = \mathscr{L}\{g(t)\}. \qquad \textbf{(1)}$$

From Theorem 7.8, (1) becomes

$$a_n[s^n Y(s) - s^{n-1} y(0) - \cdots - y^{(n-1)}(0)] +$$

$$a_{n-1}[s^{n-1} Y(s) - s^{n-2} y(0) - \cdots - y^{(n-2)}(0)] + \cdots + a_0 Y(s) = G(s)$$

or

$$[a_n s^n + a_{n-1} s^{n-1} + \cdots + a_0] Y(s) = a_n[s^{n-1} y_0 + \cdots + y_0^{(n-1)}]$$

$$+ a_{n-1}[s^{n-2} y_0 + \cdots + y_0^{(n-2)}] + \cdots + G(s), \qquad \textbf{(2)}$$

where $Y(s) = \mathscr{L}\{y(t)\}$ and $G(s) = \mathscr{L}\{g(t)\}$. By solving (2) for $Y(s)$, we find $y(t)$ by determining the inverse transform

$$y(t) = \mathscr{L}^{-1}\{Y(s)\}.$$

Figure 7.35

The procedure is outlined in Figure 7.35. Note that this method incorporates the prescribed initial conditions directly into the solution. Hence there is no need for the separate operations of determining constants in the general solution of the differential equation.

EXAMPLE I

Solve
$$\frac{dy}{dt} - 3y = e^{2t}, \qquad y(0) = 1.$$

Solution   We first take the transform of each member of the given differential equation:

$$\mathscr{L}\left\{\frac{dy}{dt}\right\} - 3\mathscr{L}\{y\} = \mathscr{L}\{e^{2t}\}.$$

We then use $\mathscr{L}\{dy/dt\} = sY(s) - y(0) = sY(s) - 1$ and $\mathscr{L}\{e^{2t}\} = 1/(s-2)$.

Solving
$$sY(s) - 1 - 3Y(s) = \frac{1}{s-2}$$

for $Y(s)$ gives
$$Y(s) = \frac{s-1}{(s-2)(s-3)}.$$

By partial fractions:

$$\frac{s-1}{(s-2)(s-3)} = \frac{A}{s-2} + \frac{B}{s-3},$$

which yields
$$s - 1 = A(s-3) + B(s-2).$$

Setting $s = 2$ and $s = 3$ in the last equation, we obtain $A = -1$ and $B = 2$, respectively. Consequently

$$Y(s) = \frac{-1}{s-2} + \frac{2}{s-3}$$

and
$$y(t) = -\mathscr{L}^{-1}\left\{\frac{1}{s-2}\right\} + 2\mathscr{L}^{-1}\left\{\frac{1}{s-3}\right\}.$$

From part (c) of Theorem 7.3 it follows that

$$y(t) = -e^{2t} + 2e^{3t}. \qquad \blacksquare$$

EXAMPLE 2

Solve
$$y'' - 6y' + 9y = t^2 e^{3t}, \qquad y(0) = 2, \quad y'(0) = 6.$$

Solution
$$\mathcal{L}\{y''\} - 6\mathcal{L}\{y'\} + 9\mathcal{L}\{y\} = \mathcal{L}\{t^2 e^{3t}\}$$

$$\underbrace{s^2 Y(s) - sy(0) - y'(0)}_{\mathcal{L}\{y''\}} - \underbrace{6[sY(s) - y(0)]}_{\mathcal{L}\{y'\}} + \underbrace{9Y(s)}_{\mathcal{L}\{y\}} = \underbrace{\frac{2}{(s-3)^3}}_{\mathcal{L}\{t^2 e^{3t}\}}.$$

Using the initial conditions and simplifying give

$$(s^2 - 6s + 9)Y(s) = 2s - 6 + \frac{2}{(s-3)^3}$$

$$(s - 3)^2 Y(s) = 2(s - 3) + \frac{2}{(s-3)^3}$$

$$Y(s) = \frac{2}{s-3} + \frac{2}{(s-3)^5}$$

and so
$$y(t) = 2\mathcal{L}^{-1}\left\{\frac{1}{s-3}\right\} + \frac{2}{4!}\mathcal{L}^{-1}\left\{\frac{4!}{(s-3)^5}\right\}.$$

Recall from the first translation theorem that

$$\mathcal{L}^{-1}\left\{\frac{4!}{s^5}\bigg|_{s \to s-3}\right\} = t^4 e^{3t}.$$

Hence we have
$$y(t) = 2e^{3t} + \frac{1}{12} t^4 e^{3t}. \qquad \blacksquare$$

EXAMPLE 3

Solve $y'' + 4y' + 6y = 1 + e^{-t}, \qquad y(0) = 0, \quad y'(0) = 0.$

Solution
$$\mathcal{L}\{y''\} + 4\mathcal{L}\{y'\} + 6\mathcal{L}\{y\} = \mathcal{L}\{1\} + \mathcal{L}\{e^{-t}\}$$

$$s^2 Y(s) - sy(0) - y'(0) + 4[sY(s) - y(0)] + 6Y(s) = \frac{1}{s} + \frac{1}{s+1}$$

$$(s^2 + 4s + 6)Y(s) = \frac{2s+1}{s(s+1)}$$

$$Y(s) = \frac{2s+1}{s(s+1)(s^2 + 4s + 6)}.$$

By partial fractions:

$$\frac{2s+1}{s(s+1)(s^2 + 4s + 6)} = \frac{A}{s} + \frac{B}{s+1} + \frac{Cs+D}{s^2 + 4s + 6},$$

which implies

$$2s + 1 = A(s + 1)(s^2 + 4s + 6) + Bs(s^2 + 4s + 6) + (Cs + D)s(s + 1).$$

Setting $s = 0$ and $s = -1$ gives, respectively, $A = 1/6$ and $B = 1/3$. Equating the coefficients of s^3 and s gives

$$A + B + C = 0$$

$$10A + 6B + D = 2,$$

so it follows that $C = -1/2$ and $D = -5/3$. Thus

$$Y(s) = \frac{1/6}{s} + \frac{1/3}{s + 1} + \frac{-s/2 - 5/3}{s^2 + 4s + 6}$$

$$= \frac{1/6}{s} + \frac{1/3}{s + 1} + \frac{(-1/2)(s + 2) - 2/3}{(s + 2)^2 + 2}$$

$$= \frac{1/6}{s} + \frac{1/3}{s + 1} - \frac{1}{2}\frac{s + 2}{(s + 2)^2 + 2} - \frac{2}{3}\frac{1}{(s + 2)^2 + 2}.$$

Finally, from parts (a) and (c) of Theorem 7.3 and the first translation theorem, we obtain

$$y(t) = \frac{1}{6}\mathscr{L}^{-1}\left\{\frac{1}{s}\right\} + \frac{1}{3}\mathscr{L}^{-1}\left\{\frac{1}{s + 1}\right\} - \frac{1}{2}\mathscr{L}^{-1}\left\{\frac{s + 2}{(s + 2)^2 + 2}\right\}$$

$$- \frac{2}{3\sqrt{2}}\mathscr{L}^{-1}\left\{\frac{\sqrt{2}}{(s + 2)^2 + 2}\right\}$$

$$= \frac{1}{6} + \frac{1}{3}e^{-t} - \frac{1}{2}e^{-2t}\cos\sqrt{2}t - \frac{\sqrt{2}}{3}e^{-2t}\sin\sqrt{2}t. \quad\blacksquare$$

EXAMPLE 4

Solve $x'' + 16x = \cos 4t, \qquad x(0) = 0, \quad x'(0) = 1.$

Solution Recall that this initial-value problem could describe the forced, undamped, and resonant motion of a mass on a spring. The mass starts with an initial velocity of 1 foot per second in the downward direction from the equilibrium position.

Transforming the equation gives

$$(s^2 + 16)X(s) = 1 + \frac{s}{s^2 + 16}$$

$$X(s) = \frac{1}{s^2 + 16} + \frac{s}{(s^2 + 16)^2}.$$

With the aid of part (d) of Theorem 7.3 and Theorem 7.7, we find

$$x(t) = \frac{1}{4}\mathcal{L}^{-1}\left\{\frac{4}{s^2 + 16}\right\} + \frac{1}{8}\mathcal{L}^{-1}\left\{\frac{8s}{(s^2 + 16)^2}\right\}$$

$$= \frac{1}{4}\sin 4t + \frac{1}{8}t \sin 4t.$$

∎

EXAMPLE 5

Solve

$$x'' + 16x = f(t), \qquad x(0) = 0, \quad x'(0) = 1,$$

where

$$f(t) = \begin{cases} \cos 4t, & 0 \leq t < \pi \\ 0, & t \geq \pi. \end{cases}$$

f(t)

$\pi/2$ π

Figure 7.36

Solution The function $f(t)$ can be interpreted as an external force that is acting on a mechanical system for only a short period of time and then is removed. See Figure 7.36. Although this problem could be solved by conventional means, the procedure is not at all convenient when $f(t)$ is defined in a piecewise manner. With the aid of (2) and (3) of Section 7.3 and the periodicity of the cosine, we can rewrite f in terms of the unit step function as

$$f(t) = \cos 4t - \cos 4t\,\mathcal{U}(t - \pi) = \cos 4t - \cos 4(t - \pi)\mathcal{U}(t - \pi).$$

The second translation theorem then yields

$$\mathcal{L}\{x''\} + 16\mathcal{L}\{x\} = \mathcal{L}\{f(t)\}$$

$$s^2 X(s) - sx(0) - x'(0) + 16X(s) = \frac{s}{s^2 + 16} - \frac{s}{s^2 + 16}e^{-\pi s}$$

$$(s^2 + 16)X(s) = 1 + \frac{s}{s^2 + 16} - \frac{s}{s^2 + 16}e^{-\pi s}$$

$$X(s) = \frac{1}{s^2 + 16} + \frac{s}{(s^2 + 16)^2} - \frac{s}{(s^2 + 16)^2}e^{-\pi s}.$$

From part (b) of Example 12 in Section 7.3 (with $k = 4$) along with (8) of that section, we find

$$x(t) = \frac{1}{4}\mathcal{L}^{-1}\left\{\frac{4}{s^2 + 16}\right\} + \frac{1}{8}\mathcal{L}^{-1}\left\{\frac{8s}{(s^2 + 16)^2}\right\}$$

$$- \frac{1}{8}\mathcal{L}^{-1}\left\{\frac{8s}{(s^2 + 16)^2}e^{-\pi s}\right\}$$

$$= \frac{1}{4}\sin 4t + \frac{1}{8}t \sin 4t - \frac{1}{8}(t - \pi)\sin 4(t - \pi)\mathcal{U}(t - \pi).$$

The foregoing solution is the same as

$$
x(t) = \begin{cases}
\dfrac{1}{4} \sin 4t + \dfrac{1}{8} t \sin 4t, & 0 \le t < \pi \\[2ex]
\dfrac{2 + \pi}{8} \sin 4t, & t \ge \pi.
\end{cases}
$$

Observe from the graph of $x(t)$ in Figure 7.37 that the amplitudes of vibration become steady as soon as the external force is turned off.

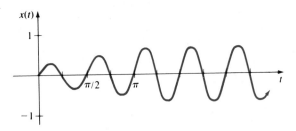

Figure 7.37   ■

EXAMPLE 6

Solve

$$y'' + 2y' + y = f(t), \qquad y(0) = 0, \quad y'(0) = 0,$$

where

$$f(t) = \mathscr{U}(t - 1) - 2\mathscr{U}(t - 2) + \mathscr{U}(t - 3).$$

Solution   By the second translation theorem and simplification, the transform of the differential equation is

$$(s + 1)^2 Y(s) = \frac{e^{-s}}{s} - 2\frac{e^{-2s}}{s} + \frac{e^{-3s}}{s}$$

or

$$Y(s) = \frac{e^{-s}}{s(s + 1)^2} - 2\frac{e^{-2s}}{s(s + 1)^2} + \frac{e^{-3s}}{s(s + 1)^2}.$$

With the aid of partial fractions, the last equation becomes

$$Y(s) = \left[\frac{1}{s} - \frac{1}{s + 1} - \frac{1}{(s + 1)^2} \right] e^{-s} - 2\left[\frac{1}{s} - \frac{1}{s + 1} - \frac{1}{(s + 1)^2} \right] e^{-2s}$$

$$+ \left[\frac{1}{s} - \frac{1}{s + 1} - \frac{1}{(s + 1)^2} \right] e^{-3s}.$$

Again using the inverse form of the second translation theorem, we find

$$y(t) = [1 - e^{-(t-1)} - (t - 1)e^{-(t-1)}]\mathscr{U}(t - 1) - 2[1 - e^{-(t-2)} - (t - 2)e^{-(t-2)}]\mathscr{U}(t - 2)$$

$$+ [1 - e^{-(t-3)} - (t - 3)e^{-(t-3)}]\mathscr{U}(t - 3). \qquad ■$$

Volterra Integral Equation

The convolution theorem is useful in solving other types of equations in which an unknown function appears under an integral sign. In the next example we solve a **Volterra integral equation**

$$f(t) = g(t) + \int_0^t f(\tau)h(t - \tau)\, d\tau$$

for $f(t)$. The functions $g(t)$ and $h(t)$ are known.

EXAMPLE 7

Solve

$$f(t) = 3t^2 - e^{-t} - \int_0^t f(\tau)e^{t-\tau}\, d\tau$$

for $f(t)$.

Solution It follows from Theorem 7.9 that

$$\mathscr{L}\{f(t)\} = 3\mathscr{L}\{t^2\} - \mathscr{L}\{e^{-t}\} - \mathscr{L}\{f(t)\}\mathscr{L}\{e^t\}$$

$$F(s) = 3 \cdot \frac{2}{s^3} - \frac{1}{s + 1} - F(s) \cdot \frac{1}{s - 1}$$

$$\left(1 + \frac{1}{s - 1}\right) F(s) = \frac{6}{s^3} - \frac{1}{s + 1}$$

$$\frac{s}{s - 1} F(s) = \frac{6}{s^3} - \frac{1}{s + 1}$$

$$F(s) = \frac{6(s - 1)}{s^4} - \frac{s - 1}{s(s + 1)}$$

$$= \frac{6}{s^3} - \frac{6}{s^4} + \frac{1}{s} - \frac{2}{s + 1}. \quad \Longleftarrow \boxed{\text{termwise division and partial fractions}}$$

Therefore

$$f(t) = 3\mathscr{L}^{-1}\left\{\frac{2!}{s^3}\right\} - \mathscr{L}^{-1}\left\{\frac{3!}{s^4}\right\} + \mathscr{L}^{-1}\left\{\frac{1}{s}\right\} - 2\mathscr{L}^{-1}\left\{\frac{1}{s + 1}\right\}$$

$$= 3t^2 - t^3 + 1 - 2e^{-t}. \qquad\qquad \blacksquare$$

Series Circuits

In a single loop or series circuit, Kirchhoff's second law states that the sum of the voltage drops across an inductor, resistor, and capacitor is equal to the impressed voltage $E(t)$. Now it is known that

$$\text{the voltage drop across the inductor} = L\frac{di}{dt},$$

$$\text{the voltage drop across the resistor} = Ri(t),$$

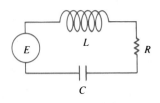

Figure 7.38

and $\quad$ the voltage drop across the capacitor $= \dfrac{1}{C}\displaystyle\int_0^t i(\tau)\,d\tau,$

where $i(t)$ is the current and L, R, and C are constants. See Section 1.2. It follows that the current in a circuit, such as that shown in Figure 7.38, is governed by the **integrodifferential equation**

$$L\frac{di}{dt} + Ri + \frac{1}{C}\int_0^t i(\tau)\,d\tau = E(t). \tag{3}$$

EXAMPLE 8

Figure 7.39

Determine the current $i(t)$ in a single loop L-R-C circuit when $L = 0.1$ henry, $R = 20$ ohms, $C = 10^{-3}$ farad, $i(0) = 0$, and the impressed voltage $E(t)$ is as given in Figure 7.39.

Solution Since the voltage is off for $t \geq 1$, we can write

$$E(t) = 120t - 120t\,\mathscr{U}(t-1). \tag{4}$$

But in order to use the second translation theorem we must rewrite (4) as

$$E(t) = 120t - 120(t-1)\mathscr{U}(t-1) - 120\mathscr{U}(t-1).$$

Equation (3) then becomes

$$0.1\frac{di}{dt} + 20i + 10^3\int_0^t i(\tau)\,d\tau = 120t - 120(t-1)\mathscr{U}(t-1) - 120\mathscr{U}(t-1). \tag{5}$$

Now recall from Theorem 7.9 that

$$\mathscr{L}\left\{\int_0^t i(\tau)\,d\tau\right\} = \frac{I(s)}{s},$$

where $I(s) = \mathscr{L}\{i(t)\}$. Thus the transform of equation (5) is

$$0.1sI(s) + 20I(s) + 10^3\frac{I(s)}{s} = 120\left[\frac{1}{s^2} - \frac{1}{s^2}e^{-s} - \frac{1}{s}e^{-s}\right]$$

or, after multiplying by $10s$,

$$(s+100)^2I(s) = 1200\left[\frac{1}{s} - \frac{1}{s}e^{-s} - e^{-s}\right]$$

$$I(s) = 1200\left[\frac{1}{s(s+100)^2} - \frac{1}{s(s+100)^2}e^{-s} - \frac{1}{(s+100)^2}e^{-s}\right].$$

By partial fractions we can write

$$I(s) = 1200 \left[\frac{1/10{,}000}{s} - \frac{1/10{,}000}{s+100} - \frac{1/100}{(s+100)^2} - \frac{1/10{,}000}{s} e^{-s} \right.$$

$$\left. + \frac{1/10{,}000}{s+100} e^{-s} + \frac{1/100}{(s+100)^2} e^{-s} - \frac{1}{(s+100)^2} e^{-s} \right].$$

Employing the inverse form of the second translation theorem, we obtain

$$i(t) = \frac{3}{25} [1 - \mathcal{U}(t-1)] - \frac{3}{25} [e^{-100t} - e^{-100(t-1)}\mathcal{U}(t-1)]$$

$$- 12te^{-100t} - 1188(t-1)e^{-100(t-1)}\mathcal{U}(t-1). \qquad \blacksquare$$

EXAMPLE 9

The differential equation for the current $i(t)$ in a single loop L-R series circuit is

$$L\frac{di}{dt} + Ri = E(t). \qquad (6)$$

Determine the current $i(t)$ when $i(0) = 0$ and $E(t)$ is the square wave function shown in Figure 7.40.

Solution The Laplace transform of the equation is

$$LsI(s) + RI(s) = \mathcal{L}\{E(t)\}. \qquad (7)$$

Since $E(t)$ is periodic with period $T = 2$, we use (8) of Section 7.4:

$$\mathcal{L}\{E(t)\} = \frac{1}{1 - e^{-2s}} \int_0^2 e^{-st} f(t)\, dt$$

$$= \frac{1}{1 - e^{-2s}} \left(\int_0^1 1 \cdot e^{-st}\, dt + \int_1^2 0 \cdot e^{-st}\, dt \right)$$

$$= \frac{1}{1 - e^{-2s}} \frac{1 - e^{-s}}{s} \qquad \Longleftarrow \boxed{1 - e^{-2s} = (1 + e^{-s})(1 - e^{-s})}$$

$$= \frac{1}{s(1 + e^{-s})}.$$

Hence from (7) we find

$$I(s) = \frac{1/L}{s(s + R/L)(1 + e^{-s})}. \qquad (8)$$

To find the inverse Laplace transform of this function we first make use of a geometric series. Recall, for $|x| < 1$,

$$\frac{1}{1 + x} = 1 - x + x^2 - x^3 + \cdots.$$

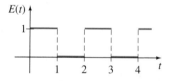

$E(t)$

Figure 7.40

With the identification $x = e^{-s}$ we then have for $s > 0$

$$\frac{1}{1 + e^{-s}} = 1 - e^{-s} + e^{-2s} - e^{-3s} + \cdots .$$

By writing

$$\frac{1}{s(s + R/L)} = \frac{L/R}{s} - \frac{L/R}{s + R/L}$$

(8) becomes

$$I(s) = \frac{1}{R}\left(\frac{1}{s} - \frac{1}{s + R/L}\right)(1 - e^{-s} + e^{-2s} - e^{-3s} + \cdots)$$

$$= \frac{1}{R}\left(\frac{1}{s} - \frac{e^{-s}}{s} + \frac{e^{-2s}}{s} - \frac{e^{-3s}}{s} + \cdots\right) - \frac{1}{R}\left(\frac{1}{s + R/L} - \frac{e^{-s}}{s + R/L} + \frac{e^{-2s}}{s + R/L} - \frac{e^{-3s}}{s + R/L} + \cdots\right).$$

By applying the inverse form of the second translation theorem to each term of both series, we obtain

$$i(t) = \frac{1}{R}(1 - \mathcal{U}(t - 1) + \mathcal{U}(t - 2) - \mathcal{U}(t - 3) + \cdots)$$

$$- \frac{1}{R}(e^{-Rt/L} - e^{-R(t-1)/L}\mathcal{U}(t - 1) + e^{-R(t-2)/L}\mathcal{U}(t - 2) - e^{-R(t-3)/L}\mathcal{U}(t - 3) + \cdots)$$

or equivalently

$$i(t) = \frac{1}{R}(1 - e^{-Rt/L}) + \frac{1}{R}\sum_{n=1}^{\infty}(-1)^n(1 - e^{-R(t-n)/L})\mathcal{U}(t - n). \quad \blacksquare$$

To interpret the solution in Example 9 let us assume for the sake of illustration that $R = 1$, $L = 1$, and $0 \leq t < 4$. In this case

$$i(t) = 1 - e^{-t} - (1 - e^{t-1})\mathcal{U}(t - 1) + (1 - e^{-(t-2)})\mathcal{U}(t - 2) - (1 - e^{-(t-3)})\mathcal{U}(t - 3);$$

in other words,

$$i(t) = \begin{cases} 1 - e^{-t}, & 0 \leq t < 1 \\ -e^{-t} + e^{-(t-1)}, & 1 \leq t < 2 \\ 1 - e^{-t} + e^{-(t-1)} - e^{-(t-2)}, & 2 \leq t < 3 \\ -e^{-t} + e^{-(t-1)} - e^{-(t-2)} + e^{-(t-3)}, & 3 \leq t < 4. \end{cases}$$

Figure 7.41

The graph of $i(t)$ on the interval $0 \leq t < 4$ is given in Figure 7.41.

Beams

In Example 9 of Section 1.2 we saw that the static deflection $y(x)$ of a uniform beam of length L carrying a load $w(x)$ per unit length is found from the fourth-order differential equation

$$EI\frac{d^4y}{dx^4} = w(x), \qquad (9)$$

where E is Young's modulus of elasticity and I is a moment of inertia of a cross section of the beam. To apply the Laplace transform to (9) we tacitly assume that $w(x)$ and $y(x)$ are defined on $(0, \infty)$ rather than on $(0, L)$. Note too that the next example is a boundary-value problem rather than an initial-value problem.

EXAMPLE 10

wall

w_0

L

Figure 7.42

A beam of length L is embedded at both ends. See Figure 7.42. In this case the deflection $y(x)$ must satisfy (9) and the conditions

$$y(0) = 0, \quad y(L) = 0, \quad y'(0) = 0, \quad \text{and} \quad y'(L) = 0.$$

The first two conditions indicate that there is no vertical deflection at the ends; the last two conditions mean that the line of deflection is horizontal (zero slope) at the ends. Find the deflection of the beam when a constant load w_0 is uniformly distributed along its length, that is, when $w(x) = w_0, 0 < x < L$.

Solution Transforming (9) with respect to the variable x gives

$$EI(s^4Y(s) - s^3y(0) - s^2y'(0) - sy''(0) - y'''(0)) = \frac{w_0}{s}$$

$$s^4Y(s) - sy''(0) - y'''(0) = \frac{w_0}{EIs}.$$

If we let $c_1 = y''(0)$ and $c_2 = y'''(0)$, then

$$Y(s) = \frac{c_1}{s^3} + \frac{c_2}{s^4} + \frac{w_0}{EIs^5}$$

and consequently

$$y(x) = \frac{c_1}{2!}\mathscr{L}^{-1}\left\{\frac{2!}{s^3}\right\} + \frac{c_2}{3!}\mathscr{L}^{-1}\left\{\frac{3!}{s^4}\right\} + \frac{w_0}{4!EI}\mathscr{L}^{-1}\left\{\frac{4!}{s^5}\right\}$$

$$= \frac{c_1}{2}x^2 + \frac{c_2}{6}x^3 + \frac{w_0}{24EI}x^4.$$

Applying the given conditions $y(L) = 0$ and $y'(L) = 0$ to the last equation yields the system

$$\frac{c_1}{2}L^2 + \frac{c_2}{6}L^3 + \frac{w_0}{24EI}L^4 = 0$$

$$c_1L + \frac{c_2}{2}L^2 + \frac{w_0}{6EI}L^3 = 0.$$

Solving, we find $c_1 = w_0 L^2/12EI$ and $c_2 = -w_0 L/2EI$. Thus the deflection is given by

$$y(x) = \frac{w_0 L^2}{24EI} x^2 - \frac{w_0 L}{12EI} x^3 + \frac{w_0}{24EI} x^4 = \frac{w_0}{24EI} x^2 (x - L)^2. \qquad \blacksquare$$

7.5 EXERCISES

Answers to odd-numbered problems begin on page A–20.

A table of the transforms of some basic functions is given in Appendix II.

In Problems 1–26 use the Laplace transform to solve the given differential equation subject to the indicated initial conditions. Where appropriate, write f in terms of unit step functions.

1. $\dfrac{dy}{dt} - y = 1, \quad y(0) = 0$

2. $\dfrac{dy}{dt} + 2y = t, \quad y(0) = -1$

3. $y' + 4y = e^{-4t}, \quad y(0) = 2$

4. $y' - y = \sin t, \quad y(0) = 0$

5. $y'' + 5y' + 4y = 0, \quad y(0) = 1, y'(0) = 0$

6. $y'' - 6y' + 13y = 0, \quad y(0) = 0, y'(0) = -3$

7. $y'' - 6y' + 9y = t, \quad y(0) = 0, y'(0) = 1$

8. $y'' - 4y' + 4y = t^3, \quad y(0) = 1, y'(0) = 0$

9. $y'' - 4y' + 4y = t^3 e^{2t}, \quad y(0) = 0, y'(0) = 0$

10. $y'' - 2y' + 5y = 1 + t, \quad y(0) = 0, y'(0) = 4$

11. $y'' + y = \sin t, \quad y(0) = 1, y'(0) = -1$

12. $y'' + 16y = 1, \quad y(0) = 1, y'(0) = 2$

13. $y'' - y' = e^t \cos t, \quad y(0) = 0, y'(0) = 0$

14. $y'' - 2y' = e^t \sinh t, \quad y(0) = 0, y'(0) = 0$

15. $2y''' + 3y'' - 3y' - 2y = e^{-t}, \quad y(0) = 0, y'(0) = 0, y''(0) = 1$

16. $y''' + 2y'' - y' - 2y = \sin 3t, \quad y(0) = 0, y'(0) = 0, y''(0) = 1$

17. $y^{(4)} - y = 0, \quad y(0) = 1, y'(0) = 0, y''(0) = -1, y'''(0) = 0$

18. $y^{(4)} - y = t, \quad y(0) = 0, y'(0) = 0, y''(0) = 0, y'''(0) = 0$

19. $y' + y = f(t)$, where $f(t) = \begin{cases} 0, & 0 \le t < 1 \\ 5, & t \ge 1 \end{cases} \quad y(0) = 0$

20. $y' + y = f(t)$, where $f(t) = \begin{cases} 1, & 0 \le t < 1 \\ -1, & t \ge 1 \end{cases} \quad y(0) = 0$

21. $y' + 2y = f(t)$, where $f(t) = \begin{cases} t, & 0 \le t < 1 \\ 0, & t \ge 1 \end{cases} \quad y(0) = 0$

22. $y'' + 4y = f(t)$, where $f(t) = \begin{cases} 1, & 0 \le t < 1 \\ 0, & t \ge 1 \end{cases}$ $y(0) = 0$, $y'(0) = -1$

23. $y'' + 4y = \sin t \, \mathcal{U}(t - 2\pi)$, $y(0) = 1$, $y'(0) = 0$

24. $y'' - 5y' + 6y = \mathcal{U}(t - 1)$, $y(0) = 0$, $y'(0) = 1$

25. $y'' + y = f(t)$, where $f(t) = \begin{cases} 0, & 0 \le t < \pi \\ 1, & \pi \le t < 2\pi \\ 0, & t \ge 2\pi \end{cases}$ $y(0) = 0$, $y'(0) = 1$

26. $y'' + 4y' + 3y = 1 - \mathcal{U}(t - 2) - \mathcal{U}(t - 4) + \mathcal{U}(t - 6)$;
    $y(0) = 0$, $y'(0) = 0$

In Problems 27 and 28 use the Laplace transform to solve the given differential equation subject to the indicated boundary conditions.

27. $y'' + 2y' + y = 0$, $y'(0) = 2$, $y(1) = 2$

28. $y'' - 9y' + 20y = 1$, $y(0) = 0$, $y'(1) = 0$

In Problems 29–38 use the Laplace transform to solve the given integral equation or integrodifferential equation.

29. $f(t) + \displaystyle\int_0^t (t - \tau) f(\tau) \, d\tau = t$

30. $f(t) = 2t - 4 \displaystyle\int_0^t \sin \tau f(t - \tau) \, d\tau$

31. $f(t) = t e^t + \displaystyle\int_0^t \tau f(t - \tau) \, d\tau$

32. $f(t) + 2 \displaystyle\int_0^t f(\tau) \cos(t - \tau) \, d\tau = 4e^{-t} + \sin t$

33. $f(t) + \displaystyle\int_0^t f(\tau) \, d\tau = 1$

34. $f(t) = \cos t + \displaystyle\int_0^t e^{-\tau} f(t - \tau) \, d\tau$

35. $f(t) = 1 + t - \dfrac{8}{3} \displaystyle\int_0^t (\tau - t)^3 f(\tau) \, d\tau$

36. $t - 2f(t) = \displaystyle\int_0^t (e^\tau - e^{-\tau}) f(t - \tau) \, d\tau$

37. $y'(t) = 1 - \sin t - \displaystyle\int_0^t y(\tau) \, d\tau$, $y(0) = 0$

38. $\dfrac{dy}{dt} + 6y(t) + 9 \displaystyle\int_0^t y(\tau) \, d\tau = 1$, $y(0) = 0$

39. Use equation (3) to determine the current $i(t)$ in a single loop L-R-C

Figure 7.43

Figure 7.44

Figure 7.47

Figure 7.48

circuit when $L = 0.005$ henry, $R = 1$ ohm, $C = 0.02$ farad, $E(t) = 100[1 - \mathcal{U}(t-1)]$ volts, and $i(0) = 0$.

40. Solve Problem 39 when $E(t) = 100[t - (t-1)\mathcal{U}(t-1)]$.

41. Recall that the differential equation for the charge $q(t)$ on the capacitor in an R-C series circuit is

$$R\frac{dq}{dt} + \frac{1}{C}q = E(t),$$

where $E(t)$ is the impressed voltage. See Section 3.2. Use the Laplace transform to determine the charge $q(t)$ when $q(0) = 0$ and $E(t) = E_0 e^{-kt}$, $k > 0$. Consider two cases: $k \neq 1/RC$ and $k = 1/RC$.

42. Use the Laplace transform to determine the charge on the capacitor in an R-C series circuit if $q(0) = q_0$, $R = 10$ ohms, $C = 0.1$ farad, and $E(t)$ is as given in Figure 7.43.

43. Use the Laplace transform to determine the charge on the capacitor in an R-C series circuit if $q(0) = 0$, $R = 2.5$ ohms, $C = 0.08$ farad, and $E(t)$ is as given in Figure 7.44.

44. Use the Laplace transform to determine the charge $q(t)$ on the capacitor in an R-C series circuit when $q(0) = 0$, $R = 50$ ohms, $C = 0.01$ farad, and $E(t)$ is as given in Figure 7.45.

Figure 7.45

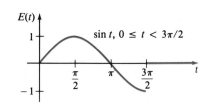

Figure 7.46

45. Use the Laplace transform to determine the current $i(t)$ in a single loop L-R series circuit when $i(0) = 0$, $L = 1$ henry, $R = 10$ ohms, and $E(t)$ is as given in Figure 7.46.

46. Solve equation (6) subject to $i(0) = 0$, where $E(t)$ is as given in Figure 7.47. [*Hint:* See Problem 31, Exercises 7.4.]

47. Solve equation (6) subject to $i(0) = 0$, where $E(t)$ is as given in Figure 7.48. Specify the solution for $0 \leq t < 2$. [*Hint:* See Problem 33, Exercises 7.4.]

48. Recall that the differential equation for the instantaneous charge $q(t)$ on the capacitor in an L-R-C series circuit is given by

$$L\frac{d^2q}{dt^2} + R\frac{dq}{dt} + \frac{1}{C}q = E(t). \tag{10}$$

See Section 5.4. Use the Laplace transform to determine $q(t)$ when $L = 1$ henry, $R = 20$ ohms, $C = 0.005$ farad, $E(t) = 150$ volts, $t > 0$, and $q(0) = 0$, $i(0) = 0$. What is the current $i(t)$? What is the charge $q(t)$ if the same constant voltage is turned off for $t \geq 2$?

49. Determine the charge $q(t)$ and current $i(t)$ for a series circuit in which $L = 1$ henry, $R = 20$ ohms, $C = 0.01$ farad, $E(t) = 120 \sin 10t$ volts, $q(0) = 0$, and $i(0) = 0$. What is the steady-state current?

50. Consider the battery of constant voltage, E_0 that charges the capacitor shown in Figure 7.49. If we divide by L and define $\lambda = R/2L$ and $\omega^2 = 1/LC$, then (10) becomes

Figure 7.49

$$\frac{d^2q}{dt^2} + 2\lambda\frac{dq}{dt} + \omega^2 q = \frac{E_0}{L}.$$

Use the Laplace transform to show that the solution of this equation, subject to $q(0) = 0$ and $i(0) = 0$, is

$$q(t) = \begin{cases} E_0C[1 - e^{-\lambda t}(\cosh\sqrt{\lambda^2 - \omega^2}\,t + \dfrac{\lambda}{\sqrt{\lambda^2 - \omega^2}}\sinh\sqrt{\lambda^2 - \omega^2}\,t)], & \lambda > \omega \\[2mm] E_0C[1 - e^{-\lambda t}(1 + \lambda t)], & \lambda = \omega \\[2mm] E_0C[1 - e^{-\lambda t}(\cos\sqrt{\omega^2 - \lambda^2}\,t + \dfrac{\lambda}{\sqrt{\omega^2 - \lambda^2}}\sin\sqrt{\omega^2 - \lambda^2}\,t)], & \lambda < \omega. \end{cases}$$

51. Use the Laplace transform to determine the charge $q(t)$ on the capacitor in an L-C series circuit when $q(0) = 0$, $i(0) = 0$, and $E(t) = E_0 e^{-kt}$, $k > 0$.

52. Suppose a 32-lb weight stretches a spring 2 ft. If the weight is released from rest at the equilibrium position, determine the equation of motion if an impressed force $f(t) = \sin t$ acts on the system for $0 \leq t < 2\pi$ and is then removed. Ignore any damping forces. [*Hint:* Write the impressed force in terms of the unit step function.]

53. A 4-lb weight stretches a spring 2 ft. The weight is released from rest 18 in. above the equilibrium position, and the resulting motion takes place in a medium offering a damping force numerically equal to 7/8 times the instantaneous velocity. Use the Laplace transform to determine the equation of motion.

54. A 16-lb weight is attached to a spring whose constant is $k = 4.5$ lb/ft. Beginning at $t = 0$, a force equal to $f(t) = 4 \sin 3t + 2 \cos 3t$ acts on the system. Assuming that no damping forces are present, use the Laplace transform to find the equation of motion if the weight is released from rest from the equilibrium position.

55. For a cantilever beam embedded at its left end ($x = 0$) and free at its right end ($x = L$), the deflection $y(x)$ must satisfy (9) and

$$y(0) = 0, \quad y'(0) = 0, \quad y''(L) = 0, \quad y'''(L) = 0. \tag{11}$$

Figure 7.50

The first two conditions state that the deflection and slope are zero at $x = 0$, and the last two conditions state that the bending moment and shear force are zero at $x = L$. Use the Laplace transform to solve equation (9) subject to (11) when a constant load w_0 is uniformly distributed along the length of the beam. See Figure 7.50. Find the deflection at the midpoint of the beam. Find the maximum deflection of the beam.

56. Solve Problem 55 when the load is given by

$$w(x) = \begin{cases} 0, & 0 < x < L/3 \\ w_0, & L/3 < x < 2L/3 \\ 0, & 2L/3 < x < L. \end{cases}$$

Write $w(x)$ in terms of unit step functions.

57. Solve Problem 55 when the load is given by

$$w(x) = \begin{cases} w_0, & 0 < x < L/2 \\ 0, & L/2 < x < L. \end{cases}$$

58. The static deflection $y(x)$ of a beam that is hinged at both ends must satisfy the differential equation (9) and the conditions

$$y(0) = 0, \quad y''(0) = 0, \quad y(L) = 0, \quad y''(L) = 0. \tag{12}$$

Use the Laplace transform to solve (9) subject to (12) when $w(x) = w_0$, $0 < x < L$. See Figure 7.51.

Figure 7.51

In Problems 59 and 60 use the Laplace transform and Theorem 7.7 to find a solution of the given equation.

59. $ty'' - y' = t^2$, $y(0) = 0$ **60.** $ty'' + 2ty' + 2y = 0$, $y(0) = 0$

61. In this problem we show how the convolution integral can be used to find a solution of an initial-value problem of the type

$$ay'' + by' + cy = g(t), \quad y(0) = 0, \quad y'(0) = 0. \tag{13}$$

(a) Show that the solution $y_1(t)$ of the initial-value problem

$$ay'' + by' + cy = 0, \quad y(0) = 0, \quad y'(0) = 1,$$

is $$y_1(t) = \mathscr{L}^{-1}\left\{\frac{a}{as^2 + bs + c}\right\}.$$

(b) Use the result in part (a) to show that the solution $y_2(t)$ of the initial-value problem in (13) is given by

$$y_2(t) = \frac{1}{a} g * y_1.$$

62. Use the procedure outlined in Problem 61 to find a solution of the initial-value problem

$$y'' + y = \sec t, \quad y(0) = 0, \quad y'(0) = 0.$$

D] 7.6 DIRAC DELTA FUNCTION

Unit Impulse

Mechanical systems are often acted upon by an external force (or emf in an electrical circuit) of large magnitude that acts only for a very short period of time. For example, a vibrating airplane wing could be struck by lightning, a mass on a spring could be given a sharp blow by a ball peen hammer, a ball (baseball, golf ball, tennis ball) could be sent soaring when struck violently by some kind of club (baseball bat, golf club, tennis racket). The function

$$\delta_a(t - t_0) = \begin{cases} 0, & 0 \le t < t_0 - a \\ \dfrac{1}{2a}, & t_0 - a \le t < t_0 + a \\ 0, & t \ge t_0 + a, \end{cases} \tag{1}$$

$a > 0$, $t_0 > 0$, shown in Figure 7.52(a), could serve as a mathematical model for such a force. For a small value of a, $\delta_a(t - t_0)$ is essentially a constant function of large magnitude that is "on" for just a very short period of time around t_0. The behavior of $\delta_a(t - t_0)$ as $a \to 0$ is illustrated in Figure 7.52(b). The function $\delta_a(t - t_0)$ is called a **unit impulse** since it possesses the integration property

$$\int_0^\infty \delta_a(t - t_0)\, dt = 1.$$

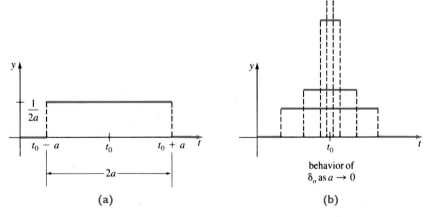

(a)   (b)

Figure 7.52

Dirac Delta Function

In practice it is convenient to work with another type of unit impulse, a "function" that approximates $\delta_a(t - t_0)$ and is defined by the limit

$$\delta(t - t_0) = \lim_{a \to 0} \delta_a(t - t_0). \tag{2}$$

The latter expression, which is not a function at all, can be characterized by the two properties

(i) $\delta(t - t_0) = \begin{cases} \infty, & t = t_0 \\ 0, & t \neq t_0 \end{cases}$,   and   **(ii)** $\int_0^\infty \delta(t - t_0)\, dt = 1.$

The expression $\delta(t - t_0)$ is called the **Dirac delta function**.*

It is possible to obtain the Laplace transform of the Dirac delta function by the formal assumption that

$$\mathscr{L}\{\delta(t - t_0)\} = \lim_{a \to 0} \mathscr{L}\{\delta_a(t - t_0)\}.$$

THEOREM 7.11   **Transform of the Dirac Delta Function**

For $t_0 > 0$,

$$\mathscr{L}\{\delta(t - t_0)\} = e^{-st_0}. \tag{3}$$

Proof   To begin, we can write $\delta_a(t - t_0)$ in terms of the unit step function by virtue of (4) and (5) of Section 7.3:

$$\delta_a(t - t_0) = \frac{1}{2a} [\mathscr{U}(t - (t_0 - a)) - \mathscr{U}(t - (t_0 + a))].$$

By linearity and (7) of Section 7.3 the Laplace transform of this last expression is

$$\mathscr{L}\{\delta_a(t - t_0)\} = \frac{1}{2a} \left[\frac{e^{-s(t_0 - a)}}{s} - \frac{e^{-s(t_0 + a)}}{s} \right]$$

or $$\mathscr{L}\{\delta_a(t - t_0)\} = e^{-st_0} \left(\frac{e^{sa} - e^{-sa}}{2sa} \right). \tag{4}$$

Since (4) has the indeterminate form $0/0$ as $a \to 0$, we apply L'Hôpital's rule:

$$\mathscr{L}\{\delta(t - t_0)\} = \lim_{a \to 0} \mathscr{L}\{\delta_a(t - t_0)\} = e^{-st_0} \lim_{a \to 0} \left(\frac{e^{sa} - e^{-sa}}{2sa} \right) = e^{-st_0}. \quad \square$$

*****Paul Adrian Maurice Dirac** (1902–1984)   The delta function was the invention of the contemporary British physicist P. A. M. Dirac. Along with Max Planck, Werner Heisenberg, Erwin Schrödinger, and Albert Einstein, Dirac was one of the founding fathers, in the era 1900–1930, of a new way of describing the behavior of atoms, molecules, and elementary particles called *quantum mechanics*. For their pioneering work in this field, Dirac and Schrödinger shared the 1933 Nobel Prize in physics. The Dirac delta function was used extensively throughout his 1932 classic treatise, *The Principles of Quantum Mechanics*.

Now when $t_0 = 0$, it seems plausible to conclude from (3) that

$$\mathscr{L}\{\delta(t)\} = 1.$$

The last result emphasizes the fact that $\delta(t)$ is not the usual type of function that we have been considering since we expect from Theorem 7.4 that $\mathscr{L}\{f(t)\} \to 0$ as $s \to \infty$.

EXAMPLE I

Solve $\qquad\qquad y'' + y = 4\delta(t - 2\pi)$

subject to **(a)** $y(0) = 1$, $y'(0) = 0$, and **(b)** $y(0) = 0$, $y'(0) = 0$.
The two initial-value problems could serve as models for describing the motion of a mass on a spring moving in a medium in which damping is negligible. At $t = 2\pi$ seconds the mass is given a sharp blow. In (a) the mass is released from rest 1 unit below the equilibrium position. In (b) the mass is at rest in the equilibrium position.

Solution **(a)** From (3) the Laplace transform of the differential equation is

$$s^2 Y(s) - s + Y(s) = 4e^{-2\pi s}$$

or $\qquad\qquad\qquad\qquad Y(s) = \dfrac{s}{s^2 + 1} + \dfrac{4e^{-2\pi s}}{s^2 + 1}.$

Utilizing the inverse form of the second translation theorem, we find

$$y(t) = \cos t + 4 \sin(t - 2\pi)\mathscr{U}(t - 2\pi):$$

Since $\sin(t - 2\pi) = \sin t$, the foregoing solution can be written as

$$y(t) = \begin{cases} \cos t, & 0 \leq t < 2\pi \\ \cos t + 4 \sin t, & t \geq 2\pi. \end{cases} \qquad\qquad \textbf{(5)}$$

In Figure 7.53 we see from the graph of (5) that the mass is exhibiting simple harmonic motion until it is struck at $t = 2\pi$. The influence of the unit impulse is to increase the amplitude of vibration to $\sqrt{17}$ for $t > 2\pi$.

Figure 7.53

(b) In this case the transform of the equation is simply

$$Y(s) = \frac{4e^{-2\pi s}}{s^2 + 1}$$

and so

$$y(t) = 4\sin(t - 2\pi)\mathscr{U}(t - 2\pi)$$

$$= \begin{cases} 0 & 0 \le t < 2\pi \\ 4\sin t, & t \ge 2\pi. \end{cases} \tag{6}$$

The graph of (6) in Figure 7.54 shows, as we would expect from the initial conditions, that the mass exhibits no motion until it is struck at $t = 2\pi$.

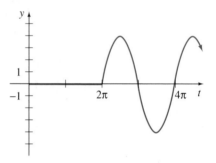

Figure 7.54 ∎

Remark If $\delta(t - t_0)$ were a function in the usual sense, then property (i) on page 380 would imply $\int_0^\infty \delta(t - t_0)\,dt = 0$ rather than $\int_0^\infty \delta(t - t_0)\,dt = 1$. Since the Dirac delta function did not "behave" like an ordinary function, even though its users produced correct results, it was met initially with great scorn by mathematicians. However, in the 1940s Dirac's controversial function was put on a rigorous footing by the French mathematician Laurent Schwartz in his book *La Théorie de distribution* and this, in turn, led to an entirely new branch of mathematics known as the *theory of distributions*, or *generalized functions*. In the modern theory of generalized functions, (2) is not an accepted definition of $\delta(t - t_0)$, nor does one speak of a function whose values are either ∞ or 0. Although this theory is much beyond the level of this text, it suffices for our purposes to say that the Dirac delta function is defined in terms of its effect or action on other functions. To see this let us suppose that f is a continuous function on $[0, \infty)$. Then, by the mean value theorem for integrals, it follows that

$$\int_0^\infty f(t)\delta_a(t - t_0)\,dt = \int_{t_0 - a}^{t_0 + a} f(t)\left(\frac{1}{2a}\right)dt = \frac{1}{2a}(2af(c)) = f(c),$$

where c is some number in the interval $t_0 - a < t < t_0 + a$. As $a \to 0$ we must have $c \to t_0$ so that

$$\int_0^\infty f(t)\delta(t - t_0)\, dt = \lim_{a \to 0} \int_0^\infty f(t)\delta_a(t - t_0)\, dt = \lim_{a \to 0} f(c)$$

implies
$$\int_0^\infty f(t)\delta(t - t_0)\, dt = f(t_0). \qquad (7)$$

Although we have used the intuitive definition (2) to arrive at (7), the result is nevertheless valid and can be obtained in a rigorous fashion. The result in (7) can be taken as the *definition* of $\delta(t - t_0)$. It is known as the *sifting property* since $\delta(t - t_0)$ has the effect of sifting the value $f(t_0)$ from the values of f. Note that property (ii) on page 380 is consistent with (7) when $f(t) = 1$, $0 \le t < \infty$. The integral operation (7) that corresponds a number $f(t_0)$ with a function f leads to the notion of a *linear functional*. We stop at this point and urge the curious reader to consult an advanced text.*

7.6 EXERCISES

Answers to odd-numbered problems begin on page A–20.

In Problems 1–12 use the Laplace transform to solve the given differential equation subject to the indicated initial conditions.

1. $y' - 3y = \delta(t - 2), \quad y(0) = 0$

2. $y' + y = \delta(t - 1), \quad y(0) = 2$

3. $y'' + y = \delta(t - 2\pi), \quad y(0) = 0, \, y'(0) = 1$

4. $y'' + 16y = \delta(t - 2\pi), \quad y(0) = 0, \, y'(0) = 0$

5. $y'' + y = \delta\left(t - \dfrac{\pi}{2}\right) + \delta\left(t - \dfrac{3\pi}{2}\right), \quad y(0) = 0, \, y'(0) = 0$

6. $y'' + y = \delta(t - 2\pi) + \delta(t - 4\pi), \quad y(0) = 1, \, y'(0) = 0$

7. $y'' + 2y' = \delta(t - 1), \quad y(0) = 0, \, y'(0) = 1$

8. $y'' - 2y' = 1 + \delta(t - 2), \quad y(0) = 0, \, y'(0) = 1$

9. $y'' + 4y' + 5y = \delta(t - 2\pi), \quad y(0) = 0, \, y'(0) = 0$

10. $y'' + 2y' + y = \delta(t - 1), \quad y(0) = 0, \, y'(0) = 0$

11. $y'' + 4y' + 13y = \delta(t - \pi) + \delta(t - 3\pi), \quad y(0) = 1, \, y'(0) = 0$

---

* See M. J. Lighthill, *Introduction to Fourier Analysis and Generalized Functions* (New York: Cambridge University Press, 1958).

12. $y'' - 7y' + 6y = e^t + \delta(t - 2) + \delta(t - 4)$; $y(0) = 0, y'(0) = 0$

13. A uniform beam of length L carries a concentrated load w_0 at $x = L/2$. The beam is embedded at its left end and is free at its right end. Use the Laplace transform to determine the deflection $y(x)$ from

$$EI \frac{d^4 y}{dx^4} = w_0 \delta\left(x - \frac{L}{2}\right),$$

where $y(0) = 0$, $y'(0) = 0$, $y''(L) = 0$, and $y'''(L) = 0$.

14. Solve the differential equation in Problem 13 subject to $y(0) = 0$, $y'(0) = 0$, $y(L) = 0$, $y'(L) = 0$. In this case the beam is embedded at both ends. See Figure 7.55.

Figure 7.55

15. Use (7) to obtain (3).

16. Use (7) to evaluate $\displaystyle\int_0^\infty t^2 e^{-3t} \delta(t - 4)\, dt$.

17. Use the Laplace transform and (7) to solve

$$y'' + 2y' + 2y = \cos t\, \delta(t - 3\pi)$$

subject to $y(0) = 1$ and $y'(0) = -1$.

18. To emphasize the unusual nature of the Dirac delta function, show that the "solution" of the initial-value problem $y'' + \omega^2 y = \delta(t)$, $y(0) = 0$, $y'(0) = 0$, does *not* satisfy the initial condition $y'(0) = 0$.

19. Solve the initial-value problem

$$L \frac{di}{dt} + Ri = \delta(t), \qquad i(0) = 0,$$

where L and R are constants. Does the solution satisfy the condition at $t = 0$?

20. If $\delta'(t - t_0)$ is the derivative of the Dirac delta function, then it is known that $\mathscr{L}\{\delta'(t - t_0)\} = se^{-st_0}, t_0 \geq 0$. Use this result to solve $y' + 5y = \delta'(t)$ subject to $y(0) = 0$.

Chapter 7   *REVIEW*

The **Laplace transform** of a function $f(t)$, $t \geq 0$, is defined by the integral

$$\mathscr{L}\{f(t)\} = \int_0^\infty e^{-st} f(t)\, dt = F(s).$$

The parameter s is usually restricted in such a manner that convergence of the integral is guaranteed. When it is applied to a linear differential equation with constant coefficients such as $ay'' + by' + cy = g(t)$, there results an algebraic equation

$$a[s^2Y(s) - sy(0) - y'(0)] + b[sY(s) - y(0)] + cY(s) = G(s),$$

which depends on the initial conditions $y(0)$ and $y'(0)$. When these values are known, we determine $y(t)$ by evaluating $y(t) = \mathcal{L}^{-1}\{Y(s)\}$.

Chapter 7   *REVIEW EXERCISES*

Answers to odd-numbered problems begin on page A–21.

In Problems 1 and 2 use the definition of the Laplace transform to find $\mathcal{L}\{f(t)\}$.

1. $f(t) = \begin{cases} t, & 0 \le t < 1 \\ 2 - t, & t \ge 1 \end{cases}$
    2. $f(t) = \begin{cases} 0, & 0 \le t < 2 \\ 1, & 2 \le t < 4 \\ 0, & t \ge 4 \end{cases}$

In Problems 3–24 fill in the blanks or answer true/false.

3. If f is not piecewise continuous on $[0, \infty)$, then $\mathcal{L}\{f(t)\}$ will not exist. _____

4. The function $f(t) = (e^t)^{10}$ is not of exponential order. _____

5. $F(s) = s^2/(s^2 + 4)$ is not the Laplace transform of a function that is piecewise continuous and of exponential order. _____

6. If $\mathcal{L}\{f(t)\} = F(s)$ and $\mathcal{L}\{g(t)\} = G(s)$, then $\mathcal{L}^{-1}\{F(s)G(s)\} = f(t)g(t)$. _____

7. $\mathcal{L}\{e^{-7t}\} = $ _____   **8.** $\mathcal{L}\{te^{-7t}\} = $ _____

9. $\mathcal{L}\{\sin 2t\} = $ _____   **10.** $\mathcal{L}\{e^{-3t} \sin 2t\} = $ _____

11. $\mathcal{L}\{t \sin 2t\} = $ _____   **12.** $\mathcal{L}\{\sin 2t\,\mathcal{U}(t - \pi)\} = $ _____

13. $\mathcal{L}^{-1}\left\{\dfrac{20}{s^6}\right\} = $ _____   **14.** $\mathcal{L}^{-1}\left\{\dfrac{1}{3s - 1}\right\} = $ _____

15. $\mathcal{L}^{-1}\left\{\dfrac{1}{(s - 5)^3}\right\} = $ _____   **16.** $\mathcal{L}^{-1}\left\{\dfrac{1}{s^2 - 5}\right\} = $ _____

17. $\mathcal{L}^{-1}\left\{\dfrac{s}{s^2 - 10s + 29}\right\} = $ _____   **18.** $\mathcal{L}^{-1}\left\{\dfrac{e^{-5s}}{s^2}\right\} = $ _____

19. $\mathcal{L}^{-1}\left\{\dfrac{s + \pi}{s^2 + \pi^2} e^{-s}\right\} = $ _____   **20.** $\mathcal{L}^{-1}\left\{\dfrac{1}{L^2s^2 + n^2\pi^2}\right\} = $ _____

21. $\mathcal{L}\{e^{-5t}\}$ exists for $s > $ _____.

22. If $\mathcal{L}\{f(t)\} = F(s)$, then $\mathcal{L}\{te^{8t}f(t)\} = $ _____.

23. If $\mathcal{L}\{f(t)\} = F(s)$ and $k > 0$, then $\mathcal{L}\{e^{a(t-k)}f(t - k)\mathcal{U}(t - k)\} = $ _____.

24. $1 * 1 = $ _____.

In Problems 25–28, **(a)** express f in terms of unit step functions, **(b)** find $\mathcal{L}\{f(t)\}$, and **(c)** find $\mathcal{L}\{e^t f(t)\}$.

25.

Figure 7.56

26.

Figure 7.57

27.

Figure 7.58

28.

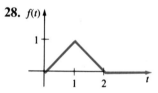

Figure 7.59

In Problems 29–36 use the Laplace transform to solve the given equation.

29. $y'' - 2y' + y = e^t, \quad y(0) = 0, y'(0) = 5$

30. $y'' - 8y' + 20y = te^t, \quad y(0) = 0, y'(0) = 0$

31. $y'' - 4y' + 6y = 30\,\mathcal{U}(t - \pi), \quad y(0) = 0, y'(0) = 0$

32. $y'' + 6y' + 5y = t - t\mathcal{U}(t - 2), \quad y(0) = 1, y'(0) = 0$

33. $y' - 5y = f(t)$, where $f(t) = \begin{cases} t^2, & 0 \le t < 1 \\ 0, & t \ge 1 \end{cases} \quad y(0) = 1.$

34. $f(t) = 1 - 2\displaystyle\int_0^t e^{-3\tau} f(t - \tau)\, d\tau$

35. $y'(t) = \cos t + \displaystyle\int_0^t y(\tau)\cos(t - \tau)\, d\tau, \quad y(0) = 1$

36. $\displaystyle\int_0^t f(\tau) f(t - \tau)\, d\tau = 6t^3$

37. The current $i(t)$ in an R-C series circuit can be determined from the integral equation

$$Ri + \frac{1}{C} \int_0^t i(\tau)\, d\tau = E(t),$$

where $E(t)$ is the impressed voltage. Determine $i(t)$ when $R = 10$ ohms, $C = 0.5$ farad, and $E(t) = 2(t^2 + t)$.

38. A series circuit contains an inductor, a resistor, and a capacitor for which $L = 1/2$ henry, $R = 10$ ohms, and $C = 0.01$ farad, respectively. The voltage

$$E(t) = \begin{cases} 10, & 0 \le t < 5 \\ 0, & t \ge 5 \end{cases}$$

is applied to the circuit. Determine the instantaneous charge $q(t)$ on the capacitor for $t > 0$ if $q(0) = 0$ and $q'(0) = 0$.

39. A uniform cantilever beam of length L is embedded at its left end ($x = 0$) and free at its right end. Find the deflection $y(x)$ if the load per unit length is given by

$$w(x) = \frac{2w_0}{L} \left[\frac{L}{2} - x + \left(x - \frac{L}{2} \right) \mathscr{U} \left(x - \frac{L}{2} \right) \right].$$

40. When a uniform beam is supported by an elastic foundation, the differential equation for its deflection $y(x)$ is

$$\frac{d^4 y}{dx^4} + 4a^4 y = \frac{w(x)}{EI},$$

where a is a constant. In the case when $a = 1$, find the deflection $y(x)$ of an elastically supported beam of length π that is embedded in concrete at both ends when a concentrated load w_0 is applied at $x = \pi/2$. [*Hint:* Use the table of Laplace transforms in Appendix II.]

CHAPTER 8

SYSTEMS OF LINEAR DIFFERENTIAL EQUATIONS

8.1 Operator Method
8.2 Laplace Transform Method
8.3 Systems of Linear First-Order Equations
8.4 Introduction to Matrices
8.5 Matrices and Systems of Linear First-Order Equations

8.6 Homogeneous Linear Systems
[O] **8.7** Undetermined Coefficients
8.8 Variation of Parameters
[O] **8.9** Matrix Exponential

Chapter 8 Review
Chapter 8 Review Exercises

Important Concepts

Systematic elimination
Normal form
Homogeneous system
Nonhomogeneous system
Matrix
Multiplicative inverse
Nonsingular matrix
Singular matrix
Augmented matrix
Elementary row operations
Gaussian elimination
Gauss-Jordan elimination
Reduced row-echelon form
Eigenvalue
Eigenvector
Characteristic equation
Solution vector
Superposition principle
Linear dependence
Linear independence
Wronskian
Fundamental set
General solution
Fundamental matrix

In applications considered in previous chapters, a physical system could be described by a single differential equation. For example, in Chapter 5 we saw that the mathematical model for the motion of a mass attached to a spring or for the response of a series electrical circuit was a relatively simple differential equation of the form $ay'' + by' + cy = f(t)$. However, if we attached two (or more) springs together or formed a parallel circuit or network, as shown in the accompanying figures, we would need two (or more)

Continues

Undetermined coefficients
Variation of parameters
Matrix exponential

Continued

coupled, or simultaneous, differential equations to describe the motion of the masses or the response of the network.

In this chapter we confine our attention to the theory and solution of systems of simultaneous linear differential equations or, simply, linear systems in which all the coefficients are constants.

8.1   OPERATOR METHOD

Simultaneous ordinary differential equations involve two or more equations that contain derivatives of two or more unknown functions of a single independent variable. If x, y, and z are functions of the variable t, then

$$4\frac{d^2x}{dt^2} = -5x + y$$

$$2\frac{d^2y}{dt^2} = 3x - y$$

and

$$x' - 3x + y' + z' = 5$$
$$x' \qquad - y' + 2z' = t^2$$
$$x + y' - 6z' = t - 1$$

are two examples of systems of simultaneous differential equations.

Solution of a System

A **solution** of a system of differential equations is a set of differentiable functions $x = f(t)$, $y = g(t)$, $z = h(t)$, and so on, that satisfies each equation of the system on some interval I.

Systematic Elimination

The first technique we shall consider for solving systems is based on the fundamental principle of **systematic algebraic elimination** of variables. We shall see that the analogue of multiplying an algebraic equation by a constant is operating on a differential equation with some combination of derivatives. Recall that a linear differential equation

$$a_n y^{(n)} + a_{n-1} y^{(n-1)} + \cdots + a_1 y' + a_0 y = g(t),$$

where the a_i, $i = 0, 1, \ldots, n$, are constants, can be written in terms of differential operators as

$$(a_n D^n + a_{n-1} D^{n-1} + \cdots + a_1 D + a_0)y = g(t).$$

EXAMPLE 1

Write the system of differential equations

$$x'' + 2x' + y'' = x + 3y + \sin t$$

$$x' + y' = -4x + 2y + e^{-t}$$

in operator notation.

Solution Rewrite the given system as

$$x'' + 2x' - x + y'' - 3y = \sin t$$
$$x' + 4x + y' - 2y = e^{-t}$$

so that

$$(D^2 + 2D - 1)x + (D^2 - 3)y = \sin t$$
$$(D + 4)x + (D - 2)y = e^{-t}.$$

∎

Method of Solution

Consider the simple system of linear first-order equations

$$Dy = 2x$$
$$Dx = 3y$$

(1)

or equivalently,

$$2x - Dy = 0$$
$$Dx - 3y = 0.$$

(2)

Operating on the first equation in (2) by D while multiplying the second by 2 and then subtracting eliminate x from the system. It follows that

$$-D^2 y + 6y = 0 \quad \text{or} \quad D^2 y - 6y = 0.$$

Since the roots of the auxiliary equation are $m_1 = \sqrt{6}$ and $m_2 = -\sqrt{6}$, we obtain

$$y(t) = c_1 e^{\sqrt{6}t} + c_2 e^{-\sqrt{6}t}.$$

(3)

Multiplying the first equation by -3 while operating on the second by D and then adding give the differential equation for x, $D^2 x - 6x = 0$. It follows immediately that

$$x(t) = c_3 e^{\sqrt{6}t} + c_4 e^{-\sqrt{6}t}.$$

(4)

Now (3) and (4) do not satisfy the system (1) for every choice of $c_1, c_2, c_3,$ and c_4. Substituting $x(t)$ and $y(t)$ into the first equation of the original system (1) gives, after we simplify,

$$(\sqrt{6}c_1 - 2c_3)e^{\sqrt{6}t} + (-\sqrt{6}c_2 - 2c_4)e^{-\sqrt{6}t} = 0.$$

Since the latter expression is to be zero for all values of t, we must have

$$\sqrt{6}c_1 - 2c_3 = 0 \quad \text{and} \quad -\sqrt{6}c_2 - 2c_4 = 0$$

or

$$c_3 = \frac{\sqrt{6}}{2}c_1, \quad c_4 = -\frac{\sqrt{6}}{2}c_2. \tag{5}$$

Hence we conclude that a solution of the system must be

$$x(t) = \frac{\sqrt{6}}{2}c_1 e^{\sqrt{6}t} - \frac{\sqrt{6}}{2}c_2 e^{-\sqrt{6}t}$$

$$y(t) = c_1 e^{\sqrt{6}t} + c_2 e^{-\sqrt{6}t}.$$

You are urged to substitute (3) and (4) into the second equation of (1) and verify that the same relationship (5) holds between the constants.

EXAMPLE 2

Solve

$$Dx + (D+2)y = 0$$

$$(D-3)x - \quad 2y = 0. \tag{6}$$

Solution Operating on the first equation by $D - 3$ and on the second by D and subtracting eliminate x from the system. It follows that the differential equation for y is

$$[(D-3)(D+2) + 2D]y = 0 \quad \text{or} \quad (D^2 + D - 6)y = 0.$$

Since the characteristic equation of this last differential equation is $m^2 + m - 6 = (m-2)(m+3) = 0$, we obtain the solution

$$y(t) = c_1 e^{2t} + c_2 e^{-3t}. \tag{7}$$

Eliminating y in a similar manner yields $(D^2 + D - 6)x = 0$, from which we find

$$x(t) = c_3 e^{2t} + c_4 e^{-3t}. \tag{8}$$

As we noted in the foregoing discussion, a solution of (6) does not contain four independent constants since the system itself puts a constraint on the actual number that can be chosen arbitrarily. Substituting (7) and (8) into the first equation of (6) gives

$$(4c_1 + 2c_3)e^{2t} + (-c_2 - 3c_4)e^{-3t} = 0$$

and so

$$4c_1 + 2c_3 = 0 \quad \text{and} \quad -c_2 - 3c_4 = 0.$$

Therefore

$$c_3 = -2c_1 \quad \text{and} \quad c_4 = -\tfrac{1}{3}c_2.$$

Accordingly, a solution of the system is

$$x(t) = -2c_1e^{2t} - \tfrac{1}{3}c_2e^{-3t}$$

$$y(t) = \quad c_1e^{2t} + c_2e^{-3t}. \qquad\qquad \blacksquare$$

Since we could just as easily solve for c_3 and c_4 in terms of c_1 and c_2, the solution in Example 2 can be written in the alternative form

$$x(t) = c_3e^{2t} + c_4e^{-3t}$$

$$y(t) = -\tfrac{1}{2}c_3e^{2t} - 3c_4e^{-3t}.$$

Also, it sometimes pays to keep one's eyes open when solving systems. Had we solved for x first, then y could be found, along with the relationship between the constants, by simply using the last equation of (6):

$$y = \frac{1}{2}(Dx - 3x) = \frac{1}{2}[2c_3e^{2t} - 3c_4e^{-3t} - 3c_3e^{2t} - 3c_4e^{-3t}]$$

or $\qquad y = -\dfrac{1}{2}c_3e^{2t} - 3c_4e^{-3t}.$

EXAMPLE 3

Solve $\qquad\qquad\qquad\qquad x' - 4x + y'' = t^2$

$$x' + x + y' = 0. \qquad\qquad (9)$$

Solution First we write the system in differential operator notation:

$$(D - 4)x + D^2y = t^2$$

$$(D + 1)x + \quad Dy = 0. \qquad\qquad (10)$$

Then, by eliminating x, we obtain

$$[(D + 1)D^2 - (D - 4)D]y = (D + 1)t^2 - (D - 4)0$$

or $\qquad\qquad\qquad\qquad (D^3 + 4D)y = t^2 + 2t.$

Since the roots of the auxiliary equation $m(m^2 + 4) = 0$ are $m_1 = 0$, $m_2 = 2i$, and $m_3 = -2i$, the complementary function is

$$y_c = c_1 + c_2 \cos 2t + c_3 \sin 2t.$$

To determine the particular solution y_p we use undetermined coefficients by assuming $y_p = At^3 + Bt^2 + Ct$. Therefore

$$y_p' = 3At^2 + 2Bt + C, \qquad y_p'' = 6At + 2B, \qquad y_p''' = 6A,$$

$$y_p''' + 4y_p' = 12At^2 + 8Bt + 6A + 4C = t^2 + 2t.$$

The last equality implies

$$12A = 1, \qquad 8B = 2, \qquad 6A + 4C = 0,$$

and hence $A = 1/12$, $B = 1/4$, $C = -1/8$. Thus

$$y = y_c + y_p$$

$$= c_1 + c_2 \cos 2t + c_3 \sin 2t + \frac{1}{12} t^3 + \frac{1}{4} t^2 - \frac{1}{8} t. \qquad \textbf{(11)}$$

Eliminating y from the system (10) leads to

$$[(D - 4) - D(D + 1)]x = t^2 \quad \text{or} \quad (D^2 + 4)x = -t^2.$$

It should be obvious that

$$x_c = c_4 \cos 2t + c_5 \sin 2t$$

and that undetermined coefficients can be applied to obtain a particular solution of the form $x_p = At^2 + Bt + C$. In this case the usual differentiations and algebra yield

$$x_p = -\frac{1}{4} t^2 + \frac{1}{8}$$

and so

$$x = x_c + x_p = c_4 \cos 2t + c_5 \sin 2t - \frac{1}{4} t^2 + \frac{1}{8}. \qquad \textbf{(12)}$$

Now c_4 and c_5 can be expressed in terms of c_2 and c_3 by substituting (11) and (12) into either equation of (9). By using the second equation, we find, after combining terms,

$$(c_5 - 2c_4 - 2c_2) \sin 2t + (2c_5 + c_4 + 2c_3) \cos 2t = 0$$

so that

$$c_5 - 2c_4 - 2c_2 = 0 \quad \text{and} \quad 2c_5 + c_4 + 2c_3 = 0.$$

Solving for c_4 and c_5 in terms of c_2 and c_3 gives

$$c_4 = -\frac{1}{5}(4c_2 + 2c_3) \quad \text{and} \quad c_5 = \frac{1}{5}(2c_2 - 4c_3).$$

Finally, a solution of (9) is found to be

$$x(t) = -\frac{1}{5}(4c_2 + 2c_3) \cos 2t + \frac{1}{5}(2c_2 - 4c_3) \sin 2t - \frac{1}{4} t^2 + \frac{1}{8}$$

$$y(t) = c_1 + c_2 \cos 2t + c_3 \sin 2t + \frac{1}{12} t^3 + \frac{1}{4} t^2 - \frac{1}{8} t. \qquad \blacksquare$$

Use of Determinants

Symbolically, if L_1, L_2, L_3, and L_4 denote linear differential operators with constant coefficients, then a system of linear differential equations in two variables x and y can be written as

$$L_1 x + L_2 y = g_1(t)$$

$$L_3 x + L_4 y = g_2(t).$$

(13)

Eliminating variables, as we would for algebraic equations, leads to

$$(L_1 L_4 - L_2 L_3)x = f_1(t) \quad \text{and} \quad (L_1 L_4 - L_2 L_3)y = f_2(t), \qquad (14)$$

where

$$f_1(t) = L_4 g_1(t) - L_2 g_2(t) \quad \text{and} \quad f_2(t) = L_1 g_2(t) - L_3 g_1(t).$$

Formally the results in (14) can be written in terms of determinants similar to those used in Cramer's rule:

$$\begin{vmatrix} L_1 & L_2 \\ L_3 & L_4 \end{vmatrix} x = \begin{vmatrix} g_1 & L_2 \\ g_2 & L_4 \end{vmatrix} \quad \text{and} \quad \begin{vmatrix} L_1 & L_2 \\ L_3 & L_4 \end{vmatrix} y = \begin{vmatrix} L_1 & g_1 \\ L_3 & g_2 \end{vmatrix}. \qquad (15)$$

The left-hand determinant in each equation in (15) can be expanded in the usual algebraic sense, with the result then operating on the functions $x(t)$ and $y(t)$. However, some care should be exercised in the expansion of the right-hand determinants in (15). We must expand these determinants in the sense of the internal differential operators actually operating upon the functions $g_1(t)$ and $g_2(t)$.

If
$$\begin{vmatrix} L_1 & L_2 \\ L_3 & L_4 \end{vmatrix} \neq 0$$

in (15) and is a differential operator of order n, then:

- The system (13) can be decoupled into two nth-order differential equations in x and y.
- The characteristic equation and hence the complementary function of each of these differential equations are the same.
- Since x and y both contain n constants, there is a total of $2n$ constants appearing.
- The total number of *independent* constants in the solution of the system is n.

If
$$\begin{vmatrix} L_1 & L_2 \\ L_3 & L_4 \end{vmatrix} = 0$$

in (13), then the system may have a solution containing any number of independent constants or may have no solution at all. Similar remarks hold for systems larger than indicated in (13).

EXAMPLE **4**

Solve
$$x' = 3x - y - 1$$

$$y' = x + y + 4e^t.$$

(16)

Solution Write the system in terms of differential operators:

$$(D - 3)x + \qquad y = -1$$

$$-x + (D - 1)y = 4e^t$$

and then use determinants:

$$\begin{vmatrix} D - 3 & 1 \\ -1 & D - 1 \end{vmatrix} x = \begin{vmatrix} -1 & 1 \\ 4e^t & D - 1 \end{vmatrix}$$

$$\begin{vmatrix} D - 3 & 1 \\ -1 & D - 1 \end{vmatrix} y = \begin{vmatrix} D - 3 & -1 \\ -1 & 4e^t \end{vmatrix}.$$

After expanding, we find that

$$(D - 2)^2 x = 1 - 4e^t$$

$$(D - 2)^2 y = -1 - 8e^t.$$

By the usual methods it follows that

$$x = x_c + x_p = c_1 e^{2t} + c_2 t e^{2t} + \frac{1}{4} - 4e^t \tag{17}$$

$$y = y_c + y_p = c_3 e^{2t} + c_4 t e^{2t} - \frac{1}{4} - 8e^t. \tag{18}$$

Substituting (17) and (18) into the second equation of (16) gives

$$(c_3 - c_1 + c_4)e^{2t} + (c_4 - c_2)t e^{2t} = 0,$$

which then implies

$$c_4 = c_2 \quad \text{and} \quad c_3 = c_1 - c_4 = c_1 - c_2.$$

Thus a solution of (16) is

$$x(t) = c_1 e^{2t} + c_2 t e^{2t} + \frac{1}{4} - 4e^t$$

$$y(t) = (c_1 - c_2)e^{2t} + c_2 t e^{2t} - \frac{1}{4} - 8e^t. \qquad \blacksquare$$

EXAMPLE 5

Given the system

$$Dx + \qquad\qquad Dz = t^2$$

$$2x + D^2 y \qquad\qquad = e^t$$

$$-2Dx - \quad 2y + (D + 1)z = 0,$$

find the differential equation for the variable y.

Solution With determinants we can write

$$\begin{vmatrix} D & 0 & D \\ 2 & D^2 & 0 \\ -2D & -2 & D+1 \end{vmatrix} y = \begin{vmatrix} D & t^2 & D \\ 2 & e^t & 0 \\ -2D & 0 & D+1 \end{vmatrix}.$$

In turn, expanding each determinant by cofactors of the first row gives

$$\left(D \begin{vmatrix} D^2 & 0 \\ -2 & D+1 \end{vmatrix} + D \begin{vmatrix} 2 & D^2 \\ -2D & -2 \end{vmatrix} \right) y = D \begin{vmatrix} e^t & 0 \\ 0 & D+1 \end{vmatrix} - \begin{vmatrix} 2 & 0 \\ -2D & D+1 \end{vmatrix} t^2 + D \begin{vmatrix} 2 & e^t \\ -2D & 0 \end{vmatrix}$$

or $\qquad\qquad\qquad D(3D^3 + D^2 - 4)y = 4e^t - 2t^2 - 4t.$

Again we remind you that the D symbol on the left-hand side is to be treated as an algebraic quantity, but this is not the case on the right-hand side. ∎

8.1 EXERCISES

Answers to odd-numbered problems begin on page A–21.

In Problems 1–22 solve, if possible, the given system of differential equations by either systematic elimination or determinants.

1. $\dfrac{dx}{dt} = 2x - y$

$\dfrac{dy}{dt} = x$

2. $\dfrac{dx}{dt} = 4x + 7y$

$\dfrac{dy}{dt} = x - 2y$

3. $\dfrac{dx}{dt} = -y + t$

$\dfrac{dy}{dt} = x - t$

4. $\dfrac{dx}{dt} - 4y = 1$

$x + \dfrac{dy}{dt} = 2$

5. $(D^2 + 5)x - \qquad 2y = 0$

$\quad -2x + (D^2 + 2)y = 0$

6. $(D + 1)x + (D - 1)y = 2$

$\quad 3x + (D + 2)y = -1$

7. $\dfrac{d^2x}{dt^2} = 4y + e^t$

$\dfrac{d^2y}{dt^2} = 4x - e^t$

8. $\dfrac{d^2x}{dt^2} + \dfrac{dy}{dt} = -5x$

$\dfrac{dx}{dt} + \dfrac{dy}{dt} = -x + 4y$

9. $\quad Dx + \qquad D^2y = e^{3t}$

$\quad (D + 1)x + (D - 1)y = 4e^{3t}$

10. $\quad D^2x - \qquad Dy = t$

$\quad (D + 3)x + (D + 3)y = 2$

11. $(D^2 - 1)x - \quad y = 0$

$\quad (D - 1)x + Dy = 0$

12. $(2D^2 - D - 1)x - (2D + 1)y = 1$

$\qquad (D - 1)x + \qquad Dy = -1$

13. $2\dfrac{dx}{dt} - 5x + \dfrac{dy}{dt} = e^t$

$\dfrac{dx}{dt} - \quad x + \dfrac{dy}{dt} = 5e^t$

14. $\quad \dfrac{dx}{dt} + \dfrac{dy}{dt} \qquad = e^t$

$\quad -\dfrac{d^2x}{dt^2} + \dfrac{dx}{dt} + x + y = 0$

15. $(D-1)x + (D^2+1)y = 1$
$(D^2-1)x + (D+1)y = 2$

16. $D^2x - 2(D^2+D)y = \sin t$
$x + \qquad Dy = 0$

17. $Dx = y$
$Dy = z$
$Dz = x$

18. $\qquad Dx + \qquad z = e^t$
$(D-1)x + Dy + Dz = 0$
$x + 2y + Dz = e^t$

19. $\dfrac{dx}{dt} - 6y \qquad = 0$

$x - \dfrac{dy}{dt} + z = 0$

$x + y - \dfrac{dz}{dt} = 0$

20. $\dfrac{dx}{dt} = -x + z$

$\dfrac{dy}{dt} = -y + z$

$\dfrac{dz}{dt} = -x + y$

21. $2Dx + (D-1)y = t$
$Dx + \qquad Dy = t^2$

22. $\qquad Dx - \qquad 2Dy = t^2$
$(D+1)x - 2(D+1)y = 1$

In Problems 23 and 24 solve the given system subject to the indicated initial conditions.

23. $\dfrac{dx}{dt} = -5x - y$

$\dfrac{dy}{dt} = 4x - y$

$x(1) = 0, \; y(1) = 1$

24. $\dfrac{dx}{dt} = y - 1$

$\dfrac{dy}{dt} = -3x + 2y$

$x(0) = 0, \; y(0) = 0$

25. Determine, if possible, a system of differential equations having

$$x(t) = c_1 + c_2 e^{2t}$$

$$y(t) = -c_1 + c_2 e^{2t}$$

as its solution.

8.2 LAPLACE TRANSFORM METHOD

When initial conditions are specified, the Laplace transform reduces a system of linear differential equations with constant coefficients to a set of simultaneous algebraic equations in the transformed functions.

EXAMPLE I

Solve
$$2x' + y' - y = t$$
$$x' + y' \quad = t^2 \tag{1}$$

subject to $x(0) = 1$, $y(0) = 0$.

Solution If $X(s) = \mathscr{L}\{x(t)\}$ and $Y(s) = \mathscr{L}\{y(t)\}$, then after transforming each equation, we obtain

$$2[sX(s) - x(0)] + sY(s) - y(0) - Y(s) = \frac{1}{s^2}$$

$$sX(s) - x(0) + sY(s) - y(0) = \frac{2}{s^3}$$

or

$$2sX(s) + (s - 1)Y(s) = 2 + \frac{1}{s^2}$$

(2)

$$sX(s) + \qquad sY(s) = 1 + \frac{2}{s^3}.$$

Multiplying the second equation of (2) by 2 and substracting yield

$$(-s - 1)Y(s) = \frac{1}{s^2} - \frac{4}{s^3}$$

$$Y(s) = \frac{4 - s}{s^3(s + 1)}.$$

(3)

Now by partial fractions

$$\frac{4 - s}{s^3(s + 1)} = \frac{A}{s} + \frac{B}{s^2} + \frac{C}{s^3} + \frac{D}{s + 1}$$

so that

$$4 - s = As^2(s + 1) + Bs(s + 1) + C(s + 1) + Ds^3.$$

Setting $s = 0$ and $s = -1$ in the last line gives $C = 4$ and $D = -5$, respectively, whereas equating the coefficients of s^3 and s^2 on each side of the equality yields

$$A + D = 0 \quad \text{and} \quad A + B = 0.$$

It follows that $A = 5$, $B = -5$. Thus (3) becomes

$$Y(s) = \frac{5}{s} - \frac{5}{s^2} + \frac{4}{s^3} - \frac{5}{s + 1}$$

and so

$$y(t) = 5\mathscr{L}^{-1}\left\{\frac{1}{s}\right\} - 5\mathscr{L}^{-1}\left\{\frac{1}{s^2}\right\} + 2\mathscr{L}^{-1}\left\{\frac{2!}{s^3}\right\} - 5\mathscr{L}^{-1}\left\{\frac{1}{s + 1}\right\}$$

$$= 5 - 5t + 2t^2 - 5e^{-t}.$$

By the second equation of (2),

$$X(s) = -Y(s) + \frac{1}{s} + \frac{2}{s^4},$$

from which it follows that

$$x(t) = -\mathscr{L}^{-1}\{Y(s)\} + \mathscr{L}^{-1}\left\{\frac{1}{s}\right\} + \frac{2}{3!}\,\mathscr{L}^{-1}\left\{\frac{3!}{s^4}\right\}$$

$$= -4 + 5t - 2t^2 + \frac{1}{3}t^3 + 5e^{-t}.$$

Hence we conclude that the solution of the given system (1) is

$$x(t) = -4 + 5t - 2t^2 + \frac{1}{3}t^3 + 5e^{-t}$$

$$y(t) = 5 - 5t + 2t^2 - 5e^{-t}. \tag{4}$$

∎

Applications

Let us turn now to some elementary applications involving systems of differential equations. The solutions of the problems that we shall consider can be obtained either by the method of the preceding section or through the use of the Laplace transform.

Coupled Springs

Two masses m_1 and m_2 are connected to two springs A and B of negligible mass having spring constants k_1 and k_2, respectively. In turn, the two springs are attached as shown in Figure 8.1. Let $x_1(t)$ and $x_2(t)$ denote the vertical displacements of the masses from their equilibrium positions. When the

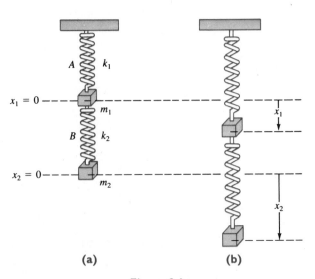

(a) (b)

Figure 8.1

system is in motion, spring B is subject to both an elongation and a compression; hence its net elongation is $x_2 - x_1$. Therefore it follows from Hooke's law that springs A and B exert forces

$$-k_1 x_1 \quad \text{and} \quad k_2(x_2 - x_1),$$

respectively, on m_1. If no external force is impressed on the system and if no damping force is present, then the net force on m_1 is

$$-k_1 x_1 + k_2(x_2 - x_1).$$

By Newton's second law we can write

$$m_1 \frac{d^2 x_1}{dt^2} = -k_1 x_1 + k_2(x_2 - x_1).$$

Similarly, the net force exerted on mass m_2 is due solely to the net elongation of B; that is,

$$-k_2(x_2 - x_1).$$

Thus it follows that

$$m_2 \frac{d^2 x_2}{dt^2} = -k_2(x_2 - x_1).$$

In other words, the motion of the coupled system is represented by the system of simultaneous second-order differential equations

$$m_1 x_1'' = -k_1 x_1 + k_2(x_2 - x_1)$$

$$m_2 x_2'' = -k_2(x_2 - x_1). \tag{5}$$

In the next example we shall solve (5) under the assumptions that

$$k_1 = 6, \qquad k_2 = 4, \qquad m_1 = 1, \qquad m_2 = 1$$

and that the masses start from their equilibrium positions with opposite unit velocities.

EXAMPLE 2

Solve
$$x_1'' + 10x_1 \qquad -4x_2 = 0$$

$$-4x_1 + x_2'' + 4x_2 = 0 \tag{6}$$

subject to $x_1(0) = 0$, $x_1'(0) = 1$, $x_2(0) = 0$, $x_2'(0) = -1$.

Solution The Laplace transform of each equation is

$$s^2 X_1(s) - s x_1(0) - x_1'(0) + 10 X_1(s) - 4 X_2(s) = 0$$

$$-4 X_1(s) + s^2 X_2(s) - s x_2(0) - x_2'(0) + 4 X_2(s) = 0,$$

where $X_1(s) = \mathscr{L}\{x_1(t)\}$ and $X_2(s) = \mathscr{L}\{x_2(t)\}$. The preceding system is the same as

$$(s^2 + 10)X_1(s) - \qquad 4X_2(s) = 1$$

$$-4X_1(s) + (s^2 + 4)X_2(s) = -1. \qquad (7)$$

Eliminating X_2 gives

$$X_1(s) = \frac{s^2}{(s^2 + 2)(s^2 + 12)}.$$

By partial fractions we can write

$$\frac{s^2}{(s^2 + 2)(s^2 + 12)} = \frac{As + B}{s^2 + 2} + \frac{Cs + D}{s^2 + 12}$$

and $\qquad s^2 = (As + B)(s^2 + 12) + (Cs + D)(s^2 + 2).$

Comparing the coefficients of s on each side of the last equality gives

$$A + C = 0$$

$$B + D = 1$$

$$12A + 2C = 0$$

$$12B + 2D = 0$$

so that $\qquad A = 0, \qquad C = 0, \qquad B = -\frac{1}{5}, \qquad D = \frac{6}{5}.$

Hence $\qquad X_1(s) = -\frac{1/5}{s^2 + 2} + \frac{6/5}{s^2 + 12}$

and therefore

$$x_1(t) = -\frac{1}{5\sqrt{2}} \mathscr{L}^{-1}\left\{\frac{\sqrt{2}}{s^2 + 2}\right\} + \frac{6}{5\sqrt{12}} \mathscr{L}^{-1}\left\{\frac{\sqrt{12}}{s^2 + 12}\right\}$$

$$= -\frac{\sqrt{2}}{10} \sin \sqrt{2}\,t + \frac{\sqrt{3}}{5} \sin 2\sqrt{3}\,t.$$

From the first equation of (7) it follows that

$$X_2(s) = -\frac{s^2 + 6}{(s^2 + 2)(s^2 + 12)}.$$

Proceeding as before with partial fractions, we obtain

$$X_2(s) = -\frac{2/5}{s^2 + 2} - \frac{3/5}{s^2 + 12}$$

and
$$x_2(t) = -\frac{2}{5\sqrt{2}}\, \mathscr{L}^{-1}\left\{\frac{\sqrt{2}}{s^2 + 2}\right\} - \frac{3}{5\sqrt{12}}\, \mathscr{L}^{-1}\left\{\frac{\sqrt{12}}{s^2 + 12}\right\}$$

$$= -\frac{\sqrt{2}}{5} \sin\sqrt{2}\,t - \frac{\sqrt{3}}{10} \sin 2\sqrt{3}\,t.$$

Finally the solution to the given system (6) is

$$x_1(t) = -\frac{\sqrt{2}}{10} \sin\sqrt{2}\,t + \frac{\sqrt{3}}{5} \sin 2\sqrt{3}\,t$$

$$x_2(t) = -\frac{\sqrt{2}}{5} \sin\sqrt{2}\,t - \frac{\sqrt{3}}{10} \sin 2\sqrt{3}\,t. \qquad \blacksquare$$

(8)

Networks

An electrical network having more than one loop also gives rise to simultaneous differential equations. As shown in Figure 8.2, the current $i_1(t)$ splits in the directions shown at point B_1 called a *branch point* of the network. By Kirchhoff's first law we can write

$$i_1(t) = i_2(t) + i_3(t). \tag{9}$$

In addition we can also apply **Kirchhoff's second law** to each loop. For loop $A_1 B_1 B_2 A_2 A_1$, summing the voltage drops across each part of the loop gives

$$E(t) = i_1 R_1 + L_1 \frac{di_2}{dt} + i_2 R_2. \tag{10}$$

Similarly, for loop $A_1 B_1 C_1 C_2 B_2 A_2 A_1$ we find

$$E(t) = i_1 R_1 + L_2 \frac{di_3}{dt}. \tag{11}$$

Using (9) to eliminate i_1 in (10) and (11) yields two first-order equations for the currents $i_2(t)$ and $i_3(t)$:

$$L_1 \frac{di_2}{dt} + (R_1 + R_2)i_2 + R_1 i_3 = E(t)$$

$$L_2 \frac{di_3}{dt} + \qquad R_1 i_2 + R_1 i_3 = E(t). \tag{12}$$

Given the natural initial conditions $i_2(0) = 0$, $i_3(0) = 0$, the system (12) is amenable to solution by the Laplace transform.

We leave it as an exercise (see Problem 18) to show that the system of differential equations describing the currents $i_1(t)$ and $i_2(t)$ in the network

Figure 8.2

Figure 8.3

containing a resistor, an inductor, and a capacitor shown in Figure 8.3 is

$$L\frac{di_1}{dt} + Ri_2 = E(t)$$

$$RC\frac{di_2}{dt} + i_2 - i_1 = 0.$$

(13)

EXAMPLE 3

Solve the system (13) under the conditions $E = 60$ volts, $L = 1$ henry, $R = 50$ ohms, $C = 10^{-4}$ farad, and i_1 and i_2 are initially zero.

Solution   We must solve

$$\frac{di_1}{dt} + 50i_2 = 60$$

$$50(10^{-4})\frac{di_2}{dt} + i_2 - i_1 = 0$$

subject to $i_1(0) = 0$, $i_2(0) = 0$.

Applying the Laplace transform to each equation of the system and simplifying give

$$sI_1(s) + 50I_2(s) = \frac{60}{s}$$

$$-200I_1(s) + (s + 200)I_2(s) = 0,$$

where $I_1(s) = \mathscr{L}\{i_1(t)\}$ and $I_2(s) = \mathscr{L}\{i_2(t)\}$. Solving the system for I_1 and I_2 yields

$$I_1(s) = \frac{60s + 12{,}000}{s(s + 100)^2}$$

$$I_2(s) = \frac{12{,}000}{s(s + 100)^2}.$$

By partial fractions we can write

$$I_1(s) = \frac{6/5}{s} - \frac{6/5}{s + 100} - \frac{60}{(s + 100)^2}$$

$$I_2(s) = \frac{6/5}{s} - \frac{6/5}{s + 100} - \frac{120}{(s + 100)^2},$$

from which it follows that

$$i_1(t) = \frac{6}{5} - \frac{6}{5}e^{-100t} - 60te^{-100t}$$

$$i_2(t) = \frac{6}{5} - \frac{6}{5}e^{-100t} - 120te^{-100t}.$$

■

Note that both $i_1(t)$ and $i_2(t)$ in Example 3 tend toward the value $E/R = 6/5$ as $t \to \infty$. Furthermore since the current through the capacitor is $i_3(t) = i_1(t) - i_2(t) = 60te^{-100t}$, we observe that $i_3(t) \to 0$ as $t \to \infty$.

8.2 EXERCISES

Answers to odd-numbered problems begin on page A–22.

In Problems 1–12 use the Laplace transform to solve the given system of differential equations.

1. $\dfrac{dx}{dt} = -x + y$

$\dfrac{dy}{dt} = 2x$

$x(0) = 0, \; y(0) = 1$

2. $\dfrac{dx}{dt} = 2y + e^t$

$\dfrac{dy}{dt} = 8x - t$

$x(0) = 1, \; y(0) = 1$

3. $\dfrac{dx}{dt} = x - 2y$

$\dfrac{dy}{dt} = 5x - y$

$x(0) = -1, \; y(0) = 2$

4. $\dfrac{dx}{dt} + 3x + \dfrac{dy}{dt} = 1$

$\dfrac{dx}{dt} - x + \dfrac{dy}{dt} - y = e^t$

$x(0) = 0, \; y(0) = 0$

5. $2\dfrac{dx}{dt} + \dfrac{dy}{dt} - 2x = 1$

$\dfrac{dx}{dt} + \dfrac{dy}{dt} - 3x - 3y = 2$

$x(0) = 0, \; y(0) = 0$

6. $\dfrac{dx}{dt} + x - \dfrac{dy}{dt} + y = 0$

$\dfrac{dx}{dt} + \dfrac{dy}{dt} + 2y = 0$

$x(0) = 0, \; y(0) = 1$

7. $\dfrac{d^2x}{dt^2} + x - y = 0$

$\dfrac{d^2y}{dt^2} + y - x = 0$

$x(0) = 0, \; x'(0) = -2,$

$y(0) = 0, \; y'(0) = 1$

8. $\dfrac{d^2x}{dt^2} + \dfrac{dx}{dt} + \dfrac{dy}{dt} = 0$

$\dfrac{d^2y}{dt^2} + \dfrac{dy}{dt} - 4\dfrac{dx}{dt} = 0$

$x(0) = 1, \; x'(0) = 0,$

$y(0) = -1, \; y'(0) = 5$

9. $\dfrac{d^2x}{dt^2} + \dfrac{d^2y}{dt^2} = t^2$

$\dfrac{d^2x}{dt^2} - \dfrac{d^2y}{dt^2} = 4t$

$x(0) = 8, \; x'(0) = 0,$

$y(0) = 0, \; y'(0) = 0$

10. $\dfrac{dx}{dt} - 4x + \dfrac{d^3y}{dt^3} = 6 \sin t$

$\dfrac{dx}{dt} + 2x - 2\dfrac{d^3y}{dt^3} = 0$

$x(0) = 0, \; y(0) = 0,$

$y'(0) = 0, \; y''(0) = 0$

11. $\dfrac{d^2x}{dt^2} + 3\dfrac{dy}{dt} + 3y = 0$

$\dfrac{d^2x}{dt^2} + 3y = te^{-t}$

$x(0) = 0, \; x'(0) = 2, \; y(0) = 0$

12. $\dfrac{dx}{dt} = 4x - 2y + 2\mathcal{U}(t - 1)$

$\dfrac{dy}{dt} = 3x - y + \mathcal{U}(t - 1)$

$x(0) = 0, \; y(0) = \frac{1}{2}$

Figure 8.4

Figure 8.5

Figure 8.6

13. Solve system (5) when

$$k_1 = 3, \qquad k_2 = 2, \qquad m_1 = 1, \qquad m_2 = 1$$

and
$$x_1(0) = 0, \qquad x_1'(0) = 1$$
$$x_2(0) = 1, \qquad x_2'(0) = 0.$$

14. Derive the system of differential equations describing the straight-line vertical motion of the coupled springs shown in Figure 8.4. Use the Laplace transform to solve the system when $k_1 = 1$, $k_2 = 1$, $k_3 = 1$, $m_1 = 1$, $m_2 = 1$, and $x_1(0) = 0$, $x_1'(0) = -1$, $x_2(0) = 0$, $x_2'(0) = 1$.

15. **(a)** Show that the system of differential equations for the currents $i_2(t)$ and $i_3(t)$ in the electrical network shown in Figure 8.5 is

$$L_1 \frac{di_2}{dt} + Ri_2 + Ri_3 = E(t)$$

$$L_2 \frac{di_3}{dt} + Ri_2 + Ri_3 = E(t).$$

(b) Solve the system in part (a) if $R = 5$ ohms, $L_1 = 0.01$ henry, $L_2 = 0.0125$ henry, $E = 100$ volts, $i_2(0) = 0$, and $i_3(0) = 0$.

(c) Determine the current $i_1(t)$.

16. **(a)** Show that the system of differential equations for the currents $i_2(t)$ and $i_3(t)$ in the electrical network shown in Figure 8.6 is

$$L \frac{di_2}{dt} + L \frac{di_3}{dt} + R_1 i_2 = E(t)$$

$$-R_1 \frac{di_2}{dt} + R_2 \frac{di_3}{dt} + \frac{1}{C} i_3 = 0.$$

(b) Solve the system in part (a) if $R_1 = 10$ ohms, $R_2 = 5$ ohms, $L = 1$ henry, $C = 0.2$ farad,

$$E(t) = \begin{cases} 120, & 0 \le t < 2 \\ 0, & t \ge 2, \end{cases}$$

$i_2(0) = 0$, and $i_3(0) = 0$.

(c) Determine the current $i_1(t)$.

17. Solve the system given in (12) when $R_1 = 6$ ohms, $R_2 = 5$ ohms, $L_1 = 1$ henry, $L_2 = 1$ henry, $E(t) = 50 \sin t$ volts.

18. Derive the system of equations (13).

19. Solve (13) when $E = 60$ volts, $L = 1/2$ henry, $R = 50$ ohms, $C = 10^{-4}$ farad, $i_1(0) = 0$, $i_2(0) = 0$.

20. Solve (13) when $E = 60$ volts, $L = 2$ henrys, $R = 50$ ohms, $C = 10^{-4}$ farad, $i_1(0) = 0$, $i_2(0) = 0$.

Figure 8.7

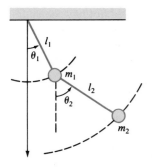

Figure 8.8

21. (a) Show that the system of differential equations for the charge on the capacitor $q(t)$ and the current $i_3(t)$ in the electrical network shown in Figure 8.7 is

$$R_1 \frac{dq}{dt} + \frac{1}{C} q + R_1 i_3 = E(t)$$

$$L \frac{di_3}{dt} + R_2 i_3 - \frac{1}{C} q = 0.$$

(b) Find the charge on the capacitor when $L = 1$ henry, $R_1 = 1$ ohm, $R_2 = 1$ ohm, $C = 1$ farad,

$$E(t) = \begin{cases} 0, & 0 < t < 1 \\ 50e^{-t}, & t \geq 1, \end{cases}$$

$i_3(0) = 0$, and $q(0) = 0$.

22. A double pendulum oscillates in a vertical plane under the influence of gravity (see Figure 8.8). For small displacements $\theta_1(t)$ and $\theta_2(t)$, it can be shown that the differential equations of motion are

$$(m_1 + m_2)l_1^2 \theta_1'' + m_2 l_1 l_2 \theta_2'' + (m_1 + m_2)l_1 g \theta_1 = 0$$

$$m_2 l_2^2 \theta_2'' + m_2 l_1 l_2 \theta_1'' + m_2 l_2 g \theta_2 = 0.$$

Use the Laplace transform to solve the system when $m_1 = 3$, $m_2 = 1$, $l_1 = l_2 = 16$, $\theta_1(0) = 1$, $\theta_2(0) = -1$, $\theta_1'(0) = 0$, and $\theta_2'(0) = 0$.

8.3 SYSTEMS OF LINEAR FIRST-ORDER EQUATIONS

In the preceding two sections we dealt with linear systems that were of the form

$$P_{11}(D)x_1 + P_{12}(D)x_2 + \cdots + P_{1n}(D)x_n = b_1(t)$$

$$P_{21}(D)x_1 + P_{22}(D)x_2 + \cdots + P_{2n}(D)x_n = b_2(t)$$

$$\vdots \qquad \vdots \qquad \vdots \tag{1}$$

$$P_{n1}(D)x_1 + P_{n2}(D)x_2 + \cdots + P_{nn}(D)x_n = b_n(t),$$

where the P_{ij} were polynomials in the differential operator D. However, the study of systems of *first-order* differential equations

$$\frac{dx_1}{dt} = g_1(t, x_1, x_2, \ldots, x_n)$$

$$\frac{dx_2}{dt} = g_2(t, x_1, x_2, \ldots, x_n) \tag{2}$$

$$\vdots$$

$$\frac{dx_n}{dt} = g_n(t, x_1, x_2, \ldots, x_n)$$

is particularly important in advanced mathematics since every nth-order differential equation

$$y^{(n)} = F(t, y, y', \ldots, y^{(n-1)}),$$

as well as most systems of differential equations, can be reduced to form (2). System (2) of n first-order equations is called an **nth-order system**.

Linear Normal Form

Of course, a system such as (2) need not be linear and need not have constant coefficients. Consequently, the system may not be readily solvable, if at all. In the remaining sections of this chapter we shall be interested only in a particular, but important, case of (2), namely, those systems having the linear **normal**, or **canonical**, form

$$
\begin{aligned}
\frac{dx_1}{dt} &= a_{11}(t)x_1 + a_{12}(t)x_2 + \cdots + a_{1n}(t)x_n + f_1(t) \\[2mm]
\frac{dx_2}{dt} &= a_{21}(t)x_1 + a_{22}(t)x_2 + \cdots + a_{2n}(t)x_n + f_2(t) \\
&\ \vdots \qquad\qquad \vdots \qquad\qquad \vdots \\
\frac{dx_n}{dt} &= a_{n1}(t)x_1 + a_{n2}(t)x_2 + \cdots + a_{nn}(t)x_n + f_n(t),
\end{aligned}
\tag{3}
$$

where the coefficients a_{ij} and the f_i are functions continuous on a common interval I. When $f_i(t) = 0$, $i = 1, 2, \ldots, n$, the system (3) is said to be **homogeneous**; otherwise it is called **nonhomogeneous**.

We shall now show that every linear nth-order differential equation can be reduced to a linear system having the normal form (3).

Equation to a System

Suppose a linear nth-order differential equation is first written as

$$\frac{d^n y}{dt^n} = -\frac{a_0}{a_n} y - \frac{a_1}{a_n} y' - \cdots - \frac{a_{n-1}}{a_n} y^{(n-1)} + f(t). \tag{4}$$

If we then introduce the variables

$$y = x_1, \quad y' = x_2, \quad y'' = x_3, \ldots, \quad y^{(n-1)} = x_n, \tag{5}$$

it follows that $y' = x_1' = x_2$, $y'' = x_2' = x_3, \ldots, y^{(n-1)} = x_{n-1}' = x_n$, and $y^{(n)} = x_n'$. Hence from (4) and (5) we find that a linear nth-order differential equation can be expressed as an nth-order system:

$$x'_1 = x_2$$

$$x'_2 = x_3$$

$$x'_3 = x_4$$

$$\vdots$$

$$x'_{n-1} = x_n$$

(6)

$$x'_n = -\frac{a_0}{a_n} x_1 - \frac{a_1}{a_n} x_2 - \cdots - \frac{a_{n-1}}{a_n} x_n + f(t).$$

Inspection of (6) reveals that it has the same form as (3).

EXAMPLE 1

Reduce the third-order equation

$$2y''' - 6y'' + 4y' + y = \sin t$$

to the normal form (3).

Solution Write the differential equation as

$$y''' = -\frac{1}{2} y - 2y' + 3y'' + \frac{1}{2} \sin t$$

and then let $y = x_1$, $y' = x_2$, $y'' = x_3$. Since

$$x'_1 = y' = x_2$$

$$x'_2 = y'' = x_3$$

$$x'_3 = y''',$$

we find

$$x'_1 = x_2$$

$$x'_2 = x_3$$

$$x'_3 = -\frac{1}{2} x_1 - 2x_2 + 3x_3 + \frac{1}{2} \sin t.$$

∎

Systems Reduced to Normal Form

Using a procedure similar to that just outlined, we can reduce *most* systems of the linear form (1) to the linear normal form (3). To accomplish this it is necessary to first solve the system for the highest-order derivative of each dependent variable. As we shall see, this may not always be possible.

EXAMPLE 2

Reduce
$$(D^2 - D + 5)x + \quad 2D^2 y = e^t$$
$$-2x + (D^2 + 2)y = 3t^2$$

to the normal form (3).

Solution Write the system as

$$D^2 x + 2D^2 y = \quad e^t - 5x + Dx$$

$$D^2 y = 3t^2 + 2x - 2y$$

and then eliminate $D^2 y$ by multiplying the second equation by 2 and subtracting. We have

$$D^2 x = e^t - 6t^2 - 9x + 4y + Dx.$$

Since the second equation of the system already expresses the highest-order derivative of y in terms of the remaining functions, we are now in a position to introduce new variables. If we let

$$Dx = u \quad \text{and} \quad Dy = v,$$

the expressions for $D^2 x$ and $D^2 y$ become, respectively,

$$Du = e^t - 6t^2 - 9x + 4y + u$$

$$Dv = 3t^2 + 2x - 2y.$$

Thus the original system can be written in the normal form

$$Dx = u$$

$$Dy = v$$

$$Du = -9x + 4y + u + e^t - 6t^2$$

$$Dv = \quad 2x - 2y + 3t^2. \qquad \blacksquare$$

Degenerate Systems

Those systems of differential equations of form (1) that cannot be reduced to a linear system in normal form are said to be **degenerate**. For example, it is a straightforward matter to show that it is impossible to solve the system

$$(D + 1)x + (D + 1)y = 0$$

$$2Dx + (2D + 1)y = 0 \qquad (7)$$

for the highest derivative of each variable, and hence the system is degenerate.*

---

* This does *not* mean that the system does not have a solution (see Problem 21).

You may be wondering why anyone would want to convert a single differential equation to a system of equations, or for that matter a system of differential equations to an even larger system. While we are not in a position to completely justify their importance, suffice it to say that these procedures are more than a theoretical exercise. There are times when it is actually desirable to work with a system rather than with one equation. In the numerical analysis of differential equations, almost all computational algorithms are established for first-order equations. Since these algorithms can be generalized directly to systems, to compute numerically, say, a second-order equation, we could reduce it to a system of two first-order equations (see Chapter 9).

A linear system such as (3) also arises naturally in some physical applications. The following example illustrates a homogeneous system in two dependent variables.

EXAMPLE 3

Tank A contains 50 gallons of water in which 25 pounds of salt are dissolved. A second tank, B, contains 50 gallons of pure water. Liquid is pumped in and out of the tanks at the rates shown in Figure 8.9. Derive the differential equations that describe the number of pounds $x_1(t)$ and $x_2(t)$ of salt at any time in tanks A and B, respectively.

Figure 8.9

Solution By an analysis similar to that used in Section 3.2, we see that the net rate of change in $x_1(t)$ in lb/min is

$$\frac{dx_1}{dt} = \overbrace{(3 \text{ gal/min}) \cdot (0 \text{ lb/gal}) + (1 \text{ gal/min}) \cdot \left(\frac{x_2}{50} \text{ lb/gal}\right)}^{\text{input}} - \overbrace{(4 \text{ gal/min}) \cdot \left(\frac{x_1}{50} \text{ lb/gal}\right)}^{\text{output}}$$

$$= -\frac{2}{25} x_1 + \frac{1}{50} x_2.$$

In addition we find that the net rate of change in $x_2(t)$ is

$$\frac{dx_2}{dt} = 4 \cdot \frac{x_1}{50} - 3 \cdot \frac{x_2}{50} - 1 \cdot \frac{x_2}{50}$$

$$= \frac{2}{25} x_1 - \frac{2}{25} x_2.$$

Thus we obtain the first-order system

$$\frac{dx_1}{dt} = -\frac{2}{25} x_1 + \frac{1}{50} x_2$$

$$\frac{dx_2}{dt} = \frac{2}{25} x_1 - \frac{2}{25} x_2.$$

(8)

Observe that the foregoing system is accompanied by the initial conditions $x_1(0) = 25$, $x_2(0) = 0$. ∎

It is left as an exercise to solve (8) by the Laplace transform. See Problem 17.

8.3 EXERCISES

Answers to odd-numbered problems begin on page A–22.

In Problems 1–8 rewrite the given differential equation as a system in normal form (3).

1. $y'' - 3y' + 4y = \sin 3t$

2. $2\dfrac{d^2y}{dt^2} + 4\dfrac{dy}{dt} - 5y = 0$

3. $y''' - 3y'' + 6y' - 10y = t^2 + 1$

4. $4y''' + y = e^t$

5. $\dfrac{d^4y}{dt^4} - 2\dfrac{d^2y}{dt^2} + 4\dfrac{dy}{dt} + y = t$

6. $2\dfrac{d^4y}{dt^4} + \dfrac{d^3y}{dt^3} - 8y = 10$

7. $(t + 1)y'' = ty$

8. $t^2y'' + ty' + (t^2 - 4)y = 0$

In Problems 9–16 rewrite, if possible, the given system in the normal form (3).

9. $\begin{aligned} x' + 4x - y' &= 7t \\ x' + \quad y' - 2y &= 3t \end{aligned}$

10. $\begin{aligned} x'' + y' &= 1 \\ x'' + y' &= -1 \end{aligned}$

11. $\begin{aligned} (D - 1)x - Dy &= t^2 \\ x + Dy &= 5t - 2 \end{aligned}$

12. $\begin{aligned} x'' - 2y'' &= \sin t \\ x'' + \quad y'' &= \cos t \end{aligned}$

13. $\begin{aligned} (2D + 1)x - 2Dy &= 4 \\ Dx - \quad Dy &= e^t \end{aligned}$

14. $\begin{aligned} m_1 x_1'' &= -k_1 x_1 + k_2(x_2 - x_1) \\ m_2 x_2'' &= -k_2(x_2 - x_1) \end{aligned}$

15. $\begin{aligned} \dfrac{d^3x}{dt^3} &= 4x - 3\dfrac{d^2x}{dt^2} + 4\dfrac{dy}{dt} \\ \dfrac{d^2y}{dt^2} &= 10t^2 - 4\dfrac{dx}{dt} + 3\dfrac{dy}{dt} \end{aligned}$

16. $\begin{aligned} D^2x + \quad\quad Dy &= 4t \\ -D^2x + (D + 1)y &= 6t^2 + 10 \end{aligned}$

17. Use the Laplace transform to solve system (8) subject to $x_1(0) = 25$ and $x_2(0) = 0$.

18. Consider two tanks A and B with liquid being pumped in and out at the same rates as given in Example 3. What is the system of differential equations if, instead of pure water, a brine solution containing 2 lb of salt per gallon is pumped into tank A?

19. Using the information given in Figure 8.10, derive the system of differential equations describing the number of pounds of salt x_1, x_2, and x_3 at any time in tanks A, B, and C, respectively.

pure water
4 gal/min

mixture
2 gal/min

mixture
1 gal/min

A
100 gal

B
100 gal

C
100 gal

mixture
6 gal/min

mixture
5 gal/min

mixture
4 gal/min

Figure 8.10

20. Consider the first-order system

$$(a_1 D - b_1)x + (a_2 D - b_2)y = 0$$

$$(a_3 D - b_3)x + (a_4 D - b_4)y = 0,$$

where the a_i are nonzero constants. Determine a condition on the a_i such that the system is degenerate.

21. Verify that the degenerate system (7) possesses the solution $x(t) = c_1 e^{-t}$, $y(t) = -2c_1 e^{-t}$.

8.4 INTRODUCTION TO MATRICES

8.4.1 Basic Definitions and Theory

Before examining a systematic procedure for solving linear first-order systems in normal form, we need the new and useful concept of a **matrix**.

DEFINITION 8.1 Matrix

A **matrix A** is any rectangular array of numbers or functions:

$$\mathbf{A} = \begin{pmatrix} a_{11} & a_{12} & \cdots & a_{1n} \\ a_{21} & a_{22} & \cdots & a_{2n} \\ \vdots & & & \vdots \\ a_{m1} & a_{m2} & \cdots & a_{mn} \end{pmatrix}. \tag{1}$$

If a matrix has m rows and n columns, we say that its **size** is m by n (written $m \times n$). An $n \times n$ matrix is called a **square** matrix of order n.

The element, or entry, in the ith row and jth column of an $m \times n$ matrix **A** is written a_{ij}. An $m \times n$ matrix **A** is then abbreviated as $\mathbf{A} = (a_{ij})_{m \times n}$, or simply $\mathbf{A} = (a_{ij})$. A 1×1 matrix is simply one constant or function.

DEFINITION 8.2 Equality of Matrices

Two $m \times n$ matrices **A** and **B** are **equal** if $a_{ij} = b_{ij}$ for each i and j.

DEFINITION 8.3 Column Matrix

A **column matrix X** is any matrix having n rows and one column:

$$\mathbf{X} = \begin{pmatrix} b_{11} \\ b_{21} \\ \vdots \\ b_{n1} \end{pmatrix} = (b_{i1})_{n \times 1}.$$

A column matrix is also called a **column vector** or simply a **vector**.

DEFINITION 8.4 Multiples of Matrices

A **multiple** of a matrix **A** is defined to be

$$k\mathbf{A} = \begin{pmatrix} ka_{11} & ka_{12} & \cdots & ka_{1n} \\ ka_{21} & ka_{22} & \cdots & ka_{2n} \\ \vdots & & & \vdots \\ ka_{m1} & ka_{m2} & \cdots & ka_{mn} \end{pmatrix} = (ka_{ij})_{m \times n},$$

where k is a constant or a function.

EXAMPLE 1

(a) $5\begin{pmatrix} 2 & -3 \\ 4 & -1 \\ \frac{1}{5} & 6 \end{pmatrix} = \begin{pmatrix} 10 & -15 \\ 20 & -5 \\ 1 & 30 \end{pmatrix}$ **(b)** $e^t\begin{pmatrix} 1 \\ -2 \\ 4 \end{pmatrix} = \begin{pmatrix} e^t \\ -2e^t \\ 4e^t \end{pmatrix}$ ∎

We note in passing that for any matrix **A**, the product $k\mathbf{A}$ is the same as $\mathbf{A}k$. For example,

$$e^{-3t}\begin{pmatrix} 2 \\ 5 \end{pmatrix} = \begin{pmatrix} 2e^{-3t} \\ 5e^{-3t} \end{pmatrix} = \begin{pmatrix} 2 \\ 5 \end{pmatrix}e^{-3t}.$$

DEFINITION 8.5   Addition of Matrices

The **sum** of two $m \times n$ matrices **A** and **B** is defined to be the matrix

$$\mathbf{A} + \mathbf{B} = (a_{ij} + b_{ij})_{m \times n}.$$

In other words, when adding two matrices of the same size, we add the corresponding elements.

EXAMPLE 2

The sum of

$$\mathbf{A} = \begin{pmatrix} 2 & -1 & 3 \\ 0 & 4 & 6 \\ -6 & 10 & -5 \end{pmatrix} \quad \text{and} \quad \mathbf{B} = \begin{pmatrix} 4 & 7 & -8 \\ 9 & 3 & 5 \\ 1 & -1 & 2 \end{pmatrix}$$

is

$$\mathbf{A} + \mathbf{B} = \begin{pmatrix} 2+4 & -1+7 & 3+(-8) \\ 0+9 & 4+3 & 6+5 \\ -6+1 & 10+(-1) & -5+2 \end{pmatrix}$$

$$= \begin{pmatrix} 6 & 6 & -5 \\ 9 & 7 & 11 \\ -5 & 9 & -3 \end{pmatrix}.$$ ∎

EXAMPLE 3

The single matrix

$$\begin{pmatrix} 3t^2 - 2e^t \\ t^2 + 7t \\ 5t \end{pmatrix}$$

can be written as the sum of three column vectors

$$\begin{pmatrix} 3t^2 - 2e^t \\ t^2 + 7t \\ 5t \end{pmatrix} = \begin{pmatrix} 3t^2 \\ t^2 \\ 0 \end{pmatrix} + \begin{pmatrix} 0 \\ 7t \\ 5t \end{pmatrix} + \begin{pmatrix} -2e^t \\ 0 \\ 0 \end{pmatrix} = \begin{pmatrix} 3 \\ 1 \\ 0 \end{pmatrix} t^2 + \begin{pmatrix} 0 \\ 7 \\ 5 \end{pmatrix} t + \begin{pmatrix} -2 \\ 0 \\ 0 \end{pmatrix} e^t. \quad \blacksquare$$

The **difference** of two $m \times n$ matrices is defined in the usual manner: $\mathbf{A} - \mathbf{B} = \mathbf{A} + (-\mathbf{B})$, where $-\mathbf{B} = (-1)\mathbf{B}$.

DEFINITION 8.6   **Multiplication of Matrices**

Let $\mathbf{A}$ be a matrix having m rows and n columns and $\mathbf{B}$ be a matrix having n rows and p columns. We define the **product AB** to be the $m \times p$ matrix

$$\mathbf{AB} = \begin{pmatrix} a_{11} & a_{12} & \cdots & a_{1n} \\ a_{21} & a_{22} & \cdots & a_{2n} \\ \vdots & \vdots & & \vdots \\ a_{m1} & a_{m2} & \cdots & a_{mn} \end{pmatrix} \begin{pmatrix} b_{11} & b_{12} & \cdots & b_{1p} \\ b_{21} & b_{22} & \cdots & b_{2p} \\ \vdots & \vdots & & \vdots \\ b_{n1} & b_{n2} & \cdots & b_{np} \end{pmatrix}$$

$$= \begin{pmatrix} a_{11}b_{11} + a_{12}b_{21} + \cdots + a_{1n}b_{n1} & \cdots & a_{11}b_{1p} + a_{12}b_{2p} + \cdots + a_{1n}b_{np} \\ a_{21}b_{11} + a_{22}b_{21} + \cdots + a_{2n}b_{n1} & \cdots & a_{21}b_{1p} + a_{22}b_{2p} + \cdots + a_{2n}b_{np} \\ \vdots & & \vdots \\ a_{m1}b_{11} + a_{m2}b_{21} + \cdots + a_{mn}b_{n1} & \cdots & a_{m1}b_{1p} + a_{m2}b_{2p} + \cdots + a_{mn}b_{np} \end{pmatrix}$$

$$= \left(\sum_{k=1}^{n} a_{ik}b_{kj} \right)_{m \times p}$$

Note carefully in Definition 8.6 that the product $\mathbf{AB} = \mathbf{C}$ is defined only when the number of columns in the matrix $\mathbf{A}$ is the same as the number of rows in $\mathbf{B}$. The size of the product can be determined from

$$\mathbf{A}_{m \times n}\mathbf{B}_{n \times p} = \mathbf{C}_{m \times p}.$$

Also, you might recognize that the entries in, say, the ith row of the final matrix $\mathbf{AB}$ are formed by using the component definition of the inner or dot product of the ith row of $\mathbf{A}$ with each of the columns of $\mathbf{B}$.

EXAMPLE 4

(a) For $\mathbf{A} = \begin{pmatrix} 4 & 7 \\ 3 & 5 \end{pmatrix}$ and $\mathbf{B} = \begin{pmatrix} 9 & -2 \\ 6 & 8 \end{pmatrix}$,

$$\mathbf{AB} = \begin{pmatrix} 4 \cdot 9 + 7 \cdot 6 & 4 \cdot (-2) + 7 \cdot 8 \\ 3 \cdot 9 + 5 \cdot 6 & 3 \cdot (-2) + 5 \cdot 8 \end{pmatrix} = \begin{pmatrix} 78 & 48 \\ 57 & 34 \end{pmatrix}.$$

(b) For $\mathbf{A} = \begin{pmatrix} 5 & 8 \\ 1 & 0 \\ 2 & 7 \end{pmatrix}$ and $\mathbf{B} = \begin{pmatrix} -4 & -3 \\ 2 & 0 \end{pmatrix}$,

$$\mathbf{AB} = \begin{pmatrix} 5 \cdot (-4) + 8 \cdot 2 & 5 \cdot (-3) + 8 \cdot 0 \\ 1 \cdot (-4) + 0 \cdot 2 & 1 \cdot (-3) + 0 \cdot 0 \\ 2 \cdot (-4) + 7 \cdot 2 & 2 \cdot (-3) + 7 \cdot 0 \end{pmatrix} = \begin{pmatrix} -4 & -15 \\ -4 & -3 \\ 6 & -6 \end{pmatrix}. \quad \blacksquare$$

In general, matrix multiplication is not commutative; that is, $\mathbf{AB} \neq \mathbf{BA}$. Observe that in part (a) of Example 4

$$\mathbf{BA} = \begin{pmatrix} 30 & 53 \\ 48 & 82 \end{pmatrix},$$

whereas in part (b) the product $\mathbf{BA}$ is *not defined* since Definition 8.6 requires that the first matrix (in this case $\mathbf{B}$) have the same number of columns as the second matrix has rows.

We are particularly interested in the product of a square matrix and a column vector.

E X A M P L E **5**

(a) $\begin{pmatrix} 2 & -1 & 3 \\ 0 & 4 & 5 \\ 1 & -7 & 9 \end{pmatrix} \begin{pmatrix} -3 \\ 6 \\ 4 \end{pmatrix} = \begin{pmatrix} 2 \cdot (-3) + (-1) \cdot 6 + 3 \cdot 4 \\ 0 \cdot (-3) + 4 \cdot 6 + 5 \cdot 4 \\ 1 \cdot (-3) + (-7) \cdot 6 + 9 \cdot 4 \end{pmatrix} = \begin{pmatrix} 0 \\ 44 \\ -9 \end{pmatrix}$

(b) $\begin{pmatrix} -4 & 2 \\ 3 & 8 \end{pmatrix} \begin{pmatrix} x \\ y \end{pmatrix} = \begin{pmatrix} -4x + 2y \\ 3x + 8y \end{pmatrix}$ $\quad \blacksquare$

Multiplicative Identity

For a given positive integer n, the $n \times n$ matrix

$$\mathbf{I} = \begin{pmatrix} 1 & 0 & 0 & \cdots & 0 \\ 0 & 1 & 0 & \cdots & 0 \\ \vdots & & & & \vdots \\ 0 & 0 & 0 & \cdots & 1 \end{pmatrix}$$

is called the **multiplicative identity matrix**. It follows from Definition 8.6 that for any $n \times n$ matrix $\mathbf{A}$,

$$\mathbf{AI} = \mathbf{IA} = \mathbf{A}.$$

Also, it is readily verified that if $\mathbf{X}$ is an $n \times 1$ column matrix, then $\mathbf{IX} = \mathbf{X}$.

Zero Matrix

A matrix consisting of all zero entries is called a **zero matrix** and is denoted by **0**. For example,

$$\mathbf{0} = \begin{pmatrix} 0 \\ 0 \end{pmatrix}, \quad \mathbf{0} = \begin{pmatrix} 0 & 0 \\ 0 & 0 \end{pmatrix}, \quad \mathbf{0} = \begin{pmatrix} 0 & 0 \\ 0 & 0 \\ 0 & 0 \end{pmatrix},$$

and so on. If **A** and **0** are $m \times n$ matrices, then

$$\mathbf{A} + \mathbf{0} = \mathbf{0} + \mathbf{A} = \mathbf{A}.$$

Associative Law

Although we shall not prove it, matrix multiplication is **associative**. If **A** is an $m \times p$ matrix, **B** a $p \times r$ matrix, and **C** an $r \times n$ matrix, then

$$\mathbf{A}(\mathbf{BC}) = (\mathbf{AB})\mathbf{C}$$

is an $m \times n$ matrix.

Distributive Law

If **B** and **C** are $r \times n$ matrices and **A** is an $m \times r$ matrix, then the **distributive law** is

$$\mathbf{A}(\mathbf{B} + \mathbf{C}) = \mathbf{AB} + \mathbf{AC}.$$

Furthermore, if the product $(\mathbf{B} + \mathbf{C})\mathbf{A}$ is defined, then

$$(\mathbf{B} + \mathbf{C})\mathbf{A} = \mathbf{BA} + \mathbf{CA}.$$

Determinant of a Matrix

Associated with every *square* matrix **A** of constants, there is a number called the **determinant of the matrix**, which is denoted by det **A** or $|\mathbf{A}|$.

EXAMPLE 6

For

$$\mathbf{A} = \begin{pmatrix} 3 & 6 & 2 \\ 2 & 5 & 1 \\ -1 & 2 & 4 \end{pmatrix}$$

we expand det **A** by cofactors. Using the first row, we obtain

$$\det \mathbf{A} = \begin{vmatrix} 3 & 6 & 2 \\ 2 & 5 & 1 \\ -1 & 2 & 4 \end{vmatrix} = 3 \begin{vmatrix} 5 & 1 \\ 2 & 4 \end{vmatrix} - 6 \begin{vmatrix} 2 & 1 \\ -1 & 4 \end{vmatrix} + 2 \begin{vmatrix} 2 & 5 \\ -1 & 2 \end{vmatrix}$$

$$= 3(20 - 2) - 6(8 + 1) + 2(4 + 5) = 18. \quad \blacksquare$$

See Appendix III for a brief review of the properties of determinants.

DEFINITION 8.7   Transpose of a Matrix

The **transpose** of the $m \times n$ matrix (1) is the $n \times m$ matrix $\mathbf{A}^T$ given by

$$\mathbf{A}^T = \begin{pmatrix} a_{11} & a_{21} & \cdots & a_{m1} \\ a_{12} & a_{22} & \cdots & a_{m2} \\ \vdots & & & \vdots \\ a_{1n} & a_{2n} & \cdots & a_{mn} \end{pmatrix}.$$

In other words, the rows of a matrix $\mathbf{A}$ become the columns of its transpose $\mathbf{A}^T$.

EXAMPLE 7

(a) The transpose of matrix $\mathbf{A}$ in Example 6 is

$$\mathbf{A}^T = \begin{pmatrix} 3 & 2 & -1 \\ 6 & 5 & 2 \\ 2 & 1 & 4 \end{pmatrix}.$$

(b) If $\mathbf{X} = \begin{pmatrix} 5 \\ 0 \\ 3 \end{pmatrix}$, then $\mathbf{X}^T = (5 \quad 0 \quad 3)$. ∎

DEFINITION 8.8   Multiplicative Inverse of a Matrix

Let $\mathbf{A}$ be an $n \times n$ matrix. If there exists an $n \times n$ matrix $\mathbf{B}$ such that

$$\mathbf{AB} = \mathbf{BA} = \mathbf{I},$$

where $\mathbf{I}$ is the multiplicative identity, then $\mathbf{B}$ is said to be the **multiplicative inverse of** $\mathbf{A}$ and is denoted by $\mathbf{B} = \mathbf{A}^{-1}$.

DEFINITION 8.9   Nonsingular/Singular Matrices

Let $\mathbf{A}$ be an $n \times n$ matrix. If det $\mathbf{A} \neq 0$, then $\mathbf{A}$ is said to be **nonsingular**. If det $\mathbf{A} = 0$, then $\mathbf{A}$ is said to be **singular**.

The following gives a necessary and sufficient condition for a square matrix to have a multiplicative inverse.

THEOREM 8.1 **Nonsingularity Implies A has an Inverse**

An $n \times n$ matrix $\mathbf{A}$ has a multiplicative inverse $\mathbf{A}^{-1}$ if and only if $\mathbf{A}$ is nonsingular.

The following theorem gives one way of finding the multiplicative inverse for a nonsingular matrix.

THEOREM 8.2 **A Formula for the Inverse of a Matrix**

Let $\mathbf{A}$ be an $n \times n$ nonsingular matrix and let $C_{ij} = (-1)^{i+j}M_{ij}$, where M_{ij} is the determinant of the $(n-1) \times (n-1)$ matrix obtained by deleting the ith row and jth column from $\mathbf{A}$. Then

$$\mathbf{A}^{-1} = \frac{1}{\det \mathbf{A}}(C_{ij})^T. \tag{2}$$

Each C_{ij} in Theorem 8.2 is simply the **cofactor** (signed minor) of the corresponding entry a_{ij} in $\mathbf{A}$. Note that the transpose is utilized in formula (2).

For future reference we observe in the case of a 2×2 nonsingular matrix

$$\mathbf{A} = \begin{pmatrix} a_{11} & a_{12} \\ a_{21} & a_{22} \end{pmatrix}$$

that $C_{11} = a_{22}$, $C_{12} = -a_{21}$, $C_{21} = -a_{12}$, and $C_{22} = a_{11}$. Thus

$$\mathbf{A}^{-1} = \frac{1}{\det \mathbf{A}} \begin{pmatrix} a_{22} & -a_{21} \\ -a_{12} & a_{11} \end{pmatrix}^T = \frac{1}{\det \mathbf{A}} \begin{pmatrix} a_{22} & -a_{12} \\ -a_{21} & a_{11} \end{pmatrix}. \tag{3}$$

For a 3×3 nonsingular matrix

$$\mathbf{A} = \begin{pmatrix} a_{11} & a_{12} & a_{13} \\ a_{21} & a_{22} & a_{23} \\ a_{31} & a_{32} & a_{33} \end{pmatrix},$$

$$C_{11} = \begin{vmatrix} a_{22} & a_{23} \\ a_{32} & a_{33} \end{vmatrix}, \qquad C_{12} = -\begin{vmatrix} a_{21} & a_{23} \\ a_{31} & a_{33} \end{vmatrix}, \qquad C_{13} = \begin{vmatrix} a_{21} & a_{22} \\ a_{31} & a_{32} \end{vmatrix},$$

and so on. Carrying out the transposition gives

$$\mathbf{A}^{-1} = \frac{1}{\det \mathbf{A}} \begin{pmatrix} C_{11} & C_{21} & C_{31} \\ C_{12} & C_{22} & C_{32} \\ C_{13} & C_{23} & C_{33} \end{pmatrix}. \tag{4}$$

EXAMPLE 8

Find the multiplicative inverse for $\mathbf{A} = \begin{pmatrix} 1 & 4 \\ 2 & 10 \end{pmatrix}$.

Solution Since det $\mathbf{A} = 10 - 8 = 2 \neq 0$, $\mathbf{A}$ is nonsingular. It follows from Theorem 8.1 that $\mathbf{A}^{-1}$ exists. From (3) we find

$$\mathbf{A}^{-1} = \frac{1}{2}\begin{pmatrix} 10 & -4 \\ -2 & 1 \end{pmatrix} = \begin{pmatrix} 5 & -2 \\ -1 & \frac{1}{2} \end{pmatrix}.$$

Check:

$$\mathbf{A}\mathbf{A}^{-1} = \begin{pmatrix} 1 & 4 \\ 2 & 10 \end{pmatrix}\begin{pmatrix} 5 & -2 \\ -1 & \frac{1}{2} \end{pmatrix} = \begin{pmatrix} 5-4 & -2+2 \\ 10-10 & -4+5 \end{pmatrix} = \begin{pmatrix} 1 & 0 \\ 0 & 1 \end{pmatrix}.$$

$$\mathbf{A}^{-1}\mathbf{A} = \begin{pmatrix} 5 & -2 \\ -1 & \frac{1}{2} \end{pmatrix}\begin{pmatrix} 1 & 4 \\ 2 & 10 \end{pmatrix} = \begin{pmatrix} 5-4 & 20-20 \\ -1+1 & -4+5 \end{pmatrix} = \begin{pmatrix} 1 & 0 \\ 0 & 1 \end{pmatrix}. \qquad \blacksquare$$

EXAMPLE 9

The matrix $\mathbf{A} = \begin{pmatrix} 2 & 2 \\ 3 & 3 \end{pmatrix}$ is singular since det $\mathbf{A} = 2(3) - 2(3) = 0$. We conclude that $\mathbf{A}^{-1}$ does not exist. $\qquad \blacksquare$

EXAMPLE 10

Find the multiplicative inverse for

$$\mathbf{A} = \begin{pmatrix} 2 & 2 & 0 \\ -2 & 1 & 1 \\ 3 & 0 & 1 \end{pmatrix}.$$

Solution Since det $\mathbf{A} = 12 \neq 0$, the given matrix is nonsingular. The cofactors corresponding to the entries in each row of det $\mathbf{A}$ are

$$C_{11} = \begin{vmatrix} 1 & 1 \\ 0 & 1 \end{vmatrix} = 1 \qquad C_{12} = -\begin{vmatrix} -2 & 1 \\ 3 & 1 \end{vmatrix} = 5 \qquad C_{13} = \begin{vmatrix} -2 & 1 \\ 3 & 0 \end{vmatrix} = -3$$

$$C_{21} = -\begin{vmatrix} 2 & 0 \\ 0 & 1 \end{vmatrix} = -2 \qquad C_{22} = \begin{vmatrix} 2 & 0 \\ 3 & 1 \end{vmatrix} = 2 \qquad C_{23} = -\begin{vmatrix} 2 & 2 \\ 3 & 0 \end{vmatrix} = 6$$

$$C_{31} = \begin{vmatrix} 2 & 0 \\ 1 & 1 \end{vmatrix} = 2 \qquad C_{32} = -\begin{vmatrix} 2 & 0 \\ -2 & 1 \end{vmatrix} = -2 \qquad C_{33} = \begin{vmatrix} 2 & 2 \\ -2 & 1 \end{vmatrix} = 6$$

It follows from (4) that

$$\mathbf{A}^{-1} = \frac{1}{12} \begin{pmatrix} 1 & -2 & 2 \\ 5 & 2 & -2 \\ -3 & 6 & 6 \end{pmatrix} = \begin{pmatrix} \frac{1}{12} & -\frac{1}{6} & \frac{1}{6} \\ \frac{5}{12} & \frac{1}{6} & -\frac{1}{6} \\ -\frac{1}{4} & \frac{1}{2} & \frac{1}{2} \end{pmatrix}.$$

You are urged to verify that $\mathbf{A}^{-1}\mathbf{A} = \mathbf{A}\mathbf{A}^{-1} = \mathbf{I}$. ∎

Formula (2) presents obvious difficulties for nonsingular matrices larger than 3×3. For example, to apply (2) to a 4×4 matrix we would have to calculate *sixteen* 3×3 determinants.* In the case of a large matrix, there are more efficient ways of finding $\mathbf{A}^{-1}$. The curious reader is referred to any text in linear algebra.

Since our goal is to apply the concept of a matrix to systems of linear differential equations in normal form, we need the following definitions.

DEFINITION 8.10   **Derivative of a Matrix of Functions**

If $\mathbf{A}(t) = (a_{ij}(t))_{m \times n}$ is a matrix whose entries are functions differentiable on a common interval, then

$$\frac{d\mathbf{A}}{dt} = \left(\frac{d}{dt} a_{ij} \right)_{m \times n}.$$

DEFINITION 8.11   **Integral of a Matrix of Functions**

If $\mathbf{A}(t) = (a_{ij}(t))_{m \times n}$ is a matrix whose entries are functions continuous on a common interval containing t and t_0, then

$$\int_{t_0}^{t} \mathbf{A}(s)\, ds = \left(\int_{t_0}^{t} a_{ij}(s)\, ds \right)_{m \times n}.$$

To differentiate (integrate) a matrix of functions we simply differentiate (integrate) each entry. The derivative of a matrix is also denoted by $\mathbf{A}'(t)$.

---

* Strictly speaking, a determinant is a number, but it is sometimes convenient to refer to a determinant as if it were an array.

EXAMPLE II

If $\quad \mathbf{X}(t) = \begin{pmatrix} \sin 2t \\ e^{3t} \\ 8t - 1 \end{pmatrix}, \quad$ then $\quad \mathbf{X}'(t) = \begin{pmatrix} \dfrac{d}{dt} \sin 2t \\[2mm] \dfrac{d}{dt} e^{3t} \\[2mm] \dfrac{d}{dt} (8t - 1) \end{pmatrix} = \begin{pmatrix} 2 \cos 2t \\ 3e^{3t} \\ 8 \end{pmatrix}$

and $\quad \displaystyle\int_0^t \mathbf{X}(s)\, ds = \begin{pmatrix} \displaystyle\int_0^t \sin 2s\, ds \\[3mm] \displaystyle\int_0^t e^{3s}\, ds \\[3mm] \displaystyle\int_0^t (8s - 1)\, ds \end{pmatrix} = \begin{pmatrix} -\dfrac{1}{2} \cos 2t + \dfrac{1}{2} \\[2mm] \dfrac{1}{3} e^{3t} - \dfrac{1}{3} \\[2mm] 4t^2 - t \end{pmatrix}. \qquad \blacksquare$

8.4.2   Gaussian and Gauss-Jordan Elimination Methods

In preparation for Section 8.6, we need to know more about solving algebraic systems of n linear equations in n unknowns

$$
\begin{aligned}
a_{11}x_1 + a_{12}x_2 + \cdots + a_{1n}x_n &= b_1 \\
a_{21}x_1 + a_{22}x_2 + \cdots + a_{2n}x_n &= b_2 \\
\vdots \qquad\qquad\qquad\quad &\ \ \vdots \\
a_{n1}x_1 + a_{n2}x_2 + \cdots + a_{nn}x_n &= b_n.
\end{aligned}
\tag{5}
$$

If $\mathbf{A}$ denotes the matrix of coefficients in (5), we know that Cramer's rule (see Appendix III) could be used to solve the system whenever $\det \mathbf{A} \neq 0$. However, that rule requires a herculean effort if $\mathbf{A}$ is larger than 3×3. The procedure that we shall now consider has the distinct advantage of being not only an efficient way of handling large systems but also a means of solving consistent systems (5) in which $\det \mathbf{A} = 0$ and a means of solving m linear equations in n unknowns.

DEFINITION 8.12   **Augmented Matrix**

The **augmented matrix** of the system (5) is the $n \times (n + 1)$ matrix

$$
\begin{pmatrix}
a_{11} & a_{12} & \cdots & a_{1n} & b_1 \\
a_{21} & a_{22} & \cdots & a_{2n} & b_2 \\
\vdots & & & & \vdots \\
a_{n1} & a_{n2} & \cdots & a_{nn} & b_n
\end{pmatrix}.
$$

If **B** is the column matrix of the b_i, $i = 1, 2, \ldots, n$, the augmented matrix of (5) is denoted by $(\mathbf{A} \,|\, \mathbf{B})$.

Elementary Row Operations

Recall from algebra that we can transform an algebraic system of equations into an equivalent system (that is, one having the same solution) by multiplying an equation by a nonzero constant, interchanging the positions of any two equations in a system, and adding a nonzero constant multiple of an equation to another equation. These operations on equations in a system are, in turn, equivalent to **elementary row operations** on an augmented matrix:

   (i) Multiply a row by a nonzero constant.

   (ii) Interchange any two rows.

   (iii) Add a nonzero constant multiple of one row to any other row.

Elimination Methods

To solve a system such as (5) using an augmented matrix we use either **Gaussian elimination** or the **Gauss-Jordan elimination method.*** In the former method we carry out a succession of elementary row operations until we

---

*Karl Friedrich Gauss (1777–1855)   Gauss was the first of a new breed of precise and demanding mathematicians—the "rigorists." As a child, Gauss was a prodigy in mathematics. As an adult, he often remarked that he could calculate or "reckon" before he could talk. However, as a college student, Gauss was torn between two loves: philology and mathematics. But he was inspired by some original mathematical achievements as a teenager and encouraged by the mathematician Wolfgang Bolyai, so the choice was not too difficult. At the age of twenty Gauss settled on a career in mathematics. At the age of twenty-two he completed a book on number theory, *Disquisitiones Arithmeticae*. Published in 1801, this text was recognized as a masterpiece, and even today remains a classic in its field. Gauss' doctoral dissertation of 1799 also remains a memorable document. Using the theory of functions of a complex variable, he was the first to prove the so-called *fundamental theorem of algebra*: Every polynomial equation has at least one root.

   Although Gauss was certainly recognized and respected as an outstanding mathematician during his lifetime, the full extent of his genius was not realized until the publication of his scientific diary in 1898, forty-four years after his death. Much to the chagrin of some nineteenth-century mathematicians, the diary revealed that Gauss had foreseen, sometimes by decades, many of their discoveries or, perhaps more accurately, rediscoveries. He was oblivious to fame; his mathematical researches were often pursued, like a child playing on a beach, simply for pleasure and self-satisfaction and not for the instruction that could be given to others through publication.

   On any list of "Greatest Mathematicians Who Ever Lived," Karl Friedrich Gauss must surely rank near or at the top. For his profound impact on so many branches of mathematics, Gauss is sometimes referred to as "the prince of mathematicians."

Wilhelm Jordan (1842–1899)   Jordan, a German engineer, used this method to solve linear systems in his 1888 text, *Handbook of Geodesy*.

arrive at an augmented matrix in **row-echelon form**:

(i) The first nonzero entry in a nonzero row is 1.

(ii) In consecutive nonzero rows, the first entry 1 in the lower row appears to the right of the first 1 in the higher row.

(iii) Rows consisting of all 0's are at the bottom of the matrix.

In the Gauss-Jordan method the row operations are continued until we obtain an augmented matrix that is in **reduced row-echelon form**. A reduced row-echelon matrix has the same three properties listed above in addition to

(iv) A column containing a first entry 1 has 0's everywhere else.

EXAMPLE 12

(a) The augmented matrices

$$\begin{pmatrix} 1 & 5 & 0 & | & 2 \\ 0 & 1 & 0 & | & -1 \\ 0 & 0 & 0 & | & 0 \end{pmatrix} \quad \text{and} \quad \begin{pmatrix} 0 & 0 & 1 & -6 & 2 & | & 2 \\ 0 & 0 & 0 & 0 & 1 & | & 4 \end{pmatrix}$$

are in row-echelon form. You should verify that the three criteria are satisfied.

(b) The augmented matrices

$$\begin{pmatrix} 1 & 0 & 0 & | & 7 \\ 0 & 1 & 0 & | & -1 \\ 0 & 0 & 0 & | & 0 \end{pmatrix} \quad \text{and} \quad \begin{pmatrix} 0 & 0 & 1 & -6 & 0 & | & -6 \\ 0 & 0 & 0 & 0 & 1 & | & 4 \end{pmatrix}$$

are in reduced row-echelon form. Note that the remaining entries in the columns containing a leading entry 1 are all 0's. ∎

It should be noted that in Gaussian elimination, we stop once we have obtained *an* augmented matrix in row-echelon form. In other words, by using different sequences of row operations we may arrive at different row-echelon forms. This method then requires the use of back-substitution. In Gauss-Jordan elimination we stop when we have obtained *the* augmented matrix in reduced row-echelon form. Any sequence of row operations will lead to the same augmented matrix in reduced row-echelon form. This method does not require back-substitution; the solution of the system will be apparent by inspection of the final matrix. In terms of the equations of the original system, our goal in both methods is simply to make the coefficient of x_1 in the first equation* equal to 1 and then use multiples of that equation to eliminate x_1 from other equations. The process is repeated on the other variables.

---

* We can always interchange equations so that the first equation contains the variable x_1.

To keep track of the row operations on an augmented matrix, we utilize the following notation:

Symbol	Meaning
R_{ij}	Interchange rows i and j
cR_i	Multiply the ith row by the nonzero constant c
$cR_i + R_j$	Multiply the ith row by c and add to the jth row

EXAMPLE 13

Solve

$$2x_1 + 6x_2 + x_3 = 7$$

$$x_1 + 2x_2 - x_3 = -1$$

$$5x_1 + 7x_2 - 4x_3 = 9$$

using **(a)** Gaussian elimination and **(b)** Gauss-Jordan elimination.

Solution **(a)** Using row operations on the augmented matrix of the system, we obtain

$$\begin{pmatrix} 2 & 6 & 1 & | & 7 \\ 1 & 2 & -1 & | & -1 \\ 5 & 7 & -4 & | & 9 \end{pmatrix} \xrightarrow{R_{12}} \begin{pmatrix} 1 & 2 & -1 & | & -1 \\ 2 & 6 & 1 & | & 7 \\ 5 & 7 & -4 & | & 9 \end{pmatrix}$$

$$\xrightarrow[-5R_1 + R_3]{-2R_1 + R_2} \begin{pmatrix} 1 & 2 & -1 & | & -1 \\ 0 & 2 & 3 & | & 9 \\ 0 & -3 & 1 & | & 14 \end{pmatrix} \xrightarrow{\frac{1}{2}R_2} \begin{pmatrix} 1 & 2 & -1 & | & -1 \\ 0 & 1 & \frac{3}{2} & | & \frac{9}{2} \\ 0 & -3 & 1 & | & 14 \end{pmatrix}$$

$$\xrightarrow{3R_2 + R_3} \begin{pmatrix} 1 & 2 & -1 & | & -1 \\ 0 & 1 & \frac{3}{2} & | & \frac{9}{2} \\ 0 & 0 & \frac{11}{2} & | & \frac{55}{2} \end{pmatrix} \xrightarrow{\frac{2}{11}R_3} \begin{pmatrix} 1 & 2 & -1 & | & -1 \\ 0 & 1 & \frac{3}{2} & | & \frac{9}{2} \\ 0 & 0 & 1 & | & 5 \end{pmatrix}.$$

The last matrix is in row-echelon form and represents the system

$$x_1 + 2x_2 - x_3 = -1$$

$$x_2 + \frac{3}{2}x_3 = \frac{9}{2}$$

$$x_3 = 5.$$

Substituting $x_3 = 5$ into the second equation then gives $x_2 = -3$. Substituting both these values back into the first equation finally yields $x_1 = 10$.

(b) We start with the last matrix above. Since the first entries in the second and third rows are 1's, we must, in turn, make the remaining entries in the second and third columns 0's:

$$\begin{pmatrix} 1 & 2 & -1 & | & -1 \\ 0 & 1 & \frac{3}{2} & | & \frac{9}{2} \\ 0 & 0 & 1 & | & 5 \end{pmatrix} \overset{-2R_2 + R_1}{\Longrightarrow} \begin{pmatrix} 1 & 0 & -4 & | & -10 \\ 0 & 1 & \frac{3}{2} & | & \frac{9}{2} \\ 0 & 0 & 1 & | & 5 \end{pmatrix} \overset{\substack{4R_3 + R_1 \\ -\frac{3}{2}R_3 + R_2}}{\Longrightarrow} \begin{pmatrix} 1 & 0 & 0 & | & 10 \\ 0 & 1 & 0 & | & -3 \\ 0 & 0 & 1 & | & 5 \end{pmatrix}.$$

The last matrix is now in reduced row-echelon form. Because of what the matrix means in terms of equations, it evident that the solution of the system is $x_1 = 10$, $x_2 = -3$, $x_3 = 5$. ∎

EXAMPLE 14

Use Gauss-Jordan elimination to solve

$$x + 3y - 2z = -7$$

$$4x + y + 3z = 5$$

$$2x - 5y + 7z = 19.$$

Solution

$$\begin{pmatrix} 1 & 3 & -2 & | & -7 \\ 4 & 1 & 3 & | & 5 \\ 2 & -5 & 7 & | & 19 \end{pmatrix} \overset{\substack{-4R_1 + R_2 \\ -2R_1 + R_3}}{\Longrightarrow} \begin{pmatrix} 1 & 3 & -2 & | & -7 \\ 0 & -11 & 11 & | & 33 \\ 0 & -11 & 11 & | & 33 \end{pmatrix}$$

$$\overset{\substack{-\frac{1}{11}R_2 \\ -\frac{1}{11}R_3}}{\Longrightarrow} \begin{pmatrix} 1 & 3 & -2 & | & -7 \\ 0 & 1 & -1 & | & -3 \\ 0 & 1 & -1 & | & -3 \end{pmatrix} \overset{\substack{-3R_2 + R_1 \\ -R_2 + R_3}}{\Longrightarrow} \begin{pmatrix} 1 & 0 & 1 & | & 2 \\ 0 & 1 & -1 & | & -3 \\ 0 & 0 & 0 & | & 0 \end{pmatrix}.$$

In this case the last matrix in reduced row-echelon form implies that the original system of three equations in three unknowns is really equivalent to two equations in three unknowns. Since only z is common to both equations (the nonzero rows), we can assign its values arbitrarily. If we let $z = t$, where t represents any real number, then we see that the system has infinitely many solutions: $x = 2 - t$, $y = -3 + t$, $z = t$. Geometrically, these equations are the parametric equations for the line of intersection of the planes $x + 0y + z = 2$ and $0x + y - z = -3$. ∎

8.4.3 The Eigenvalue Problem

Eigenvalues and Eigenvectors

Gauss-Jordan elimination can be used to find the **eigenvectors** for a square matrix.

DEFINITION 8.13 **Eigenvalues and Eigenvectors**

Let $\mathbf{A}$ be an $n \times n$ matrix. A number λ is said to be an **eigenvalue** of $\mathbf{A}$ if there exists a *nonzero* solution vector $\mathbf{K}$ of the linear system

$$\mathbf{AK} = \lambda\mathbf{K}. \tag{6}$$

The solution vector $\mathbf{K}$ is said to be an **eigenvector** corresponding to the eigenvalue λ.

The word *eigenvalue* is a combination of German and English terms adapted from the German word *eigenwert*, which, translated literally, is "proper value." Eigenvalues and eigenvectors are also called **characteristic values** and **characteristic vectors**, respectively.

EXAMPLE 15

Verify that $\mathbf{K} = \begin{pmatrix} 1 \\ -1 \\ 1 \end{pmatrix}$ is an eigenvector of the matrix $\mathbf{A} = \begin{pmatrix} 0 & -1 & -3 \\ 2 & 3 & 3 \\ -2 & 1 & 1 \end{pmatrix}$.

Solution By carrying out the multiplication $\mathbf{AK}$, we see that

$$\mathbf{AK} = \begin{pmatrix} 0 & -1 & -3 \\ 2 & 3 & 3 \\ -2 & 1 & 1 \end{pmatrix} \begin{pmatrix} 1 \\ -1 \\ 1 \end{pmatrix} = \begin{pmatrix} -2 \\ 2 \\ -2 \end{pmatrix} = (-2)\begin{pmatrix} 1 \\ -1 \\ 1 \end{pmatrix} = \overset{\text{eigenvalue}}{(-2)}\mathbf{K}.$$

We see from the preceding line and Definition 8.13 that $\lambda = -2$ is an eigenvalue of $\mathbf{A}$. ■

Using properties of matrix algebra, we can write (6) in the alternative form

$$(\mathbf{A} - \lambda\mathbf{I})\mathbf{K} = \mathbf{0}, \tag{7}$$

where $\mathbf{I}$ is the multiplicative identity. If we let

$$\mathbf{K} = \begin{pmatrix} k_1 \\ k_2 \\ \vdots \\ k_n \end{pmatrix},$$

then (7) is the same as

$$
\begin{aligned}
(a_{11} - \lambda)k_1 + \quad & a_{12}k_2 + \cdots + \quad a_{1n}k_n = 0 \\
a_{21}k_1 + (a_{22} - \lambda)k_2 + \cdots + \quad & a_{2n}k_n = 0 \\
& \vdots \\
a_{n1}k_1 + \quad & a_{n2}k_2 + \cdots + (a_{nn} - \lambda)k_n = 0.
\end{aligned}
\tag{8}
$$

Although an obvious solution of (8) is $k_1 = 0$, $k_2 = 0, \ldots, k_n = 0$, we are seeking only nontrivial solutions. Now it is known that a homogeneous system of n linear equations in n unknowns (that is, $b_i = 0$, $i = 1, 2, \ldots, n$, in (5)) has a nontrivial solution if and only if the determinant of the coefficient matrix is equal to zero. Thus to find a nonzero solution **K** for (7), we must have

$$
\det(\mathbf{A} - \lambda\mathbf{I}) = 0.
\tag{9}
$$

Inspection of (8) shows that the expansion of $\det(\mathbf{A} - \lambda\mathbf{I})$ by cofactors results in an nth-degree polynomial in λ. The equation (9) is called the **characteristic equation** of **A**. Thus *the eigenvalues of* **A** *are the roots of the characteristic equation.* To find an eigenvector corresponding to an eigenvalue λ we simply solve the system of equations $(\mathbf{A} - \lambda\mathbf{I})\mathbf{K} = \mathbf{0}$ by applying Gauss-Jordan elimination to the augmented matrix $(\mathbf{A} - \lambda\mathbf{I} \,|\, \mathbf{0})$.

EXAMPLE 16

Find the eigenvalues and eigenvectors of

$$
\mathbf{A} = \begin{pmatrix} 1 & 2 & 1 \\ 6 & -1 & 0 \\ -1 & -2 & -1 \end{pmatrix}.
$$

Solution   To expand the determinant in the characteristic equation

$$
\det(\mathbf{A} - \lambda\mathbf{I}) = \begin{vmatrix} 1 - \lambda & 2 & 1 \\ 6 & -1 - \lambda & 0 \\ -1 & -2 & -1 - \lambda \end{vmatrix} = 0
$$

we use the cofactors of the second row. It follows that

$$
-\lambda^3 - \lambda^2 + 12\lambda = 0 \quad \text{or} \quad \lambda(\lambda + 4)(\lambda - 3) = 0.
$$

Hence the eigenvalues are $\lambda_1 = 0$, $\lambda_2 = -4$, $\lambda_3 = 3$. To find the eigenvectors we must now reduce $(\mathbf{A} - \lambda\mathbf{I} \,|\, \mathbf{0})$ three times corresponding to the three distinct eigenvalues.

For $\lambda_1 = 0$ we have

$$(\mathbf{A} - 0\mathbf{I}\,|\,\mathbf{0}) = \begin{pmatrix} 1 & 2 & 1 & | & 0 \\ 6 & -1 & 0 & | & 0 \\ -1 & -2 & -1 & | & 0 \end{pmatrix} \xRightarrow[R_1 + R_3]{-6R_1 + R_2} \begin{pmatrix} 1 & 2 & 1 & | & 0 \\ 0 & -13 & -6 & | & 0 \\ 0 & 0 & 0 & | & 0 \end{pmatrix}$$

$$\xRightarrow{-\frac{1}{13}R_2} \begin{pmatrix} 1 & 2 & 1 & | & 0 \\ 0 & 1 & \frac{6}{13} & | & 0 \\ 0 & 0 & 0 & | & 0 \end{pmatrix} \xRightarrow{-2R_2 + R_1} \begin{pmatrix} 1 & 0 & \frac{1}{13} & | & 0 \\ 0 & 1 & \frac{6}{13} & | & 0 \\ 0 & 0 & 0 & | & 0 \end{pmatrix}.$$

Thus we see that $k_1 = (-1/13)k_3$ and $k_2 = (-6/13)k_3$. Choosing $k_3 = -13$, we get the eigenvector*

$$\mathbf{K}_1 = \begin{pmatrix} 1 \\ 6 \\ -13 \end{pmatrix}.$$

For $\lambda_2 = -4$,

$$(\mathbf{A} + 4\mathbf{I}\,|\,\mathbf{0}) = \begin{pmatrix} 5 & 2 & 1 & | & 0 \\ 6 & 3 & 0 & | & 0 \\ -1 & -2 & 3 & | & 0 \end{pmatrix} \xRightarrow[R_{31}]{-R_3} \begin{pmatrix} 1 & 2 & -3 & | & 0 \\ 6 & 3 & 0 & | & 0 \\ 5 & 2 & 1 & | & 0 \end{pmatrix}$$

$$\xRightarrow[-5R_1 + R_3]{-6R_1 + R_2} \begin{pmatrix} 1 & 2 & -3 & | & 0 \\ 0 & -9 & 18 & | & 0 \\ 0 & -8 & 16 & | & 0 \end{pmatrix} \xRightarrow[-\frac{1}{8}R_3]{-\frac{1}{9}R_2} \begin{pmatrix} 1 & 2 & -3 & | & 0 \\ 0 & 1 & -2 & | & 0 \\ 0 & 1 & -2 & | & 0 \end{pmatrix}$$

$$\xRightarrow[-R_2 + R_3]{-2R_2 + R_1} \begin{pmatrix} 1 & 0 & 1 & | & 0 \\ 0 & 1 & -2 & | & 0 \\ 0 & 0 & 0 & | & 0 \end{pmatrix}$$

implies $k_1 = -k_3$ and $k_2 = 2k_3$. Choosing $k_3 = 1$ then yields the second eigenvector

$$\mathbf{K}_2 = \begin{pmatrix} -1 \\ 2 \\ 1 \end{pmatrix}.$$

Finally, for $\lambda_3 = 3$ Gauss-Jordan elimination gives

$$(\mathbf{A} - 3\mathbf{I}\,|\,\mathbf{0}) = \begin{pmatrix} -2 & 2 & 1 & | & 0 \\ 6 & -4 & 0 & | & 0 \\ -1 & -2 & -4 & | & 0 \end{pmatrix} \xRightarrow{\text{row operations}} \begin{pmatrix} 1 & 0 & 1 & | & 0 \\ 0 & 1 & \frac{3}{2} & | & 0 \\ 0 & 0 & 0 & | & 0 \end{pmatrix}$$

---

*Of course k_3 could be chosen as any nonzero number. In other words, a nonzero constant multiple of an eigenvector is also an eigenvector.

and so $k_1 = -k_3$ and $k_2 = (-3/2)k_3$. The choice of $k_3 = -2$ leads to the third eigenvector:

$$\mathbf{K}_3 = \begin{pmatrix} 2 \\ 3 \\ -2 \end{pmatrix}.$$ ∎

When an $n \times n$ matrix $\mathbf{A}$ possesses n distinct eigenvalues $\lambda_1, \lambda_2, \ldots, \lambda_n$, it can be proved that a set of n linearly independent* eigenvectors $\mathbf{K}_1$, $\mathbf{K}_2, \ldots, \mathbf{K}_n$ can be found. However, when the characteristic equation has repeated roots, it may not be possible to find n linearly independent eigenvectors for $\mathbf{A}$.

E X A M P L E 17

Find the eigenvalues and eigenvectors of

$$\mathbf{A} = \begin{pmatrix} 3 & 4 \\ -1 & 7 \end{pmatrix}.$$

Solution From the characteristic equation

$$\det(\mathbf{A} - \lambda\mathbf{I}) = \begin{vmatrix} 3 - \lambda & 4 \\ -1 & 7 - \lambda \end{vmatrix} = (\lambda - 5)^2 = 0,$$

we see that $\lambda_1 = \lambda_2 = 5$ is an eigenvalue of multiplicity two. In the case of a 2×2 matrix there is no need to use Gauss-Jordan elimination. To find the eigenvector(s) corresponding to $\lambda_1 = 5$ we resort to the system $(\mathbf{A} - 5\mathbf{I} \,|\, \mathbf{0})$ in its equivalent form

$$-2k_1 + 4k_2 = 0$$

$$-k_1 + 2k_2 = 0.$$

It is apparent from this system that $k_1 = 2k_2$. Thus if we choose $k_2 = 1$, we find the single eigenvector

$$\mathbf{K}_1 = \begin{pmatrix} 2 \\ 1 \end{pmatrix}.$$ ∎

E X A M P L E 18

Find the eigenvalues and eigenvectors of

$$\mathbf{A} = \begin{pmatrix} 9 & 1 & 1 \\ 1 & 9 & 1 \\ 1 & 1 & 9 \end{pmatrix}.$$

---

*Linear independence of column vectors is defined in exactly the same manner as for functions.

Solution The characteristic equation

$$\det(\mathbf{A} - \lambda\mathbf{I}) = \begin{vmatrix} 9 - \lambda & 1 & 1 \\ 1 & 9 - \lambda & 1 \\ 1 & 1 & 9 - \lambda \end{vmatrix} = -(\lambda - 11)(\lambda - 8)^2 = 0$$

shows that $\lambda_1 = 11$ and that $\lambda_2 = \lambda_3 = 8$ is an eigenvalue of multiplicity two. For $\lambda_1 = 11$ Gauss-Jordan elimination gives

$$(\mathbf{A} - 11\mathbf{I}\,|\,\mathbf{0}) = \begin{pmatrix} -2 & 1 & 1 & | & 0 \\ 1 & -2 & 1 & | & 0 \\ 1 & 1 & -2 & | & 0 \end{pmatrix} \overset{\text{row}}{\underset{\text{operations}}{\Longrightarrow}} \begin{pmatrix} 1 & 0 & -1 & | & 0 \\ 0 & 1 & -1 & | & 0 \\ 0 & 0 & 0 & | & 0 \end{pmatrix}.$$

Hence $k_1 = k_3$ and $k_2 = k_3$. If $k_3 = 1$, then

$$\mathbf{K}_1 = \begin{pmatrix} 1 \\ 1 \\ 1 \end{pmatrix}.$$

Now for $\lambda_2 = 8$ we have

$$(\mathbf{A} - 8\mathbf{I}\,|\,\mathbf{0}) = \begin{pmatrix} 1 & 1 & 1 & | & 0 \\ 1 & 1 & 1 & | & 0 \\ 1 & 1 & 1 & | & 0 \end{pmatrix} \overset{\text{row}}{\underset{\text{operations}}{\Longrightarrow}} \begin{pmatrix} 1 & 1 & 1 & | & 0 \\ 0 & 0 & 0 & | & 0 \\ 0 & 0 & 0 & | & 0 \end{pmatrix}.$$

In the equation $k_1 + k_2 + k_3 = 0$, we are free to select two of the variables arbitrarily. Choosing, on the one hand, $k_2 = 1$, $k_3 = 0$ and, on the other, $k_2 = 0$, $k_3 = 1$, we obtain two linearly independent eigenvectors

$$\mathbf{K}_2 = \begin{pmatrix} -1 \\ 1 \\ 0 \end{pmatrix} \quad \text{and} \quad \mathbf{K}_3 = \begin{pmatrix} -1 \\ 0 \\ 1 \end{pmatrix}$$

corresponding to a single eigenvalue. ∎

8.4 EXERCISES

Answers to odd-numbered problems begin on page A–22.

[8.4.1] **1.** If $\mathbf{A} = \begin{pmatrix} 4 & 5 \\ -6 & 9 \end{pmatrix}$ and $\mathbf{B} = \begin{pmatrix} -2 & 6 \\ 8 & -10 \end{pmatrix}$,

find **(a)** $\mathbf{A} + \mathbf{B}$, **(b)** $\mathbf{B} - \mathbf{A}$, **(c)** $2\mathbf{A} + 3\mathbf{B}$.

2. If $\mathbf{A} = \begin{pmatrix} -2 & 0 \\ 4 & 1 \\ 7 & 3 \end{pmatrix}$ and $\mathbf{B} = \begin{pmatrix} 3 & -1 \\ 0 & 2 \\ -4 & -2 \end{pmatrix}$,

find **(a)** $\mathbf{A} - \mathbf{B}$, **(b)** $\mathbf{B} - \mathbf{A}$, **(c)** $2(\mathbf{A} + \mathbf{B})$.

3. If $\mathbf{A} = \begin{pmatrix} 2 & -3 \\ -5 & 4 \end{pmatrix}$ and $\mathbf{B} = \begin{pmatrix} -1 & 6 \\ 3 & 2 \end{pmatrix}$,

   find **(a) AB, (b) BA, (c)** $\mathbf{A}^2 = \mathbf{AA}$, **(d)** $\mathbf{B}^2 = \mathbf{BB}$.

4. If $\mathbf{A} = \begin{pmatrix} 1 & 4 \\ 5 & 10 \\ 8 & 12 \end{pmatrix}$ and $\mathbf{B} = \begin{pmatrix} -4 & 6 & -3 \\ 1 & -3 & 2 \end{pmatrix}$,

   find **(a) AB, (b) BA**.

5. If $\mathbf{A} = \begin{pmatrix} 1 & -2 \\ -2 & 4 \end{pmatrix}$, $\mathbf{B} = \begin{pmatrix} 6 & 3 \\ 2 & 1 \end{pmatrix}$, and $\mathbf{C} = \begin{pmatrix} 0 & 2 \\ 3 & 4 \end{pmatrix}$,

   find **(a) BC, (b) A(BC), (c) C(BA), (d) A(B + C)**.

6. If $\mathbf{A} = (5 \quad -6 \quad 7)$, $\mathbf{B} = \begin{pmatrix} 3 \\ 4 \\ -1 \end{pmatrix}$, and $\mathbf{C} = \begin{pmatrix} 1 & 2 & 4 \\ 0 & 1 & -1 \\ 3 & 2 & 1 \end{pmatrix}$,

   find **(a) AB, (b) BA, (c) (BA)C, (d) (AB)C**.

7. If $\mathbf{A} = \begin{pmatrix} 4 \\ 8 \\ -10 \end{pmatrix}$ and $\mathbf{B} = (2 \quad 4 \quad 5)$,

   find **(a)** $\mathbf{A}^T\mathbf{A}$, **(b)** $\mathbf{B}^T\mathbf{B}$, **(c)** $\mathbf{A} + \mathbf{B}^T$.

8. If $\mathbf{A} = \begin{pmatrix} 1 & 2 \\ 2 & 4 \end{pmatrix}$ and $\mathbf{B} = \begin{pmatrix} -2 & 3 \\ 5 & 7 \end{pmatrix}$,

   find **(a)** $\mathbf{A} + \mathbf{B}^T$, **(b)** $2\mathbf{A}^T - \mathbf{B}^T$, **(c)** $\mathbf{A}^T(\mathbf{A} - \mathbf{B})$.

9. If $\mathbf{A} = \begin{pmatrix} 3 & 4 \\ 8 & 1 \end{pmatrix}$ and $\mathbf{B} = \begin{pmatrix} 5 & 10 \\ -2 & -5 \end{pmatrix}$,

   find **(a)** $(\mathbf{AB})^T$, **(b)** $\mathbf{B}^T\mathbf{A}^T$.

10. If $\mathbf{A} = \begin{pmatrix} 5 & 9 \\ -4 & 6 \end{pmatrix}$ and $\mathbf{B} = \begin{pmatrix} -3 & 11 \\ -7 & 2 \end{pmatrix}$,

    find **(a)** $\mathbf{A}^T + \mathbf{B}^T$, **(b)** $(\mathbf{A} + \mathbf{B})^T$.

In Problems 11–14 write the given sum as a single column matrix.

11. $4\begin{pmatrix} -1 \\ 2 \end{pmatrix} - 2\begin{pmatrix} 2 \\ 8 \end{pmatrix} + 3\begin{pmatrix} -2 \\ 3 \end{pmatrix}$

12. $3t\begin{pmatrix} 2 \\ t \\ -1 \end{pmatrix} + (t-1)\begin{pmatrix} -1 \\ -t \\ 3 \end{pmatrix} - 2\begin{pmatrix} 3t \\ 4 \\ -5t \end{pmatrix}$

13. $\begin{pmatrix} 2 & -3 \\ 1 & 4 \end{pmatrix}\begin{pmatrix} -2 \\ 5 \end{pmatrix} - \begin{pmatrix} -1 & 6 \\ -2 & 3 \end{pmatrix}\begin{pmatrix} -7 \\ 2 \end{pmatrix}$

14. $\begin{pmatrix} 1 & -3 & 4 \\ 2 & 5 & -1 \\ 0 & -4 & -2 \end{pmatrix}\begin{pmatrix} t \\ 2t-1 \\ -t \end{pmatrix} + \begin{pmatrix} -t \\ 1 \\ 4 \end{pmatrix} - \begin{pmatrix} 2 \\ 8 \\ -6 \end{pmatrix}$

In Problems 15–22 determine whether the given matrix is singular or nonsingular. If nonsingular, find $\mathbf{A}^{-1}$.

15. $\mathbf{A} = \begin{pmatrix} -3 & 6 \\ -2 & 4 \end{pmatrix}$

16. $\mathbf{A} = \begin{pmatrix} 2 & 5 \\ 1 & 4 \end{pmatrix}$

17. $\mathbf{A} = \begin{pmatrix} 4 & 8 \\ -3 & -5 \end{pmatrix}$

18. $\mathbf{A} = \begin{pmatrix} 7 & 10 \\ 2 & 2 \end{pmatrix}$

19. $\mathbf{A} = \begin{pmatrix} 2 & 1 & 0 \\ -1 & 2 & 1 \\ 1 & 2 & 1 \end{pmatrix}$

20. $\mathbf{A} = \begin{pmatrix} 3 & 2 & 1 \\ 4 & 1 & 0 \\ -2 & 5 & -1 \end{pmatrix}$

21. $\mathbf{A} = \begin{pmatrix} 2 & 1 & 1 \\ 1 & -2 & -3 \\ 3 & 2 & 4 \end{pmatrix}$

22. $\mathbf{A} = \begin{pmatrix} 4 & 1 & -1 \\ 6 & 2 & -3 \\ -2 & -1 & 2 \end{pmatrix}$

In Problems 23 and 24 show that the given matrix is nonsingular for every real value of t. Find $\mathbf{A}^{-1}(t)$.

23. $\mathbf{A}(t) = \begin{pmatrix} 2e^{-t} & e^{4t} \\ 4e^{-t} & 3e^{4t} \end{pmatrix}$

24. $\mathbf{A}(t) = \begin{pmatrix} 2e^{t} \sin t & -2e^{t} \cos t \\ e^{t} \cos t & e^{t} \sin t \end{pmatrix}$

In Problems 25–28 find $d\mathbf{X}/dt$.

25. $\mathbf{X} = \begin{pmatrix} 5e^{-t} \\ 2e^{-t} \\ -7e^{-t} \end{pmatrix}$

26. $\mathbf{X} = \begin{pmatrix} \frac{1}{2} \sin 2t - 4 \cos 2t \\ -3 \sin 2t + 5 \cos 2t \end{pmatrix}$

27. $\mathbf{X} = 2\begin{pmatrix} 1 \\ -1 \end{pmatrix} e^{2t} + 4\begin{pmatrix} 2 \\ 1 \end{pmatrix} e^{-3t}$

28. $\mathbf{X} = \begin{pmatrix} 5te^{2t} \\ t \sin 3t \end{pmatrix}$

29. Let $\mathbf{A}(t) = \begin{pmatrix} e^{4t} & \cos \pi t \\ 2t & 3t^{2} - 1 \end{pmatrix}$.

Find (a) $\dfrac{d\mathbf{A}}{dt}$, (b) $\displaystyle\int_{0}^{2} \mathbf{A}(t)\, dt$, (c) $\displaystyle\int_{0}^{t} \mathbf{A}(s)\, ds$.

30. Let $\mathbf{A}(t) = \begin{pmatrix} \dfrac{1}{t^{2}+1} & 3t \\ t^{2} & t \end{pmatrix}$ and $\mathbf{B}(t) = \begin{pmatrix} 6t & 2 \\ 1/t & 4t \end{pmatrix}$.

Find (a) $\dfrac{d\mathbf{A}}{dt}$, (b) $\dfrac{d\mathbf{B}}{dt}$, (c) $\displaystyle\int_{0}^{1} \mathbf{A}(t)\, dt$, (d) $\displaystyle\int_{1}^{2} \mathbf{B}(t)\, dt$, (e) $\mathbf{A}(t)\mathbf{B}(t)$,

(f) $\dfrac{d}{dt}\mathbf{A}(t)\mathbf{B}(t)$, (g) $\displaystyle\int_{1}^{t} \mathbf{A}(s)\mathbf{B}(s)\, ds$.

[8.4.2] In Problems 31–38 solve the given system of equations by either Gaussian elimination or Gauss-Jordan elimination.

31. $\begin{aligned} x + y - 2z &= 14 \\ 2x - y + z &= 0 \\ 6x + 3y + 4z &= 1 \end{aligned}$

32. $\begin{aligned} 5x - 2y + 4z &= 10 \\ x + y + z &= 9 \\ 4x - 3y + 3z &= 1 \end{aligned}$

33.
$$y + z = -5$$
$$5x + 4y - 16z = -10$$
$$x - y - 5z = 7$$

34. $3x + y + z = 4$
$$4x + 2y - z = 7$$
$$x + y - 3z = 6$$

35. $2x + y + z = 4$
$$10x - 2y + 2z = -1$$
$$6x - 2y + 4z = 8$$

36. $x + 2z = 8$
$$x + 2y - 2z = 4$$
$$2x + 5y - 6z = 6$$

37. $x_1 + x_2 - x_3 - x_4 = -1$
$$x_1 + x_2 + x_3 + x_4 = 3$$
$$x_1 - x_2 + x_3 - x_4 = 3$$
$$4x_1 + x_2 - 2x_3 + x_4 = 0$$

38. $2x_1 + x_2 + x_3 = 0$
$$x_1 + 3x_2 + x_3 = 0$$
$$7x_1 + x_2 + 3x_3 = 0$$

In Problems 39 and 40 use Gauss-Jordan elimination to demonstrate that the given system of equations has no solution.

39. $x + 2y + 4z = 2$
$$2x + 4y + 3z = 1$$
$$x + 2y - z = 7$$

40. $x_1 + x_2 - x_3 + 3x_4 = 1$
$$x_2 - x_3 - 4x_4 = 0$$
$$x_1 + 2x_2 - 2x_3 - x_4 = 6$$
$$4x_1 + 7x_2 - 7x_3 = 9$$

[8.4.3]   In Problems 41–48 find the eigenvalues and eigenvectors of the given matrix.

41. $\begin{pmatrix} -1 & 2 \\ -7 & 8 \end{pmatrix}$

42. $\begin{pmatrix} 2 & 1 \\ 2 & 1 \end{pmatrix}$

43. $\begin{pmatrix} -8 & -1 \\ 16 & 0 \end{pmatrix}$

44. $\begin{pmatrix} 1 & 1 \\ \frac{1}{4} & 1 \end{pmatrix}$

45. $\begin{pmatrix} 5 & -1 & 0 \\ 0 & -5 & 9 \\ 5 & -1 & 0 \end{pmatrix}$

46. $\begin{pmatrix} 3 & 0 & 0 \\ 0 & 2 & 0 \\ 4 & 0 & 1 \end{pmatrix}$

47. $\begin{pmatrix} 0 & 4 & 0 \\ -1 & -4 & 0 \\ 0 & 0 & -2 \end{pmatrix}$

48. $\begin{pmatrix} 1 & 6 & 0 \\ 0 & 2 & 1 \\ 0 & 1 & 2 \end{pmatrix}$

In Problems 49 and 50 show that the given matrix has complex eigenvalues. Find the eigenvectors of the matrix.

49. $\begin{pmatrix} -1 & 2 \\ -5 & 1 \end{pmatrix}$

50. $\begin{pmatrix} 2 & -1 & 0 \\ 5 & 2 & 4 \\ 0 & 1 & 2 \end{pmatrix}$

51. If $\mathbf{A}(t)$ is a 2×2 matrix of differentiable functions and $\mathbf{X}(t)$ is a 2×1 column matrix of differentiable functions, prove the product rule

$$\frac{d}{dt}[\mathbf{A}(t)\mathbf{X}(t)] = \mathbf{A}(t)\mathbf{X}'(t) + \mathbf{A}'(t)\mathbf{X}(t).$$

52. Derive formula (3). [*Hint:* Find a matrix $\mathbf{B} = \begin{pmatrix} b_{11} & b_{12} \\ b_{21} & b_{22} \end{pmatrix}$ for which $\mathbf{AB} = \mathbf{I}$. Solve for $b_{11}, b_{12}, b_{21},$ and b_{22}. Then show that $\mathbf{BA} = \mathbf{I}$.]

53. If $\mathbf{A}$ is nonsingular and $\mathbf{AB} = \mathbf{AC}$, show that $\mathbf{B} = \mathbf{C}$.

54. If $\mathbf{A}$ and $\mathbf{B}$ are nonsingular, show that $(\mathbf{AB})^{-1} = \mathbf{B}^{-1}\mathbf{A}^{-1}$.

55. Let $\mathbf{A}$ and $\mathbf{B}$ be $n \times n$ matrices. In general, is $(\mathbf{A} + \mathbf{B})^2 = \mathbf{A}^2 + 2\mathbf{AB} + \mathbf{B}^2$?

8.5 MATRICES AND SYSTEMS OF LINEAR FIRST-ORDER EQUATIONS

8.5.1 Preliminary Theory

Matrix Form of a System

If $\mathbf{X}$, $\mathbf{A}(t)$, and $\mathbf{F}(t)$ denote the respective matrices

$$\mathbf{X} = \begin{pmatrix} x_1(t) \\ x_2(t) \\ \vdots \\ x_n(t) \end{pmatrix}, \qquad \mathbf{A}(t) = \begin{pmatrix} a_{11}(t) & a_{12}(t) & \cdots & a_{1n}(t) \\ a_{21}(t) & a_{22}(t) & \cdots & a_{2n}(t) \\ \vdots & & & \vdots \\ a_{n1}(t) & a_{n2}(t) & \cdots & a_{nn}(t) \end{pmatrix}, \qquad \mathbf{F}(t) = \begin{pmatrix} f_1(t) \\ f_2(t) \\ \vdots \\ f_n(t) \end{pmatrix},$$

then the system of linear first-order differential equations

$$\frac{dx_1}{dt} = a_{11}(t)x_1 + a_{12}(t)x_2 + \cdots + a_{1n}(t)x_n + f_1(t)$$

$$\frac{dx_2}{dt} = a_{21}(t)x_1 + a_{22}(t)x_2 + \cdots + a_{2n}(t)x_n + f_2(t) \tag{1}$$

$$\vdots \qquad\qquad \vdots$$

$$\frac{dx_n}{dt} = a_{n1}(t)x_1 + a_{n2}(t)x_2 + \cdots + a_{nn}(t)x_n + f_n(t)$$

can be written as

$$\frac{d}{dt}\begin{pmatrix} x_1 \\ x_2 \\ \vdots \\ x_n \end{pmatrix} = \begin{pmatrix} a_{11}(t) & a_{12}(t) & \cdots & a_{1n}(t) \\ a_{21}(t) & a_{22}(t) & \cdots & a_{2n}(t) \\ \vdots & & & \vdots \\ a_{n1}(t) & a_{n2}(t) & \cdots & a_{nn}(t) \end{pmatrix}\begin{pmatrix} x_1 \\ x_2 \\ \vdots \\ x_n \end{pmatrix} + \begin{pmatrix} f_1(t) \\ f_2(t) \\ \vdots \\ f_n(t) \end{pmatrix}$$

or simply

$$\frac{d\mathbf{X}}{dt} = \mathbf{A}(t)\mathbf{X} + \mathbf{F}(t). \tag{2}$$

If the system is homogeneous, (2) becomes

$$\frac{d\mathbf{X}}{dt} = \mathbf{A}(t)\mathbf{X}. \tag{3}$$

Equations (2) and (3) are also written as $\mathbf{X'} = \mathbf{AX} + \mathbf{F}$ and $\mathbf{X'} = \mathbf{AX}$, respectively.

EXAMPLE I

In matrix terms the nonhomogeneous system

$$\frac{dx}{dt} = -2x + 5y + e^t - 2t$$

$$\frac{dy}{dt} = 4x - 3y + 10t$$

can be written as

$$\frac{d\mathbf{X}}{dt} = \begin{pmatrix} -2 & 5 \\ 4 & -3 \end{pmatrix}\mathbf{X} + \begin{pmatrix} e^t - 2t \\ 10t \end{pmatrix}$$

or $$\mathbf{X'} = \begin{pmatrix} -2 & 5 \\ 4 & -3 \end{pmatrix}\mathbf{X} + \begin{pmatrix} 1 \\ 0 \end{pmatrix}e^t + \begin{pmatrix} -2 \\ 10 \end{pmatrix}t,$$

where $\mathbf{X} = \begin{pmatrix} x \\ y \end{pmatrix}$. ∎

EXAMPLE 2

The matrix form of the homogeneous system

$$\frac{dx}{dt} = 2x - 3y$$
$$\frac{dy}{dt} = 6x + 5y$$

is $$\frac{d\mathbf{X}}{dt} = \begin{pmatrix} 2 & -3 \\ 6 & 5 \end{pmatrix}\mathbf{X},$$

where $\mathbf{X} = \begin{pmatrix} x \\ y \end{pmatrix}$. ∎

DEFINITION 8.14 **Solution Vector**

A **solution vector** on an interval I is any column matrix

$$\mathbf{X} = \begin{pmatrix} x_1(t) \\ x_2(t) \\ \vdots \\ x_n(t) \end{pmatrix}$$

whose entries are differentiable functions satisfying the system (2) on the interval.

EXAMPLE 3

Verify that

$$\mathbf{X}_1 = \begin{pmatrix} 1 \\ -1 \end{pmatrix} e^{-2t} = \begin{pmatrix} e^{-2t} \\ -e^{-2t} \end{pmatrix} \quad \text{and} \quad \mathbf{X}_2 = \begin{pmatrix} 3 \\ 5 \end{pmatrix} e^{6t} = \begin{pmatrix} 3e^{6t} \\ 5e^{6t} \end{pmatrix}$$

are solutions of

$$\mathbf{X}' = \begin{pmatrix} 1 & 3 \\ 5 & 3 \end{pmatrix} \mathbf{X} \qquad\qquad (4)$$

on the interval $(-\infty, \infty)$.

Solution   We have

$$\mathbf{X}_1' = \begin{pmatrix} -2e^{-2t} \\ 2e^{-2t} \end{pmatrix}$$

and

$$\mathbf{A}\mathbf{X}_1 = \begin{pmatrix} 1 & 3 \\ 5 & 3 \end{pmatrix} \begin{pmatrix} e^{-2t} \\ -e^{-2t} \end{pmatrix} = \begin{pmatrix} e^{-2t} - 3e^{-2t} \\ 5e^{-2t} - 3e^{-2t} \end{pmatrix} = \begin{pmatrix} -2e^{-2t} \\ 2e^{-2t} \end{pmatrix} = \mathbf{X}_1'.$$

Now

$$\mathbf{X}_2' = \begin{pmatrix} 18e^{6t} \\ 30e^{5t} \end{pmatrix}$$

and

$$\mathbf{A}\mathbf{X}_2 = \begin{pmatrix} 1 & 3 \\ 5 & 3 \end{pmatrix} \begin{pmatrix} 3e^{6t} \\ 5e^{6t} \end{pmatrix} = \begin{pmatrix} 3e^{6t} + 15e^{6t} \\ 15e^{6t} + 15e^{6t} \end{pmatrix} = \begin{pmatrix} 18e^{6t} \\ 30e^{6t} \end{pmatrix} = \mathbf{X}_2'. \qquad \blacksquare$$

Much of the theory of systems of n linear first-order differential equations is similar to that of linear nth-order differential equations.

Initial-Value Problem

Let t_0 denote a point on an interval I and

$$\mathbf{X}(t_0) = \begin{pmatrix} x_1(t_0) \\ x_2(t_0) \\ \vdots \\ x_n(t_0) \end{pmatrix} \quad \text{and} \quad \mathbf{X}_0 = \begin{pmatrix} \gamma_1 \\ \gamma_2 \\ \vdots \\ \gamma_n \end{pmatrix},$$

where the γ_i, $i = 1, 2, \ldots, n$, are given constants. Then the problem

$$\text{Solve:} \quad \frac{d\mathbf{X}}{dt} = \mathbf{A}(t)\mathbf{X} + \mathbf{F}(t)$$

$$\text{Subject to:} \quad \mathbf{X}(t_0) = \mathbf{X}_0 \qquad\qquad (5)$$

is an **initial-value problem** on the interval.

> **THEOREM 8.3** **Existence of a Unique Solution**
>
> Let the entries of the matrices $\mathbf{A}(t)$ and $\mathbf{F}(t)$ be functions continuous on a common interval I that contains the point t_0. Then there exists a unique solution of the initial-value problem (5) on the interval.

Homogeneous Systems

In the next several definitions and theorems, we are concerned only with homogeneous systems. Without stating it, we shall always assume that the a_{ij} and the f_i are continuous functions of t on some common interval I.

Superposition Principle

The following result is a **superposition principle** for solutions of linear systems.

> **THEOREM 8.4** **Superposition Principle**
>
> Let $\mathbf{X}_1, \mathbf{X}_2, \ldots, \mathbf{X}_k$ be a set of solution vectors of the homogeneous system (3) on an interval I. Then the linear combination
>
> $$\mathbf{X} = c_1\mathbf{X}_1 + c_2\mathbf{X}_2 + \cdots + c_k\mathbf{X}_k,$$
>
> where the c_i, $i = 1, 2, \ldots, k$, are arbitrary constants, is also a solution on the interval.

It follows from Theorem 8.4 that a constant multiple of any solution vector of a homogeneous system of linear first-order differential equations is also a solution.

EXAMPLE 4

One solution of the system

$$\mathbf{X}' = \begin{pmatrix} 1 & 0 & 1 \\ 1 & 1 & 0 \\ -2 & 0 & -1 \end{pmatrix} \mathbf{X} \tag{6}$$

is

$$\mathbf{X}_1 = \begin{pmatrix} \cos t \\ -\tfrac{1}{2}\cos t + \tfrac{1}{2}\sin t \\ -\cos t - \sin t \end{pmatrix}.$$

For any constant c_1 the vector $\mathbf{X} = c_1\mathbf{X}_1$ is also a solution since

$$\frac{d\mathbf{X}}{dt} = \begin{pmatrix} -c_1 \sin t \\ \frac{1}{2}c_1 \sin t + \frac{1}{2}c_1 \cos t \\ c_1 \sin t - c_1 \cos t \end{pmatrix}$$

and $\quad \mathbf{AX} = \begin{pmatrix} 1 & 0 & 1 \\ 1 & 1 & 0 \\ -2 & 0 & -1 \end{pmatrix} \begin{pmatrix} c_1 \cos t \\ -\frac{1}{2}c_1 \cos t + \frac{1}{2}c_1 \sin t \\ -c_1 \cos t - c_1 \sin t \end{pmatrix}$

$$= \begin{pmatrix} -c_1 \sin t \\ \frac{1}{2}c_1 \cos t + \frac{1}{2}c_1 \sin t \\ -c_1 \cos t + c_1 \sin t \end{pmatrix}.$$

Inspection of the resulting matrices shows that $\mathbf{X}' = \mathbf{AX}$. ∎

EXAMPLE 5

Consider the system (6) of Example 4. If

$$\mathbf{X}_2 = \begin{pmatrix} 0 \\ e^t \\ 0 \end{pmatrix}, \quad \text{then} \quad \mathbf{X}_2' = \begin{pmatrix} 0 \\ e^t \\ 0 \end{pmatrix}$$

and $\quad \mathbf{AX}_2 = \begin{pmatrix} 1 & 0 & 1 \\ 1 & 1 & 0 \\ -2 & 0 & -1 \end{pmatrix} \begin{pmatrix} 0 \\ e^t \\ 0 \end{pmatrix}$

$$= \begin{pmatrix} 0 \\ e^t \\ 0 \end{pmatrix} = \mathbf{X}_2'.$$

Thus we see that $\mathbf{X}_2$ is also a solution vector of (6). By the superposition principle the linear combination

$$\mathbf{X} = c_1\mathbf{X}_1 + c_2\mathbf{X}_2 = c_1 \begin{pmatrix} \cos t \\ -\frac{1}{2}\cos t + \frac{1}{2}\sin t \\ -\cos t - \sin t \end{pmatrix} + c_2 \begin{pmatrix} 0 \\ e^t \\ 0 \end{pmatrix}$$

is yet another solution of the system. ∎

Linear Independence

We are primarily interested in linearly independent solutions of the homogeneous system (3).

DEFINITION 8.15   Linear Dependence/Independence

Let $\mathbf{X}_1, \mathbf{X}_2, \ldots, \mathbf{X}_k$ be a set of solution vectors of the homogeneous system (3) on an interval I. We say that the set is **linearly dependent** on the interval if there exist constants $c_1, c_2, \ldots, c_k$, not all zero, such that

$$c_1 \mathbf{X}_1 + c_2 \mathbf{X}_2 + \cdots + c_k \mathbf{X}_k = \mathbf{0}$$

for every t in the interval. If the set of vectors is not linearly dependent on the interval, it is said to be **linearly independent**.

The case when $k = 2$ should be clear; two solution vectors $\mathbf{X}_1$ and $\mathbf{X}_2$ are linearly dependent if one is a constant multiple of the other, and conversely. For $k > 2$ a set of solution vectors is linearly dependent if we can express at least one solution vector as a linear combination of the remaining vectors.

EXAMPLE 6

It can be verified that

$$\mathbf{X}_1 = \binom{3}{1} e^t \quad \text{and} \quad \mathbf{X}_2 = \binom{1}{1} e^{-t}$$

are solution vectors of the system

$$\mathbf{X}' = \begin{pmatrix} 2 & -3 \\ 1 & -2 \end{pmatrix} \mathbf{X}. \tag{7}$$

Now $\mathbf{X}_1$ and $\mathbf{X}_2$ are linearly independent on the interval $(-\infty, \infty)$ since

$$c_1 \mathbf{X}_1 + c_2 \mathbf{X}_2 = \mathbf{0} \quad \text{or} \quad c_1 \binom{3}{1} e^t + c_2 \binom{1}{1} e^{-t} = \binom{0}{0}$$

is equivalent to
$$3 c_1 e^t + c_2 e^{-t} = 0$$

$$c_1 e^t + c_2 e^{-t} = 0.$$

Solving this system for c_1 and c_2 immediately yields $c_1 = 0$ and $c_2 = 0$. ∎

EXAMPLE 7

The vector $\mathbf{X}_3 = \begin{pmatrix} e^t + \cosh t \\ \cosh t \end{pmatrix}$ is also a solution of the system (7) given in Example 6. However, $\mathbf{X}_1$, $\mathbf{X}_2$, and $\mathbf{X}_3$ are linearly dependent since

$$\mathbf{X}_3 = \frac{1}{2} \mathbf{X}_1 + \frac{1}{2} \mathbf{X}_2.$$ ∎

Wronskian

As in our earlier consideration of the theory of a single ordinary differential equation, we can introduce the concept of the **Wronskian** determinant as a test for linear independence. We state the following theorem without proof.

THEOREM 8.5   **Criterion for Linearly Independent Solutions**

Let

$$\mathbf{X}_1 = \begin{pmatrix} x_{11} \\ x_{21} \\ \vdots \\ x_{n1} \end{pmatrix}, \quad \mathbf{X}_2 = \begin{pmatrix} x_{12} \\ x_{22} \\ \vdots \\ x_{n2} \end{pmatrix}, \ldots, \quad \mathbf{X}_n = \begin{pmatrix} x_{1n} \\ x_{2n} \\ \vdots \\ x_{nn} \end{pmatrix}$$

be n solution vectors of the homogeneous system (3) on an interval I. A necessary and sufficient condition that the set of solutions be linearly independent is that the Wronskian

$$W(\mathbf{X}_1, \mathbf{X}_2, \ldots, \mathbf{X}_n) = \begin{vmatrix} x_{11} & x_{12} & \cdots & x_{1n} \\ x_{21} & x_{22} & \cdots & x_{2n} \\ \vdots & \vdots & & \vdots \\ x_{n1} & x_{n2} & \cdots & x_{nn} \end{vmatrix} \neq 0 \qquad (8)$$

for every t in I.

In fact it can be shown that if $\mathbf{X}_1, \mathbf{X}_2, \ldots, \mathbf{X}_n$ are solution vectors of (3), then either

$$W(\mathbf{X}_1, \mathbf{X}_2, \ldots, \mathbf{X}_n) \neq 0$$

for every t in I or

$$W(\mathbf{X}_1, \mathbf{X}_2, \ldots, \mathbf{X}_n) = 0$$

for every t in the interval. Thus if we can show that $W \neq 0$ for some t_0 in I, then $W \neq 0$ for every t and hence the solutions are linearly independent on the interval.

Notice that, unlike our previous definition of the Wronskian, the determinant (8) does not involve differentiation.

EXAMPLE 8

In Example 3 we saw that

$$\mathbf{X}_1 = \begin{pmatrix} 1 \\ -1 \end{pmatrix} e^{-2t} \quad \text{and} \quad \mathbf{X}_2 = \begin{pmatrix} 3 \\ 5 \end{pmatrix} e^{6t}$$

are solutions of the system (4). Clearly, $\mathbf{X}_1$ and $\mathbf{X}_2$ are linearly independent on $(-\infty, \infty)$ since neither vector is a constant multiple of the other. In addition, we have

$$W(\mathbf{X}_1, \mathbf{X}_2) = \begin{vmatrix} e^{-2t} & 3e^{6t} \\ -e^{-2t} & 5e^{6t} \end{vmatrix} = 8e^{4t} \neq 0$$

for all real values of t.   ∎

Fundamental Set of Solutions

DEFINITION 8.16 **Fundamental Set of Solutions**

Any set $\mathbf{X}_1, \mathbf{X}_2, \ldots, \mathbf{X}_n$ of n linearly independent solution vectors of the homogeneous system (3) on an interval I is said to be a **fundamental set of solutions** on the interval.

THEOREM 8.6 **Existence of a Fundamental Set**

There exists a fundamental set of solutions for the homogeneous system (3) on an interval I.

DEFINITION 8.17 **General Solution—Homogeneous Systems**

Let $\mathbf{X}_1, \mathbf{X}_2, \ldots, \mathbf{X}_n$ be a fundamental set of solutions of the homogeneous system (3) on an interval I. The **general solution** of the system on the interval is defined to be

$$\mathbf{X} = c_1\mathbf{X}_1 + c_2\mathbf{X}_2 + \cdots + c_n\mathbf{X}_n,$$

where the c_i, $i = 1, 2, \ldots, n$, are arbitrary constants.

Although we shall not give the proof, it can be shown that, for appropriate choices of the constants $c_1, c_2, \ldots, c_n$, *any* solution of (3) on the interval I can be obtained from the general solution.

EXAMPLE 9

From Example 8 we know that

$$\mathbf{X}_1 = \begin{pmatrix} 1 \\ -1 \end{pmatrix} e^{-2t} \quad \text{and} \quad \mathbf{X}_2 = \begin{pmatrix} 3 \\ 5 \end{pmatrix} e^{6t}$$

are linearly independent solutions of (4) on $(-\infty, \infty)$. Hence $\mathbf{X}_1$ and $\mathbf{X}_2$ form a fundamental set of solutions on the interval. The general solution of the system on the interval is then

$$\mathbf{X} = c_1\mathbf{X}_1 + c_2\mathbf{X}_2 = c_1\begin{pmatrix} 1 \\ -1 \end{pmatrix}e^{-2t} + c_2\begin{pmatrix} 3 \\ 5 \end{pmatrix}e^{6t}. \qquad (9)$$

∎

EXAMPLE 10

The vectors

$$\mathbf{X}_1 = \begin{pmatrix} \cos t \\ -\tfrac{1}{2}\cos t + \tfrac{1}{2}\sin t \\ -\cos t - \sin t \end{pmatrix}, \qquad \mathbf{X}_2 = \begin{pmatrix} 0 \\ 1 \\ 0 \end{pmatrix}e^t, \qquad \mathbf{X}_3 = \begin{pmatrix} \sin t \\ -\tfrac{1}{2}\sin t - \tfrac{1}{2}\cos t \\ -\sin t + \cos t \end{pmatrix}$$

are solutions of the system (6)* in Example 4. Now

$$W(\mathbf{X}_1, \mathbf{X}_2, \mathbf{X}_3) = \begin{vmatrix} \cos t & 0 & \sin t \\ -\tfrac{1}{2}\cos t + \tfrac{1}{2}\sin t & e^t & -\tfrac{1}{2}\sin t - \tfrac{1}{2}\cos t \\ -\cos t - \sin t & 0 & -\sin t + \cos t \end{vmatrix}$$

$$= e^t\begin{vmatrix} \cos t & \sin t \\ -\cos t - \sin t & -\sin t + \cos t \end{vmatrix} = e^t \neq 0$$

for all real values of t. We conclude that $\mathbf{X}_1$, $\mathbf{X}_2$, and $\mathbf{X}_3$ form a fundamental set of solutions on $(-\infty, \infty)$. Thus the general solution of the system on the interval is

$$\mathbf{X} = c_1\mathbf{X}_1 + c_2\mathbf{X}_2 + c_3\mathbf{X}_3$$

$$= c_1\begin{pmatrix} \cos t \\ -\tfrac{1}{2}\cos t + \tfrac{1}{2}\sin t \\ -\cos t - \sin t \end{pmatrix} + c_2\begin{pmatrix} 0 \\ 1 \\ 0 \end{pmatrix}e^t + c_3\begin{pmatrix} \sin t \\ -\tfrac{1}{2}\sin t - \tfrac{1}{2}\cos t \\ -\sin t + \cos t \end{pmatrix}. \qquad ∎$$

Nonhomogeneous Systems

For nonhomogeneous systems a **particular solution** $\mathbf{X}_p$ on an interval I is any vector, free of arbitrary parameters, whose entries are functions that satisfy the system (2).

EXAMPLE 11

Verify that the vector $\qquad \mathbf{X}_p = \begin{pmatrix} 3t - 4 \\ -5t + 6 \end{pmatrix}$

---

* On pages 439 and 440 it was verified that $\mathbf{X}_1$ and $\mathbf{X}_2$ are solutions; it is left as an exercise to demonstrate that $\mathbf{X}_3$ is also a solution. See Problems 16.

is a particular solution of the nonhomogeneous system

$$\mathbf{X}' = \begin{pmatrix} 1 & 3 \\ 5 & 3 \end{pmatrix}\mathbf{X} + \begin{pmatrix} 12t - 11 \\ -3 \end{pmatrix}$$

(10)

on the interval $(-\infty, \infty)$.

Solution   We have $\mathbf{X}'_p = \begin{pmatrix} 3 \\ -5 \end{pmatrix}$ and

$$\begin{pmatrix} 1 & 3 \\ 5 & 3 \end{pmatrix}\mathbf{X}_p + \begin{pmatrix} 12t - 11 \\ -3 \end{pmatrix} = \begin{pmatrix} 1 & 3 \\ 5 & 3 \end{pmatrix}\begin{pmatrix} 3t - 4 \\ -5t + 6 \end{pmatrix} + \begin{pmatrix} 12t - 11 \\ -3 \end{pmatrix}$$

$$= \begin{pmatrix} (3t - 4) + 3(-5t + 6) \\ 5(3t - 4) + 3(-5t + 6) \end{pmatrix} + \begin{pmatrix} 12t - 11 \\ -3 \end{pmatrix}$$

$$= \begin{pmatrix} -12t + 14 \\ -2 \end{pmatrix} + \begin{pmatrix} 12t - 11 \\ -3 \end{pmatrix}$$

$$= \begin{pmatrix} 3 \\ -5 \end{pmatrix} = \mathbf{X}'_p. \qquad \blacksquare$$

THEOREM 8.7

Let $\mathbf{X}_1, \mathbf{X}_2, \ldots, \mathbf{X}_k$ be a set of solution vectors of the homogeneous system (3) on an interval I and let $\mathbf{X}_p$ be any solution vector of the nonhomogeneous system (2) on the same interval. Then

$$\mathbf{X} = c_1\mathbf{X}_1 + c_2\mathbf{X}_2 + \cdots + c_k\mathbf{X}_k + \mathbf{X}_p$$

is also a solution of the nonhomogeneous system on the interval for any constants $c_1, c_2, \ldots, c_k$.

DEFINITION 8.18   *General Solution—Nonhomogeneous Systems*

Let $\mathbf{X}_p$ be a given solution of the nonhomogeneous system (2) on an interval I, and let

$$\mathbf{X}_c = c_1\mathbf{X}_1 + c_2\mathbf{X}_2 + \cdots + c_n\mathbf{X}_n$$

denote the general solution on the same interval of the corresponding homogeneous system (3). The **general solution** of the nonhomogeneous system on the interval is defined to be

$$\mathbf{X} = \mathbf{X}_c + \mathbf{X}_p.$$

The general solution $\mathbf{X}_c$ of the homogeneous system (3) is called the **complementary function** of the nonhomogeneous system (2).

EXAMPLE 12

In Example 11 it was verified that a particular solution of the nonhomogeneous system (10) on $(-\infty, \infty)$ is

$$\mathbf{X}_p = \begin{pmatrix} 3t - 4 \\ -5t + 6 \end{pmatrix}.$$

The complementary function of (10) on the same interval, or the general solution of

$$\mathbf{X}' = \begin{pmatrix} 1 & 3 \\ 5 & 3 \end{pmatrix}\mathbf{X},$$

was seen in Example 9 to be

$$\mathbf{X}_c = c_1 \begin{pmatrix} 1 \\ -1 \end{pmatrix} e^{-2t} + c_2 \begin{pmatrix} 3 \\ 5 \end{pmatrix} e^{6t}.$$

Hence by Definition 8.18,

$$\mathbf{X} = \mathbf{X}_c + \mathbf{X}_p = c_1 \begin{pmatrix} 1 \\ -1 \end{pmatrix} e^{-2t} + c_2 \begin{pmatrix} 3 \\ 5 \end{pmatrix} e^{6t} + \begin{pmatrix} 3t - 4 \\ -5t + 6 \end{pmatrix}$$

is the general solution of (10) on $(-\infty, \infty)$. ∎

As one might expect, if $\mathbf{X}$ is *any* solution of the nonhomogeneous system (2) on an interval I, then it is always possible to find appropriate constants $c_1, c_2, \ldots, c_n$ so that $\mathbf{X}$ can be obtained from the general solution.

8.5.2   A Fundamental Matrix

If $\mathbf{X}_1, \mathbf{X}_2, \ldots, \mathbf{X}_n$ is a fundamental set of solutions of the homogeneous system (3) on an interval I, then its general solution on the interval is

$\mathbf{X} = c_1\mathbf{X}_1 + c_2\mathbf{X}_2 + \cdots + c_n\mathbf{X}_n$

$$= c_1 \begin{pmatrix} x_{11} \\ x_{21} \\ \vdots \\ x_{n1} \end{pmatrix} + c_2 \begin{pmatrix} x_{12} \\ x_{22} \\ \vdots \\ x_{n2} \end{pmatrix} + \cdots + c_n \begin{pmatrix} x_{1n} \\ x_{2n} \\ \vdots \\ x_{nn} \end{pmatrix} = \begin{pmatrix} c_1 x_{11} + c_2 x_{12} + \cdots + c_n x_{1n} \\ c_1 x_{21} + c_2 x_{22} + \cdots + c_n x_{2n} \\ \vdots \\ c_1 x_{n1} + c_2 x_{n2} + \cdots + c_n x_{nn} \end{pmatrix}. \quad (11)$$

Observe that (11) can be written as the matrix product

$$\mathbf{X} = \begin{pmatrix} x_{11} & x_{12} & \cdots & x_{1n} \\ x_{21} & x_{22} & \cdots & x_{2n} \\ \vdots & & & \vdots \\ x_{n1} & x_{n2} & \cdots & x_{nn} \end{pmatrix} \begin{pmatrix} c_1 \\ c_2 \\ \vdots \\ c_n \end{pmatrix}. \quad (12)$$

We are led to the following definition.

DEFINITION 8.19 Fundamental Matrix

Let
$$\mathbf{X}_1 = \begin{pmatrix} x_{11} \\ x_{21} \\ \vdots \\ x_{n1} \end{pmatrix}, \quad \mathbf{X}_2 = \begin{pmatrix} x_{12} \\ x_{22} \\ \vdots \\ x_{n2} \end{pmatrix}, \quad \ldots, \quad \mathbf{X}_n = \begin{pmatrix} x_{1n} \\ x_{2n} \\ \vdots \\ x_{nn} \end{pmatrix}$$

be a fundamental set of n solution vectors of the homogeneous system (3) on an interval I. The matrix

$$\Phi(t) = \begin{pmatrix} x_{11} & x_{12} & \cdots & x_{1n} \\ x_{21} & x_{22} & \cdots & x_{2n} \\ \vdots & & & \vdots \\ x_{n1} & x_{n2} & \cdots & x_{nn} \end{pmatrix}$$

is said to be a **fundamental matrix** of the system on the interval.

EXAMPLE 13

The vectors

$$\mathbf{X}_1 = \begin{pmatrix} 1 \\ -1 \end{pmatrix} e^{-2t} = \begin{pmatrix} e^{-2t} \\ -e^{-2t} \end{pmatrix} \quad \text{and} \quad \mathbf{X}_2 = \begin{pmatrix} 3 \\ 5 \end{pmatrix} e^{6t} = \begin{pmatrix} 3e^{6t} \\ 5e^{6t} \end{pmatrix}$$

have been shown to form a fundamental set of solutions of the system (4) on $(-\infty, \infty)$. A fundamental matrix of the system on the interval is then

$$\Phi(t) = \begin{pmatrix} e^{-2t} & 3e^{6t} \\ -e^{-2t} & 5e^{6t} \end{pmatrix}. \tag{13}$$

∎

The result given in (12) states that the general solution of any homogeneous system $\mathbf{X}' = \mathbf{A}(t)\mathbf{X}$ can always be written in terms of a fundamental matrix of the system: $\mathbf{X} = \Phi(t)\mathbf{C}$, where $\mathbf{C}$ is an $n \times 1$ column vector of arbitrary constants.

EXAMPLE 14

The general solution given in (9) can be written as

$$\mathbf{X} = \begin{pmatrix} e^{-2t} & 3e^{6t} \\ -e^{-2t} & 5e^{6t} \end{pmatrix} \begin{pmatrix} c_1 \\ c_2 \end{pmatrix}.$$

∎

Furthermore, to say that $\mathbf{X} = \Phi(t)\mathbf{C}$ is a solution of $\mathbf{X}' = \mathbf{A}(t)\mathbf{X}$ means

$$\Phi'(t)\mathbf{C} = \mathbf{A}(t)\Phi(t)\mathbf{C}$$

or $$(\mathbf{\Phi}'(t) - \mathbf{A}(t)\mathbf{\Phi}(t))\mathbf{C} = \mathbf{0}.$$

Since the last equation is to hold for every t in the interval I and for every possible column matrix of constants $\mathbf{C}$, we must have

$$\mathbf{\Phi}'(t) - \mathbf{A}(t)\mathbf{\Phi}(t) = \mathbf{0}$$

or $$\mathbf{\Phi}'(t) = \mathbf{A}(t)\mathbf{\Phi}(t). \tag{14}$$

This result will be useful in Section 8.8.

Fundamental Matrix Is Nonsingular

Comparison of Theorem 8.5 and Definition 8.19 shows that det $\mathbf{\Phi}(t)$ is the same as the Wronskian $W(\mathbf{X}_1, \mathbf{X}_2, \ldots, \mathbf{X}_n)$.* Hence the linear independence of the columns of $\mathbf{\Phi}(t)$ on an interval I guarantees that det $\mathbf{\Phi}(t) \neq 0$ for every t in the interval; that is, $\mathbf{\Phi}(t)$ is nonsingular on the interval.

THEOREM 8.8   A Fundamental Matrix Has an Inverse

Let $\mathbf{\Phi}(t)$ be a fundamental matrix of the homogeneous system (3) on an interval I. Then $\mathbf{\Phi}^{-1}(t)$ exists for every value of t in the interval.

EXAMPLE 15

For the fundamental matrix given in (13) we see that det $\mathbf{\Phi}(t) = 8e^{4t}$. It then follows from (3) of Section 8.4 that

$$\mathbf{\Phi}^{-1}(t) = \frac{1}{8e^{4t}} \begin{pmatrix} 5e^{6t} & -3e^{6t} \\ e^{-2t} & e^{-2t} \end{pmatrix} = \begin{pmatrix} \frac{5}{8}e^{2t} & -\frac{3}{8}e^{2t} \\ \frac{1}{8}e^{-6t} & \frac{1}{8}e^{-6t} \end{pmatrix}. \qquad \blacksquare$$

Special Matrix

In some instances it is convenient to form another special $n \times n$ matrix, a matrix in which the column vectors $\mathbf{V}_i$ are solutions of $\mathbf{X}' = \mathbf{A}(t)\mathbf{X}$ that satisfy the conditions

$$\mathbf{V}_1(t_0) = \begin{pmatrix} 1 \\ 0 \\ \vdots \\ 0 \end{pmatrix}, \quad \mathbf{V}_2(t_0) = \begin{pmatrix} 0 \\ 1 \\ \vdots \\ 0 \end{pmatrix}, \ldots, \quad \mathbf{V}_n(t_0) = \begin{pmatrix} 0 \\ 0 \\ \vdots \\ 1 \end{pmatrix}. \tag{15}$$

Here t_0 is an arbitrarily chosen point in the interval on which the general solution of the system is defined. We denote this special matrix by the symbol

---

* For this reason some texts call $\mathbf{\Phi}(t)$ a *Wronski matrix*.

$\Psi(t)$. Observe that $\Psi(t)$ has the property

$$\Psi(t_0) = \begin{pmatrix} 1 & 0 & 0 & \cdots & 0 \\ 0 & 1 & 0 & \cdots & 0 \\ \vdots & & & & \vdots \\ 0 & 0 & 0 & \cdots & 1 \end{pmatrix} = I, \tag{16}$$

where I is the $n \times n$ multiplicative identity.

EXAMPLE 16

Find the matrix $\Psi(t)$ satisfying $\Psi(0) = I$ for the system given in (4).

Solution   From (9) we know that the general solution of (4) is given by

$$X = c_1 \begin{pmatrix} 1 \\ -1 \end{pmatrix} e^{-2t} + c_2 \begin{pmatrix} 3 \\ 5 \end{pmatrix} e^{6t}.$$

When $t = 0$, we first solve for constants c_1 and c_2 such that

$$c_1 \begin{pmatrix} 1 \\ -1 \end{pmatrix} + c_2 \begin{pmatrix} 3 \\ 5 \end{pmatrix} = \begin{pmatrix} 1 \\ 0 \end{pmatrix} \quad \text{or} \quad \begin{array}{r} c_1 + 3c_2 = 1 \\ -c_1 + 5c_2 = 0. \end{array}$$

We find that $c_1 = 5/8$ and $c_2 = 1/8$. Hence we define the vector V_1 to be the linear combination

$$V_1 = \frac{5}{8} \begin{pmatrix} 1 \\ -1 \end{pmatrix} e^{-2t} + \frac{1}{8} \begin{pmatrix} 3 \\ 5 \end{pmatrix} e^{6t}.$$

Again when $t = 0$ we wish to find another pair of constants c_1 and c_2 for which

$$c_1 \begin{pmatrix} 1 \\ -1 \end{pmatrix} + c_2 \begin{pmatrix} 3 \\ 5 \end{pmatrix} = \begin{pmatrix} 0 \\ 1 \end{pmatrix} \quad \text{or} \quad \begin{array}{r} c_1 + 3c_2 = 0 \\ -c_1 + 5c_2 = 1. \end{array}$$

In this case we find $c_1 = -3/8$ and $c_2 = 1/8$. We then define

$$V_2 = -\frac{3}{8} \begin{pmatrix} 1 \\ -1 \end{pmatrix} e^{-2t} + \frac{1}{8} \begin{pmatrix} 3 \\ 5 \end{pmatrix} e^{6t}.$$

Hence

$$\Psi(t) = \begin{pmatrix} \frac{5}{8} e^{-2t} + \frac{3}{8} e^{6t} & -\frac{3}{8} e^{-2t} + \frac{3}{8} e^{6t} \\ -\frac{5}{8} e^{-2t} + \frac{5}{8} e^{6t} & \frac{3}{8} e^{-2t} + \frac{5}{8} e^{6t} \end{pmatrix}. \tag{17}$$

Observe that $\Psi(0) = \begin{pmatrix} 1 & 0 \\ 0 & 1 \end{pmatrix} = I.$ ■

Note in Example 16 that since the columns of $\Psi(t)$ are linear combinations of the solutions Ψ of $X' = A(t)X$, we know from the superposition principle that each column is a solution of the system.

$\Psi(t)$ Is a Fundamental Matrix

From (16) it is seen that det $\Psi(t_0) \neq 0$, and hence we conclude from Theorem 8.5 that the columns of $\Psi(t)$ are linearly independent on the interval under consideration. Therefore $\Psi(t)$ is a fundamental matrix. Also, it follows from Theorem 8.3 that $\Psi(t)$ is the unique matrix that satisfies the condition $\Psi(t_0) = \mathbf{I}$. Last, the fundamental matrices $\Phi(t)$ and $\Psi(t)$ are related by

$$\Psi(t) = \Phi(t)\Phi^{-1}(t_0). \tag{18}$$

Equation (18) provides an alternative method for determining $\Psi(t)$ (see Problem 37).

The answer to why anyone would want to form an obviously complicated looking fundamental matrix such as (17) will be answered in Sections 8.8 and 8.9.

8.5 EXERCISES

Answers to odd-numbered problems begin on page A–23.

[8.5.1] In Problems 1–6 write the given system in matrix form.

1. $\dfrac{dx}{dt} = 3x - 5y$

$\dfrac{dy}{dt} = 4x + 8y$

2. $\dfrac{dx}{dt} = 4x - 7y$

$\dfrac{dy}{dt} = 5x$

3. $\dfrac{dx}{dt} = -3x + 4y - 9z$

$\dfrac{dy}{dt} = 6x - y$

$\dfrac{dz}{dt} = 10x + 4y + 3z$

4. $\dfrac{dx}{dt} = x - y$

$\dfrac{dy}{dt} = x + 2z$

$\dfrac{dz}{dt} = -x + z$

5. $\dfrac{dx}{dt} = x - y + z + t - 1$

$\dfrac{dy}{dt} = 2x + y - z - 3t^2$

$\dfrac{dz}{dt} = x + y + z + t^2 - t + 2$

6. $\dfrac{dx}{dt} = -3x + 4y + e^{-t}\sin 2t$

$\dfrac{dy}{dt} = 5x + 9y + 4e^{-t}\cos 2t$

In Problems 7–10 write the given system without the use of matrices.

7. $\mathbf{X}' = \begin{pmatrix} 4 & 2 \\ -1 & 3 \end{pmatrix}\mathbf{X} + \begin{pmatrix} 1 \\ -1 \end{pmatrix}e^{t}$

8. $\mathbf{X}' = \begin{pmatrix} 7 & 5 & -9 \\ 4 & 1 & 1 \\ 0 & -2 & 3 \end{pmatrix}\mathbf{X} + \begin{pmatrix} 0 \\ 2 \\ 1 \end{pmatrix}e^{5t} - \begin{pmatrix} 8 \\ 0 \\ 3 \end{pmatrix}e^{-2t}$

9. $\dfrac{d}{dt}\begin{pmatrix} x \\ y \\ z \end{pmatrix} = \begin{pmatrix} 1 & -1 & 2 \\ 3 & -4 & 1 \\ -2 & 5 & 6 \end{pmatrix}\begin{pmatrix} x \\ y \\ z \end{pmatrix} + \begin{pmatrix} 1 \\ 2 \\ 2 \end{pmatrix}e^{-t} - \begin{pmatrix} 3 \\ -1 \\ 1 \end{pmatrix}t$

10. $\dfrac{d}{dt}\begin{pmatrix} x \\ y \end{pmatrix} = \begin{pmatrix} 3 & -7 \\ 1 & 1 \end{pmatrix}\begin{pmatrix} x \\ y \end{pmatrix} + \begin{pmatrix} 4 \\ 8 \end{pmatrix}\sin t + \begin{pmatrix} t-4 \\ 2t+1 \end{pmatrix}e^{4t}$

In Problems 11–16 verify that the vector **X** is a solution of the given system.

11. $\dfrac{dx}{dt} = 3x - 4y$

$\dfrac{dy}{dt} = 4x - 7y; \quad \mathbf{X} = \begin{pmatrix} 1 \\ 2 \end{pmatrix}e^{-5t}$

12. $\dfrac{dx}{dt} = -2x + 5y$

$\dfrac{dy}{dt} = -2x + 4y; \quad \mathbf{X} = \begin{pmatrix} 5\cos t \\ 3\cos t - \sin t \end{pmatrix}e^{t}$

13. $\mathbf{X}' = \begin{pmatrix} -1 & \frac{1}{4} \\ 1 & -1 \end{pmatrix}\mathbf{X}; \quad \mathbf{X} = \begin{pmatrix} -1 \\ 2 \end{pmatrix}e^{-3t/2}$

14. $\mathbf{X}' = \begin{pmatrix} 2 & 1 \\ -1 & 0 \end{pmatrix}\mathbf{X}; \quad \mathbf{X} = \begin{pmatrix} 1 \\ 3 \end{pmatrix}e^{t} + \begin{pmatrix} 4 \\ -4 \end{pmatrix}te^{t}$

15. $\dfrac{d\mathbf{X}}{dt} = \begin{pmatrix} 1 & 2 & 1 \\ 6 & -1 & 0 \\ -1 & -2 & -1 \end{pmatrix}\mathbf{X}; \quad \mathbf{X} = \begin{pmatrix} 1 \\ 6 \\ -13 \end{pmatrix}$

16. $\mathbf{X}' = \begin{pmatrix} 1 & 0 & 1 \\ 1 & 1 & 0 \\ -2 & 0 & -1 \end{pmatrix}\mathbf{X}; \quad \mathbf{X} = \begin{pmatrix} \sin t \\ -\frac{1}{2}\sin t - \frac{1}{2}\cos t \\ -\sin t + \cos t \end{pmatrix}$

In Problems 17–20 the given vectors are solutions of a system $\mathbf{X}' = \mathbf{AX}$. Determine whether the vectors form a fundamental set on $(-\infty, \infty)$.

17. $\mathbf{X}_1 = \begin{pmatrix} 1 \\ 1 \end{pmatrix}e^{-2t},\ \mathbf{X}_2 = \begin{pmatrix} 1 \\ -1 \end{pmatrix}e^{-6t}$

18. $\mathbf{X}_1 = \begin{pmatrix} 1 \\ -1 \end{pmatrix}e^{t},\ \mathbf{X}_2 = \begin{pmatrix} 2 \\ 6 \end{pmatrix}e^{t} + \begin{pmatrix} 8 \\ -8 \end{pmatrix}te^{t}$

19. $\mathbf{X}_1 = \begin{pmatrix} 1 \\ -2 \\ 4 \end{pmatrix} + t\begin{pmatrix} 1 \\ 2 \\ 2 \end{pmatrix},\ \mathbf{X}_2 = \begin{pmatrix} 1 \\ -2 \\ 4 \end{pmatrix},\ \mathbf{X}_3 = \begin{pmatrix} 3 \\ -6 \\ 12 \end{pmatrix} + t\begin{pmatrix} 2 \\ 4 \\ 4 \end{pmatrix}$

20. $\mathbf{X}_1 = \begin{pmatrix} 1 \\ 6 \\ -13 \end{pmatrix},\ \mathbf{X}_2 = \begin{pmatrix} 1 \\ -2 \\ -1 \end{pmatrix}e^{-4t},\ \mathbf{X}_3 = \begin{pmatrix} 2 \\ 3 \\ -2 \end{pmatrix}e^{3t}$

In Problems 21–24 verify that the vector $\mathbf{X}_p$ is a particular solution of the given system.

21. $\dfrac{dx}{dt} = x + 4y + 2t - 7$

$\dfrac{dy}{dt} = 3x + 2y - 4t - 18;$ $\mathbf{X}_p = \begin{pmatrix} 2 \\ -1 \end{pmatrix} t + \begin{pmatrix} 5 \\ 1 \end{pmatrix}$

22. $\mathbf{X}' = \begin{pmatrix} 2 & 1 \\ 1 & -1 \end{pmatrix} \mathbf{X} + \begin{pmatrix} -5 \\ 2 \end{pmatrix};$ $\mathbf{X}_p = \begin{pmatrix} 1 \\ 3 \end{pmatrix}$

23. $\mathbf{X}' = \begin{pmatrix} 2 & 1 \\ 3 & 4 \end{pmatrix} \mathbf{X} - \begin{pmatrix} 1 \\ 7 \end{pmatrix} e^t;$ $\mathbf{X}_p = \begin{pmatrix} 1 \\ 1 \end{pmatrix} e^t + \begin{pmatrix} 1 \\ -1 \end{pmatrix} t e^t$

24. $\mathbf{X}' = \begin{pmatrix} 1 & 2 & 3 \\ -4 & 2 & 0 \\ -6 & 1 & 0 \end{pmatrix} \mathbf{X} + \begin{pmatrix} -1 \\ 4 \\ 3 \end{pmatrix} \sin 3t;$ $\mathbf{X}_p = \begin{pmatrix} \sin 3t \\ 0 \\ \cos 3t \end{pmatrix}$

25. Prove that the general solution of

$$\mathbf{X}' = \begin{pmatrix} 0 & 6 & 0 \\ 1 & 0 & 1 \\ 1 & 1 & 0 \end{pmatrix} \mathbf{X}$$

on the interval $(-\infty, \infty)$ is

$$\mathbf{X} = c_1 \begin{pmatrix} 6 \\ -1 \\ -5 \end{pmatrix} e^{-t} + c_2 \begin{pmatrix} -3 \\ 1 \\ 1 \end{pmatrix} e^{-2t} + c_3 \begin{pmatrix} 2 \\ 1 \\ 1 \end{pmatrix} e^{3t}.$$

26. Prove that the general solution of

$$\mathbf{X}' = \begin{pmatrix} -1 & -1 \\ -1 & 1 \end{pmatrix} \mathbf{X} + \begin{pmatrix} 1 \\ 1 \end{pmatrix} t^2 + \begin{pmatrix} 4 \\ -6 \end{pmatrix} t + \begin{pmatrix} -1 \\ 5 \end{pmatrix}$$

on the interval $(-\infty, \infty)$ is

$$\mathbf{X} = c_1 \begin{pmatrix} 1 \\ -1 - \sqrt{2} \end{pmatrix} e^{\sqrt{2}t} + c_2 \begin{pmatrix} 1 \\ -1 + \sqrt{2} \end{pmatrix} e^{-\sqrt{2}t} + \begin{pmatrix} 1 \\ 0 \end{pmatrix} t^2 + \begin{pmatrix} -2 \\ 4 \end{pmatrix} t + \begin{pmatrix} 1 \\ 0 \end{pmatrix}.$$

[8.5.2]   In Problems 27–30 the indicated column vectors form a fundamental set of solutions for the given system on $(-\infty, \infty)$. Form a fundamental matrix $\mathbf{\Phi}(t)$ and compute $\mathbf{\Phi}^{-1}(t)$.

27. $\mathbf{X}' = \begin{pmatrix} 4 & 1 \\ 6 & 5 \end{pmatrix} \mathbf{X};$ $\mathbf{X}_1 = \begin{pmatrix} 1 \\ -2 \end{pmatrix} e^{2t},$ $\mathbf{X}_2 = \begin{pmatrix} 1 \\ 3 \end{pmatrix} e^{7t}$

28. $\mathbf{X}' = \begin{pmatrix} 2 & 3 \\ 3 & 2 \end{pmatrix} \mathbf{X};$ $\mathbf{X}_1 = \begin{pmatrix} -1 \\ 1 \end{pmatrix} e^{-t},$ $\mathbf{X}_2 = \begin{pmatrix} 1 \\ 1 \end{pmatrix} e^{5t}$

29. $\mathbf{X}' = \begin{pmatrix} 4 & 1 \\ -9 & -2 \end{pmatrix} \mathbf{X}; \quad \mathbf{X}_1 = \begin{pmatrix} -1 \\ 3 \end{pmatrix} e^t, \mathbf{X}_2 = \begin{pmatrix} -1 \\ 3 \end{pmatrix} te^t + \begin{pmatrix} 0 \\ -1 \end{pmatrix} e^t$

30. $\mathbf{X}' = \begin{pmatrix} 3 & -2 \\ 5 & -3 \end{pmatrix} \mathbf{X}; \mathbf{X}_1 = \begin{pmatrix} 2\cos t \\ 3\cos t + \sin t \end{pmatrix}, \mathbf{X}_2 = \begin{pmatrix} -2\sin t \\ \cos t - 3\sin t \end{pmatrix}$

31. Find the fundamental matrix $\boldsymbol{\Psi}(t)$ satisfying $\boldsymbol{\Psi}(0) = \mathbf{I}$ for the system given in Problem 27.

32. Find the fundamental matrix $\boldsymbol{\Psi}(t)$ satisfying $\boldsymbol{\Psi}(0) = \mathbf{I}$ for the system given in Problem 28.

33. Find the fundamental matrix $\boldsymbol{\Psi}(t)$ satisfying $\boldsymbol{\Psi}(0) = \mathbf{I}$ for the system given in Problem 29.

34. Find the fundamental matrix $\boldsymbol{\Psi}(t)$ satisfying $\boldsymbol{\Psi}(\pi/2) = \mathbf{I}$ for the system given in Problem 30.

35. If $\mathbf{X} = \boldsymbol{\Phi}(t)\mathbf{C}$ is the general solution of $\mathbf{X}' = \mathbf{A}\mathbf{X}$, show that the solution of the initial-value problem $\mathbf{X}' = \mathbf{A}\mathbf{X}, \mathbf{X}(t_0) = \mathbf{X}_0$, is $\mathbf{X} = \boldsymbol{\Phi}(t)\boldsymbol{\Phi}^{-1}(t_0)\mathbf{X}_0$.

36. Show that the solution of the initial-value problem given in Problem 35 is also given by $\mathbf{X} = \boldsymbol{\Psi}(t)\mathbf{X}_0$.

37. Show that $\boldsymbol{\Psi}(t) = \boldsymbol{\Phi}(t)\boldsymbol{\Phi}^{-1}(t_0)$. [*Hint:* Compare Problems 35 and 36.]

8.6 HOMOGENEOUS LINEAR SYSTEMS

8.6.1   *Distinct Real Eigenvalues*   •

For the remainder of this chapter we shall be concerned only with linear systems with real constant coefficients.

We saw in Example 9 of Section 8.5 that the general solution of the homogeneous system

$$\frac{dx}{dt} = x + 3y$$

$$\frac{dy}{dt} = 5x + 3y$$

is
$$\mathbf{X} = c_1 \begin{pmatrix} 1 \\ -1 \end{pmatrix} e^{-2t} + c_2 \begin{pmatrix} 3 \\ 5 \end{pmatrix} e^{6t}.$$

Since both solution vectors have the basic form

$$\mathbf{X}_i = \begin{pmatrix} k_1 \\ k_2 \end{pmatrix} e^{\lambda_i t}, \quad i = 1, 2.$$

k_1 and k_2 constants, we are prompted to ask whether we can always find a solution of the form

$$\mathbf{X} = \begin{pmatrix} k_1 \\ k_2 \\ \vdots \\ k_n \end{pmatrix} e^{\lambda t} = \mathbf{K}e^{\lambda t} \tag{1}$$

for the general homogeneous linear first-order system

$$\mathbf{X}' = \mathbf{AX}, \tag{2}$$

where $\mathbf{A}$ is an $n \times n$ matrix of constants.

Eigenvalues and Eigenvectors

If (1) is to be a solution vector of (2), then $\mathbf{X}' = \mathbf{K}\lambda e^{\lambda t}$ so that the system becomes

$$\mathbf{K}\lambda e^{\lambda t} = \mathbf{AK}e^{\lambda t}.$$

After dividing out $e^{\lambda t}$ and rearranging, we obtain

$$\mathbf{AK} = \lambda\mathbf{K}$$

or
$$(\mathbf{A} - \lambda\mathbf{I})\mathbf{K} = \mathbf{0}. \tag{3}$$

Equation (3) is equivalent to the simultaneous algebraic equations (8) of Section 8.4. To find a nontrivial solution $\mathbf{X}$ of (2) we must find a nontrivial vector $\mathbf{K}$ satisfying (3). But in order that (3) have nontrivial solutions we must have

$$\det(\mathbf{A} - \lambda\mathbf{I}) = 0.$$

The latter equation is recognized as the characteristic equation of the matrix $\mathbf{A}$. In other words, $\mathbf{X} = \mathbf{K}e^{\lambda t}$ will be a solution of the system of differential equations (2) if and only if λ is an **eigenvalue** of $\mathbf{A}$ and $\mathbf{K}$ is an **eigenvector** corresponding to λ.

When the $n \times n$ matrix $\mathbf{A}$ possesses n distinct real eigenvalues λ_1, $\lambda_2, \ldots, \lambda_n$, then a set of n linearly independent eigenvectors $\mathbf{K}_1, \mathbf{K}_2, \ldots, \mathbf{K}_n$ can always be found and

$$\mathbf{X}_1 = \mathbf{K}_1 e^{\lambda_1 t}, \quad \mathbf{X}_2 = \mathbf{K}_2 e^{\lambda_2 t}, \quad \ldots, \quad \mathbf{X}_n = \mathbf{K}_n e^{\lambda_n t}$$

is a fundamental set of solutions of (2) on $(-\infty, \infty)$.

THEOREM 8.9   *General Solution—Homogeneous Systems*

Let $\lambda_1, \lambda_2, \ldots, \lambda_n$ be n distinct real eigenvalues of the coefficient matrix $\mathbf{A}$ of the homogeneous system (2) and let $\mathbf{K}_1, \mathbf{K}_2, \ldots, \mathbf{K}_n$ be the corresponding eigenvectors. Then the **general solution** of (2) on the interval $(-\infty, \infty)$ is given by

$$\mathbf{X} = c_1\mathbf{K}_1 e^{\lambda_1 t} + c_2\mathbf{K}_2 e^{\lambda_2 t} + \cdots + c_n\mathbf{K}_n e^{\lambda_n t}.$$

EXAMPLE I

Solve
$$\frac{dx}{dt} = 2x + 3y$$

$$\frac{dy}{dt} = 2x + y. \tag{4}$$

Solution We first find the eigenvalues and eigenvectors of the matrix of coefficients.

The characteristic equation is

$$\det(\mathbf{A} - \lambda\mathbf{I}) = \begin{vmatrix} 2 - \lambda & 3 \\ 2 & 1 - \lambda \end{vmatrix} = \lambda^2 - 3\lambda - 4 = 0.$$

Since $\lambda^2 - 3\lambda - 4 = (\lambda + 1)(\lambda - 4)$, we see that the eigenvalues are $\lambda_1 = -1$ and $\lambda_2 = 4$.

Now for $\lambda_1 = -1$, (3) is equivalent to

$$3k_1 + 3k_2 = 0$$

$$2k_1 + 2k_2 = 0.$$

Thus $k_1 = -k_2$. When $k_2 = -1$ the related eigenvector is

$$\mathbf{K}_1 = \begin{pmatrix} 1 \\ -1 \end{pmatrix}.$$

For $\lambda_2 = 4$ we have
$$-2k_1 + 3k_2 = 0$$

$$2k_1 - 3k_2 = 0$$

so that $k_1 = 3k_2/2$, and therefore with $k_2 = 2$, the corresponding eigenvector is

$$\mathbf{K}_2 = \begin{pmatrix} 3 \\ 2 \end{pmatrix}.$$

Since the matrix of coefficients $\mathbf{A}$ is a 2×2 matrix and since we have found two linearly independent solutions of (4),

$$\mathbf{X}_1 = \begin{pmatrix} 1 \\ -1 \end{pmatrix} e^{-t} \quad \text{and} \quad \mathbf{X}_2 = \begin{pmatrix} 3 \\ 2 \end{pmatrix} e^{4t},$$

we conclude that the general solution of the system is

$$\mathbf{X} = c_1\mathbf{X}_1 + c_2\mathbf{X}_2 = c_1 \begin{pmatrix} 1 \\ -1 \end{pmatrix} e^{-t} + c_2 \begin{pmatrix} 3 \\ 2 \end{pmatrix} e^{4t}. \tag{5}$$

■

For the sake of review, you should keep firmly in mind that a solution of a system of first-order differential equations, when written in terms of matrices, is simply an alternative to the method that we employed in Section 8.1, namely, listing the individual functions and the relationships between the constants. By adding the vectors given in (5), we obtain

$$\begin{pmatrix} x(t) \\ y(t) \end{pmatrix} = \begin{pmatrix} c_1 e^{-t} + 3c_2 e^{4t} \\ -c_1 e^{-t} + 2c_2 e^{4t} \end{pmatrix}$$

and this in turn yields the more familiar statement

$$x(t) = c_1 e^{-t} + 3c_2 e^{4t}$$

$$y(t) = -c_1 e^{-t} + 2c_2 e^{4t}.$$

EXAMPLE **2**

Solve

$$\frac{dx}{dt} = -4x + y + z$$

$$\frac{dy}{dt} = x + 5y - z \quad \text{(6)}$$

$$\frac{dz}{dt} = y - 3z.$$

Solution Using the cofactors of the third row, we find

$$\det(\mathbf{A} - \lambda\mathbf{I}) = \begin{vmatrix} -4 - \lambda & 1 & 1 \\ 1 & 5 - \lambda & -1 \\ 0 & 1 & -3 - \lambda \end{vmatrix} = -(\lambda + 3)(\lambda + 4)(\lambda - 5) = 0$$

and so the eigenvalues are $\lambda_1 = -3$, $\lambda_2 = -4$, $\lambda_3 = 5$.
Now for $\lambda_1 = -3$, Gauss-Jordan elimination gives

$$(\mathbf{A} + 3\mathbf{I}\,|\,\mathbf{0}) = \begin{pmatrix} -1 & 1 & 1 & | & 0 \\ 1 & 8 & -1 & | & 0 \\ 0 & 1 & 0 & | & 0 \end{pmatrix} \overset{\text{row operations}}{\Longrightarrow} \begin{pmatrix} 1 & 0 & -1 & | & 0 \\ 0 & 1 & 0 & | & 0 \\ 0 & 0 & 0 & | & 0 \end{pmatrix}.$$

Therefore $k_1 = k_3$, $k_2 = 0$. The choice $k_3 = 1$ gives the eigenvector

$$\mathbf{K}_1 = \begin{pmatrix} 1 \\ 0 \\ 1 \end{pmatrix}. \quad \text{(7)}$$

Similarly, for $\lambda_2 = -4$,

$$(\mathbf{A} + 4\mathbf{I}\,|\,\mathbf{0}) = \begin{pmatrix} 0 & 1 & 1 & | & 0 \\ 1 & 9 & -1 & | & 0 \\ 0 & 1 & 1 & | & 0 \end{pmatrix} \overset{\text{row operations}}{\Longrightarrow} \begin{pmatrix} 1 & 0 & -10 & | & 0 \\ 0 & 1 & 1 & | & 0 \\ 0 & 0 & 0 & | & 0 \end{pmatrix}$$

implies $k_1 = 10k_3, k_2 = -k_3$. Choosing $k_3 = 1$, we get the second eigenvector

$$\mathbf{K}_2 = \begin{pmatrix} 10 \\ -1 \\ 1 \end{pmatrix}. \tag{8}$$

Finally, when $\lambda_3 = 5$, the augmented matrices

$$(\mathbf{A} - 5\mathbf{I} \,|\, \mathbf{0}) = \begin{pmatrix} -9 & 1 & 1 & | & 0 \\ 1 & 0 & -1 & | & 0 \\ 0 & 1 & -8 & | & 0 \end{pmatrix} \xrightarrow[\text{operations}]{\text{row}} \begin{pmatrix} 1 & 0 & -1 & | & 0 \\ 0 & 1 & -8 & | & 0 \\ 0 & 0 & 0 & | & 0 \end{pmatrix}$$

yield

$$\mathbf{K}_3 = \begin{pmatrix} 1 \\ 8 \\ 1 \end{pmatrix}. \tag{9}$$

Multiplying the vectors (7), (8), and (9) by e^{-3t}, e^{-4t}, and e^{5t}, respectively, gives three solutions of (6):

$$\mathbf{X}_1 = \begin{pmatrix} 1 \\ 0 \\ 1 \end{pmatrix} e^{-3t}, \qquad \mathbf{X}_2 = \begin{pmatrix} 10 \\ -1 \\ 1 \end{pmatrix} e^{-4t}, \qquad \mathbf{X}_3 = \begin{pmatrix} 1 \\ 8 \\ 1 \end{pmatrix} e^{5t}.$$

The general solution of the system is then

$$\mathbf{X} = c_1 \begin{pmatrix} 1 \\ 0 \\ 1 \end{pmatrix} e^{-3t} + c_2 \begin{pmatrix} 10 \\ -1 \\ 1 \end{pmatrix} e^{-4t} + c_3 \begin{pmatrix} 1 \\ 8 \\ 1 \end{pmatrix} e^{5t}. \qquad \blacksquare$$

8.6.2   Complex Eigenvalues

If
$$\lambda_1 = \alpha + i\beta \quad \text{and} \quad \lambda_2 = \alpha - i\beta, \qquad i^2 = -1,$$

are complex eigenvalues of the coefficient matrix $\mathbf{A}$, we can then certainly expect their corresponding eigenvectors to also have complex entries.*

For example, the characteristic equation of the system

$$\frac{dx}{dt} = 6x - y$$

$$\frac{dy}{dt} = 5x + 4y \tag{10}$$

---

* When the characteristic equation has real coefficients, complex eigenvalues always appear in conjugate pairs.

is $\qquad \det(\mathbf{A} - \lambda\mathbf{I}) = \begin{vmatrix} 6 - \lambda & -1 \\ 5 & 4 - \lambda \end{vmatrix} = \lambda^2 - 10\lambda + 29 = 0.$

From the quadratic formula we find

$$\lambda_1 = 5 + 2i, \qquad \lambda_2 = 5 - 2i.$$

Now for $\lambda_1 = 5 + 2i$ we must solve

$$(1 - 2i)k_1 - \qquad\qquad k_2 = 0$$

$$5k_1 - (1 + 2i)k_2 = 0.$$

Since $k_2 = (1 - 2i)k_1,$* it follows, after we choose $k_1 = 1$, that one eigenvector is

$$\mathbf{K}_1 = \begin{pmatrix} 1 \\ 1 - 2i \end{pmatrix}.$$

Similarly, for $\lambda_2 = 5 - 2i$ we find the other eigenvector to be

$$\mathbf{K}_2 = \begin{pmatrix} 1 \\ 1 + 2i \end{pmatrix}.$$

Consequently two solutions of (10) are

$$\mathbf{X}_1 = \begin{pmatrix} 1 \\ 1 - 2i \end{pmatrix} e^{(5 + 2i)t} \quad \text{and} \quad \mathbf{X}_2 = \begin{pmatrix} 1 \\ 1 + 2i \end{pmatrix} e^{(5 - 2i)t}.$$

By the superposition principle another solution is

$$\mathbf{X} = c_1 \begin{pmatrix} 1 \\ 1 - 2i \end{pmatrix} e^{(5 + 2i)t} + c_2 \begin{pmatrix} 1 \\ 1 + 2i \end{pmatrix} e^{(5 - 2i)t}. \tag{11}$$

Note that the entries in $\mathbf{K}_2$ corresponding to λ_2 are the conjugates of the entries in $\mathbf{K}_1$ corresponding to λ_1. The conjugate of λ_1 is, of course, λ_2. We write this as $\lambda_2 = \bar{\lambda}_1$ and $\mathbf{K}_2 = \bar{\mathbf{K}}_1$. We have illustrated the following general result.

THEOREM 8.10 **Solutions Corresponding to a Complex Eigenvalue**

Let $\mathbf{A}$ be the coefficient matrix having real entries of the homogeneous system (2), and let $\mathbf{K}$ be an eigenvector corresponding to the complex eigenvalue $\lambda_1 = \alpha + i\beta$, α and β real. Then

$$\mathbf{X}_1 = \mathbf{K}_1 e^{\lambda_1 t} \quad \text{and} \quad \mathbf{X}_2 = \bar{\mathbf{K}}_1 e^{\bar{\lambda}_1 t}$$

are solutions of (2).

---

* Note that the second equation is simply $(1 + 2i)$ times the first.

It is desirable and relatively easy to rewrite a solution such as (11) in terms of real functions. Since

$$x = c_1 e^{(5+2i)t} + c_2 e^{(5-2i)t}$$

$$y = c_1(1 - 2i)e^{(5+2i)t} + c_2(1 + 2i)e^{(5-2i)t},$$

it follows from Euler's formula that

$$x = e^{5t}[c_1 e^{2it} + c_2 e^{-2it}]$$

$$= e^{5t}[(c_1 + c_2)\cos 2t + (c_1 i - c_2 i)\sin 2t]$$

$$y = e^{5t}[(c_1(1 - 2i) + c_2(1 + 2i))\cos 2t + (c_1 i(1 - 2i) - c_2 i(1 + 2i))\sin 2t]$$

$$= e^{5t}[(c_1 + c_2) - 2(c_1 i - c_2 i)]\cos 2t + e^{5t}[2(c_1 + c_2) + (c_1 i - c_2 i)]\sin 2t.$$

If we replace $c_1 + c_2$ by C_1 and $c_1 i - c_2 i$ by C_2, then

$$x = e^{5t}[C_1 \cos 2t + C_2 \sin 2t]$$

$$y = e^{5t}[C_1 - 2C_2]\cos 2t + e^{5t}[2C_1 + C_2]\sin 2t$$

or, in terms of vectors,

$$\mathbf{X} = \begin{pmatrix} x \\ y \end{pmatrix} = C_1 \begin{pmatrix} \cos 2t \\ \cos 2t + 2\sin 2t \end{pmatrix} e^{5t} + C_2 \begin{pmatrix} \sin 2t \\ -2\cos 2t + \sin 2t \end{pmatrix} e^{5t}. \quad (12)$$

Here, of course, it can be verified that each vector in (12) is a solution of (10). In addition the solutions are linearly independent on the interval $(-\infty, \infty)$. We may further assume that C_1 and C_2 are completely arbitrary and real. Thus (12) is the general solution of (10).

The foregoing process can be generalized. Let $\mathbf{K}_1$ be an eigenvector of the matrix $\mathbf{A}$ corresponding to the complex eigenvalue $\lambda_1 = \alpha + i\beta$. Then $\mathbf{X}_1$ and $\mathbf{X}_2$ in Theorem 8.10 can be written as

$$\mathbf{K}_1 e^{\lambda_1 t} = \mathbf{K}_1 e^{\alpha t} e^{i\beta t} = \mathbf{K}_1 e^{\alpha t}(\cos \beta t + i \sin \beta t)$$

$$\bar{\mathbf{K}}_1 e^{\bar{\lambda}_1 t} = \bar{\mathbf{K}}_1 e^{\alpha t} e^{-i\beta t} = \bar{\mathbf{K}}_1 e^{\alpha t}(\cos \beta t - i \sin \beta t).$$

The foregoing equations then yield

$$\frac{1}{2}(\mathbf{K}_1 e^{\lambda_1 t} + \bar{\mathbf{K}}_1 e^{\bar{\lambda}_1 t}) = \frac{1}{2}(\mathbf{K}_1 + \bar{\mathbf{K}}_1)e^{\alpha t}\cos \beta t - \frac{i}{2}(-\mathbf{K}_1 + \bar{\mathbf{K}}_1)e^{\alpha t}\sin \beta t$$

$$\frac{i}{2}(-\mathbf{K}_1 e^{\lambda_1 t} + \bar{\mathbf{K}}_1 e^{\bar{\lambda}_1 t}) = \frac{i}{2}(-\mathbf{K}_1 + \bar{\mathbf{K}}_1)e^{\alpha t}\cos \beta t + \frac{1}{2}(\mathbf{K}_1 + \bar{\mathbf{K}}_1)e^{\alpha t}\sin \beta t.$$

For *any* complex number $z = a + ib$, we note that $\frac{1}{2}(z + \bar{z}) = a$ and $\frac{i}{2}(-z + \bar{z}) = b$ are *real* numbers. Therefore, the entries in the column vectors $\frac{1}{2}(\mathbf{K}_1 + \bar{\mathbf{K}}_1)$ and $\frac{i}{2}(-\mathbf{K}_1 + \bar{\mathbf{K}}_1)$ are real numbers. By defining

$$\mathbf{B}_1 = \frac{1}{2}[\mathbf{K}_1 + \bar{\mathbf{K}}_1] \quad \text{and} \quad \mathbf{B}_2 = \frac{i}{2}[-\mathbf{K}_1 + \bar{\mathbf{K}}_1], \quad (13)$$

we are led to the following theorem.

THEOREM 8.11 **Real Solutions Corresponding to a Complex Eigenvalue**

Let $\lambda_1 = \alpha + i\beta$ be a complex eigenvalue of the coefficient matrix $\mathbf{A}$ in the homogeneous system (2) and let $\mathbf{B}_1$ and $\mathbf{B}_2$ denote the column vectors defined in (13). Then

$$\mathbf{X}_1 = (\mathbf{B}_1 \cos \beta t - \mathbf{B}_2 \sin \beta t)e^{\alpha t}$$

$$\mathbf{X}_2 = (\mathbf{B}_2 \cos \beta t + \mathbf{B}_1 \sin \beta t)e^{\alpha t}$$

(14)

are linearly independent solutions of (2) on $(-\infty, \infty)$.

The matrices $\mathbf{B}_1$ and $\mathbf{B}_2$ in (13) are often denoted by

$$\mathbf{B}_1 = \text{Re}(\mathbf{K}_1) \quad \text{and} \quad \mathbf{B}_2 = \text{Im}(\mathbf{K}_1) \tag{15}$$

since these vectors are, in turn, the *real* and *imaginary* parts of the eigenvector $\mathbf{K}_1$. For example, (12) follows from (14) with

$$\mathbf{K}_1 = \begin{pmatrix} 1 \\ 1 - 2i \end{pmatrix} = \begin{pmatrix} 1 \\ 1 \end{pmatrix} + i\begin{pmatrix} 0 \\ -2 \end{pmatrix}$$

$$\mathbf{B}_1 = \text{Re}(\mathbf{K}_1) = \begin{pmatrix} 1 \\ 1 \end{pmatrix} \quad \text{and} \quad \mathbf{B}_2 = \text{Im}(\mathbf{K}_1) = \begin{pmatrix} 0 \\ -2 \end{pmatrix}.$$

EXAMPLE 3

Solve

$$\mathbf{X}' = \begin{pmatrix} 2 & 8 \\ -1 & -2 \end{pmatrix}\mathbf{X}.$$

Solution First we obtain the eigenvalues from

$$\det(\mathbf{A} - \lambda\mathbf{I}) = \begin{vmatrix} 2 - \lambda & 8 \\ -1 & -2 - \lambda \end{vmatrix} = \lambda^2 + 4 = 0.$$

Thus the eigenvalues are $\lambda_1 = 2i$ and $\lambda_2 = \bar{\lambda}_1 = -2i$. For λ_1 we see that the system

$$(2 - 2i)k_1 + \qquad\qquad 8k_2 = 0$$

$$-k_1 + (-2 - 2i)k_2 = 0$$

gives $k_1 = -(2 + 2i)k_2$. By choosing $k_2 = -1$, we get

$$\mathbf{K}_1 = \begin{pmatrix} 2 + 2i \\ -1 \end{pmatrix} = \begin{pmatrix} 2 \\ -1 \end{pmatrix} + i\begin{pmatrix} 2 \\ 0 \end{pmatrix}.$$

Now from (15) we form

$$\mathbf{B}_1 = \text{Re}(\mathbf{K}_1) = \begin{pmatrix} 2 \\ -1 \end{pmatrix} \quad \text{and} \quad \mathbf{B}_2 = \text{Im}(\mathbf{K}_1) = \begin{pmatrix} 2 \\ 0 \end{pmatrix}.$$

Since $\alpha = 0$, it follows from (14) that the general solution of the system is

$$\mathbf{X} = c_1 \left[\begin{pmatrix} 2 \\ -1 \end{pmatrix} \cos 2t - \begin{pmatrix} 2 \\ 0 \end{pmatrix} \sin 2t \right] + c_2 \left[\begin{pmatrix} 2 \\ 0 \end{pmatrix} \cos 2t + \begin{pmatrix} 2 \\ -1 \end{pmatrix} \sin 2t \right]$$

$$= c_1 \begin{pmatrix} 2 \cos 2t - 2 \sin 2t \\ -\cos 2t \end{pmatrix} + c_2 \begin{pmatrix} 2 \cos 2t + 2 \sin 2t \\ -\sin 2t \end{pmatrix}. \qquad \blacksquare$$

EXAMPLE 4

Solve

$$\mathbf{X}' = \begin{pmatrix} 1 & 2 \\ -\frac{1}{2} & 1 \end{pmatrix} \mathbf{X}.$$

Solution The solutions of the characteristic equation

$$\det(\mathbf{A} - \lambda \mathbf{I}) = \begin{vmatrix} 1 - \lambda & 2 \\ -\frac{1}{2} & 1 - \lambda \end{vmatrix} = \lambda^2 - 2\lambda + 2 = 0$$

are

$$\lambda_1 = 1 + i \quad \text{and} \quad \lambda_2 = \bar{\lambda}_1 = 1 - i.$$

Now an eigenvector associated with λ_1 is

$$\mathbf{K}_1 = \begin{pmatrix} 2 \\ i \end{pmatrix} = \begin{pmatrix} 2 \\ 0 \end{pmatrix} + i \begin{pmatrix} 0 \\ 1 \end{pmatrix}.$$

From (15) we find

$$\mathbf{B}_1 = \begin{pmatrix} 2 \\ 0 \end{pmatrix} \quad \text{and} \quad \mathbf{B}_2 = \begin{pmatrix} 0 \\ 1 \end{pmatrix}.$$

Thus (14) gives

$$\mathbf{X} = c_1 \left[\begin{pmatrix} 2 \\ 0 \end{pmatrix} \cos t - \begin{pmatrix} 0 \\ 1 \end{pmatrix} \sin t \right] e^t + c_2 \left[\begin{pmatrix} 0 \\ 1 \end{pmatrix} \cos t + \begin{pmatrix} 2 \\ 0 \end{pmatrix} \sin t \right] e^t$$

$$= c_1 \begin{pmatrix} 2 \cos t \\ -\sin t \end{pmatrix} e^t + c_2 \begin{pmatrix} 2 \sin t \\ \cos t \end{pmatrix} e^t. \qquad \blacksquare$$

Alternative Method

When $\mathbf{A}$ is a 2×2 matrix having a complex eigenvalue $\lambda = \alpha + i\beta$, the general solution of the system can also be obtained from the assumption

$$\mathbf{X} = \begin{pmatrix} c_1 \\ c_2 \end{pmatrix} e^{\alpha t} \sin \beta t + \begin{pmatrix} c_3 \\ c_4 \end{pmatrix} e^{\alpha t} \cos \beta t$$

and then the substitution of $x(t)$ and $y(t)$ into one of the equations of the original system. This procedure is basically that of Section 8.1.

8.6.3 Repeated Eigenvalues

Up to this point we have not considered the case in which some of the n eigenvalues $\lambda_1, \lambda_2, \ldots, \lambda_n$ of an $n \times n$ matrix are repeated. For example, the characteristic equation of the coefficient matrix in

$$\mathbf{X}' = \begin{pmatrix} 3 & -18 \\ 2 & -9 \end{pmatrix}\mathbf{X} \tag{16}$$

is readily shown to be $(\lambda + 3)^2 = 0$, and therefore $\lambda_1 = \lambda_2 = -3$ is a root of *multiplicity two*. Now for this value we find the single eigenvector

$$\mathbf{K}_1 = \begin{pmatrix} 3 \\ 1 \end{pmatrix}$$

and so one solution of (16) is

$$\mathbf{X}_1 = \begin{pmatrix} 3 \\ 1 \end{pmatrix}e^{-3t}. \tag{17}$$

But since we are obviously interested in forming the general solution of the system, we need to pursue the question of finding a second solution.

In general, if m is a positive integer and $(\lambda - \lambda_1)^m$ is a factor of the characteristic equation, while $(\lambda - \lambda_1)^{m+1}$ is not a factor, then λ_1 is said to be an **eigenvalue of multiplicity m**. We distinguish two possibilities:

(i) For some $n \times n$ matrices $\mathbf{A}$ it may be possible to find m linearly independent eigenvectors $\mathbf{K}_1, \mathbf{K}_2, \ldots, \mathbf{K}_m$ corresponding to an eigenvalue λ_1 of multiplicity $m \leq n$. In this case the general solution of the system contains the linear combination

$$c_1\mathbf{K}_1 e^{\lambda_1 t} + c_2\mathbf{K}_2 e^{\lambda_1 t} + \cdots + c_m\mathbf{K}_m e^{\lambda_1 t}.$$

(ii) If there is only one eigenvector corresponding to the eigenvalue λ_1 of multiplicity m, then m linearly independent solutions of the form

$$\mathbf{X}_1 = \mathbf{K}_{11} e^{\lambda_1 t}$$

$$\mathbf{X}_2 = \mathbf{K}_{21} t e^{\lambda_1 t} + \mathbf{K}_{22} e^{\lambda_1 t}$$

$$\vdots$$

$$\mathbf{X}_m = \mathbf{K}_{m1}\frac{t^{m-1}}{(m-1)!}e^{\lambda_1 t} + \mathbf{K}_{m2}\frac{t^{m-2}}{(m-2)!}e^{\lambda_1 t} + \cdots + \mathbf{K}_{mm}e^{\lambda_1 t},$$

where $\mathbf{K}_{ij}$ are column vectors, can always be found.

Eigenvalue of Multiplicity Two

We begin by considering eigenvalues of multiplicity two. In the first example we illustrate a matrix for which we can find two distinct eigenvectors corresponding to a double eigenvalue.

EXAMPLE 5

Solve
$$\mathbf{X}' = \begin{pmatrix} 1 & -2 & 2 \\ -2 & 1 & -2 \\ 2 & -2 & 1 \end{pmatrix}\mathbf{X}.$$

Solution Expanding the determinant in the characteristic equation

$$\det(\mathbf{A} - \lambda\mathbf{I}) = \begin{vmatrix} 1-\lambda & -2 & 2 \\ -2 & 1-\lambda & -2 \\ 2 & -2 & 1-\lambda \end{vmatrix} = 0$$

yields $-(\lambda + 1)^2(\lambda - 5) = 0$. We see that $\lambda_1 = \lambda_2 = -1$ and $\lambda_3 = 5$.

For $\lambda_1 = -1$, Gauss-Jordan elimination gives immediately

$$(\mathbf{A} + \mathbf{I}\,|\,\mathbf{0}) = \begin{pmatrix} 2 & -2 & 2 & | & 0 \\ -2 & 2 & -2 & | & 0 \\ 2 & -2 & 2 & | & 0 \end{pmatrix} \xrightarrow{\substack{\text{row} \\ \text{operations}}} \begin{pmatrix} 1 & -1 & 1 & | & 0 \\ 0 & 0 & 0 & | & 0 \\ 0 & 0 & 0 & | & 0 \end{pmatrix}.$$

From $k_1 - k_2 + k_3 = 0$ we can express, say, k_1 in terms of k_2 and k_3. By choosing $k_2 = 1$ and $k_3 = 0$ in $k_1 = k_2 - k_3$, we obtain $k_1 = 1$ and so one eigenvector is

$$\mathbf{K}_1 = \begin{pmatrix} 1 \\ 1 \\ 0 \end{pmatrix}.$$

But the choice $k_2 = 1$, $k_3 = 1$ implies $k_1 = 0$. Hence a second eigenvector is

$$\mathbf{K}_2 = \begin{pmatrix} 0 \\ 1 \\ 1 \end{pmatrix}.$$

Since neither eigenvector is a constant multiple of the other, we have found, corresponding to the same eigenvalue, two linearly independent solutions

$$\mathbf{X}_1 = \begin{pmatrix} 1 \\ 1 \\ 0 \end{pmatrix}e^{-t} \quad \text{and} \quad \mathbf{X}_2 = \begin{pmatrix} 0 \\ 1 \\ 1 \end{pmatrix}e^{-t}.$$

Last, for $\lambda_3 = 5$, the reduction

$$(\mathbf{A} - 5\mathbf{I}\,|\,\mathbf{0}) = \begin{pmatrix} -4 & -2 & 2 & | & 0 \\ -2 & -4 & -2 & | & 0 \\ 2 & -2 & -4 & | & 0 \end{pmatrix} \overset{\text{row operations}}{\Longrightarrow} \begin{pmatrix} 1 & 0 & -1 & | & 0 \\ 0 & 1 & 1 & | & 0 \\ 0 & 0 & 0 & | & 0 \end{pmatrix}$$

implies $k_1 = k_3$ and $k_2 = -k_3$. Picking $k_3 = 1$ gives $k_1 = 1$, $k_2 = -1$, and thus a third eigenvector is

$$\mathbf{K}_3 = \begin{pmatrix} 1 \\ -1 \\ 1 \end{pmatrix}.$$

We conclude that the general solution of the system is

$$\mathbf{X} = c_1 \begin{pmatrix} 1 \\ 1 \\ 0 \end{pmatrix} e^{-t} + c_2 \begin{pmatrix} 0 \\ 1 \\ 1 \end{pmatrix} e^{-t} + c_3 \begin{pmatrix} 1 \\ -1 \\ 1 \end{pmatrix} e^{5t}. \qquad \blacksquare$$

Second Solution

Now suppose λ_1 is an eigenvalue of multiplicity two and there is only one eigenvector associated with this value. A second solution can be found of the form

$$\mathbf{X}_2 = \mathbf{K}te^{\lambda_1 t} + \mathbf{P}e^{\lambda_1 t}, \qquad (18)$$

where

$$\mathbf{K} = \begin{pmatrix} k_1 \\ k_2 \\ \vdots \\ k_n \end{pmatrix} \quad \text{and} \quad \mathbf{P} = \begin{pmatrix} p_1 \\ p_2 \\ \vdots \\ p_n \end{pmatrix}.$$

To see this we substitute (18) into the system $\mathbf{X}' = \mathbf{A}\mathbf{X}$ and simplify:

$$(\mathbf{A}\mathbf{K} - \lambda_1\mathbf{K})te^{\lambda_1 t} + (\mathbf{A}\mathbf{P} - \lambda_1\mathbf{P} - \mathbf{K})e^{\lambda_1 t} = \mathbf{0}.$$

Since this last equation is to hold for all values of t, we must have

$$(\mathbf{A} - \lambda_1\mathbf{I})\mathbf{K} = \mathbf{0} \qquad (19)$$

and

$$(\mathbf{A} - \lambda_1\mathbf{I})\mathbf{P} = \mathbf{K}. \qquad (20)$$

Equation (19) simply states that $\mathbf{K}$ must be an eigenvector of $\mathbf{A}$ associated with λ_1. By solving (19), we find one solution $\mathbf{X}_1 = \mathbf{K}e^{\lambda_1 t}$. To find the second solution $\mathbf{X}_2$ we need only solve the additional system (20) for the vector $\mathbf{P}$.

EXAMPLE 6

Find the general solution of the system given in (16).

Solution From (17) we know that $\lambda_1 = -3$ and that one solution is

$$\mathbf{X}_1 = \begin{pmatrix} 3 \\ 1 \end{pmatrix} e^{-3t}.$$

Identifying $\mathbf{K} = \begin{pmatrix} 3 \\ 1 \end{pmatrix}$ and $\mathbf{P} = \begin{pmatrix} p_1 \\ p_2 \end{pmatrix}$, we find from (20) that we must now

solve $\qquad (\mathbf{A} + 3\mathbf{I})\mathbf{P} = \mathbf{K} \quad$ or $\quad \begin{pmatrix} 6 & -18 \\ 2 & -6 \end{pmatrix}\begin{pmatrix} p_1 \\ p_2 \end{pmatrix} = \begin{pmatrix} 3 \\ 1 \end{pmatrix}.$

Multiplying out this last expression gives

$$6p_1 - 18p_2 = 3$$

$$2p_1 - 6p_2 = 1.$$

Since this system is obviously equivalent to one equation, we have an infinite number of choices for p_1 and p_2. For example, by choosing $p_1 = 1$, we find $p_2 = 1/6$. However, for simplicity, we shall choose $p_1 = 1/2$ so that $p_2 = 0$. Hence $\mathbf{P} = \begin{pmatrix} \frac{1}{2} \\ 0 \end{pmatrix}$. Thus from (18) we find

$$\mathbf{X}_2 = \begin{pmatrix} 3 \\ 1 \end{pmatrix} t e^{-3t} + \begin{pmatrix} \frac{1}{2} \\ 0 \end{pmatrix} e^{-3t}.$$

The general solution of (16) is then

$$\mathbf{X} = c_1 \begin{pmatrix} 3 \\ 1 \end{pmatrix} e^{-3t} + c_2 \left[\begin{pmatrix} 3 \\ 1 \end{pmatrix} t e^{-3t} + \begin{pmatrix} \frac{1}{2} \\ 0 \end{pmatrix} e^{-3t} \right]. \qquad \blacksquare$$

Eigenvalues of Multiplicity Three

When a matrix $\mathbf{A}$ has only one eigenvector associated with an eigenvalue λ_1 of multiplicity three, we can find a second solution of form (18) and a third solution of the form

$$\mathbf{X}_3 = \mathbf{K}\frac{t^2}{2}e^{\lambda_1 t} + \mathbf{P}te^{\lambda_1 t} + \mathbf{Q}e^{\lambda_1 t}, \qquad (21)$$

where $\qquad \mathbf{K} = \begin{pmatrix} k_1 \\ k_2 \\ \vdots \\ k_n \end{pmatrix}, \quad \mathbf{P} = \begin{pmatrix} p_1 \\ p_2 \\ \vdots \\ p_n \end{pmatrix}, \quad$ and $\quad \mathbf{Q} = \begin{pmatrix} q_1 \\ q_2 \\ \vdots \\ q_n \end{pmatrix}.$

By substituting (21) into the system $\mathbf{X}' = \mathbf{A}\mathbf{X}$, we find the column vectors $\mathbf{K}$, $\mathbf{P}$, and $\mathbf{Q}$ must satisfy

$$(\mathbf{A} - \lambda_1 \mathbf{I})\mathbf{K} = \mathbf{0} \qquad (22)$$

$$(\mathbf{A} - \lambda_1 \mathbf{I})\mathbf{P} = \mathbf{K} \qquad (23)$$

and $\qquad\qquad\qquad (\mathbf{A} - \lambda_1 \mathbf{I})\mathbf{Q} = \mathbf{P}. \qquad (24)$

Of course the solutions of (22) and (23) can be utilized in the formulation of the solutions $\mathbf{X}_1$ and $\mathbf{X}_2$.

EXAMPLE 7

Solve
$$\mathbf{X}' = \begin{pmatrix} 2 & 1 & 6 \\ 0 & 2 & 5 \\ 0 & 0 & 2 \end{pmatrix} \mathbf{X}.$$

Solution The characteristic equation $(\lambda - 2)^3 = 0$ shows that $\lambda_1 = 2$ is an eigenvalue of multiplicity three. In succession we find that a solution of

$$(\mathbf{A} - 2\mathbf{I})\mathbf{K} = \mathbf{0} \quad \text{is} \quad \mathbf{K} = \begin{pmatrix} 1 \\ 0 \\ 0 \end{pmatrix};$$

a solution of

$$(\mathbf{A} - 2\mathbf{I})\mathbf{P} = \mathbf{K} \quad \text{is} \quad \mathbf{P} = \begin{pmatrix} 0 \\ 1 \\ 0 \end{pmatrix};$$

and finally a solution of

$$(\mathbf{A} - 2\mathbf{I})\mathbf{Q} = \mathbf{P} \quad \text{is} \quad \mathbf{Q} = \begin{pmatrix} 0 \\ -\frac{6}{5} \\ \frac{1}{5} \end{pmatrix}.$$

We see from (18) and (21) that the general solution of the system is

$$\mathbf{X} = c_1 \begin{pmatrix} 1 \\ 0 \\ 0 \end{pmatrix} e^{2t} + c_2 \left[\begin{pmatrix} 1 \\ 0 \\ 0 \end{pmatrix} te^{2t} + \begin{pmatrix} 0 \\ 1 \\ 0 \end{pmatrix} e^{2t} \right] + c_3 \left[\begin{pmatrix} 1 \\ 0 \\ 0 \end{pmatrix} \frac{t^2}{2} e^{2t} + \begin{pmatrix} 0 \\ 1 \\ 0 \end{pmatrix} te^{2t} + \begin{pmatrix} 0 \\ -\frac{6}{5} \\ \frac{1}{5} \end{pmatrix} e^{2t} \right]. \quad \blacksquare$$

8.6 EXERCISES

Answers to odd-numbered problems begin on page A–24.

[8.6.1] In Problems 1–12 find the general solution of the given system.

1. $\dfrac{dx}{dt} = x + 2y$

$\dfrac{dy}{dt} = 4x + 3y$

2. $\dfrac{dx}{dt} = 2y$

$\dfrac{dy}{dt} = 8x$

3. $\dfrac{dx}{dt} = -4x + 2y$

$\dfrac{dy}{dt} = -\dfrac{5}{2}x + 2y$

4. $\dfrac{dx}{dt} = \dfrac{1}{2}x + 9y$

$\dfrac{dy}{dt} = \dfrac{1}{2}x + 2y$

5. $\mathbf{X}' = \begin{pmatrix} 10 & -5 \\ 8 & -12 \end{pmatrix} \mathbf{X}$

6. $\mathbf{X}' = \begin{pmatrix} -6 & 2 \\ -3 & 1 \end{pmatrix} \mathbf{X}$

7. $\dfrac{dx}{dt} = x + y - z$

$\dfrac{dy}{dt} = \quad\; 2y$

$\dfrac{dz}{dt} = \quad\; y - z$

8. $\dfrac{dx}{dt} = 2x - 7y$

$\dfrac{dy}{dt} = 5x + 10y + 4z$

$\dfrac{dz}{dt} = \quad\;\; 5y + 2z$

9. $\mathbf{X}' = \begin{pmatrix} -1 & 1 & 0 \\ 1 & 2 & 1 \\ 0 & 3 & -1 \end{pmatrix} \mathbf{X}$

10. $\mathbf{X}' = \begin{pmatrix} 1 & 0 & 1 \\ 0 & 1 & 0 \\ 1 & 0 & 1 \end{pmatrix} \mathbf{X}$

11. $\mathbf{X}' = \begin{pmatrix} -1 & -1 & 0 \\ \frac{3}{4} & -\frac{3}{2} & 3 \\ \frac{1}{8} & \frac{1}{4} & -\frac{1}{2} \end{pmatrix} \mathbf{X}$

12. $\mathbf{X}' = \begin{pmatrix} -1 & 4 & 2 \\ 4 & -1 & -2 \\ 0 & 0 & 6 \end{pmatrix} \mathbf{X}$

In Problems 13 and 14 solve the given system subject to the indicated initial condition.

13. $\mathbf{X}' = \begin{pmatrix} \frac{1}{2} & 0 \\ 1 & -\frac{1}{2} \end{pmatrix} \mathbf{X}, \quad \mathbf{X}(0) = \begin{pmatrix} 3 \\ 5 \end{pmatrix}$

14. $\mathbf{X}' = \begin{pmatrix} 1 & 1 & 4 \\ 0 & 2 & 0 \\ 1 & 1 & 1 \end{pmatrix} \mathbf{X}, \quad \mathbf{X}(0) = \begin{pmatrix} 1 \\ 3 \\ 0 \end{pmatrix}$

[8.6.2] In Problems 15–26 find the general solution of the given system.

15. $\dfrac{dx}{dt} = 6x - y$

$\dfrac{dy}{dt} = 5x + 2y$

16. $\dfrac{dx}{dt} = \quad\; x + y$

$\dfrac{dy}{dt} = -2x - y$

17. $\dfrac{dx}{dt} = \quad\; 5x + y$

$\dfrac{dy}{dt} = -2x + 3y$

18. $\dfrac{dx}{dt} = \quad\; 4x + 5y$

$\dfrac{dy}{dt} = -2x + 6y$

19. $\mathbf{X}' = \begin{pmatrix} 4 & -5 \\ 5 & -4 \end{pmatrix} \mathbf{X}$

20. $\mathbf{X}' = \begin{pmatrix} 1 & -8 \\ 1 & -3 \end{pmatrix} \mathbf{X}$

21. $\dfrac{dx}{dt} = z$

$\dfrac{dy}{dt} = -z$

$\dfrac{dz}{dt} = y$

22. $\dfrac{dx}{dt} = \quad 2x + y + 2z$

$\dfrac{dy}{dt} = \quad 3x \quad\;\; + 6z$

$\dfrac{dz}{dt} = -4x \quad\;\; - 3z$

23. $\mathbf{X}' = \begin{pmatrix} 1 & -1 & 2 \\ -1 & 1 & 0 \\ -1 & 0 & 1 \end{pmatrix} \mathbf{X}$ **24.** $\mathbf{X}' = \begin{pmatrix} 4 & 0 & 1 \\ 0 & 6 & 0 \\ -4 & 0 & 4 \end{pmatrix} \mathbf{X}$

25. $\mathbf{X}' = \begin{pmatrix} 2 & 5 & 1 \\ -5 & -6 & 4 \\ 0 & 0 & 2 \end{pmatrix} \mathbf{X}$ **26.** $\mathbf{X}' = \begin{pmatrix} 2 & 4 & 4 \\ -1 & -2 & 0 \\ -1 & 0 & -2 \end{pmatrix} \mathbf{X}$

In Problems 27 and 28 solve the given system subject to the indicated initial condition.

27. $\mathbf{X}' = \begin{pmatrix} 1 & -12 & -14 \\ 1 & 2 & -3 \\ 1 & 1 & -2 \end{pmatrix} \mathbf{X}, \quad \mathbf{X}(0) = \begin{pmatrix} 4 \\ 6 \\ -7 \end{pmatrix}$

28. $\mathbf{X}' = \begin{pmatrix} 6 & -1 \\ 5 & 4 \end{pmatrix} \mathbf{X}, \quad \mathbf{X}(0) = \begin{pmatrix} -2 \\ 8 \end{pmatrix}$

[8.6.3]   In Problems 29–38 find the general solution of the given system.

29. $\dfrac{dx}{dt} = 3x - y$

$\dfrac{dy}{dt} = 9x - 3y$

30. $\dfrac{dx}{dt} = -6x + 5y$

$\dfrac{dy}{dt} = -5x + 4y$

31. $\dfrac{dx}{dt} = -x + 3y$

$\dfrac{dy}{dt} = -3x + 5y$

32. $\dfrac{dx}{dt} = 12x - 9y$

$\dfrac{dy}{dt} = 4x$

33. $\dfrac{dx}{dt} = 3x - y - z$

$\dfrac{dy}{dt} = x + y - z$

$\dfrac{dz}{dt} = x - y + z$

34. $\dfrac{dx}{dt} = 3x + 2y + 4z$

$\dfrac{dy}{dt} = 2x \qquad + 2z$

$\dfrac{dz}{dt} = 4x + 2y + 3z$

35. $\mathbf{X}' = \begin{pmatrix} 5 & -4 & 0 \\ 1 & 0 & 2 \\ 0 & 2 & 5 \end{pmatrix} \mathbf{X}$ **36.** $\mathbf{X}' = \begin{pmatrix} 1 & 0 & 0 \\ 0 & 3 & 1 \\ 0 & -1 & 1 \end{pmatrix} \mathbf{X}$

37. $\mathbf{X}' = \begin{pmatrix} 1 & 0 & 0 \\ 2 & 2 & -1 \\ 0 & 1 & 0 \end{pmatrix} \mathbf{X}$ **38.** $\mathbf{X}' = \begin{pmatrix} 4 & 1 & 0 \\ 0 & 4 & 1 \\ 0 & 0 & 4 \end{pmatrix} \mathbf{X}$

In Problems 39 and 40 solve the given system subject to the indicated initial condition.

39. $\mathbf{X}' = \begin{pmatrix} 2 & 4 \\ -1 & 6 \end{pmatrix} \mathbf{X}, \quad \mathbf{X}(0) = \begin{pmatrix} -1 \\ 6 \end{pmatrix}$

40. $\mathbf{X}' = \begin{pmatrix} 0 & 0 & 1 \\ 0 & 1 & 0 \\ 1 & 0 & 0 \end{pmatrix} \mathbf{X}, \quad \mathbf{X}(0) = \begin{pmatrix} 1 \\ 2 \\ 5 \end{pmatrix}$

Miscellaneous Problems

If $\boldsymbol{\Phi}(t)$ is a fundamental matrix of the system, the initial-value problem $\mathbf{X}' = \mathbf{AX}$, $\mathbf{X}(t_0) = \mathbf{X}_0$, has the solution $\mathbf{X} = \boldsymbol{\Phi}(t)\boldsymbol{\Phi}^{-1}(t_0)\mathbf{X}_0$ (see Problem 35, Exercises 8.5). In Problems 41 and 42 use this result to solve the given system subject to the indicated initial condition.

41. $\mathbf{X}' = \begin{pmatrix} 4 & 3 \\ 3 & -4 \end{pmatrix} \mathbf{X}, \quad \mathbf{X}(0) = \begin{pmatrix} 1 \\ 1 \end{pmatrix}$

42. $\mathbf{X}' = \begin{pmatrix} -\frac{2}{25} & \frac{1}{50} \\ \frac{2}{25} & -\frac{2}{25} \end{pmatrix} \mathbf{X}, \quad \mathbf{X}(0) = \begin{pmatrix} 25 \\ 0 \end{pmatrix}$

In Problems 43 and 44 find a solution of the given system of the form $\mathbf{X} = t^\lambda \mathbf{K}$, $t > 0$, where $\mathbf{K}$ is a column vector of constants.

43. $t\mathbf{X}' = \begin{pmatrix} 1 & 3 \\ -1 & 5 \end{pmatrix} \mathbf{X}$ **44.** $t\mathbf{X}' = \begin{pmatrix} 2 & -2 \\ 2 & 7 \end{pmatrix} \mathbf{X}$

8.7 UNDETERMINED COEFFICIENTS

The methods of **undetermined coefficients** and **variation of parameters** can both be adapted to the solution of a nonhomogeneous linear system $\mathbf{X}' = \mathbf{AX} + \mathbf{F}(t)$. Of these two methods, variation of parameters is the more powerful technique. However, there are a few instances when the method of undetermined coefficients gives a quick means of finding a particular solution $\mathbf{X}_p$.

EXAMPLE I

Solve the system $\quad \mathbf{X}' = \begin{pmatrix} -1 & 2 \\ -1 & 1 \end{pmatrix} \mathbf{X} + \begin{pmatrix} -8 \\ 3 \end{pmatrix} \quad$ on $(-\infty, \infty)$.

Solution We first solve the homogeneous system

$$\mathbf{X}' = \begin{pmatrix} -1 & 2 \\ -1 & 1 \end{pmatrix} \mathbf{X}.$$

The characteristic equation

$$\det(\mathbf{A} - \lambda\mathbf{I}) = \begin{vmatrix} -1 - \lambda & 2 \\ -1 & 1 - \lambda \end{vmatrix} = \lambda^2 + 1 = 0$$

yields the complex eigenvalues $\lambda_1 = i$ and $\lambda_2 = \bar{\lambda}_1 = -i$. By the procedures of the last section, we find

$$\mathbf{X}_c = c_1 \begin{pmatrix} \cos t + \sin t \\ \cos t \end{pmatrix} + c_2 \begin{pmatrix} \cos t - \sin t \\ -\sin t \end{pmatrix}.$$

Now since $\mathbf{F}(t)$ is a constant vector, we assume a constant particular solution vector $\mathbf{X}_p = \begin{pmatrix} a_1 \\ b_1 \end{pmatrix}$. Substituting this latter assumption into the original system leads to

$$0 = -a_1 + 2b_1 - 8$$

$$0 = -a_1 + b_1 + 3.$$

Solving this system of algebraic equations gives $a_1 = 14$ and $b_1 = 11$, and so $\mathbf{X}_p = \begin{pmatrix} 14 \\ 11 \end{pmatrix}$. The general solution of the system is

$$\mathbf{X} = c_1 \begin{pmatrix} \cos t + \sin t \\ \cos t \end{pmatrix} + c_2 \begin{pmatrix} \cos t - \sin t \\ -\sin t \end{pmatrix} + \begin{pmatrix} 14 \\ 11 \end{pmatrix}. \qquad \blacksquare$$

EXAMPLE 2

Solve the system
$$\frac{dx}{dt} = 6x + y + 6t \qquad \text{on } (-\infty, \infty).$$

$$\frac{dy}{dt} = 4x + 3y - 10t + 4$$

Solution We first solve the homogeneous system

$$\frac{dx}{dt} = 6x + y$$

$$\frac{dy}{dt} = 4x + 3y$$

by the method of Section 8.6. The eigenvalues are determined from

$$\det(\mathbf{A} - \lambda \mathbf{I}) = \begin{vmatrix} 6 - \lambda & 1 \\ 4 & 3 - \lambda \end{vmatrix} = \lambda^2 - 9\lambda + 14 = 0.$$

Since $\lambda^2 - 9\lambda + 14 = (\lambda - 2)(\lambda - 7)$, we have $\lambda_1 = 2$ and $\lambda_2 = 7$. It is then easily verified that the respective eigenvectors of the coefficient matrix are

$$\mathbf{K}_1 = \begin{pmatrix} 1 \\ -4 \end{pmatrix} \quad \text{and} \quad \mathbf{K}_2 = \begin{pmatrix} 1 \\ 1 \end{pmatrix}.$$

Consequently the complementary function is

$$\mathbf{X}_c = c_1 \begin{pmatrix} 1 \\ -4 \end{pmatrix} e^{2t} + c_2 \begin{pmatrix} 1 \\ 1 \end{pmatrix} e^{7t}.$$

Because $\mathbf{F}(t)$ can be written as

$$\mathbf{F}(t) = \begin{pmatrix} 6 \\ -10 \end{pmatrix} t + \begin{pmatrix} 0 \\ 4 \end{pmatrix},$$

we shall try to find a particular solution of the system possessing the *same* form:

$$\mathbf{X}_p = \begin{pmatrix} a_2 \\ b_2 \end{pmatrix} t + \begin{pmatrix} a_1 \\ b_1 \end{pmatrix}.$$

In matrix terms we must have

$$\mathbf{X}'_p = \begin{pmatrix} 6 & 1 \\ 4 & 3 \end{pmatrix} \mathbf{X}_p + \begin{pmatrix} 6 \\ -10 \end{pmatrix} t + \begin{pmatrix} 0 \\ 4 \end{pmatrix}$$

or

$$\begin{pmatrix} a_2 \\ b_2 \end{pmatrix} = \begin{pmatrix} 6 & 1 \\ 4 & 3 \end{pmatrix} \left[\begin{pmatrix} a_2 \\ b_2 \end{pmatrix} t + \begin{pmatrix} a_1 \\ b_1 \end{pmatrix} \right] + \begin{pmatrix} 6 \\ -10 \end{pmatrix} t + \begin{pmatrix} 0 \\ 4 \end{pmatrix}$$

$$\begin{pmatrix} 0 \\ 0 \end{pmatrix} = \begin{pmatrix} (6a_2 + b_2 + 6)t + 6a_1 + b_1 - a_2 \\ (4a_2 + 3b_2 - 10)t + 4a_1 + 3b_1 - b_2 + 4 \end{pmatrix}.$$

From this last identity we conclude that

$$6a_2 + b_2 + 6 = 0 \qquad\qquad 6a_1 + b_1 - a_2 \quad\ = 0$$
$$\text{and}$$
$$4a_2 + 3b_2 - 10 = 0 \qquad\qquad 4a_1 + 3b_1 - b_2 + 4 = 0.$$

Solving the first two equations simultaneously yields $a_2 = -2$ and $b_2 = 6$. Substituting these values into the last two equations and solving for a_1 and b_1 give $a_1 = -4/7$, $b_1 = 10/7$. It follows, therefore, that a particular solution vector is

$$\mathbf{X}_p = \begin{pmatrix} -2 \\ 6 \end{pmatrix} t + \begin{pmatrix} -\frac{4}{7} \\ \frac{10}{7} \end{pmatrix},$$

and so the general solution of the system on $(-\infty, \infty)$ is

$$\mathbf{X} = \mathbf{X}_c + \mathbf{X}_p$$

$$= c_1 \begin{pmatrix} 1 \\ -4 \end{pmatrix} e^{2t} + c_2 \begin{pmatrix} 1 \\ 1 \end{pmatrix} e^{7t} + \begin{pmatrix} -2 \\ 6 \end{pmatrix} t + \begin{pmatrix} -\frac{4}{7} \\ \frac{10}{7} \end{pmatrix}. \qquad \blacksquare$$

EXAMPLE 3

Determine the form of the particular solution vector $\mathbf{X}_p$ for

$$\frac{dx}{dt} = 5x + 3y - 2e^{-t} + 1$$

$$\frac{dy}{dt} = -x + y + e^{-t} - 5t + 7.$$

Solution   Proceeding in the usual manner, we find

$$\mathbf{X}_c = c_1 \begin{pmatrix} 1 \\ -1 \end{pmatrix} e^{2t} + c_2 \begin{pmatrix} 3 \\ -1 \end{pmatrix} e^{4t}.$$

Now since

$$\mathbf{F}(t) = \begin{pmatrix} -2 \\ 1 \end{pmatrix} e^{-t} + \begin{pmatrix} 0 \\ -5 \end{pmatrix} t + \begin{pmatrix} 1 \\ 7 \end{pmatrix},$$

we assume a particular solution of the form

$$\mathbf{X}_p = \begin{pmatrix} a_3 \\ b_3 \end{pmatrix} e^{-t} + \begin{pmatrix} a_2 \\ b_2 \end{pmatrix} t + \begin{pmatrix} a_1 \\ b_1 \end{pmatrix}. \qquad \blacksquare$$

Remark   The method of undetermined coefficients is not as simple as the last three examples seem to indicate. As in Section 4.4 the method can be applied only when the entries in the matrix $\mathbf{F}(t)$ are constants, polynomials, exponential functions, sines and cosines, or finite sums and products of these functions. There are further difficulties. The assumption for $\mathbf{X}_p$ is actually predicated on a prior knowledge of the complementary function $\mathbf{X}_c$. For example, if $\mathbf{F}(t)$ is a constant vector and $\lambda = 0$ is an eigenvalue, then $\mathbf{X}_c$ contains a constant vector. In this case $\mathbf{X}_p$ is *not* a constant vector as in Example 1 but rather

$$\mathbf{X}_p = \begin{pmatrix} a_2 \\ b_2 \end{pmatrix} t + \begin{pmatrix} a_1 \\ b_1 \end{pmatrix}.$$

See Problem 11.

Similarly, in Example 3, if we replace e^{-t} in $\mathbf{F}(t)$ by e^{2t} ($\lambda = 2$ is an eigenvalue), then the correct form of the particular solution is

$$\mathbf{X}_p = \begin{pmatrix} a_4 \\ b_4 \end{pmatrix} t e^{2t} + \begin{pmatrix} a_3 \\ b_3 \end{pmatrix} e^{2t} + \begin{pmatrix} a_2 \\ b_2 \end{pmatrix} t + \begin{pmatrix} a_1 \\ b_1 \end{pmatrix}.$$

Rather than pursue these difficulties, we turn our attention now to the method of variation of parameters.

8.7 EXERCISES

Answers to odd-numbered problems begin on page A–25.

In Problems 1–8 use the method of undetermined coefficients to solve the given system on $(-\infty, \infty)$.

1. $\dfrac{dx}{dt} = 2x + 3y - 7$

   $\dfrac{dy}{dt} = -x - 2y + 5$

2. $\dfrac{dx}{dt} = 5x + 9y + 2$

   $\dfrac{dy}{dt} = -x + 11y + 6$

3. $\dfrac{dx}{dt} = x + 3y - 2t^2$

$\dfrac{dy}{dt} = 3x + y + t + 5$

4. $\dfrac{dx}{dt} = x - 4y + 4t + 9e^{6t}$

$\dfrac{dy}{dt} = 4x + y - t + e^{6t}$

5. $\mathbf{X}' = \begin{pmatrix} 4 & \frac{1}{3} \\ 9 & 6 \end{pmatrix}\mathbf{X} + \begin{pmatrix} -3 \\ 10 \end{pmatrix}e^t$

6. $\mathbf{X}' = \begin{pmatrix} -1 & 5 \\ -1 & 1 \end{pmatrix}\mathbf{X} + \begin{pmatrix} \sin t \\ -2\cos t \end{pmatrix}$

7. $\mathbf{X}' = \begin{pmatrix} 1 & 1 & 1 \\ 0 & 2 & 3 \\ 0 & 0 & 5 \end{pmatrix}\mathbf{X} + \begin{pmatrix} 1 \\ -1 \\ 2 \end{pmatrix}e^{4t}$

8. $\mathbf{X}' = \begin{pmatrix} 0 & 0 & 5 \\ 0 & 5 & 0 \\ 5 & 0 & 0 \end{pmatrix}\mathbf{X} + \begin{pmatrix} 5 \\ -10 \\ 40 \end{pmatrix}$

9. Solve $\mathbf{X}' = \begin{pmatrix} -1 & -2 \\ 3 & 4 \end{pmatrix}\mathbf{X} + \begin{pmatrix} 3 \\ 3 \end{pmatrix}$ subject to $\mathbf{X}(0) = \begin{pmatrix} -4 \\ 5 \end{pmatrix}$.

10. **(a)** Show that the system of differential equations for the currents $i_2(t)$ and $i_3(t)$ in the electrical network shown in Figure 8.11 is

$$\frac{d}{dt}\begin{pmatrix} i_2 \\ i_3 \end{pmatrix} = \begin{pmatrix} -R_1/L_1 & -R_1/L_1 \\ -R_1/L_2 & -(R_1 + R_2)/L_2 \end{pmatrix}\begin{pmatrix} i_2 \\ i_3 \end{pmatrix} + \begin{pmatrix} E/L_1 \\ E/L_2 \end{pmatrix}.$$

(b) Solve the system in part (a) if $R_1 = 2$ ohms, $R_2 = 3$ ohms, $L_1 = 1$ henry, $L_2 = 1$ henry, $E = 60$ volts, $i_2(0) = 0$, and $i_3(0) = 0$.

(c) Determine the current $i_1(t)$.

11. Solve the system $\mathbf{X}' = \begin{pmatrix} 1 & -1 \\ -1 & 1 \end{pmatrix}\mathbf{X} + \begin{pmatrix} 3 \\ -5 \end{pmatrix}$ on $(-\infty, \infty)$.

[*Hint:* A particular solution may not be unique.]

Figure 8.11

8.8 VARIATION OF PARAMETERS

In Section 8.5 we saw that the general solution of a homogeneous system $\mathbf{X}' = \mathbf{AX}$ can be written as the product

$$\mathbf{X} = \mathbf{\Phi}(t)\mathbf{C},$$

where $\mathbf{\Phi}(t)$ is a fundamental matrix of the system and $\mathbf{C}$ is an $n \times 1$ column vector of constants. As in the procedure of Section 4.5, we ask whether it is possible to replace $\mathbf{C}$ by a column matrix of functions

$$\mathbf{U}(t) = \begin{pmatrix} u_1(t) \\ u_2(t) \\ \vdots \\ u_n(t) \end{pmatrix}$$

so that
$$\mathbf{X}_p = \mathbf{\Phi}(t)\mathbf{U}(t) \tag{1}$$

is a particular solution of the nonhomogeneous system

$$\mathbf{X}' = \mathbf{AX} + \mathbf{F}(t).$$ (2)

By the product rule* the derivative of (1) is

$$\mathbf{X}'_p = \boldsymbol{\Phi}(t)\mathbf{U}'(t) + \boldsymbol{\Phi}'(t)\mathbf{U}(t).$$ (3)

Substituting (3) and (1) into (2) gives

$$\boldsymbol{\Phi}(t)\mathbf{U}'(t) + \boldsymbol{\Phi}'(t)\mathbf{U}(t) = \mathbf{A}\boldsymbol{\Phi}(t)\mathbf{U}(t) + \mathbf{F}(t).$$ (4)

Now recall from (14) of Section 8.5 that $\boldsymbol{\Phi}'(t) = \mathbf{A}\boldsymbol{\Phi}(t)$. Thus (4) becomes

$$\boldsymbol{\Phi}(t)\mathbf{U}'(t) + \mathbf{A}\boldsymbol{\Phi}(t)\mathbf{U}(t) = \mathbf{A}\boldsymbol{\Phi}(t)\mathbf{U}(t) + \mathbf{F}(t)$$

or

$$\boldsymbol{\Phi}(t)\mathbf{U}'(t) = \mathbf{F}(t).$$ (5)

Multiplying both sides of equation (5) by $\boldsymbol{\Phi}^{-1}(t)$ gives

$$\mathbf{U}'(t) = \boldsymbol{\Phi}^{-1}(t)\mathbf{F}(t)$$

or

$$\mathbf{U}(t) = \int \boldsymbol{\Phi}^{-1}(t)\mathbf{F}(t)\, dt.$$

Hence by assumption (1) we conclude that a particular solution of (2) is given by

$$\mathbf{X}_p = \boldsymbol{\Phi}(t) \int \boldsymbol{\Phi}^{-1}(t)\mathbf{F}(t)\, dt.$$ (6)

To calculate the indefinite integral of the column matrix $\boldsymbol{\Phi}^{-1}(t)\mathbf{F}(t)$ in (6) we integrate each entry. Thus the general solution of the system (2) is $\mathbf{X} = \mathbf{X}_c + \mathbf{X}_p$ or

$$\mathbf{X} = \boldsymbol{\Phi}(t)\mathbf{C} + \boldsymbol{\Phi}(t) \int \boldsymbol{\Phi}^{-1}(t)\mathbf{F}(t)\, dt.$$ (7)

EXAMPLE I

Find the general solution of the nonhomogeneous system

$$\mathbf{X}' = \begin{pmatrix} -3 & 1 \\ 2 & -4 \end{pmatrix} \mathbf{X} + \begin{pmatrix} 3t \\ e^{-t} \end{pmatrix}$$ (8)

on the interval $(-\infty, \infty)$.

---

* See Problem 51, Exercises 8.4. Note that the order of the products is very important. Since $\mathbf{U}(t)$ is a column matrix, the products $\mathbf{U}'(t)\boldsymbol{\Phi}(t)$ and $\mathbf{U}(t)\boldsymbol{\Phi}'(t)$ are not defined.

Solution   We first solve the homogeneous system

$$\mathbf{X}' = \begin{pmatrix} -3 & 1 \\ 2 & -4 \end{pmatrix} \mathbf{X}. \tag{9}$$

The characteristic equation of the coefficient matrix is

$$\det(\mathbf{A} - \lambda \mathbf{I}) = \begin{vmatrix} -3 - \lambda & 1 \\ 2 & -4 - \lambda \end{vmatrix} = (\lambda + 2)(\lambda + 5) = 0$$

and so the eigenvalues are $\lambda_1 = -2$ and $\lambda_2 = -5$. By the usual method we find that the eigenvectors corresponding to λ_1 and λ_2 are, respectively,

$$\begin{pmatrix} 1 \\ 1 \end{pmatrix} \quad \text{and} \quad \begin{pmatrix} 1 \\ -2 \end{pmatrix}.$$

The solution vectors of the system (9) are then

$$\mathbf{X}_1 = \begin{pmatrix} 1 \\ 1 \end{pmatrix} e^{-2t} \quad \text{and} \quad \mathbf{X}_2 = \begin{pmatrix} 1 \\ -2 \end{pmatrix} e^{-5t}.$$

Next we form

$$\mathbf{\Phi}(t) = \begin{pmatrix} e^{-2t} & e^{-5t} \\ e^{-2t} & -2e^{-5t} \end{pmatrix} \quad \text{and} \quad \mathbf{\Phi}^{-1}(t) = \begin{pmatrix} \frac{2}{3}e^{2t} & \frac{1}{3}e^{2t} \\ \frac{1}{3}e^{5t} & -\frac{1}{3}e^{5t} \end{pmatrix}.$$

From (6) we then obtain

$$\mathbf{X}_p = \mathbf{\Phi}(t) \int \mathbf{\Phi}^{-1}(t) \mathbf{F}(t) \, dt$$

$$= \begin{pmatrix} e^{-2t} & e^{-5t} \\ e^{-2t} & -2e^{-5t} \end{pmatrix} \int \begin{pmatrix} \frac{2}{3}e^{2t} & \frac{1}{3}e^{2t} \\ \frac{1}{3}e^{5t} & -\frac{1}{3}e^{5t} \end{pmatrix} \begin{pmatrix} 3t \\ e^{-t} \end{pmatrix} dt$$

$$= \begin{pmatrix} e^{-2t} & e^{-5t} \\ e^{-2t} & -2e^{-5t} \end{pmatrix} \int \begin{pmatrix} 2te^{2t} + \frac{1}{3}e^{t} \\ te^{5t} - \frac{1}{3}e^{4t} \end{pmatrix} dt$$

$$= \begin{pmatrix} e^{-2t} & e^{-5t} \\ e^{-2t} & -2e^{-5t} \end{pmatrix} \begin{pmatrix} te^{2t} - \frac{1}{2}e^{2t} + \frac{1}{3}e^{t} \\ \frac{1}{5}te^{5t} - \frac{1}{25}e^{5t} - \frac{1}{12}e^{4t} \end{pmatrix}$$

$$= \begin{pmatrix} \frac{6}{5}t - \frac{27}{50} + \frac{1}{4}e^{-t} \\ \frac{3}{5}t - \frac{21}{50} + \frac{1}{2}e^{-t} \end{pmatrix}.$$

Hence from (7) the general solution of (8) on the interval is

$$\mathbf{X} = \begin{pmatrix} e^{-2t} & e^{-5t} \\ e^{-2t} & -2e^{-5t} \end{pmatrix} \begin{pmatrix} c_1 \\ c_2 \end{pmatrix} + \begin{pmatrix} \frac{6}{5}t - \frac{27}{50} + \frac{1}{4}e^{-t} \\ \frac{3}{5}t - \frac{21}{50} + \frac{1}{2}e^{-t} \end{pmatrix}$$

$$= c_1 \begin{pmatrix} 1 \\ 1 \end{pmatrix} e^{-2t} + c_2 \begin{pmatrix} 1 \\ -2 \end{pmatrix} e^{-5t} + \begin{pmatrix} \frac{6}{5} \\ \frac{3}{5} \end{pmatrix} t - \begin{pmatrix} \frac{27}{50} \\ \frac{21}{50} \end{pmatrix} + \begin{pmatrix} \frac{1}{4} \\ \frac{1}{2} \end{pmatrix} e^{-t}. \qquad \blacksquare$$

The general solution of (2) on an interval can be written in the alternative manner

$$\mathbf{X} = \mathbf{\Phi}(t)\mathbf{C} + \mathbf{\Phi}(t)\int_{t_0}^{t} \mathbf{\Phi}^{-1}(s)\mathbf{F}(s)\,ds, \tag{10}$$

where t and t_0 are points in the interval. This last form is useful in solving (2) subject to an initial condition $\mathbf{X}(t_0) = \mathbf{X}_0$. Substituting $t = t_0$ in (10) yields

$$\mathbf{X}_0 = \mathbf{\Phi}(t_0)\mathbf{C},$$

from which we see immediately that

$$\mathbf{C} = \mathbf{\Phi}^{-1}(t_0)\mathbf{X}_0.$$

We conclude that the solution of the initial-value problem is given by

$$\mathbf{X} = \mathbf{\Phi}(t)\mathbf{\Phi}^{-1}(t_0)\mathbf{X}_0 + \mathbf{\Phi}(t)\int_{t_0}^{t} \mathbf{\Phi}^{-1}(s)\mathbf{F}(s)\,ds. \tag{11}$$

Recall from Section 8.5 that an alternative way of forming a fundamental matrix is to choose its column vectors $\mathbf{V}_i$ in such a manner that

$$\mathbf{V}_1(t_0) = \begin{pmatrix} 1 \\ 0 \\ \vdots \\ 0 \end{pmatrix}, \quad \mathbf{V}_2(t_0) = \begin{pmatrix} 0 \\ 1 \\ \vdots \\ 0 \end{pmatrix}, \quad \ldots, \quad \mathbf{V}_n(t_0) = \begin{pmatrix} 0 \\ 0 \\ \vdots \\ 1 \end{pmatrix}. \tag{12}$$

This fundamental matrix is denoted by $\mathbf{\Psi}(t)$. As a consequence of (12) we know that $\mathbf{\Psi}(t)$ has the property

$$\mathbf{\Psi}(t_0) = \mathbf{I}. \tag{13}$$

But since $\mathbf{\Psi}(t)$ is nonsingular for all values of t in an interval, (13) implies

$$\mathbf{\Psi}^{-1}(t_0) = \mathbf{I}. \tag{14}$$

Thus when $\mathbf{\Psi}(t)$ is used rather than $\mathbf{\Phi}(t)$, it follows from (14) that (11) can be written

$$\mathbf{X} = \mathbf{\Psi}(t)\mathbf{X}_0 + \mathbf{\Psi}(t)\int_{t_0}^{t} \mathbf{\Psi}^{-1}(s)\mathbf{F}(s)\,ds. \tag{15}$$

8.8 **EXERCISES**   *Answers to odd-numbered problems begin on page A–26.*

In Problems 1–20 use variation of parameters to solve the given system.

1. $\dfrac{dx}{dt} = 3x - 3y + 4$

   $\dfrac{dy}{dt} = 2x - 2y - 1$

2. $\dfrac{dx}{dt} = 2x - y$

   $\dfrac{dy}{dt} = 3x - 2y + 4t$

3. $\mathbf{X}' = \begin{pmatrix} 3 & -5 \\ \frac{3}{4} & -1 \end{pmatrix} \mathbf{X} + \begin{pmatrix} 1 \\ -1 \end{pmatrix} e^{t/2}$

4. $\mathbf{X}' = \begin{pmatrix} 2 & -1 \\ 4 & 2 \end{pmatrix} \mathbf{X} + \begin{pmatrix} \sin 2t \\ 2\cos 2t \end{pmatrix} e^{2t}$

5. $\mathbf{X}' = \begin{pmatrix} 0 & 2 \\ -1 & 3 \end{pmatrix} \mathbf{X} + \begin{pmatrix} 1 \\ -1 \end{pmatrix} e^{t}$

6. $\mathbf{X}' = \begin{pmatrix} 0 & 2 \\ -1 & 3 \end{pmatrix} \mathbf{X} + \begin{pmatrix} 2 \\ e^{-3t} \end{pmatrix}$

7. $\mathbf{X}' = \begin{pmatrix} 1 & 8 \\ 1 & -1 \end{pmatrix} \mathbf{X} + \begin{pmatrix} 12 \\ 12 \end{pmatrix} t$

8. $\mathbf{X}' = \begin{pmatrix} 1 & 8 \\ 1 & -1 \end{pmatrix} \mathbf{X} + \begin{pmatrix} e^{-t} \\ te^{t} \end{pmatrix}$

9. $\mathbf{X}' = \begin{pmatrix} 3 & 2 \\ -2 & -1 \end{pmatrix} \mathbf{X} + \begin{pmatrix} 2e^{-t} \\ e^{-t} \end{pmatrix}$

10. $\mathbf{X}' = \begin{pmatrix} 3 & 2 \\ -2 & -1 \end{pmatrix} \mathbf{X} + \begin{pmatrix} 1 \\ 1 \end{pmatrix}$

11. $\mathbf{X}' = \begin{pmatrix} 0 & -1 \\ 1 & 0 \end{pmatrix} \mathbf{X} + \begin{pmatrix} \sec t \\ 0 \end{pmatrix}$

12. $\mathbf{X}' = \begin{pmatrix} 1 & -1 \\ 1 & 1 \end{pmatrix} \mathbf{X} + \begin{pmatrix} 3 \\ 3 \end{pmatrix} e^{t}$

13. $\mathbf{X}' = \begin{pmatrix} 1 & -1 \\ 1 & 1 \end{pmatrix} \mathbf{X} + \begin{pmatrix} \cos t \\ \sin t \end{pmatrix} e^{t}$

14. $\mathbf{X}' = \begin{pmatrix} 2 & -2 \\ 8 & -6 \end{pmatrix} \mathbf{X} + \begin{pmatrix} 1 \\ 3 \end{pmatrix} \frac{e^{-2t}}{t}$

15. $\mathbf{X}' = \begin{pmatrix} 0 & 1 \\ -1 & 0 \end{pmatrix} \mathbf{X} + \begin{pmatrix} 0 \\ \sec t \tan t \end{pmatrix}$

16. $\mathbf{X}' = \begin{pmatrix} 0 & 1 \\ -1 & 0 \end{pmatrix} \mathbf{X} + \begin{pmatrix} 1 \\ \cot t \end{pmatrix}$

17. $\mathbf{X}' = \begin{pmatrix} 1 & 2 \\ -\frac{1}{2} & 1 \end{pmatrix} \mathbf{X} + \begin{pmatrix} \csc t \\ \sec t \end{pmatrix} e^{t}$

18. $\mathbf{X}' = \begin{pmatrix} 1 & -2 \\ 1 & -1 \end{pmatrix} \mathbf{X} + \begin{pmatrix} \tan t \\ 1 \end{pmatrix}$

19. $\mathbf{X}' = \begin{pmatrix} 1 & 1 & 0 \\ 1 & 1 & 0 \\ 0 & 0 & 3 \end{pmatrix} \mathbf{X} + \begin{pmatrix} e^{t} \\ e^{2t} \\ te^{3t} \end{pmatrix}$

20. $\mathbf{X}' = \begin{pmatrix} 3 & -1 & -1 \\ 1 & 1 & -1 \\ 1 & -1 & 1 \end{pmatrix} \mathbf{X} + \begin{pmatrix} 0 \\ t \\ 2e^{t} \end{pmatrix}$

In Problems 21 and 22 use (11) to solve the given system subject to the indicated initial condition.

21. $\mathbf{X}' = \begin{pmatrix} 3 & -1 \\ -1 & 3 \end{pmatrix} \mathbf{X} + \begin{pmatrix} 4e^{2t} \\ 4e^{4t} \end{pmatrix}, \quad \mathbf{X}(0) = \begin{pmatrix} 1 \\ 1 \end{pmatrix}$

22. $\mathbf{X}' = \begin{pmatrix} 1 & -1 \\ 1 & -1 \end{pmatrix} \mathbf{X} + \begin{pmatrix} 1/t \\ 1/t \end{pmatrix}, \quad \mathbf{X}(1) = \begin{pmatrix} 2 \\ -1 \end{pmatrix}$

In Problems 23 and 24 use (15) to solve the given system subject to the indicated initial condition. Use the results of Problems 31 and 34 in Exercises 8.5.

23. $\mathbf{X}' = \begin{pmatrix} 4 & 1 \\ 6 & 5 \end{pmatrix} \mathbf{X} + \begin{pmatrix} 50e^{7t} \\ 0 \end{pmatrix}, \quad \mathbf{X}(0) = \begin{pmatrix} 5 \\ -5 \end{pmatrix}$

24. $\mathbf{X}' = \begin{pmatrix} 3 & -2 \\ 5 & -3 \end{pmatrix} \mathbf{X} + \begin{pmatrix} 2 \\ 3 \end{pmatrix}, \quad \mathbf{X}(\pi/2) = \begin{pmatrix} 0 \\ 0 \end{pmatrix}$

25. (a) Show that the system of differential equations for the currents $i_1(t)$ and $i_2(t)$ in the electrical network shown in Figure 8.12 is

$$\frac{d}{dt}\begin{pmatrix} i_1 \\ i_2 \end{pmatrix} = \begin{pmatrix} -(R_1 + R_2)/L_2 & R_2/L_2 \\ R_2/L_1 & -R_2/L_1 \end{pmatrix}\begin{pmatrix} i_1 \\ i_2 \end{pmatrix} + \begin{pmatrix} E/L_2 \\ 0 \end{pmatrix}.$$

(b) Solve the system in part (a) if $R_1 = 8$ ohms, $R_2 = 3$ ohms, $L_1 = 1$ henry, $L_2 = 1$ henry, $E(t) = 100 \sin t$ volts, $i_1(0) = 0$, and $i_2(0) = 0$.

Figure 8.12

[o] 8.9 MATRIX EXPONENTIAL

Matrices can be utilized in an entirely different manner to solve a homogeneous system of linear first-order differential equations.

Recall that the simple linear first-order differential equation

$$x' = ax,$$

where a is a constant, has the general solution

$$x = ce^{at}.$$

It seems natural then to ask whether we can define a matrix exponential $e^{t\mathbf{A}}$ so that the homogeneous system

$$\mathbf{X}' = \mathbf{A}\mathbf{X},$$

where $\mathbf{A}$ is an $n \times n$ matrix of constants, has a solution

$$\mathbf{X} = e^{t\mathbf{A}}\mathbf{C}. \tag{1}$$

Since $\mathbf{C}$ is to be an $n \times 1$ column vector of arbitrary constants, we want $e^{t\mathbf{A}}$ to be an $n \times n$ matrix. While the complete development of the meaning of the **matrix exponential** would necessitate a more thorough investigation of matrix algebra, one means of computing $e^{t\mathbf{A}}$ is given in the following definition.

DEFINITION 8.20 **Matrix Exponential**

For any $n \times n$ matrix $\mathbf{A}$,

$$e^{t\mathbf{A}} = \sum_{n=0}^{\infty} \frac{(t\mathbf{A})^n}{n!} = \mathbf{I} + t\mathbf{A} + \frac{t^2}{2!}\mathbf{A}^2 + \frac{t^2}{3!}\mathbf{A}^3 + \cdots. \tag{2}$$

It can be shown that the series given in (2) converges to an $n \times n$ matrix for every value of t. Also, $\mathbf{A}^2 = \mathbf{A}\mathbf{A}$, $\mathbf{A}^3 = \mathbf{A}(\mathbf{A}^2)$, and so on.

Now the general solution of the single differential equation

$$x' = ax + f(t),$$

where a is a constant, can be expressed as

$$x = x_c + x_p = ce^{at} + e^{at}\int_{t_0}^{t} e^{-as}f(s)\, ds.$$

For systems of linear first-order differential equations, it can be shown that the general solution of

$$\mathbf{X}' = \mathbf{A}\mathbf{X} + \mathbf{F}(t),$$

where $\mathbf{A}$ is an $n \times n$ matrix of constants, is

$$\mathbf{X} = \mathbf{X}_c + \mathbf{X}_p = e^{t\mathbf{A}}\mathbf{C} + e^{t\mathbf{A}} \int_{t_0}^{t} e^{-s\mathbf{A}}\mathbf{F}(s)\, ds. \tag{3}$$

The matrix exponential $e^{t\mathbf{A}}$ is always nonsingular and $e^{-s\mathbf{A}} = (e^{s\mathbf{A}})^{-1}$. In practice, $e^{-s\mathbf{A}}$ can be obtained from $e^{t\mathbf{A}}$ by replacing t by $-s$.

Additional Properties

From (2) it is seen that

$$e^{0} = \mathbf{I}. \tag{4}$$

Also, formal termwise differentiation of (2) shows that

$$\frac{d}{dt} e^{t\mathbf{A}} = \mathbf{A}e^{t\mathbf{A}}. \tag{5}$$

If we denote the matrix exponential by $\mathbf{\Psi}(t)$, then (5) and (4) are equivalent to

$$\mathbf{\Psi}'(t) = \mathbf{A}\mathbf{\Psi}(t) \tag{6}$$

and

$$\mathbf{\Psi}(0) = \mathbf{I}, \tag{7}$$

respectively. The notation here is chosen deliberately. Comparing (6) with (14) of Section 8.5 reveals that $e^{t\mathbf{A}}$ is a fundamental matrix of the system $\mathbf{X}' = \mathbf{A}\mathbf{X}$. It is precisely this formulation of the fundamental matrix that was discussed on pages 448 and 449 of Section 8.5.

By multiplying the series defining $e^{t\mathbf{A}}$ and $e^{-s\mathbf{A}}$, we can prove that

$$e^{t\mathbf{A}}e^{-s\mathbf{A}} = e^{(t-s)\mathbf{A}} \quad \text{or equivalently} \quad \mathbf{\Psi}(t)\mathbf{\Psi}^{-1}(s) = \mathbf{\Psi}(t-s).*$$

This last result enables us to relate (3) to (10) of the preceding section

$$\mathbf{X} = e^{t\mathbf{A}}\mathbf{C} + \int_{t_0}^{t} e^{t\mathbf{A}}e^{-s\mathbf{A}}\mathbf{F}(s)\, ds$$

$$= e^{t\mathbf{A}}\mathbf{C} + \int_{t_0}^{t} e^{(t-s)\mathbf{A}}\mathbf{F}(s)\, ds \tag{8}$$

$$\mathbf{X} = \mathbf{\Psi}(t)\mathbf{C} + \int_{t_0}^{t} \mathbf{\Psi}(t-s)\mathbf{F}(s)\, ds. \tag{9}$$

Equation (9) possesses a form simpler than (10) of Section 8.8. In other words, there is no need to compute $\mathbf{\Psi}^{-1}$; we need only replace t by $t - s$ in $\mathbf{\Psi}(t)$.

---

* Although $e^{t\mathbf{A}}e^{-s\mathbf{A}} = e^{(t-s)\mathbf{A}}$, it is interesting to note that $e^{\mathbf{A}}e^{\mathbf{B}}$ is, in general, not the same as $e^{\mathbf{A}+\mathbf{B}}$ for $n \times n$ matrices $\mathbf{A}$ and $\mathbf{B}$.

8.9 EXERCISES

Answers to odd-numbered problems begin on page A–26.

In Problems 1 and 2 use (2) to compute $e^{t\mathbf{A}}$ and $e^{-t\mathbf{A}}$.

1. $\mathbf{A} = \begin{pmatrix} 0 & 1 \\ 1 & 0 \end{pmatrix}$

2. $\mathbf{A} = \begin{pmatrix} 1 & 0 \\ 0 & 2 \end{pmatrix}$

In Problems 3 and 4 use (1) to find the general solution of each system.

3. $\mathbf{X}' = \begin{pmatrix} 0 & 1 \\ 1 & 0 \end{pmatrix}\mathbf{X}$

4. $\mathbf{X}' = \begin{pmatrix} 1 & 0 \\ 0 & 2 \end{pmatrix}\mathbf{X}$

In Problems 5–8 use (3) to find the general solution of each system.

5. $\mathbf{X}' = \begin{pmatrix} 0 & 1 \\ 1 & 0 \end{pmatrix}\mathbf{X} + \begin{pmatrix} 1 \\ 1 \end{pmatrix}$

6. $\mathbf{X}' = \begin{pmatrix} 0 & 1 \\ 1 & 0 \end{pmatrix}\mathbf{X} + \begin{pmatrix} \cosh t \\ \sinh t \end{pmatrix}$

7. $\mathbf{X}' = \begin{pmatrix} 1 & 0 \\ 0 & 2 \end{pmatrix}\mathbf{X} + \begin{pmatrix} t \\ e^{4t} \end{pmatrix}$

8. $\mathbf{X}' = \begin{pmatrix} 1 & 0 \\ 0 & 2 \end{pmatrix}\mathbf{X} + \begin{pmatrix} 3 \\ -1 \end{pmatrix}$

Miscellaneous Problems

Let $\mathbf{P}$ denote a matrix whose columns are eigenvectors $\mathbf{K}_1, \mathbf{K}_2, \ldots, \mathbf{K}_n$ corresponding to *distinct* eigenvalues $\lambda_1, \lambda_2, \ldots, \lambda_n$ of an $n \times n$ matrix $\mathbf{A}$. Then it can be shown that $\mathbf{A} = \mathbf{PDP}^{-1}$, where $\mathbf{D}$ is defined by

$$\mathbf{D} = \begin{pmatrix} \lambda_1 & 0 & \cdots & 0 \\ 0 & \lambda_2 & \cdots & 0 \\ \vdots & & & \vdots \\ 0 & 0 & \cdots & \lambda_n \end{pmatrix}. \tag{10}$$

In Problems 9 and 10 verify the above result for the given matrix.

9. $\mathbf{A} = \begin{pmatrix} 2 & 1 \\ -3 & 6 \end{pmatrix}$

10. $\mathbf{A} = \begin{pmatrix} 2 & 1 \\ 1 & 2 \end{pmatrix}$

11. Suppose $\mathbf{A} = \mathbf{PDP}^{-1}$, where $\mathbf{D}$ is defined in (10). Use (2) to show that $e^{t\mathbf{A}} = \mathbf{P}e^{t\mathbf{D}}\mathbf{P}^{-1}$.

12. Use (2) to show that

$$e^{t\mathbf{D}} = \begin{pmatrix} e^{\lambda_1 t} & 0 & \cdots & 0 \\ 0 & e^{\lambda_2 t} & \cdots & 0 \\ \vdots & & & \vdots \\ 0 & 0 & \cdots & e^{\lambda_n t} \end{pmatrix},$$

where $\mathbf{D}$ is defined in (10).

In Problems 13 and 14 use the results of Problems 9–12 to solve the given system.

13. $\mathbf{X}' = \begin{pmatrix} 2 & 1 \\ -3 & 6 \end{pmatrix} \mathbf{X}$ **14.** $\mathbf{X}' = \begin{pmatrix} 2 & 1 \\ 1 & 2 \end{pmatrix} \mathbf{X}$

Chapter 8 *REVIEW*

Throughout this chapter we have considered **systems of linear differential equations**.

For linear systems probably the most basic technique of solution consists of rewriting the entire system in **operator notation** and then using **systematic elimination** to obtain single differential equations in one dependent variable that can be solved by the usual procedures. Also, **determinants** can usually be utilized to accomplish the same result. Once all dependent variables have been determined, it is necessary to use the system itself to find various relationships between the parameters.

When initial conditions are specified, the **Laplace transform** can be used to reduce the system to simultaneous algebraic equations in the transformed functions.

A first-order system in **normal form** in *two* dependent variables is any system

$$\frac{dx}{dt} = a_{11}(t)x + a_{12}(t)y + f_1(t)$$

$$\frac{dy}{dt} = a_{21}(t)x + a_{22}(t)y + f_2(t), \tag{1}$$

where the coefficients $a_{ij}(t)$, $f_1(t)$, and $f_2(t)$ are continuous on some common interval I. When $f_1(t) = 0$, $f_2(t) = 0$, the system is said to be **homogeneous**; otherwise it is **nonhomogeneous**. Any linear second-order differential equation can be expressed in this form.

Using matrices, we can write the system (1) compactly as

$$\frac{d\mathbf{X}}{dt} = \mathbf{A}(t)\mathbf{X} + \mathbf{F}(t), \tag{2}$$

where

$$\mathbf{X} = \begin{pmatrix} x \\ y \end{pmatrix}, \quad \mathbf{A}(t) = \begin{pmatrix} a_{11}(t) & a_{12}(t) \\ a_{21}(t) & a_{22}(t) \end{pmatrix}, \quad \text{and} \quad \mathbf{F}(t) = \begin{pmatrix} f_1(t) \\ f_2(t) \end{pmatrix}.$$

The **general solution of the homogeneous system** in two dependent variables

$$\frac{d\mathbf{X}}{dt} = \mathbf{A}(t)\mathbf{X} \tag{3}$$

is defined to be the linear combination

$$\mathbf{X} = c_1\mathbf{X}_1 + c_2\mathbf{X}_2, \tag{4}$$

where $\mathbf{X}_1$ and $\mathbf{X}_2$ form a **fundamental set of solutions** of (3) on I. The **general solution of the nonhomogeneous system** (2) is defined to be

$$\mathbf{X} = \mathbf{X}_c + \mathbf{X}_p,$$

where $\mathbf{X}_c$ is defined by (4) and $\mathbf{X}_p$ is *any* solution vector of (2).

To solve a homogeneous system (3) we determine the **eigenvalues** of the coefficient matrix $\mathbf{A}$ and then find the corresponding **eigenvectors**.

To solve a nonhomogeneous system we first solve the associated homogeneous system. A particular solution vector $\mathbf{X}_p$ of the nonhomogeneous system is found by either **undetermined coefficients** or **variation of parameters**.

A **fundamental matrix** of a homogeneous system (3) in two dependent variables is defined to be

$$\mathbf{\Phi}(t) = \begin{pmatrix} x_1 & x_2 \\ y_1 & y_2 \end{pmatrix}. \tag{5}$$

The columns in (5) are obtained from two linearly independent solution vectors $\mathbf{X}_1$ and $\mathbf{X}_2$ of (3). In terms of matrices, the method of variation of parameters leads to a particular solution given by

$$\mathbf{X}_p = \mathbf{\Phi}(t) \int \mathbf{\Phi}^{-1}(t)\mathbf{F}(t)\, dt.$$

The general solution of (2) on an interval is

$$\mathbf{X} = \mathbf{\Phi}(t)\mathbf{C} + \mathbf{\Phi}(t) \int \mathbf{\Phi}^{-1}(t)\mathbf{F}(t)\, dt,$$

where $\mathbf{C}$ is a column matrix containing two arbitrary constants. The matrix $\mathbf{\Phi}^{-1}(t)$ is called the **multiplicative inverse** of $\mathbf{\Phi}(t)$; the multiplicative inverse satisfies $\mathbf{\Phi}(t)\mathbf{\Phi}^{-1}(t) = \mathbf{\Phi}^{-1}(t)\mathbf{\Phi}(t) = \mathbf{I}$, where $\mathbf{I}$ is the 2×2 **multiplicative identity**.

Chapter 8   *REVIEW EXERCISES*

Answers to odd-numbered problems begin on page A–26.

Answer Problems 1–12 without referring back to the text. Fill in the blank or answer true/false.

1. Every second-order linear differential equation can be expressed as a system of two linear first-order differential equations. _____

2. If $\mathbf{A} = \begin{pmatrix} 1 \\ 2 \end{pmatrix}$ and $\mathbf{B} = (3 \quad 4)$, then $\mathbf{AB} = $ _____ and $\mathbf{BA} = $ _____.

3. If $\mathbf{A} = \begin{pmatrix} 1 & 2 \\ 3 & 4 \end{pmatrix}$, then $\mathbf{A}^{-1} = $ _____.

4. If $\mathbf{A}$ is a nonsingular matrix for which $\mathbf{AB} = \mathbf{AC}$, then $\mathbf{B} = \mathbf{C}$. _____

5. If $\mathbf{X}_1$ is a solution of $\mathbf{X}' = \mathbf{AX}$ and $\mathbf{X}_2$ is a solution of $\mathbf{X}' = \mathbf{AX} + \mathbf{F}$, then $\mathbf{X} = \mathbf{X}_1 + \mathbf{X}_2$ is a solution of $\mathbf{X}' = \mathbf{AX} + \mathbf{F}$. _____

6. A fundamental matrix $\mathbf{\Phi}$ of a system $\mathbf{X}' = \mathbf{AX}$ is always nonsingular. _____

7. Let $\mathbf{A}$ be an $n \times n$ matrix. The eigenvalues of $\mathbf{A}$ are the nonzero solutions of $\det(\mathbf{A} - \lambda\mathbf{I}) = 0$. _____

8. A nonzero constant multiple of an eigenvector is an eigenvector corresponding to the same eigenvalue. _____

9. An $n \times 1$ column vector $\mathbf{K}$ with all zero entries is never an eigenvector of an $n \times n$ matrix $\mathbf{A}$. _____

10. Let $\mathbf{A}$ be an $n \times n$ matrix with real entries. If λ is a complex eigenvalue, then $\bar{\lambda}$ is also an eigenvalue of $\mathbf{A}$. _____

11. An $n \times n$ matrix $\mathbf{A}$ always possesses n linearly independent eigenvectors. _____

12. The augmented matrix

$$\begin{pmatrix} 1 & 1 & 1 & | & 2 \\ 0 & 1 & 0 & | & 3 \\ 0 & 0 & 0 & | & 0 \end{pmatrix}$$

is in reduced row-echelon form. _____

In Problems 13–16 use systematic elimination or determinants to solve the given system.

13. $x' + \ y' = 2x + 2y + 1$
$\quad\ x' + 2y' = \qquad y + 3$

14. $\dfrac{dx}{dt} = 2x + \ y + t - 2$

$\quad\ \dfrac{dy}{dt} = 3x + 4y - 4t$

15. $(D - 2)x - \qquad\quad y = -e^t$
$\qquad -3x + (D - 4)y = -7e^t$

16. $(D + 2)x + (D + 1)y = \sin 2t$
$\qquad\ 5x + (D + 3)y = \cos 2t$

In Problems 17 and 18 use the Laplace transform to solve each system.

17. $\ x' + y = t$
$\quad 4x + y' = 0$
$\quad x(0) = 1, \ y(0) = 2$

18. $\ x'' + y'' = \ e^{2t}$
$\quad 2x' + y'' = -e^{2t}$
$\quad x(0) = 0, \ y(0) = 0$
$\quad x'(0) = 0, \ y'(0) = 0$

19. (a) Write as one column matrix $\mathbf{X}$:

$$\begin{pmatrix} 3 & 1 & 1 \\ -1 & 2 & -1 \\ 0 & -2 & 4 \end{pmatrix}\begin{pmatrix} t \\ t^2 \\ t^3 \end{pmatrix} - \begin{pmatrix} 2 \\ -2 \\ -1 \end{pmatrix} + 2t\begin{pmatrix} 1 \\ 0 \\ 4 \end{pmatrix} + 2t^2\begin{pmatrix} 1 \\ -1 \\ 7 \end{pmatrix}.$$

(b) Find $d\mathbf{X}/dt$.

20. Write the differential equation

$$3y^{(4)} - 5y'' + 9y = 6e^t - 2t$$

as a system of first-order equations in linear normal form.

21. Write the system

$$(2D^2 + D)y - \qquad D^2 x = \ln t$$

$$D^2 y + (D + 1)x = 5t - 2$$

as a system of first-order equations in linear normal form.

22. Verify that the general solution of the system

$$\frac{dx}{dt} = y$$

$$\frac{dy}{dt} = -x + 2y - 2\cos t$$

on the interval $(-\infty, \infty)$ is

$$\mathbf{X} = c_1\begin{pmatrix} 1 \\ 1 \end{pmatrix}e^t + c_2\left[\begin{pmatrix} 1 \\ 1 \end{pmatrix}te^t + \begin{pmatrix} 0 \\ 1 \end{pmatrix}e^t\right] + \begin{pmatrix} \sin t \\ \cos t \end{pmatrix}.$$

In Problems 23–28 use the concept of eigenvalues and eigenvectors to solve each system.

23. $\dfrac{dx}{dt} = 2x + y$

$\dfrac{dy}{dt} = -x$

24. $\dfrac{dx}{dt} = -4x + 2y$

$\dfrac{dy}{dt} = 2x - 4y$

25. $\mathbf{X}' = \begin{pmatrix} 1 & 2 \\ -2 & 1 \end{pmatrix}\mathbf{X}$

26. $\mathbf{X}' = \begin{pmatrix} -2 & 5 \\ -2 & 4 \end{pmatrix}\mathbf{X}$

27. $\mathbf{X}' = \begin{pmatrix} 1 & 1 & 1 \\ 1 & 1 & 1 \\ 1 & 1 & 1 \end{pmatrix}\mathbf{X}$

28. $\mathbf{X}' = \begin{pmatrix} 1 & -1 & 1 \\ 0 & 1 & 3 \\ 4 & 3 & 1 \end{pmatrix}\mathbf{X}$

In Problems 29–32 use either undetermined coefficients or variation of parameters to solve the given system.

29. $X' = \begin{pmatrix} 2 & 8 \\ 0 & 4 \end{pmatrix} X + \begin{pmatrix} 2 \\ 16t \end{pmatrix}$

30. $\dfrac{dx}{dt} = \quad x + 2y$

$\dfrac{dy}{dt} = -\dfrac{1}{2}x + \ y + e^t \tan t$

31. $X' = \begin{pmatrix} -1 & 1 \\ -2 & 1 \end{pmatrix} X + \begin{pmatrix} 1 \\ \cot t \end{pmatrix}$

32. $X' = \begin{pmatrix} 3 & 1 \\ -1 & 1 \end{pmatrix} X + \begin{pmatrix} -2 \\ 1 \end{pmatrix} e^{2t}$

C H 9 A P T E R

NUMERICAL METHODS FOR ORDINARY DIFFERENTIAL EQUATIONS

9.1 Direction Fields
9.2 The Euler Methods
9.3 The Three-Term Taylor Method
9.4 The Runge-Kutta Method
9.5 Multistep Methods
9.6 Errors and Stability
9.7 Higher-Order Equations and Systems

9.8 Second-Order Boundary-Value Problems

Chapter 9 Review
Chapter 9 Review Exercises
Essay: Nerve Impulse Models

Important Concepts

Lineal elements
Isocline
Slope field
Euler's method
Step size
Absolute error
Relative error
Percentage relative error
Improved Euler's method
Three-term Taylor method
Fourth-order Runge-Kutta method
Single-step methods
Multistep methods
Adams-Bashforth/ Adams-Moulton method
Milne's method
Stable method
Round-off error
Local truncation error
Global truncation error

A differential equation does not have to have a solution, and even if a solution exists, we may not always be able to find (that is, exhibit) an explicit or implicit solution of the equation. In many instances, particularly in the study of nonlinear equations, we may have to be content with an approximation to the solution.

If a solution of a differential equation exists, it represents a locus of points (points connected by a smooth curve) in the Cartesian plane. Beginning in Section 9.2 we shall consider numerical procedures that utilize the differential equation to obtain a sequence of distinct points whose coordinates, as shown in the accompanying figure, approximate the coordinates of the points on the actual solution curve.

Our primary focus in this chapter is on first-order initial-value problems: $dy/dx = f(x, y)$, $y(x_0) = y_0$. We shall see that the numerical procedures developed for first-order equations may be adapted to systems of first-order equations in a very natural manner. As a consequence, we can approximate solutions of higher-

Continues

Difference quotient
Finite difference
Finite difference equation

Continues

order initial-value problems by simply reducing the differential equation to a system of first-order equations (see Section 8.3). In Section 9.8 we shall turn our attention to a numerical procedure for linear second-order boundary-value problems.

We begin with the study of direction fields. Although it is not a numerical method, the concept of a direction field enables us to obtain a rough sketch of a solution of a first-order differential equation without actually solving it.

9.1 DIRECTION FIELDS

Lineal Elements

Suppose for the moment that we do not know the general solution of the simple equation $y' = y$. Specifically, the differential equation implies that the slopes of tangent lines to a solution curve are given by the function $f(x, y) = y$. When $f(x, y)$ is held constant, that is, when

$$y = c, \tag{1}$$

where c is any constant, we are in effect stating that the slope of the tangents to the solution curves is the same constant value along a horizontal line. For example, for $y = 2$ let us draw a sequence of short line segments, or **lineal elements**, each having slope 2 and its midpoint on the line. As shown in Figure 9.1, the solution curves pass through this horizontal line at every point tangent to the lineal elements.

solution curves

Figure 9.1

Isoclines and Direction Fields

Equation (1) represents a one-parameter family of horizontal lines. In general, any member of the family $f(x, y) = c$ is called an **isocline**, which literally means a curve along which the inclination (of the tangents) is the same. As the parameter c is varied, we obtain a collection of isoclines on which the lineal elements are judiciously constructed. The totality of these lineal elements is called a **direction field**, **slope field**, or **lineal element field** of the differential equation $y' = f(x, y)$. As we see in Figure 9.2(a), the direction field suggests the "flow pattern" for the family of solution curves of the differential equation $y' = y$. In particular, if we want the one solution that passes through the point $(0, 1)$, then, as indicated in Figure 9.2(b), we construct a curve through this point and passing through the isoclines with the appropriate slopes.

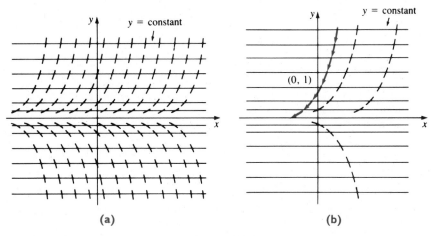

(a)   (b)

Figure 9.2

EXAMPLE 1

Determine the isoclines for the differential equation

$$\frac{dy}{dx} = 4x^2 + 9y^2.$$

Solution   For $c > 0$ the isoclines are the curves

$$4x^2 + 9y^2 = c.$$

As Figure 9.3 shows, the curves are a concentric family of ellipses with major axis along the x-axis.   ∎

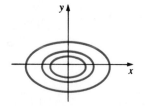

Figure 9.3

EXAMPLE 2

Sketch the direction field and indicate several possible members of the family of solution curves for

$$\frac{dy}{dx} = \frac{x}{y}.$$

Solution   Before sketching the direction field corresponding to the isoclines $x/y = c$ or $y = x/c$, we note that the differential equation gives the following information:

(a) If a solution curve crosses the x-axis ($y = 0$), it does so tangent to a vertical lineal element at every point except possibly $(0, 0)$.

(b) If a solution curve crosses the y-axis ($x = 0$), it does so tangent to a horizontal lineal element at every point except possibly $(0, 0)$.

(c) The lineal elements corresponding to the isoclines $c = 1$ and $c = -1$ are collinear with the lines $y = x$ and $y = -x$, respectively. Indeed, it is easily verified that these isoclines are both particular solutions of the given differential equation. However, it should be noted that *in general* isoclines are themselves not solutions to a differential equation. See Example 4.

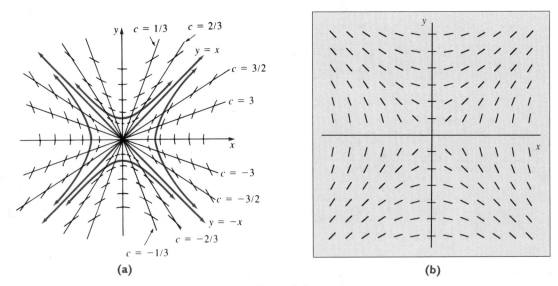

(a)

(b)

Figure 9.4

Figure 9.4(a) shows the direction field and several possible solution curves. Remember, on any particular isocline all the lineal elements are parallel. Also, the lineal elements may be drawn in such a manner as to suggest the flow of a particular curve. In other words, imagine the isoclines so close together that if the lineal elements were connected, we would have a polygonal curve suggestive of the shape of a smooth curve. Alternatively, the lineal elements can be drawn uniformly spaced on their isoclines as shown in the computer-generated version of the same direction field given in Figure 9.4(b). Note, however, the isoclines themselves are not drawn in part (b).   ∎

EXAMPLE 3

In Section 2.1 we indicated that the differential equation

$$\frac{dy}{dx} = x^2 + y^2$$

cannot be solved in terms of elementary functions. Use a direction field to locate an approximate solution satisfying $y(0) = 1$.

Solution   The isoclines are concentric circles defined by

$$x^2 + y^2 = c, \qquad c > 0.$$

By choosing $c = 1/4$, $c = 1$, $c = 9/4$, and $c = 4$, we obtain the circles with radii 1/2, 1, 3/2, and 2 shown in Figure 9.5(a). The lineal elements superimposed on each circle have slope corresponding to the particular value of c. It seems plausible from inspection of Figure 9.5(a) that a solution curve of the given initial-value problem might have the shape given in Figure 9.5(b). Unfortunately, we are not able to obtain any formula that describes this curve.

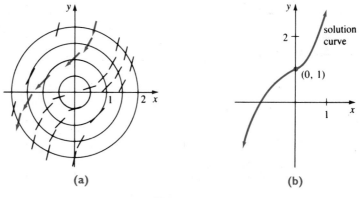

Figure 9.5

The concept of the direction field is used primarily to establish the existence of, and possibly to locate an approximate solution curve for, a first-order differential equation that cannot be solved by the usual standard techniques. However, the preceding discussion is of little value in determining specific values of a solution $y(x)$ at given points. For example, if we want to know the approximate value of $y(0.5)$ for the solution of

$$\frac{dy}{dx} = x^2 + y^2, \qquad y(0) = 1,$$

then Figure 9.5(b) can do nothing more than indicate that $y(0.5)$ may be in the same "ball park" as $y = 2$.

When the isoclines are straight lines it is easy to determine which, if any, of these isoclines are also particular solutions of the differential equation.

EXAMPLE 4

The isoclines of the differential equation

$$y' = 2x + y \tag{2}$$

are the straight lines

$$2x + y = c. \tag{3}$$

A line in this latter family will be a solution of the differential equation whenever its slope is the same as c. In other words, both the original equation and the line satisfy $y' = c$. Since the slope of (3) is -2 if we choose $c = -2$, then $2x + y = -2$ is a solution of (2). ∎

Remark Sketching a direction field by hand is a straightforward but time-consuming task. If available to you, computer software such as *The Grapher* supplement to this text is recommended.

9.1 **EXERCISES**   *Answers to odd-numbered problems begin on page A–27.*

In Problems 1–4 use the given computer-generated direction field to sketch several possible solution curves for the indicated differential equation.

1. $y' = xy$ **2.** $y' = 1 - xy$

Figure 9.6

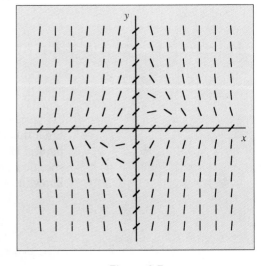

Figure 9.7

3. $y' = y - x$ **4.** $y' = \cos x / \sin y$

Figure 9.8

Figure 9.9

In Problem 5–14 identify the isoclines for the given differential equation.

5. $\dfrac{dy}{dx} = x + 4$

6. $\dfrac{dy}{dx} = 2x + y$

7. $\dfrac{dy}{dx} = x^2 - y^2$

8. $\dfrac{dy}{dx} = y - x^2$

9. $y' = \sqrt{x^2 + y^2 + 2y + 1}$

10. $y' = (x^2 + y^2)^{-1}$

11. $\dfrac{dy}{dx} = y(x + y)$

12. $\dfrac{dy}{dx} = y + e^x$

13. $\dfrac{dy}{dx} = \dfrac{y - 1}{x - 2}$

14. $\dfrac{dy}{dx} = \dfrac{x - y}{x + y}$

In Problems 15–22 sketch, or use a computer to obtain, the direction field for the given differential equation. Indicate several possible solution curves.

15. $y' = x$

16. $y' = x + y$

17. $y\dfrac{dy}{dx} = -x$

18. $\dfrac{dy}{dx} = \dfrac{1}{y}$

19. $\dfrac{dy}{dx} = 0.2x^2 + y$

20. $\dfrac{dy}{dx} = xe^y$

21. $y' = y - \cos\dfrac{\pi}{2} x$

22. $y' = 1 - \dfrac{y}{x}$

23. Formally show that the isoclines for the differential equation

$$\frac{dy}{dx} = \frac{\alpha x + \beta y}{\gamma x + \delta y}$$

are straight lines through the origin.

24. Show that $y = cx$ is a solution of the differential equation in Problem 23 if and only if $(\beta - \gamma)^2 + 4\alpha\delta \geq 0$.

In Problems 25–30 find those isoclines that are also solutions of the given differential equation. See Problems 23 and 24 and Example 4.

25. $y' = 3x + 2y$

26. $y' = 6x - 2y$

27. $y' = \dfrac{2x}{y}$

28. $y' = \dfrac{2y}{x + y}$

29. $\dfrac{dy}{dx} = \dfrac{4x + 3y}{y}$

30. $\dfrac{dy}{dx} = \dfrac{5x + 10y}{-4x + 3y}$

9.2 THE EULER METHODS

Euler's Method

One of the simplest techniques for approximating solutions of differential equations is known as **Euler's method**, or the method of **tangent lines**. Suppose we wish to approximate the solution of the initial-value problem

$$y' = f(x, y), \qquad y(x_0) = y_0.$$

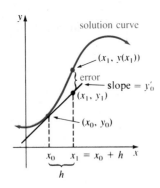

If h is a positive increment on the x-axis, then as Figure 9.10 shows, we can find a point $(x_1, y_1) = (x_0 + h, y_1)$ on the line tangent to the unknown solution curve at (x_0, y_0).

By the point-slope form of the equation of a line, we have

$$\frac{y_1 - y_0}{(x_0 + h) - x_0} = y_0' \quad \text{or} \quad y_1 = y_0 + hy_0',$$

Figure 9.10

where $y_0' = f(x_0, y_0)$. If we label $x_0 + h$ by x_1, then the point (x_1, y_1) on the tangent line is an approximation to the point $(x_1, y(x_1))$ on the solution curve; that is, $y_1 \approx y(x_1)$. Of course the accuracy of the approximation depends heavily on the size of the increment h. Usually we must choose this **step size** to be "reasonably small."

Assuming a uniform (constant) value of h, we can obtain a succession of points $(x_1, y_1), (x_2, y_2), \ldots, (x_n, y_n)$, which we hope are close to the points $(x_1, y(x_1)), (x_2, y(x_2)), \ldots, (x_n, y(x_n))$. See Figure 9.11. Now using (x_1, y_1), we can obtain the value of y_2, which is the ordinate of a point on a new "tangent"

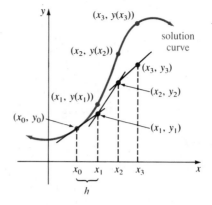

Figure 9.11

line. We have

$$\frac{y_2 - y_1}{h} = y'_1 \quad \text{or} \quad y_2 = y_1 + hy'_1 = y_1 + hf(x_1, y_1).$$

In general it follows that

$$y_{n+1} = y_n + hy'_n$$
$$= y_n + hf(x_n, y_n), \tag{1}$$

where $x_n = x_0 + nh$.

As an example, suppose we try the iteration scheme (1) on a differential equation for which we know the explicit solution; in this way we can compare the estimated values y_n and the true values $y(x_n)$.

EXAMPLE 1

Consider the initial-value problem

$$y' = 0.2xy, \qquad y(1) = 1.$$

Use the Euler method to obtain an approximation to $y(1.5)$ using first $h = 0.1$ and then $h = 0.05$.

Solution We first identify $f(x, y) = 0.2xy$ so that (1) becomes

$$y_{n+1} = y_n + h(0.2x_n y_n).$$

Then for $h = 0.1$ we find

$$y_1 = y_0 + (0.1)(0.2x_0y_0)$$
$$= 1 + (0.1)[0.2(1)(1)] = 1.02,$$

which is an estimate to the value of $y(1.1)$. However, if we use $h = 0.05$, it takes *two* iterations to reach $x = 1.1$. We have

$$y_1 = 1 + (0.05)[0.2(1)(1)] = 1.01$$
$$y_2 = 1.01 + (0.05)[0.2(1.05)(1.01)] = 1.020605.$$

Here we note that $y_1 \approx y(1.05)$ and $y_2 \approx y(1.1)$. The remainder of the calculations are summarized in Tables 9.1 and 9.2. Each entry is rounded to four decimal places.

TABLE 9.1 Euler's Method with $h = 0.1$

x_n	y_n	True Value	Abs. Error	% Rel. Error
1.00	1.0000	1.0000	0.0000	0.00
1.10	1.0200	1.0212	0.0012	0.12
1.20	1.0424	1.0450	0.0025	0.24
1.30	1.0675	1.0714	0.0040	0.37
1.40	1.0952	1.1008	0.0055	0.50
1.50	1.1259	1.1331	0.0073	0.64

TABLE 9.2 Euler's Method with $h = 0.05$

x_n	y_n	True Value	Abs. Error	% Rel. Error
1.00	1.0000	1.0000	0.0000	0.00
1.05	1.0100	1.0103	0.0003	0.03
1.10	1.0206	1.0212	0.0006	0.06
1.15	1.0318	1.0328	0.0009	0.09
1.20	1.0437	1.0450	0.0013	0.12
1.25	1.0562	1.0579	0.0016	0.16
1.30	1.0694	1.0714	0.0020	0.19
1.35	1.0833	1.0857	0.0024	0.22
1.40	1.0980	1.1008	0.0028	0.25
1.45	1.1133	1.1166	0.0032	0.29
1.50	1.1295	1.1331	0.0037	0.32

In Example 1 the true values were calculated from the known solution $y = e^{0.1(x^2 - 1)}$. Also, the **absolute error** is defined to be

$$|true\ value - approximation|.$$

The **relative error** and the **percentage relative error** are, in turn,

$$\frac{|true\ value - approximation|}{|true\ value|}$$

and
$$\frac{|true\ value - approximation|}{|true\ value|} \times 100 = \frac{absolute\ error}{|true\ value|} \times 100.$$

Computer software enables us to examine approximations to the graph of the solution $y(x)$ of an initial-value problem by plotting straight lines through the points (x_n, y_n) generated by Euler's method. In Figure 9.12 we have compared, on the interval $[1, 3]$, the graph of the exact solution of the initial-value problem in Example 1 with the graphs obtained from Euler's method using the step sizes $h = 1$, $h = 0.5$, and $h = 0.1$. It is apparent from the figure that the approximation improves as the step size decreases.

Figure 9.12

Although we see that the percentage relative error in Tables 9.1 and 9.2 is growing, it does not appear to be that bad. But you should not be deceived by Example 1 and Figure 9.12. Watch what happens in the next example when we simply change the coefficient 0.2 of the differential equation in Example 1 to the number 2.

EXAMPLE 2

Use the Euler method to obtain the approximate value of $y(1.5)$ for the solution of

$$y' = 2xy, \qquad y(1) = 1.$$

Solution You should verify that the exact or analytic solution is now $y = e^{x^2 - 1}$. Proceeding as in Example 1, we obtain the following results.

TABLE 9.3 Euler's Method with $h = 0.1$

x_n	y_n	True Value	Abs. Error	% Rel. Error
1.00	1.0000	1.0000	0.0000	0.00
1.10	1.2000	1.2337	0.0337	2.73
1.20	1.4640	1.5527	0.0887	5.71
1.30	1.8154	1.9937	0.1784	8.95
1.40	2.2874	2.6117	0.3244	12.42
1.50	2.9278	3.4904	0.5625	16.12

TABLE 9.4 Euler's Method with $h = 0.05$

x_n	y_n	True Value	Abs. Error	% Rel. Error
1.00	1.0000	1.0000	0.0000	0.00
1.05	1.1000	1.1079	0.0079	0.72
1.10	1.2155	1.2337	0.0182	1.47
1.15	1.3492	1.3806	0.0314	2.27
1.20	1.5044	1.5527	0.0483	3.11
1.25	1.6849	1.7551	0.0702	4.00
1.30	1.8955	1.9937	0.0982	4.93
1.35	2.1419	2.2762	0.1343	5.90
1.40	2.4311	2.6117	0.1806	6.92
1.45	2.7714	3.0117	0.2403	7.98
1.50	3.1733	3.4904	0.3171	9.08

∎

In this case, with a step size $h = 0.1$, a 16% relative error in the calculation of the approximation to $y(1.5)$ is totally unacceptable. At the expense of doubling the number of calculations, a slight improvement in accuracy is obtained by halving the step size to $h = 0.05$.

We will discuss the errors for Euler's method in detail in Section 9.6.

Of course in many instances we may not know the solution of a particular differential equation, or for that matter whether a solution of an initial-value problem actually exists. The following nonlinear equation does possess a solution in closed form, but we leave it as an exercise for you to find it. (See Problem 1.)

EXAMPLE 3

Use the Euler method to obtain the approximate value of $y(0.5)$ for the solution of

$$y' = (x + y - 1)^2, \qquad y(0) = 2.$$

Solution For $n = 0$ and $h = 0.1$, we have

$$y_1 = y_0 + (0.1)(x_0 + y_0 - 1)^2$$
$$= 2 + (0.1)(1)^2 = 2.1.$$

The remaining calculations are summarized in Tables 9.5 and 9.6 for $h = 0.1$ and $h = 0.05$, respectively.

TABLE 9.5 Euler's Method with $h = 0.1$	
x_n	y_n
0.00	2.0000
0.10	2.1000
0.20	2.2440
0.30	2.4525
0.40	2.7596
0.50	3.2261

TABLE 9.6 Euler's Method with $h = 0.05$	
x_n	y_n
0.00	2.0000
0.05	2.0500
0.10	2.1105
0.15	2.1838
0.20	2.2727
0.25	2.3812
0.30	2.5142
0.35	2.6788
0.40	2.8845
0.45	3.1455
0.50	3.4823

We may want greater accuracy than that displayed, say, in Table 9.4, and so we could try a step size even smaller than $h = 0.05$. However, rather than resorting to this extra labor, it probably would be more advantageous to employ an alternative numerical procedure. The Euler formula by itself, though attractive in its simplicity, is seldom used in serious calculations.

Improved Euler's Method

The formula

$$y_{n+1} = y_n + h \frac{f(x_n, y_n) + f(x_{n+1}, y_{n+1}^*)}{2},$$

$$\text{where} \qquad y_{n+1}^* = y_n + hf(x_n, y_n) \tag{2}$$

is known as the **improved Euler formula**, or **Heun's formula**. The values $f(x_n, y_n)$ and $f(x_{n+1}, y_{n+1}^*)$ are approximations to the slope of the curve at $(x_n, y(x_n))$ and $(x_{n+1}, y(x_{n+1}))$ and, consequently, the quotient

$$\frac{f(x_n, y_n) + f(x_{n+1}, y_{n+1}^*)}{2}$$

can be interpreted as an average slope on the interval between x_n and x_{n+1}.

The equations in (2) can be readily visualized. In Figure 9.13 we show the case in which $n = 0$. Note that

$$f(x_0, y_0) \quad \text{and} \quad f(x_1, y_1^*)$$

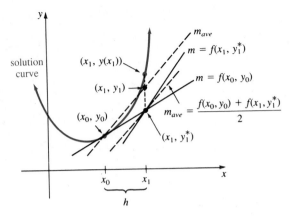

Figure 9.13

are slopes of the indicated straight lines passing through the points (x_0, y_0) and (x_1, y_1^*), respectively. By taking an average of these slopes, we obtain the slope of the dashed skew lines. Rather than advancing along the line with slope $m = f(x_0, y_0)$ to the point with ordinate y_1^* obtained by the usual Euler method, we advance along the line through (x_0, y_0) with slope m_{ave} until we reach x_1. It seems plausible from inspection of the figure that y_1 is an improvement over y_1^*. In fact, in Section 9.6 we will see that the improved Euler method is more accurate than the Euler method.

We might also say that the value of

$$y_1^* = y_0 + hf(x_0, y_0)$$

predicts a value of $y(x_1)$, whereas

$$y_1 = y_0 + h\frac{f(x_0, y_0) + f(x_1, y_1^*)}{2}$$

corrects this estimate.

EXAMPLE 4

Use the improved Euler formula to obtain the approximate value of $y(1.5)$ for the solution of the initial-value problem in Example 2. Compare the results for $h = 0.1$ and $h = 0.05$.

Solution   For $n = 0$ and $h = 0.1$, we first compute

$$y_1^* = y_0 + (0.1)(2x_0y_0) = 1.2.$$

Then from (2)

$$y_1 = y_0 + (0.1)\frac{2x_0 y_0 + 2x_1 y_1^*}{2}$$

$$= 1 + (0.1)\frac{2(1)(1) + 2(1.1)(1.2)}{2} = 1.232.$$

The comparative values of the calculations for $h = 0.1$ and $h = 0.5$ are given in Tables 9.7 and 9.8, respectively.

TABLE 9.7 Improved Euler's Method with $h = 0.1$

x_n	y_n	True Value	Abs. Error	% Rel. Error
1.00	1.0000	1.0000	0.0000	0.00
1.10	1.2320	1.2337	0.0017	0.14
1.20	1.5479	1.5527	0.0048	0.31
1.30	1.9832	1.9937	0.0106	0.53
1.40	2.5908	2.6117	0.0209	0.80
1.50	3.4509	3.4904	0.0394	1.13

TABLE 9.8 Improved Euler's Method with $h = 0.05$

x_n	y_n	True Value	Abs. Error	% Rel. Error
1.00	1.0000	1.0000	0.0000	0.00
1.05	1.1077	1.1079	0.0002	0.02
1.10	1.2332	1.2337	0.0004	0.04
1.15	1.3798	1.3806	0.0008	0.06
1.20	1.5514	1.5527	0.0013	0.08
1.25	1.7531	1.7551	0.0020	0.11
1.30	1.9909	1.9937	0.0029	0.14
1.35	2.2721	2.2762	0.0041	0.18
1.40	2.6060	2.6117	0.0057	0.22
1.45	3.0038	3.0117	0.0079	0.26
1.50	3.4795	3.4904	0.0108	0.31

■

A brief word of caution is in order here. We cannot compute all the values of y_n^* first and then substitute these values in the first formula of (2). In other words, we cannot use the data in Table 9.3 to help construct the values in Table 9.7. Why not?

EXAMPLE 5

Use the improved Euler formula to obtain the approximate value of $y(0.5)$ for the solution of the initial-value problem in Example 3.

Solution For $n = 0$ and $h = 0.1$, we have

$$y_1^* = y_0 + (0.1)(x_0 + y_0 - 1)^2 = 2.1$$

and so
$$y_1 = y_0 + (0.1)\frac{(x_0 + y_0 - 1)^2 + (x_1 + y_1^* - 1)^2}{2}$$

$$= 2 + (0.1)\frac{1 + 1.44}{2} = 2.122.$$

The remaining calculations are summarized in Tables 9.9 and 9.10 for $h = 0.1$ and $h = 0.05$, respectively.

TABLE 9.9 Improved Euler's Method with $h = 0.1$		**TABLE 9.10** Improved Euler's Method with $h = 0.05$	
x_n	y_n	x_n	y_n
0.00	2.0000	0.00	2.0000
0.10	2.1220	0.05	2.0553
0.20	2.3049	0.10	2.1228
0.30	2.5858	0.15	2.2056
0.40	3.0378	0.20	2.3075
0.50	3.8254	0.25	2.4342
		0.30	2.5931
		0.35	2.7953
		0.40	3.0574
		0.45	3.4057
		0.50	3.8840

■

Computer Programs

BASIC, Fortran, and Pascal programs for the Euler methods, as well as for other numerical techniques, can be found in the software supplement that accompanies this text.

9.2 EXERCISES

Answers to odd-numbered problems begin on page A–28.

1. Solve the initial-value problem

$$y' = (x + y - 1)^2, \qquad y(0) = 2,$$

in terms of elementary functions.

2. Let $y(x)$ be the solution of the initial-value problem given in Problem 1. Round to four decimal places as you compute the exact values of $y(0.1)$, $y(0.2)$, $y(0.3)$, $y(0.4)$, and $y(0.5)$. Compare these values with the entries in Tables 9.5, 9.6, 9.9, and 9.10.

Given the initial-value problems in Problems 3–12, use the Euler formula to obtain a four-decimal approximation to the indicated value. First use **(a)** $h = 0.1$ and then **(b)** $h = 0.05$.

3. $y' = 2x - 3y + 1$, $y(1) = 5$; $y(1.5)$

4. $y' = 4x - 2y$, $y(0) = 2$; $y(0.5)$

5. $y' = 1 + y^2$, $y(0) = 0$; $y(0.5)$

6. $y' = x^2 + y^2$, $y(0) = 1$; $y(0.5)$

7. $y' = e^{-y}$, $y(0) = 0$; $y(0.5)$

8. $y' = x + y^2$, $y(0) = 0$; $y(0.5)$

9. $y' = (x - y)^2$, $y(0) = 0.5$; $y(0.5)$

10. $y' = xy + \sqrt{y}$, $y(0) = 1$; $y(0.5)$

11. $y' = xy^2 - \dfrac{y}{x}$, $y(1) = 1$; $y(1.5)$

12. $y' = y - y^2$, $y(0) = 0.5$; $y(0.5)$

13. As parts **(a)**–**(e)** of this problem, repeat the calculations of Problems 3, 5, 7, 9, and 11 using the improved Euler formula.

14. As parts **(a)**–**(e)** of this problem, repeat the calculations of Problems 4, 6, 8, 10, and 12 using the improved Euler formula.

15. Although it may not be obvious from the differential equation, its solution could "behave badly" near a point x at which we wish to approximate $y(x)$. Numerical procedures may then give widely differing results near this point. Let $y(x)$ be the solution of the initial-value problem

$$y' = x^2 + y^3, \qquad y(1) = 1.$$

Using the step size $h = 0.1$, compare the results obtained from the Euler formula with the results from the improved Euler formula in the approximation of $y(1.4)$.

16. Derive the basic Euler formula by integrating both sides of the equation $y' = f(x, y)$ on the interval, $x_n \le x \le x_{n+1}$. Approximate the integral of the right side by replacing the function $f(x, y)$ by its value at the left endpoint of the interval of integration.

17. By following the procedure outlined in Problem 16, derive the improved Euler formula. [*Hint:* Replace the integrand of the right side by the average of its values at the endpoints of the interval of integration.]

9.3 THE THREE-TERM TAYLOR METHOD

The numerical method considered in this section, the **three-term Taylor method**, is more of theoretical interest than of practical importance since the results obtained using the following formula (5) do not differ substantially from those obtained using the improved Euler method.

In this study of numerical solutions of differential equations, many computational algorithms can be derived from a Taylor series expansion. Recall from calculus that the form of this expansion centered at a point $x = a$ is

$$y(x) = y(a) + y'(a)\frac{(x - a)}{1!} + y''(a)\frac{(x - a)^2}{2!} + \cdots. \tag{1}$$

It is understood that the function $y(x)$ possesses derivatives of all orders and that the series (1) converges in some interval defined by $|x - a| < R$. Notice, in particular, that if we set $a = x_n$ and $x = x_n + h$, then (1) becomes

$$y(x_n + h) = y(x_n) + y'(x_n)h + y''(x_n)\frac{h^2}{2} + \cdots. \tag{2}$$

Euler's Method Revisited

Furthermore, let us now assume that the function $y(x)$ is a solution of the first-order differential equation

$$y' = f(x, y).$$

If we then truncate the series (2) after, say, two terms, we obtain the approximation

$$y(x_n + h) \approx y(x_n) + y'(x_n)h$$

or

$$y(x_n + h) \approx y(x_n) + f(x_n, y(x_n))h. \tag{3}$$

Observe that we can obtain the Euler formula

$$y_{n+1} = y_n + hf(x_n, y_n) \tag{4}$$

of the preceding section by replacing $y(x_n + h)$ and $y(x_n)$ in (3) by their approximations y_{n+1} and y_n, respectively. The approximation symbol $\approx$ is replaced by an equality since we are defining the left side of (4) by the numbers obtained from the right-hand member.

Taylor's Method

By retaining three terms in the series (2), we can write

$$y(x_n + h) \approx y(x_n) + y'(x_n)h + y''(x_n)\frac{h^2}{2}.$$

After using the replacements noted in the preceding material, it follows that

$$y_{n+1} = y_n + y'_n h + y''_n \frac{h^2}{2}. \tag{5}$$

The second derivative y'' can be obtained by differentiating $y' = f(x, y)$.

At this point let us reexamine two initial-value problems from Section 9.2.

EXAMPLE I

Use the three-term Taylor formula to obtain the approximate value of $y(1.5)$ for the solution of

$$y' = 2xy, \qquad y(1) = 1.$$

Compare the results for $h = 0.1$ and $h = 0.05$.

Solution   Since $y' = 2xy$, it follows by the product rule that $y'' = 2xy' + 2y$. Thus, for example, when $h = 0.1$, $n = 0$, we can first calculate

$$y_0' = 2x_0y_0 = 2(1)(1) = 2$$

and then $$y_0'' = 2x_0y_0' + 2y_0 = 2(1)(2) + 2(1) = 6.$$

Hence (5) becomes

$$y_1 = y_0 + y_0'(0.1) + y_0'' \frac{(0.1)^2}{2}$$

$$= 1 + 2(0.1) + 6(0.005) = 1.23.$$

The results of the iteration, along with the comparative exact values, are summarized in Tables 9.11 and 9.12.

TABLE 9.11   Three-Term Taylor Method with $h = 0.1$

x_n	y_n	True Value	Abs. Error	% Rel. Error
1.00	1.0000	1.0000	0.0000	0.00
1.10	1.2300	1.2337	0.0037	0.30
1.20	1.5427	1.5527	0.0100	0.65
1.30	1.9728	1.9937	0.0210	1.05
1.40	2.5721	2.6117	0.0396	1.52
1.50	3.4188	3.4904	0.0715	2.05

TABLE 9.12   Three-Term Taylor Method with $h = 0.05$

x_n	y_n	True Value	Abs. Error	% Rel. Error
1.00	1.0000	1.0000	0.0000	0.00
1.05	1.1075	1.1079	0.0004	0.04
1.10	1.2327	1.2337	0.0010	0.08
1.15	1.3788	1.3806	0.0018	0.13
1.20	1.5499	1.5527	0.0028	0.18
1.25	1.7509	1.7551	0.0041	0.23
1.30	1.9879	1.9937	0.0059	0.29
1.35	2.2681	2.2762	0.0081	0.36
1.40	2.6006	2.6117	0.0111	0.43
1.45	2.9967	3.0117	0.0150	0.50
1.50	3.4702	3.4904	0.0202	0.58

■

EXAMPLE 2

Use the three-term Taylor formula to obtain the approximate value of $y(0.5)$ for the solution of

$$y' = (x + y - 1)^2, \qquad y(0) = 2.$$

Solution   In this case we compute y'' by the power rule. We have

$$y'' = 2(x + y - 1)(1 + y').$$

The results are summarized in Tables 9.13 and 9.14 for $h = 0.1$ and $h = 0.05$, respectively.

TABLE 9.13 Three-Term Taylor's Method with $h = 0.1$

x_n	y_n
0.00	2.0000
0.10	2.1200
0.20	2.2992
0.30	2.5726
0.40	3.0077
0.50	3.7511

TABLE 9.14 Three-Term Taylor's Method with $h = 0.05$

x_n	y_n
0.00	2.0000
0.05	2.0550
0.10	2.1222
0.15	2.2045
0.20	2.3058
0.25	2.4315
0.30	2.5890
0.35	2.7889
0.40	3.0475
0.45	3.3898
0.50	3.8574

A comparison of the last two examples with the corresponding results obtained from the improved Euler method shows no startling dissimilarities. In fact, when $f(x, y)$ is linear in both variables x and y, the Taylor method gives the *same* values of y_n as the improved Euler method for a given value of h. See Problems 12 and 13. In the error analysis given in Section 9.6 we will see that these two methods have the same accuracy.

9.3 EXERCISES

Answers to odd-numbered problems begin on page A–31.

Given the initial-value problems in Problems 1–10, use the three-term Taylor formula to obtain a four-decimal approximation to the indicated value. First use **(a)** $h = 0.1$ and then **(b)** $h = 0.05$.

1. $y' = 2x - 3y + 1$, $y(1) = 5$; $y(1.5)$

2. $y' = 4x - 2y$, $y(0) = 2$; $y(0.5)$

3. $y' = 1 + y^2$, $y(0) = 0$; $y(0.5)$

4. $y' = x^2 + y^2$, $y(0) = 1$; $y(0.5)$

5. $y' = e^{-y}$, $y(0) = 0$; $y(0.5)$

6. $y' = x + y^2$, $y(0) = 0$; $y(0.5)$

7. $y' = (x - y)^2$, $y(0) = 0.5$; $y(0.5)$

8. $y' = xy + \sqrt{y}$, $y(0) = 1$; $y(0.5)$

9. $y' = xy^2 - \dfrac{y}{x}$, $y(1) = 1$; $y(1.5)$

10. $y' = y - y^2$, $y(0) = 0.5$; $y(0.5)$

11. Let $y(x)$ be the solution of the initial-value problem

$$y' = x^2 + y^3, \qquad y(1) = 1.$$

Use $h = 0.1$ and the three-term Taylor formula to obtain an approximation to $y(1.4)$. Compare your answer with the results obtained in Problem 15 of Exercises 9.2.

12. Consider the differential equation $y' = f(x, y)$, where f is linear in x and y. In this case prove that the improved Euler formula is the same as the three-term Taylor formula. [*Hint:* Recall from calculus that a Taylor series for a function g of two variables is

$$g(a + h, b + k) = g(a, b) + g_x(a, b)h + g_y(a, b)k$$

$$+ \frac{1}{2} (h^2 g_{xx} + 2hk g_{xy} + k^2 g_{yy}) \Big|_{(a, b)}$$

$$+ \ terms \ involving \ higher\text{-}order \ derivatives.$$

Apply this result to $f(x_n + h, y_n + hf(x_n, y_n))$ in the improved Euler formula. Also use the fact that $y''(x) = (d/dx)y'(x) = f_x + f_y y'$.]

13. Compare the approximate values of $y(1.5)$ for

$$y' = x + y - 1, \qquad y(1) = 5,$$

using the three-term Taylor method and the improved Euler method with $h = 0.1$. Solve the initial-value problem and compute the true values $y(x_n)$, $n = 0, 1, \ldots, 5$.

9.4 THE RUNGE-KUTTA METHOD

Probably one of the most popular as well as accurate numerical procedures used in obtaining approximate solutions to differential equations is the

fourth-order Runge-Kutta method.* As the name suggests, there are Runge-Kutta methods of different orders.

For the moment let us consider a **second-order** procedure. This consists of finding constants a, b, α, and β such that the formula

$$y_{n+1} = y_n + ak_1 + bk_2, \tag{1}$$

where
$$k_1 = hf(x_n, y_n)$$
$$\tag{2}$$
$$k_2 = hf(x_n + \alpha h, y_n + \beta k_1)$$

agrees with a Taylor series expansion to as many terms as possible. The obvious purpose is to achieve the accuracy of the Taylor method without having to compute higher-order derivatives. Now it can be shown that whenever the constants satisfy

$$a + b = 1, \qquad b\alpha = \frac{1}{2}, \qquad b\beta = \frac{1}{2},$$

(1) agrees with a Taylor expansion out to the h^2, or third term. It should be of interest to observe that when $a = 1/2$, $b = 1/2$, $\alpha = 1$, $\beta = 1$, then (1) reduces to the improved Euler method. Thus we can conclude that the three-term Taylor formula is essentially equivalent to the improved Euler formula. Also, the basic Euler method is a **first-order** Runge-Kutta procedure.

Notice too that the sum $ak_1 + bk_2$, $a + b = 1$, in equation (1) is simply a *weighted average* of k_1 and k_2. The numbers k_1 and k_2 are multiples of approximations to the slope at two different points.

Fourth-Order Runge-Kutta Formula

The **fourth-order** Runge-Kutta method consists of determining appropriate constants so that a formula such as

$$y_{n+1} = y_n + ak_1 + bk_2 + ck_3 + dk_4$$

agrees with a Taylor expansion out to h^4, or the fifth term. As in (2) the k_i are constant multiples of $f(x, y)$ evaluated at select points. The derivation of the actual method is tedious, to say the least, so we state the results:

---

*Carl D. T. Runge (1856–1927)** A German professor of applied mathematics at the University of Göttingen, Runge devised this numerical method around 1895. (M. W. Kutta expanded on this work in 1901.) In mathematics Runge also did notable work in the field of Diophantine equations. In physics Runge is remembered for his work on the Zeeman effect (the spectral lines in the emission spectrum of an element are affected by the presence of a magnetic field).

Martin W. Kutta (1867–1944) Also a German applied mathematician, Kutta made significant contributions to the field of aerodynamics.

$$y_{n+1} = y_n + \tfrac{1}{6}(k_1 + 2k_2 + 2k_3 + k_4)$$

$$k_1 = hf(x_n, y_n)$$

$$k_2 = hf(x_n + \tfrac{1}{2}h, y_n + \tfrac{1}{2}k_1) \qquad (3)$$

$$k_3 = hf(x_n + \tfrac{1}{2}h, y_n + \tfrac{1}{2}k_2)$$

$$k_4 = hf(x_n + h, y_n + k_3).$$

You are advised to look carefully at the formulas in (3); note that k_2 depends on k_1, k_3 depends on k_2, and so on. Also, k_2 and k_3 involve approximations to the slope at the midpoint of the interval between x_n and $x_{n+1} = x_n + h$.

EXAMPLE I

Use the Runge-Kutta method with $h = 0.1$ to obtain an approximation to $y(1.5)$ for the solution of

$$y' = 2xy, \qquad y(1) = 1.$$

Solution For the sake of illustration let us compute the case when $n = 0$. From (3) we find

$$k_1 = (0.1)f(x_0, y_0) = (0.1)(2x_0 y_0) = 0.2$$

$$k_2 = (0.1)f\left(x_0 + \frac{1}{2}(0.1), y_0 + \frac{1}{2}(0.2)\right)$$

$$= (0.1)2\left(x_0 + \frac{1}{2}(0.1)\right)\left(y_0 + \frac{1}{2}(0.2)\right) = 0.231$$

$$k_3 = (0.1)f\left(x_0 + \frac{1}{2}(0.1), y_0 + \frac{1}{2}(0.231)\right)$$

$$= (0.1)2\left(x_0 + \frac{1}{2}(0.1)\right)\left(y_0 + \frac{1}{2}(0.231)\right) = 0.234255$$

$$k_4 = (0.1)f(x_0 + 0.1, y_0 + 0.234255)$$

$$= (0.1)2(x_0 + 0.1)(y_0 + 0.234255) = 0.2715361$$

and therefore

$$y_1 = y_0 + \frac{1}{6}(k_1 + 2k_2 + 2k_3 + k_4)$$

$$= 1 + \frac{1}{6}(0.2 + 2(0.231) + 2(0.234255) + 0.2715361) = 1.23367435.$$

The accompanying table, whose entries are rounded to four decimal places, should convince you why the Runge-Kutta method is so popular. Of course there is no need to use any smaller step size. The reason for the high accuracy of the Runge-Kutta method will be seen in Section 9.6.

TABLE 9.15 Runge-Kutta Method with $h = 0.1$

x_n	y_n	True Value	Abs. Error	% Rel. Error
1.00	1.0000	1.0000	0.0000	0.00
1.10	1.2337	1.2337	0.0000	0.00
1.20	1.5527	1.5527	0.0000	0.00
1.30	1.9937	1.9937	0.0000	0.00
1.40	2.6116	2.6117	0.0001	0.00
1.50	3.4902	3.4904	0.0001	0.00

∎

You might be interested in inspecting Tables 9.16 and 9.17 at this point. These tables compare the results obtained from the various formulas that we have examined applied to the two specific problems

$$y' = 2xy, \qquad y(1) = 1$$

$$y' = (x + y - 1)^2, \qquad y(0) = 2,$$

that we have considered throughout the last three sections.

TABLE 9.16 $y' = 2xy$, $y(1) = 1$

	Comparison of Numerical Methods with $h = 0.1$				
x_n	Euler	Improved Euler	Three-Term Taylor	Runge-Kutta	True Value
1.00	1.0000	1.0000	1.0000	1.0000	1.0000
1.10	1.2000	1.2320	1.2300	1.2337	1.2337
1.20	1.4640	1.5479	1.5427	1.5527	1.5527
1.30	1.8154	1.9832	1.9728	1.9937	1.9937
1.40	2.2874	2.5908	2.5721	2.6116	2.6117
1.50	2.9278	3.4509	3.4188	3.4902	3.4904

TABLE 9.16 *Continued*

			Three-		
		Improved	Term	Runge-	True
x_n	Euler	Euler	Taylor	Kutta	Value

Comparison of Numerical Methods with $h = 0.05$

x_n	Euler	Improved Euler	Three-Term Taylor	Runge-Kutta	True Value
1.00	1.0000	1.0000	1.0000	1.0000	1.0000
1.05	1.1000	1.1077	1.1075	1.1079	1.1079
1.10	1.2155	1.2332	1.2327	1.2337	1.2337
1.15	1.3492	1.3798	1.3788	1.3806	1.3806
1.20	1.5044	1.5514	1.5499	1.5527	1.5527
1.25	1.6849	1.7531	1.7509	1.7551	1.7551
1.30	1.8955	1.9909	1.9879	1.9937	1.9937
1.35	2.1419	2.2721	2.2681	2.2762	2.2762
1.40	2.4311	2.6060	2.6006	2.6117	2.6117
1.45	2.7714	3.0038	2.9967	3.0117	3.0117
1.50	3.1733	3.4795	3.4702	3.4903	3.4904

TABLE 9.17 $y' = (x + y - 1)^2$, $y(0) = 2$

Comparison of Numerical Methods with $h = 0.1$

x_n	Euler	Improved Euler	Three-Term Taylor	Runge-Kutta	True Value
0.00	2.0000	2.0000	2.0000	2.0000	2.0000
0.10	2.1000	2.1220	2.1200	2.1230	2.1230
0.20	2.2440	2.3049	2.2992	2.3085	2.3085
0.30	2.4525	2.5858	2.5726	2.5958	2.5958
0.40	2.7596	3.0378	3.0077	3.0649	3.0650
0.50	3.2261	3.8254	3.7511	3.9078	3.9082

Comparison of Numerical Methods with $h = 0.05$

x_n	Euler	Improved Euler	Three-Term Taylor	Runge-Kutta	True Value
0.00	2.0000	2.0000	2.0000	2.0000	2.0000
0.05	2.0500	2.0553	2.0550	2.0554	2.0554
0.10	2.1105	2.1228	2.1222	2.1230	2.1230
0.15	2.1838	2.2056	2.2045	2.2061	2.2061
0.20	2.2727	2.3075	2.3058	2.3085	2.3085
0.25	2.3812	2.4342	2.4315	2.4358	2.4358
0.30	2.5142	2.5931	2.5890	2.5958	2.5958
0.35	2.6788	2.7953	2.7889	2.7998	2.7997
0.40	2.8845	3.0574	3.0475	3.0650	3.0650
0.45	3.1455	3.4057	3.3898	3.4189	3.4189
0.50	3.4823	3.8840	3.8574	3.9082	3.9082

9.4 EXERCISES

Answers to odd-numbered problems begin on page A–32.

Given the initial-value problems in Problems 1–10, use the Runge-Kutta method with $h = 0.1$ to obtain a four-decimal approximation to the indicated value.

1. $y' = 2x - 3y + 1$, $y(1) = 5$, $y(1.5)$

2. $y' = 4x - 2y$, $y(0) = 2$, $y(0.5)$

3. $y' = 1 + y^2$, $y(0) = 0$, $y(0.5)$

4. $y' = x^2 + y^2$, $y(0) = 1$, $y(0.5)$

5. $y' = e^{-y}$, $y(0) = 0$, $y(0.5)$

6. $y' = x + y^2$, $y(0) = 0$, $y(0.5)$

7. $y' = (x - y)^2$, $y(0) = 0.5$, $y(0.5)$

8. $y' = xy + \sqrt{y}$, $y(0) = 1$, $y(0.5)$

9. $y' = xy^2 - \dfrac{y}{x}$, $y(1) = 1$, $y(1.5)$

10. $y' = y - y^2$, $y(0) = 0.5$, $y(0.5)$

11. If air resistance is proportional to the square of the instantaneous velocity, then the velocity v of a mass m dropped from a height h is determined from

$$m\frac{dv}{dt} = mg - kv^2, \qquad k > 0$$

(see Problem 8, Chapter 3 Review Exercises). If $v(0) = 0$, $k = 0.125$, $m = 5$ slugs, and $g = 32$ ft/s², use the Runge-Kutta method to find an approximation to the velocity of the falling mass at $t = 5$ seconds. Use $h = 1$.

12. Solve the initial-value problem in Problem 11 by one of the methods of Chapter 2. Find the true value of $v(5)$.

13. A mathematical model for the area A (in cm²) that a colony of bacteria (*B. dendroides*) occupies is given by*

$$\frac{dA}{dt} = A(2.128 - 0.0432A).$$

If $A(0) = 0.24$ cm², use the Runge-Kutta method to complete the following table. Use $h = 0.5$.

---

* See. V. A. Kostitzin, *Mathematical Biology* (London: Harrap, 1939).

t (days)	1	2	3	4	5
A (observed)	2.78	13.53	36.30	47.50	49.40
A (approximated)					

14. Solve the initial-value problem in Problem 13. Compute the values $A(1)$, $A(2)$, $A(3)$, $A(4)$, and $A(5)$. [*Hint:* See Section 3.3.]

15. Let $y(x)$ be the solution of the initial-value problem

$$y' = x^2 + y^3, \qquad y(1) = 1.$$

Determine whether the Runge-Kutta formula can be used to obtain an approximation for $y(1.4)$. Use $h = 0.1$.

16. Consider the differential equation $y' = f(x)$. In this case show that the fourth-order Runge-Kutta method reduces to Simpson's rule for the integral of $f(x)$ on the interval $x_n \le x \le x_{n+1}$.

9.5 MULTISTEP METHODS

Adams-Bashforth/Adams-Moulton Method

There are many additional formulas that can be applied to obtain approximations to solutions of differential equations. Although it is not our intention to survey the vast field of numerical methods, several additional formulas deserve mention. The **Adams-Bashforth/Adams-Moulton method**, like the improved Euler formula, is a predictor-corrector method. By using the Adams-Bashforth formula

$$y^*_{n+1} = y_n + \frac{h}{24}(55y'_n - 59y'_{n-1} + 37y'_{n-2} - 9y'_{n-3}), \tag{1}$$

$$y'_n = f(x_n, y_n)$$

$$y'_{n-1} = f(x_{n-1}, y_{n-1})$$

$$y'_{n-2} = f(x_{n-2}, y_{n-2})$$

$$y'_{n-3} = f(x_{n-3}, y_{n-3}),$$

for $n \geq 3$, as a predictor, we are able to substitute the value of y_{n+1}^* into the Adams-Moulton corrector

$$y_{n+1} = y_n + \frac{h}{24}(9y_{n+1}' + 19y_n' - 5y_{n-1}' + y_{n-2}'),$$ (2)

$$y_{n+1}' = f(x_{n+1}, y_{n+1}^*).$$

Notice formula (1) requires that we know the values of y_0, y_1, y_2, and y_3 in order to obtain y_4. The value of y_0 is, of course, the given initial condition; the values of y_1, y_2, and y_3 are computed by an accurate method such as the Runge-Kutta formula.

Since the Adams-Bashforth/Adams-Moulton formulas demand that we know more than just y_n to compute y_{n+1}, the procedure is called a **multistep**, or **continuing**, method. The Euler formulas, the three-term Taylor formula, and the Runge-Kutta formulas are examples of **single-step**, or **starting**, methods.

EXAMPLE I

Use the Adams-Bashforth/Adams-Moulton method with $h = 0.2$ to obtain an approximation to $y(0.8)$ for the solution of

$$y' = x + y - 1, \qquad y(0) = 1.$$

Solution With a step size of $h = 0.2$, $y(0.8)$ will be approximated by y_4. To get started we use the Runge-Kutta method with $x_0 = 0$, $y_0 = 1$, and $h = 0.2$ to obtain

$$y_1 = 1.02140000$$

$$y_2 = 1.09181796$$

$$y_3 = 1.22210646.$$

Now with the identifications $x_0 = 0$, $x_1 = 0.2$, $x_2 = 0.4$, $x_3 = 0.6$, and $f(x, y) = x + y - 1$, we find

$$y_0' = f(x_0, y_0) = (0) + (1) - 1 = 0$$

$$y_1' = f(x_1, y_1) = (0.2) + (1.02140000) - 1 = 0.22140000$$

$$y_2' = f(x_2, y_2) = (0.4) + (1.09181796) - 1 = 0.49181796$$

$$y_3' = f(x_3, y_3) = (0.6) + (1.22210646) - 1 = 0.82210646.$$

With the foregoing values the predictor (1) then gives

$$y_4^* = y_3 + \frac{0.2}{24}(55y_3' - 59y_2' + 37y_1' - 9y_0') = 1.42535975.$$

To use the corrector (2) we first need

$$y'_4 = f(x_4, y_4^*) = (0.8) + (1.42535975) - 1 = 1.22535975.$$

Finally, (2) yields

$$y_4 = y_3 + \frac{0.2}{24}(9y'_4 + 19y'_3 - 5y'_2 + y'_1) = 1.42552788. \qquad \blacksquare$$

You should verify that the exact value of $y(0.8)$ in Example 1 is $y(0.8) = 1.42554093$.

Milne's Method

Another multistep method, which admittedly has limited use (for a reason explained in Section 9.6), is called **Milne's method**. In this method the predictor is

$$y_{n+1}^* = y_{n-3} + \frac{4h}{3}(2y'_n - y'_{n-1} + 2y'_{n-2}), \qquad (3)$$

where

$$y'_n = f(x_n, y_n)$$

$$y'_{n-1} = f(x_{n-1}, y_{n-1})$$

$$y'_{n-2} = f(x_{n-2}, y_{n-2})$$

for $n \geq 3$. The corrector for this formula, based on Simpson's rule of integration, is given by

$$y_{n+1} = y_{n-1} + \frac{h}{3}(y'_{n+1} + 4y'_n + y'_{n-1}), \qquad (4)$$

where

$$y'_{n+1} = f(x_{n+1}, y_{n+1}^*).$$

As in the Adams-Bashforth/Adams-Moulton method, we generally use the Runge-Kutta method to compute y_1, y_2, and y_3.

Choice of Starting Methods

As stated earlier in this section, a multistep method requires the use of a single-step method or starting method. Some care should be taken in the selection of this latter method. In practice, starting methods are usually chosen to have the same accuracy as the multistep method with which they are used. By *accuracy* we mean the same order of error (see Section 9.6). Using a less accurate starting method with a highly accurate multistep method is not advisable because then the values used in the multistep method will be unreliable. On the other hand, a highly accurate starting method is complicated and expensive to calculate and this effort will be wasted if the multistep method is of low accuracy.

Advantages and Disadvantages of Multistep Methods

Many considerations enter into the choice of a method to solve a differential equation numerically. Single-step methods, particularly the Runge-Kutta method, are often chosen because of their accuracy and the fact that they are easy to program. However, a major drawback to them is that the right-hand side of the differential equation must be evaluated many times at each step. For instance, the fourth-order Runge-Kutta method requires four function evaluations for each step. On the other hand, if the function evaluations in the previous step have been calculated and stored, a multistep method requires only one new function evaluation for each step. This can lead to a great savings in time and expense.

As an example, to solve $y' = f(x, y)$, $y(x_0) = y_0$, numerically using n steps by the fourth-order Runge-Kutta method requires $4n$ function evaluations. The Adams-Bashforth multistep method requires 16 function evaluations for the Runge-Kutta fourth-order starter and $n - 4$ for the n Adams-Bashforth steps, giving a total of $n + 12$ function evaluations for this method. In general the Adams-Bashforth multistep method requires slightly more than a quarter of the number of function evaluations required for the fourth-order Runge-Kutta method. If the evaluation of $f(x, y)$ is complicated, the multistep method will be more efficient.

Another issue involved with multistep methods is how many times the corrector formula should be repeated in each step. Each time the corrector is used, another function evaluation is done, and so the accuracy is increased at the expense of losing an advantage of the multistep method. In practice, the corrector is calculated once, and if the value of y_{n+1} is changed by a large amount, the entire problem is restarted using a smaller step size. This is often the basis of the variable step size methods, whose discussion is beyond the scope of this text.

9.5   EXERCISES

Answers to odd-numbered problems begin on page A–32.

1. Find the exact solution of the initial-value problem in Example 1. Compare the exact values of $y(0.2)$, $y(0.4)$, $y(0.6)$, and $y(0.8)$ with the approximations y_1, y_2, y_3, and y_4.

2. Write a computer program for the Adams-Bashforth/Adams-Moulton method.

In Problems 3 and 4 use the Adams-Bashforth/Adams-Moulton method to approximate $y(0.8)$, where $y(x)$ is the solution of the given initial-value problem. Use $h = 0.2$ and the Runge-Kutta method to compute y_1, y_2, and y_3.

3. $y' = 2x - 3y + 1$, $y(0) = 1$ 4. $y' = 4x - 2y$, $y(0) = 2$

In Problems 5–8 use the Adams-Bashforth/Adams-Moulton method to approximate $y(1.0)$, where $y(x)$ is the solution of the given initial-value problem.

Use $h = 0.2$ and $h = 0.1$ and the Runge-Kutta method to compute y_1, y_2, and y_3.

5. $y' = 1 + y^2$, $y(0) = 0$ **6.** $y' = y + \cos x$, $y(0) = 1$

7. $y' = (x - y)^2$, $y(0) = 0$ **8.** $y' = xy + \sqrt{y}$, $y(0) = 1$

9. Use Milne's method to approximate the value of $y(0.4)$, where $y(x)$ is the solution of the initial-value problem in Example 1. Use $h = 0.1$ and the Runge-Kutta method to compute y_1, y_2, and y_3.

9.6 ERRORS AND STABILITY

Errors in Numerical Methods

In choosing and using a numerical method for the solution of an initial-value problem, we must be aware of the various sources of errors. For some kinds of computation, the accumulation of errors might reduce the accuracy of an approximation to the point of being useless. On the other hand, depending upon the use to which a numerical solution may be put, extreme accuracy may not be worth the added expense and complication.

One source of error always present in calculations is **round-off error**. This is caused by the fact that any calculator or computer can compute to only, at most, a finite number of decimal places. Suppose for the sake of illustration that we have a calculator that can display six digits while carrying eight digits internally. If we multiply two numbers, each having six decimals, then the product actually contains 12 decimal places. But the number that we see is rounded to six decimal places, while the machine has stored a number rounded to eight decimal places. In one calculation such as this, the round-off error may not be significant, but a problem could arise if many calculations are performed with rounded numbers. One way to reduce the effect of round-off error is to minimize the number of calculations. Another technique on a computer is to use double precision capabilities. Since round-off error is unpredictable and difficult to analyze, we will neglect it in the error analysis that follows. We will concentrate on investigating the error made by using a formula or algorithm to approximate the values of the solution.

Truncation Errors for Euler's Method

When iterating Euler's formula

$$y_{n+1} = y_n + hf(x_n, y_n)$$

we obtain a sequence of values y_1, y_2, y_3, ... The value y_1 will not agree with $y(x_1)$, the actual solution evaluated at x_1, because the algorithm gives only a straight-line approximation to the solution. See Figure 9.10 in Section 9.2. This error is called the **local truncation error**, **formula error**, or **discretization error**. It occurs at each step; that is, if we assume that y_n is accurate, then y_{n+1} will contain local truncation error.

To derive a formula for the local truncation error for Euler's method we will use Taylor's formula with remainder. This is very similar to the Taylor series expansion (1) in Section 9.3. If a function $y(x)$ possesses $k + 1$ continuous derivatives on an open interval containing x and a, then

$$y(x) = y(a) + y'(a)\frac{(x - a)}{1!} + \cdots + y^{(k)}(a)\frac{(x - a)^k}{k!} + y^{(k+1)}(c)\frac{(x - a)^{k+1}}{(k + 1)!}, \quad \textbf{(1)}$$

where c is some point between a and x. Setting $k = 1$, $a = x_n$, and $x = x_{n+1} = x_n + h$, we get

$$y(x_{n+1}) = y(x_n) + y'(x_n)\frac{h}{1!} + y''(c)\frac{h^2}{2!}$$

or

$$y_{n+1} = y_n + hf(x_n, y_n) + y''(c)\frac{h^2}{2!}.$$

Euler's method is this formula without the last term; hence, the local truncation error in y_{n+1} is

$$y''(c)\frac{h^2}{2!}, \qquad \text{where} \quad x_n < c < x_{n+1}.$$

Unfortunately, the value of c is usually unknown (it exists theoretically) and so the exact error cannot be calculated, but an upper bound on the absolute value of the error is

$$M\frac{h^2}{2}, \qquad \text{where} \quad M = \max_{x_n < x < x_{n+1}} |y''(x)|.$$

The following terminology is used when discussing errors of numerical methods. A calculation is said to be of **order** α, denoted $O(h^\alpha)$, if the error behaves like Ch^α, where C is a constant. From this we see that if the step size is halved, then the new error should be approximately $C(h^\alpha/2^\alpha)$; that is, the error is reduced by a factor of $1/2^\alpha$.

Using this definition, we see that the local truncation error for the Euler method is of order two, or $O(h^2)$.

EXAMPLE I

Find a bound for the local truncation errors for Euler's method applied to

$$y' = 2xy, \qquad y(1) = 1.$$

Solution This differential equation was studied in Example 2 of Section 9.2 and its analytic solution is $y(x) = e^{x^2 - 1}$.

The local truncation error is

$$y''(c)\frac{h^2}{2} = (2 + 4c^2)e^{(c^2 - 1)}\frac{h^2}{2},$$

where c is between x_n and $x_n + h$. In particular, for $h = 0.1$, we can get an upper bound on the local truncation error for y_1 by replacing c by 1.1:

$$[2 + (4)(1.1)^2]e^{((1.1)^2 - 1)}\frac{(0.1)^2}{2} = 0.0422.$$

From Table 9.3 we see that the error after the first step is 0.0337, less than the value given by the bound.

Similarly we can get a bound for the local truncation error for any of the five steps given in Table 9.3 by replacing c by 1.5 (this value of c gives the largest value of $y''(c)$ for any of the steps and may be too generous for the first few steps). Doing this gives

$$[2 + (4)(1.5)^2]e^{((1.5)^2 - 1)}\frac{(0.1)^2}{2} = 0.1920 \tag{2}$$

as an upper bound for the local truncation error in each step. ∎

Note in Example 1 that if h is halved to 0.05, then the error bound (2) would be 0.0480, about one fourth as much. This is expected since the local truncation error for Euler's method is $O(h^2)$.

In the above analysis we assumed that the value of y_n is correct in the calculation of y_{n+1}, but it is not because it contains local truncation errors from previous steps. The total error in y_{n+1} is an accumulation of the errors in each of the previous steps. See Figure 9.11 in Section 9.2. This total error is called the **global truncation error**. A complete analysis of the global truncation error is beyond the scope of this text, but it can be shown that the global truncation error for the Euler method is $O(h)$.

We expect that for Euler's method if the step size is halved, then the error would be approximately halved as well. This is borne out in Example 2 of Section 9.2, where the absolute error at $x = 1.50$ with $h = 0.1$ is 0.5625 and the absolute error with $h = 0.05$ is 0.3171, approximately half as large. See Tables 9.3 and 9.4 in Section 9.2.

In general it can be shown that if a method for the numerical solution of a differential equation has local truncation error $O(h^{\alpha+1})$, then the global truncation error is $O(h^\alpha)$. Hence we can say that the order of global truncation error for any method is one less than the order of the local truncation error.

Truncation Errors for Other Methods

The local and global truncation errors for the other methods we have studied can be analyzed similarly. For instance, it was pointed out in Section 9.4 that the improved Euler method is a second-order Runge-Kutta method and thus one condition it satisfies is that it agrees with the Taylor polynomial through $k = 2$. Thus the local truncation error is

$$y^{(3)}(c)\frac{h^3}{3!}.$$

The derivation of this result is left for you. (See Problem 1.)

Since the local truncation error is $O(h^3)$, we know that the global truncation error is $O(h^2)$. This can be seen in Tables 9.7 and 9.8 of Section 9.2; when the step size is halved from $h = 0.1$ to $h = 0.05$, the global truncation error at $x = 1.50$ is reduced from 0.0394 to 0.0108. This is a reduction of approximately one fourth.

Similarly, a condition in the derivation of the fourth-order Runge-Kutta method is that it should agree with the Taylor polynomial through $k = 4$. Hence the local truncation error is

$$y^{(5)}(c) \frac{h^5}{5!} \quad \text{or} \quad O(h^5)$$

and the global truncation error is $O(h^4)$. It is now obvious why this is called the fourth-order Runge-Kutta method.

EXAMPLE 2

Analyze the local and global truncation errors for the fourth-order Runge-Kutta method applied to

$$y' = 2xy, \qquad y(1) = 1.$$

Solution By differentiating the known solution $y(x) = e^{x^2 - 1}$, we get

$$y^{(5)}(c) \frac{h^5}{5!} = (120c + 160c^3 + 32c^5)e^{c^2 - 1} \frac{h^5}{5!}. \tag{3}$$

Thus with $c = 1.5$, (3) yields a bound of 0.00028 on the local discretization error for each of the five steps when $h = 0.1$. Note that in Table 9.15 the error in y_1 is much less than this bound.

In Table 9.18 we summarize the approximations to the solution of the initial-value problem at $x = 1.5$ given by the fourth-order Runge-Kutta method with $h = 0.1$ and $h = 0.05$. Because the method is so exact, many decimal places must be kept in the numerical solution to see the effect of halving the step size. Note that when we halve h, the error is divided by a factor of about $2^4 = 16$, as expected.

TABLE 9.18 Runge-Kutta Method

h	Approximation	Error
0.1	3.49021064	$1.323210889 \times 10^{-4}$
0.05	3.49033382	$9.137760898 \times 10^{-6}$

A similar analysis can be done for the errors in each of the methods we have studied. Table 9.19 gives the results.

TABLE 9.19 Orders of Error for Different Methods

Method	Order of Global Truncation Error
Euler	1
Improved Euler	2
Three-term Taylor	2
Fourth-order Runge-Kutta	4
Adams-Bashforth	4
Adams-Moulton	4
Milne	4

Stability of Numerical Methods

In accepting the results of a numerical method, a user should always be concerned with the stability of the method. Put simply, a method is **stable** if small changes in the data make only small changes in the results. This is important since the numbers used in a calculator or computer are not exact. Also, in physical applications the data are often obtained by imprecise measurements. If a method is not stable, it is possible to have an error greater than would be expected from global truncation or round-off error. All single-step methods are stable if the step size taken is small enough; however, some multistep methods, particularly Milne's method, are not stable.

EXAMPLE 3

Consider the initial-value problem

$$y' = -10y + 5, \qquad y(1) = 1.$$

Use Milne's method to obtain an approximation to $y(2.0)$ using $h = 0.1$.

Solution The instability of the method for this relatively simple problem is apparent from inspection of the results given in Table 9.20. The starting method is the fourth-order Runge-Kutta method, which is very accurate and stable, but nevertheless the very large errors at the end of the calculations make the approximation for $y(2.0)$ quite useless.

TABLE 9.20   Instability of Milne's Method

x	Actual	Milne	Error
1.0	1.00000	1.00000	
1.1	0.68393	0.68750	0.00357
1.2	0.56767	0.57031	0.00264
1.3	0.52489	0.52637	0.00148
1.4	0.50917	0.52344	0.01427
1.5	0.50337	0.47266	−0.03071
1.6	0.50125	0.60417	0.10292
1.7	0.50045	0.14963	−0.35082
1.8	0.50017	1.66956	1.16939
1.9	0.50005	−3.39111	−3.89116
2.0	0.50002	13.47418	12.97416

Milne's method is the only numerical procedure we have studied that has serious stability problems, but stability must be considered when any numerical method is used.

9.6 EXERCISES

Answers to odd-numbered problems begin on page A–33.

1. Use the Taylor polynomial with remainder (1) to find the formula for the local truncation error for the improved Euler method (recall that it is a second-order Runge-Kutta method).

2. Repeat Problem 1 for the three-term Taylor method.

3. Repeat Problem 1 for the fourth-order Runge-Kutta method.

4. Consider the initial-value problem $y' = 2y$, $y(0) = 1$. The analytic solution is $y(x) = e^{2x}$.

   (a) Approximate $y(0.1)$ using one step and the Euler method.

   (b) Find a bound for the local truncation error in y_1.

   (c) Compare the actual error in y_1 with your error bound.

   (d) Approximate $y(0.1)$ using two steps and the Euler method.

   (e) Verify that the global truncation error for Euler's method is $O(h)$ by comparing the errors in parts (a) and (d).

5. Repeat Problem 4 using the improved Euler method. Its global truncation error is $O(h^2)$.

6. Repeat Problem 4 using the three-term Taylor method. Its global truncation error is $O(h^2)$.

7. Repeat Problem 4 using the fourth-order Runge-Kutta method. Its global truncation error is $O(h^4)$. Keep eight or nine decimal places in your calculations.

8.–11. Repeat Problems 4–7 using the initial-value problem $y' = -2y + x$, $y(0) = 1$. The analytic solution is $y(x) = \frac{1}{2}x - \frac{1}{4} + \frac{5}{4}e^{-2x}$.

12. Consider the initial-value problem $y' = 2x - 3y + 1$, $y(1) = 5$. The analytic solution is $y(x) = \frac{1}{9} + \frac{2}{3}x + \frac{38}{9}e^{-3(x-1)}$.

   (a) Find a formula involving c and h for the local truncation error in the nth step if the Euler method is used.

   (b) Find a bound for the local truncation error in each step if $h = 0.1$ is used to approximate $y(1.5)$.

   (c) Approximate $y(1.5)$ using $h = 0.1$ and $h = 0.05$ with the Euler method. See Problem 3, Exercises 9.2.

   (d) Calculate the errors in part (c) and verify that the global truncation error of the Euler method is $O(h)$.

13. Repeat Problem 12 using the improved Euler method, which has global truncation error $O(h^2)$. See Problem 13(a), Exercises 9.2. You may need to keep more than four decimal places to see the effect of reducing the order of error.

14. Repeat Problem 12 using the three-term Taylor series method, which has global truncation error $O(h^2)$. See Problem 1, Exercises 9.3. You may need to keep more than four decimal places to see the effect of reducing the order of error.

15. Repeat Problem 12 using the fourth-order Runge-Kutta method, which has global truncation error $O(h^4)$. See Problem 1, Exercise 9.4. You will need to keep more than six decimal places to see the effect of reducing the order of error.

16.–19. Repeat Problems 12–15 for the initial-value problem $y' = e^{-y}$, $y(0) = 0$. The analytic solution is $y(x) = \ln(x + 1)$. Approximate $y(0.5)$. See Problems 7 and 13(c) in Exercises 9.2, Problem 5 in Exercises 9.3, and Problem 5 in Exercises 9.4.

9.7 HIGHER-ORDER EQUATIONS AND SYSTEMS

Second-Order Initial-Value Problem

The numerical procedures we have discussed so far in this chapter applied to only a first-order equation $dy/dx = f(x, y)$ subject to an initial condition $y(x_0) = y_0$. To approximate the solution to, say, a second-order initial-value problem

$$\frac{d^2y}{dx^2} = f(x, y, y'), \qquad y(x_0) = y_0, \quad y'(x_0) = y_0', \qquad (1)$$

we reduce the second-order differential equation to a second-order system. (See Section 8.3.) If we let $y' = u$, then the equation in (1) becomes

$$y' = u$$

$$u' = f(x, y, u). \qquad (2)$$

We now apply a particular method to each equation in the resulting system. For example, the basic **Euler method** for the system (2) would be

$$y_{n+1} = y_n + hu_n$$

$$u_{n+1} = u_n + hf(x_n, y_n, u_n). \tag{3}$$

EXAMPLE 1

Use the Euler method to obtain the approximate value of $y(0.2)$, where $y(x)$ is the solution of the initial-value problem

$$y'' + xy' + y = 0, \qquad y(0) = 1, \quad y'(0) = 2.$$

Solution   In terms of the substitution $y' = u$, the equation is equivalent to the system

$$y' = u$$

$$u' = -xu - y.$$

Thus from (3) we obtain

$$y_{n+1} = y_n + hu_n$$

$$u_{n+1} = u_n + h[-x_n u_n - y_n].$$

Using the step size $h = 0.1$ and $y_0 = 1$, $u_0 = 2$, we find

$$y_1 = y_0 + (0.1)u_0 = 1 + (0.1)2 = 1.2$$

$$u_1 = u_0 + (0.1)[-x_0 u_0 - y_0] = 2 + (0.1)[-(0)(2) - 1] = 1.9$$

$$y_2 = y_1 + (0.1)u_1 = 1.2 + (0.1)(1.9) = 1.39$$

$$u_2 = u_1 + (0.1)[-x_1 u_1 - y_1] = 1.9 + (0.1)[-(0.1)(1.9) - 1.2] = 1.761.$$

In other words, $y(0.2) \approx 1.39$ and $y'(0.2) \approx 1.761$.   ∎

Systems

An initial-value problem for a second-order system, that is, a system of two first-order differential equations, is

$$x' = f(t, x, y)$$

$$y' = g(t, x, y) \tag{4}$$

$$x(t_0) = x_0, \qquad y(t_0) = y_0.$$

As we did in (3), to approximate a solution of this problem we apply a numerical method to each equation in (4). The **fourth-order Runge-Kutta method** looks like this:

$$x_{n+1} = x_n + \frac{1}{6}(m_1 + 2m_2 + 2m_3 + m_4)$$

$$y_{n+1} = y_n + \frac{1}{6}(k_1 + 2k_2 + 2k_3 + k_4),$$

(5)

where

$$m_1 = hf(t_n, x_n, y_n) \qquad\qquad k_1 = hg(t_n, x_n, y_n)$$

$$m_2 = hf\left(t_n + \frac{1}{2}h, x_n + \frac{1}{2}m_1, y_n + \frac{1}{2}k_1\right) \qquad k_2 = hg\left(t_n + \frac{1}{2}h, x_n + \frac{1}{2}m_1, y_n + \frac{1}{2}k_1\right)$$

$$m_3 = hf\left(t_n + \frac{1}{2}h, x_n + \frac{1}{2}m_2, y_n + \frac{1}{2}k_2\right) \qquad k_3 = hg\left(t_n + \frac{1}{2}h, x_n + \frac{1}{2}m_2, y_n + \frac{1}{2}k_2\right)$$

$$m_4 = hf(t_n + h, x_n + m_3, y_n + k_3) \qquad k_4 = hg(t_n + h, x_n + m_3, y_n + k_3).$$

(6)

EXAMPLE 2

Consider the initial-value problem

$$x' = 2x + 4y$$

$$y' = -x + 6y$$

$$x(0) = -1, \qquad y(0) = 6.$$

Use the fourth-order Runge-Kutta method to approximate $x(0.6)$ and $y(0.6)$. Compare the results for $h = 0.2$ and $h = 0.1$.

Solution We illustrate the computations of x_1 and y_1 with the step size $h = 0.2$. With the identifications $f(t, x, y) = 2x + 4y$, $g(t, x, y) = -x + 6y$, $t_0 = 0$, $x_0 = -1$, and $y_0 = 6$, we see from (6) that

$$m_1 = hf(t_0, x_0, y_0) = 0.2f(0, -1, 6) = 0.2[2(-1) + 4(6)] = 4.4000$$

$$k_1 = hg(t_0, x_0, y_0) = 0.2g(0, -1, 6) = 0.2[-1(-1) + 6(6)] = 7.4000$$

$$m_2 = hf\left(t_0 + \frac{1}{2}h, x_0 + \frac{1}{2}m_1, y_0 + \frac{1}{2}k_1\right) = 0.2f(0.1, 1.2, 9.7) = 8.2400$$

$$k_2 = hg\left(t_0 + \frac{1}{2}h, x_0 + \frac{1}{2}m_1, y_0 + \frac{1}{2}k_1\right) = 0.2g(0.1, 1.2, 9.7) = 11.4000$$

$$m_3 = hf\left(t_0 + \frac{1}{2}h, x_0 + \frac{1}{2}m_2, y_0 + \frac{1}{2}k_2\right) = 0.2f(0.1, 3.12, 11.7) = 10.6080$$

$$k_3 = hg\left(t_0 + \frac{1}{2}h, x_0 + \frac{1}{2}m_2, y_0 + \frac{1}{2}k_2\right) = 0.2g(0.1, 3.12, 11.7) = 13.4160$$

$$m_4 = hf(t_0 + h, x_0 + m_3, y_0 + k_3) = 0.2f(0.2, 8, 20.216) = 19.3760$$

$$k_4 = hg(t_0 + h, x_0 + m_3, y_0 + k_3) = 0.2g(0.2, 8, 20.216) = 21.3776.$$

Therefore, from (5) we get

$$x_1 = x_0 + \frac{1}{6}(m_1 + 2m_2 + 2m_3 + m_4)$$

$$= -1 + \frac{1}{6}(4.4 + 2(8.24) + 2(10.608) + 19.3760) = 9.2453$$

$$y_1 = y_0 + \frac{1}{6}(k_1 + 2k_2 + 2k_3 + k_4)$$

$$= 6 + \frac{1}{6}(7.4 + 2(11.4) + 2(13.416) + 21.3776) = 19.0683,$$

where, as usual, the computed values are rounded to four decimal places. These numbers give us the approximations $x_1 \approx x(0.2)$ and $y_1 \approx y(0.2)$. The subsequent values, obtained with the aid of a computer, are summarized in Tables 9.21 and 9.22.

TABLE 9.21   Runge-Kutta Method with $h = 0.2$

m_1	m_2	m_3	m_4	k_1	k_2	k_3	k_4	t_n	x_n	y_n
								0.00	−1.0000	6.0000
4.4000	8.2400	10.6080	19.3760	7.4000	11.4000	13.4160	21.3776	0.20	9.2453	19.0683
18.9527	31.1564	37.8870	63.6848	21.0329	31.7573	36.9716	57.8214	0.40	46.0327	55.1203
62.5093	97.7863	116.0063	187.3669	56.9378	84.8495	98.0688	151.4191	0.60	158.9430	150.8192

TABLE 9.22   Runge-Kutta Method with $h = 0.1$

m_1	m_2	m_3	m_4	k_1	k_2	k_3	k_4	t_n	x_n	y_n
								0.00	−1.0000	6.0000
2.2000	3.1600	3.4560	4.8720	3.7000	4.7000	4.9520	6.3256	0.10	2.3840	10.8883
4.8321	6.5742	7.0778	9.5870	6.2946	7.9413	8.3482	10.5957	0.20	9.3379	19.1332
9.5208	12.5821	13.4258	17.7609	10.5461	13.2339	13.8872	17.5358	0.30	22.5541	32.8539
17.6524	22.9090	24.3055	31.6554	17.4569	21.8114	22.8549	28.7393	0.40	46.5103	55.4420
31.4788	40.3496	42.6387	54.9202	28.6141	35.6245	37.2840	46.7207	0.50	88.5729	93.3006
54.6348	69.4029	73.1247	93.4107	46.5231	57.7482	60.3774	75.4370	0.60	160.7563	152.0025

You should verify that the solution of the initial-value problem in Example 2 (see Problem 39, Exercises 8.6) is given by $x(t) = (26t - 1)e^{4t}$, $y(t) = (13t + 6)e^{4t}$. From these equations we see that the exact values are $x(0.6) = 160.9384$ and $y(0.6) = 152.1198$.

In conclusion, we state **Euler's method** for the general system (4):

$$x_{n+1} = x_n + hf(t_n, x_n, y_n)$$

$$y_{n+1} = y_n + hg(t_n, x_n, y_n).$$

9.7 EXERCISES

Answers to odd-numbered problems begin on page A−34.

1. Use the Euler method to approximate $y(0.2)$, where $y(x)$ is the solution of the initial-value problem

$$y'' - 4y' + 4y = 0, \qquad y(0) = -2, \quad y'(0) = 1.$$

Use $h = 0.1$. Find the exact solution of the problem and compare the exact value of $y(0.2)$ with y_2.

2. Use the Euler method to approximate $y(1.2)$, where $y(x)$ is the solution of the initial-value problem

$$x^2 y'' - 2xy' + 2y = 0, \qquad y(1) = 4, \quad y'(1) = 9,$$

where $x > 0$. Use $h = 0.1$. Find the exact solution of the problem and compare the exact value of $y(1.2)$ with y_2.

3. Repeat Problem 1 using the Runge-Kutta method with $h = 0.2$ and $h = 0.1$.

4. Repeat Problem 2 using the Runge-Kutta method with $h = 0.2$ and $h = 0.1$.

5. Use the Runge-Kutta method to obtain the approximate value of $y(0.2)$, where $y(x)$ is a solution of the initial-value problem

$$y'' - 2y' + 2y = e^t \cos t, \qquad y(0) = 1, \quad y'(0) = 2.$$

Use $h = 0.2$ and $h = 0.1$.

6. When $E = 100$ volts, $R = 10$ ohms, and $L = 1$ henry, the system of differential equations for the currents $i_1(t)$ and $i_3(t)$ in the electrical network given in Figure 9.14 is

$$\frac{di_1}{dt} = -20i_1 + 10i_3 + 100$$

$$\frac{di_3}{dt} = 10i_1 - 20i_3,$$

where $i_1(0) = 0$ and $i_3(0) = 0$. Use the Runge-Kutta method to approximate $i_1(t)$ and $i_3(t)$ at $t = 0.1, 0.2, 0.3, 0.4,$ and 0.5. Use $h = 0.1$.

Figure 9.14

In Problems 7–10 use the Runge-Kutta method to approximate $x(0.2)$ and $y(0.2)$. Compare the results for $h = 0.2$ and $h = 0.1$.

7. $x' = 2x - y$
  $y' = x$
  $x(0) = 6, y(0) = 2$

8. $x' = x + 2y$
  $y' = 4x + 3y$
  $x(0) = 1, y(0) = 1$

9. $x' = -y + t$
  $y' = x - t$
  $x(0) = -3, y(0) = 5$

10. $x' = 6x + y + 6t$
  $y' = 4x + 3y - 10t + 4$
  $x(0) = 0.5, y(0) = 0.2$

9.8 SECOND-ORDER BOUNDARY-VALUE PROBLEMS

In Sections 9.2–9.5 we focused on techniques that yield an approximation to a solution of a first-order initial-value problem $y' = f(x, y)$, $y(x_0) = y_0$. In addition, we saw in Section 9.7 that we can adapt the approximation techniques to a second-order *initial-value problem* $y'' = f(x, y, y')$, $y(x_0) = y_0$, $y'(x_0) = y'_0$, by reducing the second-order differential equation to a system of first-order equations. In this section we are going to examine a method for approximating a solution to a second-order *boundary-value problem* $y'' = f(x, y, y')$, $y(a) = \alpha$, $y(b) = \beta$. We note at the outset that this method does not require reducing the second-order differential equation to a system of equations.

Finite Difference Approximations

Recall from (2) of Section 9.3 that a Taylor series expansion of a function $y(x)$ can be written as

$$y(x + h) = y(x) + y'(x)h + y''(x)\frac{h^2}{2} + y'''(x)\frac{h^3}{6} + \cdots \quad \text{(1)}$$

and so

$$y(x - h) = y(x) - y'(x)h + y''(x)\frac{h^2}{2} - y'''(x)\frac{h^3}{6} + \cdots \quad \text{(2)}$$

If h is small we can ignore terms involving $h^4, h^5, \ldots$ since these values are negligible. Indeed, if we ignore all terms involving h^2 and higher, then (1) and

(2) yield, in turn, the following approximations for the first derivative $y'(x)$:

$$y'(x) \approx \frac{1}{h} [y(x + h) - y(x)] \tag{3}$$

$$y'(x) \approx \frac{1}{h} [y(x) - y(x - h)]. \tag{4}$$

Subtracting (1) and (2) also gives

$$y'(x) \approx \frac{1}{2h} [y(x + h) - y(x - h)]. \tag{5}$$

On the other hand, if we ignore terms involving h^3 and higher, then by adding (1) and (2) we obtain an approximation for the second derivative $y''(x)$:

$$y''(x) \approx \frac{1}{h^2} [y(x + h) - 2y(x) + y(x - h)]. \tag{6}$$

The right sides of (3), (4), (5), and (6) are called **difference quotients**. The expressions

$$y(x + h) - y(x) \qquad\qquad y(x) - y(x - h)$$

$$y(x + h) - y(x - h) \quad \text{and} \quad y(x + h) - 2y(x) + y(x - h)$$

are called **finite differences**. Specifically, $y(x + h) - y(x)$ is called a **forward difference**, $y(x) - y(x - h)$ is a **backward difference**, and both $y(x + h) - y(x - h)$ and $y(x + h) - 2y(x) + y(x - h)$ are called **central differences**. The results given in (5) and (6) are referred to as **central difference approximations** for the derivatives y' and y''.

Finite Difference Equation

Consider now a linear second-order boundary-value problem

$$y'' + P(x)y' + Q(x)y = f(x), \qquad y(a) = \alpha, \quad y(b) = \beta. \tag{7}$$

Suppose $a = x_0 < x_1 < x_2 < \cdots < x_{n-1} < x_n = b$ represents a regular partition of the interval $[a, b]$; that is, $x_i = a + ih$, where $i = 0, 1, 2, \ldots, n$ and $h = (b - a)/n$. The points

$$x_1 = a + h, \quad x_2 = a + 2h, \quad \ldots, \quad x_{n-1} = a + (n - 1)h$$

are called **interior mesh points** of the interval $[a, b]$. If we let

$$y_i = y(x_i), \quad P_i = P(x_i), \quad Q_i = Q(x_i), \quad \text{and} \quad f_i = f(x_i)$$

and if y'' and y' in (7) are replaced by the central difference approximations (5) and (6), we get

$$\frac{y_{i+1} - 2y_i + y_{i-1}}{h^2} + P_i \frac{y_{i+1} - y_{i-1}}{2h} + Q_i y_i = f_i$$

or, after simplifying,

$$\left(1 + \frac{h}{2} P_i\right) y_{i+1} + (-2 + h^2 Q_i) y_i + \left(1 - \frac{h}{2} P_i\right) y_{i-1} = h^2 f_i. \qquad \textbf{(8)}$$

The last equation, known as a **finite difference equation**, is an approximation to the differential equation. It enables us to approximate the solution $y(x)$ of (7) at the interior mesh points $x_1, x_2, \ldots, x_{n-1}$ of the interval $[a, b]$. By letting i take on the values $1, 2, \ldots, n - 1$ in (8), we obtain $n - 1$ equations in the $n - 1$ unknowns $y_1, y_2, \ldots, y_{n-1}$. Bear in mind that we know y_0 and y_n since these are the prescribed boundary conditions $y_0 = y(x_0) = y(a) = \alpha$ and $y_n = y(x_n) = y(b) = \beta$.

In Example 1 we consider a boundary-value problem for which we can compare the approximate values found with the exact values of an explicit solution.

EXAMPLE 1

Use the difference equation (8) with $n = 4$ to approximate the solution of the boundary-value problem

$$y'' - 4y = 0, \qquad y(0) = 0, \quad y(1) = 5.$$

Solution To use (8) we identify $P(x) = 0$, $Q(x) = -4$, $f(x) = 0$, and $h = (1 - 0)/4 = 1/4$. Hence the difference equation is

$$y_{i+1} - 2.25 y_i + y_{i-1} = 0. \qquad \textbf{(9)}$$

Now, the interior points are $x_1 = 0 + 1/4$, $x_2 = 0 + 2/4$, $x_3 = 0 + 3/4$, and so for $i = 1, 2,$ and 3, (9) yields the following system for the corresponding $y_1, y_2,$ and y_3:

$$y_2 - 2.25 y_1 + y_0 = 0$$

$$y_3 - 2.25 y_2 + y_1 = 0$$

$$y_4 - 2.25 y_3 + y_2 = 0.$$

With the boundary conditions $y_0 = 0$ and $y_4 = 5$, the foregoing system becomes

$$-2.25 y_1 + \quad y_2 \qquad\qquad = 0$$

$$y_1 - 2.25 y_2 + \quad y_3 = 0$$

$$y_2 - 2.25 y_3 = -5.$$

Solving the system gives $y_1 = 0.7256$, $y_2 = 1.6327$, and $y_3 = 2.9479$.

Now the general solution of the given differential equation is $y = c_1 \cosh 2x + c_2 \sinh 2x$. The condition $y(0) = 0$ implies $c_1 = 0$. The other boundary condition gives c_2. In this way we see that an explicit solution of the boundary-value problem is $y(x) = (5 \sinh 2x)/\sinh 2$. Thus the exact values (rounded to four decimal places) of this solution at the interior points are $y(0.25) = 0.7184$, $y(0.5) = 1.6201$, and $y(0.75) = 2.9354$. ∎

The accuracy of the approximations in Example 1 can be improved by using a smaller value of h. Of course, the trade-off here is that a smaller value of h necessitates solving a larger systems of equations. It is left as an exercise to show that with $h = 1/8$, approximations to $y(0.25)$, $y(0.5)$, and $y(0.75)$ are 0.7202, 1.6233, and 2.9386, respectively. See Problem 11.

EXAMPLE 2

Use the difference equation (8) with $n = 10$ to approximate the solution of

$$y'' + 3y' + 2y = 4x^2, \qquad y(1) = 1, \quad y(2) = 6.$$

Solution In this case we identify $P(x) = 3$, $Q(x) = 2$, $f(x) = 4x^2$, and $h = (2 - 1)/10 = 0.1$, and so (8) becomes

$$1.15y_{i+1} - 1.98y_i + 0.85y_{i-1} = 0.04x_i^2. \qquad \textbf{(10)}$$

Now, the interior points are $x_1 = 1.1$, $x_2 = 1.2$, $x_3 = 1.3$, $x_4 = 1.4$, $x_5 = 1.5$, $x_6 = 1.6$, $x_7 = 1.7$, $x_8 = 1.8$, and $x_9 = 1.9$. Thus, for $i = 1, 2, \ldots, 9$ and $y_0 = 1$, $y_{10} = 6$, (10) gives a system of nine equations and nine unknowns:

$$1.15y_2 - 1.98y_1 \qquad\qquad = -0.8016$$

$$1.15y_3 - 1.98y_2 + 0.85y_1 = 0.0576$$

$$1.15y_4 - 1.98y_3 + 0.85y_2 = 0.0676$$

$$1.15y_5 - 1.98y_4 + 0.85y_3 = 0.0784$$

$$1.15y_6 - 1.98y_5 + 0.85y_4 = 0.0900$$

$$1.15y_7 - 1.98y_6 + 0.85y_5 = 0.1024$$

$$1.15y_8 - 1.98y_7 + 0.85y_6 = 0.1156$$

$$1.15y_9 - 1.98y_8 + 0.85y_7 = 0.1296$$

$$- 1.98y_9 + 0.85y_8 = -6.7556.$$

We can solve this large system using Gaussian elimination or, with relative ease, by means of a computer algebra system such as Mathematica. The result is found to be $y_1 = 2.4047$, $y_2 = 3.4432$, $y_3 = 4.2010$, $y_4 = 4.7469$, $y_5 = 5.1359$, $y_6 = 5.4124$, $y_7 = 5.6117$, $y_8 = 5.7620$, and $y_9 = 5.8855$.

Remark The approximation method that we have just considered can be extended to boundary-value problems in which the first derivative is specified at a boundary, for example, a problem such as $y'' = f(x, y, y')$, $y'(a) = \alpha$, $y(b) = \beta$. See Problem 13.

9.8 EXERCISES

Answers to odd-numbered problems begin on page A–34.

In Problems 1–10 use the finite difference method and the indicated value of n to approximate the solution of the given boundary-value problem.

1. $y'' + 9y = 0$, $y(0) = 4$, $y(2) = 1$; $n = 4$

2. $y'' - y = x^2$, $y(0) = 0$, $y(1) = 0$; $n = 4$

3. $y'' + 2y' + y = 5x$, $y(0) = 0$, $y(1) = 0$; $n = 5$

4. $y'' - 10y' + 25y = 1$, $y(0) = 1$, $y(1) = 0$; $n = 5$

5. $y'' - 4y' + 4y = (x + 1)e^{2x}$, $y(0) = 3$, $y(1) = 0$; $n = 6$

6. $y'' + 5y' = 4\sqrt{x}$, $y(1) = 1$, $y(2) = -1$; $n = 6$

7. $x^2 y'' + 3xy' + 3y = 0$, $y(1) = 5$, $y(2) = 0$; $n = 8$

8. $x^2 y'' - xy' + y = \ln x$, $y(1) = 0$, $y(2) = -2$; $n = 8$

9. $y'' + (1 - x)y' + xy = x$, $y(0) = 0$, $y(1) = 2$; $n = 10$

10. $y'' + xy' + y = x$, $y(0) = 1$, $y(1) = 0$; $n = 10$

11. Rework Example 1 using $n = 8$.

12. The electrostatic potential u between two concentric spheres of radius $r = 1$ and $r = 4$ is determined from

$$\frac{d^2 u}{dr^2} + \frac{2}{r}\frac{du}{dr} = 0, \qquad u(1) = 50, \quad u(4) = 100.$$

Use the method of this section with $n = 6$ to approximate the solution of this boundary-value problem.

13. Consider the boundary-value problem $y'' + xy = 0$, $y'(0) = 1$, $y(1) = -1$.

   (a) Find the difference equation corresponding to the differential equation. Show that for $i = 0, 1, 2, \ldots, n - 1$ the difference equation yields n equations in $n + 1$ unknowns $y_{-1}, y_0, y_1, y_2, \ldots, y_{n-1}$. Here y_{-1} and y_0 are unknowns since y_{-1} represents an approximation to y at the exterior point $x = -h$ and y_0 is not specified at $x = 0$.

   (b) Use the central difference approximation (5) to show that $y_1 - y_{-1} = 2h$. Use this equation to eliminate y_{-1} from the system in part (a).

   (c) Use $n = 5$ and the system of equations found in parts (a) and (b) to approximate the solution of the original boundary-value problem.

Chapter 9   *REVIEW*

A solution of a differential equation may exist and yet we may not be able to determine it in terms of the familiar elementary functions. A way of convincing oneself that a first-order equation $y' = f(x, y)$ possesses a solution passing through a specific point (x_0, y_0) is to sketch the **direction field** associated with the equation. The equation $f(x, y) = c$ determines the **isoclines**, or curves of constant inclination. This means that every solution curve that passes through a particular isocline does so with the same slope. The direction field is the totality of short line segments throughout two-dimensional space that have midpoints on the isoclines and that possess slope equal to the value of the parameter c. A carefully plotted sequence of these **lineal elements** can suggest the shape of a solution curve passing through the given point (x_0, y_0).

At best, a direction field can give only the crudest form of an approximation to a numerical value of the solution $y(x)$ of the initial-value problem when x is close to x_0.

To obtain the approximate values of $y(x)$, we used **Euler's formula**:

$$y_{n+1} = y_n + hf(x_n, y_n);$$

the **improved Euler formula**:

$$y_{n+1} = y_n + h\,\frac{f(x_n, y_n) + f(x_{n+1}, y_{n+1}^*)}{2},$$

where

$$y_{n+1}^* = y_n + hf(x_n, y_n);$$

the **three-term Taylor formula**:

$$y_{n+1} = y_n + y_n'h + y_n''\frac{h^2}{2};$$

the **Runge-Kutta formula**:

$$y_{n+1} = y_n + \frac{1}{6}(k_1 + 2k_2 + 2k_3 + k_4),$$

where

$$k_1 = hf(x_n, y_n)$$

$$k_2 = hf\left(x_n + \frac{1}{2}h, y_n + \frac{1}{2}k_1\right)$$

$$k_3 = hf\left(x_n + \frac{1}{2}h, y_n + \frac{1}{2}k_2\right)$$

$$k_4 = hf(x_n + h, y_n + k_3);$$

and the **Adams-Bashforth/Adams-Moulton formulas**:

$$y^*_{n+1} = y_n + \frac{h}{24}(55y'_n - 59y'_{n-1} + 37y'_{n-2} - 9y'_{n-3})$$

$$y_{n+1} = y_n + \frac{h}{24}(9y'_{n+1} + 19y'_n - 5y'_{n-1} + y'_{n-2}),$$

where $$y'_{n+1} = f(x_{n+1}, y^*_{n+1}).$$

In each of the foregoing formulas, the number h is the length of a uniform step. In other words, $x_1 = x_0 + h, x_2 = x_1 + h = x_0 + 2h, \ldots, x_n = x_{n-1} + h = x_0 + nh$.

The first four methods are known as **single-step, or starting, methods**. The **Adams-Bashforth/Adams-Moulton** and **Milne methods** are examples of **multistep**, or **continuing, methods**. To use these multistep methods we need to first compute y_1, y_2, and y_3 by some starting method. The **Adams-Bashforth/Adams-Moulton** and **Milne methods** are examples of a class of approximating formulas known as **predictor-corrector formulas**.

An important question in using numerical techniques is the size of error. The order of the global truncation error is a good way to measure the accuracy of a method.

To obtain numerical approximations to higher-order differential equations, we can reduce the differential equation to a system of first-order equations. We then apply a particular numerical technique to each equation of the system.

We can approximate a solution of a second-order boundary-value problem by replacing the differential equation by a **finite difference equation**.

Chapter 9   *REVIEW EXERCISES*

Answers to odd-numbered problems begin on page A–34.

In Problems 1 and 2 sketch the direction field for the given differential equation. Indicate several possible solution curves.

1. $y\,dx - x\,dy = 0$ **2.** $y' = 2x - y$

In Problems 3–6 construct a table comparing the indicated values of $y(x)$ using the Euler, improved Euler, three-term Taylor, and Runge-Kutta methods. Compute to four rounded decimal places. Use $h = 0.1$ and $h = 0.05$.

3. $y' = 2 \ln xy,\ y(1) = 2$
  $y(1.1), y(1.2), y(1.3), y(1.4), y(1.5)$

4. $y' = \sin x^2 + \cos y^2,\ y(0) = 0$
  $y(0.1), y(0.2), y(0.3), y(0.4), y(0.5)$

5. $y' = \sqrt{x + y}$, $y(0.5) = 0.5$
   $y(0.6)$, $y(0.7)$, $y(0.8)$, $y(0.9)$, $y(1.0)$

6. $y' = xy + y^2$, $y(1) = 1$
   $y(1.1)$, $y(1.2)$, $y(1.3)$, $y(1.4)$, $y(1.5)$

7. Use the Euler method to obtain the approximate value of $y(0.2)$, where $y(x)$ is the solution of the initial-value problem

$$y'' - (2x + 1)y = 1, \qquad y(0) = 3, \quad y'(0) = 1.$$

First use one step with $h = 0.2$ and then repeat the calculations using $h = 0.1$.

8. Use the Adams-Bashforth/Adams-Moulton method to approximate the value of $y(0.4)$, where $y(x)$ is the solution of

$$y' = 4x - 2y, \qquad y(0) = 2.$$

Use the Runge-Kutta formula and $h = 0.1$ to obtain the values of y_1, y_2, and y_3.

9. Use the Euler method and $h = 0.1$ to approximate the values of $x(0.2)$, $y(0.2)$, where $x(t)$ and $y(t)$ are solutions of

$$x' = x + y$$

$$y' = x - y,$$

$$x(0) = 1, \qquad y(0) = 2.$$

10. Use the finite difference method with $n = 10$ to approximate the solution of the boundary-value problem

$$y'' + 6.55(1 + x)y = 1, \qquad y(0) = 0, \quad y(1) = 0.$$

NERVE IMPULSE MODELS

C. J. Knickerbocker
Department of Mathematics
St. Lawrence University

A biologically interesting and mathematically rich problem is the transmission of an electrical impulse down a nerve axon. A pictorial representation of the impulse, or **action potential**, is represented in Figure 9.15 as a voltage change over time at some point on the nerve axon.

Figure 9.15

Figure 9.16 Nerve Axon

Physically, a nerve axon can be thought of as a cylinder surrounded by a thin permeable membrane with sodium ions outside the axon membrane and potassium ions inside the membrane. See Figure 9.16.

Pretend we are standing at a point on the axon as a nerve impulse approaches. This "disturbance" induces a change in the voltage near us by having the sodium ions rush through the membrane, making the inside of the axon more positively charged. In Figure 9.15 this corresponds to the voltage rise at the left of the pulse. At this stage the potassium ions inside react to the voltage change by rushing out through the membrane, trying to re-equalize the potential across the membrane. This is seen as the voltage decline in Figure 9.15. But the potassium ions overcompensate and the charge becomes negative relative to the undisturbed case. A slow recovery period, the long tail in Figure 9.15, now occurs as the ions move back to their original sides of the membrane.

In about 1950, A. L. Hodgkin and A. F. Huxley developed a system of nonlinear partial differential equations that model the action potential in a nerve axon. The Hodgkin-Huxley system is similar to a nonlinear heat equation in four dependent variables and two independent variables. In general, the solution to this type of equation is difficult, if not impossible, to find, and in particular the Hodgkin-Huxley equation cannot be solved analytically. Therefore, we need some alternatives to finding an analytical solution. The first possibility is to find an analytical solution to a simplified version of the Hodgkin-Huxley equation. Much research has been done on this idea, but in order to find an exact solution to the simplified equations, the Hodgkin-Huxley equation must be reduced to the point where most of the interesting physical properties are lost. A second possibility is to find a numerical solution to a simplified Hodgkin-Huxley equation; that is, we approximate the Hodgkin-Huxley equation and then find a numerical solution to this approximate version. The simplified model we will consider is called the **Fitzhugh-Nagumo equation**, given by

$$\frac{\partial^2 u}{\partial x^2} = \frac{\partial u}{\partial t} + f(u) + z$$

$$\frac{\partial z}{\partial t} = \varepsilon u$$

$$f(u) = u(1 - u)(a - u),$$

where ε is very small and $0 \leq a \leq 1$. The Hodgkin-Huxley equation is simplified by scaling variables, noting that two of the dependent variables are roughly time independent and that a certain nonlinear term can be simulated by a cubic function of the voltage. This equation was first described by Fitzhugh and was extensively studied by Nagumo. This simplified equation does not yield any real quantitative information about the action potential, but it is rich in the qualitative aspects of the physical problem.

We can make another simplification by noting that the action potential moves at a constant velocity θ and does not change its shape as it moves; that is, we can transfer the equation from an x, t-coordinate system to a traveling coordinate system in ξ by assuming a solution of the form $u = u(\xi) = u(x - \theta t)$.

This yields the following ordinary differential equation in ξ:

$$\frac{d^2u}{d\xi^2} - \theta\,\frac{du}{d\xi} - f(u) - z = 0$$

$$\theta\,\frac{dz}{d\xi} = \varepsilon u$$

$$f(u) = u(1-u)(a-u).$$

Some analytical results are known for this equation when specific values are chosen for the parameters. For example, when $\theta = 0$, $\varepsilon = 0$, and u is assumed to be small (so that $f(u) \approx au$), an analytical solution is easily determined.

Before any attempt is made to solve this equation numerically, we should look for any analytic "clues" that may help guide us in the choice and implementation of a numerical scheme. An analysis of the Fitzhugh-Nagumo equation produces two important pieces of information. First, using a technique called perturbation analysis, we anticipate a solution whose shape is similar to that of the action potential (see Figure 9.15). Second, using phase-plane analysis, we find that the ordinary differential equation is very "sensitive" to the choices of the parameters θ, ε, and a. That is to say that the numerical solutions might vary dramatically with small changes in any of these parameters.

The sensitive nature of the equation can be seen in the numerical results. Using the fourth-order Runge-Kutta method, we begin by choosing $\varepsilon = 0.005$, $a = 0.1$, $u(0) = 0.001$, and $\theta = 0.5184627034239536284$. This yields the solution given in Figure 9.17.

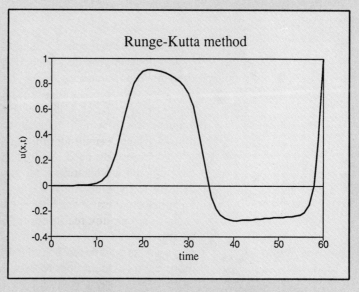

Figure 9.17 Runge-Kutta Solution 1

Note that all the qualitative features of the action potential are represented: a sharp rise in voltage followed by a sharp decline followed by a long slow recovery period. But, note that the numerical solution "blows up" to positive infinity on the right-hand side. Now by changing the last digit of θ, yielding $\theta = 0.5184627034239536283$, we get the solution shown· in Figure 9.18.

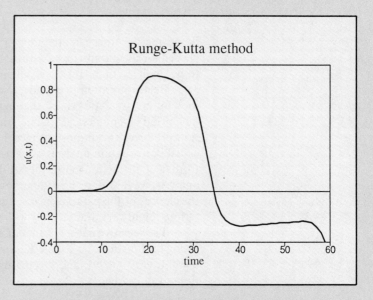

Figure 9.18   Runge-Kutta Solution 2

This solution appears identical to the first solution except on the right-hand side where it goes to negative infinity. This clearly shows how sensitive the numerical solution is to the choice of parameters.

A careful application of the fourth-order Runge-Kutta method also generates other possible solutions. For example, a numerical solution can be found in the form of a small slow pulse followed by a slightly larger pulse, both of which are considerably smaller than the "normal" solution.

A more advanced numerical method that can be applied to this ordinary differential equation is a technique called the **finite elements method**. In contrast to the finite difference approach, where an exact solution is found to an approximation of the differential equation, the finite element approach is to find an approximate solution to the exact equation. This is accomplished by subdividing the ξ-axis with mesh width $\Delta \xi$. Over each interval a set of polynomials is defined. The higher the order of the set of polynomials, the more accurate the numerical solution will be. The polynomials are selected so that the values of the functions and the corresponding derivatives have certain properties at the boundaries of the interval. For the problem at hand we have selected over each interval six fifth-degree polynomials, called the Hermite quintics. The polynomials are then "plugged into" the differential equation. Clearly these polynomials cannot solve the differential equation on the complete interval, so we "force" the polynomials to work at a few points (called the Gaussian points) on each interval. Each Gaussian point now yields a nonlinear algebraic equation to solve, and therefore the method generates a system of coupled nonlinear algebraic equations to solve. The results of this type of approach used with the parameters stated previously are given in Figure 9.19.

Figure 9.19   Finite Element Solution

The advantage to this method is that it invokes conditions at both ends of the domain, thereby treating the problem as a boundary-value problem as opposed to the fourth-order Runge-Kutta method, which treats the problem as an initial-value problem. The disadvantage to the method is the complexity of the technique and difficulties in programming.

The interested reader can find a survey on research that has been done on the various aspects of the Hodgkin-Huxley equation in an article entitled "The Electrophysics of a Nerve Fiber" by A. C. Scott in *Reviews of Modern Physics* (Volume 47, No. 2, April 1975). This article contains a very extensive list of references.

C H A P T E R 10

PLANE AUTONOMOUS SYSTEMS AND STABILITY

10.1 Autonomous Systems, Critical Points, and Periodic Solutions

10.2 Stability of Linear Systems

10.3 Linearization and Local Stability

10.4 Applications of Autonomous Systems

Chapter 10 Review
Chapter 10 Review Exercises

Important Concepts

Autonomous systems
Critical point
Arc
Periodic solution or cycle
Stable node
Unstable node
Saddle point
Degenerate node
Center
Spiral points
Stable critical point
Unstable critical point
Linearization
Jacobian matrix
Phase-plane method

The principal focus in Chapter 8 was on techniques for solving systems of linear first-order differential equations of the form $\mathbf{X}' = \mathbf{AX} + \mathbf{F}(t)$. When the system of differential equations is not linear, it is usually not possible to find solutions in terms of elementary functions. In this chapter we demonstrate that valuable information on the geometric nature of

Continues

Continued

solutions can be obtained by first analyzing special constant solutions called critical points and by searching for periodic solutions. These special solutions are further classified as stable or unstable according to the behavior of nearby solutions. This important concept of stability will be introduced and illustrated with examples from physics and ecology.

10.1 AUTONOMOUS SYSTEMS, CRITICAL POINTS, AND PERIODIC SOLUTIONS

Terminology and Notation

A system of first-order differential equations is called **autonomous** when the system can be written in the form

$$\frac{dx_1}{dt} = g_1(x_1, x_2, \ldots, x_n)$$

$$\frac{dx_2}{dt} = g_2(x_1, x_2, \ldots, x_n)$$

$$\vdots \qquad\qquad \vdots$$

$$\frac{dx_n}{dt} = g_n(x_1, x_2, \ldots, x_n).$$

(1)

Thus the independent variable t does *not* appear explicitly on the right-hand side of each differential equation (compare with (2) in Section 8.3).

EXAMPLE 1

The system of differential equations

$$\frac{dx_1}{dt} = x_1 - 3x_2 + t^2$$

$$\frac{dx_2}{dt} = x_1 \sin(x_2 t)$$

is *not* autonomous because of the presence of t^2 and $\sin(x_2 t)$ on the right-hand sides. ∎

If $\mathbf{X}(t)$ and $\mathbf{g}(\mathbf{X})$ denote the respective column vectors

$$\mathbf{X}(t) = \begin{pmatrix} x_1(t) \\ x_2(t) \\ \vdots \\ x_n(t) \end{pmatrix}, \qquad \mathbf{g}(\mathbf{X}) = \begin{pmatrix} g_1(x_1, x_2, \ldots, x_n) \\ g_2(x_1, x_2, \ldots, x_n) \\ \vdots \\ g_n(x_1, x_2, \ldots, x_n) \end{pmatrix},$$

then the autonomous system (1) can be written in the compact **column vector form** $\mathbf{X}' = \mathbf{g}(\mathbf{X})$. The homogeneous linear system $\mathbf{X}' = \mathbf{A}\mathbf{X}$ studied in Sections 8.5 and 8.6 is an important special case.

In this chapter it is also convenient to write (1) using row vectors. If we let

$$\mathbf{X}(t) = (x_1(t), x_2(t), \ldots, x_n(t))$$

and $\mathbf{g}(\mathbf{X}) = (g_1(x_1, x_2, \ldots, x_n), g_2(x_1, x_2, \ldots, x_n), \ldots, g_n(x_1, x_2, \ldots, x_n)),$

then the autonomous system (1) can be written in the compact **row vector form** $\mathbf{X}' = \mathbf{g}(\mathbf{X})$. It should be clear from the context whether we are using column or row vector form, and therefore we will not distinguish between $\mathbf{X}$ and $\mathbf{X}^T$, the transpose of $\mathbf{X}$. In particular, when $n = 2$, it is convenient to use row vector form and write an initial condition as $\mathbf{X}(0) = (x_0, y_0)$.

When the variable t is interpreted as time, we can refer to a solution $\mathbf{X}(t)$ as the **state of the system** at time t. Using this terminology, we say a system of differential equations is autonomous when the rate $\mathbf{X}'(t)$ at which the system changes depends on only the system's present state $\mathbf{X}(t)$. The linear system $\mathbf{X}' = \mathbf{A}\mathbf{X} + \mathbf{F}(t)$ studied in Chapter 8 is then autonomous when $\mathbf{F}(t)$ is constant.

Note that when $n = 1$, an autonomous differential equation takes the simple form $dx/dt = g(x)$. Explicit solutions can be constructed since this differential equation is separable, and we will make use of this fact to give illustrations of the concepts in this chapter.

Vector Field Interpretation

When $n = 2$, the system is called a **plane autonomous system**, and we write the system as

$$\frac{dx}{dt} = P(x, y)$$

$$\frac{dy}{dt} = Q(x, y).$$

The vector $\mathbf{V}(x, y) = (P(x, y), Q(x, y))$ defines a **vector field** in a region of the plane, and a solution to the system may be interpreted as the resulting path of a particle as it moves through the region. To be more specific, let $\mathbf{V}(x, y) = (P(x, y), Q(x, y))$ denote the velocity of a stream at position (x, y), and suppose that a small particle (such as a cork) is released at a position

(x_0, y_0) in the stream. If $\mathbf{X}(t) = (x(t), y(t))$ denotes the position of the particle at time t, then $\mathbf{X}'(t) = (x'(t), y'(t))$ is the velocity vector $\mathbf{v}$. When external forces are not present and frictional forces are neglected, the velocity of the particle at time t is the velocity of the stream at position $\mathbf{X}(t)$:

$$\mathbf{X}'(t) = \mathbf{V}(x(t), y(t));$$

that is,

$$\frac{dx}{dt} = P(x(t), y(t))$$

$$\frac{dy}{dt} = Q(x(t), y(t)).$$

Thus the path of the particle is the solution to a system that satisfies the initial condition $\mathbf{X}(0) = (x_0, y_0)$. We will frequently call on this simple interpretation of a plane autonomous system to illustrate new concepts.

EXAMPLE 2

A vector field for the steady-state flow of a fluid around a cylinder of radius 1 is given by

$$\mathbf{V}(x, y) = V_0\left(1 - \frac{x^2 - y^2}{(x^2 + y^2)^2}, \frac{-2xy}{(x^2 + y^2)^2}\right),$$

where V_0 is the speed of the fluid far from the cylinder. If a small cork is released at $(-3, 1)$, the path $\mathbf{X}(t) = (x(t), y(t))$ of the cork satisfies the plane autonomous system

$$\frac{dx}{dt} = V_0\left(1 - \frac{x^2 - y^2}{(x^2 + y^2)^2}\right)$$

$$\frac{dy}{dt} = V_0\left(\frac{-2xy}{(x^2 + y^2)^2}\right)$$

subject to the initial condition $\mathbf{X}(0) = (-3, 1)$. See Figure 10.1.

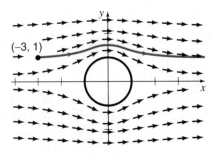

Figure 10.1

Any second-order nonlinear differential equation $x'' = g(x, x')$ can be written as a plane autonomous system. With the introduction of $y = x'$, the equation becomes

$$x' = y$$

$$y' = g(x, y).$$

EXAMPLE 3

In Example 3 in Section 1.2, we showed that the displacement angle θ for a pendulum satisfies the second-order nonlinear differential equation

$$\frac{d^2\theta}{dt^2} + \frac{g}{l}\sin\theta = 0.$$

If we let $x = \theta$ and $y = \theta'$, this second-order differential equation may be rewritten as the plane autonomous system

$$x' = y$$

$$y' = -\frac{g}{l}\sin x. \qquad \blacksquare$$

Types of Solutions

If $P(x, y)$, $Q(x, y)$, and the first-order partial derivatives $\partial P/\partial x$, $\partial P/\partial y$, $\partial Q/\partial x$, and $\partial Q/\partial y$ are continuous in a region R of the plane, then the solutions to the plane autonomous system

$$\frac{dx}{dt} = P(x, y)$$

$$\frac{dy}{dt} = Q(x, y)$$

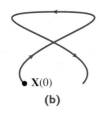

Figure 10.2

are of three basic types:

(i) A **constant solution** $x(t) = x_0$, $y(t) = y_0$ (or $\mathbf{X}(t) = \mathbf{X}_0$ for all t). A constant solution is called a **critical** or **stationary point**. When the particle is placed at a critical point $\mathbf{X}_0$ (that is, $\mathbf{X}(0) = \mathbf{X}_0$), it remains there indefinitely. Note that since $\mathbf{X}'(t) = \mathbf{0}$, a critical point is a solution of the system of algebraic equations

$$P(x, y) = 0$$

$$Q(x, y) = 0.$$

(ii) A solution $x = x(t)$, $y = y(t)$ that defines an **arc**, a plane curve that does *not* cross itself. Thus the curve in Figure 10.2(a) can be a

Figure 10.3

solution to a plane autonomous system, whereas the curve in Figure 10.2(b) cannot be a solution.

(iii) A **periodic solution** $x = x(t)$, $y = y(t)$. A periodic solution is called a **cycle**. If p is the period of the solution, then $\mathbf{X}(t + p) = \mathbf{X}(t)$ and a particle placed on the curve at $\mathbf{X}_0$ will cycle around the curve and return to $\mathbf{X}_0$ in p units of time (see Figure 10.3).

EXAMPLE 4

Find all critical points of each of the following plane autonomous systems:

(a) $x' = -x + y$
$\quad\;\; y' = x - y$

(b) $x' = x^2 + y^2 - 6$
$\quad\;\; y' = x^2 - y$

(c) $x' = 0.01x(100 - x - y)$
$\quad\;\; y' = 0.05y(60 - y - 0.2x)$

Solution We find the critical points by setting the right-hand sides of the differential equations equal to zero.

(a) The solution to the system

$$-x + y = 0$$

$$x - y = 0$$

consists of all points on the line $y = x$. Thus there are infinitely many critical points.

(b) To solve the system

$$x^2 + y^2 - 6 = 0$$

$$x^2 - y = 0$$

we substitute the second equation $x^2 = y$ into the first equation to obtain $y^2 + y - 6 = (y + 3)(y - 2) = 0$. If $y = -3$, then $x^2 = -3$, and so there are no real solutions. If $y = 2$, then $x = \pm\sqrt{2}$, and so the critical points are $(\sqrt{2}, 2)$ and $(-\sqrt{2}, 2)$.

(c) Finding the critical points in part (c) requires a careful consideration of cases. The equation $0.01x(100 - x - y) = 0$ implies $x = 0$ or $x + y = 100$. If $x = 0$, then, substituting in $0.05y(60 - y - 0.2x) = 0$, we have $y(60 - y) = 0$. Thus $y = 0$ or 60, and so $(0, 0)$ and $(0, 60)$ are critical points. If $x + y = 100$, then $0 = y(60 - y - 0.2(100 - y)) = y(40 - 0.8y)$. It follows that $y = 0$ or 50, and so $(100, 0)$ and $(50, 50)$ are critical points. ∎

When the plane autonomous system is linear, we can use the methods in Chapter 8 to investigate solutions.

EXAMPLE 5

Determine whether the given linear system possesses a periodic solution.

(a) $x' = 2x + 8y$
$y' = -x - 2y$

(b) $x' = x + 2y$
$y' = -\dfrac{1}{2}x + y$

In each case sketch the graph of the solution that satisfies $\mathbf{X}(0) = (2, 0)$.

Solution (a) In Example 3 in Section 8.6, we used the eigenvalue-eigenvector method to show that

$$x = c_1(2 \cos 2t - 2 \sin 2t) + c_2(2 \cos 2t + 2 \sin 2t)$$

$$y = c_1(-\cos 2t) - c_2 \sin 2t.$$

Thus every solution is periodic with period $p = \pi$. The solution satisfying $\mathbf{X}(0) = (2, 0)$ is

$$x = 2 \cos 2t + 2 \sin 2t$$

$$y = -\sin 2t.$$

This solution generates the ellipse shown in Figure 10.4(a).

(b) In Example 4 in Section 8.6, we used the eigenvalue-eigenvector method to show that

$$x = c_1(2e^t \cos t) + c_2(2e^t \sin t)$$

$$y = c_1(-e^t \sin t) + c_2(e^t \cos t).$$

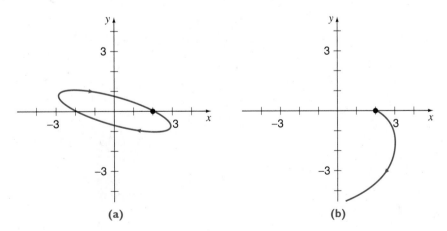

(a)   (b)

Figure 10.4

Because of the presence of e^t in the general solution, there are no periodic solutions (that is, cycles). The solution satisfying $\mathbf{X}(0) = (2, 0)$ is

$$x = 2e^t \cos t$$

$$y = -e^t \sin t$$

and this curve is shown in Figure 10.4(b). ∎

Changing to Polar Coordinates

Except for the case of constant solutions, it is usually not possible to find explicit expressions for the solutions of a nonlinear autonomous system. We can solve some nonlinear systems, however, by changing to polar coordinates. From the formulas $r^2 = x^2 + y^2$ and $\theta = \tan^{-1}(y/x)$, we obtain

$$\frac{dr}{dt} = \frac{1}{r}\left(x\frac{dx}{dt} + y\frac{dy}{dt} \right)$$

$$\frac{d\theta}{dt} = \frac{1}{r^2}\left(-y\frac{dx}{dt} + x\frac{dy}{dt} \right). \tag{2}$$

EXAMPLE 6

Find the solution of the nonlinear plane autonomous system

$$x' = -y - x\sqrt{x^2 + y^2}$$

$$y' = x - y\sqrt{x^2 + y^2}$$

satisfying the initial condition $\mathbf{X}(0) = (3, 3)$.

Solution From the expressions for dr/dt and $d\theta/dt$ in (2) we obtain

$$\frac{dr}{dt} = \frac{1}{r}[x(-y - xr) + y(x - yr)] = -r^2$$

$$\frac{d\theta}{dt} = \frac{1}{r^2}[-y(-y - xr) + x(x - yr)] = 1$$

with $r(0) = 3\sqrt{2}$ and $\theta(0) = \pi/4$. We can use separation of variables to show that the general solution of the system is

$$r = \frac{1}{t + c_1}$$

$$\theta = t + c_2$$

for $r \neq 0$. (Check this!) Applying the initial conditions then gives $r = 1/(t + \sqrt{2}/6)$ and $\theta = t + \pi/4$. The spiral $r = 1/(\theta + \sqrt{2}/6 - \pi/4)$ is sketched in Figure 10.5.

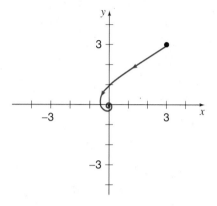

Figure 10.5   ■

EXAMPLE 7

When expressed in polar coordinates, a plane autonomous system takes the form

$$\frac{dr}{dt} = 0.5(3 - r)$$

$$\frac{d\theta}{dt} = 1.$$

Find and sketch the solutions satisfying $\mathbf{X}(0) = (3, 0)$ and $\mathbf{X}(0) = (0, 1)$ in rectangular coordinates.

Solution Applying separation of variables to $dr/dt = 0.5(3 - r)$ and integrating $d\theta/dt = 1$ lead to the general solution

$$r = 3 + c_1 e^{-0.5t}$$

$$\theta = t + c_2.$$

If $\mathbf{X}(0) = (3, 0)$, then $r(0) = 3$ and $\theta(0) = 0$. It follows that $c_1 = c_2 = 0$, and so $r = 3$ and $\theta = t$. Hence $x = r \cos \theta = 3 \cos t$ and $y = r \sin \theta = 3 \sin t$, and so the solution is periodic.

If $\mathbf{X}(0) = (0, 1)$, then $r(0) = 1$ and $\theta(0) = \pi/2$, and so $c_1 = -2$ and $c_2 = \pi/2$. The solution curve is the spiral $r = 3 - 2e^{-0.5(\theta - \pi/2)}$. Note that as

$t \to \infty$, θ increases without bound and r approaches 3. Both solutions are shown in Figure 10.6.

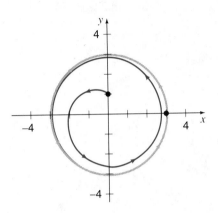

Figure 10.6

■

10.1 EXERCISES

Answers to odd-numbered problems begin on page A–35.

In Problems 1–6 write the given nonlinear second-order differential equation as a plane autonomous system. Find all critical points of the resulting system.

1. $x'' + 9 \sin x = 0$

2. $x'' + (x')^2 + 2x = 0$

3. $x'' + x'(1 - x^3) - x^2 = 0$

4. $x'' + 4\dfrac{x}{1 + x^2} + 2x' = 0$

5. $x'' + x = \varepsilon x^3$ for $\varepsilon > 0$

6. $x'' + x - \varepsilon x|x| = 0$ for $\varepsilon > 0$

In Problems 7–16 find all critical points of the given plane autonomous system.

7. $x' = x + xy$
  $y' = -y - xy$

8. $x' = y^2 - x$
  $y' = x^2 - y$

9. $x' = 3x^2 - 4y$
  $y' = x - y$

10. $x' = x^3 - y$
  $y' = x - y^3$

11. $x' = x(10 - x - \frac{1}{2}y)$
  $y' = y(16 - y - x)$

12. $x' = -2x + y + 10$
  $y' = 2x - y - 15\dfrac{y}{y + 5}$

13. $x' = x^2 e^y$
  $y' = y(e^x - 1)$

14. $x' = \sin y$
  $y' = e^{x-y} - 1$

15. $x' = x(1 - x^2 - 3y^2)$
  $y' = y(3 - x^2 - 3y^2)$

16. $x' = -x(4 - y^2)$
  $y' = 4y(1 - x^2)$

In Problems 17–22 for the given linear plane autonomous system (taken from Exercises 8.6):

(a) find the general solution and determine whether there are periodic solutions,

(b) find the solution satisfying the given initial condition, and

(c) with the aid of a graphics calculator or graphing software, sketch the solution in part (b) and indicate the direction in which the curve is traversed.

17. $\begin{aligned} x' &= x + 2y \\ y' &= 4x + 3y \end{aligned}$, $\mathbf{X}(0) = (2, -2)$ (Problem 1, Exercises 8.6)

18. $\begin{aligned} x' &= -6x + 2y \\ y' &= -3x + y \end{aligned}$, $\mathbf{X}(0) = (3, 4)$ (Problem 6, Exercises 8.6)

19. $\begin{aligned} x' &= 4x - 5y \\ y' &= 5x - 4y \end{aligned}$, $\mathbf{X}(0) = (4, 5)$ (Problem 19, Exercises 8.6)

20. $\begin{aligned} x' &= x + y \\ y' &= -2x - y \end{aligned}$, $\mathbf{X}(0) = (-2, 2)$ (Problem 16, Exercises 8.6)

21. $\begin{aligned} x' &= 5x + y \\ y' &= -2x + 3y \end{aligned}$, $\mathbf{X}(0) = (-1, 2)$ (Problem 17, Exercises 8.6)

22. $\begin{aligned} x' &= x - 8y \\ y' &= x - 3y \end{aligned}$, $\mathbf{X}(0) = (2, 1)$ (Problem 20, Exercises 8.6)

In Problems 23–26 solve the given nonlinear plane autonomous system by changing to polar coordinates. Describe the geometric behavior of the solution that satisfies the given initial condition(s).

23. $\begin{aligned} x' &= -y - x(x^2 + y^2)^2 \\ y' &= x - y(x^2 + y^2)^2 \end{aligned}$, $\mathbf{X}(0) = (4, 0)$

24. $\begin{aligned} x' &= y + x(x^2 + y^2) \\ y' &= -x + y(x^2 + y^2) \end{aligned}$, $\mathbf{X}(0) = (4, 0)$

25. $\begin{aligned} x' &= -y + x(1 - x^2 - y^2) \\ y' &= x + y(1 - x^2 - y^2) \end{aligned}$, $\mathbf{X}(0) = (1, 0)$ and $\mathbf{X}(0) = (2, 0)$

[*Hint:* The resulting differential equation for r is a Bernoulli differential equation. See Section 2.6.]

26. $\begin{aligned} x' &= y - \dfrac{x}{\sqrt{x^2 + y^2}} (4 - x^2 - y^2) \\ y' &= -x - \dfrac{y}{\sqrt{x^2 + y^2}} (4 - x^2 - y^2) \end{aligned}$, $\mathbf{X}(0) = (1, 0)$ and $\mathbf{X}(0) = (2, 0)$

[*Hint:* See Example 6, Section 2.2.]

27. If $z = f(x, y)$ is a function with continuous first partial derivatives in a region R, then a flow $\mathbf{V}(x, y) = (P(x, y), Q(x, y))$ in R may be defined by letting $P(x, y) = -(\partial f/\partial y)(x, y)$ and $Q(x, y) = (\partial f/\partial x)(x, y)$. Show that if

$\mathbf{X}(t) = (x(t), y(t))$ is a solution of the plane autonomous system

$$x' = P(x, y)$$

$$y' = Q(x, y),$$

then $f(x(t), y(t)) = c$ for some constant c. Thus a solution curve lies on a *level curve* of f. [*Hint:* Use the chain rule to compute $(d/dt)f(x(t), y(t))$.]

10.2 STABILITY OF LINEAR SYSTEMS

In Section 10.1 we pointed out that the plane autonomous system

$$\frac{dx}{dt} = P(x, y)$$

$$\frac{dy}{dt} = Q(x, y)$$

critical point

(a)

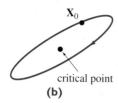

critical point

(b)

gives rise to a vector field $\mathbf{V}(x, y) = (P(x, y), Q(x, y))$, and a solution $\mathbf{X} = \mathbf{X}(t)$ may be interpreted as the resulting path of a particle that is initially placed at position $\mathbf{X}(0) = \mathbf{X}_0$. If $\mathbf{X}_0$ is a critical point, the particle remains stationary. If $\mathbf{X}_0$ is placed near a critical point $\mathbf{X}_1$, however, we ask the following questions:

(i) Will the particle return to the critical point? More precisely, if $\mathbf{X} = \mathbf{X}(t)$ is the solution that satisfies $\mathbf{X}(0) = \mathbf{X}_0$, then is $\lim_{t\to\infty} \mathbf{X}(t) = \mathbf{X}_1$?

(ii) If the particle does *not* return to the critical point, does it remain close to the critical point or move away from the critical point? It is conceivable, for example, that the particle may simply circle the critical point or even return to a different critical point. See Figure 10.7.

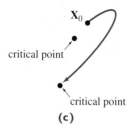

critical point

critical point

(c)

Figure 10.7

If in some neighborhood of the critical point case (a) or (b) in Figure 10.7 *always* occurs, we call the critical point **locally stable**. If, however, an initial value $\mathbf{X}_0$ that results in behavior similar to case (c) can be found in *any* given neighborhood, we call the critical point **unstable**. These concepts will be made more precise in Section 10.3, where questions (i) and (ii) are investigated for nonlinear systems.

Stability Analysis

We will first investigate these two stability questions for linear plane autonomous systems and lay the foundation for Section 10.3. The solution methods of Chapter 8 enable us to give a careful geometric analysis of the solutions to

$$x' = ax + by$$

$$y' = cx + dy$$

(1)

in terms of the eigenvalues and eigenvectors of the coefficient matrix

$$\mathbf{A} = \begin{pmatrix} a & b \\ c & d \end{pmatrix}.$$

To ensure that $\mathbf{X}_0 = (0, 0)$ is the only critical point, we assume that the determinant $\Delta = ad - bc \neq 0$. If $\tau = a + d$ is the trace* of matrix $\mathbf{A}$, then the characteristic equation $\det(\mathbf{A} - \lambda\mathbf{I}) = 0$ may be rewritten as

$$\lambda^2 - \tau\lambda + \Delta = 0.$$

Therefore the eigenvalues of $\mathbf{A}$ are $\lambda = (\tau \pm \sqrt{\tau^2 - 4\Delta})/2$, and the usual three cases for these roots occur according to whether $\tau^2 - 4\Delta$ is positive, negative, or zero.

> **CASE I   REAL DISTINCT EIGENVALUES** $(\tau^2 - 4\Delta > 0)$ According to Theorem 8.9 in Section 8.6 the general solution of (1) is given by
>
> $$\mathbf{X}(t) = c_1\mathbf{K}_1 e^{\lambda_1 t} + c_2\mathbf{K}_2 e^{\lambda_2 t}, \tag{2}$$
>
> where λ_1 and λ_2 are the eigenvalues, and $\mathbf{K}_1$ and $\mathbf{K}_2$ are the corresponding eigenvectors. Note that $\mathbf{X}(t)$ can also be written as
>
> $$\mathbf{X}(t) = e^{\lambda_1 t}[c_1\mathbf{K}_1 + c_2\mathbf{K}_2 e^{(\lambda_2 - \lambda_1)t}]. \tag{3}$$
>
> **(a) Both eigenvalues negative** $(\tau^2 - 4\Delta > 0,\ \tau < 0,\ \text{and}\ \Delta > 0)$
> *Stable Node*   From (2), it follows that $\lim_{t\to\infty} \mathbf{X}(t) = \mathbf{0}$. If we assume that $\lambda_2 < \lambda_1$, then $\lambda_2 - \lambda_1 < 0$, and so we may conclude from (3) that
>
> $$\mathbf{X}(t) \approx c_1\mathbf{K}_1 e^{\lambda_1 t}$$
>
> for large values of t. When $c_1 \neq 0$, $\mathbf{X}(t)$ approaches $\mathbf{0}$ from one of the two directions determined by the eigenvector $\mathbf{K}_1$ corresponding to λ_1. If $c_1 = 0$, $\mathbf{X}(t) = c_2\mathbf{K}_2 e^{\lambda_2 t}$ and $\mathbf{X}(t)$ approaches $\mathbf{0}$ along the line determined by the eigenvector $\mathbf{K}_2$. Figure 10.8 shows a collection of solution curves around the origin. A critical point is called a **stable node** when both eigenvalues are negative.
>
> **(b) Both eigenvalues positive** $(\tau^2 - 4\Delta > 0,\ \tau > 0,\ \text{and}\ \Delta > 0)$
> *Unstable Node*   The analysis for this case is similar to case (a). Again from (2), $\mathbf{X}(t)$ becomes unbounded as t increases. Moreover, from (3), $\mathbf{X}(t)$ becomes unbounded in one of the directions determined by the eigenvector $\mathbf{K}_1$ (when $c_1 \neq 0$) or along the line determined by the eigenvector $\mathbf{K}_2$ (when $c_1 = 0$). Figure 10.9 shows a typical collection of solution curves. This type of critical point, corresponding to the case when both eigenvalues are positive, is called an **unstable node**.

---

* In general, if $\mathbf{A}$ is an $n \times n$ matrix, then the **trace** of $\mathbf{A}$ is the sum of the main diagonal entries.

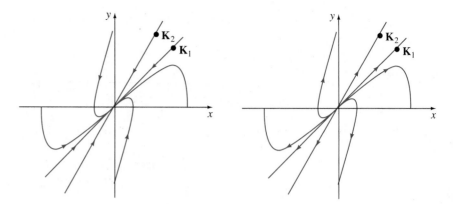

Figure 10.8 Stable Node   **Figure 10.9** Unstable Node

(c) **Eigenvalues with opposite signs** ($\tau^2 - 4\Delta > 0$ and $\Delta < 0$)

Saddle Point The analysis of the solutions is identical to case (b) with one exception. When $c_1 = 0$, $\mathbf{X}(t) = c_2 \mathbf{K}_2 e^{\lambda_2 t}$ and, since $\lambda_2 < 0$, $\mathbf{X}(t)$ will approach $\mathbf{0}$ along the line determined by the eigenvector $\mathbf{K}_2$. If $\mathbf{X}(0)$ does not lie on the line determined by $\mathbf{K}_2$, the line determined by $\mathbf{K}_1$ serves as an asymptote for $\mathbf{X}(t)$. This unstable critical point is called a **saddle point**. See Figure 10.10.

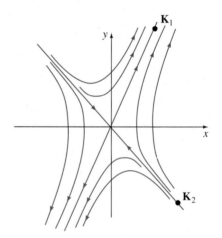

Figure 10.10 Saddle Point

EXAMPLE I

Classify the critical point (0, 0) of each of the following linear systems $\mathbf{X}' = \mathbf{AX}$ as a stable node, unstable node, or saddle point.

(a) $\mathbf{A} = \begin{pmatrix} 2 & 3 \\ 2 & 1 \end{pmatrix}$ (b) $\mathbf{A} = \begin{pmatrix} -10 & 6 \\ 15 & -19 \end{pmatrix}$.

In each case discuss the nature of the solutions in a neighborhood of $(0, 0)$.

Solution (a) Since the trace τ is 3 and the determinant $\Delta = -4$, the eigenvalues are

$$\lambda = \frac{\tau \pm \sqrt{\tau^2 - 4\Delta}}{2} = \frac{3 \pm \sqrt{3^2 - 4(-4)}}{2} = \frac{3 \pm 5}{2} = 4, -1.$$

The eigenvalues have opposite signs and so $(0, 0)$ is a saddle point. It is not hard to show (see Example 1, Section 8.6) that the eigenvectors corresponding to $\lambda_1 = 4$ and $\lambda_2 = -1$ are

$$\mathbf{K}_1 = \begin{pmatrix} 3 \\ 2 \end{pmatrix} \quad \text{and} \quad \mathbf{K}_2 = \begin{pmatrix} 1 \\ -1 \end{pmatrix},$$

respectively. If $\mathbf{X}(0) = \mathbf{X}_0$ lies on the line $y = -x$, $\mathbf{X}(t)$ approaches $\mathbf{0}$. For any other initial condition, $\mathbf{X}(t)$ becomes unbounded in the directions determined by $\mathbf{K}_1$. In other words, the line $y = \frac{2}{3}x$ serves as an asymptote for all these solution curves.

(b) From $\tau = -29$ and $\Delta = 100$, it follows that the eigenvalues of $\mathbf{A}$ are $\lambda_1 = -4$ and $\lambda_2 = -25$. Both eigenvalues are negative and so $(0, 0)$ is in this case a stable node. Since the eigenvectors corresponding to $\lambda_1 = -4$ and $\lambda_2 = -25$ are

$$\mathbf{K}_1 = \begin{pmatrix} 1 \\ 1 \end{pmatrix} \quad \text{and} \quad \mathbf{K}_2 = \begin{pmatrix} 2 \\ -5 \end{pmatrix},$$

respectively, it follows that all solutions approach $\mathbf{0}$ from the direction defined by $\mathbf{K}_1$ except those solutions for which $\mathbf{X}(0) = \mathbf{X}_0$ lies on the line $y = -\frac{5}{2}x$ determined by $\mathbf{K}_2$. These solutions approach $\mathbf{0}$ along $y = -\frac{5}{2}x$. ∎

CASE II A REPEATED REAL EIGENVALUE $(\tau^2 - 4\Delta = 0)$ *Degenerate Nodes* Recall from Section 8.6 that the general solution takes on one of two different forms depending on whether one or two linearly independent eigenvectors can be found for the repeated eigenvalue λ_1.

(a) **Two linearly independent eigenvectors.** If $\mathbf{K}_1$ and $\mathbf{K}_2$ are two linearly independent eigenvectors corresponding to λ_1, then the general solution is given by

$$\mathbf{X}(t) = c_1 \mathbf{K}_1 e^{\lambda_1 t} + c_2 \mathbf{K}_2 e^{\lambda_1 t}$$

$$= (c_1 \mathbf{K}_1 + c_2 \mathbf{K}_2) e^{\lambda_1 t}.$$

If $\lambda_1 < 0$, $\mathbf{X}(t)$ approaches $\mathbf{0}$ along the line determined by the vector $c_1 \mathbf{K}_1 + c_2 \mathbf{K}_2$, and the critical point is called a **degenerate stable**

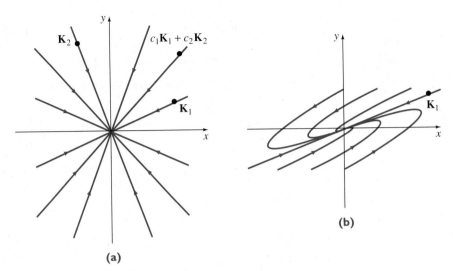

Figure 10.11 Degenerate Stable Nodes

node (see Figure 10.11(a)). The arrows in Figure 10.11(a) are reversed when $\lambda_1 > 0$, and we have a **degenerate unstable node**.

(b) A single linearly independent eigenvector. When only a single linearly independent eigenvector $\mathbf{K}_1$ exists, the general solution is given by

$$\mathbf{X}(t) = c_1 \mathbf{K}_1 e^{\lambda_1 t} + c_2 (\mathbf{K}_1 t e^{\lambda_1 t} + \mathbf{P} e^{\lambda_1 t}),$$

where $(\mathbf{A} - \lambda_1 \mathbf{I})\mathbf{P} = \mathbf{K}_1$ (see Section 8.6, equations (18)–(20)), and the solution may be rewritten as

$$\mathbf{X}(t) = t e^{\lambda_1 t}\left[c_2 \mathbf{K}_1 + \frac{c_1}{t}\mathbf{K}_1 + \frac{c_2}{t}\mathbf{P}\right].$$

If $\lambda_1 < 0$, $\lim_{t \to \infty} t e^{\lambda_1 t} = 0$ and it follows that $\mathbf{X}(t)$ approaches $\mathbf{0}$ in one of the directions determined by the vector $\mathbf{K}_1$ (see Figure 10.11(b)). The critical point is again called a **degenerate stable node**. When $\lambda_1 > 0$, the solutions look like those in Figure 10.11(b) with the arrows reversed. The line determined by $\mathbf{K}_1$ is an asymptote for *all* solutions. The critical point is again called a **degenerate unstable node**.

CASE III   COMPLEX EIGENVALUES $(\tau^2 - 4\Delta < 0)$ If $\lambda_1 = \alpha + i\beta$ and $\bar{\lambda}_1 = \alpha - i\beta$ are the complex eigenvalues and $\mathbf{K}_1 = \mathbf{B}_1 + i\mathbf{B}_2$ is a complex eigenvector corresponding to λ_1, then the general solution can be written as $\mathbf{X}(t) = c_1 \mathbf{X}_1(t) + c_2 \mathbf{X}_2(t)$, where

$$\mathbf{X}_1(t) = (\mathbf{B}_1 \cos \beta t - \mathbf{B}_2 \sin \beta t)e^{\alpha t}$$

and $$\mathbf{X}_2(t) = (\mathbf{B}_2 \cos \beta t + \mathbf{B}_1 \sin \beta t)e^{\alpha t}$$

(see Section 8.6, equations (14) and (15)). A solution can therefore be written in the form

$$x(t) = e^{\alpha t}(c_{11} \cos \beta t + c_{12} \sin \beta t)$$

$$y(t) = e^{\alpha t}(c_{21} \cos \beta t + c_{22} \sin \beta t)$$

(4)

and when $\alpha = 0$, we have

$$x(t) = c_{11} \cos \beta t + c_{12} \sin \beta t$$

$$y(t) = c_{21} \cos \beta t + c_{22} \sin \beta t.$$

(5)

(a) Pure imaginary roots ($\tau^2 - 4\Delta < 0, \tau = 0$)
  Center   When $\alpha = 0$, the eigenvalues are pure imaginary and, from (5), all solutions are periodic with period $p = 2\pi/\beta$. Notice that if both c_{12} and c_{21} happened to be zero, then (5) would reduce to

$$x(t) = c_{11} \cos \beta t$$

$$y(t) = c_{22} \sin \beta t,$$

which is a standard parametric representation for an ellipse. By solving the system of equations in (5) for $\cos \beta t$ and $\sin \beta t$ and using the identity $\sin^2 \beta t + \cos^2 \beta t = 1$, it possible to show that *all solutions are ellipses* with center at the origin. The critical point $(0, 0)$ is called a **center**, and Figure 10.12 shows a typical collection of solution curves. The ellipses are either *all* traversed in the clockwise direction or all traversed in the counterclockwise direction.

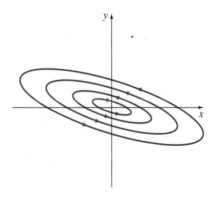

Figure 10.12   Center

(b) Nonzero real part ($\tau^2 - 4\Delta < 0, \tau \neq 0$)
  Spiral Points   When $\alpha \neq 0$, the effect of the term $e^{\alpha t}$ in (4) is similar to the effect of the exponential term in the analysis of *damped motion* given in Section 5.2. When $\alpha < 0$, $e^{\alpha t} \to 0$, and the elliptical-like solution spirals closer and closer to the origin. The critical point

Figure 10.13   Stable and Unstable Spiral Points

is called a **stable spiral point**. When $\alpha > 0$, the effect is the opposite. An elliptical-like solution is driven farther and farther from the origin, and the critical point is called an **unstable spiral point** (see Figure 10.13).

E X A M P L E   **2**

Classify the critical point $(0, 0)$ of each of the following linear systems $\mathbf{X}' = \mathbf{A}\mathbf{X}$.

(a) $\mathbf{A} = \begin{pmatrix} 3 & -18 \\ 2 & -9 \end{pmatrix}$

(b) $\mathbf{A} = \begin{pmatrix} -1 & 2 \\ -1 & 1 \end{pmatrix}$

In each case discuss the nature of the solution that satisfies $\mathbf{X}(0) = (1, 0)$. Determine parametric equations for each solution.

Solution   **(a)** Since $\tau = -6$ and $\Delta = 9$, the characteristic polynomial is

$$\lambda^2 + 6\lambda + 9 = (\lambda + 3)^2$$

and so $(0, 0)$ is a degenerate stable node. For the repeated eigenvalue $\lambda = -3$, we find a single eigenvector

$$\mathbf{K}_1 = \begin{pmatrix} 3 \\ 1 \end{pmatrix}$$

and so the solution $\mathbf{X}(t)$ that satisfies $\mathbf{X}(0) = (1, 0)$ approaches $(0, 0)$ from the direction specified by the line $y = x/3$.

(b) Since $\tau = 0$ and $\Delta = 1$, the eigenvalues are $\lambda = \pm i$, and so $(0, 0)$ is a center. The solution $\mathbf{X}(t)$ that satisfies $\mathbf{X}(0) = (1, 0)$ is an ellipse that circles the origin every 2π units of time.

From Example 6 in Section 8.6, the general solution of the system in (a) is

$$\mathbf{X}(t) = c_1 \begin{pmatrix} 3 \\ 1 \end{pmatrix} e^{-3t} + c_2 \left[\begin{pmatrix} 3 \\ 1 \end{pmatrix} te^{-3t} + \begin{pmatrix} \frac{1}{2} \\ 0 \end{pmatrix} e^{-3t} \right].$$

The initial condition gives $c_1 = 0$ and $c_2 = 2$, and so $x = (6t + 1)e^{-3t}$ and $y = 2te^{-3t}$ are parametric equations for the solution.

From Example 1 in Section 8.7, the general solution of the system in (b) is

$$\mathbf{X}(t) = c_1 \begin{pmatrix} \cos t + \sin t \\ \cos t \end{pmatrix} + c_2 \begin{pmatrix} \cos t - \sin t \\ -\sin t \end{pmatrix}.$$

The initial condition gives $c_1 = 0$ and $c_2 = 1$, and so $x = \cos t - \sin t$ and $y = -\sin t$ are parametric equations for the ellipse. Note that $y < 0$ for small positive values of t, and therefore the ellipse is traversed in the clockwise direction.

The solutions of (a) and (b) are shown in Figures 10.14(a) and (b), respectively.

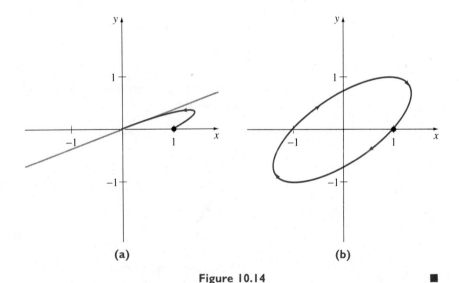

(a)   (b)

Figure 10.14   ■

Figure 10.15 conveniently summarizes the results of this section. The general geometric nature of the solutions can be determined by computing the trace and determinant of $\mathbf{A}$. In practice, graphs of the solutions are most

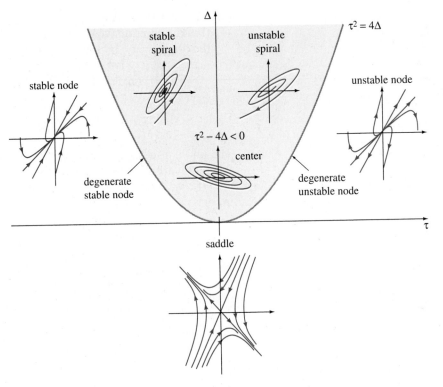

Figure 10.15

easily obtained *not* by constructing explicit eigenvalue-eigenvector solutions but rather by generating the solutions numerically using a method such as the Runge-Kutta method for first-order systems (see Section 9.7).

EXAMPLE 3

Classify the critical point $(0, 0)$ of each of the following linear systems $\mathbf{X}' = \mathbf{AX}$:

(a) $\mathbf{A} = \begin{pmatrix} 1.01 & 3.10 \\ -1.10 & -1.02 \end{pmatrix}$ 　　　　 **(b)** $\mathbf{A} = \begin{pmatrix} -a\hat{x} & -ab\hat{x} \\ -cd\hat{y} & -d\hat{y} \end{pmatrix}$

for positive constants $a, b, c, d, \hat{x},$ and $\hat{y}$.

Solution For the matrix in (a), $\tau = -0.01$, $\Delta = 2.3798$, and so $\tau^2 - 4\Delta < 0$. Using Figure 10.15, we see that $(0, 0)$ is a stable spiral point.

　　　The matrix in (b) arises from the Lotka-Volterra competition model we will study in Section 10.4. Since $\tau = -(a\hat{x} + d\hat{y})$ and all constants in the matrix are positive, $\tau < 0$. The determinant may be written as $\Delta = ad\hat{x}\hat{y}(1 - bc)$. If $bc > 1$, then $\Delta < 0$ and the critical point is a saddle. If $bc < 1$, then $\Delta > 0$ and the critical point is a stable node, a degenerate stable node, or a stable spiral point. In all three of these cases, $\lim_{t \to \infty} \mathbf{X}(t) = \mathbf{0}$. ∎

We can now answer each of the questions posed at the beginning of Section 10.2 for the linear plane autonomous system

$$x' = ax + by$$

$$y' = cx + dy$$

with $ad - bc \neq 0$. The answers are summarized in Theorem 10.1.

THEOREM 10.1 **Stability Criteria for Linear Systems**

For a linear plane autonomous system, let $\mathbf{X} = \mathbf{X}(t)$ denote the solution that satisfies the initial condition $\mathbf{X}(0) = \mathbf{X}_0$, where $\mathbf{X}_0 \neq \mathbf{0}$.

(a) $\lim_{t \to \infty} \mathbf{X}(t) = \mathbf{0}$ if and only if the eigenvalues of **A** have negative real parts. This occurs when $\Delta > 0$ and $\tau < 0$.

(b) $\mathbf{X}(t)$ is periodic if and only if the eigenvalues of **A** are pure imaginary. This occurs when $\Delta > 0$ and $\tau = 0$.

(c) In all other cases, given any neighborhood of the origin, there is at least one $\mathbf{X}_0$ in the neighborhood for which $\mathbf{X}(t)$ becomes unbounded as t increases.

10.2 EXERCISES

Answers to odd-numbered problems begin on page A–36.

In Problems 1–8 the general solution of the linear system $\mathbf{X}' = \mathbf{AX}$ is given.

(a) In each case discuss the nature of the solutions in a neighborhood of $(0, 0)$.

(b) With the aid of a graphics calculator or graphing software, sketch the solution that satisfies the initial condition $\mathbf{X}(0) = (1, 1)$.

1. $\mathbf{A} = \begin{pmatrix} -2 & -2 \\ -2 & -5 \end{pmatrix}$, $\mathbf{X}(t) = c_1 \begin{pmatrix} 2 \\ -1 \end{pmatrix} e^{-t} + c_2 \begin{pmatrix} 1 \\ 2 \end{pmatrix} e^{-6t}$

2. $\mathbf{A} = \begin{pmatrix} -1 & -2 \\ 3 & 4 \end{pmatrix}$, $\mathbf{X}(t) = c_1 \begin{pmatrix} 1 \\ -1 \end{pmatrix} e^{t} + c_2 \begin{pmatrix} -4 \\ 6 \end{pmatrix} e^{2t}$

3. $\mathbf{A} = \begin{pmatrix} 1 & -1 \\ 1 & 1 \end{pmatrix}$, $\mathbf{X}(t) = e^{t} \left[c_1 \begin{pmatrix} -\sin t \\ \cos t \end{pmatrix} + c_2 \begin{pmatrix} \cos t \\ \sin t \end{pmatrix} \right]$

4. $\mathbf{A} = \begin{pmatrix} -1 & -4 \\ 1 & -1 \end{pmatrix}$, $\mathbf{X}(t) = e^{-t} \left[c_1 \begin{pmatrix} 2 \cos 2t \\ \sin 2t \end{pmatrix} + c_2 \begin{pmatrix} -2 \sin 2t \\ \cos 2t \end{pmatrix} \right]$

5. $\mathbf{A} = \begin{pmatrix} -6 & 5 \\ -5 & 4 \end{pmatrix}$, $\mathbf{X}(t) = c_1 \begin{pmatrix} 1 \\ 1 \end{pmatrix} e^{-t} + c_2 \left[\begin{pmatrix} 1 \\ 1 \end{pmatrix} t e^{-t} + \begin{pmatrix} 0 \\ \frac{1}{5} \end{pmatrix} e^{-t} \right]$

6. $\mathbf{A} = \begin{pmatrix} 2 & 4 \\ -1 & 6 \end{pmatrix}$, $\mathbf{X}(t) = c_1 \begin{pmatrix} 2 \\ 1 \end{pmatrix} e^{4t} + c_2 \left[\begin{pmatrix} 2 \\ 1 \end{pmatrix} t e^{4t} + \begin{pmatrix} 1 \\ 1 \end{pmatrix} e^{4t} \right]$

7. $\mathbf{A} = \begin{pmatrix} 2 & -1 \\ 3 & -2 \end{pmatrix}$, $\mathbf{X}(t) = c_1 \begin{pmatrix} 1 \\ 1 \end{pmatrix} e^{t} + c_2 \begin{pmatrix} 1 \\ 3 \end{pmatrix} e^{-t}$

8. $\mathbf{A} = \begin{pmatrix} -1 & 5 \\ -1 & 1 \end{pmatrix}$, $\mathbf{X}(t) = c_1 \begin{pmatrix} 5 \cos 2t \\ \cos 2t - 2 \sin 2t \end{pmatrix} + c_2 \begin{pmatrix} 5 \sin 2t \\ 2 \cos 2t + \sin 2t \end{pmatrix}$

In Problems 9–16 classify the critical point (0, 0) of the given linear system by computing the trace τ and determinant Δ and using Figure 10.15.

9. $x' = -5x + 3y$
$y' = 2x + 7y$

10. $x' = -5x + 3y$
$y' = 2x - 7y$

11. $x' = -5x + 3y$
$y' = -2x + 5y$

12. $x' = -5x + 3y$
$y' = -7x + 4y$

13. $x' = -\frac{3}{2}x + \frac{1}{4}y$
$y' = -x - \frac{1}{2}y$

14. $x' = \frac{3}{2}x + \frac{1}{4}y$
$y' = -x + \frac{1}{2}y$

15. $x' = 0.02x - 0.11y$
$y' = 0.10x - 0.05y$

16. $x' = 0.03x + 0.01y$
$y' = -0.01x + 0.05y$

17. Determine conditions on the real constant μ so that (0, 0) is a center for the linear system

$$x' = -\mu x + y$$

$$y' = -x + \mu y.$$

18. Determine a condition on the real constant μ so that (0, 0) is a stable spiral point of the linear system

$$x' = y$$

$$y' = -x + \mu y.$$

19. Show that (0, 0) is always an unstable critical point of the linear system

$$x' = \mu x + y$$

$$y' = -x + y,$$

where μ is a real constant and $\mu \neq 1$. When is (0, 0) an unstable saddle point? When is (0, 0) an unstable spiral point?

20. Let $\mathbf{X} = \mathbf{X}(t)$ be the solution of the linear system

$$x' = \alpha x - \beta y$$

$$y' = \beta x + \alpha y$$

that satisfies the initial condition $\mathbf{X}(0) = \mathbf{X}_0$. Determine conditions on the real constants α and β that will ensure

$$\lim_{t \to \infty} \mathbf{X}(t) = (0, 0).$$

Can $(0, 0)$ be a node or a saddle point?

21. Show that the nonhomogeneous linear system $\mathbf{X}' = \mathbf{A}\mathbf{X} + \mathbf{F}$ has a unique critical point $\mathbf{X}_1$ when $\Delta = \det \mathbf{A} \neq 0$. Conclude that if $\mathbf{X} = \mathbf{X}(t)$ is a solution of the nonhomogeneous system, $\tau < 0$ and $\Delta > 0$,

$$\lim_{t \to \infty} \mathbf{X}(t) = \mathbf{X}_1.$$

[*Hint:* $\mathbf{X}(t) = \mathbf{X}_c(t) + \mathbf{X}_1$.]

22. In Example 3(b) show that $(0, 0)$ is a stable node when $bc < 1$.

10.3 LINEARIZATION AND LOCAL STABILITY

We start this section by refining the stability concepts introduced in Section 10.2 in such a way that they apply to nonlinear autonomous systems as well. While the linear system $\mathbf{X}' = \mathbf{A}\mathbf{X}$ had only one critical point when $\det \mathbf{A} \neq 0$, we saw in Section 10.1 that a nonlinear system may have many critical points. We therefore cannot expect that a particle placed initially at $\mathbf{X}_0$ will remain near a given critical point $\mathbf{X}_1$ unless $\mathbf{X}_0$ has been placed sufficiently close to $\mathbf{X}_1$ to begin with. The particle might well be driven to a second critical point $\mathbf{X}_2$.

To emphasize this idea, consider the physical system shown in Figure 10.16 in which a bead slides along the curve $z = f(x)$ under the influence of gravity alone. We will show in Section 10.4 that the x-coordinate of the bead satisfies a second-order nonlinear differential equation $x'' = g(x, x')$

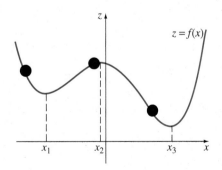

Figure 10.16

and therefore, when $y = x'$, satisfies the nonlinear autonomous system

$$x' = y$$

$$y' = g(x, y).$$

If the bead is positioned at $P = (x, f(x))$ and given zero initial velocity, the bead will remain at P provided $f'(x) = 0$. If the bead is placed near the critical point located at $x = x_1$, it will remain near $x = x_1$ only if its initial velocity does not drive it over the "hump" at $x = x_2$ toward the critical point located at $x = x_3$. Therefore $\mathbf{X}(0) = (x(0), x'(0))$ must be near $(x_1, 0)$.

In Definition 10.1 we will denote the distance between two points $\mathbf{X}$ and $\mathbf{Y}$ by $|\mathbf{X} - \mathbf{Y}|$.

DEFINITION 10.1   Stable Critical Points

Let $\mathbf{X}_1$ be a critical point of an autonomous system, and let $\mathbf{X} = \mathbf{X}(t)$ denote the solution that satisfies the initial condition $\mathbf{X}(0) = \mathbf{X}_0$, where $\mathbf{X}_0 \neq \mathbf{X}_1$. We say that $\mathbf{X}_1$ is a **stable critical point** when, given any radius $\rho > 0$, there is a corresponding radius $r > 0$ such that if the initial position $\mathbf{X}_0$ satisfies $|\mathbf{X}_0 - \mathbf{X}_1| < r$, then the corresponding solution $\mathbf{X}(t)$ satisfies $|\mathbf{X}(t) - \mathbf{X}_1| < \rho$ for all $t > 0$. If, in addition, $\lim_{t \to \infty} \mathbf{X}(t) = \mathbf{X}_1$ whenever $|\mathbf{X}_0 - \mathbf{X}_1| < r$, we call $\mathbf{X}_1$ an **asymptotically stable critical point**.

This definition is illustrated in Figure 10.17(a). Given any disk of radius ρ about the critical point $\mathbf{X}_1$, a solution will remain inside this disk provided $\mathbf{X}(0) = \mathbf{X}_0$ is selected sufficiently close to $\mathbf{X}_1$. It is *not* necessary that a solution approach the critical point in order for $\mathbf{X}_1$ to be stable. Stable nodes, stable spiral points, and centers are examples of stable critical points for linear systems. To emphasize that $\mathbf{X}_0$ must be selected close to $\mathbf{X}_1$, the terminology **locally stable critical point** is also used.

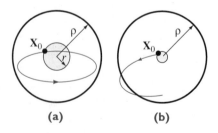

(a)   (b)

Figure 10.17

By negating Definition 10.1, we obtain the definition of an unstable critical point.

DEFINITION 10.2   Unstable Critical Points

Let X_1 be a critical point of an autonomous system, and let $X = X(t)$ denote the solution that satisfies the initial condition $X(0) = X_0$, where $X_0 \neq X_1$. We say that X_1 is an **unstable critical point** when there is a disk of radius $\rho > 0$ with the property that, for any $r > 0$, there is an initial position X_0 that satisfies $|X_0 - X_1| < r$ yet the corresponding solution $X(t)$ satisfies $|X(t) - X_1| \geq \rho$ for at least one $t > 0$.

If a critical point X_1 is unstable, no matter how small the neighborhood about X_1, then an initial position X_0 can always be found that results in the solution leaving some disk of radius ρ at some time t (see Figure 10.17(b)). Therefore unstable nodes, unstable spiral points, and saddles are all examples of unstable critical points for linear systems. In Figure 10.16 the critical point $(x_2, 0)$ is unstable. The slightest displacement or initial velocity will result in the bead sliding away from the point $(x_2, f(x_2))$.

EXAMPLE 1

Show that $(0, 0)$ is a stable critical point of the nonlinear plane autonomous system

$$x' = -y - x\sqrt{x^2 + y^2}$$
$$y' = x - y\sqrt{x^2 + y^2}$$

considered in Example 6 in Section 10.1.

Solution   In Example 6 in Section 10.1, we showed that in polar coordinates

$$r = \frac{1}{t + c_1}$$

$$\theta = t + c_2$$

is the general solution of the system. If $X(0) = (r_0, \theta_0)$ is the initial condition in polar coordinates, then $r_0 = 1/c_1$, $\theta_0 = c_2$, and so

$$r = \frac{r_0}{r_0 t + 1}$$

$$\theta = t + \theta_0.$$

Note that $r \leq r_0$ for $t \geq 0$, and r approaches $(0, 0)$ as t increases. Therefore, given $\rho > 0$, a solution that starts less than ρ units from $(0, 0)$ remains within ρ units of the origin for all $t \geq 0$. Hence the critical point $(0, 0)$ is stable and is in fact asymptotically stable. ∎

EXAMPLE 2

When expressed in polar coordinates, a plane autonomous system takes the form

$$\frac{dr}{dt} = 0.05r(3 - r)$$

$$\frac{d\theta}{dt} = -1.$$

Show that $(x, y) = (0, 0)$ is an unstable critical point.

Solution   Since $x = r \cos \theta$ and $y = r \sin \theta$, we have

$$\frac{dx}{dt} = -r \sin \theta \, \frac{d\theta}{dt} + \frac{dr}{dt} \cos \theta$$

$$\frac{dy}{dt} = r \cos \theta \, \frac{d\theta}{dt} + \frac{dr}{dt} \sin \theta.$$

From $dr/dt = 0.05r(3 - r)$ we see that $dr/dt = 0$ when $r = 0$, and we can conclude that $(x, y) = (0, 0)$ is a critical point by substituting $r = 0$ into the new system.

The differential equation $dr/dt = 0.05r(3 - r)$ is a logistic equation that can be solved using either separation of variables or equation (6) in Section 3.3. If $r(0) = r_0$ and $r_0 \neq 0$, then

$$r = \frac{3}{1 + c_0 e^{-0.15t}},$$

where $c_0 = (3 - r_0)/r_0$. Since

$$\lim_{t \to \infty} \frac{3}{1 + c_0 e^{-0.15t}} = 3,$$

it follows that no matter how close to $(0, 0)$ a solution starts, the solution will leave a disk of radius 1 about the origin. Therefore, $(0, 0)$ is an unstable critical point. ∎

Linearization

It is rarely possible to determine the stability of a critical point of a nonlinear system by finding explicit solutions as in Examples 1 and 2. Instead we replace the term $\mathbf{g}(\mathbf{X})$ in the original autonomous system $\mathbf{X}' = \mathbf{g}(\mathbf{X})$ by a linear term $\mathbf{A}(\mathbf{X} - \mathbf{X}_1)$ that most closely approximates $\mathbf{g}(\mathbf{X})$ in a neighborhood of $\mathbf{X}_1$. This replacement process, called **linearization**, will be illustrated first for the first-order differential equation $x' = g(x)$.

An equation of the tangent line to the curve $y = g(x)$ at $x = x_1$ is

$$y = g(x_1) + g'(x_1)(x - x_1)$$

and if x_1 is a critical point of $x' = g(x)$, we have

$$x' = g(x) \approx g'(x_1)(x - x_1).$$

The general solution to the linear differential equation $x' = g'(x_1)(x - x_1)$ is $x = x_1 + ce^{\lambda_1 t}$, where $\lambda_1 = g'(x_1)$, and so, if $g'(x_1) < 0$, $x(t)$ approaches x_1. Theorem 10.2 asserts that the same behavior occurs in the original differential equation provided $x(0) = x_0$ is selected close enough to x_1.

THEOREM 10.2   **Stability Criteria for $x' = g(x)$**

Let x_1 be a critical point of the autonomous differential equation $x' = g(x)$, where g is differentiable at x_1.

(a) If $g'(x_1) < 0$, then x_1 is an asymptotically stable critical point.

(b) If $g'(x_1) > 0$, then x_1 is an unstable critical point.

EXAMPLE 3

Without solving explicitly, analyze the critical points of the logistic differential equation $x' = rx\left(\dfrac{K - x}{K}\right)$, where r and K are positive constants.

Solution   The two critical points are $x = 0$ and $x = K$. Since

$$g'(x) = \frac{r}{K}(K - 2x),$$

$g'(0) = r$ and $g'(K) = -r$. We may therefore conclude that $x = 0$ is an unstable critical point and $x = K$ is an asymptotically stable critical point.   ■

Jacobian Matrix

A similar analysis may be carried out for a plane autonomous system. An equation of the tangent plane to the surface $z = g(x, y)$ at $\mathbf{X}_1 = (x_1, y_1)$ is

$$z = g(x_1, y_1) + \frac{\partial g}{\partial x}\bigg|_{(x_1, y_1)} (x - x_1) + \frac{\partial g}{\partial y}\bigg|_{(x_1, y_1)} (y - y_1)$$

and $g(x, y)$ may be approximated by its tangent plane in a neighborhood of $\mathbf{X}_1$. When $\mathbf{X}_1$ is a critical point of a plane autonomous system, $P(x_1, y_1) = Q(x_1, y_1) = 0$ and we have

$$x' = P(x, y) \approx \left.\frac{\partial P}{\partial x}\right|_{(x_1, y_1)} (x - x_1) + \left.\frac{\partial P}{\partial y}\right|_{(x_1, y_1)} (y - y_1)$$

$$y' = Q(x, y) \approx \left.\frac{\partial Q}{\partial x}\right|_{(x_1, y_1)} (x - x_1) + \left.\frac{\partial Q}{\partial y}\right|_{(x_1, y_1)} (y - y_1).$$

The original system $\mathbf{X}' = \mathbf{g}(\mathbf{X})$ may be approximated in a neighborhood of the critical point $\mathbf{X}_1$ by the linear system $\mathbf{X}' = \mathbf{A}(\mathbf{X} - \mathbf{X}_1)$, where

$$\mathbf{A} = \begin{pmatrix} \left.\dfrac{\partial P}{\partial x}\right|_{(x_1, y_1)} & \left.\dfrac{\partial P}{\partial y}\right|_{(x_1, y_1)} \\[2mm] \left.\dfrac{\partial Q}{\partial x}\right|_{(x_1, y_1)} & \left.\dfrac{\partial Q}{\partial y}\right|_{(x_1, y_1)} \end{pmatrix}.$$

This matrix is called the **Jacobian matrix** at $\mathbf{X}_1$ and is denoted by $\mathbf{g}'(\mathbf{X}_1)$. If we let $\mathbf{H} = \mathbf{X} - \mathbf{X}_1$, then the linear system $\mathbf{X}' = \mathbf{A}(\mathbf{X} - \mathbf{X}_1)$ becomes $\mathbf{H}' = \mathbf{A}\mathbf{H}$, which is the form of the linear system analyzed in Section 10.2. The critical point $\mathbf{X} = \mathbf{X}_1$ for $\mathbf{X}' = \mathbf{A}(\mathbf{X} - \mathbf{X}_1)$ now corresponds to the critical point $\mathbf{H} = \mathbf{0}$ for $\mathbf{H}' = \mathbf{A}\mathbf{H}$. If the eigenvalues of $\mathbf{A}$ have negative real parts, then $\mathbf{0}$ is an asymptotically stable critical point for $\mathbf{H}' = \mathbf{A}\mathbf{H}$. If there is an eigenvalue with a positive real part, $\mathbf{H} = \mathbf{0}$ is an unstable critical point. Theorem 10.3 asserts that the same conclusions can be made for the critical point $\mathbf{X}_1$ of the original system.

THEOREM 10.3   **Stability Criteria for Plane Autonomous Systems**

Let $\mathbf{X}_1$ be an isolated critical point of the plane autonomous system $\mathbf{X}' = \mathbf{g}(\mathbf{X})$, where $P(x, y)$ and $Q(x, y)$ have continuous first partial derivatives in a neighborhood of $\mathbf{X}_1$.

(a) If the eigenvalues of $\mathbf{A} = \mathbf{g}'(\mathbf{X}_1)$ have negative real parts, then $\mathbf{X}_1$ is an asymptotically stable critical point of $\mathbf{X}' = \mathbf{g}(\mathbf{X})$.

(b) If $\mathbf{A} = \mathbf{g}'(\mathbf{X}_1)$ has an eigenvalue with a positive real part, then $\mathbf{X}_1$ is an unstable critical point of $\mathbf{X}' = \mathbf{g}(\mathbf{X})$.

EXAMPLE 4

Classify (if possible) the critical points of each of the following plane autonomous systems as stable or unstable.

(a) $x' = x^2 + y^2 - 6$
  $y' = x^2 - y$

(b) $x' = 0.01x(100 - x - y)$
  $y' = 0.05y(60 - y - 0.2x)$

Solution The critical points of each system were determined in Example 4 in Section 10.1.

(a) The critical points are $(\sqrt{2}, 2)$ and $(-\sqrt{2}, 2)$, the Jacobian matrix is

$$\mathbf{g}'(\mathbf{X}) = \begin{pmatrix} 2x & 2y \\ 2x & -1 \end{pmatrix},$$

and so

$$\mathbf{A}_1 = \mathbf{g}'((\sqrt{2}, 2)) = \begin{pmatrix} 2\sqrt{2} & 4 \\ 2\sqrt{2} & -1 \end{pmatrix}, \qquad \mathbf{A}_2 = \mathbf{g}'((-\sqrt{2}, 2)) = \begin{pmatrix} -2\sqrt{2} & 4 \\ -2\sqrt{2} & -1 \end{pmatrix}.$$

Since the determinant of $\mathbf{A}_1$ is negative, $\mathbf{A}_1$ has a positive real eigenvalue. Therefore $(\sqrt{2}, 2)$ is an unstable critical point. Matrix $\mathbf{A}_2$ has a positive determinant and a negative trace and so both eigenvalues have negative real parts. It follows that $(-\sqrt{2}, 2)$ is a stable critical point.

(b) The critical points are $(0, 0)$, $(0, 60)$, $(100, 0)$, and $(50, 50)$, the Jacobian matrix is

$$\mathbf{g}'(\mathbf{X}) = \begin{pmatrix} 0.01(100 - 2x - y) & -0.01x \\ -0.01y & 0.05(60 - 2y - 0.2x) \end{pmatrix},$$

and so

$$\mathbf{A}_1 = \mathbf{g}'((0, 0)) = \begin{pmatrix} 1 & 0 \\ 0 & 3 \end{pmatrix} \qquad \mathbf{A}_2 = \mathbf{g}'((0, 60)) = \begin{pmatrix} 0.4 & 0 \\ -0.6 & -3 \end{pmatrix}$$

$$\mathbf{A}_3 = \mathbf{g}'((100, 0)) = \begin{pmatrix} -1 & -1 \\ 0 & 2 \end{pmatrix} \qquad \mathbf{A}_4 = \mathbf{g}'((50, 50)) = \begin{pmatrix} -0.5 & -0.5 \\ -0.5 & -2.5 \end{pmatrix}.$$

Since the matrix $\mathbf{A}_1$ has a positive determinant and a positive trace, both eigenvalues have positive real parts. Therefore $(0, 0)$ is an unstable critical point. The determinants of matrices $\mathbf{A}_2$ and $\mathbf{A}_3$ are negative, and so in each case one of the eigenvalues is positive. Therefore both $(0, 60)$ and $(100, 0)$ are unstable critical points. Since the matrix $\mathbf{A}_4$ has a positive determinant and a negative trace, $(50, 50)$ is a stable critical point. ∎

In Example 4 we did not compute $\tau^2 - 4\Delta$ (as in Section 10.2) and attempt to further classify the critical points as stable nodes, stable spiral points, saddle points, and so on. For example, for $\mathbf{X}_1 = (-\sqrt{2}, 2)$ in Example 4(a), $\tau^2 - 4\Delta < 0$ and, if the system were linear, we would be able to conclude that $\mathbf{X}_1$ is a stable spiral point. Figure 10.18 shows several solution curves near $\mathbf{X}_1$ that were obtained numerically, and each solution does *appear* to spiral in toward the critical point.

Classifying Critical Points

It is natural to ask whether we can infer more geometric information about the solutions near a critical point $\mathbf{X}_1$ of a nonlinear autonomous system from

Figure 10.18

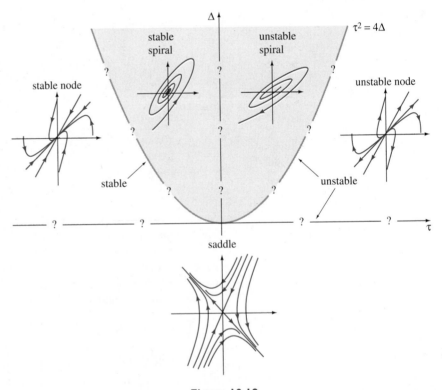

Figure 10.19

an analysis of the critical point of the corresponding linear system. The answer is summarized in Figure 10.19, but you should note the following comments:

(i) In five separate cases (stable node, stable spiral point, unstable spiral point, unstable node, and saddle), the critical point may be categorized like the critical point in the corresponding linear system. The solutions have the *same general geometric features* as the solutions to the linear system, and the smaller the neighborhood about X_1, the closer the resemblance.

(ii) If $\tau^2 = 4\Delta$ and $\tau > 0$, the critical point X_1 is unstable but, in this borderline case, *we are not yet able to decide whether X_1 is an unstable spiral, unstable node, or degenerate unstable node.* Likewise, if $\tau^2 = 4\Delta$ and $\tau < 0$, the critical point X_1 is stable but may be a stable spiral, a stable node, or a degenerate stable node.

(iii) If $\tau = 0$ and $\Delta > 0$, the eigenvalues of $A = g'(X)$ are pure imaginary and, in this borderline case, X_1 may be a stable spiral, an unstable spiral, or a center. *It is therefore not yet possible to determine whether X_1 is stable or unstable.*

EXAMPLE 5

Classify each critical point of the plane autonomous system in Example 4(b) as a stable node, a stable spiral point, an unstable spiral point, an unstable node, or a saddle.

Solution For the matrix A_1 corresponding to $(0, 0)$, $\Delta = 3$, $\tau = 4$, and so $\tau^2 - 4\Delta = 4$. Therefore $(0, 0)$ is an unstable node. The critical points $(0, 60)$ and $(100, 0)$ are saddles since $\Delta < 0$ in both cases. For matrix $A_4, \Delta > 0, \tau < 0$, and $\tau^2 - 4\Delta > 0$. It follows that $(50, 50)$ is a stable node. ∎

EXAMPLE 6

The plane autonomous system corresponding to the second-order nonlinear differential equation $x'' + x - x^3 = 0$ is

$$x' = y$$

$$y' = x^3 - x.$$

Find and classify (if possible) the critical points.

Solution Since $x^3 - x = x(x^2 - 1)$, the critical points are $(0, 0)$, $(1, 0)$, and $(-1, 0)$. The corresponding Jacobian matrices are

$$A_1 = g'((0, 0)) = \begin{pmatrix} 0 & 1 \\ -1 & 0 \end{pmatrix}, \qquad A_2 = g'((1, 0)) = g'((-1, 0)) = \begin{pmatrix} 0 & 1 \\ 2 & 0 \end{pmatrix}.$$

Since $\det A_2 < 0$, the critical points $(1, 0)$ and $(-1, 0)$ are both saddle points. The eigenvalues of matrix A_1 are $\pm i$, and according to comment (iii), the status of the critical point at $(0, 0)$ remains in doubt. It may be a stable spiral, an unstable spiral, or a center. ∎

Phase-Plane Method

The linearization method, when successful, can provide useful information on the local behavior of solutions near critical points. It is of little help if we are interested in solutions whose initial position $X(0) = X_0$ is not close to a critical point or if we wish to obtain a global view of the family of solution curves. The **phase-plane method** is based on the fact that

$$\frac{dy}{dx} = \frac{dy/dt}{dx/dt} = \frac{Q(x, y)}{P(x, y)}$$

and attempts to find y as a function of x using one of many methods available for solving first-order differential equations (see Chapter 2). As we show in Example 7, the method can sometimes be used to decide whether a critical point like $(0, 0)$ in Example 6 is a stable spiral, an unstable spiral, or a center.

EXAMPLE 7

Use the phase-plane method to determine the nature of the solutions to $x'' + x - x^3 = 0$ in a neighborhood of $(0, 0)$.

Solution   If we let $dx/dt = y$, then $dy/dt = x^3 - x$. From this we obtain the first-order differential equation

$$\frac{dy}{dx} = \frac{dy/dt}{dx/dt} = \frac{x^3 - x}{y},$$

which can be solved by separation of variables:

$$\int y \, dy = \int (x^3 - x) \, dx$$

$$\frac{y^2}{2} = \frac{x^4}{4} - \frac{x^2}{2} + c = \frac{x^4 - 2x^2 + 1}{4} + \left(c - \frac{1}{4} \right).$$

The solution may be written as $y^2 = (x^2 - 1)^2/2 + c_0$. If $X(0) = (x_0, 0)$, where $0 < x_0 < 1$, then $c_0 = -(x_0^2 - 1)^2/2$ and so

$$y^2 = \frac{(x^2 - 1)^2}{2} - \frac{(x_0^2 - 1)^2}{2}$$

$$= \frac{(2 - x^2 - x_0^2)(x_0^2 - x^2)}{2}.$$

Note that $y = 0$ when $x = -x_0$. In addition, the right-hand side is positive when $-x_0 < x < x_0$, and so each x therefore has *two* corresponding values of y. The solution $X = X(t)$ that satisfies $X(0) = (x_0, 0)$ is therefore periodic and so $(0, 0)$ is a center.

Figure 10.20 shows a family of solution curves. The original plane autonomous system was used to determine the directions of the solutions.

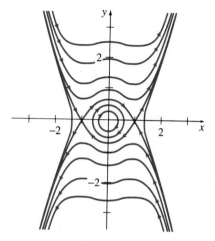

Figure 10.20

10.3 EXERCISES

Answers to odd-numbered problems begin on page A–37.

1. Show that $(0, 0)$ is an asymptotically stable point of the nonlinear autonomous system

$$x' = \alpha x - \beta y + y^2$$

$$y' = \beta x + \alpha y - xy$$

when $\alpha < 0$ and an unstable critical point when $\alpha > 0$. [*Hint:* Switch to polar coordinates.]

2. When expressed in polar coordinates, a plane autonomous system takes the form

$$\frac{dr}{dt} = \alpha r(5 - r)$$

$$\frac{d\theta}{dt} = -1.$$

Show that $(0, 0)$ is an asymptotically stable critical point if and only if $\alpha < 0$.

In Problems 3–10, without solving explicitly, classify the critical points of the given first-order autonomous differential equation as either asymptotically stable or unstable. All constants are assumed to be positive.

3. $\dfrac{dx}{dt} = kx(n + 1 - x)$ **4.** $\dfrac{dx}{dt} = -kx \ln(x/K), \quad x > 0$

5. $\dfrac{dT}{dt} = k(T - T_0)$

6. $m\dfrac{dv}{dt} = mg - kv$

7. $\dfrac{dx}{dt} = k(\alpha - x)(\beta - x), \quad \alpha > \beta$

8. $\dfrac{dx}{dt} = k(\alpha - x)(\beta - x)(\gamma - x), \quad \alpha > \beta > \gamma$

9. $\dfrac{dP}{dt} = P(a - bP)(1 - cP^{-1}), \quad P > 0, a < bc$

10. $\dfrac{dA}{dt} = k\sqrt{A}(K - \sqrt{A}), \quad A > 0$

In Problems 11–20 classify (if possible) each critical point of the given plane autonomous system as a stable node, a stable spiral point, an unstable spiral point, an unstable node, or a saddle point.

11. $x' = 1 - 2xy$
    $y' = 2xy - y$

12. $x' = x^2 - y^2 - 1$
    $y' = 2y$

13. $x' = y - x^2 + 2$
    $y' = x^2 - xy$

14. $x' = 2x - y^2$
    $y' = -y + xy$

15. $x' = -3x + y^2 + 2$
    $y' = x^2 - y^2$

16. $x' = xy - 3y - 4$
    $y' = y^2 - x^2$

17. $x' = -2xy$
    $y' = y - x + xy - y^3$

18. $x' = x(1 - x^2 - 3y^2)$
    $y' = y(3 - x^2 - 3y^2)$

19. $x' = x(10 - x - 1/2y)$
    $y' = y(16 - y - x)$

20. $x' = -2x + y + 10$
    $y' = 2x - y - 15\dfrac{y}{y + 5}$

In Problems 21–26 classify (if possible) each critical point of the given second-order differential equation as a stable node, a stable spiral point, an unstable spiral point, an unstable node, or a saddle.

21. $\theta'' = (\cos\theta - 0.5)\sin\theta, \quad |\theta| < \pi$

22. $x'' + x = [1/2 - 3(x')^2]x' - x^2$

23. $x'' + x'(1 - x^3) - x^2 = 0$

24. $x'' + 4\dfrac{x}{1 + x^2} + 2x' = 0$

25. $x'' + x = \varepsilon x^3 \quad$ for $\varepsilon > 0$

26. $x'' + x - \varepsilon x|x| = 0 \quad$ for $\varepsilon > 0$
    $\left[Hint: \dfrac{d}{dx}x|x| = 2|x|.\right]$

27. Show that the second-order nonlinear differential equation

$$(1 + \alpha^2 x^2)x'' + (\beta + \alpha^2(x')^2)x = 0$$

has a saddle point at $(0, 0)$ when $\beta < 0$.

28. Show that the plane autonomous system

$$x' = -\alpha x + xy$$

$$y' = 1 - \beta y - x^2$$

has a unique critical point when $\alpha\beta > 1$ and that this critical point is stable when $\beta > 0$.

29. (a) Show that the plane autonomous system

$$x' = -x + y - x^3$$

$$y' = -x - y + y^2$$

has two critical points by sketching the graphs of $-x + y - x^3 = 0$ and $-x - y + y^2 = 0$. Classify the critical point at $(0, 0)$.

(b) Show that the second critical point $\mathbf{X}_1 = (0.88054, 1.56327)$ is a saddle.

30. (a) Show that $(0, 0)$ is the only critical point of the Rayleigh differential equation

$$x'' + \varepsilon\left[\frac{1}{3}(x')^3 - x'\right] + x = 0.$$

(b) Show that $(0, 0)$ is unstable when $\varepsilon > 0$. When is $(0, 0)$ an unstable spiral point?

(c) Show that $(0, 0)$ is stable when $\varepsilon < 0$. When is $(0, 0)$ a stable spiral point?

(d) Show that $(0, 0)$ is a center when $\varepsilon = 0$.

31. Use the phase-plane method to show that $(0, 0)$ is a center of the second-order nonlinear differential equation $x'' + 2x^3 = 0$.

32. Use the phase-plane method to show that the solution to the second-order nonlinear differential equation $x'' + 2x - x^2 = 0$ that satisfies $x(0) = 1$ and $x'(0) = 0$ is periodic.

33. (a) Find the critical points of the plane autonomous system

$$x' = 2xy$$

$$y' = 1 - x^2 + y^2$$

and show that linearization gives no information about the nature of these critical points.

(b) Use the phase-plane method to show that the critical points in part (a) are both centers. [*Hint:* Let $u = y^2/x$ and show that $(x - c)^2 + y^2 = c^2 - 1$.]

34. The origin is the only critical point of the second-order nonlinear differential equation $x'' + (x')^2 + x = 0$.

    (a) Show that the phase-plane method leads to the Bernoulli differential equation $\dfrac{dy}{dx} = -y - xy^{-1}$.

    (b) Show that the solution satisfying $x(0) = 1/2$ and $x'(0) = 0$ is *not* periodic.

35. A solution of the second-order nonlinear differential equation $x'' + x - x^3 = 0$ satisfies $x(0) = 0$ and $x'(0) = v_0$. Use the phase-plane method to determine when the resulting solution is periodic. [*Hint:* See Example 7.]

36. The nonlinear differential equation $x'' + x = 1 + \varepsilon x^2$ arises in the analysis of planetary motion using relativity theory. Classify (if possible) all critical points of the corresponding plane autonomous system.

37. When a nonlinear capacitor is present in an *L-R-C* circuit, the voltage drop is no longer given by q/C but is more accurately described by $\alpha q + \beta q^3$, where α and β are constants and $\alpha > 0$. Differential equation (3) of Section 5.4 for the free circuit is then replaced by

$$L\frac{d^2q}{dt^2} + R\frac{dq}{dt} + \alpha q + \beta q^3 = 0.$$

Find and classify all critical points of this nonlinear differential equation. [*Hint:* Consider the two cases $\beta > 0$ and $\beta < 0$.]

10.4 APPLICATIONS OF AUTONOMOUS SYSTEMS

A plane autonomous system can serve as a mathematical model for diverse physical phenomena. These include problems in steady-state fluid dynamics, mechanical oscillations, circuits containing nonlinear capacitors, predator-prey and competition interactions in ecology, and the kinetics of medical drugs.

    Many applications from physics give rise to nonlinear second-order differential equations of the form $x'' = g(x, x')$. For example, in the analysis of free, damped motion in Section 5.2, we assumed that the damping force was proportional to the velocity x'. Frequently the damping force is proportional to the square of the velocity, and the new differential equation becomes

$$x'' = -\frac{\beta}{m}x'|x'| - \frac{k}{m}x.$$

The corresponding plane autonomous system is therefore

$$x' = y$$

$$y' = -\frac{\beta}{m}y|y| - \frac{k}{m}x.$$

In this section, we will use the results of Section 10.3 to analyze the nonlinear pendulum, the motion of a bead on a curve, the Lotka-Volterra predator-prey model, and the Lotka-Volterra competition model. Additional models are presented in the exercises.

Nonlinear Pendulum

In Example 3 in Section 1.2, we showed that the displacement angle θ for a pendulum satisfies the second-order nonlinear differential equation

$$\frac{d^2\theta}{dt^2} + \frac{g}{l}\sin\theta = 0.$$

When we let $x = \theta$ and $y = \theta'$, this second-order differential equation may be rewritten as the plane autonomous system

$$x' = y$$

$$y' = -\frac{g}{l}\sin x.$$

The critical points are $(\pm k\pi, 0)$ and the Jacobian matrix is easily shown to be

$$\mathbf{g}'((\pm k\pi, 0)) = \begin{pmatrix} 0 & 1 \\ (-1)^{k+1}\dfrac{g}{l} & 0 \end{pmatrix}.$$

If $k = 2n + 1$, then $\Delta < 0$, and so all critical points $(\pm(2n + 1)\pi, 0)$ are saddles. In particular, the critical point at $(\pi, 0)$ is unstable as expected. See Figure 10.21b. When $k = 2n$, the eigenvalues are pure imaginary and so the nature of these critical points remains in doubt. Since we have assumed that there are no damping forces acting on the pendulum, we expect that all critical points $(\pm 2n\pi, 0)$ are centers. This can be verified using the phase-plane method.

From

$$\frac{dy}{dx} = \frac{dy/dt}{dx/dt} = -\frac{g}{l}\frac{\sin x}{y},$$

it follows that

$$y^2 = \frac{2g}{l}\cos x + c.$$

If $\mathbf{X}(0) = (x_0, 0)$, then

$$y^2 = \frac{2g}{l}(\cos x - \cos x_0).$$

Note that $y = 0$ when $x = -x_0$, and $(2g/l)(\cos x - \cos x_0) > 0$ for $|x| < |x_0| < \pi$. Thus each such x has two corresponding values of y and so

(a) $\theta = 0, \theta' = 0$

(b) $\theta = \pi, \theta' = 0$

Figure 10.21

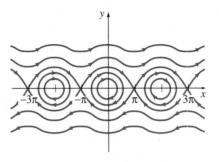

Figure 10.22

the solution $\mathbf{X} = \mathbf{X}(t)$ that satisfies $\mathbf{X}(0) = (x_0, 0)$ is periodic. We may conclude that $(0, 0)$ is a center. Figure 10.22 shows a family of solution curves.

EXAMPLE 1

A pendulum in an equilibrium position with $\theta = 0$ is given an initial angular velocity of ω_0 radians/s. Determine under what conditions the resulting motion is periodic.

Solution   We are asked to examine the solution of the plane autonomous system that satisfies $\mathbf{X}(0) = (0, \omega_0)$. From $y^2 = (2g/l) \cos x + c$, it follows that

$$y^2 = \frac{2g}{l} \left(\cos x - 1 + \frac{l}{2g} \omega_0^2 \right).$$

To establish that the solution $\mathbf{X}(t)$ is periodic, it is sufficient to show that there are two x-intercepts $x = \pm x_0$ between $-\pi$ and π, and that the right-hand side is positive for $|x| < |x_0|$. Each such x then has two corresponding values of y.

If $y = 0$, $\cos x = 1 - (l/2g)\omega_0^2$, and this equation has two solutions $x = \pm x_0$ between $-\pi$ and π provided $1 - (l/2g)\omega_0^2 > -1$. Note that $(2g/l)(\cos x - \cos x_0)$ is then positive for $|x| < |x_0|$. This restriction on the initial angular velocity may be written as

$$|\omega_0| < 2\sqrt{\frac{g}{l}}.$$ ∎

Nonlinear Oscillations: The Sliding Bead

Suppose, as shown in Figure 10.23, a bead with mass m slides along a thin wire whose shape is described by the function $z = f(x)$. A wide variety of non-linear oscillations can be obtained by changing the shape of the wire and making different assumptions about the forces acting on the bead.

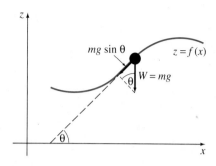

Figure 10.23

The tangential force $\mathbf{F}$ due to the weight $W = mg$ has magnitude $mg \sin \theta$ and therefore the x-component of $\mathbf{F}$ is $F_x = -mg \sin \theta \cos \theta$. Since $\tan \theta = f'(x)$, we may use the identities $1 + \tan^2\theta = \sec^2\theta$ and $\sin^2\theta = 1 - \cos^2\theta$ to conclude

$$F_x = -mg \sin \theta \cos \theta = -mg \frac{f'(x)}{1 + [f'(x)]^2}.$$

We assume (as in Section 5.2) that a damping force $\mathbf{D}$, acting in the direction opposite to the motion, is a constant multiple of the velocity of the bead. The x-component of $\mathbf{D}$ is therefore

$$D_x = -\beta \frac{dx}{dt}.$$

If we ignore the frictional force between the wire and the bead and assume that no other external forces are impressed on the system, it follows from Newton's second law that

$$mx'' = -mg \frac{f'(x)}{1 + [f'(x)]^2} - \beta x'$$

and the corresponding plane autonomous system is

$$x' = y$$

$$y' = -g \frac{f'(x)}{1 + [f'(x)]^2} - \frac{\beta}{m} y.$$

If $\mathbf{X}_1 = (x_1, y_1)$ is a critical point of the system, then $y_1 = 0$ and therefore $f'(x_1) = 0$. The bead must therefore be at rest at a point on the wire where the tangent line is horizontal. When f is twice differentiable, the Jacobian matrix at $\mathbf{X}_1$ is

$$\mathbf{g}'(\mathbf{X}_1) = \begin{pmatrix} 0 & 1 \\ -gf''(x_1) & -\beta/m \end{pmatrix}$$

and so $\tau = -\beta/m$, $\Delta = gf''(x_1)$, and $\tau^2 - 4\Delta = \beta^2/m^2 - 4gf''(x_1)$. Using the results of Section 10.3, we can make the following conclusions:

(i) $f''(x_1) < 0$

A relative maximum therefore occurs at $x = x_1$ and since $\Delta < 0$, an *unstable saddle point* occurs at $\mathbf{X}_1 = (x_1, 0)$.

(ii) $f''(x_1) > 0$ and $\beta > 0$

A relative minimum therefore occurs at $x = x_1$, and since $\tau < 0$ and $\Delta > 0$, $\mathbf{X}_1 = (x_1, 0)$ is a *stable critical point*. If $\beta^2 > 4gm^2f''(x_1)$, the system is **overdamped** and the critical point is a *stable node*. If $\beta^2 < 4gm^2f''(x_1)$, the system is **underdamped** and the critical point is a *stable spiral point*. The exact nature of the stable critical point is still in doubt if $\beta^2 = 4gm^2f''(x_1)$.

(iii) $f''(x_1) > 0$ and the system is undamped ($\beta = 0$)

In this case the eigenvalues are pure imaginary, but the phase-plane method can be used to show that the critical point is a *center*. Therefore solutions with $\mathbf{X}(0) = (x(0), x'(0))$ near $\mathbf{X}_1 = (x_1, 0)$ are periodic.

EXAMPLE 2

A 10-g bead slides along the graph of $z = \sin x$. According to conclusion (ii), the relative minima at $x_1 = -\pi/2$ and $3\pi/2$ give rise to stable critical points. See Figure 10.24. Since $f''(-\pi/2) = f''(3\pi/2) = 1$, the system will be underdamped provided $\beta^2 < 4gm^2$. If we use SI units, $m = 0.01$ kg, $g = 9.8$ m/s^2, and so the condition for an underdamped system becomes $\beta^2 < 3.92 \times 10^{-3}$.

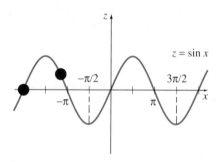

Figure 10.24

If $\beta = 0.01$ is the damping constant, both of these critical points are *stable spiral points*. The two solutions corresponding to initial conditions $\mathbf{X}(0) = ((x(0), x'(0)) = (-2\pi, 10)$ and $\mathbf{X}(0) = (-2\pi, 15)$, respectively, were obtained numerically and are shown in Figure 10.25. When $x'(0) = 10$, the bead

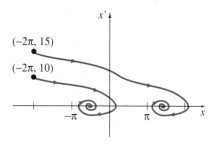

Figure 10.25 $\beta = 0.01$

has enough momentum to make it over the hill at $x = -3\pi/2$ but not over the hill at $x = \pi/2$. The bead then approaches the relative minimum based at $x = -\pi/2$. If $x'(0) = 15$, the bead has the momentum to make it over both hills but then rocks back and forth in the valley based at $x = 3\pi/2$ and approaches the point $(3\pi/2, -1)$ on the wire.

Figure 10.26 shows a collection of solution curves for the undamped case. Since $\beta = 0$, the critical points corresponding to $x_1 = -\pi/2$ and $3\pi/2$ are now centers. When $\mathbf{X}(0) = (-2\pi, 10)$, the bead has sufficient momentum to move over all hills. The figure also indicates that when the bead is released from rest at a position on the wire between $x = -3\pi/2$ and $x = \pi/2$, the resulting motion is periodic.

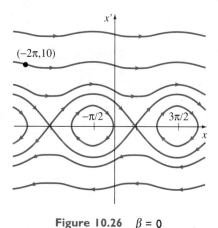

Figure 10.26 $\beta = 0$ ∎

Lotka-Volterra Predator-Prey Model

A **predator-prey interaction** between two species occurs when one species (the predator) feeds on a second species (the prey). For example, the snowy owl feeds almost exclusively on a common arctic rodent called a lemming, while a lemming uses arctic tundra plants as its food supply. Interest in using mathematics to help explain predator-prey interactions has been stimulated by the observation of population cycles in many arctic mammals. In the MacKenzie River district of Canada, for example, the principal prey of the lynx is the snowshoe hare, and both populations cycle with a period of about 10 years.

There are many predator-prey models that lead to plane autonomous systems with at least one periodic solution. The first such model was constructed independently by pioneer biomathematicians A. Lotka (1925) and V. Volterra (1926).* If x denotes the number of predators and y denotes the number of prey, then the Lotka-Volterra model takes the form

$$x' = -ax + bxy = x(-a + by)$$

$$y' = -cxy + dy = y(-cx + d),$$

where a, b, c, and d are positive constants.

Note that in the absence of predators $(x = 0)$, $y' = dy$, and so the number of prey grows exponentially. In the absence of prey, $x' = -ax$, and so the predator population becomes extinct. The term $-cxy$ represents the death rate due to predation. The model therefore assumes that this death rate is directly proportional to the number of possible encounters xy between predator and prey at a particular time t, and the term bxy represents the resulting positive contribution to the predator population.

The critical points of this plane autonomous system are $(0, 0)$ and $(d/c, a/b)$, and the corresponding Jacobian matrices are

$$\mathbf{A}_1 = \mathbf{g}'((0, 0)) = \begin{pmatrix} -a & 0 \\ 0 & d \end{pmatrix}, \qquad \mathbf{A}_2 = \mathbf{g}'((d/c, a/b)) = \begin{pmatrix} 0 & bd/c \\ -ac/b & 0 \end{pmatrix}$$

The critical point at $(0, 0)$ is a saddle, and Figure 10.27 shows a typical profile of solutions that are in the first quadrant and near $(0, 0)$.

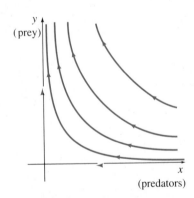

Figure 10.27   Solutions near $(0, 0)$

---

* See page 119 for biographical information on Lotka and Volterra.

Since the matrix $\mathbf{A}_2$ has pure imaginary eigenvalues $\lambda = \pm\sqrt{ad}\,i$, the critical point $(d/c, a/b)$ *may* be a center. This possibility can be investigated using the phase-plane method. Since

$$\frac{dy}{dx} = \frac{y(-cx+d)}{x(-a+by)},$$

we can separate variables and obtain

$$\int \frac{-a+by}{y}\,dy = \int \frac{-cx+d}{x}\,dx$$

so that
$$-a\ln y + by = -cx + d\ln x + c_1$$

or
$$(x^d e^{-cx})(y^a e^{-by}) = c_0.$$

The following argument establishes that all solution curves that originate in the first quadrant are periodic.

Typical graphs of the nonnegative functions $F(x) = x^d e^{-cx}$ and $G(y) = y^a e^{-by}$ are shown in Figure 10.28. It is not hard to show that $F(x)$ has an absolute maximum at $x = d/c$, while $G(y)$ has an absolute maximum at $y = a/b$. Note that, with the exception of zero and the absolute maximum, F and G each take on all values in their range exactly twice.

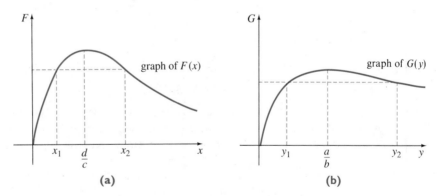

(a)   (b)

Figure 10.28

These graphs can be used to establish the following properties of a solution curve that originates at a noncritical point (x_0, y_0) in the first quadrant.

(i) If $y = a/b$, the equation $F(x)G(y) = c_0$ has exactly two solutions x_m and x_M that satisfy $x_m < d/c < x_M$.

(ii) If $x_m < x_1 < x_M$ and $x = x_1$, then $F(x)G(y) = c_0$ has exactly two solutions y_1 and y_2 that satisfy $y_1 < a/b < y_2$.

(iii) If x is outside the interval $[x_m, x_M]$, then $F(x)G(y) = c_0$ has no solutions.

We will give the demonstration of (i) and outline properties (ii) and (iii) in the exercises. Since $(x_0, y_0) \neq (d/c, a/b)$, $F(x_0)G(y_0) < F(d/c)G(a/b)$. If $y = a/b$, then

$$0 < \frac{c_0}{G(a/b)} = \frac{F(x_0)G(y_0)}{G(a/b)} < \frac{F(d/c)G(a/b)}{G(a/b)} = F(d/c).$$

Therefore $F(x) = c_0/G(a/b)$ has precisely two solutions x_m and x_M that satisfy $x_m < d/c < x_M$. The graph of a typical periodic solution is shown in Figure 10.29.

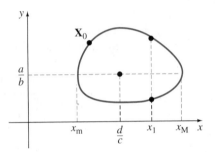

Figure 10.29

EXAMPLE 3

If we let $a = 0.1$, $b = 0.002$, $c = 0.0025$, and $d = 0.2$ in the Lotka-Volterra predator-prey model, the critical point in the first quadrant is $(d/c, a/b) = (80, 50)$, and we know that this critical point is a center. See Figure 10.30. The closer the initial condition X_0 is to $(80, 50)$, the more the periodic solutions resemble the elliptical solutions to the corresponding linear system. The eigenvalues of $g'((80, 50))$ are $\pm\sqrt{ad}\,i = \pm(\sqrt{2}/10)i$, and so the solutions near the critical point have period $p \approx 10\sqrt{2}\,\pi$, or about 44.4. ∎

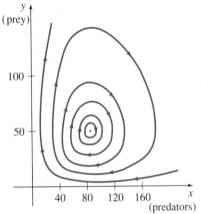

Figure 10.30

Lotka-Volterra Competition Model

A **competitive interaction** occurs when two or more species compete for the food, water, light, and space resources of an ecosystem. The use of one of these resources by one population therefore inhibits the ability of another population to survive and grow. Under what conditions can two competing species coexist? A number of mathematical models have been constructed that offer insights into conditions that permit *coexistence*. If x denotes the number in species I and y denotes the number in species II, then the Lotka-Volterra model takes the form

$$x' = \frac{r_1}{K_1} x(K_1 - x - \alpha_{12}y)$$

$$y' = \frac{r_2}{K_2} y(K_2 - y - \alpha_{21}x).$$

Note that in the absence of species II ($y = 0$), $x' = \frac{r_1}{K_1} x(K_1 - x)$, and so the first population grows logistically and approaches the steady-state population K_1 (see Section 3.3 and Example 3, Section 10.3). A similar statement holds for species II growing in the absence of species I. The term $-\alpha_{21}xy$ in the second equation stems from the competitive effect of species I on species II. The model therefore assumes that this rate of inhibition is directly proportional to the number of possible competitive pairs xy at a particular time t.

This plane autonomous system has critical points at $(0, 0)$, $(K_1, 0)$, and $(0, K_2)$. When $\alpha_{12}\alpha_{21} \neq 0$, the lines $K_1 - x - \alpha_{12}y = 0$ and $K_2 - y - \alpha_{21}x = 0$ intersect to produce a fourth critical point $\hat{\mathbf{X}} = (\hat{x}, \hat{y})$. Figure 10.31 shows the two conditions under which $(\hat{x}, \hat{y})$ is in the first quadrant.

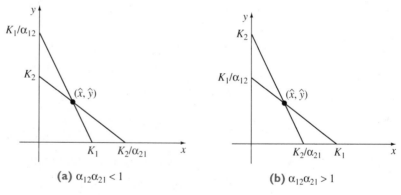

(a) $\alpha_{12}\alpha_{21} < 1$ **(b)** $\alpha_{12}\alpha_{21} > 1$

Figure 10.31

The trace and determinant of the Jacobian matrix at $(\hat{x}, \hat{y})$ are

$$\tau = -\hat{x}\frac{r_1}{K_1} - \hat{y}\frac{r_2}{K_2}$$

$$\Delta = (1 - \alpha_{12}\alpha_{21})\hat{x}\hat{y}\frac{r_1 r_2}{K_1 K_2}.$$

In case (a) in Figure 10.31, $\dfrac{K_1}{\alpha_{12}} > K_2$ and $\dfrac{K_2}{\alpha_{21}} > K_1$. It follows that $\alpha_{12}\alpha_{21} < 1$, $\tau < 0$, and $\Delta > 0$. Since

$$\tau^2 - 4\Delta = \left(\hat{x}\,\frac{r_1}{K_1} + \hat{y}\,\frac{r_2}{K_2}\right)^2 + 4(\alpha_{12}\alpha_{21} - 1)\hat{x}\hat{y}\,\frac{r_1 r_2}{K_1 K_2}$$

$$= \left[\left(\hat{x}\,\frac{r_1}{K_1} + \hat{y}\,\frac{r_2}{K_2}\right)^2 - 4\hat{x}\hat{y}\,\frac{r_1 r_2}{K_1 K_2}\right] + 4\alpha_{12}\alpha_{21}\hat{x}\hat{y}\,\frac{r_1 r_2}{K_1 K_2}$$

$$= \left(\hat{x}\,\frac{r_1}{K_1} - \hat{y}\,\frac{r_2}{K_2}\right)^2 + 4\alpha_{12}\alpha_{21}\hat{x}\hat{y}\,\frac{r_1 r_2}{K_1 K_2},$$

$\tau^2 - 4\Delta > 0$ and so $(\hat{x}, \hat{y})$ is a stable node. Therefore, if $\mathbf{X}(0) = \mathbf{X}_0$ is sufficiently close to $\hat{\mathbf{X}} = (\hat{x}, \hat{y})$, then

$$\lim_{t \to \infty} \mathbf{X}(t) = \hat{\mathbf{X}}$$

and we may conclude that coexistence is possible. The demonstration that case (b) leads to a saddle point and the investigation of the nature of critical points at $(0, 0)$, $(K_1, 0)$, and $(0, K_2)$ are left to the exercises.

When the competitive interactions between two species are weak, both of the coefficients α_{12} and α_{21} are small and so the conditions $K_1/\alpha_{12} > K_2$ and $K_2/\alpha_{21} > K_1$ may be satisfied. This might occur when there is a small overlap in the ranges of two predator species that hunt for a common prey.

EXAMPLE 4

A competitive interaction is described by the Lotka-Volterra competition model

$$x' = 0.004x(50 - x - 0.75y)$$

$$y' = 0.001y(100 - y - 3.0x).$$

Find and classify all critical points of the system.

Solution Critical points occur at $(0, 0)$, $(50, 0)$, $(0, 100)$, and the solution $(20, 40)$ of the system

$$x + 0.75y = 50$$

$$3.0x + y = 100.$$

Since $\alpha_{12}\alpha_{21} = 2.25 > 1$, we have case (b) in Figure 10.31, and so the critical point at $(20, 40)$ is a saddle point. All four critical points can be classified using the Jacobian matrix

$$\mathbf{g}'(\mathbf{X}) = \begin{pmatrix} 0.2 - 0.008x - 0.003y & -0.003x \\ -0.003y & 0.1 - 0.002y - 0.003x \end{pmatrix}$$

and we obtain

$$A_1 = g'((0, 0)) = \begin{pmatrix} 0.2 & 0 \\ 0 & 0.1 \end{pmatrix}, \qquad A_2 = g'((50, 0)) = \begin{pmatrix} -0.2 & -0.15 \\ 0 & -0.05 \end{pmatrix}$$

$$A_3 = g'((20, 40)) = \begin{pmatrix} -0.08 & -0.12 \\ -0.06 & -0.04 \end{pmatrix}, \qquad A_4 = g'((0, 100)) = \begin{pmatrix} -0.1 & 0 \\ -0.3 & -0.1 \end{pmatrix}.$$

Therefore $(0, 0)$ is an unstable node, while both $(50, 0)$ and $(0, 100)$ are stable nodes. (Check this!) Since $\det A_3 < 0$, we have a second demonstration that $(20, 40)$ is a saddle point. ∎

Coexistence can also occur in the Lotka-Volterra competition model if there is at least one periodic solution lying entirely in the first quadrant. It is possible to show, however, that this model has no periodic solutions.

10.4 EXERCISES

Answers to odd-numbered problems begin on page A–37.

Nonlinear Pendulum

1. A pendulum is released at $\theta = \pi/3$ and is given an initial angular velocity of ω_0 radians/s. Determine under what conditions the resulting motion is periodic.

2. (a) If a pendulum is released from rest at $\theta = \theta_0$, show that the angular velocity is again zero when $\theta = -\theta_0$.

   (b) The period T of the pendulum is the amount of time needed for θ to change from θ_0 to $-\theta_0$ and back to θ_0. Show that

   $$T = \sqrt{\frac{2l}{g}} \int_{-\theta_0}^{\theta_0} \frac{1}{\sqrt{\cos\theta - \cos\theta_0}} \, d\theta.$$

Nonlinear Oscillations — The Sliding Bead

3. A bead with mass m slides along a thin wire whose shape is described by the function $z = f(x)$. If $X_1 = (x_1, y_1)$ is a critical point of the plane autonomous system associated with the sliding bead, verify that the Jacobian matrix at X_1 is

   $$g'(X_1) = \begin{pmatrix} 0 & 1 \\ -gf''(x_1) & -\beta/m \end{pmatrix}.$$

4. A bead with mass m slides along a thin wire whose shape is described by the function $z = f(x)$. When $f'(x_1) = 0$, $f''(x_1) > 0$, and the system is undamped, the critical point $X_1 = (x_1, 0)$ is a center. Estimate the period of the bead when $x(0)$ is near x_1 and $x'(0) = 0$.

5. A bead is released from the position $x(0) = x_0$ on the curve $z = x^2/2$ with initial velocity $x'(0) = v_0$ cm/s.

   (a) Use the phase-plane method to show that the resulting solution is periodic when the system is undamped.

   (b) Show that the maximum height z_{max} to which the bead rises is given by

$$z_{max} = \frac{1}{2} [e^{v_0^2/g}(1 + x_0^2) - 1].$$

6. Rework Problem 5 with $z = \cosh x$.

Lotka-Volterra Predator-Prey Model

7. (Refer to Figure 10.29.) If $x_m < x_1 < x_M$ and $x = x_1$, show that

$$F(x)G(y) = c_0$$

has exactly two solutions y_1 and y_2 that satisfy $y_1 < a/b < y_2$. [*Hint:* First show that $G(y) = \dfrac{c_0}{F(x_1)} < G(a/b).$]

8. From (i) and (iii) on page 585, conclude that the maximum number of predators occurs when $y = a/b$.

9. In many fishery science models, the rate at which a species is caught is assumed to be directly proportional to its abundance. If both predator and prey are being exploited in this manner, the Lotka-Volterra differential equations take the form

$$x' = -ax + bxy - \varepsilon_1 x$$

$$y' = -cxy + dy - \varepsilon_2 y,$$

where ε_1 and ε_2 are positive constants.

   (a) When $\varepsilon_2 < d$, show that there is a new critical point in the first quadrant that is a center.

   (b) **Volterra's principle** states that a moderate amount of exploitation increases the average number of prey and decreases the average number of predators. Is this fisheries model consistent with Volterra's principle?

Lotka-Volterra Competition Model

10. Show that $(0, 0)$ is always an unstable node.

11. Show that $(K_1, 0)$ is a stable node when $K_1 > K_2/\alpha_{21}$ and a saddle point when $K_1 < K_2/\alpha_{21}$.

12. Use Problems 10 and 11 to establish that $(0, 0)$, $(K_1, 0)$, and $(0, K_2)$ are unstable when $\hat{\mathbf{X}} = (\hat{x}, \hat{y})$ is a stable node.

13. Show that $\hat{X} = (\hat{x}, \hat{y})$ is a saddle point when $\dfrac{K_1}{\alpha_{12}} < K_2$ and $\dfrac{K_2}{\alpha_{21}} < K_1$.

Additional Applications

14. If we assume that a damping force acts in a direction opposite to the motion of a pendulum and with a magnitude directly proportional to the angular velocity $d\theta/dt$, then the displacement angle θ for the pendulum satisfies the second-order nonlinear differential equation

$$ml\frac{d^2\theta}{dt^2} = -mg \sin\theta - \beta\frac{d\theta}{dt}.$$

(a) Write the second-order differential equation as a plane autonomous system and find all critical points.

(b) Find a condition on m, l, and β that will make $(0, 0)$ a stable spiral point.

15. In the analysis of free, damped motion in Section 5.2, we assumed that the damping force was proportional to the velocity x'. Frequently this damping force is proportional to the square of the velocity, and the new differential equation is

$$x'' = -\frac{\beta}{m}x'|x'| - \frac{k}{m}x.$$

(a) Write the second-order differential equation as a plane autonomous system and find all critical points.

(b) The system is called *overdamped* when $(0, 0)$ is a stable node and is called *underdamped* when $(0, 0)$ is a stable spiral point. Physical considerations suggest that $(0, 0)$ must be an asymptotically stable critical point. Show that the system is necessarily underdamped.

$$\left[\text{Hint: } \frac{d}{dy}y|y| = 2|y|.\right]$$

16. A bead with mass m slides along a thin wire whose shape may be described by the function $z = f(x)$. Small stretches of the wire act like an inclined plane, and in mechanics it is assumed that the magnitude of the frictional force between the bead and wire is directly proportional to $mg \cos\theta$ (see Figure 10.23).

(a) Show that the new differential equation for the x-coordinate of the bead is

$$x'' = g\frac{\mu - f'(x)}{1 + [f'(x)]^2} - \frac{\beta}{m}x'$$

for some positive constant μ.

(b) Characterize the critical points of the corresponding plane autonomous system. Under what conditions is a critical point a saddle? a stable spiral point?

17. An undamped oscillation satisfies a second-order nonlinear differential equation of the form $x'' + f(x) = 0$, where $f(0) = 0$ and $xf(x) > 0$ for $x \neq 0$ and $-d < x < d$. Use the phase-plane method to show that it is not possible for the critical point $(0, 0)$ to be a stable spiral. [*Hint:* Let $F(x) = \int_0^x f(u)\, du$ and show that $y^2 + 2F(x) = c$.]

18. The Lotka-Volterra predator-prey model assumes that, in the absence of predators, the number of prey grows exponentially. If we make the alternative assumption that the prey population grows logistically, the new plane autonomous system is

$$x' = -ax + bxy$$

$$y' = -cxy + \frac{r}{K} y(K - y),$$

where a, b, c, r, and K are positive and $K > a/b$.

(a) Show that the system has critical points at $(0, 0)$, $(0, K)$, and $(\hat{x}, \hat{y})$, where $\hat{y} = a/b$ and $c\hat{x} = \frac{r}{K}(K - \hat{y})$.

(b) Show that the critical points at $(0, 0)$, and $(0, K)$ are saddles, whereas the critical point at $(\hat{x}, \hat{y})$ is either a stable node or a stable spiral point.

(c) Show that $(\hat{x}, \hat{y})$ is a stable spiral point if $\hat{y} < \frac{4bK^2}{r + 4bK}$. Explain why this case occurs when the carrying capacity K of the prey is large.

19. The plane autonomous system

$$x' = \alpha \frac{y}{1 + y} x - x$$

$$y' = -\frac{y}{1 + y} x - y + \beta$$

arises in a model for the growth of microorganisms in a *chemostat*, a simple laboratory device in which a nutrient from a supply source flows into a growth chamber. In the system, x denotes the concentration of the microorganisms in the growth chamber, y denotes the concentration of nutrients, and α and β are positive constants that can be adjusted by the experimenter. Show that when $\alpha > 1$ and $\beta(\alpha - 1) > 1$, the system has a unique critical point $(\hat{x}, \hat{y})$ in the first quadrant, and demonstrate that this critical point is a stable node.

ILLUSTRATED APPLICATIONS

The instability inherent in the circus balancing act illustrated in the accompanying photograph is similar to the unstable critical point $(\pi, 0)$ for a nonlinear pendulum. See Section 10.4.

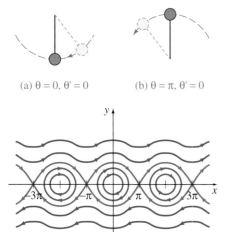

(a) $\theta = 0$, $\theta' = 0$ (b) $\theta = \pi$, $\theta' = 0$

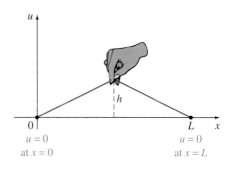

Many vibrational phenomena are modeled by a linear partial differential equation $a^2 u_{xx} = u_{tt}$, called the *one-dimensional wave equation*. For example, a solution $u(x, t)$ of this equation could represent the small vertical displacements of an ideal vibrating string such as a guitar or violin string. The strings of a guitar or violin can be set into motion by plucking them. See Section 12.2.

Musical instruments such as a guitar or a violin produce sound when someone causes their strings to vibrate either by plucking or bowing them. The piano is also a stringed instrument but it is a stringed percussion instrument. Vibrations occur when the player presses keys connected to felt covered hammers that strike the strings. See the essay at the end of Chapter 12.

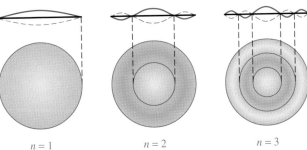

$n = 1$ $n = 2$ $n = 3$

It is not always convenient to use rectangular coordinates in the construction of a mathematical model. The description of a problem that involves, say, a circular or a spherical region is often best carried out by the use of polar and spherical coordinates, respectively. A solution $u(r, t)$ of the one-dimensional wave equation $a^2(u_{rr} + u/r) = u_{tt}$ in polar coordinates could represent the vertical displacements of an ideal vibrating circular drumhead. See Section 13.2.

Chapter 10   *REVIEW*

Solutions $\mathbf{X}(t) = (x(t), y(t))$ of a **plane autonomous system**

$$\frac{dx}{dt} = P(x, y)$$

$$\frac{dy}{dt} = Q(x, y)$$

are either **critical points** (corresponding to constant solutions), **arcs** (in which the solution curve does not cross itself), or **cycles** (which correspond to periodic solutions). Critical points can be found by solving the system of algebraic equations

$$P(x, y) = 0,$$

$$Q(x, y) = 0,$$

whereas periodic solutions can sometimes be found by changing to polar coordinates.

When the plane autonomous system is the linear system $\mathbf{X}' = \mathbf{AX}$ with det $\mathbf{A} \neq 0$, a precise geometric description of solutions near the critical point $(0, 0)$ can be given once the eigenvalues and eigenvectors are known. The critical point is classified as a **stable or unstable node, stable or unstable spiral point, center,** or **saddle**.

When the plane autonomous system is nonlinear, an isolated critical point $\mathbf{X}_1 = (x_1, y_1)$ can be classified by finding the eigenvalues of the Jacobian matrix

$$\mathbf{A} = \begin{pmatrix} \dfrac{\partial P}{\partial x}\Big|_{(x_1, y_1)} & \dfrac{\partial P}{\partial y}\Big|_{(x_1, y_1)} \\[2ex] \dfrac{\partial Q}{\partial x}\Big|_{(x_1, y_1)} & \dfrac{\partial Q}{\partial y}\Big|_{(x_1, y_1)} \end{pmatrix}.$$

If the eigenvalues of $\mathbf{A}$ have negative real parts, then $\mathbf{X}_1$ is an **asymptotically stable** critical point. If $\mathbf{A}$ has an eigenvalue with a positive real part, then $\mathbf{X}_1$ is an **unstable** critical point.

In five separate cases (stable node, stable spiral point, unstable spiral point, unstable node, and saddle), the critical point of the nonlinear system can be categorized like the critical point in the corresponding linear system $\mathbf{X}' = \mathbf{AX}$, and solutions close to the critical point have the same general geometric features as the solutions to the linear system. When the eigenvalues of $\mathbf{A}$ are pure imaginary, no conclusions can be made and further analysis is necessary.

The **phase-plane method** attempts to find y as a function of x by solving the first-order differential equation

$$\frac{dy}{dx} = \frac{Q(x, y)}{P(x, y)}.$$

The method can sometimes be used to sketch solution curves and to decide whether a critical point is a stable spiral, an unstable spiral, or a center.

Plane autonomous systems can serve as mathematical models for such diverse physical phenomena as mechanical oscillations and predator-prey or competition interactions in ecology.

Chapter 10   *REVIEW EXERCISES*

Answers to odd-numbered problems begin on page A–38.

Answer Problems 1–8 without referring back to the text. Fill in the blank or answer true/false.

1. The second-order differential equation $x'' + f(x') + g(x) = 0$ can be written as a plane autonomous system. _____

2. If $X = X(t)$ is a solution of a plane autonomous system and $X(t_1) = X(t_2)$ for $t_1 \neq t_2$, then $X(t)$ is a periodic solution. _____

3. If the trace of the matrix A is zero and det $A \neq 0$, then the critical point $(0, 0)$ of the linear system $X' = AX$ may be classified as _____.

4. If the critical point $(0, 0)$ of the linear system $X' = AX$ is a stable spiral point, then the eigenvalues of A are _____.

5. If the critical point $(0, 0)$ of the linear system $X' = AX$ is a saddle point and $X = X(t)$ is a solution, then $\lim_{t \to \infty} X(t)$ does not exist. _____

6. If the Jacobian matrix $A = g'(X_1)$ at a critical point of a plane autonomous system has a positive trace and determinant, then the critical point X_1 is unstable. _____

7. It is possible to show that a nonlinear plane autonomous system has periodic solutions using linearization. _____

8. All solutions of the pendulum equation $\dfrac{d^2\theta}{dt^2} + \dfrac{g}{l} \sin \theta = 0$ are periodic.
_____

9. Solve the following nonlinear plane autonomous system by switching to polar coordinates, and describe the geometric behavior of the solution that satisfies the given initial condition:

$$\begin{array}{l} x' = -y - x(\sqrt{x^2 + y^2})^3 \\ y' = x - y(\sqrt{x^2 + y^2})^3 \end{array}, \quad X(0) = (1, 0).$$

10. Discuss the geometric nature of the solutions to the linear system $\mathbf{X}' = \mathbf{AX}$ given that the general solution is:

(a) $\mathbf{X}(t) = c_1 \begin{pmatrix} 1 \\ 1 \end{pmatrix} e^{-t} + c_2 \begin{pmatrix} 1 \\ -2 \end{pmatrix} e^{-2t}$

(b) $\mathbf{X}(t) = c_1 \begin{pmatrix} 1 \\ -1 \end{pmatrix} e^{-t} + c_2 \begin{pmatrix} 1 \\ 2 \end{pmatrix} e^{2t}$

11. Classify the critical point $(0, 0)$ of the given linear system by computing the trace τ and determinant Δ

(a) $x' = -3x + 4y$
$\quad y' = -5x + 3y$

(b) $x' = -3x + 2y$
$\quad y' = -2x + \ y$

12. Find and classify (if possible) the critical points of the plane autonomous system

$$x' = x + xy - 3x^2$$

$$y' = 4y - 2xy - y^2.$$

13. Classify the critical point $(0, 0)$ of the plane autonomous system corresponding to the second-order nonlinear differential equation

$$x'' + \mu(x^2 - 1)x' + x = 0,$$

where μ is a real constant.

14. Without solving explicitly, classify (if possible) the critical points of the first-order autonomous differential equation $x' = (x^2 - 1)e^{-x/2}$ as asymptotically stable or unstable.

15. Use the phase-plane method to show that the solutions to the second-order nonlinear differential equation $x'' = -2x\sqrt{(x')^2 + 1}$ that satisfy $x(0) = x_0$ and $x'(0) = 0$ are periodic.

16. (Hard spring) In Section 5.2 we assumed that the restoring force F of the spring satisfied Hooke's law $F = ks$, where s is the elongation of the spring and k is a positive constant of proportionality. If we replace this assumption with the nonlinear law $F = ks^3$, the new differential equation for damped motion becomes $mx'' = -\beta x' - k(s + x)^3 + mg$, where $ks^3 = mg$. The system is called overdamped when $(0, 0)$ is a stable node and is called underdamped when $(0, 0)$ is a stable spiral point. Find new conditions on m, k, and β that will lead to overdamping and underdamping.

17. (The rotating pendulum) The rod of a pendulum is attached to a movable joint at point P and rotates at an angular speed of ω (radians/s) in the plane perpendicular to the rod. (See Figure 10.32.) As a result, the bob of the pendulum experiences an additional centripetal force and the new differential equation for θ becomes

$$ml\frac{d^2\theta}{dt^2} = \omega^2\, ml \sin\theta \cos\theta - mg \sin\theta - \beta\frac{d\theta}{dt}.$$

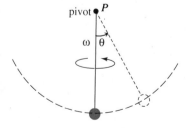

Figure 10.32

(a) If $\omega^2 < g/l$, show that $(0, 0)$ is a stable critical point and is the only critical point in the domain $-\pi < \theta < \pi$. Describe what occurs physically when $\theta(0) = \theta_0$, $\theta'(0) = 0$, and θ_0 is small.

(b) If $\omega^2 > g/l$, show that $(0, 0)$ is unstable and that there are two additional stable critical points $(\pm\hat{\theta}, 0)$ in the domain $-\pi < \theta < \pi$. Describe what occurs physically when $\theta(0) = \theta_0$, $\theta'(0) = 0$, and θ_0 is small.

C H A P T E R 11

ORTHOGONAL FUNCTIONS AND FOURIER SERIES

11.1 Orthogonal Functions
11.2 Fourier Series
11.3 Fourier Cosine and Sine Series
11.4 Sturm-Liouville Problem

11.5 Bessel and Legendre Series

Chapter 11 Review
Chapter 11 Review Exercises

Important Concepts

Orthogonal functions
Orthogonal set
Norm
Square norm
Orthogonality with respect to a weight function
Generalized Fourier series
Fourier series
Even function
Odd function
Cosine series
Sine series
Half-range expansion
Eigenvalues
Eigenfunctions
Sturm-Liouville problem
Fourier-Bessel series
Fourier-Legendre series

In calculus you studied vectors in 2- and 3-space and saw that two nonzero vectors are orthogonal when their dot or inner product is zero. Beyond the level of calculus, the notions of vectors, orthogonality, and inner product often lose their geometric interpretation. These concepts have been generalized; it is perfectly normal to think of a function as a vector. We can then say that two different functions are orthogonal when their inner product is zero. In this case, we shall see that the inner product of these vectors is actually a definite integral. The concept of orthogonal functions is fundamental to the material covered in Chapters 12 and 13.

Another concept encountered in calculus was the expansion of a function in an infinite series called a power series. In this chapter we shall see how to expand a given function in terms of an infinite set of orthogonal functions.

11.1 ORTHOGONAL FUNCTIONS

In advanced mathematics a function is considered to be a generalization of a vector. In this section we shall see how the two vector concepts of inner (dot) product and orthogonality can be extended to functions.

Suppose $\mathbf{u}$ and $\mathbf{v}$ are vectors in 3-space. The inner product $(\mathbf{u}, \mathbf{v})$ of the vectors, also written as $\mathbf{u} \cdot \mathbf{v}$, possesses the following properties:

(i) $(\mathbf{u}, \mathbf{v}) = (\mathbf{v}, \mathbf{u})$

(ii) $(k\mathbf{u}, \mathbf{v}) = k(\mathbf{u}, \mathbf{v})$, $\quad k$ a scalar

(iii) $(\mathbf{u}, \mathbf{u}) = 0$ if $\mathbf{u} = \mathbf{0}$ $\quad$ and $\quad$ $(\mathbf{u}, \mathbf{u}) > 0$ if $\mathbf{u} \neq \mathbf{0}$

(iv) $(\mathbf{u} + \mathbf{v}, \mathbf{w}) = (\mathbf{u}, \mathbf{w}) + (\mathbf{v}, \mathbf{w})$

We expect that a generalization of the inner product concept should have these same properties.

Inner Product

Now suppose f_1 and f_2 are functions defined on an interval $[a, b]$.* Since a definite integral on the interval of the product $f_1(x)f_2(x)$ also possesses properties (i)–(iv) whenever the integrals exist, we are prompted to make the following definition.

DEFINITION 11.1 **Inner Product**

The **inner product** of two functions f_1 and f_2 on an interval $[a, b]$ is the number

$$(f_1, f_2) = \int_a^b f_1(x)f_2(x)\, dx.$$

Orthogonal Functions

Motivated by the fact that two vectors $\mathbf{u}$ and $\mathbf{v}$ are orthogonal whenever their inner product is zero, we define **orthogonal functions** in a similar manner.

DEFINITION 11.2 **Orthogonal Functions**

Two functions f_1 and f_2 are said to be **orthogonal** on an interval $[a, b]$ if

$$(f_1, f_2) = \int_a^b f_1(x)f_2(x)\, dx = 0. \tag{1}$$

---

* The interval could also be $(-\infty, \infty)$, $[0, \infty)$, and so on.

Unlike in vector analysis, where the word *orthogonal* is a synonym for "perpendicular," in this present context the term *orthogonal* and condition (1) have no geometric significance.

EXAMPLE I

The functions $f_1(x) = x^2$ and $f_2(x) = x^3$ are orthogonal on the interval $[-1, 1]$ since

$$(f_1, f_2) = \int_{-1}^{1} f_1(x) f_2(x)\, dx$$

$$= \int_{-1}^{1} x^2 \cdot x^3\, dx = \frac{1}{6} x^6 \Big|_{-1}^{1}$$

$$= \frac{1}{6}[1 - (-1)^6] = 0.$$ ∎

Orthogonal Sets

We are primarily interested in infinite sets of orthogonal functions.

DEFINITION 11.3   Orthogonal Sets

A set of real-valued functions

$$\{\phi_0(x), \phi_1(x), \phi_2(x), \ldots\}$$

is said to be **orthogonal** on an interval $[a, b]$ if

$$(\phi_m, \phi_n) = \int_{a}^{b} \phi_m(x) \phi_n(x)\, dx = 0, \qquad m \neq n. \tag{2}$$

The norm, or length $\|\mathbf{u}\|$, of a vector $\mathbf{u}$ can be expressed in terms of the inner product; namely, $(\mathbf{u}, \mathbf{u}) = \|\mathbf{u}\|^2$ or $\|\mathbf{u}\| = \sqrt{(\mathbf{u}, \mathbf{u})}$. The **norm**, or generalized length, of a function ϕ_n is $\|\phi_n(x)\| = \sqrt{(\phi_n, \phi_n)}$; that is,

$$\|\phi_n(x)\| = \sqrt{\int_{a}^{b} \phi_n^2(x)\, dx}.$$

The number $$\|\phi_n(x)\|^2 = \int_{a}^{b} \phi_n^2(x)\, dx \tag{3}$$

is called the **square norm** of ϕ_n. If $\{\phi_n(x)\}$ is an orthogonal set of functions on the interval $[a, b]$ with the property that $\|\phi_n(x)\| = 1$ for $n = 0, 1, 2, \ldots$, then $\{\phi_n(x)\}$ is said to be an **orthonormal set** on the interval.

EXAMPLE 2

Show that the set $\{1, \cos x, \cos 2x, \ldots\}$ is orthogonal on the interval $[-\pi, \pi]$.

Solution If we make the identification $\phi_0(x) = 1$ and $\phi_n(x) = \cos nx$, we must then show that $\int_{-\pi}^{\pi} \phi_0(x)\phi_n(x)\, dx = 0, n \neq 0$, and $\int_{-\pi}^{\pi} \phi_m(x)\phi_n(x)\, dx = 0$, $m \neq n$. We have, in the first case,

$$
\begin{aligned}
(\phi_0, \phi_n) &= \int_{-\pi}^{\pi} \phi_0(x)\phi_n(x)\, dx \\
&= \int_{-\pi}^{\pi} \cos nx\, dx \\
&= \frac{1}{n} \sin nx \Big|_{-\pi}^{\pi} \\
&= \frac{1}{n} [\sin n\pi - \sin(-n\pi)] = 0, \qquad n \neq 0,
\end{aligned}
$$

and in the second,

$$
\begin{aligned}
(\phi_m, \phi_n) &= \int_{-\pi}^{\pi} \phi_m(x)\phi_n(x)\, dx \\
&= \int_{-\pi}^{\pi} \cos mx \cos nx\, dx \qquad \boxed{\text{trig identity}} \\
&= \frac{1}{2} \int_{-\pi}^{\pi} [\cos(m + n)x + \cos(m - n)x]\, dx \\
&= \frac{1}{2} \left[\frac{\sin(m + n)x}{m + n} + \frac{\sin(m - n)x}{m - n} \right]_{-\pi}^{\pi} = 0, \qquad m \neq n. \quad \blacksquare
\end{aligned}
$$

EXAMPLE 3

Find the norms of each function in the orthogonal set given in Example 2.

Solution For $\phi_0(x) = 1$ we have from (3)

$$
\|\phi_0(x)\|^2 = \int_{-\pi}^{\pi} dx = 2\pi
$$

so that $\|\phi_0(x)\| = \sqrt{2\pi}$. For $\phi_n(x) = \cos nx, n > 0$, it follows that

$$
\begin{aligned}
\|\phi_n(x)\|^2 &= \int_{-\pi}^{\pi} \cos^2 nx\, dx \\
&= \frac{1}{2} \int_{-\pi}^{\pi} [1 + \cos 2nx]\, dx = \pi.
\end{aligned}
$$

Thus for $n > 0$, $\|\phi_n(x)\| = \sqrt{\pi}$. $\quad \blacksquare$

Any orthogonal set of nonzero functions $\{\phi_n(x)\}, n = 0, 1, 2, \ldots$, can be *normalized*, that is, made into an orthonormal set, by dividing each function by its norm.

EXAMPLE 4

It follows from Examples 2 and 3 that the set

$$\left\{ \frac{1}{\sqrt{2\pi}}, \frac{\cos x}{\sqrt{\pi}}, \frac{\cos 2x}{\sqrt{\pi}}, \ldots \right\}$$

is orthonormal on $[-\pi, \pi]$.   ∎

We shall make one more analogy between vectors and functions. Suppose $\mathbf{v}_1$, $\mathbf{v}_2$, and $\mathbf{v}_3$ are three mutually orthogonal nonzero vectors in 3-space. Such an orthogonal set can be used as a basis for 3-space; that is, any three-dimensional vector can be written as a linear combination

$$\mathbf{u} = c_1 \mathbf{v}_1 + c_2 \mathbf{v}_2 + c_3 \mathbf{v}_3, \tag{4}$$

where the c_i, $i = 1, 2, 3$, are scalars called the components of the vector. Each component c_i can be expressed in terms of $\mathbf{u}$ and the corresponding vector $\mathbf{v}_i$. To see this we take the inner product of (4) with $\mathbf{v}_1$:

$$(\mathbf{u}, \mathbf{v}_1) = c_1(\mathbf{v}_1, \mathbf{v}_1) + c_2(\mathbf{v}_2, \mathbf{v}_1) + c_3(\mathbf{v}_3, \mathbf{v}_1)$$

$$= c_1 \|\mathbf{v}_1\|^2 + c_2 \cdot 0 + c_3 \cdot 0.$$

Hence $c_1 = \dfrac{(\mathbf{u}, \mathbf{v}_1)}{\|\mathbf{v}_1\|^2}.$

In like manner we find that the components c_2 and c_3 are given by

$$c_2 = \frac{(\mathbf{u}, \mathbf{v}_2)}{\|\mathbf{v}_2\|^2} \quad \text{and} \quad c_3 = \frac{(\mathbf{u}, \mathbf{v}_3)}{\|\mathbf{v}_3\|^2}.$$

Hence (4) can be expressed as

$$\mathbf{u} = \frac{(\mathbf{u}, \mathbf{v}_1)}{\|\mathbf{v}_1\|^2} \mathbf{v}_1 + \frac{(\mathbf{u}, \mathbf{v}_2)}{\|\mathbf{v}_2\|^2} \mathbf{v}_2 + \frac{(\mathbf{u}, \mathbf{v}_3)}{\|\mathbf{v}_3\|^2} \mathbf{v}_3 = \sum_{n=1}^{3} \frac{(\mathbf{u}, \mathbf{v}_n)}{\|\mathbf{v}_n\|^2} \mathbf{v}_n. \tag{5}$$

Generalized Fourier Series

Suppose $\{\phi_n(x)\}$ is an infinite orthogonal set of functions on an interval $[a, b]$. We ask: If $y = f(x)$ is a function defined on the interval $[a, b]$, is it possible to determine a set of coefficients c_n, $n = 0, 1, 2, \ldots$, for which

$$f(x) = c_0 \phi_0(x) + c_1 \phi_1(x) + \cdots + c_n \phi_n(x) + \cdots ? \tag{6}$$

As in the foregoing discussion on finding components of a vector, we can find the coefficients c_n by utilizing the inner product. Multiplying (6) by $\phi_m(x)$ and integrating over the interval $[a, b]$ give

$$\int_a^b f(x)\phi_m(x)\,dx = c_0 \int_a^b \phi_0(x)\phi_m(x)\,dx + c_1 \int_a^b \phi_1(x)\phi_m(x)\,dx + \cdots + c_n \int_a^b \phi_n(x)\phi_m(x)\,dx + \cdots$$

$$= c_0(\phi_0, \phi_m) + c_1(\phi_1, \phi_m) + \cdots + c_n(\phi_n, \phi_m) + \cdots.$$

By orthogonality each term on the right-hand side of the last equation is zero *except* when $m = n$. In this case we have

$$\int_a^b f(x)\phi_n(x)\,dx = c_n \int_a^b \phi_n^2(x)\,dx.$$

It follows that the required coefficients are

$$c_n = \frac{\int_a^b f(x)\phi_n(x)\,dx}{\int_a^b \phi_n^2(x)\,dx}, \qquad n = 0, 1, 2, \ldots .$$

In other words,
$$f(x) = \sum_{n=0}^{\infty} c_n\phi_n(x), \tag{7}$$

where
$$c_n = \frac{\int_a^b f(x)\phi_n(x)\,dx}{\|\phi_n(x)\|^2}. \tag{8}$$

With inner product notation, (7) becomes

$$f(x) = \sum_{n=0}^{\infty} \frac{(f, \phi_n)}{\|\phi_n(x)\|^2}\,\phi_n(x). \tag{9}$$

Thus (9) is seen to be the functional analogue of the vector result given in (5).

DEFINITION 11.4 **Orthogonal Set / Weight Function**

A set of functions $\{\phi_n(x)\}$, $n = 0, 1, 2, \ldots$, is said to be **orthogonal with respect to a weight function** $w(x)$ on an interval $[a, b]$ if

$$\int_a^b w(x)\phi_m(x)\phi_n(x)\,dx = 0, \qquad m \neq n.$$

The usual assumption is that $w(x) > 0$ on the interval of orthogonality $[a, b]$.

EXAMPLE 5

The set $\{1, \cos x, \cos 2x, \ldots\}$ is orthogonal with respect to the constant weight function $w(x) = 1$ on the interval $[-\pi, \pi]$. ∎

If $\{\phi_n(x)\}$ is orthogonal with respect to a weight function $w(x)$ on $[a, b]$, then multiplying (6) by $w(x)\phi_m(x)$ and integrating yield

$$c_n = \frac{\displaystyle\int_a^b f(x)w(x)\phi_n(x)\,dx}{\|\phi_n(x)\|^2}, \tag{10}$$

where

$$\|\phi_n(x)\|^2 = \int_a^b w(x)\phi_n^2(x)\,dx. \tag{11}$$

The series (7) with coefficients given by either (8) or (10) is called a **generalized Fourier series**.

Complete Sets

We note that the procedure outlined for determining the c_n was *formal*; that is, basic questions on whether or not a series expansion such as (7) is actually possible were ignored. Also, to expand f in a series of orthogonal functions, it is certainly necessary that f not be orthogonal to each ϕ_n of the orthogonal set $\{\phi_n(x)\}$. (If f were orthogonal to every ϕ_n, then $c_n = 0, n = 0, 1, 2, \ldots$.) To avoid the latter problem we shall assume, for the remainder of the chapter, that an orthogonal set is **complete**. This means that the only function orthogonal to each member of the set is the zero function.

11.1 EXERCISES

Answers to odd-numbered problems begin on page A–38.

In Problems 1–6 show that the given functions are orthogonal on the indicated interval.

1. $f_1(x) = x,\ f_2(x) = x^2,\quad [-2, 2]$

2. $f_1(x) = x^3,\ f_2(x) = x^2 + 1,\quad [-1, 1]$

3. $f_1(x) = e^x,\ f_2(x) = xe^{-x} - e^{-x},\quad [0, 2]$

4. $f_1(x) = \cos x,\ f_2(x) = \sin^2 x,\quad [0, \pi]$

5. $f_1(x) = x,\ f_2(x) = \cos 2x,\quad [-\pi/2, \pi/2]$

6. $f_1(x) = e^x,\ f_2(x) = \sin x,\quad [\pi/4, 5\pi/4]$

In Problems 7–12 show that the given set of functions is orthogonal on the indicated interval. Find the norm of each function in the set.

7. $\{\sin x, \sin 3x, \sin 5x, \ldots\},\quad [0, \pi/2]$

8. $\{\cos x, \cos 3x, \cos 5x, \ldots\},\quad [0, \pi/2]$

9. $\{\sin nx\}, n = 1, 2, 3, \ldots,\quad [0, \pi]$

10. $\left\{\sin \dfrac{n\pi}{p} x\right\}, n = 1, 2, 3, \ldots,\quad [0, p]$

11. $\left\{1, \cos\dfrac{n\pi}{p}x\right\}, n = 1, 2, 3, \ldots,\quad [0, p]$

12. $\left\{1, \cos\dfrac{n\pi}{p}x, \sin\dfrac{m\pi}{p}x\right\}, n = 1, 2, 3, \ldots, m = 1, 2, 3, \ldots,\quad [-p, p]$

In Problems 13 and 14 verify by direct integration that the functions are orthogonal with respect to the indicated weight function on the given interval.

13. $H_0(x) = 1,\ H_1(x) = 2x,\ H_2(x) = 4x^2 - 2;\quad w(x) = e^{-x^2}, (-\infty, \infty)$

14. $L_0(x) = 1,\ L_1(x) = -x + 1,\ L_2(x) = \frac{1}{2}x^2 - 2x + 1;\quad w(x) = e^{-x}, [0, \infty)$

15. Let $\{\phi_n(x)\}$ be an orthogonal set of functions on $[a, b]$ such that $\phi_0(x) = 1$. Show that $\int_a^b \phi_n(x)\,dx = 0$ for $n = 1, 2, \ldots$.

16. Let $\{\phi_n(x)\}$ be an orthogonal set of functions on $[a, b]$ such that $\phi_0(x) = 1$ and $\phi_1(x) = x$. Show that $\int_a^b (\alpha x + \beta)\phi_n(x)\,dx = 0$ for $n = 2, 3, \ldots$ and any constants α and β.

17. Let $\{\phi_n(x)\}$ be an orthogonal set of functions on $[a, b]$. Show that $\|\phi_m(x) + \phi_n(x)\|^2 = \|\phi_m(x)\|^2 + \|\phi_n(x)\|^2, m \neq n$.

18. From Problem 1 we know that $f_1(x) = x$ and $f_2(x) = x^2$ are orthogonal on $[-2, 2]$. Find constants c_1 and c_2 such that $f_3(x) = x + c_1 x^2 + c_2 x^3$ is orthogonal to both f_1 and f_2 on the same interval.

19. The set of functions $\{\sin nx\}, n = 1, 2, 3, \ldots$, is orthogonal on the interval $[-\pi, \pi]$. Show that the set is not complete.

20. Suppose f_1, f_2, and f_3 are functions continuous on the interval $[a, b]$. Show that $(f_1 + f_2, f_3) = (f_1, f_3) + (f_2, f_3)$.

11.2 FOURIER SERIES

The set of functions

$$\left\{1, \cos\frac{\pi}{p}x, \cos\frac{2\pi}{p}x, \ldots, \sin\frac{\pi}{p}x, \sin\frac{2\pi}{p}x, \sin\frac{3\pi}{p}x, \ldots\right\} \quad (1)$$

is orthogonal on the interval $[-p, p]$ (see Problem 12, Exercises 11.1). Suppose f is a function defined on the interval $[-p, p]$ that can be expanded in the trigonometric series

$$f(x) = \frac{a_0}{2} + \sum_{n=1}^{\infty}\left(a_n \cos\frac{n\pi}{p}x + b_n \sin\frac{n\pi}{p}x\right). \quad (2)$$

Then the coefficients $a_0, a_1, a_2, \ldots, b_1, b_2, \ldots$ can be determined in exactly the same manner as in the discussion of generalized Fourier series in Section 11.1.*

---

* We have chosen to write the coefficient of 1 in the series (2) as $a_0/2$ rather than a_0. This is for convenience only; the formula for a_n will then reduce to a_0 when $n = 0$.

Integrating both sides of (2) from $-p$ to p gives

$$\int_{-p}^{p} f(x)\,dx = \frac{a_0}{2}\int_{-p}^{p} dx + \sum_{n=1}^{\infty}\left(a_n \int_{-p}^{p}\cos\frac{n\pi}{p}x\,dx + b_n \int_{-p}^{p}\sin\frac{n\pi}{p}x\,dx\right). \quad (3)$$

Since each function $\cos(n\pi x/p)$, $\sin(n\pi x/p)$, $n > 1$, is orthogonal to 1 on the interval, the right side of (3) reduces to a single term and, consequently,

$$\int_{-p}^{p} f(x)\,dx = \frac{a_0}{2}\int_{-p}^{p} dx = \frac{a_0}{2} x\Big|_{-p}^{p} = pa_0.$$

Solving for a_0 yields

$$a_0 = \frac{1}{p}\int_{-p}^{p} f(x)\,dx. \quad (4)$$

Now we multiply (2) by $\cos(m\pi x/p)$ and integrate:

$$\int_{-p}^{p} f(x)\cos\frac{m\pi}{p}x\,dx = \frac{a_0}{2}\int_{-p}^{p}\cos\frac{m\pi}{p}x\,dx$$

$$+ \sum_{n=1}^{\infty}\left(a_n \int_{-p}^{p}\cos\frac{m\pi}{p}x\cos\frac{n\pi}{p}x\,dx + b_n \int_{-p}^{p}\cos\frac{m\pi}{p}x\sin\frac{n\pi}{p}x\,dx\right). \quad (5)$$

By orthogonality we have

$$\int_{-p}^{p}\cos\frac{m\pi}{p}x\,dx = 0, \qquad m > 0$$

$$\int_{-p}^{p}\cos\frac{m\pi}{p}x\cos\frac{n\pi}{p}x\,dx \begin{cases} = 0, & m \neq n \\ = p, & m = n \end{cases}$$

and

$$\int_{-p}^{p}\cos\frac{m\pi}{p}x\sin\frac{n\pi}{p}x\,dx = 0.$$

Thus (5) reduces to

$$\int_{-p}^{p} f(x)\cos\frac{n\pi}{p}x\,dx = a_n p$$

and so

$$a_n = \frac{1}{p}\int_{-p}^{p} f(x)\cos\frac{n\pi}{p}x\,dx. \quad (6)$$

Finally, if we multiply (2) by $\sin(m\pi x/p)$, integrate, and make use of the results

$$\int_{-p}^{p}\sin\frac{m\pi}{p}x\,dx = 0, \qquad m > 0$$

$$\int_{-p}^{p}\sin\frac{m\pi}{p}x\cos\frac{n\pi}{p}x\,dx = 0$$

$$\int_{-p}^{p}\sin\frac{m\pi}{p}x\sin\frac{n\pi}{p}x\,dx \begin{cases} = 0, & m \neq n \\ = p, & m = n, \end{cases}$$

we find that
$$b_n = \frac{1}{p} \int_{-p}^{p} f(x) \sin \frac{n\pi}{p} x \, dx. \tag{7}$$

The trigonometric series (2) with coefficients a_0, a_n, and b_n defined by (4), (6), and (7), respectively, is said to be the **Fourier series** of the function f.* The coefficients obtained from (4), (6), and (7) are referred to as **Fourier coefficients** of f.

In finding the coefficients a_0, a_n, and b_n, we assumed that f was integrable on the interval and that (2), as well as the series obtained by multiplying (2) by $\cos(m\pi x/p)$, converged in such a manner as to permit term-by-term integration. Until (2) is shown to be convergent for a given function f, the equality sign is not to be taken in a strict or literal sense. Some texts use the symbol $\sim$ in place of $=$. In view of the fact that most functions in applications are of a type that guarantee convergence of the series, we shall use the equality symbol. We summarize the results:

DEFINITION 11.5 **Fourier Series**

The **Fourier series** of a function f defined on the interval $(-p, p)$ is given by

$$f(x) = \frac{a_0}{2} + \sum_{n=1}^{\infty} \left(a_n \cos \frac{n\pi}{p} x + b_n \sin \frac{n\pi}{p} x \right), \tag{8}$$

where
$$a_0 = \frac{1}{p} \int_{-p}^{p} f(x) \, dx \tag{9}$$

$$a_n = \frac{1}{p} \int_{-p}^{p} f(x) \cos \frac{n\pi}{p} x \, dx \tag{10}$$

$$b_n = \frac{1}{p} \int_{-p}^{p} f(x) \sin \frac{n\pi}{p} x \, dx. \tag{11}$$

---

*Jean-Baptiste Joseph Fourier (1766–1830) A French mathematical physicist, Fourier used such trigonometric series in his investigations into the theory of heat, and they appear throughout his 1822 treatise *Théorie analytique de la chaleur*. However, Fourier did not "invent" Fourier series. The development of the theory of expanding functions in trigonometric series was due principally to Daniel Bernoulli and Leonhard Euler. The integral formulas that define the coefficients a_0, a_n, and b_n were discovered by Euler in 1777. Today, Fourier series, the Fourier integral, and the Fourier transform constitute a branch of mathematical analysis that is invaluable in the study of wave phenomena.

A friend and confidant of Napoleon, Fourier served in the emperor's retinue during Napoleon's 1798 campaign to "civilize" Egypt. Fourier is also remembered for his patronage of the young Jean François Champollion, who was the first to decipher Egyptian hieroglyphics through his work on the Rosetta stone.

EXAMPLE I

Expand

$$f(x) = \begin{cases} 0, & -\pi < x < 0 \\ \pi - x, & 0 < x < \pi, \end{cases} \tag{12}$$

in a Fourier series.

Solution The graph of f is given in Figure 11.1. With $p = \pi$ we have from (9) and (10) that

$$a_0 = \frac{1}{\pi} \int_{-\pi}^{\pi} f(x)\, dx = \frac{1}{\pi}\left[\int_{-\pi}^{0} 0\, dx + \int_{0}^{\pi} (\pi - x)\, dx \right]$$

$$= \frac{1}{\pi}\left[\pi x - \frac{x^2}{2} \right]_{0}^{\pi} = \frac{\pi}{2}$$

$$a_n = \frac{1}{\pi} \int_{-\pi}^{\pi} f(x) \cos nx\, dx = \frac{1}{\pi}\left[\int_{-\pi}^{0} 0\, dx + \int_{0}^{\pi} (\pi - x) \cos nx\, dx \right]$$

$$= \frac{1}{\pi}\left[(\pi - x) \frac{\sin nx}{n}\Big|_{0}^{\pi} + \frac{1}{n} \int_{0}^{\pi} \sin nx\, dx \right]$$

$$= -\frac{1}{n\pi} \frac{\cos nx}{n}\Big|_{0}^{\pi}$$

$$= \frac{-\cos n\pi + 1}{n^2 \pi} \qquad \boxed{\cos n\pi = (-1)^n}$$

$$= \frac{1 - (-1)^n}{n^2 \pi}.$$

In like manner we find from (11) that

$$b_n = \frac{1}{\pi} \int_{0}^{\pi} (\pi - x) \sin nx\, dx = \frac{1}{n}.$$

Therefore $\quad f(x) = \frac{\pi}{4} + \sum_{n=1}^{\infty} \left\{ \frac{1 - (-1)^n}{n^2 \pi} \cos nx + \frac{1}{n} \sin nx \right\}. \tag{13}$

∎

Note that a_n defined by (10) reduces to a_0 given by (9) when we set $n = 0$. But as Example 1 shows, this may not be the case *after* the integral for a_n is evaluated.

Convergence of a Fourier Series

The following theorem gives sufficient conditions for convergence of a Fourier series at a point.

Figure 11.1

THEOREM 11.1 **Conditions for Convergence**

Let f and f' be piecewise continuous on the interval $(-p, p)$; that is, let f and f' be continuous except at a finite number of points in the interval and have only finite discontinuities at these points. Then the Fourier series of f on the interval converges to $f(x)$ at a point of continuity. At a point of discontinuity, the Fourier series converges to the average

$$\frac{f(x+) + f(x-)}{2},$$

where $f(x+)$ and $f(x-)$ denotes the limit of f at x from the right and from the left, respectively.*

For a proof of this theorem, you are referred to the classic text by Churchill and Brown.[†]

EXAMPLE 2

The function (12) in Example 1 satisfies the conditions of Theorem 11.1. Thus for every x in the interval $(-\pi, \pi)$, except at $x = 0$, the series (13) will converge to $f(x)$. At $x = 0$ the function is discontinuous and so the series (13) will converge to

$$\frac{f(0+) + f(0-)}{2} = \frac{\pi + 0}{2} = \frac{\pi}{2}.$$ ∎

Periodic Extension

Observe that the functions in the basic set (1) have a common period $2p$. Hence the right side of (2) is periodic. We conclude that a Fourier series not only represents the function on the interval $(-p, p)$ but also gives the **periodic extension** of f outside this interval. We can now apply Theorem 11.1 to the periodic extension of f, or we may assume from the outset that the given function is periodic with period $2p$ (that is, $f(x + 2p) = f(x)$). When f is piecewise continuous and the right- and left-hand derivatives exist at $x = -p$ and $x = p$, respectively, then the series (8) converges to the average $[f(p-) + f(-p+)]/2$ at these endpoints and to this value extended periodically to $\pm 3p, \pm 5p, \pm 7p$, and so on.

---

* In other words, for x a point in the interval and $h > 0$,

$$f(x+) = \lim_{h \to 0} f(x + h), \qquad f(x-) = \lim_{h \to 0} f(x - h).$$

[†] *Fourier Series and Boundary Value Problems* by Ruel V. Churchill and James Ward Brown (New York: McGraw-Hill Book Co.)

EXAMPLE 3

The Fourier series (13) converges to the periodic extension of (12) on the entire x-axis. The solid dots in Figure 11.2 represent the value

$$\frac{f(0+) + f(0-)}{2} = \frac{\pi}{2}$$

at $0, \pm 2\pi, \pm 4\pi, \ldots$. At $\pm \pi, \pm 3\pi, \pm 5\pi, \ldots$, the series converges to the value

$$\frac{f(\pi-) + f(-\pi+)}{2} = 0.$$

Figure 11.2   ∎

11.2 EXERCISES

Answers to odd-numbered problems begin on page A–39.

In Problems 1–16 find the Fourier series of f on the given interval.

1. $f(x) = \begin{cases} 0, & -\pi < x < 0 \\ 1, & 0 \le x < \pi \end{cases}$

2. $f(x) = \begin{cases} -1, & -\pi < x < 0 \\ 2, & 0 \le x < \pi \end{cases}$

3. $f(x) = \begin{cases} 1, & -1 < x < 0 \\ x, & 0 \le x < 1 \end{cases}$

4. $f(x) = \begin{cases} 0, & -1 < x < 0 \\ x, & 0 \le x < 1 \end{cases}$

5. $f(x) = \begin{cases} 0, & -\pi < x < 0 \\ x^2, & 0 \le x < \pi \end{cases}$

6. $f(x) = \begin{cases} \pi^2, & -\pi < x < 0 \\ \pi^2 - x^2, & 0 \le x < \pi \end{cases}$

7. $f(x) = x + \pi, \quad -\pi < x < \pi$

8. $f(x) = 3 - 2x, \quad -\pi < x < \pi$

9. $f(x) = \begin{cases} 0, & -\pi < x < 0 \\ \sin x, & 0 \le x < \pi \end{cases}$

10. $f(x) = \begin{cases} 0, & -\pi/2 < x < 0 \\ \cos x, & 0 \le x < \pi/2 \end{cases}$

11. $f(x) = \begin{cases} 0, & -2 < x < -1 \\ -2, & -1 \le x < 0 \\ 1, & 0 \le x < 1 \\ 0, & 1 \le x < 2 \end{cases}$

12. $f(x) = \begin{cases} 0, & -2 < x < 0 \\ x, & 0 \le x < 1 \\ 1, & 1 \le x < 2 \end{cases}$

13. $f(x) = \begin{cases} 1, & -5 < x < 0 \\ 1 + x, & 0 \le x < 5 \end{cases}$

14. $f(x) = \begin{cases} 2 + x, & -2 < x < 0 \\ 2, & 0 \le x < 2 \end{cases}$

15. $f(x) = e^x$, $-\pi < x < \pi$

16. $f(x) = \begin{cases} 0, & -\pi < x < 0 \\ e^x - 1, & 0 \le x < \pi \end{cases}$

17. Use the result of Problem 5 to show

$$\frac{\pi^2}{6} = 1 + \frac{1}{2^2} + \frac{1}{3^2} + \frac{1}{4^2} + \cdots \quad \text{and} \quad \frac{\pi^2}{12} = 1 - \frac{1}{2^2} + \frac{1}{3^2} - \frac{1}{4^2} + \cdots.$$

18. Use Problem 17 to find a series that gives the numerical value of $\pi^2/8$.

19. Use the result of Problem 7 to show

$$\frac{\pi}{4} = 1 - \frac{1}{3} + \frac{1}{5} - \frac{1}{7} + \cdots.$$

20. Use the result of Problem 9 to show

$$\frac{\pi}{4} = \frac{1}{2} + \frac{1}{1 \cdot 3} - \frac{1}{3 \cdot 5} + \frac{1}{5 \cdot 7} - \frac{1}{7 \cdot 9} + \cdots.$$

21. **(a)** Use the complex exponential form of the cosine and sine:

$$\cos \frac{n\pi}{p} x = \frac{e^{in\pi x/p} + e^{-in\pi x/p}}{2}, \qquad \sin \frac{n\pi}{p} x = \frac{e^{in\pi x/p} - e^{-in\pi x/p}}{2i}$$

to show that (8) can be written in the **complex form**

$$f(x) = \sum_{n=-\infty}^{\infty} c_n e^{in\pi x/p},$$

where $c_0 = a_0/2$, $c_n = (a_n - ib_n)/2$, and $c_{-n} = (a_n + ib_n)/2$, where $n = 1, 2, 3, \ldots$.

(b) Show that c_0, c_n, and c_{-n} of part (a) can be written as one integral

$$c_n = \frac{1}{2p} \int_{-p}^{p} f(x) e^{-in\pi x/p} \, dx, \qquad n = 0, \pm 1, \pm 2, \ldots.$$

22. Use the results of Problem 21 to find the complex form of the Fourier series of $f(x) = e^{-x}$ on the interval $-\pi < x < \pi$.

11.3 FOURIER COSINE AND SINE SERIES

Even and Odd Functions

You may recall that a function f is said to be

$$\textbf{even if} \quad f(-x) = f(x), \quad \text{and}$$

$$\textbf{odd if} \quad f(-x) = -f(x).$$

Figure 11.3

Figure 11.4

EXAMPLE 1

(a) $f(x) = x^2$ is even since

$$f(-x) = (-x)^2 = x^2 = f(x).$$

See Figure 11.3.

(b) $f(x) = x^3$ is odd since

$$f(-x) = (-x)^3 = -x^3 = -f(x)$$

See Figure 11.4. ∎

   As illustrated in Figures 11.3 and 11.4, the graph of an even function is symmetric with respect to the y-axis, and the graph of an odd function is symmetric with respect to the origin.

EXAMPLE 2

Since $\cos(-x) = \cos x$ and $\sin(-x) = -\sin x$, the cosine and sine are even and odd functions, respectively. ∎

Properties of Even and Odd Functions

The following theorem lists some properties of even and odd functions.

THEOREM 11.2   Properties of Even/Odd Functions

(a) The product of two even functions is even.

(b) The product of two odd functions is even.

(c) The product of an even function and an odd function is odd.

(d) The sum (difference) of two even functions is even.

(e) The sum (difference) of two odd functions is odd.

(f) If f is even, then $\int_{-a}^{a} f(x)\, dx = 2 \int_{0}^{a} f(x)\, dx$.

(g) If f is odd, then $\int_{-a}^{a} f(x)\, dx = 0$.

   Proof of (b)   Let us suppose that f and g are odd functions. Then we have $f(-x) = -f(x)$ and $g(-x) = -g(x)$. If we define the product of f and g as $F(x) = f(x)g(x)$, then

$$F(-x) = f(-x)g(-x) = (-f(x))(-g(x)) = f(x)g(x) = F(x).$$

This shows that the product F of two odd functions is an even function. The proofs of the remaining properties are left as exercises. (See Problems 45–49.) □

Cosine and Sine Series

If f is an even function on $(-p, p)$, then in view of the foregoing properties, the coefficients (9), (10), and (11) of Section 11.2 become

$$a_0 = \frac{1}{p} \int_{-p}^{p} f(x)\, dx = \frac{2}{p} \int_{0}^{p} f(x)\, dx$$

$$a_n = \frac{1}{p} \int_{-p}^{p} \underbrace{f(x) \cos \frac{n\pi}{p} x}_{\text{even}}\, dx = \frac{2}{p} \int_{0}^{p} f(x) \cos \frac{n\pi}{p} x\, dx$$

$$b_n = \frac{1}{p} \int_{-p}^{p} \underbrace{f(x) \sin \frac{n\pi}{p} x}_{\text{odd}}\, dx = 0.$$

Similarly, when f is odd on the interval $(-p, p)$,

$$a_n = 0, \quad n = 0, 1, 2, \ldots, \qquad b_n = \frac{2}{p} \int_{0}^{p} f(x) \sin \frac{n\pi}{p} x\, dx.$$

We summarize the results.

DEFINITION 11.6 **Fourier Cosine and Sine Series**

(i) The Fourier series of an even function on the interval $(-p, p)$ is the **cosine series**

$$f(x) = \frac{a_0}{2} + \sum_{n=1}^{\infty} a_n \cos \frac{n\pi}{p} x, \tag{1}$$

where

$$a_0 = \frac{2}{p} \int_{0}^{p} f(x)\, dx \tag{2}$$

$$a_n = \frac{2}{p} \int_{0}^{p} f(x) \cos \frac{n\pi}{p} x\, dx. \tag{3}$$

(ii) The Fourier series of an odd function on the interval $(-p, p)$ is the **sine series**

$$f(x) = \sum_{n=1}^{\infty} b_n \sin \frac{n\pi}{p} x, \tag{4}$$

where

$$b_n = \frac{2}{p} \int_{0}^{p} f(x) \sin \frac{n\pi}{p} x\, dx. \tag{5}$$

$y = x, \; -2 < x < 2$

Figure 11.5

EXAMPLE 3

Expand $f(x) = x$, $-2 < x < 2$, in a Fourier series.

Solution We expand f in a sine series, since inspection of Figure 11.5 shows that the function is odd on the interval $(-2, 2)$.

With the identification $2p = 4$, or $p = 2$, we can write (5) as

$$b_n = \int_0^2 x \sin \frac{n\pi}{2} x \, dx.$$

Integration by parts then yields

$$b_n = \frac{4(-1)^{n+1}}{n\pi}.$$

Therefore

$$f(x) = \frac{4}{\pi} \sum_{n=1}^{\infty} \frac{(-1)^{n+1}}{n} \sin \frac{n\pi}{2} x. \tag{6}$$

∎

EXAMPLE 4

The function in Example 3 satisfies the conditions of Theorem 11.1. Hence the series (6) converges to the function on $(-2, 2)$ and the periodic extension (of period 4) given in Figure 11.6.

Figure 11.6 ∎

EXAMPLE 5

The function

$$f(x) = \begin{cases} -1, & -\pi < x < 0 \\ 1, & 0 \le x < \pi \end{cases}$$

shown in Figure 11.7 is odd on the interval $(-\pi, \pi)$. With $p = \pi$ we have from (5)

$$b_n = \frac{2}{\pi} \int_0^\pi (1) \sin nx \, dx = \frac{2}{\pi} \frac{1 - (-1)^n}{n}$$

Figure 11.7

and so
$$f(x) = \frac{2}{\pi} \sum_{n=1}^{\infty} \frac{1 - (-1)^n}{n} \sin nx.$$
(7)

∎

Sequence of Partial Sums

It is interesting to see how the sequence of partial sums of a Fourier series approximates a function. In Figure 11.8 the graph of the function f in Example 5 is compared with the graphs of the first three partial sums of (7):

$$S_1 = \frac{4}{\pi} \sin x, \qquad S_2 = \frac{4}{\pi} \left(\sin x + \frac{\sin 3x}{3} \right), \qquad S_3 = \frac{4}{\pi} \left(\sin x + \frac{\sin 3x}{3} + \frac{\sin 5x}{5} \right).$$

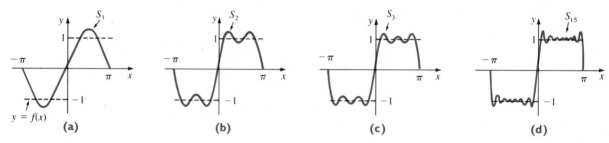

Figure 11.8

As seen in Figure 11.8(d), the graph of the partial sum S_{15} has pronounced spikes near the discontinuities at $x = 0$, $x = \pi$, $x = -\pi$, and so on. This "overshooting" of the partial sums S_N from the functional values near a point of discontinuity does not smooth out but remains fairly constant, even by taking the value of N to be large. This behavior of a Fourier series near a point at which f is discontinuous is known as the **Gibbs phenomenon.**[*]

Half-Range Expansions

Throughout the preceding discussion it was understood that a function f was defined on an interval with the origin as midpoint, that is, $-p < x < p$. However, in many instances we are interested in representing a function that

---

[*]**Josiah Willard Gibbs** (1839–1903) This "overshooting" phenomenon was discovered by the American physicist and physical chemist Josiah Gibbs in 1899. He is also remembered for his pioneering work in three-dimensional vector analysis. In 1881 Gibbs, for the benefit of his students at Yale, published a pamphlet entitled *Elements of Vector Analysis*. E. B. Wilson incorporated this pamphlet and Gibbs' lectures in the text *Vector Analysis*, which was published in 1901. In these works Gibbs introduced much of the notation that is still used in modern vector analysis.

is defined only for $0 < x < L$ by a trigonometric series. This can be done in many different ways by supplying an arbitrary *definition* of the function on the interval $-L < x < 0$. For brevity we consider the three most important cases.

If $y = f(x)$ is defined on the interval $0 < x < L$,

(i) reflect the graph of the function about the y-axis onto $-L < x < 0$; the function is now even on $-L < x < L$ (see Figure 11.9); or

(ii) reflect the graph of the function through the origin onto $-L < x < 0$; the function is now odd on $-L < x < L$ (see Figure 11.10); or

(iii) define f on $-L < x < 0$ by $f(x) = f(x + L)$ (see Figure 11.11).

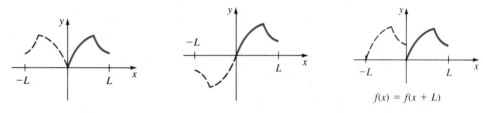

$$f(x) = f(x + L)$$

| **Figure 11.9** | **Figure 11.10** | **Figure 11.11** |

Note that the coefficients of the series (1) and (4) utilize only the definition of the function on $0 < x < p$ (that is, half of the interval $-p < x < p$). Hence in practice there is no actual need to make the reflections described in (i) and (ii). If f is defined on $0 < x < L$, we simply identify the half-period as the length of the interval $p = L$. The coefficient formulas (2), (3), and (5) and the corresponding series yield either an even or an odd periodic extension of period $2L$ of the original function. The cosine and sine series obtained in this manner are known as **half-range expansions**. Last, in case (iii) we are defining the functional values on the interval $-L < x < 0$ to be the same as the values on $0 < x < L$. As in the previous two cases, there is no real need to do this. It can be shown that the set of functions in (1) of Section 11.2 is orthogonal on $a \le x \le a + 2p$ for any real number a. By choosing $a = -p$, we obtain the limits of integration in (9), (10), and (11) of that section. But for $a = 0$ the limits of integration are from $x = 0$ to $x = 2p$. Thus if f is defined over the interval $0 < x < L$, we identify $2p = L$ or $p = L/2$. The resulting Fourier series will give the periodic extension of f with period L. In this manner the values to which the series converges on $-L < x < 0$ will be the same as on $0 < x < L$.

EXAMPLE 6

Expand $$f(x) = x^2, \qquad 0 < x < L,$$

(a) in a cosine series, **(b)** in a sine series, and **(c)** in a Fourier series.

Figure 11.12

Solution The graph of the function is given in Figure 11.12.

(a) We have
$$a_0 = \frac{2}{L} \int_0^L x^2 \, dx = \frac{2}{3} L^2$$

and, integrating by parts,

$$a_n = \frac{2}{L} \int_0^L x^2 \cos \frac{n\pi}{L} x \, dx = \frac{2}{L} \left[\frac{Lx^2 \sin \frac{n\pi}{L} x}{n\pi} \bigg|_0^L - \frac{2L}{n\pi} \int_0^L x \sin \frac{n\pi}{L} x \, dx \right]$$

$$= -\frac{4}{n\pi} \left[-\frac{Lx \cos \frac{n\pi}{L} x}{n\pi} \bigg|_0^L + \frac{L}{n\pi} \int_0^L \cos \frac{n\pi}{L} x \, dx \right]$$

$$= \frac{4L^2 (-1)^n}{n^2 \pi^2}.$$

Thus
$$f(x) = \frac{L^2}{3} + \frac{4L^2}{\pi^2} \sum_{n=1}^{\infty} \frac{(-1)^n}{n^2} \cos \frac{n\pi}{L} x. \tag{8}$$

(b) In this case
$$b_n = \frac{2}{L} \int_0^L x^2 \sin \frac{n\pi}{L} x \, dx.$$

After integrating by parts we find

$$b_n = \frac{2L^2 (-1)^{n+1}}{n\pi} + \frac{4L^2}{n^3 \pi^3} [(-1)^n - 1].$$

Thus
$$f(x) = \frac{2L^2}{\pi} \sum_{n=1}^{\infty} \left\{ \frac{(-1)^{n+1}}{n} + \frac{2}{n^3 \pi^2} [(-1)^n - 1] \right\} \sin \frac{n\pi}{L} x. \tag{9}$$

(c) With $p = L/2$, $1/p = 2/L$, and $n\pi/p = 2n\pi/L$, we have

$$a_0 = \frac{2}{L} \int_0^L x^2 \, dx = \frac{2}{3} L^2$$

$$a_n = \frac{2}{L} \int_0^L x^2 \cos \frac{2n\pi}{L} x \, dx = \frac{L^2}{n^2 \pi^2}$$

$$b_n = \frac{2}{L} \int_0^L x^2 \sin \frac{2n\pi}{L} x \, dx = -\frac{L^2}{n\pi}.$$

Therefore

$$f(x) = \frac{L^2}{3} + \frac{L^2}{\pi} \sum_{n=1}^{\infty} \left\{ \frac{1}{n^2 \pi} \cos \frac{2n\pi}{L} x - \frac{1}{n} \sin \frac{2n\pi}{L} x \right\}. \tag{10}$$

The series (8), (9), and (10) converge to the $2L$-periodic even extension of f, to the $2L$-periodic odd extension of f, and the L-periodic extension

(a) cosine series

(b) sine series

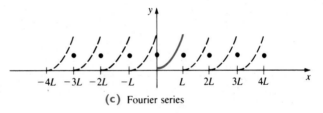

(c) Fourier series

Figure 11.13

of f, respectively. The graphs of these periodic extensions are shown in Figure 11.13. ∎

Periodic Driving Force

Fourier series are sometimes useful in determining a particular solution of a differential equation describing a physical system in which the input or driving force $f(t)$ is periodic. In the next example we find a particular solution of the differential equation

$$m\frac{d^2x}{dt^2} + kx = f(t) \tag{11}$$

by first representing f by a half-range sine expansion

$$f(t) = \sum_{n=1}^{\infty} b_n \sin\frac{n\pi}{p}t$$

and then assuming a particular solution of the form

$$x_p(t) = \sum_{n=1}^{\infty} B_n \sin \frac{n\pi}{p} t. \tag{12}$$

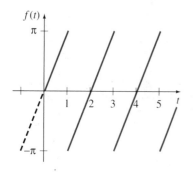

$f(t)$

Figure 11.14

EXAMPLE 7

An undamped spring-mass system, in which the mass $m = 1/16$ slug and the spring constant $k = 4$ lb/ft, is driven by the 2-periodic external force $f(t)$ shown in Figure 11.14. Although the force $f(t)$ acts on the system for $t > 0$, note that if we extend the graph of the function in a 2-periodic manner to the negative t-axis we obtain an odd function. In practical terms this means that we need only find the half-range sine expansion of $f(t) = \pi t, 0 < t < 1$. With $p = 1$ it follows from (5) and integration by parts that

$$b_n = 2 \int_0^1 \pi t \sin n\pi t \, dt = \frac{2(-1)^{n+1}}{n}.$$

From (11) the differential equation of motion is seen to be

$$\frac{1}{16} \frac{d^2x}{dt^2} + 4x = \sum_{n=1}^{\infty} \frac{2(-1)^{n+1}}{n} \sin n\pi t. \tag{13}$$

To find a particular solution $x_p(t)$ of (13) we substitute (12) into the equation and equate coefficients of $\sin n\pi t$. This yields

$$\left(-\frac{1}{16} n^2\pi^2 + 4 \right) B_n = \frac{2(-1)^{n+1}}{n} \quad \text{or} \quad B_n = \frac{32(-1)^{n+1}}{n(64 - n^2\pi^2)}.$$

Thus

$$x_p(t) = \sum_{n=1}^{\infty} \frac{32(-1)^{n+1}}{n(64 - n^2\pi^2)} \sin n\pi t. \tag{14}$$

■

Observe in the solution (14), there is no integer $n \geq 1$ for which the denominator $64 - n^2\pi^2$ of B_n is zero. In general, if there *is* a value of n, say N, for which $N\pi/p = \omega$, where $\omega = \sqrt{k/m}$, then the system described by (11) would be in a state of pure resonance. In other words, we have pure resonance if the Fourier series expansion of the driving force $f(t)$ contains a term $\sin(N\pi/L)t$ (or $\cos(N\pi/L)t$) that has the same frequency as the frequency of free vibrations.

Of course, if the $2p$-periodic extension of the driving force f onto the negative t-axis yields an even function, then we expand f in a cosine series.

11.3 EXERCISES

Answers to odd-numbered problems begin on page A–39.

In Problems 1–10 determine whether the function is even, odd, or neither.

1. $f(x) = \sin 3x$

2. $f(x) = x \cos x$

3. $f(x) = x^2 + x$

4. $f(x) = x^3 - 4x$

5. $f(x) = e^{|x|}$

6. $f(x) = |x^5|$

7. $f(x) = \begin{cases} x^2, & -1 < x < 0 \\ -x^2, & 0 < x < 1 \end{cases}$

8. $f(x) = \begin{cases} x + 5, & -2 < x < 0 \\ -x + 5, & 0 \le x < 2 \end{cases}$

9. $f(x) = x^3, \quad 0 \le x \le 2$

10. $f(x) = 2|x| - 1$

In Problems 11–24 expand the given function in an appropriate cosine or sine series.

11. $f(x) = \begin{cases} -1, & -\pi < x < 0 \\ 1, & 0 \le x < \pi \end{cases}$

12. $f(x) = \begin{cases} 1, & -2 < x < -1 \\ 0, & -1 < x < 1 \\ 1, & 1 < x < 2 \end{cases}$

13. $f(x) = |x|, \quad -\pi < x < \pi$

14. $f(x) = x, \quad -\pi < x < \pi$

15. $f(x) = x^2, \quad -1 < x < 1$

16. $f(x) = x|x|, \quad -1 < x < 1$

17. $f(x) = \pi^2 - x^2, \quad -\pi < x < \pi$

18. $f(x) = x^3, \quad -\pi < x < \pi$

19. $f(x) = \begin{cases} x - 1, & -\pi < x < 0 \\ x + 1, & 0 \le x < \pi \end{cases}$

20. $f(x) = \begin{cases} x + 1, & -1 < x < 0 \\ x - 1, & 0 \le x < 1 \end{cases}$

21. $f(x) = \begin{cases} 1, & -2 < x < -1 \\ -x, & -1 \le x < 0 \\ x, & 0 \le x < 1 \\ 1, & 1 \le x < 2 \end{cases}$

22. $f(x) = \begin{cases} -\pi, & -2\pi < x < -\pi \\ x, & -\pi \le x < \pi \\ \pi, & \pi \le x < 2\pi \end{cases}$

23. $f(x) = |\sin x|, \quad -\pi < x < \pi$

24. $f(x) = \cos x, \quad -\pi/2 < x < \pi/2$

In Problems 25–34 find the half-range cosine and sine expansions of the given function.

25. $f(x) = \begin{cases} 1, & 0 < x < \frac{1}{2} \\ 0, & \frac{1}{2} \le x < 1 \end{cases}$

26. $f(x) = \begin{cases} 0, & 0 < x < \frac{1}{2} \\ 1, & \frac{1}{2} \le x < 1 \end{cases}$

27. $f(x) = \cos x, \quad 0 < x < \pi/2$

28. $f(x) = \sin x, \quad 0 < x < \pi$

29. $f(x) = \begin{cases} x, & 0 < x < \pi/2 \\ \pi - x, & \pi/2 \le x < \pi \end{cases}$

30. $f(x) = \begin{cases} 0, & 0 < x < \pi \\ x - \pi, & \pi \le x < 2\pi \end{cases}$

31. $f(x) = \begin{cases} x, & 0 < x < 1 \\ 1, & 1 \le x < 2 \end{cases}$

32. $f(x) = \begin{cases} 1, & 0 < x < 1 \\ 2 - x, & 1 \le x < 2 \end{cases}$

33. $f(x) = x^2 + x, \quad 0 < x < 1$

34. $f(x) = x(2 - x), \quad 0 < x < 2$

In Problems 35–38 expand the given function in a Fourier series.

35. $f(x) = x^2, \quad 0 < x < 2\pi$

36. $f(x) = x, \quad 0 < x < \pi$

37. $f(x) = x + 1, \quad 0 < x < 1$

38. $f(x) = 2 - x, \quad 0 < x < 2$

In Problems 39 and 40 proceed as in Example 7 to find a particular solution of equation (11) when $m = 1$, $k = 10$, and the driving force $f(t)$ is as given. Assume that when $f(t)$ is extended to the negative t-axis in a periodic manner, the resulting function is odd.

39. $f(t) = \begin{cases} 5, & 0 < t < \pi \\ -5, & \pi < t < 2\pi \end{cases}; \quad f(t + 2\pi) = f(t)$

40. $f(t) = 1 - t, 0 < t < 2; \quad f(t + 2) = f(t)$

In Problems 41 and 42 find a particular solution of equation (11) when $m = 1/4$, $k = 12$, and the driving force $f(t)$ is as given. Assume that when $f(t)$ is extended to the negative t-axis in a periodic manner, the resulting function is even.

41. $f(t) = 2\pi t - t^2, 0 < t < 2\pi; \quad f(t + 2\pi) = f(t)$

42. $f(t) = \begin{cases} t, & 0 < t < \frac{1}{2} \\ 1 - t, & \frac{1}{2} \le t < 1 \end{cases}; \quad f(t + 1) = f(t)$

43. Suppose a uniform beam of length L is simply supported at $x = 0$ and $x = L$. If the load per unit length is given by $w(x) = w_0 x/L, 0 < x < L$, then the differential equation for the deflection $y(x)$ is

$$EI \frac{d^4 y}{dx^4} = \frac{w_0 x}{L},$$

where E, I, and w_0 are constants. (See equation (20), Section 1.2.)

(a) Expand $w(x)$ in a half-range sine series.

(b) Use the method of Example 7 to find a particular solution $y(x)$ of the differential equation.

44. Proceed as in Problem 43 to find the deflection $y(x)$ when $w(x)$ is as given in Figure 11.15.

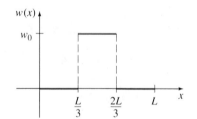

Figure 11.15

Miscellaneous Problems

45. Prove Property (a) in Theorem 11.2.

46. Prove Property (c).

47. Prove Property (d).

48. Prove Property (f).

49. Prove Property (g).

50. Prove that any function f can be written as a sum of an even and an odd function. $\left[\right.$ *Hint:* Use the identity

$$f(x) = \frac{f(x) + f(-x)}{2} + \frac{f(x) - f(-x)}{2}. \left.\right]$$

51. Find the Fourier series of

$$f(x) = \begin{cases} 0, & -\pi < x < 0 \\ x, & 0 \le x < \pi \end{cases}$$

using the identity $f(x) = (|x| + x)/2, \, -\pi < x < \pi$, and the results of Problems 13 and 14. Observe that $|x|/2$ and $x/2$ are even and odd, respectively, on the interval (see Problem 50).

52. The **double sine series** for a function $f(x, y)$ defined over a rectangular region $0 \le x \le b, 0 \le y \le c$ is given by

$$f(x, y) = \sum_{m=1}^{\infty} \sum_{n=1}^{\infty} A_{mn} \sin \frac{m\pi}{b} x \sin \frac{n\pi}{c} y,$$

where

$$A_{mn} = \frac{4}{bc} \int_0^c \int_0^b f(x, y) \sin \frac{m\pi}{b} x \sin \frac{n\pi}{c} y \, dx \, dy.$$

Find the double sine series of $f(x, y) = 1, 0 \le x \le \pi, 0 \le y \le \pi$.

53. The **double cosine series** of a function $f(x, y)$ defined over a rectangular region $0 \le x \le b, 0 \le y \le c$ is given by

$$f(x, y) = A_{00} + \sum_{m=1}^{\infty} A_{m0} \cos \frac{m\pi}{b} x + \sum_{n=1}^{\infty} A_{0n} \cos \frac{n\pi}{c} y$$

$$+ \sum_{m=1}^{\infty} \sum_{n=1}^{\infty} A_{mn} \cos \frac{m\pi}{b} x \cos \frac{n\pi}{c} y,$$

where

$$A_{00} = \frac{1}{bc} \int_0^c \int_0^b f(x, y) \, dx \, dy$$

$$A_{m0} = \frac{2}{bc} \int_0^c \int_0^b f(x, y) \cos \frac{m\pi}{b} x \, dx \, dy$$

$$A_{0n} = \frac{2}{bc} \int_0^c \int_0^b f(x, y) \cos \frac{n\pi}{c} y \, dx \, dy$$

$$A_{mn} = \frac{4}{bc} \int_0^c \int_0^b f(x, y) \cos \frac{m\pi}{b} x \cos \frac{n\pi}{c} y \, dx \, dy.$$

Find the double cosine series of $f(x, y) = xy, 0 \le x \le 1, 0 \le y \le 1$.

11.4 STURM-LIOUVILLE PROBLEM

REVIEW   For convenience we present here a brief review of some of the differential equations and their general solutions that will be of importance in the sections and chapters that follow.

(i) Linear first-order equation: $y' + ky = 0,$ k a constant
General solution: $y = c_1 e^{-kx}$

(ii) Linear second-order equation $y'' + \lambda y = 0,$ $\lambda > 0$
General solution: $y = c_1 \cos \sqrt{\lambda}\, x + c_2 \sin \sqrt{\lambda}\, x$

(iii) Linear second-order equation $y'' - \lambda y = 0,$ $\lambda > 0$
The general solution of this differential equation has two real forms:

$$y = c_1 \cosh \sqrt{\lambda}\, x + c_2 \sinh \sqrt{\lambda}\, x$$

$$y = c_1 e^{-\sqrt{\lambda}\, x} + c_2 e^{\sqrt{\lambda}\, x}$$

It should be noted that often in practice the exponential form is used when the domain of x is an infinite or semi-infinite interval* and the hyperbolic form is used when the domain of x is a finite interval.

(iv) Cauchy-Euler equation: $x^2 y'' + xy' - \lambda^2 y = 0$
General solution: $\lambda \neq 0:$ $y = c_1 x^{\lambda} + c_2 x^{-\lambda}$
  $\lambda = 0:$ $y = c_1 + c_2 \ln x$

(v) Parametric Bessel equation:

$$x^2 y'' + xy' + (\lambda^2 x^2 - n^2) y = 0, \qquad n = 0, 1, 2, \dots$$

General solution: $y = c_1 J_n(\lambda x) + c_2 Y_n(\lambda x)$
It is important that you recognize Bessel's differential equation when $n = 0$:

$$xy'' + y' + \lambda^2 x y = 0$$

General solution: $y = c_1 J_0(\lambda x) + c_2 Y_0(\lambda x)$
Recall, $Y_n(\lambda x) \to -\infty$ as $x \to 0^+$.

(vi) Legendre's differential equation:

$$(1 - x^2) y'' - 2xy' + n(n + 1) y = 0, \qquad n = 0, 1, 2, \dots$$

Particular solutions are the Legendre polynomials $y = P_n(x)$, where

$$y = P_0(x) = 1, \quad y = P_1(x) = x, \quad y = P_2(x) = \frac{1}{2}(3x^2 - 1), \quad \dots$$

---

* Infinite: $(-\infty, \infty)$; semi-infinite: $(-\infty, 0]$, $(1, \infty)$, and so on.

Orthogonal functions often arise in the solution of differential equations. More to the point, an orthogonal set of functions can be generated by solving a two-point boundary-value problem involving a linear second-order differential equation containing a parameter λ.

EXAMPLE 1

Solve the boundary-value problem

$$y'' + \lambda y = 0, \qquad y(0) = 0, \quad y(L) = 0.$$

Solution   We consider three cases: $\lambda = 0$, $\lambda < 0$, and $\lambda > 0$.

CASE I   For $\lambda = 0$ the solution of $y'' = 0$ is $y = c_1 x + c_2$. The conditions $y(0) = 0$ and $y(L) = 0$ imply, in turn, $c_2 = 0$ and $c_1 = 0$. Hence for $\lambda = 0$ the only solution of the boundary-value problem is the trivial solution $y = 0$.

CASE II   For $\lambda < 0$ we have $y = c_1 \cosh \sqrt{-\lambda}\, x + c_2 \sinh \sqrt{-\lambda}\, x$.* Again, $y(0) = 0$ gives $c_1 = 0$, and so $y = c_2 \sinh \sqrt{-\lambda}\, x$. The second condition $y(L) = 0$ dictates that $c_2 \sinh \sqrt{-\lambda}\, L = 0$. Since $\sinh \sqrt{-\lambda}\, L \neq 0$, we must have $c_2 = 0$. Thus $y = 0$.

CASE III   For $\lambda > 0$ the general solution of $y'' + \lambda y = 0$ is $y = c_1 \cos \sqrt{\lambda}\, x + c_2 \sin \sqrt{\lambda}\, x$. As before, $y(0) = 0$ yields $c_1 = 0$, but $y(L) = 0$ implies

$$c_2 \sin \sqrt{\lambda}\, L = 0. \tag{1}$$

If $c_2 = 0$, then necessarily $y = 0$. However, if $c_2 \neq 0$, then (1) is satisfied if $\sin \sqrt{\lambda}\, L = 0$. The last condition implies that the argument of the sine function must be an integer multiple of π:

$$\sqrt{\lambda}\, L = n\pi \quad \text{or} \quad \lambda = \frac{n^2 \pi^2}{L^2}, \qquad n = 1, 2, 3, \ldots. \tag{2}$$

Therefore for any real nonzero c_2, $y = c_2 \sin(n\pi x/L)$ is a solution of the problem for each n. Since the differential equation is homogeneous, we may, if desired, not write c_2. In other words, for a given number in the sequence

$$\frac{\pi^2}{L^2}, \frac{4\pi^2}{L^2}, \frac{9\pi^2}{L^2}, \ldots,$$

the *corresponding* function in the sequence

$$\sin \frac{\pi}{L}\, x, \sin \frac{2\pi}{L}\, x, \sin \frac{3\pi}{L}\, x, \ldots$$

---

*$\sqrt{-\lambda}$ looks a little strange, but bear in mind that $\lambda < 0$ is equivalent to $-\lambda > 0$.

is a nontrivial solution of the original problem. You should recognize that $\{\sin(n\pi x/L)\}$, $n = 1, 2, 3, \ldots$, is the orthogonal set of functions on $[0, L]$ used as the basis for the Fourier sine series. ∎

Eigenvalues and Eigenfunctions

The numbers $\lambda = n^2\pi^2/L^2$, $n = 1, 2, 3, \ldots$, for which the boundary-value problem in Example 1 has a nontrivial solution are known as **characteristic values** or, more commonly, **eigenvalues**. The solutions depending on these values of λ, $y = c_2 \sin(n\pi x/L)$ (or simply $y = \sin(n\pi x/L)$, are called **characteristic functions**, or **eigenfunctions**.* For example,

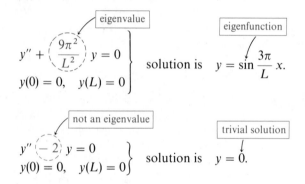

EXAMPLE 2

The eigenvalues and eigenfunctions of the boundary-value problem

$$y'' + \lambda y = 0, \qquad y'(0) = 0, \quad y'(L) = 0,$$

are, respectively, $\lambda = n^2\pi^2/L^2$, $n = 0, 1, 2, \ldots$, and $y = c_1 \cos(n\pi x/L)$, $c_1 \neq 0$. In contrast to Example 1, $\lambda = 0$ is an eigenvalue of this problem and $y = 1$ is the corresponding eigenfunction. The latter comes from solving $y'' = 0$, $y'(0) = 0$, $y'(L) = 0$ but can be incorporated in $y = \cos(n\pi x/L)$ by permitting $n = 0$ and $c_1 = 1$. The set $\{\cos(n\pi x/L)\}$, $n = 0, 1, 2, \ldots$, is orthogonal on the interval $[0, L]$ and is used in Fourier cosine expansions. ∎

---

* Notice in (2) we need only take n to be a positive integer. If n is a negative integer, $n = -k$, $k = 1, 2, 3, \ldots$, we can use the trigonometric identity $\sin(-\theta) = -\sin\theta$ to rewrite $\sin(-k\pi x/L)$ as $-\sin(k\pi x/L)$. The factor -1 can be absorbed in the coefficient c_2. If $n = 0$, then $\lambda = 0$ and, from case I, $y = 0$.

Regular Sturm-Liouville Problem

The boundary-value problems in Examples 1 and 2 are special cases of the following general problem:

Let p, q, r, and r' be real-valued functions continuous on an interval $[a, b]$ and let $r(x) > 0$ and $p(x) > 0$ for every x in the interval. The two-point boundary-value problem:

Solve:
$$\frac{d}{dx}[r(x)y'] + (q(x) + \lambda p(x))y = 0 \tag{3}$$

subject to the homogeneous boundary conditions:

$$\alpha_1 y(a) + \beta_1 y'(a) = 0 \tag{4}$$

$$\alpha_2 y(b) + \beta_2 y'(b) = 0 \tag{5}$$

is said to be a **regular Sturm-Liouville problem.*** The coefficients in (4) and (5) are assumed to be real and independent of λ. In addition α_1 and β_1 are not both zero, and α_2 and β_2 are not both zero.

Since the differential equation (3) is also homogeneous, the Sturm-Liouville problem always has the trivial solution $y = 0$. However, this solution is of no interest to us. As in Examples 1 and 2, in solving such a problem we seek to find numbers λ (eigenvalues) and nontrivial solutions y that depend on λ (eigenfunctions).

Properties

Theorem 11.3 is a list of some of the more important of the many properties of the regular Sturm-Liouville problem. We shall prove only the last property.

***Charles Jacques François Sturm** (1803–1855)   Born in Switzerland, Sturm spent his adult life in Paris. His primary interests were fluid mechanics and differential equations. Sturm was the first to accurately determine the speed of sound in water. He also won a prize for an article on compressible fluids. In mathematics he won the coveted Grand Prix des Sciences Mathématiques for his work in differential equations. Sturm held the chair of mechanics at the Sorbonne and was elected a member of the French Academy of Sciences. The work done with Liouville on differential equations and boundary-value problems appeared in the *Journal des mathématiques pures et appliquées* during the period 1836 to 1838.

Joseph Liouville (1809–1882)   Besides his work in differential equations, the Frenchman Joseph Liouville is remembered for his work in number theory and complex variables. His research in number theory led to the discovery of transcendental numbers, that is, numbers that are not roots of any polynomial equation with integer coefficients. The numbers e and π are examples of transcendental numbers, although Liouville was not able to show this. In complex analysis his notable contribution was Liouville's theorem that states that every bounded entire function must be a constant. Liouville, like Sturm, was a member of the French Academy of Sciences and a professor at the Sorbonne. He was the founder and editor of the *Journal des mathématiques pures et appliquées*.

THEOREM 11.3 **Properties of the Regular Sturm-Liouville Problem**

(a) There exists an infinite number of real eigenvalues that can be arranged in increasing order $\lambda_1 < \lambda_2 < \lambda_3 < \cdots \lambda_n < \cdots$ such that $\lambda_n \to \infty$ as $n \to \infty$.

(b) For each eigenvalue there is only one eigenfunction (except for nonzero multiples).

(c) Eigenfunctions corresponding to different eigenvalues are linearly independent.

(d) The set of eigenfunctions corresponding to the set of eigenvalues is orthogonal with respect to the weight function $p(x)$ on the interval $[a, b]$.

Proof of (d)   Let y_m and y_n be eigenfunctions corresponding to eigenvalues λ_m and λ_n, respectively. Then

$$\frac{d}{dx}[r(x)y_m'] + (q(x) + \lambda_m p(x))y_m = 0 \tag{6}$$

$$\frac{d}{dx}[r(x)y_n'] + (q(x) + \lambda_n p(x))y_n = 0. \tag{7}$$

Multiplying (6) by y_n and (7) by y_m and subtracting the two equations give

$$(\lambda_m - \lambda_n)p(x)y_m y_n = y_m \frac{d}{dx}[r(x)y_n'] - y_n \frac{d}{dx}[r(x)y_m'].$$

Integrating this last result by parts from $x = a$ to $x = b$ then yields

$$(\lambda_m - \lambda_n)\int_a^b p(x)y_m y_n\, dx = r(b)[y_m(b)y_n'(b) - y_n(b)y_m'(b)]$$

$$-r(a)[y_m(a)y_n'(a) - y_n(a)y_m'(a)]. \tag{8}$$

Now the eigenfunctions y_m and y_n must both satisfy the boundary conditions (4) and (5). In particular, from (4) we have

$$\alpha_1 y_m(a) + \beta_1 y_m'(a) = 0$$

$$\alpha_1 y_n(a) + \beta_1 y_n'(a) = 0.$$

In order that this system be satisfied by α_1 and β_1, not both zero, the determinant of the coefficients must be zero:

$$y_m(a)y_n'(a) - y_n(a)y_m'(a) = 0.$$

A similar argument applied to (5) also gives

$$y_m(b)y_n'(b) - y_n(b)y_m'(b) = 0.$$

Since both members of the right side of (8) are zero, we have established the orthogonality relation

$$\int_a^b p(x)y_m y_n \, dx = 0, \qquad \lambda_m \neq \lambda_n. \tag{9}$$

□

EXAMPLE 3

Solve the boundary-value problem

$$y'' + \lambda y = 0, \qquad y(0) = 0, \quad y(1) + y'(1) = 0. \tag{10}$$

Solution (10) is a regular Sturm-Liouville problem with $r(x) = 1$, $q(x) = 0$, $p(x) = 1$, $\alpha_1 = 1$, $\beta_1 = 0$, $\alpha_2 = 1$, $\beta_2 = 1$. Now for $\lambda > 0$ the condition $y(0) = 0$ implies $c_1 = 0$ in the general solution $y = c_1 \cos \sqrt{\lambda}x + c_2 \sin \sqrt{\lambda}x$. When $y = c_2 \sin \sqrt{\lambda}x$, the boundary condition at $x = 1$ is satisfied if

$$c_2 \sin \sqrt{\lambda} + c_2\sqrt{\lambda} \cos \sqrt{\lambda} = 0.$$

Choosing $c_2 \neq 0$, we see that the last equation is equivalent to

$$\tan \sqrt{\lambda} = -\sqrt{\lambda}. \tag{11}$$

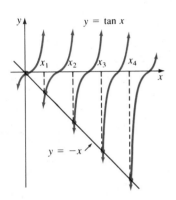

Figure 11.16

Equation (11) can be solved only by a numerical procedure such as Newton's method (see Problem 21). Nonetheless, if we let $x = \sqrt{\lambda}$, then Figure 11.16 shows the plausibility that there exists an infinite number of roots of $\tan x = -x$. The eigenvalues of the problem are then $\lambda_n = x_n^2$, where the $x_n, n = 1, 2, 3, \ldots$, are the consecutive positive roots. The corresponding eigenfunctions are $y = \sin \sqrt{\lambda_n}x$. Furthermore, $\{\sin \sqrt{\lambda_n}x\}, n = 1, 2, 3, \ldots$, is an orthogonal set with respect to the weight function $p(x) = 1$ on the interval $[0, 1]$.

Observe that even though $\lambda = 0$ satisfies (11), $\lambda = 0$ is not an eigenvalue since the solution of the problem $y'' = 0$, $y(0) = 0$, $y(1) + y'(1) = 0$ is $y = 0$. Similarly, for $\lambda < 0$ the boundary-value problem has only the trivial solution. ∎

In some circumstances we can prove the orthogonality of the solutions of (3) without the necessity of specifying a boundary condition at $x = a$ and at $x = b$.

Singular Sturm-Liouville Problem

If $r(a) = 0$, or $r(b) = 0$, or $r(a) = r(b) = 0$, the Sturm-Liouville problem consisting of the differential equation (3) and boundary conditions (4) and (5) is said to be **singular**. In the first case, $x = a$ may be a singular point of the differential equation and consequently a solution of (3) may become unbounded as $x \to a$. However, we see from (8) that if $r(a) = 0$, then no boundary condition is required at $x = a$ to prove orthogonality of the eigenfunctions provided these solutions are bounded at that point. This latter requirement guarantees the existence of the integrals involved. By assuming the solutions of (3) are bounded on the closed interval $[a, b]$, we can say that,

    (i) if $r(a) = 0$, then the orthogonality relation (9) holds with no boundary condition at $x = a$;

    (ii) if $r(b) = 0$, then the orthogonality relation (9) holds with no boundary condition at $x = b$;* and

    (iii) if $r(a) = r(b) = 0$, then the orthogonality relation (9) holds with no boundary condition at either $x = a$ or $x = b$.

We note in passing that a Sturm-Liouville problem is singular when the interval under consideration is infinite. See Problems 19 and 20.

Self-Adjoint Form

If the coefficients are continuous and $a(x) \neq 0$ for all x in some interval, then any second-order differential equation

$$a(x)y'' + b(x)y' + (c(x) + \lambda d(x))y = 0$$

can be put into the so-called **self-adjoint form** (3) by multiplying the equation by the integrating factor

$$\frac{1}{a(x)} e^{\int (b(x)/a(x)) \, dx}. \tag{12}$$

To see this, we observe that the differential equation

$$e^{\int (b/a) \, dx} y'' + \frac{b(x)}{a(x)} e^{\int (b/a) \, dx} y' + \left(\frac{c(x)}{a(x)} e^{\int (b/a) \, dx} + \lambda \frac{d(x)}{a(x)} e^{\int (b/a) \, dx} \right) y = 0$$

is the same as

$$\frac{d}{dx} [\underbrace{e^{\int (b/a) \, dx} y'}_{r(x)}] + \left(\underbrace{\frac{c(x)}{a(x)} e^{\int (b/a) \, dx}}_{q(x)} + \lambda \underbrace{\frac{d(x)}{a(x)} e^{\int (b/a) \, dx}}_{p(x)} \right) y = 0.$$

---

*Conditions (i) and (ii) are equivalent to choosing $\alpha_1 = 0$, $\beta_1 = 0$ and $\alpha_2 = 0$, $\beta_2 = 0$, respectively.

EXAMPLE 4

In Section 6.5 we saw that the general solution of the parametric Bessel equation $x^2y'' + xy' + (\lambda^2x^2 - n^2)y = 0$, $n = 0, 1, 2, \ldots$, was $y = c_1J_n(\lambda x) + c_2Y_n(\lambda x)$. From (12) we find that $1/x$ is an integrating factor. Multiplying Bessel's equation by this factor yields the self-adjoint form

$$\frac{d}{dx}[xy'] + \left(\lambda^2x - \frac{n^2}{x}\right)y = 0,$$

where we identify $r(x) = x$, $q(x) = -n^2/x$, $p(x) = x$, and λ is replaced with λ^2. Now $r(0) = 0$, and of the two solutions $J_n(\lambda x)$ and $Y_n(\lambda x)$, only $J_n(\lambda x)$ is bounded at $x = 0$. Thus in view of (i), the set $\{J_n(\lambda_i x)\}$, $i = 1, 2, 3, \ldots$, is orthogonal with respect to the weight function $p(x) = x$ on an interval $[0, b]$,

$$\int_0^b xJ_n(\lambda_i x)J_n(\lambda_j x)\, dx = 0, \qquad \lambda_i \neq \lambda_j,$$

provided the eigenvalues λ_i, $i = 1, 2, 3, \ldots$, are defined by means of a boundary condition at $x = b$ of the type

$$\alpha_2J_n(\lambda b) + \beta_2\lambda J_n'(\lambda b) = 0.* \tag{13}$$

For any choice of α_2 and β_2, not both zero, it is known that (13) has an infinite number of roots $x_i = \lambda_i b$ from which we get $\lambda_i = x_i/b$. More will be said about the eigenvalues in the next chapter. ∎

EXAMPLE 5

The Legendre polynomials $P_n(x)$ are bounded solutions of Legendre's differential equation $(1 - x^2)y'' - 2xy' + n(n + 1)y = 0$, $n = 0, 1, 2, \ldots$, on the interval $[-1, 1]$. From the self-adjoint form

$$\frac{d}{dx}[(1 - x^2)y'] + n(n + 1)y = 0,$$

we read off $r(x) = 1 - x^2$, $q(x) = 0$, $p(x) = 1$, and $\lambda = n(n + 1)$. Observing that $r(-1) = r(1) = 0$, it follows from (iii) that the set $\{P_n(x)\}$, $n = 0, 1, 2, \ldots$, is orthogonal with respect to the weight function $p(x) = 1$ on $[-1, 1]$:

$$\int_{-1}^1 P_m(x)P_n(x)\, dx = 0, \qquad m \neq n.$$

∎

---

* The extra factor of λ comes from the chain rule: $(d/dx)J_n(\lambda x) = \lambda J_n'(\lambda x)$.

11.4 EXERCISES

Answers to odd-numbered problems begin on page A–40.

In Problems 1–12 find the eigenvalues and eigenfunctions of the given boundary-value problem.

1. $y'' + \lambda y = 0$
$y(0) = 0, \quad y(\pi) = 0$

2. $y'' + \lambda y = 0$
$y(0) = 0, \quad y(\pi/4) = 0$

3. $y'' + \lambda y = 0$
$y'(0) = 0, \quad y(L) = 0$

4. $y'' + \lambda y = 0$
$y(0) = 0, \quad y'(\pi/2) = 0$

5. $y'' + \lambda y = 0$
$y'(0) = 0, \quad y'(\pi) = 0$

6. $y'' + \lambda y = 0$
$y(-\pi) = 0, \quad y(\pi) = 0$

7. $y'' + \lambda y = 0$
$y'(0) = 0, \quad y(1) + y'(1) = 0$

8. $y'' + \lambda y = 0$
$y(0) + y'(0) = 0, \quad y(1) = 0$

9. $y'' + 2y' + (\lambda + 1)y = 0$
$y(0) = 0, \quad y(5) = 0$

10. $y'' + (\lambda + 1)y = 0$
$y'(0) = 0, \quad y'(1) = 0$

11. $y'' + \lambda^2 y = 0$
$y(0) = 0, \quad y(L) = 0$

12. $y'' + \lambda^2 y = 0$
$y(0) = 0, \quad y'(3\pi) = 0$

13. When a constant load P is applied to a long slender column of uniform cross section and length L, the y-coordinate of a point on its deflection curve is given by

$$EI\frac{d^2y}{dx^2} + Py = 0,$$

where the constants E and I are, respectively, Young's modulus of elasticity and the moment of inertia of the cross section. See Figure 11.17. If the column is constrained from rotating at its top and bottom, the boundary conditions are $y(0) = 0$ and $y(L) = 0$.

(a) Show that this boundary-value problem indicates that the column does not bend at all unless the load P has one of the values $P_n = (n^2\pi^2/L^2)EI, n = 1, 2, 3, \ldots$. These different loads are called **critical loads**.

(b) Show that the deflection curve corresponding to the lowest critical load $P_1 = (\pi^2/L^2)EI$, called the **Euler load**, is $y_1(x) = c_2 \sin(\pi x/L)$. Then $y_1(x)$ is called the **first buckling mode**.

14. In (9) of Section 1.2 we saw that the differential equation for the displacement $y(x)$ of a rotating string in which the tension T is not constant is given by

$$\frac{d}{dx}\left[T(x)\frac{dy}{dx}\right] + \rho\omega^2 y = 0.$$

Suppose $1 < x < e$ and the tension is given by $T(x) = x^2$. If $y(1) = 0$, $y(e) = 0$, and $\rho\omega^2 > 0.25$, show that the **critical speeds** of angular rota-

Figure 11.17

tion are $\omega_n = \frac{1}{2}\sqrt{(4n^2\pi^2 + 1)/\rho}$ and that the corresponding deflection curves are $y_n(x) = c_2\, x^{-1/2}\sin(n\pi\ln x)$, $n = 1, 2, 3, \ldots$.

In Problems 15 and 16 find the eigenvalues and eigenfunctions of the given boundary-value problem. Put the Cauchy-Euler equation in self-adjoint form. Give an orthogonality relation.

15. $x^2y'' + xy' + \lambda y = 0$
$\quad y(1) = 0, \quad y(e^\pi) = 0$

16. $x^2y'' + 2xy' + \lambda y = 0$
$\quad y(1) = 0, \quad y(e^2) = 0$

17. Find the square norm of each eigenfunction in Problem 7.

18. Show that for the eigenfunctions in Example 3,

$$\|\sin\sqrt{\lambda_n}\,x\|^2 = \frac{1}{2}\left[1 + \cos^2\sqrt{\lambda_n}\,\right].$$

19. Laguerre's differential equation* $xy'' + (1 - x)y' + ny = 0$, $n = 0$, $1, 2, \ldots$, has polynomial solutions $L_n(x)$. Put the equation in self-adjoint form and give an orthogonality relation.

20. Hermite's differential equation† $y'' - 2xy' + 2ny = 0$, $n = 0, 1, 2, \ldots$, has polynomial solutions $H_n(x)$. Put the equation in self-adjoint form and give an orthogonality relation.

21. Find the first four eigenvalues in Example 3 to four decimal places.

22. Find the first four eigenvalues in Problem 7 to four decimal places.

23. Show that the orthogonality relation (9) holds if $r(a) = r(b)$ and the eigenfunctions satisfy the **periodic boundary conditions** $y(a) = y(b)$, $y'(a) = y'(b)$.

24. Consider $y'' + \lambda y = 0$ subject to the periodic boundary conditions $y(-L) = y(L)$, $y'(-L) = y'(L)$. Show that the eigenfunctions are
$$\left\{1, \cos\frac{\pi}{L}x, \cos\frac{2\pi}{L}x, \ldots, \sin\frac{\pi}{L}x, \sin\frac{2\pi}{L}x, \sin\frac{3\pi}{L}x, \ldots\right\}.$$ This set, which is orthogonal on $[-L, L]$, is the basis for the Fourier series.

11.5 BESSEL AND LEGENDRE SERIES

Fourier series, Fourier cosine series, and Fourier sine series are three ways of expanding a function in terms of an orthogonal set of functions. But such expansions are by no means limited to orthogonal sets of trigonometric functions. We saw in Section 11.1 that a function f defined on an interval (a, b) could be expanded at least formally in terms of any set of functions $\{\phi_n(x)\}$ that is orthogonal with respect to a weight function on $[a, b]$. Many of these

---

* Named after French mathematician **Edmond Laguerre** (1834–1886).

† Named after French mathematician **Charles Hermite** (1822–1901). Hermite polynomials are important in some aspects of quantum mechanics.

so-called generalized Fourier series come from Sturm-Liouville problems arising in physical applications of linear partial differential equations. Fourier series, generalized Fourier series, as well as the two series considered in this section, will appear again in the subsequent consideration of these applications.

11.5.1 Fourier-Bessel Series

In Example 4 of Section 11.4 it was seen that the set of Bessel functions $J_n(\lambda_i x)$ is orthogonal with respect to the weight function $p(x) = x$ on an interval $[0, b]$ when the eigenvalues are defined by means of a boundary condition

$$\alpha_2 J_n(\lambda b) + \beta_2 \lambda J_n'(\lambda b) = 0. \tag{1}$$

From (7) and (8) of Section 11.1 the generalized Fourier expansion of a function f defined on $(0, b)$ in terms of this orthogonal set is

$$f(x) = \sum_{i=1}^{\infty} c_i J_n(\lambda_i x), \tag{2}$$

where

$$c_i = \frac{\int_0^b x J_n(\lambda_i x) f(x)\, dx}{\int_0^b x J_n^2(\lambda_i x)\, dx}. \tag{3}$$

The series (2) with coefficients (3) is called a **Fourier-Bessel series**.

Differential Recurrence Relations

The **differential recurrence relations** that were discussed in Section 6.5

$$\frac{d}{dx}\left[x^n J_n(x) \right] = x^n J_{n-1}(x) \tag{4}$$

$$\frac{d}{dx}\left[x^{-n} J_n(x) \right] = -x^{-n} J_{n+1}(x) \tag{5}$$

are often useful in the evaluation of the coefficients in (3).

Square Norm

The value of the square norm $\|J_n(\lambda_i x)\|^2 = \int_0^b x J_n^2(\lambda_i x)\, dx$ depends on how the eigenvalues λ_i are defined. If $y = J_n(\lambda x)$, then we know from Example 4 of Section 11.4 that

$$\frac{d}{dx}\left[xy' \right] + \left(\lambda^2 x - \frac{n^2}{x} \right) y = 0.$$

After we multiply by $2xy'$ this equation can be written as

$$\frac{d}{dx}\left[xy' \right]^2 + (\lambda^2 x^2 - n^2)\frac{d}{dx}\left[y \right]^2 = 0.$$

Integrating the last result by parts on $[0, b]$ then gives

$$2\lambda^2 \int_0^b xy^2 \, dx = \left([xy']^2 + (\lambda^2 x^2 - n^2)y^2 \right) \Big|_0^b.$$

Since $y = J_n(\lambda x)$, the lower limit is zero for $n > 0$ since $J_n(0) = 0$. For $n = 0$ the quantity $[xy']^2 + \lambda^2 x^2 y^2$ is zero at $x = 0$. Thus

$$2\lambda^2 \int_0^b xJ_n^2(\lambda x) \, dx = \lambda^2 b^2 [J_n'(\lambda b)]^2 + (\lambda^2 b^2 - n^2)[J_n(\lambda b)]^2, \qquad \text{(6)}$$

where we have used $y' = \lambda J_n'(\lambda x)$.

We now consider three cases of (1).

CASE I   If we choose $\alpha_2 = 1$ and $\beta_2 = 0$, then (1) is

$$J_n(\lambda b) = 0. \qquad \text{(7)}$$

There are an infinite number of positive roots x_i of (7) (see Figure 6.5), and so the eigenvalues are positive and are given by $\lambda_i = x_i/b$. No new eigenvalues result from the negative roots of (7) since $J_n(-x) = (-1)^n J_n(x)$. (See Problem 38 in Exercises 6.5.) The number 0 is not an eigenvalue for any n since $J_n(0) = 0$ for $n = 1, 2, 3, \ldots$ and $J_0(0) = 1$. In other words, if $\lambda = 0$, we get the trivial function (which is never an eigenfunction) for $n = 1, 2, 3, \ldots$ and for $n = 0$, $\lambda = 0$ does not satisfy equation (7). When (5) is written in the form $xJ_n'(x) = nJ_n(x) - xJ_{n+1}(x)$ it follows from (6) and (7) that the square norm of $J_n(\lambda_i x)$ is

$$\|J_n(\lambda_i x)\|^2 = \frac{b^2}{2} J_{n+1}^2(\lambda_i b). \qquad \text{(8)}$$

$\square$

CASE II   If we choose $\alpha_2 = h \geq 0$, $\beta_2 = b$, then (1) is

$$hJ_n(\lambda b) + \lambda bJ_n'(\lambda b) = 0. \qquad \text{(9)}$$

Equation (9) has an infinite number of positive roots x_i for each positive integer $n = 1, 2, 3, \ldots$. As before, the eigenvalues are obtained from $\lambda_i = x_i/b$. $\lambda = 0$ is not an eigenvalue for $n = 1, 2, 3, \ldots$. Substituting $\lambda_i bJ_n'(\lambda_i b) = -hJ_n(\lambda_i b)$ in (6), we find that the square norm of $J_n(\lambda_i x)$ is now

$$\|J_n(\lambda_i x)\|^2 = \frac{\lambda_i^2 b^2 - n^2 + h^2}{2\lambda_i^2} J_n^2(\lambda_i b). \qquad \text{(10)}$$

$\square$

CASE III   If $h = 0$ and $n = 0$ in (9), the eigenvalues λ_i are defined from the roots of

$$J_0'(\lambda b) = 0. \qquad \text{(11)}$$

Even though (11) is just a special case of (9), it is the only situation for which $\lambda = 0$ is an eigenvalue. To see this observe that for $n = 0$, the result in (5) implies that

$$J_0'(\lambda b) = 0 \quad \text{is equivalent to} \quad J_1(\lambda b) = 0. \tag{12}$$

Since $x_1 = 0$ is a root of this equation and since $J_0(0) = 1$ is nontrivial, we conclude that $\lambda_1 = 0$ is an eigenvalue. But obviously we cannot use (10) when $h = 0$, $n = 0$, and $\lambda_1 = 0$. However, by definition of the square norm, we have

$$\|1\|^2 = \int_0^b x \, dx = \frac{b^2}{2}. \tag{13}$$

For $\lambda_i > 0$ we can use (10) with $h = 0$ and $n = 0$:

$$\|J_0(\lambda_i x)\|^2 = \frac{b^2}{2} J_0^2(\lambda_i b). \tag{14}$$

$\square$

The following summary gives the corresponding three forms of the series (2).

DEFINITION 11.7 **Fourier-Bessel Series**

The **Fourier-Bessel series** of a function f defined on the interval $(0, b)$ is given by

(i)
$$f(x) = \sum_{i=1}^{\infty} c_i J_n(\lambda_i x) \tag{15}$$

$$c_i = \frac{2}{b^2 J_{n+1}^2(\lambda_i b)} \int_0^b x J_n(\lambda_i x) f(x) \, dx, \tag{16}$$

where the λ_i are defined by $J_n(\lambda b) = 0$.

(ii)
$$f(x) = \sum_{i=1}^{\infty} c_i J_n(\lambda_i x) \tag{17}$$

$$c_i = \frac{2 \lambda_i^2}{(\lambda_i^2 b^2 - n^2 + h^2) J_n^2(\lambda_i b)} \int_0^b x J_n(\lambda_i x) f(x) \, dx, \tag{18}$$

where the λ_i are defined by $h J_n(\lambda b) + \lambda b J_n'(\lambda b) = 0$.

(iii)
$$f(x) = c_1 + \sum_{i=2}^{\infty} c_i J_0(\lambda_i x) \tag{19}$$

$$c_1 = \frac{2}{b^2} \int_0^b x f(x) \, dx$$

$$c_i = \frac{2}{b^2 J_0^2(\lambda_i b)} \int_0^b x J_0(\lambda_i x) f(x) \, dx, \tag{20}$$

where the λ_i are defined by $J_0'(\lambda b) = 0$.

Convergence of a Fourier-Legendre Series

Sufficient conditions for the convergence of a Fourier-Bessel series are not particularly restrictive.

THEOREM 11.4   Conditions for Convergence

If f and f' are piecewise continuous on the open interval $(0, b)$, then a Fourier-Bessel expansion of f converges to $f(x)$ at any point where f is continuous and to the average $[f(x-) + f(x+)]/2$ at a point where f is discontinuous.

EXAMPLE 1

Expand $f(x) = x$, $0 < x < 3$, in a Fourier-Bessel series using Bessel functions of order one that satisfy the boundary condition $J_1(3\lambda) = 0$.

Solution   We use (15) where the coefficients c_i are given by (16):

$$c_i = \frac{2}{3^2 J_2^2(3\lambda_i)} \int_0^3 x^2 J_1(\lambda_i x)\, dx.$$

To evaluate this integral we let $t = \lambda_i x$ and use (4) in the form $(d/dt)[t^2 J_2(t)] = t^2 J_1(t)$:

$$c_i = \frac{2}{9\lambda_i^3 J_2^2(3\lambda_i)} \int_0^{3\lambda_i} \frac{d}{dt}[t^2 J_2(t)]\, dt = \frac{2}{\lambda_i J_2(3\lambda_i)}.$$

Therefore the desired expansion is

$$f(x) = 2 \sum_{i=1}^{\infty} \frac{1}{\lambda_i J_2(3\lambda_i)} J_1(\lambda_i x). \qquad \blacksquare$$

EXAMPLE 2

If the eigenvalues λ_i in Example 1 are defined by $J_1(3\lambda) + \lambda J_1'(3\lambda) = 0$, then the only thing that changes in the expansion is the value of the square norm. Multiplying the boundary condition by 3 gives $3J_1(3\lambda) + 3\lambda J_1'(3\lambda) = 0$, which now matches (9) when $h = 3$, $b = 3$, and $n = 1$. Thus (18) and (17) yield, in turn,

$$c_i = \frac{18\lambda_i J_2(3\lambda_i)}{(9\lambda_i^2 + 8)J_1^2(3\lambda_i)}$$

and

$$f(x) = 18 \sum_{i=1}^{\infty} \frac{\lambda_i J_2(3\lambda_i)}{(9\lambda_i^2 + 8)J_1^2(3\lambda_i)} J_1(\lambda_i x). \qquad \blacksquare$$

11.5.2 Fourier–Legendre Series

From Example 5 of Section 11.4 we know that the set of Legendre polynomials $\{P_n(x)\}$, $n = 0, 1, 2, \ldots$, is orthogonal with respect to the weight function $p(x) = 1$ on $[-1, 1]$. Furthermore, it can be proved that

$$\|P_n(x)\|^2 = \int_{-1}^{1} P_n^2(x)\,dx = \frac{2}{2n + 1}.$$

The generalized Fourier series of Legendre polynomials is summarized as follows.

DEFINITION 11.8   **Fourier–Legendre Series**

The **Fourier–Legendre series** of a function f defined on the interval $(-1, 1)$ is given by

$$f(x) = \sum_{n=0}^{\infty} c_n P_n(x), \tag{21}$$

where

$$c_n = \frac{2n + 1}{2} \int_{-1}^{1} f(x) P_n(x)\,dx. \tag{22}$$

Convergence of a Fourier–Legendre Series

THEOREM 11.5   **Conditions for Convergence**

If f and f' are piecewise continuous on $(-1, 1)$, then the Fourier–Legendre series (21) converges to $f(x)$ at a point of continuity and to $[f(x+) + f(x-)]/2$ at a point of discontinuity.

EXAMPLE 3

Write out the first four nonzero terms in the Fourier–Legendre expansion of

$$f(x) = \begin{cases} 0, & -1 < x < 0 \\ 1, & 0 < x < 1. \end{cases}$$

Solution   The first several Legendre polynomials are listed on page 316. From these and (22) we find

$$c_0 = \frac{1}{2} \int_{-1}^{1} f(x) P_0(x)\,dx = \frac{1}{2} \int_{0}^{1} 1 \cdot 1\,dx = \frac{1}{2}$$

$$c_1 = \frac{3}{2} \int_{-1}^{1} f(x) P_1(x)\, dx = \frac{3}{2} \int_{0}^{1} 1 \cdot x\, dx = \frac{3}{4}$$

$$c_2 = \frac{5}{2} \int_{-1}^{1} f(x) P_2(x)\, dx = \frac{5}{2} \int_{0}^{1} 1 \cdot \frac{1}{2}(3x^2 - 1)\, dx = 0$$

$$c_3 = \frac{7}{2} \int_{-1}^{1} f(x) P_3(x)\, dx = \frac{7}{2} \int_{0}^{1} 1 \cdot \frac{1}{2}(5x^3 - 3x)\, dx = -\frac{7}{16}$$

$$c_4 = \frac{9}{2} \int_{-1}^{1} f(x) P_4(x)\, dx = \frac{9}{2} \int_{0}^{1} 1 \cdot \frac{1}{8}(35x^4 - 30x^2 + 3)\, dx = 0$$

$$c_5 = \frac{11}{2} \int_{-1}^{1} f(x) P_5(x)\, dx = \frac{11}{2} \int_{0}^{1} 1 \cdot \frac{1}{8}(63x^5 - 70x^3 + 15x)\, dx = \frac{11}{32}.$$

Hence $$f(x) = \frac{1}{2} P_0(x) + \frac{3}{4} P_1(x) - \frac{7}{16} P_3(x) + \frac{11}{32} P_5(x) + \cdots .$$ ∎

Alternative Form

In applications the Fourier-Legendre series appears in an alternative form. If we let $x = \cos\theta$, then $dx = -\sin\theta\, d\theta$ and (21) and (22) become

$$F(\theta) = \sum_{n=0}^{\infty} c_n P_n(\cos\theta) \tag{23}$$

$$c_n = \frac{2n + 1}{2} \int_{0}^{\pi} F(\theta) P_n(\cos\theta) \sin\theta\, d\theta, \tag{24}$$

where $f(\cos\theta)$ has been replaced by $F(\theta)$.

11.5   EXERCISES

Answers to odd-numbered problems begin on page A–41.

[11.5.1]   **1.** Find the first four eigenvalues λ_k defined by $J_1(3\lambda) = 0$. [*Hint:* See Table 6.1 on page 311.]

2. Find the first four eigenvalues λ_k defined by $J'_0(2\lambda) = 0$.

In Problems 3–6 expand $f(x) = 1$, $0 < x < 2$, in a Fourier-Bessel series using Bessel functions of order zero that satisfy the given boundary condition.

3. $J_0(2\lambda) = 0$ **4.** $J'_0(2\lambda) = 0$

5. $J_0(2\lambda) + 2\lambda J'_0(2\lambda) = 0$ **6.** $J_0(2\lambda) + \lambda J'_0(2\lambda) = 0$

In Problems 7–10 expand the given function in a Fourier-Bessel series using Bessel functions of the same order as in the indicated boundary condition.

7. $f(x) = 5x$, $0 < x < 4$
    $3J_1(4\lambda) + 4\lambda J'_1(4\lambda) = 0$

8. $f(x) = x^2$, $0 < x < 1$
    $J_2(\lambda) = 0$

9. $f(x) = x^2$, $0 < x < 3$
   $J_0'(3\lambda) = 0$ [*Hint*: $t^3 = t^2 \cdot t$.]

10. $f(x) = 1 - x^2$, $0 < x < 1$
    $J_0(\lambda) = 0$

[11.5.2] A Fourier-Legendre expansion of a polynomial function defined on the interval $(-1, 1)$ is necessarily a finite series. (Why?) In Problems 11 and 12 find the Fourier-Legendre expansion of the given function.

11. $f(x) = x^2$ 12. $f(x) = x^3$

In Problems 13 and 14 write out the first four nonzero terms in the Fourier-Legendre expansion of the given function.

13. $f(x) = \begin{cases} 0, & -1 < x < 0 \\ x, & 0 < x < 1 \end{cases}$ 14. $f(x) = e^x$, $-1 < x < 1$

15. The first three Legendre polynomials are $P_0(x) = 1$, $P_1(x) = x$, and $P_2(x) = \frac{1}{2}(3x^2 - 1)$. If $x = \cos \theta$, then $P_0(\cos \theta) = 1$ and $P_1(\cos \theta) = \cos \theta$. Show that $P_2(\cos \theta) = \frac{1}{4}(3 \cos 2\theta + 1)$.

16. Use the results given in Problem 15 to find a Fourier-Legendre expansion (23) of $F(\theta) = 1 - \cos 2\theta$.

17. A Legendre polynomial $P_n(x)$ is an even or odd function, depending on whether n is even or odd. Show that if f is an even function on $(-1, 1)$, then (21) and (22) become, respectively,

$$f(x) = \sum_{n=0}^{\infty} c_{2n} P_{2n}(x) \tag{25}$$

$$c_{2n} = (4n + 1) \int_0^1 f(x) P_{2n}(x)\, dx. \tag{26}$$

18. Show that if f is an odd function on the interval $(-1, 1)$, then (21) and (22) become, respectively,

$$f(x) = \sum_{n=0}^{\infty} c_{2n+1} P_{2n+1}(x) \tag{27}$$

$$c_{2n+1} = (4n + 3) \int_0^1 f(x) P_{2n+1}(x)\, dx. \tag{28}$$

The series (25) and (27) can also be used when f is defined on only $(0, 1)$. Both series represent f on $(0, 1)$; but on $(-1, 0)$, (25) represents its even extension, whereas (27) represents its odd extension. In Problems 19 and 20 write out the first three nonzero terms in the indicated expansion of the given function. What function does the series represent on $(-1, 1)$?

19. $f(x) = x$, $0 < x < 1$; (25)   20. $f(x) = 1$, $0 < x < 1$; (27)

Chapter 11   REVIEW

A set of functions $\{\phi_n(x)\}$, $n = 0, 1, 2, \ldots$, is said to be **orthogonal** on an interval $[a, b]$ if

$$\int_a^b \phi_m(x)\phi_n(x)\, dx \begin{cases} = 0, & m \neq n \\ \neq 0, & m = n. \end{cases}$$

A function f defined on the interval $(-p, p)$ can be expanded in terms of the trigonometric functions in the orthogonal set

$$\left\{ 1, \cos\frac{\pi}{p}x, \cos\frac{2\pi}{p}x, \ldots, \sin\frac{\pi}{p}x, \sin\frac{2\pi}{p}x, \ldots \right\}.$$

We say

$$f(x) = \frac{a_0}{2} + \sum_{n=1}^{\infty} \left(a_n \cos\frac{n\pi}{p}x + b_n \sin\frac{n\pi}{p}x \right),$$

where

$$a_0 = \frac{1}{p}\int_{-p}^{p} f(x)\, dx, \qquad a_n = \frac{1}{p}\int_{-p}^{p} f(x) \cos\frac{n\pi}{p}x\, dx,$$

$$b_n = \frac{1}{p}\int_{-p}^{p} f(x) \sin\frac{n\pi}{p}x\, dx$$

is the **Fourier series** corresponding to f. The coefficients are obtained using the concept of orthogonality. If f is an even function on the interval, then $b_n = 0$, $n = 1, 2, 3, \ldots$, and

$$a_0 = \frac{2}{p}\int_0^p f(x)\, dx \quad \text{and} \quad a_n = \frac{2}{p}\int_0^p f(x) \cos\frac{n\pi}{p}x\, dx. \tag{1}$$

Similarly, if f is odd on the interval, then $a_n = 0$, $n = 0, 1, 2, \ldots$,

and

$$b_n = \frac{2}{p}\int_0^p f(x) \sin\frac{n\pi}{p}x\, dx. \tag{2}$$

When f is defined only on an interval $(0, L)$, it can be neither even nor odd. Nonetheless we can expand f in a cosine or a sine series. By defining $p = L$, we obtain a cosine series by using the coefficients (1) and a sine series by using (2). Such series are known as **half-range expansions**.

The **Fourier-Bessel series** of a function f defined on the interval $(0, b)$ is

$$f(x) = \sum_{i=1}^{\infty} c_i J_n(\lambda_i x),$$

where

$$c_i = \frac{2}{b^2 J_{n+1}^2(\lambda_i b)} \int_0^b x J_n(\lambda_i x) f(x)\, dx, \quad \lambda_i \text{ defined by } J_n(\lambda b) = 0;$$

$$c_i = \frac{2\lambda_i^2}{(\lambda_i^2 b^2 - n^2 + h^2)J_n^2(\lambda_i b)} \int_0^b x J_n(\lambda_i x) f(x)\, dx, \quad \lambda_i \text{ defined by } h J_n(\lambda b) + \lambda b J_n'(\lambda b) = 0;$$

$$c_1 = \frac{2}{b^2} \int_0^b x f(x)\, dx, \quad c_i = \frac{2}{b^2 J_0^2(\lambda_i b)} \int_0^b x J_0(\lambda_i x) f(x)\, dx, \quad \lambda_i \text{ defined by } J_0'(\lambda b) = 0.$$

The **Fourier-Legendre series** of a function f defined on $(-1, 1)$ is

$$f(x) = \sum_{n=0}^{\infty} c_n P_n(x), \quad \text{where} \quad c_n = \frac{2n+1}{2} \int_{-1}^{1} f(x) P_n(x)\, dx.$$

Chapter 11   REVIEW EXERCISES

Answers to odd-numbered problems begin on page A–41.

Answer Problems 1–10 without referring back to the text. Fill in the blank or answer true/false.

1. The functions $f(x) = x^2 - 1$ and $f(x) = x^5$ are orthogonal on $[-\pi, \pi]$. _____

2. The product of an odd function with an odd function is _____ .

3. To expand $f(x) = |x| + 1, -\pi < x < \pi$, in an appropriate Fourier series we would use a _____ series.

4. Since $f(x) = x^2, 0 < x < 2$, is not an even function, it cannot be expanded in a Fourier cosine series. _____

5. The Fourier series of $f(x) = \begin{cases} 3, & -\pi < x < 0 \\ 0, & 0 < x < \pi \end{cases}$ will converge to _____ at $x = 0$.

6. $y = 0$ is never an eigenfunction of a Sturm-Liouville problem. _____

7. $\lambda = 0$ is never an eigenvalue of a Sturm-Liouville problem. _____

8. For $\lambda = 25$ the corresponding eigenfunction for the boundary-value problem $y'' + \lambda y = 0, y'(0) = 0, y(\pi/2) = 0$, is _____ .

9. **Chebyshev's differential equation** $(1 - x^2)y'' - xy' + n^2 y = 0$ has a polynomial solution $y = T_n(x)$ for $n = 0, 1, 2, \ldots$. The set of Chebyshev polynomials $\{T_n(x)\}$ is orthogonal with respect to the weight function _____ on the interval _____ .

10. The set $\{P_n(x)\}$ of Legendre polynomials is orthogonal with respect to the weight function $p(x) = 1$ on $[-1, 1]$ and $P_0(x) = 1$. Hence $\int_{-1}^{1} P_n(x)\, dx = $ _____ for $n > 0$.

11. Show that the set

$$\left\{ \sin \frac{\pi}{2L} x, \sin \frac{3\pi}{2L} x, \sin \frac{5\pi}{2L} x, \ldots \right\}$$

is orthogonal on the interval $0 \leq x \leq L$.

12. Find the norm of each function in Problem 11. Construct an orthonormal set.

13. Expand $f(x) = |x| - x$, $-1 < x < 1$, in a Fourier series.

14. Expand $f(x) = 2x^2 - 1$, $-1 < x < 1$, in a Fourier series.

15. Expand $f(x) = e^{-x}$, $0 < x < 1$, in a cosine series.

16. Expand the function given in Problem 15 in a sine series.

17. Find the eigenvalues and eigenfunctions of the boundary-value problem

$$x^2 y'' + xy' + 9\lambda y = 0$$

$$y'(1) = 0, \qquad y(e) = 0.$$

18. State an orthogonality relation for the eigenfunctions in Problem 17.

19. Expand $f(x) = \begin{cases} 1, & 0 < x < 2 \\ 0, & 2 < x < 4 \end{cases}$ in a Fourier-Bessel series using Bessel functions of order zero that satisfy the boundary condition $J_0(4\lambda) = 0$.

20. Expand $f(x) = x^4$, $-1 < x < 1$, in a Fourier-Legendre series.

CHAPTER 12

BOUNDARY-VALUE PROBLEMS IN RECTANGULAR COORDINATES

12.1 Separable Partial Differential Equations

12.2 Classical Equations and Boundary-Value Problems

12.3 Heat Equation

12.4 Wave Equation

12.5 Laplace's Equation

12.6 Nonhomogeneous Equations and Boundary Conditions

12.7 Use of Generalized Fourier Series

[O] **12.8** Boundary-Value Problems Involving Fourier Series in Two Variables

Chapter 12 Review

Chapter 12 Review Exercises

Essay: Where Is Middle C?

Important Concepts

Linear partial differential equation

Heat equation

Wave equation

Laplace's equation

Separation of variables

Boundary-value problem

Boundary conditions

Initial conditions

Standing waves

Dirichlet problem

Superposition principle

Double sine series

In this and the next two chapters, the emphasis will be on two procedures that are used in solving partial differential equations that occur frequently in problems involving temperature distributions, vibrations, and potentials. These problems, called boundary-value problems, are described by relatively simple linear second-order partial differential equations. The thrust of both these procedures is to find particular solutions of a partial differential equation by reducing it to one or more ordinary differential equations.

We begin with the method of separation of variables for partial differential equations. The application of this method leads back to the material covered in Chapter 11, namely, to eigenvalues, eigenfunctions, and the expansion of a function in a series of orthogonal functions.

12.1 SEPARABLE PARTIAL DIFFERENTIAL EQUATIONS

Linear Equations

The general form of a **linear second-order partial differential equation (PDE)** in two independent variables x and y is given by

$$A \frac{\partial^2 u}{\partial x^2} + B \frac{\partial^2 u}{\partial x \, \partial y} + C \frac{\partial^2 u}{\partial y^2} + D \frac{\partial u}{\partial x} + E \frac{\partial u}{\partial y} + Fu = G,$$

where A, B, C, . . . , G are functions of x and y. When $G(x, y) = 0$, the equation is said to be **homogeneous**; otherwise, it is **nonhomogeneous**.

EXAMPLE 1

The equation $\dfrac{\partial^2 u}{\partial x^2} + \dfrac{\partial^2 u}{\partial y^2} - u = 0$ is homogeneous, whereas the equation

$\dfrac{\partial^2 u}{\partial x^2} - \dfrac{\partial u}{\partial y} = x^2$ is nonhomogeneous. ∎

Solutions

A **solution** of a partial differential equation in two independent variables x and y is a function $u(x, y)$ that possesses all partial derivatives occurring in the equation and that satisfies the equation in some region of the xy-plane.

As suggested in the introduction to this chapter, it is not our intention to focus on procedures for finding general solutions of partial differential equations. Unfortunately, for most second-order linear equations, even those with constant coefficients, a general solution cannot readily be obtained. But this is not as bad as it may sound since it is usually possible, and indeed easy, to find **particular solutions** of the important linear equations that appear in many applications.

Separation of Variables

Although there are several methods that may be tried to find particular solutions (see Problems 28 and 29), we shall be interested in just one: the so-called **method of separation of variables**. When we seek a particular solution in the form of a product of a function of x and a function of y,

$$u(x, y) = X(x)Y(y),$$

it is *sometimes* possible to reduce a linear partial differential equation in two variables to two *ordinary* differential equations. To this end we note

$$\frac{\partial u}{\partial x} = X'Y, \qquad \frac{\partial u}{\partial y} = XY'$$

and

$$\frac{\partial^2 u}{\partial x^2} = X''Y, \quad \frac{\partial^2 u}{\partial y^2} = XY'',$$

where the primes denote ordinary differentiation.

EXAMPLE 2

Find product solutions of $\dfrac{\partial^2 u}{\partial x^2} = 4\dfrac{\partial u}{\partial y}$. $\hspace{2cm}$ **(1)**

Solution If $u(x, y) = X(x)Y(y)$, the equation becomes

$$X''Y = 4XY'.$$

After dividing both sides by $4XY$, we have separated the variables:

$$\frac{X''}{4X} = \frac{Y'}{Y}.$$

Since the left-hand side of the last equation is independent of y and is equal to the right-hand side, which is independent of x, we conclude that both sides of the equation are independent of x *and* y. In other words, each side of the equation must be a constant. In practice it is convenient to write this real *separation constant* as either λ^2 or $-\lambda^2$. We distinguish the following three cases.

CASE I If $\lambda^2 > 0$, the two equalities

$$\frac{X''}{4X} = \frac{Y'}{Y} = \lambda^2$$

lead to $\hspace{1cm} X'' - 4\lambda^2 X = 0 \quad\text{and}\quad Y' - \lambda^2 Y = 0.$

These latter equations have the solutions

$$X = c_1 \cosh 2\lambda x + c_2 \sinh 2\lambda x \quad\text{and}\quad Y = c_3 e^{\lambda^2 y},$$

respectively. Thus a particular solution of the partial differential equation is

$$u = XY$$

$$= (c_1 \cosh 2\lambda x + c_2 \sinh 2\lambda x)(c_3 e^{\lambda^2 y})$$

$$= A_1 e^{\lambda^2 y} \cosh 2\lambda x + B_1 e^{\lambda^2 y} \sinh 2\lambda x, \hspace{1cm} \textbf{(2)}$$

where $A_1 = c_1 c_3$ and $B_1 = c_2 c_3$.

CASE II If $-\lambda^2 < 0$, the two equalities

$$\frac{X''}{4X} = \frac{Y'}{Y} = -\lambda^2$$

give
$$X'' + 4\lambda^2 X = 0 \quad \text{and} \quad Y' + \lambda^2 Y = 0.$$

Since the solutions of these equations are
$$X = c_4 \cos 2\lambda x + c_5 \sin 2\lambda x \quad \text{and} \quad Y = c_6 e^{-\lambda^2 y},$$

respectively, another particular solution is
$$u = A_2 e^{-\lambda^2 y} \cos 2\lambda x + B_2 e^{-\lambda^2 y} \sin 2\lambda x, \tag{3}$$

where $A_2 = c_4 c_6$ and $B_2 = c_5 c_6$.

CASE III If $\lambda^2 = 0$, it follows that
$$X'' = 0 \quad \text{and} \quad Y' = 0.$$

In this case
$$X = c_7 x + c_8 \quad \text{and} \quad Y = c_9$$

so that
$$u = A_3 x + B_3, \tag{4}$$

where $A_3 = c_7 c_9$ and $B_3 = c_8 c_9$. ∎

It is left as an exercise to verify that (2), (3), and (4) satisfy the given equation. See Problem 30.

Separation of variables is not a general method for finding particular solutions; some linear partial differential equations are simply not separable. You should verify that the assumption $u = XY$ does not lead to a solution for $\partial^2 u / \partial x^2 - \partial u / \partial y = x$.

Superposition Principle

The following theorem is analogous to Theorem 4.3 and is known as the **superposition principle**.

THEOREM 12.1   Superposition Principle

If $u_1, u_2, \ldots, u_k$ are solutions of a homogeneous linear partial differential equation, then the linear combination
$$u = c_1 u_1 + c_2 u_2 + \cdots + c_k u_k,$$

where the c_i, $i = 1, 2, \ldots, k$, are constants, is also a solution.

Throughout the remainder of the chapter, we shall assume that, whenever we have an infinite set
$$u_1, u_2, u_3, \ldots$$

of solutions of a homogeneous linear equation, we can construct yet another solution u by forming the infinite series

$$u = \sum_{k=1}^{\infty} u_k.$$

Classification of Equations

A linear second-order partial differential equation in two independent variables with constant coefficients can be classified as one of three types. This classification depends only on the coefficients of the second-order derivatives. Of course, we assume that at least one of the coefficients A, B, and C is not zero.

DEFINITION 12.1   Classification of Equations

The linear second-order partial differential equation

$$A\frac{\partial^2 u}{\partial x^2} + B\frac{\partial^2 u}{\partial x\,\partial y} + C\frac{\partial^2 u}{\partial y^2} + D\frac{\partial u}{\partial x} + E\frac{\partial u}{\partial y} + Fu = 0,$$

where A, B, C, D, E, and F are real constants, is said to be

hyperbolic if	$B^2 - 4AC > 0,$
parabolic if	$B^2 - 4AC = 0,$
elliptic if	$B^2 - 4AC < 0.$

EXAMPLE 3

Classify the equations.

(a) $3\dfrac{\partial^2 u}{\partial x^2} = \dfrac{\partial u}{\partial y}$ **(b)** $\dfrac{\partial^2 u}{\partial x^2} = \dfrac{\partial^2 u}{\partial y^2}$ **(c)** $\dfrac{\partial^2 u}{\partial x^2} + \dfrac{\partial^2 u}{\partial y^2} = 0$

Solution   **(a)** By rewriting the given equation as

$$3\frac{\partial^2 u}{\partial x^2} - \frac{\partial u}{\partial y} = 0,$$

we can make the identifications $A = 3$, $B = 0$, and $C = 0$. Since $B^2 - 4AC = 0$, the equation is parabolic.

(b) By rewriting the equation as

$$\frac{\partial^2 u}{\partial x^2} - \frac{\partial^2 u}{\partial y^2} = 0,$$

we see that $A = 1$, $B = 0$, $C = -1$, and $B^2 - 4AC = -4(1)(-1) > 0$. The equation is hyperbolic.

(c) With $A = 1$, $B = 0$, $C = 1$, and $B^2 - 4AC = -4(1)(1) < 0$. The equation is elliptic. ∎

A detailed explanation of why we would want to classify a second-order partial differential equation is beyond the scope of this text. But the answer lies in the fact that we wish to solve partial differential equations subject to certain side conditions known as boundary and initial conditions. The kinds of side conditions appropriate for a given equation depend on whether the equation is hyperbolic, parabolic, or elliptic.

12.1 EXERCISES

Answers to odd-numbered problems begin on page A–41.

In Problems 1–16 use separation of variables to find, if possible, product solutions for the given partial differential equation.

1. $\dfrac{\partial u}{\partial x} = \dfrac{\partial u}{\partial y}$

2. $\dfrac{\partial u}{\partial x} + 3\dfrac{\partial u}{\partial y} = 0$

3. $u_x + u_y = u$

4. $u_x = u_y + u$

5. $x\dfrac{\partial u}{\partial x} = y\dfrac{\partial u}{\partial y}$

6. $y\dfrac{\partial u}{\partial x} + x\dfrac{\partial u}{\partial y} = 0$

7. $\dfrac{\partial^2 u}{\partial x^2} + \dfrac{\partial^2 u}{\partial x\, \partial y} + \dfrac{\partial^2 u}{\partial y^2} = 0$

8. $y\dfrac{\partial^2 u}{\partial x\, \partial y} + u = 0$

9. $k\dfrac{\partial^2 u}{\partial x^2} - u = \dfrac{\partial u}{\partial t}$, $k > 0$

10. $k\dfrac{\partial^2 u}{\partial x^2} = \dfrac{\partial u}{\partial t}$, $k > 0$

11. $a^2\dfrac{\partial^2 u}{\partial x^2} = \dfrac{\partial^2 u}{\partial t^2}$

12. $a^2\dfrac{\partial^2 u}{\partial x^2} = \dfrac{\partial^2 u}{\partial t^2} + 2k\dfrac{\partial u}{\partial t}$, $k > 0$

13. $\dfrac{\partial^2 u}{\partial x^2} + \dfrac{\partial^2 u}{\partial y^2} = 0$

14. $x^2\dfrac{\partial^2 u}{\partial x^2} + \dfrac{\partial^2 u}{\partial y^2} = 0$

15. $u_{xx} + u_{yy} = u$

16. $a^2 u_{xx} - g = u_{tt}$, g a constant

In Problems 17–26 classify the given partial differential equation as hyperbolic, parabolic, or elliptic.

17. $\dfrac{\partial^2 u}{\partial x^2} + \dfrac{\partial^2 u}{\partial x\, \partial y} + \dfrac{\partial^2 u}{\partial y^2} = 0$

18. $3\dfrac{\partial^2 u}{\partial x^2} + 5\dfrac{\partial^2 u}{\partial x\, \partial y} + \dfrac{\partial^2 u}{\partial y^2} = 0$

19. $\dfrac{\partial^2 u}{\partial x^2} + 6\dfrac{\partial^2 u}{\partial x\, \partial y} + 9\dfrac{\partial^2 u}{\partial y^2} = 0$

20. $\dfrac{\partial^2 u}{\partial x^2} - \dfrac{\partial^2 u}{\partial x\, \partial y} - 3\dfrac{\partial^2 u}{\partial y^2} = 0$

21. $\dfrac{\partial^2 u}{\partial x^2} = 9\dfrac{\partial^2 u}{\partial x\, \partial y}$

22. $\dfrac{\partial^2 u}{\partial x\, \partial y} - \dfrac{\partial^2 u}{\partial y^2} + 2\dfrac{\partial u}{\partial x} = 0$

23. $\dfrac{\partial^2 u}{\partial x^2} + 2\dfrac{\partial^2 u}{\partial x\,\partial y} + \dfrac{\partial^2 u}{\partial y^2} + \dfrac{\partial u}{\partial x} - 6\dfrac{\partial u}{\partial y} = 0$

24. $\dfrac{\partial^2 u}{\partial x^2} + \dfrac{\partial^2 u}{\partial y^2} = u$

25. $a^2 \dfrac{\partial^2 u}{\partial x^2} = \dfrac{\partial^2 u}{\partial t^2}$ **26.** $k\dfrac{\partial^2 u}{\partial x^2} = \dfrac{\partial u}{\partial t}, \quad k > 0$

27. Show that the equation

$$k\left(\frac{\partial^2 u}{\partial r^2} + \frac{1}{r}\frac{\partial u}{\partial r}\right) = \frac{\partial u}{\partial t}$$

possesses the product solution

$$u = e^{-k\lambda^2 t}(AJ_0(\lambda r) + BY_0(\lambda r)).$$

28. (a) Show that the equation

$$a^2 \frac{\partial^2 u}{\partial x^2} = \frac{\partial^2 u}{\partial t^2}$$

can be put into the form $\partial^2 u/\partial \eta\,\partial\xi = 0$ by means of the substitutions $\xi = x + at$, $\eta = x - at$.

(b) Show that a solution of the equation is

$$u = F(x + at) + G(x - at),$$

where F and G are arbitrary, twice-differentiable functions.

29. Find solutions of the form $u = e^{mx + ny}$ for $\dfrac{\partial^2 u}{\partial x^2} + \dfrac{\partial^2 u}{\partial x\,\partial y} - 6\dfrac{\partial^2 u}{\partial y^2} = 0.$

30. Verify that the products (2), (3), and (4) satisfy equation (1).

31. Definition 12.1 generalizes to linear equations with coefficients that are functions of x and y. Determine the regions in the xy-plane for which the equation

$$(xy + 1)\frac{\partial^2 u}{\partial x^2} + (x + 2y)\frac{\partial^2 u}{\partial x\,\partial y} + \frac{\partial^2 u}{\partial y^2} + xy^2 u = 0$$

is hyperbolic, parabolic, or elliptic.

12.2 CLASSICAL EQUATIONS AND BOUNDARY-VALUE PROBLEMS

For the remainder of this and the next chapter, we shall be concerned principally with finding product solutions of the partial differential equations

$$k\frac{\partial^2 u}{\partial x^2} = \frac{\partial u}{\partial t}, \qquad k > 0 \tag{1}$$

$$a^2 \frac{\partial^2 u}{\partial x^2} = \frac{\partial^2 u}{\partial t^2} \tag{2}$$

$$\frac{\partial^2 u}{\partial x^2} + \frac{\partial^2 u}{\partial y^2} = 0 \tag{3}$$

or slight variations of these equations. These classical equations of mathematical physics are known, respectively, as the **one-dimensional heat equation**, the **one-dimensional wave equation**, and **Laplace's equation in two dimensions**. "One-dimensional" refers to the fact that x denotes a spatial dimension, whereas t represents time. Laplace's equation is abbreviated as $\nabla^2 u = 0$, where

$$\nabla^2 u = \frac{\partial^2 u}{\partial x^2} + \frac{\partial^2 u}{\partial y^2}$$

is called the **two-dimensional Laplacian** of the function u. In three dimensions the **Laplacian** of u is

$$\nabla^2 u = \frac{\partial^2 u}{\partial x^2} + \frac{\partial^2 u}{\partial y^2} + \frac{\partial^2 u}{\partial z^2}.$$

Note that the heat equation (1) is parabolic, the wave equation (2) is hyperbolic, and Laplace's equation (3) is elliptic.

Heat Equation

Equation (1) occurs in the theory of heat flow, that is, heat transferred by conduction in a rod or thin wire. The function $u(x, t)$ is temperature. Problems in mechanical vibrations often lead to the wave equation (2). For purposes of discussion, a solution $u(x, t)$ of (2) will represent the displacement of an idealized string. Last, a solution $u(x, y)$ of Laplace's equation (3) can be interpreted as the steady-state (that is, time-independent) temperature distribution throughout a thin, two-dimensional plate.

Even though we have to make many simplifying assumptions, it is worthwhile to see how equations such as (1) and (2) arise.

Suppose a thin circular rod of length L has a cross-sectional area A and coincides with the x-axis on the interval $[0, L]$. See Figure 12.1. Let us suppose:

cross section of area A

Figure 12.1

- The flow of heat within the rod takes place only in the x-direction.
- The lateral, or curved, surface of the rod is insulated; that is, no heat escapes from this surface.
- No heat is being generated within the rod.
- The rod is homogeneous; that is, its mass per unit volume ρ is a constant.
- The specific heat γ and thermal conductivity K of the material of the rod are constants.

To derive the partial differential equation satisfied by the temperature $u(x, t)$ we need two empirical laws of heat conduction:

(i) *The quantity of heat Q in an element of mass m is*

$$Q = \gamma m u, \tag{4}$$

where u is the temperature of the element.

(ii) *The rate of heat flow Q_t through the cross section indicated in Figure 12.1 is proportional to the area A of the cross section and the partial derivative with respect to x of the temperature:*

$$Q_t = -KAu_x. \tag{5}$$

Since heat flows in the direction of decreasing temperature, the minus sign in (5) is used to ensure that Q_t is positive for $u_x < 0$ (heat flow to the right) and negative for $u_x > 0$ (heat flow to the left). If the circular slice of the rod shown in Figure 12.1 between x and $x + \Delta x$ is very thin, then $u(x, t)$ can be taken as the approximate temperature at each point in the interval. Now the mass of the slice is $m = \rho(A \, \Delta x)$, and so it follows from (4) that the quantity of heat in it is

$$Q = \gamma \rho A \, \Delta x \, u. \tag{6}$$

Furthermore, when heat flows in the positive x-direction, we see from (5) that heat builds up in the slice at the net rate

$$-KAu_x(x, t) - (-KAu_x(x + \Delta x, t)) = KA[u_x(x + \Delta x, t) - u_x(x, t)]. \tag{7}$$

By differentiating (6) with respect to t, we see that this net rate is also given by

$$Q_t = \gamma \rho A \, \Delta x \, u_t. \tag{8}$$

Equating (7) and (8) gives

$$\frac{K}{\gamma \rho} \frac{u_x(x + \Delta x, t) - u_x(x, t)}{\Delta x} = u_t. \tag{9}$$

Taking the limit of (9) as $\Delta x \to 0$ finally yields (1) in the form*

$$\frac{K}{\gamma \rho} u_{xx} = u_t.$$

It is customary to let $k = K/\gamma \rho$ and call this positive constant the **thermal diffusivity**.

Wave Equation

Consider a string of length L, such as a guitar string, stretched taut between two points on the x-axis, say, $x = 0$ and $x = L$. When the string starts to

---

* Recall from calculus that $u_{xx} = \lim\limits_{\Delta x \to 0} \dfrac{u_x(x + \Delta x, t) - u_x(x, t)}{\Delta x}$.

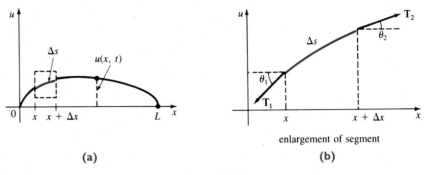

enlargement of segment

(a)   **(b)**

Figure 12.2

vibrate, assume that the motion takes place in the xy-plane in such a manner that each point on the string moves in a direction perpendicular to the x-axis (transverse vibrations). As shown in Figure 12.2(a), let $u(x, t)$ denote the vertical displacement of any point on the string measured from the x-axis for $t > 0$. We further assume that:

- The string is perfectly flexible.
- The string is homogeneous; that is, its mass per unit length ρ is a constant.
- The displacements u are small compared to the length of the string.
- The slope of the curve is small at all points.
- The tension $\mathbf{T}$ acts tangent to the string and its magnitude T is the same at all points.
- The tension is large compared with the force of gravity.
- No other external forces act on the string.

Now in Figure 12.2(b) the tensions $\mathbf{T}_1$ and $\mathbf{T}_2$ are tangent to the ends of the curve on the interval $[x, x + \Delta x]$. For small θ_1 and θ_2 the net vertical force acting on the corresponding element Δs of the string is then

$$T \sin \theta_2 - T \sin \theta_1 \approx T \tan \theta_2 - T \tan \theta_1$$

$$= T[u_x(x + \Delta x, t) - u_x(x, t)],^*$$

where $T = |\mathbf{T}_1| = |\mathbf{T}_2|$. Now $\rho\, \Delta s \approx \rho\, \Delta x$ is the mass of the string on $[x, x + \Delta x]$ and so Newton's second law gives

$$T[u_x(x + \Delta x, t) - u_x(x, t)] = \rho\, \Delta x\, u_{tt}$$

or

$$\frac{u_x(x + \Delta x, t) - u_x(x, t)}{\Delta x} = \frac{\rho}{T}\, u_{tt}.$$

---

* $\tan \theta_2 = u_x(x + \Delta x, t)$ and $\tan \theta_1 = u_x(x, t)$ are equivalent expressions for slope.

temperature as a function of position on the hot plate

thermometer

(x, y)

Figure 12.3

By taking the limit as $\Delta x \to 0$, the last equation becomes $u_{xx} = (\rho/T)u_{tt}$. This of course is (2) with $a^2 = T/\rho$.

Laplace's Equation

Although we shall not present its derivation, Laplace's equation in two and three dimensions occurs in time-independent problems involving potentials such as electrostatic, gravitational, and velocity in fluid mechanics. Moreover, a solution of Laplace's equation can also be interpreted as a steady-state temperature distribution. As illustrated in Figure 12.3, a solution $u(x, y)$ of (3) could represent the temperature that varies from point to point, but not with time, of a rectangular plate.

We often wish to find solutions of equations (1), (2), and (3) that satisfy certain side conditions.

Initial Conditions

Since solutions of (1) and (2) depend on time t, we can prescribe what happens at $t = 0$; that is, we can give **initial conditions (IC)**. If $f(x)$ denotes the initial temperature distribution throughout the rod in Figure 12.1, then a solution $u(x, t)$ of (1) must satisfy the single initial condition $u(x, 0) = f(x), 0 < x < L$. On the other hand, for a vibrating string, we can specify its initial displacement (or shape) $f(x)$ as well as its initial velocity $g(x)$. In mathematical terms we seek a function $u(x, t)$ satisfying (2) and the two initial conditions:

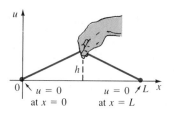

$u = 0$ at $x = 0$ $u = 0$ at $x = L$

Figure 12.4

$$u(x, 0) = f(x), \qquad \frac{\partial u}{\partial t}\bigg|_{t=0} = g(x), \qquad 0 < x < L. \tag{10}$$

For example, the string could be plucked, as shown in Figure 12.4, and released from rest ($g(x) = 0$).

Boundary Conditions

The string in Figure 12.4 is secured to the x-axis at $x = 0$ and $x = L$ for all time. We interpret this by the two **boundary conditions (BC)**:

$$u(0, t) = 0, \qquad u(L, t) = 0, \qquad t > 0.$$

Note that in this context the function f in (10) is continuous and, consequently, $f(0) = 0$ and $f(L) = 0$. In general, there are three types of boundary conditions associated with equations (1), (2), and (3). On a boundary we can specify the values of *one* of the following:

$$\textbf{(i)} \ u, \quad \textbf{(ii)} \ \frac{\partial u}{\partial n}, \quad \text{or} \quad \textbf{(iii)} \ \frac{\partial u}{\partial n} + hu, \qquad h \text{ a constant.}$$

Here $\partial u/\partial n$ denotes the normal derivative of u (the directional derivative of u in the direction perpendicular to the boundary). A boundary condition of the first type (i) is called a **Dirichlet condition**; the second type (ii) is called a **Neumann condition**; and a boundary condition of the third type (iii) is known as a **Robin condition**.* For example, for $t > 0$, a typical condition at the right-hand end of the rod in Figure 12.1 can be

(i)' $u(L, t) = u_0,$ u_0 a constant,

(ii)' $\dfrac{\partial u}{\partial x}\bigg|_{x=L} = 0,$ or

(iii)' $\dfrac{\partial u}{\partial x}\bigg|_{x=L} = -h(u(L, t) - u_m),$ $h > 0$ and u_m constants.

Condition (i)' simply states that the boundary $x = L$ is held by some means at a constant *temperature* u_0 for all time $t > 0$. Condition (ii)' indicates that the boundary $x = L$ is *insulated*. From the empirical law of heat transfer, the flux of heat across a boundary (that is, the amount of heat per unit area per unit time conducted across the boundary) is proportional to the value of the normal derivative $\partial u / \partial n$ of the temperature u. Thus when the boundary $x = L$ is thermally insulated, no heat flows into or out of the rod and so

$$\frac{\partial u}{\partial x}\bigg|_{x=L} = 0.$$

We can interpret (iii)' to mean that *heat is lost* from the right-hand end of the rod by being in contact with a medium, such as air or water, that is held at a constant temperature. From Newton's law of cooling, the outward flux of heat from the rod is proportional to the difference between the temperature $u(L, t)$ at the boundary and the temperature u_m of the surrounding medium. We note that if heat is lost from the left-hand end of the rod, the boundary condition is

$$\frac{\partial u}{\partial x}\bigg|_{x=0} = h(u(0, t) - u_m).$$

The change in algebraic sign is consistent with the assumption that the rod is at a higher temperature than the medium surrounding the ends so that $u(0, t) > u_m$ and $u(L, t) > u_m$. At $x = 0$ and $x = L$, the slopes $u_x(0, t)$ and $u_x(L, t)$ must be positive and negative, respectively.

Of course, at the ends of the rod we can specify different conditions at the same time. For example, we could have

$$\frac{\partial u}{\partial x}\bigg|_{x=0} = 0 \quad \text{and} \quad u(L, t) = u_0, \qquad t > 0.$$

---

* Named after the French mathematician **Victor G. Robin** (1855–1897) and the German mathematicians **Peter G. L. Dirichlet** (1805–1859) and **Carl G. Neumann** (1832–1925).

Boundary-Value Problems

Problems such as

$$\text{Solve:} \quad a^2 \frac{\partial^2 u}{\partial x^2} = \frac{\partial^2 u}{\partial t^2}, \qquad 0 < x < L, \quad t > 0$$

subject to:
(BC) $\quad u(0, t) = 0, \qquad\qquad u(L, t) = 0, \qquad\qquad t > 0$ $\qquad$ **(11)**

(IC) $\quad u(x, 0) = f(x), \qquad \left.\frac{\partial u}{\partial t}\right|_{t=0} = g(x), \qquad 0 < x < L$

and

$$\text{Solve:} \quad \frac{\partial^2 u}{\partial x^2} + \frac{\partial^2 u}{\partial y^2} = 0, \qquad 0 < x < a, \quad 0 < y < b$$

subject to:

(BC) $\begin{cases} \left.\dfrac{\partial u}{\partial x}\right|_{x=0} = 0, \quad \left.\dfrac{\partial u}{\partial x}\right|_{x=a} = 0, \qquad 0 < y < b \\[2mm] u(x, 0) = 0, \qquad u(x, b) = f(x), \qquad 0 < x < a \end{cases}$ $\qquad$ **(12)**

are called **boundary-value problems**.

Modifications

The partial differential equations (1), (2), and (3) must be modified to take into consideration internal or external influences acting on the physical system. More general forms of the one-dimensional heat and wave equations are, respectively,

$$k \frac{\partial^2 u}{\partial x^2} + G(x, t, u) = \frac{\partial u}{\partial t} \qquad\qquad \textbf{(13)}$$

and

$$a^2 \frac{\partial^2 u}{\partial x^2} + F(x, t, u, u_t) = \frac{\partial^2 u}{\partial t^2}. \qquad\qquad \textbf{(14)}$$

For example, if there is heat transfer from the lateral surface of a rod into a surrounding medium that is held at a constant temperature u_m, then the heat equation (13) is

$$k \frac{\partial^2 u}{\partial x^2} - h(u - u_m) = \frac{\partial u}{\partial t}.$$

In (14) the function F could represent the various forces acting on the string. For example, when external, damping, and elastic restoring forces are taken into account, (14) assumes the form

$$a^2 \frac{\partial^2 u}{\partial x^2} + f(x, t) = \frac{\partial^2 u}{\partial t^2} + c\,\frac{\partial u}{\partial t} + ku. \tag{15}$$

external force   damping force   restoring force

Remark   The analysis of a wide variety of diverse phenomena yields equation (1), (2), or (3), or their generalizations involving a greater number of spatial variables. For example, (1) is sometimes called the **diffusion equation** since the diffusion of dissolved substances in solution is analogous to the flow of heat in a solid. The function $u(x, t)$ satisfying the partial differential equation in this case represents the concentration of the dissolved substance. Similarly, equation (1) arises in the study of the flow of electricity in a long cable or a transmission line. In this setting, (1) is known as a **telegraph equation**. It can be shown that under certain assumptions the current and the voltage in the line are functions satisfying two equations identical in form with equation (1). The wave equation (2) also appears in the theory of high-frequency transmission lines, fluid mechanics, acoustics, and elasticity. Laplace's equation (3) is encountered in engineering problems in static displacements of membranes.

12.2 EXERCISES

Answers to odd-numbered problems begin on page A–42.

In Problems 1–4 a rod of length L coincides with the interval $[0, L]$ on the x-axis. Set up the boundary-value problem for the temperature $u(x, t)$.

1. The left end is held at temperature zero and the right end is insulated. The initial temperature is $f(x)$ throughout.

2. The left end is held at temperature u_0 and the right end is held at temperature u_1. The initial temperature is zero throughout.

3. The left end is held at temperature 100 and there is heat transfer from the right end into the surrounding medium at temperature zero. The initial temperature is $f(x)$ throughout.

4. The ends are insulated and there is heat transfer from the lateral surface into the surrounding medium at temperature 50. The initial temperature is 100 throughout.

In Problems 5–8 a string of length L coincides with the interval $[0, L]$ on the x-axis. Set up the boundary-value problem for the displacement $u(x, t)$.

5. The ends are secured to the x-axis. The string is released from rest from the initial displacement $x(L - x)$.

6. The ends are secured to the x-axis. Initially the string is undisplaced but has the initial velocity $\sin(\pi x/L)$.

7. The left end is secured to the x-axis but the right end moves in a transverse manner according to $\sin \pi t$. The string is released from rest from the initial displacement $f(x)$. For $t > 0$ the transverse vibrations are damped with a force proportional to the instantaneous velocity.

8. The ends are secured to the x-axis and the string is initially at rest on that axis. An external vertical force proportional to the horizontal distance from the left end acts on the string for $t > 0$.

In Problems 9 and 10 set up the boundary-value problem for the steady-state temperature $u(x, y)$.

9. A thin rectangular plate coincides with the region defined by $0 \leq x \leq 4$, $0 \leq y \leq 2$. The left end and the bottom of the plate are insulated. The top of the plate is held at temperature zero and the right end of the plate is held at temperature $f(y)$.

10. A semi-infinite plate coincides with the region defined by $0 \leq x \leq \pi$, $y \geq 0$. The left end is held at temperature e^{-y} and the right end is held at temperature 100 for $0 < y \leq 1$ and temperature zero for $y > 1$. The bottom of the plate is held at temperature $f(x)$.

12.3 HEAT EQUATION

Figure 12.5

Consider a thin rod of length L with an initial temperature $f(x)$ throughout and whose ends are held at temperature zero for all time $t > 0$. If the rod shown in Figure 12.5 satisfies the assumptions given on page 650, the temperature $u(x, t)$ in the rod is determined from the boundary-value problem:

$$k \frac{\partial^2 u}{\partial x^2} = \frac{\partial u}{\partial t}, \qquad 0 < x < L, \quad t > 0 \tag{1}$$

$$u(0, t) = 0, \qquad u(L, t) = 0, \qquad t > 0 \tag{2}$$

$$u(x, 0) = f(x), \qquad 0 < x < L. \tag{3}$$

Solution

Using the product $u = X(x)T(t)$ and $-\lambda^2$ as a separation constant leads to

$$\frac{X''}{X} = \frac{T'}{kT} = -\lambda^2 \tag{4}$$

and

$$X'' + \lambda^2 X = 0 \tag{5}$$

$$T' + k\lambda^2 T = 0$$

$$X = c_1 \cos \lambda x + c_2 \sin \lambda x \tag{6}$$

$$T = c_3 e^{-k\lambda^2 t}. \tag{7}$$

Now since
$$u(0, t) = X(0)T(t) = 0$$

$$u(L, t) = X(L)T(t) = 0,$$

we must have $X(0) = 0$ and $X(L) = 0$. These boundary conditions together with equation (5) constitute a Sturm-Liouville problem. Applying the first of these conditions in (6) immediately gives $c_1 = 0$. Therefore

$$X = c_2 \sin \lambda x.$$

The second boundary condition now implies

$$X(L) = c_2 \sin \lambda L = 0.$$

If $c_2 = 0$, then $X = 0$ so that $u = 0$. To obtain a nontrivial solution u, we must have $c_2 \neq 0$, and so the last equation is satisfied when

$$\sin \lambda L = 0.$$

This implies that $\lambda L = n\pi$ or $\lambda = n\pi/L$, where $n = 1, 2, 3, \ldots$. The values

$$\lambda = \frac{n\pi}{L}, \qquad n = 1, 2, 3, \ldots, \tag{8}$$

and the corresponding solutions

$$X = c_2 \sin \frac{n\pi}{L} x, \qquad n = 1, 2, 3, \ldots, \tag{9}$$

are the **eigenvalues** and **eigenfunctions**, respectively, of the problem.
From (7) we have $T = c_3 e^{-k(n^2\pi^2/L^2)t}$ and so

$$u_n = XT = A_n e^{-k(n^2\pi^2/L^2)t} \sin \frac{n\pi}{L} x, \tag{10}$$

where we have replaced the constant $c_2 c_3$ by A_n. The products $u_n(x, t)$ satisfy the partial differential equation (1) and the boundary conditions (2) for each value of the positive integer n.[*] However, in order that the functions given in (10) satisfy the initial condition (3), we would have to choose the coefficient A_n in such a manner that

$$u_n(x, 0) = f(x) = A_n \sin \frac{n\pi}{L} x. \tag{11}$$

In general, we would not expect condition (11) to be satisfied for an arbitrary, but reasonable, choice of f. Therefore we are forced to admit that $u_n(x, t)$ is *not a solution of the given problem*. Now by the superposition principle the

---

[*] You are urged to verify that if the separation constant is chosen as either $\lambda^2 = 0$ or $\lambda^2 > 0$, then the only solution of (1) satisfying (2) is $u = 0$.

function

$$u(x, t) = \sum_{n=1}^{\infty} u_n = \sum_{n=1}^{\infty} A_n e^{-k(n^2\pi^2/L^2)t} \sin \frac{n\pi}{L} x \qquad \textbf{(12)}$$

must also, though formally, satisfy (1) and (2). Substituting $t = 0$ in (12) implies

$$u(x, 0) = f(x) = \sum_{n=1}^{\infty} A_n \sin \frac{n\pi}{L} x.$$

This last expression is recognized as the half-range expansion of f in a sine series. Thus if we make the identification $A_n = b_n$, $n = 1, 2, 3, \ldots$, it follows from (5) of Section 11.3 that

$$A_n = \frac{2}{L} \int_0^L f(x) \sin \frac{n\pi}{L} x \, dx.$$

We conclude that a solution of the boundary-value problem described in (1), (2), and (3) is given by the infinite series

$$u(x, t) = \frac{2}{L} \sum_{n=1}^{\infty} \left(\int_0^L f(x) \sin \frac{n\pi}{L} x \, dx \right) e^{-k(n^2\pi^2/L^2)t} \sin \frac{n\pi}{L} x.$$

In the special case when $u(x, 0) = 100$, $L = \pi$, and $k = 1$, you should verify that the coefficients A_n are given by

$$A_n = \frac{200}{\pi} \left[\frac{1 - (-1)^n}{n} \right]$$

and so
$$u(x, t) = \frac{200}{\pi} \sum_{n=1}^{\infty} \left[\frac{1 - (-1)^n}{n} \right] e^{-n^2 t} \sin nx. \qquad \textbf{(13)}$$

With the aid of a computer algebra system we show in Figure 12.6 the graphs of the solution $u(x, t)$ on the interval $[0, \pi]$ for various values of time t. The graphs verify that which is apparent in (13), namely, $u(x, t) \to 0$ as $t \to \infty$.

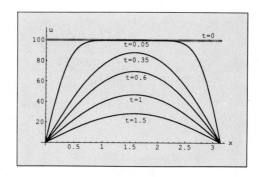

Figure 12.6

12.3 **EXERCISES**

Answers to odd-numbered problems begin on page A-42.

In Problems 1 and 2 solve the heat equation (1) subject to the given conditions. Assume a rod of length L.

1. $u(0, t) = 0,$ $u(L, t) = 0$

$$u(x, 0) = \begin{cases} 1, & 0 < x < L/2 \\ 0, & L/2 < x < L \end{cases}$$

2. $u(0, t) = 0,$ $u(L, t) = 0$

$u(x, 0) = x(L - x)$

3. Find the temperature $u(x, t)$ in a rod of length L if the initial temperature is $f(x)$ throughout and if the ends $x = 0$ and $x = L$ are insulated.

4. Solve Problem 3 if $L = 2$ and $f(x) = \begin{cases} x, & 0 < x < 1 \\ 0, & 1 < x < 2. \end{cases}$

5. Suppose heat is lost from the lateral surface of a thin rod of length L into a surrounding medium at temperature zero. If the linear law of heat transfer applies, then the heat equation takes on the form $k\partial^2 u/\partial x^2 - hu = \partial u/\partial t$, $0 < x < L$, $t > 0$, h a constant. Find the temperature $u(x, t)$ if the initial temperature is $f(x)$ throughout and the ends $x = 0$ and $x = L$ are insulated. See Figure 12.7.

6. Solve Problem 5 if the ends $x = 0$ and $x = L$ are held at temperature zero.

insulated $0°$ insulated

0 $0°$ L x

heat transfer from
lateral surface of
the rod

Figure 12.7

12.4 **WAVE EQUATION**

We are now in a position to solve the boundary-value problem (11) of Section 12.2. The vertical displacement of $u(x, t)$ of the vibrating string of length L shown in Figure 12.2(a) is determined from

$$a^2 \frac{\partial^2 u}{\partial x^2} = \frac{\partial^2 u}{\partial t^2}, \qquad 0 < x < L, \quad t > 0 \tag{1}$$

$$u(0, t) = 0, \qquad u(L, t) = 0, \qquad t > 0 \tag{2}$$

$$u(x, 0) = f(x), \qquad \frac{\partial u}{\partial t}\bigg|_{t=0} = g(x), \qquad 0 < x < L. \tag{3}$$

Solution

Separating variables in (1) gives

$$\frac{X''}{X} = \frac{T''}{a^2 T} = -\lambda^2$$

so that

$$X'' + \lambda^2 X = 0$$

$$T'' + \lambda^2 a^2 T = 0$$

and therefore
$$X = c_1 \cos \lambda x + c_2 \sin \lambda x$$

$$T = c_3 \cos \lambda at + c_4 \sin \lambda at.$$

As before, the boundary conditions (2) translate into $X(0) = 0$ and $X(L) = 0$. In turn we find

$$c_1 = 0 \quad \text{and} \quad c_2 \sin \lambda L = 0.$$

This last equation yields the eigenvalues $\lambda = n\pi/L$, where $n = 1, 2, 3, \ldots$. The corresponding eigenfunctions are

$$X = c_2 \sin \frac{n\pi}{L} x, \qquad n = 1, 2, 3, \ldots.$$

Thus solutions of equation (1) satisfying the boundary conditions (2) are

$$u_n = \left(A_n \cos \frac{n\pi a}{L} t + B_n \sin \frac{n\pi a}{L} t \right) \sin \frac{n\pi}{L} x \tag{4}$$

and
$$u(x, t) = \sum_{n=1}^{\infty} \left(A_n \cos \frac{n\pi a}{L} t + B_n \sin \frac{n\pi a}{L} t \right) \sin \frac{n\pi}{L} x. \tag{5}$$

Setting $t = 0$ in (5) gives

$$u(x, 0) = f(x) = \sum_{n=1}^{\infty} A_n \sin \frac{n\pi}{L} x,$$

which is a half-range expansion for f in a sine series. As in the discussion of the heat equation, we can write $A_n = b_n$:

$$A_n = \frac{2}{L} \int_0^L f(x) \sin \frac{n\pi}{L} x \, dx. \tag{6}$$

To determine B_n we differentiate (5) with respect to t and then set $t = 0$:

$$\frac{\partial u}{\partial t} = \sum_{n=1}^{\infty} \left(-A_n \frac{n\pi a}{L} \sin \frac{n\pi a}{L} t + B_n \frac{n\pi a}{L} \cos \frac{n\pi a}{L} t \right) \sin \frac{n\pi}{L} x$$

$$\frac{\partial u}{\partial t}\bigg|_{t=0} = g(x) = \sum_{n=1}^{\infty} \left(B_n \frac{n\pi a}{L} \right) \sin \frac{n\pi}{L} x.$$

In order for this last series to be the half-range sine expansion of g on the interval, the *total* coefficient $B_n n\pi a/L$ must be given by the form of (5) in Section 11.3—that is,

$$B_n \frac{n\pi a}{L} = \frac{2}{L} \int_0^L g(x) \sin \frac{n\pi}{L} x \, dx,$$

from which we obtain

$$B_n = \frac{2}{n\pi a} \int_0^L g(x) \sin \frac{n\pi}{L} x \, dx. \tag{7}$$

The solution of the problem consists of the series (5) with A_n and B_n defined by (6) and (7), respectively.

We note that when the string is released from *rest*, then $g(x) = 0$ for every x in $0 \le x \le L$ and, consequently, $B_n = 0$.

Standing Waves

Recall from the derivation of the wave equation in Section 12.2 that the constant a appearing in the solution of the boundary-value problem in (1), (2), and (3) is given by $\sqrt{T/\rho}$, where ρ is mass per unit length and T is the magnitude of the tension in the string. When T is large enough, the vibrating string produces a musical sound. This sound is the result of standing waves. The solution (5) is a superposition of product solutions called **standing waves**, or **normal modes**:

$$u(x, t) = u_1(x, t) + u_2(x, t) + u_3(x, t) + \cdots.$$

In view of (8) and (9) of Section 5.1, the product solutions (4) can be written as

$$u_n(x, t) = C_n \sin\left(\frac{n\pi a}{L} t + \phi_n\right) \sin \frac{n\pi}{L} x, \qquad (8)$$

where $C_n = \sqrt{A_n^2 + B_n^2}$ and ϕ_n is defined by $\sin \phi_n = A_n/C_n$ and $\cos \phi_n = B_n/C_n$. For $n = 1, 2, 3, \ldots$, the standing waves are essentially the graphs of $\sin(n\pi x/L)$, with a time-varying amplitude given by

$$C_n \sin\left(\frac{n\pi a}{L} t + \phi_n\right).$$

Alternatively, we see from (8) that at a fixed value of x, each product function $u_n(x, t)$ represents simple harmonic motion with amplitude $C_n|\sin(n\pi x/L)|$ and frequency $f_n = na/2L$. In other words, each point on a standing wave vibrates with a different amplitude but with the same frequency. When $n = 1$,

$$u_1(x, t) = C_1 \sin\left(\frac{\pi a}{L} t + \phi_1\right) \sin \frac{\pi}{L} x$$

is called the **first standing wave**, the **first normal mode**, or the **fundamental mode of vibration**. The first three standing waves, or normal modes, are shown in Figure 12.8. The dashed graphs represent the standing waves at various values of time. The points in the interval $(0, L)$, for which $\sin(n\pi/L)x = 0$, correspond to points on a standing wave where there is no motion. These points are called **nodes**. For example, in Figures 12.8(b) and (c) we see that the second standing wave has one node at $L/2$ and the third standing wave has two nodes at $L/3$ and $2L/3$. In general, the nth normal mode of vibration has $n - 1$ nodes.

The frequency

$$f_1 = \frac{a}{2L} = \frac{1}{2L}\sqrt{\frac{T}{\rho}}$$

(a) first standing wave

(b) second standing wave

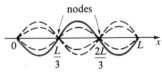

(c) third standing wave

Figure 12.8

of the first normal mode is called the **fundamental frequency**, or **first harmonic**, and is directly related to the pitch produced by a stringed instrument. It is apparent that the greater the tension on the string, the higher the pitch of the sound. The frequencies f_n of the other normal modes, which are integer multiples of the fundamental frequency, are called **overtones**. The second harmonic is the first overtone, and so on.

12.4 **EXERCISES**

Answers to odd-numbered problems begin on page A–43.

In Problems 1–8 solve the wave equation (1) subject to the given conditions.

Figure 12.9

Figure 12.9

1. $u(0, t) = 0, \quad u(L, t) = 0$

$u(x, 0) = \frac{1}{4}x(L - x), \quad \left.\dfrac{\partial u}{\partial t}\right|_{t=0} = 0$

2. $u(0, t) = 0, \quad u(L, t) = 0$

$u(x, 0) = 0, \quad \left.\dfrac{\partial u}{\partial t}\right|_{t=0} = x(L - x)$

3. $u(0, t) = 0, \quad u(L, t) = 0$

$u(x, 0)$, as specified in Figure 12.9

$\left.\dfrac{\partial u}{\partial t}\right|_{t=0} = 0$

4. $u(0, t) = 0, \quad u(\pi, t) = 0$

$u(x, 0) = \frac{1}{6}x(\pi^2 - x^2)$

$\left.\dfrac{\partial u}{\partial t}\right|_{t=0} = 0$

5. $u(0, t) = 0, \quad u(\pi, t) = 0$

$u(x, 0) = 0$

$\left.\dfrac{\partial u}{\partial t}\right|_{t=0} = \sin x$

6. $u(0, t) = 0, \quad u(1, t) = 0$

$u(x, 0) = 0.01 \sin 3\pi x$

$\left.\dfrac{\partial u}{\partial t}\right|_{t=0} = 0$

7. $u(0, t) = 0, \quad u(L, t) = 0$

$$u(x, 0) = \begin{cases} \dfrac{2hx}{L}, & 0 < x < \dfrac{L}{2} \\ 2h\left(1 - \dfrac{x}{L}\right), & \dfrac{L}{2} \le x < L \end{cases}$$

$\left.\dfrac{\partial u}{\partial t}\right|_{t=0} = 0$

The constant h is positive but small compared to L. This is referred to as the "plucked string" problem.

8. $\left.\dfrac{\partial u}{\partial x}\right|_{x=0} = 0, \quad \left.\dfrac{\partial u}{\partial x}\right|_{x=L} = 0$

$u(x, 0) = x, \quad \left.\dfrac{\partial u}{\partial t}\right|_{t=0} = 0$

This problem could describe the longitudinal displacement $u(x, t)$ of a vibrating elastic bar. The boundary conditions at $x = 0$ and $x = L$ are called **free-end conditions**. See Figure 12.10.

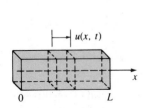

Figure 12.10

9. A string is stretched and secured on the x-axis at $x = 0$ and $x = \pi$ for

$t > 0$. If the transverse vibrations take place in a medium that imparts a resistance proportional to the instantaneous velocity, then the wave equation takes on the form

$$\frac{\partial^2 u}{\partial x^2} = \frac{\partial^2 u}{\partial t^2} + 2\beta \frac{\partial u}{\partial t}, \qquad 0 < \beta < 1, \quad t > 0.$$

Find the displacement $u(x, t)$ if the string starts from rest from the initial displacement $f(x)$.

10. Show that a solution of the boundary-value problem

$$\frac{\partial^2 u}{\partial x^2} = \frac{\partial^2 u}{\partial t^2} + u, \qquad 0 < x < \pi, \quad t > 0$$

$$u(0, t) = 0, \qquad u(\pi, t) = 0, \qquad t > 0$$

$$u(x, 0) = \begin{cases} x, & 0 < x < \pi/2 \\ \pi - x, & \pi/2 \le x < \pi \end{cases}$$

$$\frac{\partial u}{\partial t}\bigg|_{t=0} = 0, \quad 0 < x < \pi$$

is $\qquad u(x, t) = \frac{4}{\pi} \sum_{k=1}^{\infty} \frac{(-1)^{k+1}}{(2k-1)^2} \sin(2k-1)x \cos \sqrt{(2k-1)^2 + 1}\, t.$

11. The transverse displacement $u(x, t)$ of a vibrating beam of length L is determined from a fourth-order partial differential equation

$$a^2 \frac{\partial^4 u}{\partial x^4} + \frac{\partial^2 u}{\partial t^2} = 0, \qquad 0 < x < L, \quad t > 0.$$

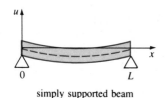

simply supported beam

Figure 12.11

If the beam is **simply supported**, as shown in Figure 12.11, the boundary and initial conditions are

$$u(0, t) = 0, \qquad u(L, t) = 0$$
$$\frac{\partial^2 u}{\partial x^2}\bigg|_{x=0} = 0, \qquad \frac{\partial^2 u}{\partial x^2}\bigg|_{x=L} = 0 \quad\Bigg\} \, t > 0$$

$$u(x, 0) = f(x), \qquad \frac{\partial u}{\partial t}\bigg|_{t=0} = g(x), \qquad 0 < x < L.$$

Solve for $u(x, t)$. [*Hint:* For convenience use λ^4 instead of λ^2 when separating variables.]

12. What are the boundary conditions when the ends of the beam in Problem 11 are embedded at $x = 0$ and $x = L$?

13. Consider the boundary-value problem given in (1), (2), and (3) of this section. If $g(x) = 0$ on $0 < x < L$, show that the solution of the problem can be written as

$$u(x, t) = \frac{1}{2} \left[f(x + at) + f(x - at) \right].$$

[*Hint:* Use the identity

$$2 \sin \theta_1 \cos \theta_2 = \sin(\theta_1 + \theta_2) + \sin(\theta_1 - \theta_2).]$$

14. The vertical displacement $u(x, t)$ of an infinitely long string is determined from the initial-value problem

$$a^2 \frac{\partial^2 u}{\partial x^2} = \frac{\partial^2 u}{\partial t^2}, \qquad -\infty < x < \infty, \quad t > 0$$

$$u(x, 0) = f(x), \qquad \frac{\partial u}{\partial t}\bigg|_{t=0} = g(x).$$

This problem can be solved without separating variables.

(a) Recall from Problem 28 of Exercises 12.1 that the wave equation can be put into the form $\partial^2 u/\partial \eta \, \partial \xi = 0$ by means of the substitutions $\xi = x + at$ and $\eta = x - at$. Integrating the last partial differential equation with respect to η and then with respect to ξ shows that $u(x, t) = F(x + at) + G(x - at)$, where F and G are arbitrary twice-differentiable functions, is a solution of the wave equation. Use this solution and the given initial conditions to show that

$$F(x) = \frac{1}{2} f(x) + \frac{1}{2a} \int_{x_0}^{x} g(s) \, ds + c$$

and

$$G(x) = \frac{1}{2} f(x) - \frac{1}{2a} \int_{x_0}^{x} g(s) \, ds - c,$$

where x_0 is arbitrary and c is a constant of integration.

(b) Use the results in part (a) to show that

$$u(x, t) = \frac{1}{2} \left[f(x + at) + f(x - at) \right] + \frac{1}{2a} \int_{x - at}^{x + at} g(s) \, ds. \qquad \textbf{(9)}$$

Note that when the initial velocity $g(x) = 0$, we obtain

$$u(x, t) = \frac{1}{2} \left[f(x + at) + f(x - at) \right], \qquad -\infty < x < \infty.$$

The last solution can be interpreted as a superposition of two **traveling waves**, one moving to the right (that is, $\frac{1}{2} f(x - at)$) and one moving to the left ($\frac{1}{2} f(x + at)$). Both waves travel with speed a and have the same basic shape as the initial displacement $f(x)$. The form of $u(x, t)$ given in (9) is called **d'Alembert's solution.** *

---

*Jean le Rond d'Alembert (1717–1783)   Educated in law and medicine, the Frenchman d'Alembert devoted his life to the study of physics and mathematics. D'Alembert was also a collaborator with the famous philosopher Denis Diderot on the latter's *Encyclopédie*.

In Problems 15–17 use d'Alembert's solution (9) to solve the initial-value problem in Problem 14 subject to the given initial conditions.

15. $f(x) = \sin x$, $g(x) = 1$

16. $f(x) = \sin x$, $g(x) = \cos x$

17. $f(x) = 0$, $g(x) = \sin 2x$

18. Suppose $f(x) = 1/(1 + x^2)$, $g(x) = 0$, and $a = 1$ for the initial-value problem stated in Problem 14. Graph d'Alembert's solution in this case at the times $t = 0$, $t = 1$, and $t = 3$.

12.5 LAPLACE'S EQUATION

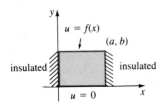

Figure 12.12

Suppose we wish to find the steady-state temperature $u(x, y)$ in a rectangular plate with insulated boundaries, as shown in Figure 12.12. When no heat escapes from the lateral faces of the plate, we solve Laplace's equation

$$\frac{\partial^2 u}{\partial x^2} + \frac{\partial^2 u}{\partial y^2} = 0, \qquad 0 < x < a, \qquad 0 < y < b$$

subject to

$$\frac{\partial u}{\partial x}\bigg|_{x=0} = 0, \qquad \frac{\partial u}{\partial x}\bigg|_{x=a} = 0, \qquad 0 < y < b$$

$$u(x, 0) = 0, \qquad u(x, b) = f(x), \qquad 0 < x < a.$$

Solution

Separation of variables leads to

$$\frac{X''}{X} = -\frac{Y''}{Y} = -\lambda^2$$

$$X'' + \lambda^2 X = 0 \tag{1}$$

$$Y'' - \lambda^2 Y = 0 \tag{2}$$

$$X = c_1 \cos \lambda x + c_2 \sin \lambda x \tag{3}$$

and, since $0 < y < b$ is a finite interval, we use the solution

$$Y = c_3 \cosh \lambda y + c_4 \sinh \lambda y. \tag{4}$$

Now the first three boundary conditions translate into $X'(0) = 0$, $X'(a) = 0$, and $Y(0) = 0$. Differentiating X and setting $x = 0$ imply $c_2 = 0$ and, therefore, $X = c_1 \cos \lambda x$. Differentiating this last expression and then setting $x = a$ give $-c_1 \lambda \sin \lambda a = 0$. This last condition is satisfied when $\lambda = 0$ or when $\lambda a = n\pi$, or $\lambda = n\pi/a$, $n = 1, 2, \ldots$. Observe that $\lambda = 0$ implies (1) is $X'' = 0$. The general solution of this equation is given by the linear function $X = c_1 + c_2 x$ and *not* by (3). In this case the boundary conditions $X'(0) = 0$, $X'(a) = 0$ demand that $X = c_1$. In this example, unlike the previous two examples, we are forced to conclude that $\lambda = 0$ is an eigenvalue. Corresponding $\lambda = 0$ with $n = 0$, we get the eigenfunctions

$$X = c_1, \quad n = 0, \quad \text{and} \quad X = c_1 \cos \frac{n\pi}{a} x, \quad n = 1, 2, \dots .$$

Finally, the condition that $Y(0) = 0$ dictates that $c_3 = 0$ in (4) when $\lambda > 0$. However, when $\lambda = 0$, equation (2) becomes $Y'' = 0$ and thus the solution is given by $Y = c_3 + c_4 y$ rather than by (4). But $Y(0) = 0$ implies again that $c_3 = 0$ and so $Y = c_4 y$. Thus product solutions of the equation satisfying the first three boundary conditions are

$$A_0 y, \quad n = 0, \quad \text{and} \quad A_n \sinh \frac{n\pi}{a} y \cos \frac{n\pi}{a} x, \quad n = 1, 2, \dots .$$

The superposition principle yields another solution

$$u(x, y) = A_0 y + \sum_{n=1}^{\infty} A_n \sinh \frac{n\pi}{a} y \cos \frac{n\pi}{a} x. \tag{5}$$

Substituting $y = b$ in (5) gives

$$u(x, b) = f(x) = A_0 b + \sum_{n=1}^{\infty} \left(A_n \sinh \frac{n\pi}{a} b \right) \cos \frac{n\pi}{a} x,$$

which, in this case, is a half-range expansion of f in a cosine series. If we make the identifications $A_0 b = a_0/2$ and $A_n \sinh(n\pi b/a) = a_n$, $n = 1, 2, 3, \dots$, it follows from (2) and (3) of Section 11.3 that

$$2A_0 b = \frac{2}{a} \int_0^a f(x)\, dx$$

$$A_0 = \frac{1}{ab} \int_0^a f(x)\, dx \tag{6}$$

and

$$A_n \sinh \frac{n\pi}{a} b = \frac{2}{a} \int_0^a f(x) \cos \frac{n\pi}{a} x\, dx$$

$$A_n = \frac{2}{a \sinh \dfrac{n\pi}{a} b} \int_0^a f(x) \cos \frac{n\pi}{a} x\, dx. \tag{7}$$

The solution of this problem consists of the series given in (5), where A_0 and A_n are defined by (6) and (7), respectively.

Dirichlet Problem

A boundary-value problem in which we seek a solution to an elliptic partial differential equation, such as Laplace's equation $\nabla^2 u = 0$, within a region such that u takes on prescribed values on the entire boundary of the region is called a **Dirichlet problem**.

Superposition Principle

A Dirichlet problem for a rectangle can readily be solved by separation of variables when homogeneous boundary conditions are specified on two *parallel* boundaries. However, the method of separation of variables is not applicable to a Dirichlet problem when the boundary conditions on all four sides of the rectangle are nonhomogeneous. To get around this difficulty we break the problem

$$\frac{\partial^2 u}{\partial x^2} + \frac{\partial^2 u}{\partial y^2} = 0, \qquad 0 < x < a, \;\; 0 < y < b$$

$$u(0, y) = F(y), \qquad u(a, y) = G(y), \qquad 0 < y < b \tag{8}$$

$$u(x, 0) = f(x), \qquad u(x, b) = g(x), \qquad 0 < x < a$$

into two problems, each of which has homogeneous boundary conditions on parallel boundaries, as shown:

| Problem 1 | | | Problem 2 | | |

$$\frac{\partial^2 u_1}{\partial x^2} + \frac{\partial^2 u_1}{\partial y^2} = 0, \;\; 0 < x < a, \qquad 0 < y < b \qquad\qquad \frac{\partial^2 u_2}{\partial x^2} + \frac{\partial^2 u_2}{\partial y^2} = 0, \;\; 0 < x < a, \qquad\qquad 0 < y < b$$

$$u_1(0, y) = 0, \qquad u_1(a, y) = 0, \qquad 0 < y < b \qquad\qquad u_2(0, y) = F(y), \qquad u_2(a, y) = G(y), \;\; 0 < y < b$$

$$u_1(x, 0) = f(x), \qquad u_1(x, b) = g(x), \;\; 0 < x < a \qquad\qquad u_2(x, 0) = 0, \qquad u_2(x, b) = 0, \qquad 0 < x < a$$

Suppose u_1 and u_2 are the solutions of Problems 1 and 2, respectively. If we define $u(x, y) = u_1(x, y) + u_2(x, y)$, it is seen that u satisfies all boundary conditions in the original problem (8). For example,

$$u(0, y) = u_1(0, y) + u_2(0, y) = 0 + F(y) = F(y)$$

$$u(x, b) = u_1(x, b) + u_2(x, b) = g(x) + 0 = g(x),$$

and so on. Furthermore, u is a solution of Laplace's equation by Theorem 12.1. In other words, by solving Problems 1 and 2 and adding their solutions, we have solved the original problem. This additive property of solutions is known as the superposition principle. See Figure 12.13.

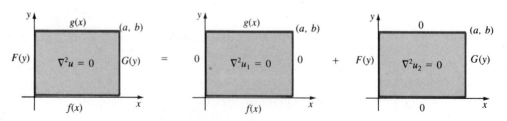

Solution u = Solution u_1 of Problem 1 + Solution u_2 of Problem 2

Figure 12.13

We leave it as exercises (see Problems 11 and 12) to show that a solution of Problem 1 is

$$u_1(x, y) = \sum_{n=1}^{\infty} \left\{ A_n \cosh \frac{n\pi}{a} y + B_n \sinh \frac{n\pi}{a} y \right\} \sin \frac{n\pi}{a} x,$$

where $\qquad A_n = \frac{2}{a} \int_0^a f(x) \sin \frac{n\pi}{a} x \, dx$

$$B_n = \frac{1}{\sinh \dfrac{n\pi}{a} b} \left(\frac{2}{a} \int_0^a g(x) \sin \frac{n\pi}{a} x \, dx - A_n \cosh \frac{n\pi}{a} b \right),$$

and that a solution of Problem 2 is

$$u_2(x, y) = \sum_{n=1}^{\infty} \left\{ A_n \cosh \frac{n\pi}{b} x + B_n \sinh \frac{n\pi}{b} x \right\} \sin \frac{n\pi}{b} y,$$

where $\qquad A_n = \frac{2}{b} \int_0^b F(y) \sin \frac{n\pi}{b} y \, dy$

$$B_n = \frac{1}{\sinh \dfrac{n\pi}{b} a} \left(\frac{2}{b} \int_0^b G(y) \sin \frac{n\pi}{b} y \, dy - A_n \cosh \frac{n\pi}{b} a \right).$$

12.5 EXERCISES

Answers to odd-numbered problems begin on page A–43.

In Problems 1–8 find the steady-state temperature for a rectangular plate with boundary conditions as given.

1. $u(0, y) = 0, \quad u(a, y) = 0$
$u(x, 0) = 0, \quad u(x, b) = f(x)$

2. $u(0, y) = 0, \quad u(a, y) = 0$
$\left. \dfrac{\partial u}{\partial y} \right|_{y=0} = 0, \quad u(x, b) = f(x)$

3. $u(0, y) = 0, \quad u(a, y) = 0$
$u(x, 0) = f(x), \quad u(x, b) = 0$

4. $\left. \dfrac{\partial u}{\partial x} \right|_{x=0} = 0, \quad \left. \dfrac{\partial u}{\partial x} \right|_{x=a} = 0$
$u(x, 0) = x, \quad u(x, b) = 0$

5. $u(0, y) = 0, \quad u(1, y) = 1 - y$
$\left. \dfrac{\partial u}{\partial y} \right|_{y=0} = 0, \quad \left. \dfrac{\partial u}{\partial y} \right|_{y=1} = 0$

6. $u(0, y) = g(y), \quad \left. \dfrac{\partial u}{\partial x} \right|_{x=1} = 0$
$\left. \dfrac{\partial u}{\partial y} \right|_{y=0} = 0, \quad \left. \dfrac{\partial u}{\partial y} \right|_{y=\pi} = 0$

7. $\left. \dfrac{\partial u}{\partial x} \right|_{x=0} = u(0, y), \quad u(\pi, y) = 1$
$u(x, 0) = 0, \quad u(x, \pi) = 0$

8. $u(0, y) = 0, \quad u(1, y) = 0$
$\left. \dfrac{\partial u}{\partial y} \right|_{y=0} = u(x, 0), \quad u(x, 1) = f(x)$

In Problems 9 and 10 find the steady-state temperature in the given semi-infinite plate extending in the positive y-direction. In each case assume $u(x, y)$ is bounded as $y \to \infty$.

9.

10.

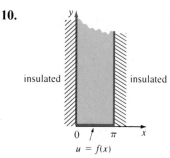

Figure 12.14   Figure 12.15

In Problems 11 and 12 find the steady-state temperature for a rectangular plate with boundary conditions as given.

11. $u(0, y) = 0, \quad u(a, y) = 0$
$u(x, 0) = f(x), \quad u(x, b) = g(x)$

12. $u(0, y) = F(y), \quad u(a, y) = G(y)$
$u(x, 0) = 0, \quad u(x, b) = 0$

In Problems 13 and 14 use the superposition principle to find the steady-state temperature for a square plate with boundary conditions as given.

13. $u(0, y) = 1, \quad u(\pi, y) = 1$
$u(x, 0) = 0, \quad u(x, \pi) = 1$

14. $u(0, y) = 0, \quad u(2, y) = y(2 - y)$
$$u(x, 0) = 0, \quad u(x, 2) = \begin{cases} x, & 0 < x < 1 \\ 2 - x, & 1 \le x < 2 \end{cases}$$

12.6 NONHOMOGENEOUS EQUATIONS AND BOUNDARY CONDITIONS

The method of separation of variables may not be applicable to a boundary-value problem when the partial differential equation or boundary conditions are nonhomogeneous. For example, when heat is generated at a constant rate r within a rod of finite length, the form of the heat equation is

$$k \frac{\partial^2 u}{\partial x^2} + r = \frac{\partial u}{\partial t}. \tag{1}$$

Equation (1) is nonhomogeneous and is readily shown to be not separable. On the other hand, suppose we wish to solve the usual heat equation $k u_{xx} = u_t$ when the boundaries $x = 0$ and $x = L$ are held at nonzero temperatures k_1 and k_2. Even though the assumption $u(x, t) = X(x)T(t)$ separates the partial differential equation, we quickly find ourselves at an impasse in determining eigenvalues and eigenfunctions since no conclusions can be obtained from $u(0, t) = X(0)T(t) = k_1$ and $u(L, t) = X(L)T(t) = k_2$.

A few problems involving nonhomogeneous equations or nonhomogeneous boundary conditions can be solved by means of a change of dependent variable.

$$u = v + \psi.$$

The basic idea is to determine ψ, a function of *one* variable, in such a manner that v, a function of *two* variables, is made to satisfy a homogeneous partial differential equation and homogeneous boundary conditions. The following example illustrates the procedure.

EXAMPLE I

Solve (1) subject to

$$u(0, t) = 0, \qquad u(1, t) = u_0, \qquad t > 0$$

$$u(x, 0) = f(x), \qquad 0 < x < 1.$$

Solution Both the equation and the boundary condition at $x = 1$ are nonhomogeneous. If we let $u(x, t) = v(x, t) + \psi(x)$, then

$$\frac{\partial^2 u}{\partial x^2} = \frac{\partial^2 v}{\partial x^2} + \psi'' \quad \text{and} \quad \frac{\partial u}{\partial t} = \frac{\partial v}{\partial t}.$$

Substituting these results into (1) gives

$$k\frac{\partial^2 v}{\partial x^2} + k\psi'' + r = \frac{\partial v}{\partial t}. \tag{2}$$

Equation (2) reduces to a homogeneous equation if we demand that ψ satisfy

$$k\psi'' + r = 0 \quad \text{or} \quad \psi'' = -\frac{r}{k}.$$

Integrating the last equation twice reveals that

$$\psi(x) = -\frac{r}{2k}x^2 + c_1 x + c_2. \tag{3}$$

Furthermore $\qquad u(0, t) = v(0, t) + \psi(0) = 0$

$$u(1, t) = v(1, t) + \psi(1) = u_0.$$

We have $v(0, t) = 0$ and $v(1, t) = 0$, provided

$$\psi(0) = 0 \quad \text{and} \quad \psi(1) = u_0.$$

Applying the latter two conditions to (3) gives, in turn, $c_2 = 0$ and $c_1 = \frac{r}{2k} + u_0$. Consequently,

$$\psi(x) = -\frac{r}{2k} x^2 + \left(\frac{r}{2k} + u_0 \right) x.$$

Last, the initial condition $u(x, 0) = v(x, 0) + \psi(x)$ implies $v(x, 0) = u(x, 0) - \psi(x) = f(x) - \psi(x)$. Thus to determine $v(x, t)$ we solve the *new* boundary-value problem

$$k \frac{\partial^2 v}{\partial x^2} = \frac{\partial v}{\partial t}, \qquad 0 < x < 1, \quad t > 0$$

$$v(0, t) = 0, \qquad v(1, t) = 0, \qquad t > 0$$

$$v(x, 0) = f(x) + \frac{r}{2k} x^2 - \left(\frac{r}{2k} + u_0 \right) x$$

by separation of variables. In the usual manner we find

$$v(x, t) = \sum_{n=1}^{\infty} A_n e^{-kn^2\pi^2 t} \sin n\pi x,$$

where $\qquad A_n = 2 \int_0^1 \left[f(x) + \frac{r}{2k} x^2 - \left(\frac{r}{2k} + u_0 \right) x \right] \sin n\pi x \, dx. \qquad$ **(4)**

Finally, a solution of the original problem is obtained by adding $\psi(x)$ and $v(x, t)$:

$$u(x, t) = -\frac{r}{2k} x^2 + \left(\frac{r}{2k} + u_0 \right) x + \sum_{n=1}^{\infty} A_n e^{-kn^2\pi^2 t} \sin n\pi x, \qquad \textbf{(5)}$$

where the A_n are defined in (4). ∎

Observe in (5) that $u(x, t) \to \psi(x)$ as $t \to \infty$. In the context of solving forms of the heat equation, ψ is called a **steady-state solution**. Since $v(x, t) \to 0$ as $t \to \infty$, v is called a **transient solution**.

The substitution $u = v + \psi$ can also be used on problems involving forms of the wave equation as well as Laplace's equation.

12.6 EXERCISES

Answers to odd-numbered problems begin on page A–43.

In Problems 1 and 2 solve the heat equation $ku_{xx} = u_t$, $0 < x < 1$, $t > 0$, subject to the given conditions.

1. $u(0, t) = 100, \quad u(1, t) = 100$
  $u(x, 0) = 0$

2. $u(0, t) = u_0, \quad u(1, t) = 0$
  $u(x, 0) = f(x)$

In Problems 3 and 4 solve the partial differential equation (1) subject to the given conditions.

3. $u(0, t) = u_0,$ $u(1, t) = u_0$
$u(x, 0) = 0$

4. $u(0, t) = u_0,$ $u(1, t) = u_1$
$u(x, 0) = f(x)$

5. Solve the boundary-value problem:

$$k \frac{\partial^2 u}{\partial x^2} + Ae^{-\beta x} = \frac{\partial u}{\partial t}, \qquad \beta > 0, \quad 0 < x < 1, \quad t > 0$$

$$u(0, t) = 0, \qquad u(1, t) = 0, \qquad t > 0$$

$$u(x, 0) = f(x), \qquad 0 < x < 1.$$

The partial differential equation is a form of the heat equation when heat is generated within a thin rod due to radioactive decay of the material.

6. Solve the boundary-value problem:

$$k \frac{\partial^2 u}{\partial x^2} - hu = \frac{\partial u}{\partial t}, \qquad 0 < x < \pi, \quad t > 0$$

$$u(0, t) = 0, \qquad u(\pi, t) = u_0, \qquad t > 0$$

$$u(x, 0) = 0, \qquad 0 < x < \pi.$$

7. Find a steady-state solution $\psi(x)$ of the boundary-value problem:

$$k \frac{\partial^2 u}{\partial x^2} - h(u - u_0) = \frac{\partial u}{\partial t}, \qquad 0 < x < 1, \quad t > 0$$

$$u(0, t) = u_0, \qquad u(1, t) = 0, \qquad t > 0$$

$$u(x, 0) = f(x), \qquad 0 < x < 1.$$

8. Find a steady-state solution $\psi(x)$ if the rod in Problem 7 is semi-infinite extending in the positive x-direction, radiates from its lateral surface into a medium at temperature zero, and

$$u(0, t) = u_0, \qquad \lim_{x \to \infty} u(x, t) = 0, \qquad t > 0$$

$$u(x, 0) = f(x), \qquad x > 0.$$

9. When a vibrating string is subjected to an external vertical force that varies with the horizontal distance from the left end, the wave equation takes on the form

$$a^2 \frac{\partial^2 u}{\partial x^2} + Ax = \frac{\partial^2 u}{\partial t^2}.$$

Solve this partial differential equation subject to

$$u(0, t) = 0, \qquad u(1, t) = 0, \qquad t > 0$$

$$u(x, 0) = 0, \qquad \frac{\partial u}{\partial t}\bigg|_{t=0} = 0, \qquad 0 < x < 1.$$

10. A string initially at rest on the x-axis is secured on the x-axis at $x = 0$ and $x = 1$. If the string is allowed to fall under its own weight for $t > 0$, the displacement $u(x, t)$ satisfies

$$a^2 \frac{\partial^2 u}{\partial x^2} - g = \frac{\partial^2 u}{\partial t^2}, \qquad 0 < x < 1, \quad t > 0,$$

where g is the acceleration of gravity. Solve for $u(x, t)$.

11. Find the steady-state temperature $u(x, y)$ in the semi-infinite plate shown in Figure 12.16. Assume that the temperature is bounded as $x \to \infty$. [*Hint:* Try $u(x, y) = v(x, y) + \psi(y)$.]

12. Poisson's equation*

$$\frac{\partial^2 u}{\partial x^2} + \frac{\partial^2 u}{\partial y^2} = -h, \qquad h > 0,$$

occurs in many problems involving electric potential. Solve the above equation subject to the conditions

$$u(0, y) = 0, \qquad u(\pi, y) = 1, \qquad y > 0$$

$$u(x, 0) = 0, \qquad 0 < x < \pi.$$

Figure 12.16

12.7 USE OF GENERALIZED FOURIER SERIES

For certain types of boundary conditions, the method of separation of variables and the superposition principle lead to an expansion of a function in a trigonometric series that is not a Fourier series. To solve the problems in this section we shall utilize the concept of generalized Fourier series developed in Section 11.1.

EXAMPLE I

The temperature in a rod of unit length in which there is heat transfer from its right boundary into a surrounding medium kept at a constant tempera-

---

* **Simeon-Denis Poisson** (1781–1840) was a French mathematician and physicist.

ture zero is determined from

$$k \frac{\partial^2 u}{\partial x^2} = \frac{\partial u}{\partial t}, \qquad 0 < x < 1, \quad t > 0$$

$$u(0, t) = 0, \qquad \frac{\partial u}{\partial x}\bigg|_{x=1} = -hu(1, t), \qquad h > 0, \quad t > 0$$

$$u(x, 0) = 1, \qquad 0 < x < 1.$$

Solve for $u(x, t)$.

Solution   The method of separation of variables gives

$$X'' + \lambda^2 X = 0$$

$$T' + k\lambda^2 T = 0 \tag{1}$$

and

$$X(x) = c_1 \cos \lambda x + c_2 \sin \lambda x$$

$$T(t) = c_3 e^{-k\lambda^2 t}.$$

Since $u = XT$, we find that the boundary conditions become

$$X(0) = 0 \quad \text{and} \quad X'(1) = -hX(1). \tag{2}$$

The first condition in (2) immediately gives $c_1 = 0$. Applying the second condition in (2) to $X(x) = c_2 \sin \lambda x$ yields

$$\lambda \cos \lambda = -h \sin \lambda \quad \text{or} \quad \tan \lambda = -\frac{\lambda}{h}. \tag{3}$$

From the analysis in Example 3 of Section 11.4, we know that the last equation in (3) has an infinite number of roots. The consecutive positive roots λ_n, $n = 1, 2, 3, \dots$, are the eigenvalues of the problem, and the corresponding eigenfunctions are $X(x) = c_2 \sin \lambda_n x$, $n = 1, 2, 3, \dots$. Thus

$$u_n = XT = A_n e^{-k\lambda_n^2 t} \sin \lambda_n x$$

and

$$u(x, t) = \sum_{n=1}^{\infty} A_n e^{-k\lambda_n^2 t} \sin \lambda_n x.$$

Now at $t = 0$, $u(x, 0) = 1$, $0 < x < 1$, so that

$$1 = \sum_{n=1}^{\infty} A_n \sin \lambda_n x. \tag{4}$$

The series in (4) is not a Fourier sine series; it is an expansion in terms of the orthogonal functions arising from the Sturm-Liouville problem consisting of the differential equation (1) and boundary conditions (2). It follows that the set of eigenfunctions $\{\sin \lambda_n x\}$, $n = 1, 2, 3, \dots$, where the λ's are defined by $\tan \lambda = -\lambda/h$, is orthogonal with respect to the weight function $p(x) = 1$ on the interval $[0, 1]$. With $f(x) = 1$ in (8) of Section 11.1, we can write

$$A_n = \frac{\int_0^1 \sin \lambda_n x \, dx}{\int_0^1 \sin^2 \lambda_n x \, dx}. \tag{5}$$

To evaluate the square norm of each of the eigenfunctions we use a trigonometric identity:

$$\int_0^1 \sin^2 \lambda_n x \, dx = \frac{1}{2} \int_0^1 [1 - \cos 2\lambda_n x] \, dx = \frac{1}{2} \left[1 - \frac{1}{2\lambda_n} \sin 2\lambda_n \right]. \tag{6}$$

Using $\sin 2\lambda_n = 2 \sin \lambda_n \cos \lambda_n$ and $\lambda \cos \lambda = -h \sin \lambda$, we simplify (6) to

$$\int_0^1 \sin^2 \lambda_n x \, dx = \frac{1}{2h} [h + \cos^2 \lambda_n].$$

Also $\qquad \int_0^1 \sin \lambda_n x \, dx = -\frac{1}{\lambda_n} \cos \lambda_n x \Big|_0^1 = \frac{1}{\lambda_n} [1 - \cos \lambda_n].$

Consequently (5) becomes

$$A_n = \frac{2h(1 - \cos \lambda_n)}{\lambda_n(h + \cos^2 \lambda_n)}.$$

Finally a solution of the boundary-value problem is

$$u(x, t) = 2h \sum_{n=1}^{\infty} \frac{1 - \cos \lambda_n}{\lambda_n(h + \cos^2 \lambda_n)} e^{-k\lambda_n^2 t} \sin \lambda_n x. \qquad \blacksquare$$

EXAMPLE 2

The twist angle $\theta(x, t)$ of a torsionally vibrating shaft of unit length is determined from

$$a^2 \frac{\partial^2 \theta}{\partial x^2} = \frac{\partial^2 \theta}{\partial t^2}, \qquad 0 < x < 1, \quad t > 0$$

$$\theta(0, t) = 0, \qquad \frac{\partial \theta}{\partial x}\Big|_{x=1} = 0, \qquad t > 0$$

$$\theta(x, 0) = x, \qquad \frac{\partial \theta}{\partial t}\Big|_{t=0} = 0, \qquad 0 < x < 1.$$

twisted shaft

Figure 12.17

See Figure 12.17. The boundary condition at $x = 1$ is called a free-end condition. Solve for $\theta(x, t)$.

Solution   Using $\theta = XT$, we find

$$X'' + \lambda^2 X = 0$$

$$T'' + a^2 \lambda^2 T = 0$$

and

$$X(x) = c_1 \cos \lambda x + c_2 \sin \lambda x$$

$$T(t) = c_3 \cos a\lambda t + c_4 \sin a\lambda t.$$

The boundary conditions $X(0) = 0$ and $X'(1) = 0$ give

$$c_1 = 0 \quad \text{and} \quad c_2 \cos \lambda = 0,$$

respectively. Since the cosine function is zero at odd multiples of $\pi/2$, the eigenvalues of the problem are $\lambda = (2n - 1)(\pi/2)$, $n = 1, 2, 3, \ldots$. The initial condition $T'(0) = 0$ gives $c_4 = 0$ so

$$\theta_n = XT = A_n \cos a\left(\frac{2n-1}{2}\right)\pi t \sin\left(\frac{2n-1}{2}\right)\pi x.$$

In order to satisfy the remaining initial condition, we form

$$\theta(x, t) = \sum_{n=1}^{\infty} A_n \cos a\left(\frac{2n-1}{2}\right)\pi t \sin\left(\frac{2n-1}{2}\right)\pi x. \tag{7}$$

When $t = 0$, we must have for $0 < x < 1$,

$$\theta(x, 0) = x = \sum_{n=1}^{\infty} A_n \sin\left(\frac{2n-1}{2}\right)\pi x. \tag{8}$$

As in Example 1, the set of eigenfunctions $\left\{\sin\left(\frac{2n-1}{2}\right)\pi x\right\}$, $n = 1, 2, 3, \ldots$, is orthogonal with respect to the weight function $p(x) = 1$ on the interval $[0, 1]$. The series $\sum_{n=1}^{\infty} A_n \sin\left(\frac{2n-1}{2}\right)\pi x$ is not a Fourier sine series since the argument of the sine is not an integer multiple of $\pi x/L$ ($L = 1$ in this case). The series is again a generalized Fourier series. Hence from (8) of Section 11.1 the coefficients in (7) are

$$A_n = \frac{\displaystyle\int_0^1 x \sin\left(\frac{2n-1}{2}\right)\pi x \, dx}{\displaystyle\int_0^1 \sin^2\left(\frac{2n-1}{2}\right)\pi x \, dx}.$$

Carrying out the two integrations, we arrive at

$$A_n = \frac{8(-1)^{n+1}}{(2n-1)^2\pi^2}.$$

The twist angle is then

$$\theta(x, t) = \frac{8}{\pi^2} \sum_{n=1}^{\infty} \frac{(-1)^{n+1}}{(2n-1)^2} \cos a\left(\frac{2n-1}{2}\right)\pi t \sin\left(\frac{2n-1}{2}\right)\pi x. \qquad \blacksquare$$

12.7 EXERCISES

Answers to odd-numbered problems begin on page A–44.

1. In Example 1 find the temperature $u(x, t)$ when the left end of the rod is insulated.

2. Solve the boundary-value problem:

$$k\frac{\partial^2 u}{\partial x^2} = \frac{\partial u}{\partial t}, \qquad 0 < x < 1, \quad t > 0$$

$$u(0, t) = 0, \qquad \frac{\partial u}{\partial x}\bigg|_{x=1} = -h(u(1, t) - u_0), \qquad h > 0, \quad t > 0$$

$$u(x, 0) = f(x), \qquad 0 < x < 1.$$

3. Find the steady-state temperature for a rectangular plate for which the boundary conditions are

$$u(0, y) = 0, \qquad \frac{\partial u}{\partial x}\bigg|_{x=a} = -hu(a, y), \qquad 0 < y < b$$

$$u(x, 0) = 0, \qquad u(x, b) = f(x), \qquad 0 < x < a.$$

4. Solve the boundary-value problem:

$$\frac{\partial^2 u}{\partial x^2} + \frac{\partial^2 u}{\partial y^2} = 0, \qquad 0 < y < 1, \quad x > 0$$

$$u(0, y) = u_0, \qquad \lim_{x \to \infty} u(x, y) = 0, \qquad 0 < y < 1$$

$$\frac{\partial u}{\partial y}\bigg|_{y=0} = 0, \qquad \frac{\partial u}{\partial y}\bigg|_{y=1} = -hu(x, 1), \qquad h > 0, \quad x > 0.$$

5. Find the temperature $u(x, t)$ in a rod of length L if the initial temperature is $f(x)$ throughout and if the end $x = 0$ is kept at temperature zero and the end $x = L$ is insulated.

6. Solve the boundary-value problem:

$$a^2 \frac{\partial^2 u}{\partial x^2} = \frac{\partial^2 u}{\partial t^2}, \qquad 0 < x < L, \quad t > 0$$

$$u(0, t) = 0, \qquad E\frac{\partial u}{\partial x}\bigg|_{x=L} = F_0, \qquad t > 0$$

$$u(x, 0) = 0, \qquad \frac{\partial u}{\partial t}\bigg|_{t=0} = 0, \qquad 0 < x < L.$$

The solution $u(x, t)$ represents the longitudinal displacement of a vibrating elastic bar that is anchored at its left end and is subjected to a constant force of magnitude F_0 at its right end. See Figure 12.10 on page 663. E is a constant called the modulus of elasticity.

7. Solve the boundary-value problem:

$$\frac{\partial^2 u}{\partial x^2} + \frac{\partial^2 u}{\partial y^2} = 0, \qquad 0 < x < 1, \quad 0 < y < 1$$

$$\frac{\partial u}{\partial x}\bigg|_{x=0} = 0, \qquad u(1, y) = u_0, \qquad 0 < y < 1$$

$$u(x, 0) = 0, \qquad \frac{\partial u}{\partial y}\bigg|_{y=1} = 0, \qquad 0 < x < 1.$$

8. The initial temperature in a rod of unit length is $f(x)$ throughout. There is heat transfer from both ends, $x = 0$ and $x = 1$, into a surrounding medium kept at a constant temperature zero. Show that

$$u(x, t) = \sum_{n=1}^{\infty} A_n e^{-k\lambda_n^2 t}(\lambda_n \cos \lambda_n x + h \sin \lambda_n x),$$

where $\quad A_n = \dfrac{2}{(\lambda_n^2 + 2h + h^2)} \displaystyle\int_0^1 f(x)(\lambda_n \cos \lambda_n x + h \sin \lambda_n x)\, dx$

and the λ_n, $n = 1, 2, 3, \ldots$, are the consecutive positive roots of $\tan \lambda = 2\lambda h/(\lambda^2 - h^2)$.

9. A vibrating cantilever beam is embedded at its left end ($x = 0$) and free at its right end ($x = 1$). See Figure 12.18. The transverse displacement $u(x, t)$ of the beam is determined from the boundary-value problem

$$\frac{\partial^4 u}{\partial x^4} + \frac{\partial^2 u}{\partial t^2} = 0, \qquad 0 < x < 1, \quad t > 0$$

$$\left.\begin{array}{ll} u(0, t) = 0, & \dfrac{\partial u}{\partial x}\bigg|_{x=0} = 0 \\[3mm] \dfrac{\partial^2 u}{\partial x^2}\bigg|_{x=1} = 0, & \dfrac{\partial^3 u}{\partial x^3}\bigg|_{x=1} = 0 \end{array}\right\} \quad t > 0$$

$$u(x, 0) = f(x), \qquad \frac{\partial u}{\partial t}\bigg|_{t=0} = g(x), \qquad x > 0.$$

(a) Using the separation constant λ^4, show that the eigenvalues of the problem are determined from the equation $\cos \lambda \cosh \lambda = -1$.

Figure 12.18

(b) Use a calculator or computer to find approximations to the first two positive eigenvalues.

10. (a) Find an equation that defines the eigenvalues when the ends of the beam in Problem 9 are embedded at $x = 0$ and $x = 1$.

(b) Use a calculator or computer to find approximations to the first two positive eigenvalues.

[O] 12.8 BOUNDARY-VALUE PROBLEMS INVOLVING FOURIER SERIES IN TWO VARIABLES

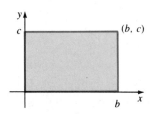

Figure 12.19

Suppose the rectangular region in Figure 12.19 is a thin plate in which the temperature u is a function of time t and position (x, y). Then, under suitable conditions, $u(x, y, t)$ can be shown to satisfy the **two-dimensional heat equation**

$$k\left(\frac{\partial^2 u}{\partial x^2} + \frac{\partial^2 u}{\partial y^2}\right) = \frac{\partial u}{\partial t}. \qquad (1)$$

On the other hand, suppose Figure 12.20 represents a rectangular frame over which a thin flexible membrane has been stretched (a rectangular drumhead). If the membrane is set in motion, then its displacement u, measured from the xy-plane (transverse vibrations), is also a function of time t and position (x, y). When the vibrations are small, free, and undamped, $u(x, y, t)$ satisfies the **two-dimensional wave equation**

$$a^2\left(\frac{\partial^2 u}{\partial x^2} + \frac{\partial^2 u}{\partial y^2}\right) = \frac{\partial^2 u}{\partial t^2}. \qquad (2)$$

As the next example shows, solutions of boundary-value problems involving (1) and (2) by the method of separation of variables lead to the notion of Fourier series in two variables.

Figure 12.20

EXAMPLE I

Find the temperature $u(x, y, t)$ in the plate shown in Figure 12.19 if the initial temperature is $f(x, y)$ throughout and if the boundaries are held at temperature zero.

Solution We must solve

$$k\left(\frac{\partial^2 u}{\partial x^2} + \frac{\partial^2 u}{\partial y^2}\right) = \frac{\partial u}{\partial t}, \qquad 0 < x < b, \quad 0 < y < c, \quad t > 0$$

subject to $u(0, y, t) = 0, \qquad u(b, y, t) = 0, \qquad 0 < y < c, \quad t > 0$

$u(x, 0, t) = 0, \qquad u(x, c, t) = 0, \qquad 0 < x < b, \quad t > 0$

$u(x, y, 0) = f(x, y), \qquad 0 < x < b, \quad 0 < y < c.$

To separate variables in the partial differential equation in three independent variables we try to find a product solution $u(x, y, t) = X(x)Y(y)T(t)$. After substitution, this assumption yields

$$k(X''YT + XY''T) = XYT'$$

or

$$\frac{X''}{X} = -\frac{Y''}{Y} + \frac{T'}{kT}. \tag{3}$$

Since the left side of (3) depends only on x and the right side depends only on y and t, we must have both sides equal to a constant $-\lambda^2$:

$$\frac{X''}{X} = -\frac{Y''}{Y} + \frac{T'}{kT} = -\lambda^2$$

and so

$$X'' + \lambda^2 X = 0 \tag{4}$$

$$\frac{Y''}{Y} = \frac{T'}{kT} + \lambda^2. \tag{5}$$

By the same reasoning, if we introduce another separation constant $-\mu^2$ in (5), then

$$\frac{Y''}{Y} = -\mu^2 \quad \text{and} \quad \frac{T'}{kT} + \lambda^2 = -\mu^2$$

$$Y'' + \mu^2 Y = 0 \quad \text{and} \quad T' + k(\lambda^2 + u^2)T = 0. \tag{6}$$

Now the solutions of the equations in (4) and (6) are, respectively,

$$X(x) = c_1 \cos \lambda x + c_2 \sin \lambda x \tag{7}$$

$$Y(y) = c_3 \cos \mu y + c_4 \sin \mu y \tag{8}$$

$$T(t) = c_5 e^{-k(\lambda^2 + \mu^2)t}. \tag{9}$$

But the boundary conditions

$$\left. \begin{array}{l} u(0, y, t) = 0, \quad u(b, y, t) = 0 \\ u(x, 0, t) = 0, \quad u(x, c, t) = 0 \end{array} \right\} \text{ imply } \begin{cases} X(0) = 0, \quad X(b) = 0 \\ Y(0) = 0, \quad Y(c) = 0. \end{cases}$$

Applying these conditions to (7) and (8) gives $c_1 = 0, c_3 = 0$, and $c_2 \sin \lambda b = 0$, $c_4 \sin \mu c = 0$. The latter equations in turn imply

$$\lambda = \frac{m\pi}{b}, \quad m = 1, 2, 3, \ldots; \qquad \mu = \frac{n\pi}{c}, \quad n = 1, 2, 3, \ldots.$$

Thus a product solution of the two-dimensional heat equation that satisfies the boundary conditions is

$$u_{mn}(x, y, t) = A_{mn} e^{-k[(m\pi/b)^2 + (n\pi/c)^2]t} \sin \frac{m\pi}{b} x \sin \frac{n\pi}{c} y,$$

where A_{mn} is an arbitrary constant. Because we have two independent sets of eigenvalues, we are prompted to try the superposition principle in the form of a double sum

$$u(x, y, t) = \sum_{m=1}^{\infty} \sum_{n=1}^{\infty} A_{mn} e^{-k[(m\pi/b)^2 + (n\pi/c)^2]t} \sin \frac{m\pi}{b} x \sin \frac{n\pi}{c} y. \qquad (10)$$

Now at $t = 0$ we must have

$$u(x, y, 0) = f(x, y) = \sum_{m=1}^{\infty} \sum_{n=1}^{\infty} A_{mn} \sin \frac{m\pi}{b} x \sin \frac{n\pi}{c} y. \qquad (11)$$

We can find the coefficients A_{mn} by multiplying the double sum (11) by the product $\sin(m\pi x/b) \sin(n\pi y/c)$ and integrating over the rectangle $0 \leq x \leq b$, $0 \leq y \leq c$. It follows that

$$A_{mn} = \frac{4}{bc} \int_0^c \int_0^b f(x, y) \sin \frac{m\pi}{b} x \sin \frac{n\pi}{c} y \, dx \, dy. \qquad (12)$$

Thus the solution of the boundary-value problem consists of (1) with the A_{mn} defined by (12). ∎

The series (11) with coefficients (12) is called a **sine series in two variables**, or a **double sine series** (see Problem 52, Exercises 11.3).

12.8 EXERCISES

Answers to odd-numbered problems begin on page A–44.

In Problems 1 and 2 solve the heat equation (1) subject to the given conditions.

1. $u(0, y, t) = 0, \quad u(\pi, y, t) = 0$
$u(x, 0, t) = 0, \quad u(x, \pi, t) = 0$
$u(x, y, 0) = u_0$

2. $\dfrac{\partial u}{\partial x}\bigg|_{x=0} = 0, \quad \dfrac{\partial u}{\partial x}\bigg|_{x=1} = 0$
$\dfrac{\partial u}{\partial y}\bigg|_{y=0} = 0, \quad \dfrac{\partial u}{\partial y}\bigg|_{y=1} = 0$
$u(x, y, 0) = xy$
[*Hint:* See Problem 53, Exercises 11.3.]

In Problems 3 and 4 solve the wave equation (2) subject to the given conditions.

3. $u(0, y, t) = 0, \quad u(\pi, y, t) = 0$
$u(x, 0, t) = 0, \quad u(x, \pi, t) = 0$
$u(x, y, 0) = xy(x - \pi)(y - \pi)$
$\dfrac{\partial u}{\partial t}\bigg|_{t=0} = 0$

4. $u(0, y, t) = 0, \quad u(b, y, t) = 0$
$u(x, 0, t) = 0, \quad u(x, c, t) = 0$
$u(x, y, 0) = f(x, y)$
$\dfrac{\partial u}{\partial t}\bigg|_{t=0} = g(x, y)$

The steady-state temperature $u(x, y, z)$ in the rectangular parallelepiped shown in Figure 12.21 satisfies Laplace's equation in three dimensions:

$$\frac{\partial^2 u}{\partial x^2} + \frac{\partial^2 u}{\partial y^2} + \frac{\partial^2 u}{\partial z^2} = 0. \qquad (13)$$

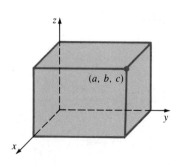

(a, b, c)

Figure 12.21

5. Solve Laplace's equation (13) if the top $(z = c)$ of the parallelepiped is kept at temperature $f(x, y)$ and the remaining sides are kept at temperature zero.

6. Solve Laplace's equation (13) if the bottom $(z = 0)$ of the parallelepiped is kept at temperature $f(x, y)$ and the remaining sides are kept at temperature zero.

Chapter 12   *REVIEW*

A particular solution of a linear partial differential equation in two variables can possibly be found by assuming a solution in the form of a product $u = XY$, where X is a function of x only and Y is a function of y only. If applicable, this **method of separation of variables** leads to two ordinary differential equations.

A **boundary-value problem** consists of finding a function that satisfies a partial differential equation as well as side conditions consisting of perhaps both boundary conditions and initial conditions. We applied the method of separation of variables to obtain solutions of certain boundary-value problems involving the heat equation, the wave equation, and Laplace's equation. The procedure consisted of five basic steps:

(i) Separate the variables.

(ii) Solve the separated ordinary differential equations and find the eigenvalues and eigenfunctions of the problem.

(iii) Form the products u_n.

(iv) Use the superposition principle to form an infinite series of the function u_n.

(v) After using a boundary, or the initial condition(s), obtain the coefficients in the series by making an appropriate identification with half-range sine or cosine expansions, Fourier series, or generalized Fourier series.

Chapter 12   *REVIEW EXERCISES*

Answers to odd-numbered problems begin on page A–44.

1. Use separation of variables to find product solutions of
$$\frac{\partial^2 u}{\partial x \, \partial y} = u.$$

2. Use separation of variables to find product solutions of
$$\frac{\partial^2 u}{\partial x^2} + \frac{\partial^2 u}{\partial y^2} + 2\frac{\partial u}{\partial x} + 2\frac{\partial u}{\partial y} = 0.$$

Is it possible to choose a constant of separation so that both X and Y are oscillatory functions?

3. Find a steady-state solution $\psi(x)$ of the boundary-value problem:

$$k \frac{\partial^2 u}{\partial x^2} = \frac{\partial u}{\partial t}, \qquad 0 < x < \pi, \quad t > 0$$

$$u(0, t) = u_0, \qquad -\frac{\partial u}{\partial x}\bigg|_{x=\pi} = u(\pi, t) - u_1, \qquad t > 0$$

$$u(x, 0) = 0, \qquad 0 < x < \pi.$$

4. Give a physical interpretation for the boundary conditions in Problem 3.

5. At $t = 0$ a string of unit length is stretched on the positive x-axis. The ends of the string $x = 0$ and $x = 1$ are secured on the x-axis for $t > 0$. Find the displacement $u(x, t)$ if the initial velocity $g(x)$ is as given in Figure 12.22.

Figure 12.22

6. The partial differential equation

$$\frac{\partial^2 u}{\partial x^2} + x^2 = \frac{\partial^2 u}{\partial t^2}$$

is a form of the wave equation when an external vertical force proportional to the square of the horizontal distance from the left end is applied to the string. The string is secured at $x = 0$ one unit above the x-axis and on the x-axis at $x = 1$ for $t > 0$. Find the displacement $u(x, t)$ if the string starts from rest from the initial displacement $f(x)$.

7. Find the steady-state temperature $u(x, y)$ in the square plate shown in Figure 12.23.

Figure 12.23

8. Find the steady-state temperature $u(x, y)$ in the semi-infinite plate shown in Figure 12.24.

Figure 12.24

9. Solve Problem 8 if the boundaries $y = 0$ and $y = \pi$ are held at temperature zero for all time.

10. Find the temperature $u(x, t)$ in the infinite plate of width $2L$ shown in

Figure 12.25

Figure 12.25 if the initial temperature is u_0 throughout [*Hint:* $u(x, 0) = u_0$, $-L < x < L$, is an even function of x.]

11. Solve the boundary-value problem:

$$\frac{\partial^2 u}{\partial x^2} = \frac{\partial u}{\partial t}, \qquad 0 < x < \pi, \quad t > 0$$

$$u(0, t) = 0, \qquad u(\pi, t) = 0, \qquad t > 0$$

$$u(x, 0) = \sin x, \qquad 0 < x < \pi.$$

12. Solve the boundary-value problem:

$$k\frac{\partial^2 u}{\partial x^2} + \sin 2\pi x = \frac{\partial u}{\partial t}, \qquad 0 < x < 1, \quad t > 0$$

$$u(0, t) = 0, \qquad u(1, t) = 0, \qquad t > 0$$

$$u(x, 0) = \sin \pi x, \qquad 0 < x < 1.$$

13. Find a formal series solution of the problem

$$\frac{\partial^2 u}{\partial x^2} + 2\frac{\partial u}{\partial x} = \frac{\partial^2 u}{\partial t^2} + 2\frac{\partial u}{\partial t} + u, \qquad 0 < x < \pi, \quad t > 0$$

$$u(0, t) = 0, \qquad u(\pi, t) = 0, \qquad t > 0$$

$$\left.\frac{\partial u}{\partial t}\right|_{t=0} = 0, \qquad 0 < x < \pi.$$

Do not attempt to evaluate the coefficients in the series.

14. The concentration $c(x, t)$ of a substance that both diffuses in a medium and is convected by currents in the medium satisfies the partial differential equation

$$k\frac{\partial^2 c}{\partial x^2} - h\frac{\partial c}{\partial x} = \frac{\partial c}{\partial t}, \qquad h \text{ a constant.}$$

Solve this equation subject to

$$c(0, t) = 0, \qquad c(1, t) = 0, \qquad t > 0$$

$$c(x, 0) = c_0, \qquad 0 < x < 1,$$

where c_0 is a constant.

WHERE IS MIDDLE C?

Ruth G. Favro

Department of Mathematics and Computer Science
Lawrence Technological University

How can we relate the Fourier series solution of the vibrating string problem to what we hear from a guitar, violin, or piano? What correspondence does mathematics have with middle C?

The sound generated by a musical instrument is transmitted to our ears by a pressure wave that travels through the air. In a stringed instrument, the vibrations created when we pluck or strike or bow the string are transmitted to the body of the instrument. This is done by stretching the strings over the bridge, a thin bar mounted on a part of the body called the sound board. The vibration of the materials in the body changes the pressure in the air inside the box; these pressure waves are then transmitted to our ears through openings in the body. The vibrations of the string are transverse waves, which move perpendicular to the direction of motion of the string. The pressure waves are longitudinal waves; they move in the same direction as the sound.

Some attributes of musical sound are pitch, timbre, and loudness. The primary attribute is pitch, which distinguishes whether a sound we hear is, say, middle C or an octave above or below. Timbre can be said to qualify the pitch (for example, by making the sound "bright" or "dull").

In order to talk about the sounds, it is convenient to introduce some standard terminology. The piano keyboard is used as a standard reference because of the wide range of notes it produces. We label the keys as shown in Figure 12.26. Twelve successive keys, seven white and five black, form an **octave**. The interval between each key is called a **semitone**. There are more

A3 B3 C4 D4 E4 F4 G4 A4 B4 C5 D5

Figure 12.26

than seven octaves on a piano. Middle C is referred to as C4, so the key that is an octave above is C5.

Mathematical Model

We want to analyze the vibrations of a string under tension, after it has been plucked or struck. The simplest problem statement is to find the vertical displacement $u(x, t)$ of an "ideal" string at any point x and at any later time t after it is set in motion. We assume it is fixed at the ends, is of small diameter, and vibrates only in a plane. It has length L and constant density ρ and is under constant tension T, all with suitable units. The mathematical model for the displacement $u(x, t)$ is

$$u_{tt} = c^2 u_{xx}, \qquad c = \sqrt{\frac{T}{\rho}}$$

$$\begin{cases} u(0, t) = u(L, t) = 0, & 0 < t \\ u(x, 0) = f(x), \quad u_t(x, 0) = g(x), & 0 < x < L. \end{cases}$$

The units of c are length per time; c represents the wave velocity. Separation of variables yields a solution

$$u(x, t) = \sum \left(A_n \cos \frac{n\pi ct}{L} + B_n \sin \frac{n\pi ct}{L} \right) \sin \frac{n\pi x}{L} \tag{1}$$

with the A_n and B_n to be determined by the initial conditions.

Each term in the series (1) is called a **standing wave** on $(0, L)$, or a **mode of vibration**. We can think of each one as a sine wave with a time-varying amplitude. Alternatively, using trigonometric identities, we can write (1) as

$$\sum C_n \sin \frac{n\pi x}{L} \sin \left(\frac{n\pi ct}{L} + \phi_n \right), \qquad C_n^2 = A_n^2 + B_n^2.$$

Then the motion looks like a traveling wave with a position-varying amplitude. See Figure 12.27. The points where each mode crosses the axis are called nodes; these are fixed points in the motion of the standing wave. It is the periodicity of the modes of vibration, all of which have at least the half-period $(0, L)$, that produces periodic pressure waves, and thus what we call musical sound.

$n = 1$ $n = 2$ $n = 3$

Figure 12.27 First Three Standing Waves

Frequency, Pitch, and Timbre

The nth mode of vibration has period p_n and frequency $f_n = 1/p_n$. Taking the "traveling wave" form of (1), we find $f_n = nc/2L$, $c = \sqrt{T/\rho}$. Frequency is measured in cycles per second; one cycle per second is called one hertz (Hz), named after Heinrich Hertz, a nineteenth-century German physicist. The lowest frequency, f_1, is called the **fundamental**, or the **first partial**. The next is the **second harmonic**, or the **second partial**, and so on. (Harmonics are also called **overtones**: the second harmonic is called the first overtone.)

The fundamental mode of vibration, f_1, determines the pitch of a musical sound. To sound the A above middle C (A4), the string must vibrate at 440 Hz, the standard frequency for concert instrument tuning. Middle C is 261.6 (~ 262) Hz. Since the frequency depends on the tension and density (mass per unit length) of the string, if our ideal string is steel with density 0.006 kg/m and tension 650 N, then $262 = (1/2L)\sqrt{(650/0.006)}$. Then a string length of $L = \sim 0.63$ m or 25 in. will sound middle C [2, p. 178].

Note that the frequencies of each term in the series (1) are integral multiples of each other: $f_n = nf_1$. The sequence $f_1, f_2, \ldots, f_n, \ldots$ is called a **harmonic series** in acoustics. The consonant intervals in music—such as thirds, fourths, fifths, and octaves—are produced by pitches whose fundamental frequencies bear certain ratios to each other. The accompanying table shows some of the basic relationships [4, p. 64]. A note an octave above a given note has a fundamental that is double the fundamental of the original. So if C4 has fundamental f (~ 262 Hz), then C5 has fundamental $2f$ (~ 524 Hz). An interval called a fifth is sounded by two notes that are seven semitones apart, for example, C4 and G4. The note G4 has a fundamental frequency that is $(3/2)f$ (~ 396 Hz). This relationship holds for the higher harmonics as well. If you have a wire that vibrates at a certain pitch, you can force its pitch up a fifth by placing a wedge at a point 2/3 of its length.

Interval Name	Notes (in C major)	Ideal Ratio	Number of Semitones
Octave	C-C	2:1	12
Minor third	E-G	6:5	3
Major third	C-E	5:4	4
Fourth	C-F	4:3	5
Fifth	C-G	3:2	7
Minor sixth	E-C	8:5	8
Major sixth	C-A	5:3	9

Effect of Initial Conditions

The initial position and velocity of the string provide the conditions that determine the constants A_n and B_n. Physically, this could mean plucking

Figure 12.28

Figure 12.29

Figure 12.30

the string in one place (as on a guitar or a harpsichord in Figure 12.28), or plucking in two places (as on a harp in Figure 12.29), or striking the string with a hammer (as on a piano or dulcimer in Figure 12.30). Let us take a very simplified model of each of these. The A_n and B_n are Fourier coefficients:

$g(x) = 0$ plucked in the middle (Figure 12.28)

$$A_n = \frac{8h}{n^2\pi^2}\sin\frac{n\pi}{2}, \quad B_n = 0 \qquad (2)$$

$g(x) = 0$ plucked in two places (harp) (Figure 12.29)

$$A_n = \frac{16h}{n^2\pi^2}\left[\sin\frac{n\pi}{4} - \sin\frac{3n\pi}{4}\right], \quad B_n = 0 \qquad (3)$$

$f(x) = 0$ struck (piano) (Figure 12.30)

$$A_n = 0, \quad B_n = \frac{2hL}{n^2\pi^2 c}\left[\cos\frac{n\pi}{4} - \cos\frac{n\pi}{2}\right] \qquad (4)$$

How do these conditions affect the sound? If some of the harmonics are missing, it affects the tone of the pitch we hear. The timbre, or tone, of a sound depends on the higher harmonics. If the higher harmonics are intense (that is, have amplitudes that do not decay too fast), the sound is called bright. If they decay very fast, the sound is called dull.

Plucking at a point where a standing wave has a node suppresses that frequency. For example, plucking at $L/2$ suppresses $f_2, f_4, \ldots$; in (2), we observe all $A_{2n} = 0$. Plucking at $L/4$ suppresses $f_4, f_8, \ldots$; in (3), all $A_{4n} = 0$.* However, plucking at a point of maximum amplitude of a standing wave (an antinode) reinforces that mode, making its contribution to the complex waveform stronger.

Loudness

Loudness, the third attribute, can be measured physically as intensity (watts/m^2), but for convenience this is usually converted to the familiar decibel (dB) scale. The loudness of the sound depends on how much energy can be efficiently transferred from the string to the instrument sound board. In a piano, the middle and upper registers of strings are triple-strung, both for sound quality and in order to increase the loudness. A string in this range is very thin and strung at maximum tension. One string is not enough to produce the loudness or the tone fullness we expect from a piano.

In real life, many other factors also affect the sound, for example, frictional forces of air resistance, stiffness of the strings (especially on the lower piano strings, which are wound to increase the diameter), and the width and hardness of the felt on the piano hammer. While consideration of these prob-

---

* It is interesting to note that many more frequencies in (3) are missing, including f_1.

lems is beyond the scope of this essay, one simple change we can make in our mathematical model is to put in a term corresponding to air resistance in the PDE. The new equation is $u_{tt} + \beta u_t = c^2 u_{xx}$ whose solution contains a time-dependent decaying exponential factor, as we would expect.

Other Instruments

We can apply our analysis to other musical instruments, such as the organ. The sound in a pipe organ is produced by directing an air flow across a hole. The same process is used in a flute and a recorder, or when we blow across the mouth of a narrow-necked bottle. The sounds are called **edgetones**.

The pitch of an organ pipe is determined by its length and by whether one end of the pipe is open or closed (stopped). The end of the pipe with the air hole is always "open." When air is forced across the edge and through the pipe, periodic oscillations are created in the air mass in the pipe. A longitudinal wave is created (rather than a transverse wave as on a string) that has both a motion and a pressure component.

Air is displaced back and forth from its equilibrium position as the wave travels down the tube and is reflected from the other end. This creates a periodic change in pressure at each point. At a closed end, the pressure is at a maximum and is reflected from the end; at an open end, the pressure should be at a minimum (and the air displacement at a maximum) in order for the wave to oscillate back.

The partial differential equations governing both the air displacement and the pressure are identical to the wave equation. If $p(x, t)$ represents the pressure at any time and position in the tube, then $p_{tt} = k^2 p_{xx}$, where k represents the speed of sound in air (344 m/s). There will be periodic standing waves, whose interference pattern forms the complex waveform of the sound. Appropriate boundary conditions for the pressure equation can be modeled by considering the standing waves at the ends of the tube.

At a closed end, the pressure wave is at a maximum and is reflected back. The pressure gradient is zero, so an appropriate condition is $p_x(L, t) = 0$. At an open end, the pressure wave must have a node in order for most of it to be reflected; the pressure just outside should be approximately the same as the pressure just inside of the pipe. The appropriate condition is $p(L, t) = c$, where c is a constant. This leads to the following boundary values for a pipe of length L:

Open at $x = 0$, open at $x = L$: $p(0, t) = c, \quad p(L, t) = c.$

Open at $x = 0$, closed at $x = L$: $p(0, t) = c, \quad p_x(L, t) = 0.$

The eigenvalues for the closed pipe are $nk/4L$ and for the open pipe, $nk/2L$.

Since the fundamental frequency of an open pipe is twice that of a closed pipe of the same length, the pitch of the open pipe is an octave higher. In a closed pipe the even harmonics are missing, which gives it a tone different from an open pipe's.

Where is middle C for the organ? The length of pipe we need to produce $f_1 = \sim 262$ Hz is [2, p. 254]:

For an open pipe, $L = \dfrac{k}{2f_1} = (344 \text{ m/s})/(524 \text{ Hz}) = \sim 0.66$ m.

For a closed pipe, $L = \dfrac{k}{4f_1} = (344 \text{ m/s})/(1048 \text{ Hz}) = \sim 0.33$ m.

REFERENCES

1. Benade, Arthur H. *Fundamentals of Musical Acoustics.* New York: Oxford University Press, 1976.

2. Hall, Donald E. *Musical Acoustics.* Belmont, CA: Wadsworth, 1980.

3. Morse, Philip M. *Vibration and Sound.* New York: McGraw-Hill, 1948.

4. Pierce, John R. *The Science of Musical Sound.* New York: Scientific American Books, 1983.

C H A P T E R 13

BOUNDARY-VALUE PROBLEMS IN OTHER COORDINATE SYSTEMS

13.1   Problems Involving
       Laplace's Equation in
       Polar Coordinates
13.2   Problems in Polar and
       Cylindrical Coordinates

13.3   Problems in Spherical
       Coordinates

       Chapter 13 Review
       Chapter 13 Review Exercises

All the boundary-value problems that we have considered so far have been expressed in terms of a rectangular coordinate system. If, however, we wish to find, say, temperatures in a circular disk, in a circular cylinder, or in a sphere, we would naturally try to

disk   cylinder   sphere

describe the problem in terms of polar coordinates, cylindrical coordinates, and spherical coordinates, respectively. It is important then that we express the Laplacian $\nabla^2 u$ in terms of these different coordinates.

In Sections 13.2 and 13.3 we will put the theory of Fourier-Bessel series and Fourier-Legendre series to practical use.

Important Concepts

   Laplacian in
      polar coordinates
      cylindrical coordinates
      spherical coordinates
   Double eigenvalues
   Radial symmetry
   Radial vibrations
   Standing waves
   Nodal lines

13.1 PROBLEMS INVOLVING LAPLACE'S EQUATION IN POLAR COORDINATES

(x, y) or
(r, θ)

Figure 13.1

The relationships between polar coordinates in the plane and rectangular coordinates are given by

$$x = r \cos \theta, \quad y = r \sin \theta, \quad \text{and} \quad r^2 = x^2 + y^2, \quad \tan \theta = \frac{y}{x}.$$

See Figure 13.1. The first pair of equations transform polar coordinates (r, θ) into rectangular coordinates (x, y); the second pair of equations enable us to transform rectangular coordinates into polar coordinates. These equations also make it possible to convert the two-dimensional Laplacian of the function u, $\nabla^2 u = \partial^2 u / \partial x^2 + \partial^2 u / \partial y^2$, into polar coordinates. You are encouraged to work through the details of the chain rule and show that

$$\frac{\partial u}{\partial x} = \frac{\partial u}{\partial r} \frac{\partial r}{\partial x} + \frac{\partial u}{\partial \theta} \frac{\partial \theta}{\partial x} = \cos \theta \, \frac{\partial u}{\partial r} - \frac{\sin \theta}{r} \frac{\partial u}{\partial \theta}$$

$$\frac{\partial u}{\partial y} = \frac{\partial u}{\partial r} \frac{\partial r}{\partial y} + \frac{\partial u}{\partial \theta} \frac{\partial \theta}{\partial y} = \sin \theta \, \frac{\partial u}{\partial r} + \frac{\cos \theta}{r} \frac{\partial u}{\partial \theta}$$

$$\frac{\partial^2 u}{\partial x^2} = \cos^2 \theta \, \frac{\partial^2 u}{\partial r^2} - \frac{2 \sin \theta \cos \theta}{r} \frac{\partial^2 u}{\partial r \, \partial \theta}$$

$$+ \frac{\sin^2 \theta}{r^2} \frac{\partial^2 u}{\partial \theta^2} + \frac{\sin^2 \theta}{r} \frac{\partial u}{\partial r} + \frac{2 \sin \theta \cos \theta}{r} \frac{\partial u}{\partial \theta} \qquad \textbf{(1)}$$

$$\frac{\partial^2 u}{\partial y^2} = \sin^2 \theta \, \frac{\partial^2 u}{\partial r^2} + \frac{2 \sin \theta \cos \theta}{r} \frac{\partial^2 u}{\partial r \, \partial \theta}$$

$$+ \frac{\cos^2 \theta}{r^2} \frac{\partial^2 u}{\partial \theta^2} + \frac{\cos^2 \theta}{r} \frac{\partial u}{\partial r} - \frac{2 \sin \theta \cos \theta}{r} \frac{\partial u}{\partial \theta}. \qquad \textbf{(2)}$$

Adding (1) and (2) and simplifying yield the Laplacian of u in polar coordinates:

$$\nabla^2 u = \frac{\partial^2 u}{\partial x^2} + \frac{\partial^2 u}{\partial y^2} = \frac{\partial^2 u}{\partial r^2} + \frac{1}{r} \frac{\partial u}{\partial r} + \frac{1}{r^2} \frac{\partial^2 u}{\partial \theta^2}.$$

In this section we shall focus only on problems involving Laplace's equation in polar coordinates

$$\frac{\partial^2 u}{\partial r^2} + \frac{1}{r} \frac{\partial u}{\partial r} + \frac{1}{r^2} \frac{\partial^2 u}{\partial \theta^2} = 0.$$

Our first example is the Dirichlet problem for a circular disk.

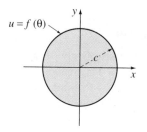

$u = f(\theta)$

Figure 13.2

EXAMPLE 1

Find the steady-state temperature $u(r, \theta)$ in the circular plate of radius c if the temperature of the circumference is $u(c, \theta) = f(\theta)$, $0 < \theta < 2\pi$. See Figure 13.2.

Solution We must solve Laplace's equation

$$\frac{\partial^2 u}{\partial r^2} + \frac{1}{r}\frac{\partial u}{\partial r} + \frac{1}{r^2}\frac{\partial^2 u}{\partial \theta^2} = 0, \qquad 0 < \theta < 2\pi, \;\; 0 < r < c$$

subject to $u(c, \theta) = f(\theta), \qquad 0 < \theta < 2\pi$.

If we define $u = R(r)\Theta(\theta)$, then separation of variables gives

$$\frac{r^2 R'' + rR'}{R} = -\frac{\Theta''}{\Theta} = \lambda^2$$

and

$$r^2 R'' + rR' - \lambda^2 R = 0 \tag{3}$$

$$\Theta'' + \lambda^2 \Theta = 0. \tag{4}$$

The solution of (4) is

$$\Theta = c_1 \cos \lambda\theta + c_2 \sin \lambda\theta. \tag{5}$$

Equation (3) is recognized as a Cauchy-Euler equation. Its general solution is

$$R = c_3 r^\lambda + c_4 r^{-\lambda}. \tag{6}$$

Now there are no explicit conditions in the statement of this problem that enable us to determine any of the coefficients or the eigenvalues. However, there are some implicit conditions.

First, our physical intuition leads us to expect that the temperature $u(r, \theta)$ should be bounded inside the circle $r = c$. Moreover, the temperature u should be the same at a specified point in the circle regardless of the polar description of that point. Since $(r, \theta + 2\pi)$ is an equivalent description of the point (r, θ), we must have

$$u(r, \theta) = u(r, \theta + 2\pi).$$

In other words, the temperature $u(r, \theta)$ must be periodic with period 2π. But the only way the solutions of (4) can be 2π-periodic is to take $\lambda = n$, where $n = 0, 1, 2, \ldots$. Thus for $n > 0$, (5) and (6) become

$$\Theta = c_1 \cos n\theta + c_2 \sin n\theta \tag{7}$$

$$R = c_3 r^n + c_4 r^{-n}. \tag{8}$$

Now observe in (8) that $r^{-n} = 1/r^n$. In order that the solution be bounded at the center of the plate, that is, at $r = 0$, we must define $c_4 = 0$. Hence for

$\lambda = n > 0,$

$$u_n = R\Theta = r^n(A_n \cos n\theta + B_n \sin n\theta), \tag{9}$$

where c_3c_1 and c_3c_2 have been replaced by A_n and B_n, respectively.

There is one last item of consideration; namely, $n = 0$ is an eigenvalue. For this value the solutions of (3) and (4) are not given by (5) and (6). The solutions of the differential equations $\Theta'' = 0$ and $r^2R'' + rR' = 0$ are

$$\Theta = c_5\theta + c_6 \quad \text{and} \quad R = c_7 + c_8 \ln r.$$

Periodicity demands that we take $c_5 = 0$, whereas the expectation that the temperature is finite at $r = 0$ demands that we take $c_8 = 0$. In this case a product solution of the partial differential equation is simply

$$u_0 = A_0, \tag{10}$$

where A_0 represents c_6c_7.

The superposition principle then gives

$$u(r, \theta) = \sum_{n=0}^{\infty} u_n$$

or

$$u(r, \theta) = A_0 + \sum_{n=1}^{\infty} r^n(A_n \cos n\theta + B_n \sin n\theta). \tag{11}$$

Finally, by applying the boundary condition at $r = c$ to the result in (11), we recognize

$$f(\theta) = A_0 + \sum_{n=1}^{\infty} c^n(A_n \cos n\theta + B_n \sin n\theta)$$

as an expansion of f in a full Fourier series. Consequently, we can make the identifications

$$A_0 = \frac{a_0}{2}, \quad c^nA_n = a_n, \quad \text{and} \quad c^nB_n = b_n.$$

That is,

$$A_0 = \frac{1}{2\pi} \int_0^{2\pi} f(\theta)\, d\theta \tag{12}$$

$$A_n = \frac{1}{c^n\pi} \int_0^{2\pi} f(\theta) \cos n\theta\, d\theta \tag{13}$$

$$B_n = \frac{1}{c^n\pi} \int_0^{2\pi} f(\theta) \sin n\theta\, d\theta. \tag{14}$$

The solution of the problem consists of the series given in (11), where A_0, A_n, and B_n are defined in (12), (13), and (14). ∎

Observe in Example 1 that corresponding to each positive eigenvalue, $n = 1, 2, 3, \ldots$, there are two different eigenfunctions, namely, $\cos n\theta$ and $\sin n\theta$. In this situation the eigenvalues are sometimes called **double eigenvalues**.

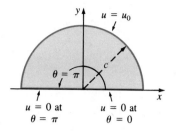

y

$u = u_0$

$\theta = \pi$

c

x

$u = 0$ at
$\theta = \pi$

$u = 0$ at
$\theta = 0$

Figure 13.3

EXAMPLE 2

Find the steady-state temperature $u(r, \theta)$ in the semi-circular plate shown in Figure 13.3.

Solution The boundary-value problem is

$$\frac{\partial^2 u}{\partial r^2} + \frac{1}{r}\frac{\partial u}{\partial r} + \frac{1}{r^2}\frac{\partial^2 u}{\partial \theta^2} = 0, \qquad 0 < \theta < \pi, \quad 0 < r < c$$

$$u(c, \theta) = u_0, \qquad 0 < \theta < \pi$$

$$u(r, 0) = 0, \qquad u(r, \pi) = 0, \qquad 0 < r < c.$$

Defining $u = R(r)\Theta(\theta)$ and separating variables give

$$\frac{r^2 R'' + rR'}{R} = -\frac{\Theta''}{\Theta} = \lambda^2$$

and

$$r^2 R'' + rR' - \lambda^2 R = 0 \tag{15}$$

$$\Theta'' + \lambda^2 \Theta = 0. \tag{16}$$

Applying the boundary conditions $\Theta(0) = 0$ and $\Theta(\pi) = 0$ to the solution $\Theta = c_1 \cos \lambda\theta + c_2 \sin \lambda\theta$ of (16) gives, in turn, $c_1 = 0$ and $\lambda = n, n = 1, 2, 3, \ldots$. Hence $\Theta = c_2 \sin n\theta$. Unlike the problem in Example 1, $n = 0$ is not an eigenvalue. With $\lambda = n$ the solution of (15) is

$$R = c_3 r^n + c_4 r^{-n}.$$

But the assumption that $u(r, \theta)$ is bounded at $r = 0$ prompts us to define $c_4 = 0$. Therefore

$$u_n = R(r)\Theta(\theta) = A_n r^n \sin n\theta$$

and

$$u(r, \theta) = \sum_{n=1}^{\infty} A_n r^n \sin n\theta.$$

The remaining boundary condition at $r = c$ gives the sine series

$$u_0 = \sum_{n=1}^{\infty} A_n c^n \sin n\theta.$$

Consequently

$$A_n c^n = \frac{2}{\pi}\int_0^{\pi} u_0 \sin n\theta \, d\theta$$

and so

$$A_n = \frac{2u_0}{\pi c^n}\frac{1 - (-1)^n}{n}.$$

Hence the solution of the problem is given by

$$u(r, \theta) = \frac{2u_0}{\pi}\sum_{n=1}^{\infty}\frac{1 - (-1)^n}{n}\left(\frac{r}{c}\right)^n \sin n\theta.$$

■

13.1 EXERCISES

Answers to odd-numbered problems begin on page A–44.

In Problems 1–4 find the steady-state temperature $u(r, \theta)$ in a circular plate of radius $r = 1$ if the temperature on the circumference is as given.

1. $u(1, \theta) = \begin{cases} u_0, & 0 < \theta < \pi \\ 0, & \pi < \theta < 2\pi \end{cases}$

2. $u(1, \theta) = \begin{cases} \theta, & 0 < \theta < \pi \\ \pi - \theta, & \pi < \theta < 2\pi \end{cases}$

3. $u(1, \theta) = 2\pi\theta - \theta^2, \quad 0 < \theta < 2\pi$

4. $u(1, \theta) = \theta, \quad 0 < \theta < 2\pi$

5. Solve the exterior Dirichlet problem for a circular disk of radius c if $u(c, \theta) = f(\theta), 0 < \theta < 2\pi$. In other words, find the steady-state temperature $u(r, \theta)$ in a plate that coincides with the entire xy-plane in which a circular hole of radius c has been cut out around the origin and the temperature on the circumference of the hole is $f(\theta)$. [*Hint:* Assume that the temperature is bounded as $r \to \infty$.]

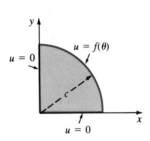

$u = 0$ $u = f(\theta)$ $u = 0$ c

Figure 13.4

6. Find the steady-state temperature in the quarter-circular plate shown in Figure 13.4.

7. If the boundaries $\theta = 0$ and $\theta = \pi/2$ in Figure 13.4 are insulated, we then have, respectively,

$$\left.\frac{\partial u}{\partial \theta}\right|_{\theta = 0} = 0, \qquad \left.\frac{\partial u}{\partial \theta}\right|_{\theta = \pi/2} = 0.$$

Find the steady-state temperature if $u(c, \theta) = \begin{cases} 1, & 0 < \theta < \pi/4, \\ 0, & \pi/4 < \theta < \pi/2. \end{cases}$

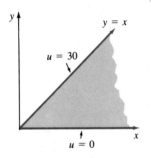

$y = x$ $u = 30$ $u = 0$

Figure 13.5

8. Find the steady-state temperature in the infinite wedge-shaped plate shown in Figure 13.5. [*Hint:* Assume that the temperature is bounded as $r \to 0$ and as $r \to \infty$.]

9. Find the steady-state temperature $u(r, \theta)$ in the circular ring shown in Figure 13.6. [*Hint:* Proceed as in Example 1.]

10. If the boundary conditions for the circular ring in Figure 13.6 are $u(a, \theta) = u_0, u(b, \theta) = u_1, 0 < \theta < 2\pi, u_0$ and u_1 constants, show that the steady-state temperature is given by $u(r, \theta) = [u_0 \ln(r/b) - u_1 \ln(r/a)]/\ln(a/b)$. [*Hint:* Try a solution of the form $u(r, \theta) = v(r, \theta) + \psi(r)$.]

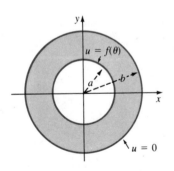

$u = f(\theta)$ a b $u = 0$

Figure 13.6

11. Find the steady-state temperature $u(r, \theta)$ in a semicircular ring if

$$u(a, \theta) = \theta(\pi - \theta), \qquad u(b, \theta) = 0, \qquad 0 < \theta < \pi$$

$$u(r, 0) = 0, \qquad u(r, \pi) = 0, \qquad a < r < b.$$

12. Find the steady-state temperature $u(r, \theta)$ in a semicircular plate of radius $r = 1$ if

$$u(1, \theta) = u_0, \qquad 0 < \theta < \pi$$

$$u(r, 0) = 0, \qquad u(r, \pi) = u_0, \qquad 0 < r < 1,$$

u_0 a constant.

13. Find the steady-state temperature $u(r, \theta)$ in a semicircular plate of radius $r = 2$ if

$$u(2, \theta) = \begin{cases} u_0, & 0 < \theta < \pi/2 \\ 0, & \pi/2 < \theta < \pi, \end{cases}$$

u_0 a constant, and the edges $\theta = 0$ and $\theta = \pi$ are insulated.

14. Show that the steady-state temperature in Example 1 can be written as the integral

$$u(r, \theta) = \frac{1}{2\pi} \int_0^{2\pi} \frac{c^2 - r^2}{c^2 - 2cr \cos(t - \theta) + r^2} f(t)\, dt.$$

This result is known as **Poisson's integral formula for a circle**. [*Hint:* First show that

$$u(r, \theta) = \frac{1}{2\pi} \int_0^{2\pi} f(t) \left[1 + 2 \sum_{n=1}^{\infty} \left(\frac{r}{c}\right)^n \cos n(t - \theta) \right] dt.$$

Then use $\cos nv = \frac{1}{2}(e^{inv} + e^{-inv})$ and geometric series to show that

$$1 + 2 \sum_{n=1}^{\infty} u^n \cos nv = \frac{1 - u^2}{1 - 2u \cos v + u^2}, \qquad |u| < 1.]$$

13.2 PROBLEMS IN POLAR AND CYLINDRICAL COORDINATES

The two-dimensional heat and wave equations:

$$k\left(\frac{\partial^2 u}{\partial x^2} + \frac{\partial^2 u}{\partial y^2}\right) = \frac{\partial u}{\partial t} \quad \text{and} \quad a^2\left(\frac{\partial^2 u}{\partial x^2} + \frac{\partial^2 u}{\partial y^2}\right) = \frac{\partial^2 u}{\partial t^2}$$

expressed in polar coordinates are, respectively,

$$k\left(\frac{\partial^2 u}{\partial r^2} + \frac{1}{r}\frac{\partial u}{\partial r} + \frac{1}{r^2}\frac{\partial^2 u}{\partial \theta^2}\right) = \frac{\partial u}{\partial t} \quad \text{and} \quad a^2\left(\frac{\partial^2 u}{\partial r^2} + \frac{1}{r}\frac{\partial u}{\partial r} + \frac{1}{r^2}\frac{\partial^2 u}{\partial \theta^2}\right) = \frac{\partial^2 u}{\partial t^2},$$

where $u = u(r, \theta, t)$. To solve a boundary-value problem involving either of these equations by separation of variables we define $u = R(r)\Theta(\theta)T(t)$. As in Section 12.8 this assumption leads to multiple infinite series. See Problem 12. In the following discussion we shall consider the simpler, but still important, problems that possess **radial symmetry**, that is, problems that are independent of the angular coordinate θ. In this case the heat and wave equations take, in turn, the forms

$$k\left(\frac{\partial^2 u}{\partial r^2} + \frac{1}{r}\frac{\partial u}{\partial r}\right) = \frac{\partial u}{\partial t} \quad \text{and} \quad a^2\left(\frac{\partial^2 u}{\partial r^2} + \frac{1}{r}\frac{\partial u}{\partial r}\right) = \frac{\partial^2 u}{\partial t^2}, \tag{1}$$

where $u = u(r, t)$. Vibrations described by the second equation in (1) are said to be **radial vibrations**.

The first example deals with the free radial vibrations of a thin circular membrane. We assume that the displacements are small and that the motion is such that each point on the membrane moves in a direction perpendicular to the xy-plane (transverse vibrations); that is, the u-axis is perpendicular to the xy-plane. A physical model to keep in mind while reading this example is a vibrating drumhead.

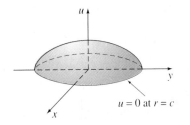

$u = 0$ at $r = c$

Figure 13.7

EXAMPLE 1

Find the displacement $u(r, t)$ of a circular membrane of radius c clamped along its circumference if its initial displacement is $f(r)$ and its initial velocity is $g(r)$. See Figure 13.7.

Solution The boundary-value problem to be solved is

$$a^2 \left(\frac{\partial^2 u}{\partial r^2} + \frac{1}{r} \frac{\partial u}{\partial r} \right) = \frac{\partial^2 u}{\partial t^2}, \qquad 0 < r < c, \quad t > 0$$

$$u(c, t) = 0, \qquad t > 0$$

$$u(r, 0) = f(r), \qquad \left. \frac{\partial u}{\partial t} \right|_{t=0} = g(r), \qquad 0 < r < c.$$

Substituting $u = R(r)T(t)$ into the partial differential equation and separating variables give

$$\frac{R'' + \dfrac{1}{r} R'}{R} = \frac{T''}{a^2 T} = -\lambda^2$$

and

$$rR'' + R' + \lambda^2 rR = 0 \tag{2}$$

$$T'' + a^2 \lambda^2 T = 0. \tag{3}$$

Now (2) is not a Cauchy-Euler equation but is the parametric Bessel differential equation with order $v = 0$. Its general solution is

$$R = c_1 J_0(\lambda r) + c_2 Y_0(\lambda r).$$

The general solution of the familiar equation (3) is

$$T = c_3 \cos a\lambda t + c_4 \sin a\lambda t.$$

Now recall $Y_0(\lambda r) \to -\infty$ as $r \to 0^+$, and so the implicit assumption that the displacement $u(r, t)$ should be bounded at $r = 0$ forces us to define $c_2 = 0$. Thus $R = c_1 J_0(\lambda r)$.

Since the boundary condition $u(c, t) = 0$ is equivalent to $R(c) = 0$, we must have $c_1 J_0(\lambda c) = 0$. We rule out $c_1 = 0$ (this would lead to a trivial solution of the partial differential equation), so consequently

$$J_0(\lambda c) = 0. \tag{4}$$

This equation defines the positive eigenvalues λ_n of the problem: If x_n are the positive roots of (4), then $\lambda_n = x_n/c$. Product solutions that satisfy the partial differential equation and the boundary condition are

$$u_n = RT = (A_n \cos a\lambda_n t + B_n \sin a\lambda_n t)J_0(\lambda_n r), \tag{5}$$

where we have done the usual relabeling of the constants. The superposition principle then gives

$$u(r, t) = \sum_{n=1}^{\infty} (A_n \cos a\lambda_n t + B_n \sin a\lambda_n t)J_0(\lambda_n r). \tag{6}$$

The given initial conditions determine the coefficients A_n and B_{n*}
Setting $t = 0$ in (6) and using $u(r, 0) = f(r)$ give

$$f(r) = \sum_{n=1}^{\infty} A_n J_0(\lambda_n r). \tag{7}$$

This last result is recognized as the Fourier-Bessel expansion of the function f on the interval $(0, c)$. Hence by a direct comparison of (4) and (7) with (7) and (15) of Section 11.5, we can identify the coefficients A_n with those given in (16) of Section 11.5:

$$A_n = \frac{2}{c^2 J_1^2(\lambda_n c)} \int_0^c r J_0(\lambda_n r)f(r)\, dr. \tag{8}$$

Next, we differentiate (6) with respect to t, set $t = 0$, and use $u_t(r, 0) = g(r)$:

$$g(r) = \sum_{n=1}^{\infty} a\lambda_n B_n J_0(\lambda_n r).$$

This is now a Fourier-Bessel expansion of the function g. By identifying the total coefficient $a\lambda_n B_n$ with (16) of Section 11.5, we can write

$$B_n = \frac{2}{a\lambda_n c^2 J_1^2(\lambda_n c)} \int_0^c r J_0(\lambda_n r)g(r)\, dr. \tag{9}$$

Finally, the solution of the given boundary-value problem is the series (6) with A_n and B_n defined in (8) and (9). ∎

Analogous to (8) of Section 12.4, the product solutions (5) are called **standing waves**. For $n = 1, 2, 3, \ldots$ the standing waves are basically the graph of $J_0(\lambda_n r)$ with the time-varying amplitude

$$A_n \cos a\lambda_n t + B_n \sin a\lambda_n t.$$

The standing waves at different values of time are represented by the dashed graphs in Figure 13.8. The zeros of each standing wave in the interval $(0, c)$

$n = 1$

(a)

$n = 2$

(b)

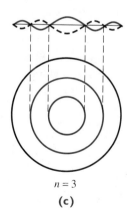

$n = 3$

(c)

Figure 13.8

are the roots of $J_0(\lambda_n r) = 0$ and correspond to the set of points on a standing wave where there is no motion. This set of points is called a **nodal line**. If (as in Example 1) the positive roots of $J_0(\lambda_n c) = 0$ are denoted by x_n, then $\lambda_n c = x_n$ implies $\lambda_n = x_n/c$ and consequently the zeros of the standing waves are determined from

$$J_0(\lambda_n r) = J_0\left(\frac{x_n}{c} r\right) = 0.$$

Now the first three positive zeros of J_0 are (approximately) $x_1 = 2.4$, $x_2 = 5.5$, and $x_3 = 8.7$. Thus for $n = 1$, the first positive root of

$$J_0\left(\frac{x_1}{c} r\right) = 0 \quad \text{is} \quad \frac{2.4}{c} r = 2.4 \quad \text{or} \quad r = c.$$

Since we are seeking zeros of the standing waves in the open interval $(0, c)$, the last result means that the first standing wave has no nodal line. For $n = 2$, the first two positive roots of

$$J_0\left(\frac{x_2}{c} r\right) = 0 \quad \text{are determined from} \quad \frac{5.5}{c} r = 2.4 \quad \text{and} \quad \frac{5.5}{c} r = 5.5.$$

Thus the second standing wave has one nodal line defined by $r = x_1 c/x_2 = 2.4c/5.5$. Note that $r \approx 0.44c < c$. For $n = 3$, a similar analysis shows that there are two nodal lines defined by $r = x_1 c/x_3 = 2.4c/8.7$ and $r = x_2 c/x_3 = 5.5c/8.7$. In general, the nth standing wave has $n - 1$ nodal lines $r = x_1 c/x_n$, $r = x_2 c/x_n, \ldots, r = x_{n-1} c/x_n$. Since $r = constant$ is an equation of a circle in polar coordinates, we see in Figure 13.8 that the nodal lines of a standing wave are concentric circles.

Laplacian in Cylindrical Coordinates

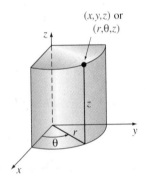

(x,y,z) or (r,θ,z)

Figure 13.9

From Figure 13.9 we see that the relationship between the cylindrical coordinates of a point in space and its rectangular coordinates is given by

$$x = r \cos \theta, \qquad y = r \sin \theta, \qquad z = z.$$

It follows immediately from the derivation of the Laplacian in polar coordinates (see Section 13.1) that the Laplacian of a function u in cylindrical coordinates is

$$\nabla^2 u = \frac{\partial^2 u}{\partial x^2} + \frac{\partial^2 u}{\partial y^2} + \frac{\partial^2 u}{\partial z^2} = \frac{\partial^2 u}{\partial r^2} + \frac{1}{r} \frac{\partial u}{\partial r} + \frac{1}{r^2} \frac{\partial^2 u}{\partial \theta^2} + \frac{\partial^2 u}{\partial z^2}.$$

$u = u_0$ at $z = 4$

$u = 0$ at $r = 2$

$u = 0$ at $z = 0$

Figure 13.10

EXAMPLE 2

Find the steady-state temperature in the circular cylinder shown in Figure 13.10.

Solution   The prescribed boundary conditions suggest that the temperature u has radial symmetry. Accordingly $u(r, z)$ is determined from:

$$\frac{\partial^2 u}{\partial r^2} + \frac{1}{r}\frac{\partial u}{\partial r} + \frac{\partial^2 u}{\partial z^2} = 0, \qquad 0 < r < 2, \quad 0 < z < 4$$

$$u(2, z) = 0, \qquad 0 < z < 4$$

$$u(r, 0) = 0, \qquad u(r, 4) = u_0, \qquad 0 < r < 2.$$

Using $u = R(r)Z(z)$ and separating variables give

$$\frac{R'' + \dfrac{1}{r}R'}{R} = -\frac{Z''}{Z} = -\lambda^2$$

and

$$rR'' + R' + \lambda^2 rR = 0 \qquad \text{(10)}$$

$$Z'' - \lambda^2 Z = 0. \qquad \text{(11)}$$

A negative separation constant is used here since there is no reason to expect the solution $u(r, z)$ to be periodic in z. The solution of (10) is

$$R = c_1 J_0(\lambda r) + c_2 Y_0(\lambda r),$$

and since the solution of (11) is defined on the finite interval $[0, 2]$, we write its general solution as

$$Z = c_3 \cosh \lambda z + c_4 \sinh \lambda z.$$

As in Example 1, the assumption that the function u is bounded at $r = 0$ demands that $c_2 = 0$. The condition $u(2, z) = 0$ implies $R(2) = 0$. This equation,

$$J_0(2\lambda) = 0, \qquad \text{(12)}$$

defines the positive eigenvalues λ_n of the problem. Last, $Z(0) = 0$ implies $c_3 = 0$. Hence we have $R = c_1 J_0(\lambda_n r)$, $Z = c_4 \sinh \lambda_n z$, and

$$u_n = RZ = A_n \sinh \lambda_n z \, J_0(\lambda_n r)$$

$$u(r, z) = \sum_{n=1}^{\infty} A_n \sinh \lambda_n z \, J_0(\lambda_n r).$$

The remaining boundary condition at $z = 4$ then gives the Fourier-Bessel series

$$u_0 = \sum_{n=1}^{\infty} A_n \sinh 4\lambda_n J_0(\lambda_n r)$$

so that in view of (12) the coefficients are defined by (16) of Section 11.5;

$$A_n \sinh 4\lambda_n = \frac{2u_0}{2^2 J_1^2(2\lambda_n)} \int_0^2 r J_0(\lambda_n r) \, dr.$$

To evaluate the last integral we first use the substitution $t = \lambda_n r$, followed by $\dfrac{d}{dt}[tJ_1(t)] = tJ_0(t)$:

$$A_n \sinh 4\lambda_n = \frac{u_0}{2\lambda_n^2 J_1^2(2\lambda_n)} \int_0^{2\lambda_n} \frac{d}{dt}[tJ_1(t)]\, dt = \frac{u_0}{\lambda_n J_1(2\lambda_n)}.$$

Finally we arrive at

$$A_n = \frac{u_0}{\lambda_n \sinh 4\lambda_n J_1(2\lambda_n)}.$$

Thus the temperature in the cylinder is given by

$$u(r, z) = u_0 \sum_{n=1}^{\infty} \frac{\sinh \lambda_n z \; J_0(\lambda_n r)}{\lambda_n \sinh 4\lambda_n \, J_1(2\lambda_n)}. \qquad \blacksquare$$

13.2 EXERCISES

Answers to odd-numbered problems begin on page A–45.

1. Find the displacement $u(r, t)$ in Example 1 if $f(r) = 0$ and the circular membrane is given an initial unit velocity in the upward direction.

2. A circular membrane of unit radius 1 is clamped along its circumference. Find the displacement $u(r, t)$ if the membrane starts from rest from the initial displacement $f(r) = 1 - r^2, 0 < r < 1$. [*Hint:* See Problem 10, Exercises 11.5.]

3. Find the steady-state temperature $u(r, z)$ in the cylinder in Example 2 if the boundary conditions are $u(2, z) = 0, 0 < z < 4, u(r, 0) = u_0, u(r, 4) = 0,$ $0 < r < 2.$

4. If the lateral side of the cylinder in Example 2 is insulated, then $\dfrac{\partial u}{\partial r}\Big|_{r=2} = 0, 0 < z < 4.$

   (a) Find the steady-state temperature $u(r, z)$ when $u(r, 4) = f(r), 0 < r < 2.$

   (b) Show that the steady-state temperature in part (a) reduces to $u(r, z) = u_0 z/4$ when $f(r) = u_0$. [*Hint:* Use (11) of Section 11.5.]

5. The temperature in a circular plate of radius c is determined from the boundary-value problem

$$k\left(\frac{\partial^2 u}{\partial r^2} + \frac{1}{r}\frac{\partial u}{\partial r}\right) = \frac{\partial u}{\partial t}, \qquad 0 < r < c, \quad t > 0$$

$$u(c, t) = 0, \qquad t > 0$$

$$u(r, 0) = f(r), \qquad 0 < r < c.$$

Solve for $u(r, t)$.

Figure 13.11

Figure 13.12

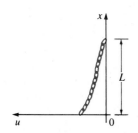

Figure 13.13

6. Solve Problem 5 if the edge $r = c$ of the plate is insulated.

7. When there is heat transfer from the lateral side of an infinite circular cylinder of unit radius (see Figure 13.11) into a surrounding medium at temperature zero, the temperature inside the cylinder is determined from

$$k\left(\frac{\partial^2 u}{\partial r^2} + \frac{1}{r}\frac{\partial u}{\partial r}\right) = \frac{\partial u}{\partial t}, \qquad 0 < r < 1, \quad t > 0$$

$$\left.\frac{\partial u}{\partial r}\right|_{r=1} = -hu(1, t), \qquad h > 0, \quad t > 0$$

$$u(r, 0) = f(r), \qquad 0 < r < 1.$$

Solve for $u(r, t)$.

8. Find the steady-state temperature $u(r, z)$ in a semi-infinite cylinder of unit radius ($z \geq 0$) if there is heat transfer from its lateral side into a surrounding medium at temperature zero and if the temperature of the base $z = 0$ is held at a constant temperature u_0.

9. A circular plate is a composite of two different materials in the form of concentric circles. See Figure 13.12. The temperature in the plate is determined from the boundary-value problem.

$$\frac{\partial^2 u}{\partial r^2} + \frac{1}{r}\frac{\partial u}{\partial r} = \frac{\partial u}{\partial t}, \qquad 0 < r < 2, \quad t > 0$$

$$u(2, t) = 100, \qquad t > 0$$

$$u(r, 0) = \begin{cases} 200, & 0 < r < 1 \\ 100, & 1 < r < 2. \end{cases}$$

Solve for $u(r, t)$. [*Hint:* Let $u(r, t) = v(r, t) + \psi(r)$.]

10. Solve the boundary-value problem

$$\frac{\partial^2 u}{\partial r^2} + \frac{1}{r}\frac{\partial u}{\partial r} + \beta = \frac{\partial u}{\partial t}, \qquad 0 < r < 1, \quad t > 0, \quad \beta \text{ a constant}$$

$$u(1, t) = 0, \qquad t > 0$$

$$u(r, 0) = 0, \qquad 0 < r < 1.$$

11. The horizontal displacement $u(x, t)$ of a heavy chain of length L oscillating in a vertical plane satisfies

$$g\frac{\partial}{\partial x}\left(x\frac{\partial u}{\partial x}\right) = \frac{\partial^2 u}{\partial t^2}, \qquad 0 < x < L, \quad t > 0.$$

See Figure 13.13.

(a) Using $-\lambda^2$ as a separation constant, show that the ordinary differential equation in the spatial variable x is $xX'' + X' + \lambda^2 X = 0$. Solve this equation by means of the substitution $x = \tau^2/4$.

(b) Use the result of part (a) to solve the given partial differential equation subject to

$$u(L, t) = 0, \qquad t > 0$$

$$u(x, 0) = f(x), \qquad \left.\frac{\partial u}{\partial t}\right|_{t=0} = 0, \qquad 0 < x < L.$$

[*Hint:* Assume the oscillations at the free end $x = 0$ are finite.]

12. In this problem we consider the general case of the vibrating circular membrane of radius c:

$$a^2\left(\frac{\partial^2 u}{\partial r^2} + \frac{1}{r}\frac{\partial u}{\partial r} + \frac{1}{r^2}\frac{\partial^2 u}{\partial \theta^2}\right) = \frac{\partial^2 u}{\partial t^2}, \qquad 0 < r < c, \quad t > 0$$

$$u(c, \theta, t) = 0, \qquad\qquad 0 < \theta < 2\pi, \quad t > 0$$

$$u(r, \theta, 0) = f(r, \theta), \qquad 0 < r < c, \quad 0 < \theta < 2\pi$$

$$\left.\frac{\partial u}{\partial t}\right|_{t=0} = g(r, \theta), \qquad 0 < r < c, \quad 0 < \theta < 2\pi.$$

(a) Assume that $u = R(r)\Theta(\theta)T(t)$ and that the separation constants are $-\lambda^2$ and $-v^2$. Show that the separated differential equations are

$$T'' + a^2\lambda^2 T = 0$$

$$\Theta'' + v^2\Theta = 0$$

$$r^2 R'' + rR' + (\lambda^2 r^2 - v^2)R = 0.$$

(b) Solve the separated equations.

(c) Show that the eigenvalues and eigenfunctions of the problem are:

Eigenvalues: $v = n$, $n = 0, 1, 2, \ldots$; eigenfunctions: 1, $\cos n\theta$, $\sin n\theta$.

Eigenvalues: $\lambda_{ni} = x_{ni}/c$, $i = 1, 2, \ldots$, where for each n, x_{ni} are the positive roots of $J_n(\lambda c) = 0$; eigenfunctions: $J_n(\lambda_{ni} r)$.

(d) Use the superposition principle to determine a multiple series solution. Do not attempt to evaluate the coefficients.

13. (a) Consider Example 1 with $a = 1$, $c = 10$, $g(r) = 0$, and $f(r) = 1 - r/10$, $0 < r < 10$. Find the numerical values of the first three eigenvalues $\lambda_1, \lambda_2, \lambda_3$ of the boundary-value problem and the first three coefficients A_1, A_2, A_3 of the series solution $u(r, t)$. Write the third partial sum S_3 of the series solution. [*Hint:* Use the table on page 311 and the reference on page 312. Also see Problem 25 in Exercises 6.5.]

(b) Use computer software to graph S_3 for $t = 0, 4, 10, 12, 20$.

13.3   PROBLEMS IN SPHERICAL COORDINATES

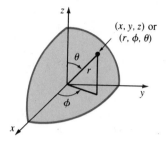

Figure 13.14

In the spherical coordinate system shown in Figure 13.14, we have
$x = r \sin \theta \cos \phi$, $y = r \sin \theta \sin \phi$, $z = r \cos \theta$.   The   Laplacian

$$\nabla^2 u = \frac{\partial^2 u}{\partial x^2} + \frac{\partial^2 u}{\partial y^2} + \frac{\partial^2 u}{\partial z^2} \quad \text{then becomes}$$

$$\nabla^2 u = \frac{\partial^2 u}{\partial r^2} + \frac{2}{r}\frac{\partial u}{\partial r} + \frac{1}{r^2 \sin^2\theta}\frac{\partial^2 u}{\partial \phi^2} + \frac{1}{r^2}\frac{\partial^2 u}{\partial \theta^2} + \frac{\cot\theta}{r^2}\frac{\partial u}{\partial \theta}. \tag{1}$$

As you might imagine, problems involving (1) can be quite formidable. Consequently we shall consider only a few of the simpler problems that are independent of the azimuthal angle ϕ.

EXAMPLE I

Find the steady-state temperature $u(r, \theta)$ in the sphere shown in Figure 13.15.

Solution   The temperature is determined from

$$\frac{\partial^2 u}{\partial r^2} + \frac{2}{r}\frac{\partial u}{\partial r} + \frac{1}{r^2}\frac{\partial^2 u}{\partial \theta^2} + \frac{\cot\theta}{r^2}\frac{\partial u}{\partial \theta} = 0, \quad 0 < r < c, \;\; 0 < \theta < \pi,$$

$$u(c, \theta) = f(\theta), \quad 0 < \theta < \pi.$$

If $u = R(r)\Theta(\theta)$, the partial differential equation separates as

$$\frac{r^2 R'' + 2rR'}{R} = -\frac{\Theta'' + \cot\theta\, \Theta'}{\Theta} = \lambda^2$$

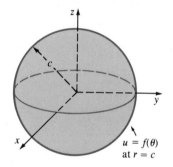

Figure 13.15

and so

$$r^2 R'' + 2rR' - \lambda^2 R = 0 \tag{2}$$

$$\sin\theta\, \Theta'' + \cos\theta\, \Theta' + \lambda^2 \sin\theta\, \Theta = 0. \tag{3}$$

After we substitute $x = \cos\theta, 0 \le \theta \le \pi$, (3) becomes

$$(1 - x^2)\frac{d^2\Theta}{dx^2} - 2x\frac{d\Theta}{dx} + \lambda^2\Theta = 0, \quad -1 \le x \le 1. \tag{4}$$

The latter equation is a form of Legendre's equation (see Problem 41, Exercises 6.5). Now the only solutions of (4) that are continuous and have continuous derivatives on the closed interval $[-1, 1]$ are the Legendre polynomials $P_n(x)$ corresponding to $\lambda^2 = n(n + 1), n = 0, 1, 2, \ldots$. Thus we take the solutions of (3) to be

$$\Theta = P_n(\cos\theta).$$

Furthermore when $\lambda^2 = n(n + 1)$, the general solution of the Cauchy-Euler equation (2) is

$$R = c_1 r^n + c_2 r^{-(n+1)}.$$

Since we again expect $u(r, \theta)$ to be bounded at $r = 0$, we define $c_2 = 0$. Hence

$$u_n = A_n r^n P_n(\cos \theta)$$

and
$$u(r, \theta) = \sum_{n=0}^{\infty} A_n r^n P_n(\cos \theta).$$

At $r = c$,
$$f(\theta) = \sum_{n=0}^{\infty} A_n c^n P_n(\cos \theta).$$

Therefore $A_n c^n$ are the coefficients of the Fourier-Legendre series (23) of Section 11.5:

$$A_n = \frac{2n + 1}{2c^n} \int_0^{\pi} f(\theta) P_n(\cos \theta) \sin \theta \, d\theta.$$

It follows that the solution is

$$u(r, \theta) = \sum_{n=0}^{\infty} \left(\frac{2n + 1}{2} \int_0^{\pi} f(\theta) P_n(\cos \theta) \sin \theta \, d\theta \right) \left(\frac{r}{c} \right)^n P_n(\cos \theta). \qquad \blacksquare$$

13.3 EXERCISES

Answers to odd-numbered problems begin on page A–45.

1. Solve the problem in Example 1 if

$$f(\theta) = \begin{cases} 50, & 0 < \theta < \pi/2 \\ 0, & \pi/2 < \theta < \pi. \end{cases}$$

Write out the first four nonzero terms of the series solution. [*Hint:* See Example 3, Section 11.5.]

2. The solution $u(r, \theta)$ in Example 1 could also be interpreted as the potential inside the sphere due to a charge distribution $f(\theta)$ on its surface. Find the potential outside the sphere.

3. Find the solution of the problem in Example 1 if $f(\theta) = \cos \theta, 0 < \theta < \pi$. [*Hint:* $P_1(\cos \theta) = \cos \theta$. Use orthogonality.]

4. Find the solution of the problem in Example 1 if $f(\theta) = 1 - \cos 2\theta$, $0 < \theta < \pi$. [*Hint:* See Problem 16, Exercises 11.5.].

5. Find the steady-state temperature $u(r, \theta)$ within a hollow sphere $a < r < b$ if its inner surface $r = a$ is kept at temperature $f(\theta)$ and its outer surface $r = b$ is kept at temperature zero. The sphere in the first octant is shown in Figure 13.16.

6. The steady-state temperature in a hemisphere of radius $r = c$ is determined from

$$\frac{\partial^2 u}{\partial r^2} + \frac{2}{r} \frac{\partial u}{\partial r} + \frac{1}{r^2} \frac{\partial^2 u}{\partial \theta^2} + \frac{\cot \theta}{r^2} \frac{\partial u}{\partial \theta} = 0, \qquad 0 < r < c, \quad 0 < \theta < \pi/2$$

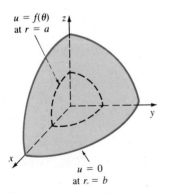

$u = f(\theta)$
at $r = a$

$u = 0$
at $r = b$

Figure 13.16

$$u(r, \pi/2) = 0, \qquad 0 < r < c$$

$$u(c, \theta) = f(\theta), \qquad 0 < \theta < \pi/2.$$

Solve for $u(r, \theta)$. [*Hint:* $P_n(0) = 0$ only if n is odd. Also see Problem 18, Exercises 11.5.]

7. Solve Problem 6 when the base of the hemisphere is insulated; that is,

$$\left. \frac{\partial u}{\partial \theta} \right|_{\theta = \pi/2} = 0, \qquad 0 < r < c.$$

8. Solve Problem 6 for $r > c$.

9. The time-dependent temperature within a sphere of unit radius is determined from

$$\frac{\partial^2 u}{\partial r^2} + \frac{2}{r} \frac{\partial u}{\partial r} = \frac{\partial u}{\partial t}, \qquad 0 < r < 1, \quad t > 0$$

$$u(1, t) = 100, \qquad t > 0$$

$$u(r, 0) = 0, \qquad 0 < r < 1.$$

Solve for $u(r, t)$. [*Hint:* Verify that the left side of the partial differential equation can be written as $\dfrac{1}{r} \dfrac{\partial^2}{\partial r^2} (ru)$. Let $ru(r, t) = v(r, t) + \psi(r)$. Use only functions that are bounded as $r \to 0$.]

10. A uniform solid sphere of radius 1 at an initial constant temperature u_0 throughout is dropped into a large container of fluid that is kept at a constant temperature u_1 ($u_1 > u_0$) for all time. See Figure 13.17. Since there is heat transfer across the boundary $r = 1$, the temperature $u(r, t)$ in the sphere is determined from the boundary-value problem

Figure 13.17

$$\frac{\partial^2 u}{\partial r^2} + \frac{2}{r} \frac{\partial u}{\partial r} = \frac{\partial u}{\partial t}, \qquad 0 < r < 1, \quad t > 0$$

$$\left. \frac{\partial u}{\partial r} \right|_{r=1} = -h(u(1, t) - u_1), \qquad 0 < h < 1$$

$$u(r, 0) = u_0, \qquad 0 < r < 1.$$

Solve for $u(r, t)$. [*Hint:* Proceed as in Problem 9.]

11. Solve the boundary-value problem involving spherical vibrations:

$$a^2 \left(\frac{\partial^2 u}{\partial r^2} + \frac{2}{r} \frac{\partial u}{\partial r} \right) = \frac{\partial^2 u}{\partial t^2}, \qquad 0 < r < c, \quad t > 0$$

$$u(c, t) = 0, \qquad t > 0$$

$$u(r, 0) = f(r), \qquad \left. \frac{\partial u}{\partial t} \right|_{t=0} = g(r), \qquad 0 < r < c.$$

$$\left[\text{Hint: Write the left side of the partial differential equation as} \right.$$

$$\left. a^2 \frac{1}{r} \frac{\partial^2}{\partial r^2} (ru). \text{ Let } v(r, t) = ru(r, t). \right]$$

12. A conducting sphere of radius $r = c$ is grounded and placed in a uniform electric field that has intensity E in the z direction. The potential $u(r, \theta)$ outside the sphere is determined from the boundary-value problem

$$\frac{\partial^2 u}{\partial r^2} + \frac{2}{r} \frac{\partial u}{\partial r} + \frac{1}{r^2} \frac{\partial^2 u}{\partial \theta^2} + \frac{\cot \theta}{r^2} \frac{\partial u}{\partial \theta} = 0, \quad r > c, \quad 0 < \theta < \pi$$

$$u(c, \theta) = 0, \quad 0 < \theta < \pi$$

$$\lim_{r \to \infty} u(r, \theta) = -Ez = -Er \cos \theta.$$

Show that

$$u(r, \theta) = -Er \cos \theta + E \frac{c^3}{r^2} \cos \theta.$$

[*Hint:* Explain why $\int_0^\pi \cos \theta \, P_n(\cos \theta) \sin \theta \, d\theta = 0$ for all nonnegative integers except $n = 1$. See (24) of Section 11.5.]

Chapter 13 *REVIEW*

Certain kinds of applied problems are best analyzed in a coordinate system other than rectangular coordinates. In this chapter we considered boundary-value problems involving the heat equation, wave equation, and Laplace's equation in **polar, cylindrical**, and **spherical coordinates**. In each of these coordinate systems, the method of separation of variables led to an ordinary differential equation with variable coefficients. We encountered the Cauchy-Euler equation in the course of solving problems involving Laplace's equation in polar coordinates, Bessel's equation arose in the solution of some radially symmetric problems in polar and cylindrical coordinates, and Legendre's equation occurred in special kinds of problems in spherical coordinates. The solution of those problems in polar or cylindrical coordinates that gave rise to Bessel's differential equation depended in the final analysis on our ability to expand functions in an infinite series of Bessel functions, that is, in a **Fourier-Bessel series**. Similarly, we used the concept of **Fourier-Legendre series** to solve problems in spherical coordinates.

Chapter 13   REVIEW EXERCISES

Answers to odd-numbered problems begin on page A-46.

1. Find the steady-state temperature $u(r, \theta)$ in a circular plate of radius c if the temperature on the circumference is given by

$$u(c, \theta) = \begin{cases} u_0, & 0 < \theta < \pi \\ -u_0, & \pi < \theta < 2\pi. \end{cases}$$

2. Find the steady-state temperature in the circular plate in Problem 1 if

$$u(c, \theta) = \begin{cases} 1, & 0 < \theta < \pi/2 \\ 0, & \pi/2 < \theta < 3\pi/2 \\ 1, & 3\pi/2 < \theta < 2\pi. \end{cases}$$

3. Find the steady-state temperature $u(r, \theta)$ in a semicircular plate of radius 1 if

$$u(1, \theta) = u_0(\pi\theta - \theta^2), \qquad 0 < \theta < \pi$$

$$u(r, 0) = 0, \qquad u(r, \pi) = 0, \qquad 0 < r < 1.$$

4. Find the steady-state temperature $u(r, \theta)$ in the semicircular plate in Problem 3 if $u(1, \theta) = \sin\theta, 0 < \theta < \pi$.

5. Find the steady-state temperature $u(r, \theta)$ in the plate shown in Figure 13.18.

Figure 13.18

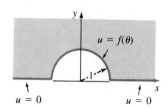

Figure 13.19

6. Find the steady-state temperature $u(r, \theta)$ in the infinite plate shown in Figure 13.19.

7. Suppose heat is lost from the flat surfaces of a very thin circular unit disk into a surrounding medium at temperature zero. If the linear law of heat transfer applies, the heat equation assumes the form

$$\frac{\partial^2 u}{\partial r^2} + \frac{1}{r}\frac{\partial u}{\partial r} - hu = \frac{\partial u}{\partial t}, \qquad h > 0, \quad 0 < r < 1, \quad t > 0.$$

See Figure 13.20. Find the temperature $u(r, t)$ if the edge $r = 1$ is kept at temperature zero and if initially the temperature of the plate is unity throughout.

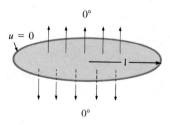

Figure 13.20

8. Suppose x_k is a positive zero of J_0. Show that a solution of the boundary-value problem

$$a^2\left(\frac{\partial^2 u}{\partial r^2} + \frac{1}{r}\frac{\partial u}{\partial r}\right) = \frac{\partial^2 u}{\partial t^2}, \qquad 0 < r < 1, \quad t > 0$$

$$u(1, t) = 0, \qquad t > 0$$

$$u(r, 0) = u_0 J_0(x_k r), \qquad \frac{\partial u}{\partial t}\bigg|_{t=0} = 0, \qquad 0 < r < 1,$$

is $u(r, t) = u_0 J_0(x_k r) \cos a x_k t$.

9. Find the steady-state temperature $u(r, z)$ in the cylinder in Figure 13.10 if the lateral side is kept at temperature zero, the top $z = 4$ is kept at temperature 50, and the base $z = 0$ is insulated.

10. Solve the boundary-value problem

$$\frac{\partial^2 u}{\partial r^2} + \frac{1}{r}\frac{\partial u}{\partial r} + \frac{\partial^2 u}{\partial z^2} = 0, \qquad 0 < r < 1, \quad 0 < z < 1$$

$$\frac{\partial u}{\partial r}\bigg|_{r=1} = 0, \qquad 0 < z < 1$$

$$u(r, 0) = f(r), \qquad u(r, 1) = g(r), \qquad 0 < r < 1.$$

11. Find the steady-state temperature $u(r, \theta)$ in a sphere of unit radius if the surface is kept at

$$u(1, \theta) = \begin{cases} 100, & 0 < \theta < \pi/2 \\ -100, & \pi/2 < \theta < \pi. \end{cases}$$

[*Hint:* See Problem 20, Exercises 11.5.]

12. Solve the boundary-value problem

$$\frac{\partial^2 u}{\partial r^2} + \frac{2}{r}\frac{\partial u}{\partial r} = \frac{\partial^2 u}{\partial t^2}, \qquad 0 < r < 1, \quad t > 0$$

$$\frac{\partial u}{\partial r}\bigg|_{r=1} = 0, \qquad t > 0$$

$$u(r, 0) = f(r), \qquad \frac{\partial u}{\partial t}\bigg|_{t=0} = g(r), \qquad 0 < r < 1.$$

[*Hint:* Proceed as in Problems 9 and 10, Exercises 13.3, but let $v(r, t) = ru(r, t)$. See Section 12.7.]

13. The function $u(x) = Y_0(\lambda a)J_0(\lambda x) - J_0(\lambda a)Y_0(\lambda x)$, $a > 0$, is a solution of the parametric Bessel equation

$$x^2 \frac{d^2 u}{dx^2} + x \frac{du}{dx} + \lambda^2 x^2 u = 0$$

on the interval $a \le x \le b$. If the eigenvalues λ_n are defined by the positive roots of the equation $Y_0(\lambda a)J_0(\lambda b) - J_0(\lambda a)Y_0(\lambda b) = 0$, show that the functions

$$u_m(x) = Y_0(\lambda_m a)J_0(\lambda_m x) - J_0(\lambda_m a)Y_0(\lambda_m x)$$

$$u_n(x) = Y_0(\lambda_n a)J_0(\lambda_n x) - J_0(\lambda_n a)Y_0(\lambda_n x)$$

are orthogonal with respect to the weight function $p(x) = x$ on the interval $[a, b]$; that is,

$$\int_a^b x u_m(x)u_n(x)\, dx = 0, \qquad m \ne n.$$

[*Hint:* Follow the procedure on page 626.]

14. Use the results of Problem 13 to solve the following boundary-value problem for the temperature $u(r, t)$ in a circular ring:

$$\frac{\partial^2 u}{\partial r^2} + \frac{1}{r}\frac{\partial u}{\partial r} = \frac{\partial u}{\partial t}, \qquad a < r < b, \quad t > 0$$

$$u(a, t) = 0, \qquad u(b, t) = 0, \qquad t > 0$$

$$u(r, 0) = f(r), \qquad a < r < b.$$

C H A P T E R 14

INTEGRAL TRANSFORM METHOD

14.1 Error Function
14.2 Applications of the Laplace Transform
14.3 Fourier Integral

14.4 Fourier Transforms

Chapter 14 Review
Chapter 14 Review Exercises

Important Concepts

Error function
Complementary error function
Fourier integral
Cosine integral
Sine integral
Transform pairs
Fourier transform
Inverse Fourier transform
Fourier sine transform
Inverse Fourier sine transform
Fourier cosine transform
Inverse Fourier cosine transform

The method of separation of variables that we employed in Chapters 12 and 13 is a powerful, but not universally applicable, method for solving boundary-value problems. If the partial differential equation is nonhomogeneous, or if the boundary conditions are time dependent, or if the domain of the spatial variable is infinite or semi-infinite, we may be able to use an integral transform to solve the problem.

In Section 14.2 we shall solve problems involving the heat and wave equations by means of the familiar Laplace transform. In Section 14.4 three new integral transforms, Fourier transforms, will be introduced and used.

14.1 ERROR FUNCTION

erf(x) and erfc(x)

Before applying the Laplace transform to boundary-value problems, we need to consider a function that plays an important role in many applications. This function, like the natural logarithm, is defined by means of an integral.

DEFINITION 14.1   Error Function

The **error function**, denoted by erf(x), is

$$\operatorname{erf}(x) = \frac{2}{\sqrt{\pi}} \int_0^x e^{-u^2}\, du. \qquad (1)$$

The **complementary error function** erfc(x) is defined in terms of the error function:

$$\operatorname{erfc}(x) = 1 - \operatorname{erf}(x). \qquad (2)$$

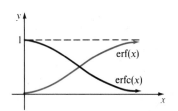

Figure 14.1

In calculus it is usually shown that $\int_0^\infty e^{-u^2}\, du = \sqrt{\pi}/2$ or $(2/\sqrt{\pi}) \int_0^\infty e^{-u^2}\, du = 1$. Hence (2) can be written as

$$\operatorname{erfc}(x) = \frac{2}{\sqrt{\pi}} \left[\int_0^\infty e^{-u^2}\, du - \int_0^x e^{-u^2}\, du \right]$$

or

$$\operatorname{erfc}(x) = \frac{2}{\sqrt{\pi}} \int_x^\infty e^{-u^2}\, du.$$

The graphs of erf(x) and erfc(x) are given in Figure 14.1. Note that erf(0) = 0, erfc(0) = 1, and erf(x) → 1 and erfc(x) → 0 as $x \to \infty$. Numerical values of erf(x) and erfc(x) have been extensively tabulated. In tables the error

$f(t),\ a > 0$	$\mathcal{L}\{f(t)\} = F(s)$	$f(t),\ a > 0$	$\mathcal{L}\{f(t)\} = F(s)$
1. $\dfrac{1}{\sqrt{\pi t}} e^{-a^2/4t}$	$\dfrac{e^{-a\sqrt{s}}}{\sqrt{s}}$	**4.** $2\sqrt{\dfrac{t}{\pi}}\, e^{-a^2/4t} - a\,\operatorname{erfc}\left(\dfrac{a}{2\sqrt{t}}\right)$	$\dfrac{e^{-a\sqrt{s}}}{s\sqrt{s}}$
2. $\dfrac{a}{2\sqrt{\pi t^3}} e^{-a^2/4t}$	$e^{-a\sqrt{s}}$	**5.** $e^{ab}e^{b^2 t}\,\operatorname{erfc}\left(b\sqrt{t} + \dfrac{a}{2\sqrt{t}}\right)$	$\dfrac{e^{-a\sqrt{s}}}{\sqrt{s}(\sqrt{s} + b)}$
3. $\operatorname{erfc}\left(\dfrac{a}{2\sqrt{t}}\right)$	$\dfrac{e^{-a\sqrt{s}}}{s}$	**6.** $-e^{ab}e^{b^2 t}\,\operatorname{erfc}\left(b\sqrt{t} + \dfrac{a}{2\sqrt{t}}\right) + \operatorname{erfc}\left(\dfrac{a}{2\sqrt{t}}\right)$	$\dfrac{be^{-a\sqrt{s}}}{s(\sqrt{s} + b)}$

function is often referred to as the **probability integral**. The preceding table of Laplace transforms will be useful in the exercises of the next section. The proofs of these results are complicated and will not be given.

14.1 EXERCISES

Answers to odd-numbered problems begin on page A–46.

1. (a) Show that $\operatorname{erf}(\sqrt{t}) = \dfrac{1}{\sqrt{\pi}} \displaystyle\int_0^t \dfrac{e^{-\tau}}{\sqrt{\tau}} \, d\tau$.

(b) Use the convolution theorem and the results of Problems 43 and 44 in Exercises 7.1 to show that $\mathscr{L}\{\operatorname{erf}(\sqrt{t})\} = \dfrac{1}{s\sqrt{s+1}}$.

2. Use the result of Problem 1 to show that

$$\mathscr{L}\{\operatorname{erfc}(\sqrt{t})\} = \frac{1}{s}\left[1 - \frac{1}{\sqrt{s+1}}\right].$$

3. Use the result of Problem 1 to show that

$$\mathscr{L}\{e^t \operatorname{erf}(\sqrt{t})\} = \frac{1}{\sqrt{s}(s-1)}.$$

4. Use the result of Problem 2 to show that

$$\mathscr{L}\{e^t \operatorname{erfc}(\sqrt{t})\} = \frac{1}{\sqrt{s}(\sqrt{s}+1)}.$$

5. Let $C, G, R,$ and x be constants. Use the table in this section to show that

$$\mathscr{L}^{-1}\left\{\frac{C}{Cs+G}(1 - e^{-x\sqrt{RCs+RG}})\right\} = e^{-Gt/C} \operatorname{erf}\left(\frac{x}{2}\sqrt{\frac{RC}{t}}\right).$$

6. Let a be a constant. Show that

$$\mathscr{L}^{-1}\left\{\frac{\sinh a\sqrt{s}}{s \sinh \sqrt{s}}\right\} = \sum_{n=0}^{\infty}\left[\operatorname{erf}\left(\frac{2n+1+a}{2\sqrt{t}}\right) - \operatorname{erf}\left(\frac{2n+1-a}{2\sqrt{t}}\right)\right].$$

[*Hint:* Use the exponential definition of the hyperbolic sine. Expand $1/(1 - e^{-2\sqrt{s}})$ in a geometric series.]

7. Use the Laplace transform and the table in this section to solve the integral equation

$$y(t) = 1 - \int_0^t \frac{y(\tau)}{\sqrt{t-\tau}} \, d\tau.$$

8. Use the third and fifth entries in the table in this section to derive the sixth entry.

9. Show that $\displaystyle\int_a^b e^{-u^2}\,du = \frac{\sqrt{\pi}}{2}\,[\mathrm{erf}(b) - \mathrm{erf}(a)]$.

10. Show that $\displaystyle\int_{-a}^a e^{-u^2}\,du = \sqrt{\pi}\,\mathrm{erf}(a)$.

14.2 APPLICATIONS OF THE LAPLACE TRANSFORM

We define the Laplace transform of a function of two variables $u(x, t)$ with respect to the variable t by

$$\mathscr{L}\{u(x, t)\} = \int_0^\infty e^{-st} u(x, t)\,dt = U(x, s),$$

where x is treated as a parameter. Throughout this section we shall assume that all the operational properties of Sections 7.3 and 7.4 apply to functions of two variables. For example, by Theorem 7.8 the transform of the partial derivative $\partial u/\partial t$ is

$$\mathscr{L}\left\{\frac{\partial u}{\partial t}\right\} = s\mathscr{L}\{u(x, t)\} - u(x, 0);$$

that is,
$$\mathscr{L}\left\{\frac{\partial u}{\partial t}\right\} = sU(x, s) - u(x, 0). \tag{1}$$

Similarly
$$\mathscr{L}\left\{\frac{\partial^2 u}{\partial t^2}\right\} = s^2 U(x, s) - su(x, 0) - u_t(x, 0). \tag{2}$$

Since we are transforming with respect to t, we further suppose that it is legitimate to interchange integration and differentiation in the transform of $\partial^2 u/\partial x^2$:

$$\mathscr{L}\left\{\frac{\partial^2 u}{\partial x^2}\right\} = \int_0^\infty e^{-st}\frac{\partial^2 u}{\partial x^2}\,dt = \int_0^\infty \frac{\partial^2}{\partial x^2}[e^{-st}u(x, t)]\,dt$$

$$= \frac{d^2}{dx^2}\int_0^\infty e^{-st}u(x, t)\,dt = \frac{d^2}{dx^2}\mathscr{L}\{u(x, t)\};$$

that is,
$$\mathscr{L}\left\{\frac{\partial^2 u}{\partial x^2}\right\} = \frac{d^2 U}{dx^2}. \tag{3}$$

In view of (1) and (2), we see that the Laplace transform is suited to problems with initial conditions, namely, those problems associated with the heat equation or the wave equation.

EXAMPLE 1

Find the Laplace transform of the wave equation $a^2 \dfrac{\partial^2 u}{\partial x^2} = \dfrac{\partial^2 u}{\partial t^2}$, $t > 0$.

Solution From (2) and (3),

$$\mathcal{L}\left\{a^2 \frac{\partial^2 u}{\partial x^2}\right\} = \mathcal{L}\left\{\frac{\partial^2 u}{\partial t^2}\right\}$$

becomes $\quad a^2 \dfrac{d^2}{dx^2} \mathcal{L}\{u(x, t)\} = s^2 \mathcal{L}\{u(x, t)\} - su(x, 0) - u_t(x, 0)$

or $\qquad a^2 \dfrac{d^2 U}{dx^2} - s^2 U = -su(x, 0) - u_t(x, 0).$ (4)

∎

The Laplace transform with respect to t of either the wave equation or the heat equation eliminates that variable and, for the one-dimensional equations, the transformed equations are then *ordinary differential equations* in the spatial variable x. In solving a transformed equation, we treat s as a parameter.

EXAMPLE 2

Solve $\qquad \dfrac{\partial^2 u}{\partial x^2} = \dfrac{\partial^2 u}{\partial t^2}, \qquad 0 < x < 1, \quad t > 0,$

subject to $\qquad u(0, t) = 0, \qquad u(1, t) = 0, \qquad t > 0$

$$u(x, 0) = 0, \qquad \frac{\partial u}{\partial t}\bigg|_{t=0} = \sin \pi x, \qquad 0 < x < 1.$$

Solution The partial differential equation is recognized as the wave equation with $a = 1$. From (4) and the given initial conditions, the transformed equation is

$$\frac{d^2 U}{dx^2} - s^2 U = -\sin \pi x,$$ (5)

where $U(x, s) = \mathcal{L}\{u(x, t)\}$. Since the boundary conditions are functions of t, we must also find their Laplace transforms:

$$\mathcal{L}\{u(0, t)\} = U(0, s) = 0 \quad \text{and} \quad \mathcal{L}\{u(1, t)\} = U(1, s) = 0.$$ (6)

The results in (6) are boundary conditions for the ordinary differential equation (5). Since (5) is defined over a finite interval, its complementary function is

$$U_c(x, s) = c_1 \cosh sx + c_2 \sinh sx.$$

The method of undetermined coefficients yields a particular solution

$$U_p(x, s) = \frac{1}{s^2 + \pi^2} \sin \pi x.$$

Hence

$$U(x, s) = c_1 \cosh sx + c_2 \sinh sx + \frac{1}{s^2 + \pi^2} \sin \pi x.$$

But the conditions $U(0, s) = 0$ and $U(1, s) = 0$ yield, in turn, $c_1 = 0$ and $c_2 = 0$. We conclude

$$U(x, s) = \frac{1}{s^2 + \pi^2} \sin \pi x$$

$$u(x, t) = \mathscr{L}^{-1}\left\{\frac{1}{s^2 + \pi^2} \sin \pi x\right\} = \frac{1}{\pi} \sin \pi x \, \mathscr{L}^{-1}\left\{\frac{\pi}{s^2 + \pi^2}\right\}.$$

Therefore

$$u(x, t) = \frac{1}{\pi} \sin \pi x \sin \pi t. \qquad \blacksquare$$

EXAMPLE 3

A very long string is initially at rest on the nonnegative x-axis. The string is secured at $x = 0$ and its distant right end slides down a frictionless vertical support. The string is set in motion by letting it fall under its own weight. Find the displacement $u(x, t)$.

Solution Since the force of gravity is taken into consideration, it can be shown that the wave equation has the form

$$a^2 \frac{\partial^2 u}{\partial x^2} - g = \frac{\partial^2 u}{\partial t^2}, \qquad x > 0, \quad t > 0.$$

The boundary and initial conditions are, respectively,

$$u(0, t) = 0, \qquad \lim_{x \to \infty} \frac{\partial u}{\partial x} = 0, \qquad t > 0$$

$$u(x, 0) = 0, \qquad \frac{\partial u}{\partial t}\bigg|_{t=0} = 0, \qquad x > 0.$$

The second boundary condition $\lim_{x \to \infty} \partial u/\partial x = 0$ indicates that the string is horizontal at a great distance from the left end. Now from (2) and (3),

$$\mathscr{L}\left\{a^2 \frac{\partial^2 u}{\partial x^2}\right\} - \mathscr{L}\{g\} = \mathscr{L}\left\{\frac{\partial^2 u}{\partial t^2}\right\}$$

becomes

$$a^2 \frac{d^2U}{dx^2} - \frac{g}{s} = s^2U - su(x, 0) - u_t(x, 0)$$

or, in view of the initial conditions,

$$\frac{d^2U}{dx^2} - \frac{s^2}{a^2}U = \frac{g}{a^2 s}.$$

The transforms of the boundary conditions are

$$\mathscr{L}\{u(0, t)\} = U(0, s) = 0 \quad \text{and} \quad \mathscr{L}\left\{\lim_{x\to\infty} \frac{\partial u}{\partial x}\right\} = \lim_{x\to\infty} \frac{dU}{dx} = 0.$$

With the aid of undetermined coefficients, the general solution of the transformed equation is found to be

$$U(x, s) = c_1 e^{-(x/a)s} + c_2 e^{(x/a)s} - \frac{g}{s^3}.$$

The boundary condition $\lim_{x\to\infty} dU/dx = 0$ implies $c_2 = 0$ and $U(0, s) = 0$ gives $c_1 = g/s^3$. Therefore

$$U(x, s) = \frac{g}{s^3} e^{-(x/a)s} - \frac{g}{s^3}.$$

Now by the second translation theorem, we have

$$u(x, t) = \mathscr{L}^{-1}\left\{\frac{g}{s^3} e^{-(x/a)s} - \frac{g}{s^3}\right\} = \frac{1}{2} g\left(t - \frac{x}{a}\right)^2 \mathscr{U}\left(t - \frac{x}{a}\right) - \frac{1}{2} gt^2$$

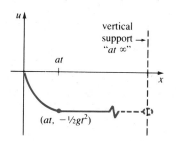

or

$$u(x, t) = \begin{cases} -\dfrac{1}{2} gt^2, & 0 \le t < x/a \\[2mm] -\dfrac{g}{2a^2}(2axt - x^2), & t \ge x/a. \end{cases}$$

To interpret the solution let us suppose $t > 0$ is fixed. For $0 \le x \le at$ the string is the shape of a parabola passing through $(0, 0)$ and $(at, -\frac{1}{2}gt^2)$. For $x > at$ the string is described by the horizontal line $u = -\frac{1}{2}gt^2$. See Figure 14.2. ∎

Figure label (left): u, vertical support →"at ∞", at, x, $(at, -\frac{1}{2}gt^2)$

Figure 14.2

Observe that the problem in the next example could be solved by the procedure in Section 12.6. The Laplace transform provides an alternative solution.

EXAMPLE 4

Solve the heat equation

$$\frac{\partial^2 u}{\partial x^2} = \frac{\partial u}{\partial t}, \qquad 0 < x < 1, \quad t > 0,$$

subject to $\qquad u(0, t) = 0, \qquad u(1, t) = u_0, \qquad t > 0$

$$u(x, 0) = 0, \qquad 0 < x < 1.$$

Solution From (1) and (3) and the given initial condition,

$$\mathscr{L}\left\{\frac{\partial^2 u}{\partial x^2}\right\} = \mathscr{L}\left\{\frac{\partial u}{\partial t}\right\}$$

becomes $\qquad \dfrac{d^2 U}{dx^2} - sU = 0. \qquad\qquad\qquad (7)$

The transforms of the boundary conditions are

$$U(0, s) = 0 \quad \text{and} \quad U(1, s) = \frac{u_0}{s}. \qquad\qquad (8)$$

Since we are concerned with a finite interval on the x-axis, we choose to write the general solution of (7) as

$$U(x, s) = c_1 \cosh(\sqrt{s}\,x) + c_2 \sinh(\sqrt{s}\,x).$$

Applying the two boundary conditions in (8) yields, in turn, $c_1 = 0$ and $c_2 = u_0/(s \sinh \sqrt{s})$. Thus

$$U(x, s) = u_0 \frac{\sinh(\sqrt{s}\,x)}{s \sinh \sqrt{s}}.$$

Now the inverse transform of the latter function cannot be found in most tables. However, by writing

$$\frac{\sinh(\sqrt{s}\,x)}{s \sinh \sqrt{s}} = \frac{e^{\sqrt{s}\,x} - e^{-\sqrt{s}\,x}}{s(e^{\sqrt{s}} - e^{-\sqrt{s}})} = \frac{e^{(x-1)\sqrt{s}} - e^{-(x+1)\sqrt{s}}}{s(1 - e^{-2\sqrt{s}})}$$

and using the geometric series

$$\frac{1}{1 - e^{-2\sqrt{s}}} = \sum_{n=0}^{\infty} e^{-2n\sqrt{s}},$$

we find

$$\frac{\sinh(\sqrt{s}\,x)}{s \sinh \sqrt{s}} = \sum_{n=0}^{\infty} \left[\frac{e^{-(2n+1-x)\sqrt{s}}}{s} - \frac{e^{-(2n+1+x)\sqrt{s}}}{s}\right].$$

If we assume that the inverse Laplace transform can be done term by term, it follows from entry 3 of the table in Section 14.1 that

$$u(x, t) = u_0 \mathscr{L}^{-1}\left\{\frac{\sinh(\sqrt{s}\,x)}{s \sinh \sqrt{s}}\right\} = u_0 \sum_{n=0}^{\infty} \left[\mathscr{L}^{-1}\left\{\frac{e^{-(2n+1-x)\sqrt{s}}}{s}\right\} - \mathscr{L}^{-1}\left\{\frac{e^{-(2n+1+x)\sqrt{s}}}{s}\right\}\right]$$

$$= u_0 \sum_{n=0}^{\infty} \left[\operatorname{erfc}\left(\frac{2n+1-x}{2\sqrt{t}}\right) - \operatorname{erfc}\left(\frac{2n+1+x}{2\sqrt{t}}\right)\right]. \qquad (9)$$

The solution (9) can be rewritten in terms of the error function using $\text{erfc}(x) = 1 - \text{erf}(x)$:

$$u(x, t) = u_0 \sum_{n=0}^{\infty} \left[\text{erf}\left(\frac{2n + 1 + x}{2\sqrt{t}}\right) - \text{erf}\left(\frac{2n + 1 - x}{2\sqrt{t}}\right) \right].$$ ■

14.2 EXERCISES

Answers to odd-numbered problems begin on page A–46.

In the following problems use tables as necessary.

1. A string is stretched along the x-axis between $(0, 0)$ and $(L, 0)$. Find the displacement $u(x, t)$ if the string starts from rest in the initial position $A \sin(\pi x/L)$.

2. Solve the boundary-value problem:

$$\frac{\partial^2 u}{\partial x^2} = \frac{\partial^2 u}{\partial t^2}, \qquad 0 < x < 1, \quad t > 0$$

$$u(0, t) = 0, \qquad u(1, t) = 0$$

$$u(x, 0) = 0, \qquad \left.\frac{\partial u}{\partial t}\right|_{t=0} = 2 \sin \pi x + 4 \sin 3\pi x.$$

3. The displacement of a semi-infinite elastic string is determined from

$$a^2 \frac{\partial^2 u}{\partial x^2} = \frac{\partial^2 u}{\partial t^2}, \qquad x > 0, \quad t > 0$$

$$u(0, t) = f(t), \qquad \lim_{x \to \infty} u(x, t) = 0, \qquad t > 0$$

$$u(x, 0) = 0, \qquad \left.\frac{\partial u}{\partial t}\right|_{t=0} = 0, \qquad x > 0.$$

Solve for $u(x, t)$.

4. Solve the boundary-value problem in Problem 3 when

$$f(t) = \begin{cases} \sin \pi t, & 0 \le t \le 1 \\ 0, & t > 1. \end{cases}$$

Sketch the displacement $u(x, t)$ for $t > 1$.

5. In Example 3 find the displacement $u(x, t)$ when the left end of the string at $x = 0$ is given an oscillatory motion described by $f(t) = A \sin \omega t$.

6. The displacement $u(x, t)$ of a string that is driven by an external force is

determined from

$$\frac{\partial^2 u}{\partial x^2} + \sin \pi x \sin \omega t = \frac{\partial^2 u}{\partial t^2}, \qquad 0 < x < 1, \quad t > 0$$

$$u(0, t) = 0, \qquad u(1, t) = 0, \qquad t > 0$$

$$u(x, 0) = 0, \qquad \frac{\partial u}{\partial t}\bigg|_{t=0} = 0, \qquad 0 < x < 1.$$

Solve for $u(x, t)$.

7. A uniform bar is clamped at $x = 0$ and is initially at rest. If a constant force F_0 is applied to the free end at $x = L$, the longitudinal displacement $u(x, t)$ of a cross section of the bar is determined from

$$a^2 \frac{\partial^2 u}{\partial x^2} = \frac{\partial^2 u}{\partial t^2}, \qquad 0 < x < L, \quad t > 0$$

$$u(0, t) = 0, \qquad E \frac{\partial u}{\partial x}\bigg|_{x=L} = F_0, \qquad E \text{ a constant}, \quad t > 0$$

$$u(x, 0) = 0, \qquad \frac{\partial u}{\partial t}\bigg|_{t=0} = 0, \qquad 0 < x < L.$$

Solve for $u(x, t)$. [*Hint:* Expand $1/(1 + e^{-2sL/a})$ in a geometric series.]

8. A uniform semi-infinite elastic beam moving along the x-axis with a constant velocity $-v_0$ is brought to a stop by hitting a wall at time $t = 0$. See Figure 14.3. The longitudinal displacement $u(x, t)$ is determined from

$$a^2 \frac{\partial^2 u}{\partial x^2} = \frac{\partial^2 u}{\partial t^2}, \qquad x > 0, \quad t > 0$$

$$u(0, t) = 0, \qquad \lim_{x \to \infty} \frac{\partial u}{\partial x} = 0, \qquad t > 0,$$

$$u(x, 0) = 0, \qquad \frac{\partial u}{\partial t}\bigg|_{t=0} = -v_0, \qquad x > 0.$$

Solve for $u(x, t)$.

wall | beam | $\leftarrow v_0$

$x = 0$

Figure 14.3

9. Solve the boundary-value problem:

$$\frac{\partial^2 u}{\partial x^2} = \frac{\partial^2 u}{\partial t^2}, \qquad x > 0, \quad t > 0$$

$$u(0, t) = 0, \qquad \lim_{x \to \infty} u(x, t) = 0, \qquad t > 0$$

$$u(x, 0) = xe^{-x}, \qquad \frac{\partial u}{\partial t}\bigg|_{t=0} = 0, \qquad x > 0.$$

10. Solve the boundary-value problem:

$$\frac{\partial^2 u}{\partial x^2} = \frac{\partial^2 u}{\partial t^2}, \qquad x > 0, \quad t > 0$$

$$u(0, t) = 1, \qquad \lim_{x \to \infty} u(x, t) = 0, \qquad t > 0$$

$$u(x, 0) = e^{-x}, \qquad \frac{\partial u}{\partial t}\bigg|_{t=0} = 0, \qquad x > 0.$$

11. The temperature in a semi-infinite solid is determined from

$$k\frac{\partial^2 u}{\partial x^2} = \frac{\partial u}{\partial t}, \qquad x > 0, \qquad t > 0$$

$$u(0, t) = u_0, \qquad \lim_{x \to \infty} u(x, t) = 0, \qquad t > 0$$

$$u(x, 0) = 0, \qquad x > 0.$$

Solve for $u(x, t)$.

12. In Problem 11 if there is a constant flux of heat into the solid at its left-hand boundary, then the boundary condition is $\dfrac{\partial u}{\partial x}\bigg|_{x=0} = -A, t > 0.$
Solve for $u(x, t)$.

In Problems 13–20 use the Laplace transform to solve the heat equation $u_{xx} = u_t$, $x > 0$, $t > 0$, subject to the given conditions.

13. $u(0, t) = u_0, \quad \lim\limits_{x \to \infty} u(x, t) = u_1$

  $u(x, 0) = u_1$

14. $u(0, t) = u_0, \quad \lim\limits_{x \to \infty} u(x, t)/x = u_1$

  $u(x, 0) = u_1 x$

15. $\dfrac{\partial u}{\partial x}\bigg|_{x=0} = u(0, t), \quad \lim\limits_{x \to \infty} u(x, t) = u_0$

  $u(x, 0) = u_0$

16. $\dfrac{\partial u}{\partial x}\bigg|_{x=0} = u(0, t) - 50, \quad \lim\limits_{x \to \infty} u(x, t) = 0$

  $u(x, 0) = 0$

17. $u(0, t) = f(t), \quad \lim\limits_{x \to \infty} u(x, t) = 0$

  $u(x, 0) = 0$

  [*Hint:* Use the convolution theorem.]

18. $\dfrac{\partial u}{\partial x}\bigg|_{x=0} = -f(t), \quad \lim\limits_{x \to \infty} u(x, t) = 0$

  $u(x, 0) = 0$

19. $u(0, t) = 60 + 40\mathcal{U}(t - 2), \quad \lim\limits_{x \to \infty} u(x, t) = 60$

  $u(x, 0) = 60$

20. $u(0, t) = \begin{cases} 20, & 0 < t < 1, \\ 0, & t \geq 1, \end{cases} \qquad \lim_{x \to \infty} u(x, t) = 100$

$u(x, 0) = 100$

21. Solve the boundary-value problem;

$$\frac{\partial^2 u}{\partial x^2} = \frac{\partial u}{\partial t}, \qquad -\infty < x < 1, \quad t > 0$$

$$\frac{\partial u}{\partial x}\bigg|_{x=1} = 100 - u(1, t), \qquad \lim_{x \to -\infty} u(x, t) = 0, \qquad t > 0$$

$$u(x, 0) = 0, \qquad -\infty < x < 1.$$

22. Show that a solution of the boundary-value problem

$$k \frac{\partial^2 u}{\partial x^2} + r = \frac{\partial u}{\partial t}, \qquad x > 0, \quad t > 0$$

$$u(0, t) = 0, \qquad \lim_{x \to \infty} \frac{\partial u}{\partial x} = 0, \qquad t > 0$$

$$u(x, 0) = 0, \qquad x > 0,$$

where r is a constant, is given by

$$u(x, t) = rt - r \int_0^t \text{erfc}\left(\frac{x}{2\sqrt{k\tau}}\right) d\tau.$$

23. A rod of length L is held at a constant temperature u_0 at its ends $x = 0$ and $x = L$. If the initial temperature of the rod is $u_0 + u_0 \sin(x\pi/L)$, solve the heat equation $u_{xx} = u_t$, $0 < x < L$, $t > 0$, for the temperature $u(x, t)$.

24. If there is a heat transfer from the lateral surface of a thin wire of length L into a medium at constant temperature u_m, then the heat equation takes on the form

$$k \frac{\partial^2 u}{\partial x^2} - h(u - u_m) = \frac{\partial u}{\partial t}, \qquad 0 < x < L, \quad t > 0,$$

h a constant. Find the temperature $u(x, t)$ if the initial temperature is a constant u_0 throughout and the ends $x = 0$ and $x = L$ are insulated.

25. A rod of unit length is insulated at $x = 0$ and is kept at temperature zero at $x = 1$. If the initial temperature of the rod is a constant u_0, solve $ku_{xx} = u_t$, $0 < x < 1$, $t > 0$, for the temperature $u(x, t)$. [*Hint:* Expand $1/(1 + e^{-2\sqrt{s/k}})$ in a geometric series.]

26. An infinite porous slab of unit width is immersed in a solution of constant concentration c_0. A dissolved substance in the solution diffuses into the slab. The concentration $c(x, t)$ in the slab is determined from

$$D \frac{\partial^2 c}{\partial x^2} = \frac{\partial c}{\partial t}, \qquad 0 < x < 1, \quad t > 0$$

$$c(0, t) = c_0, \qquad c(1, t) = c_0, \qquad t > 0$$

$$c(x, 0) = 0, \qquad 0 < x < 1.$$

Solve for $c(x, t)$.

27. A very long telephone transmission line is initially at a constant potential u_0. If the line is grounded at $x = 0$ and insulated at the distant right end, then the potential $u(x, t)$ at a point x along the line at time t is determined from

$$\frac{\partial^2 u}{\partial x^2} - RC \frac{\partial u}{\partial t} - RGu = 0, \qquad x > 0, \qquad t > 0$$

$$u(0, t) = 0, \qquad \lim_{x \to \infty} \frac{\partial u}{\partial x} = 0, \qquad t > 0$$

$$u(x, 0) = u_0, \qquad x > 0,$$

where R, C, and G are constants known as resistance, capacitance, and conductance, respectively. Solve for $u(x, t)$. [*Hint:* See Problem 5, Exercises 14.1.]

28. Show that a solution of the boundary-value problem

$$\frac{\partial^2 u}{\partial x^2} - hu = \frac{\partial u}{\partial t}, \qquad x > 0, \quad t > 0$$

$$u(0, t) = u_0, \qquad \lim_{x \to \infty} u(x, t) = 0, \qquad t > 0$$

$$u(x, 0) = 0, \qquad x > 0,$$

is $\qquad u(x, t) = \dfrac{u_0 x}{2\sqrt{\pi}} \displaystyle\int_0^t \frac{e^{-h\tau - x^2/4\tau}}{\tau^{3/2}} \, d\tau.$

29. Solve the boundary-value problem:

$$k \frac{\partial^2 u}{\partial x^2} = \frac{\partial u}{\partial t}, \qquad x > 0, \quad t > 0$$

$$\left. \frac{\partial u}{\partial x} \right|_{x=0} = -A\delta(t), \qquad \lim_{x \to \infty} u(x, t) = 0, \qquad t > 0$$

$$u(x, 0) = 0, \qquad x > 0,$$

where $\delta(t)$ is the Dirac delta function (see Section 7.6). Interpret the boundary condition at $x = 0$.

30. Starting at $t = 0$, a concentrated load of magnitude F_0 moves with a constant velocity v_0 along a semi-infinite string. In this case the wave equation becomes

$$a^2 \frac{\partial^2 u}{\partial x^2} = \frac{\partial^2 u}{\partial t^2} + F_0 \delta\left(t - \frac{x}{v_0}\right).$$

Solve the above partial differential equation subject to

$$u(0, t) = 0, \qquad \lim_{x \to \infty} u(x, t) = 0, \qquad t > 0$$

$$u(x, 0) = 0, \qquad \left.\frac{\partial u}{\partial t}\right|_{t=0} = 0, \qquad x > 0,$$

(a) when $v_0 \neq a$ and   **(b)** when $v_0 = a$.

14.3 FOURIER INTEGRAL

In Chapters 11 and 12 we used Fourier series to represent a function f defined on a *finite* interval such as $(-p, p)$ or $(0, L)$. When f and f' are piecewise continuous on such an interval, a Fourier series represents the function on the interval and converges to a periodic extension of f outside of the interval. In this way we are justified in saying that Fourier series are associated only with periodic functions. We shall now derive, in a nonrigorous fashion, a means of representing certain kinds of nonperiodic functions that are defined on either an *infinite* interval $(-\infty, \infty)$ or a *semi-infinite* interval $(0, \infty)$.

From Fourier Series to Fourier Integral

Suppose a function f is defined on $(-p, p)$ and

$$f(x) = \frac{a_0}{2} + \sum_{n=1}^{\infty} \left(a_n \cos \frac{n\pi}{p} x + b_n \sin \frac{n\pi}{p} x \right)$$

$$= \frac{1}{2p} \int_{-p}^{p} f(t)\, dt + \frac{1}{p} \sum_{n=1}^{\infty} \left[\left(\int_{-p}^{p} f(t) \cos \frac{n\pi}{p} t\, dt \right) \cos \frac{n\pi}{p} x \right.$$

$$\left. + \left(\int_{-p}^{p} f(t) \sin \frac{n\pi}{p} t\, dt \right) \sin \frac{n\pi}{p} x \right]. \tag{1}$$

If we let $\alpha_n = n\pi/p$, $\Delta\alpha = \alpha_{n+1} - \alpha_n = \pi/p$, then (1) becomes

$$f(x) = \frac{1}{2\pi} \left(\int_{-p}^{p} f(t)\, dt \right) \Delta\alpha + \frac{1}{\pi} \sum_{n=1}^{\infty} \left[\left(\int_{-p}^{p} f(t) \cos \alpha_n t\, dt \right) \cos \alpha_n x \right.$$

$$\left. + \left(\int_{-p}^{p} f(t) \sin \alpha_n t\, dt \right) \sin \alpha_n x \right] \Delta\alpha. \tag{2}$$

We now expand the interval $(-p, p)$ by letting $p \to \infty$. Since $p \to \infty$ implies $\Delta\alpha \to 0$, the limit of (2) has the form $\lim_{\Delta\alpha \to 0} \sum_{n=1}^{\infty} F(\alpha_n)\, \Delta\alpha$, which is suggestive of the definition of the integral $\int_0^\infty F(\alpha) \, d\alpha$. Thus if $\int_{-\infty}^\infty f(t) \, dt$ exists, the limit of the first term in (2) is zero and the limit of the sum becomes

$$f(x) = \frac{1}{\pi} \int_0^\infty \left[\left(\int_{-\infty}^\infty f(t) \cos \alpha t \, dt \right) \cos \alpha x + \left(\int_{-\infty}^\infty f(t) \sin \alpha t \, dt \right) \sin \alpha x \right] d\alpha. \quad (3)$$

The result given in (3) is called the **Fourier integral** of f on $(-\infty, \infty)$. As the following summary shows, the basic structure of the Fourier integral is reminiscent of that of a Fourier series.

DEFINITION 14.2   **Fourier Integral**

The **Fourier integral** of a function f defined on the interval $(-\infty, \infty)$ is given by

$$f(x) = \frac{1}{\pi} \int_0^\infty [A(\alpha) \cos \alpha x + B(\alpha) \sin \alpha x] \, d\alpha, \quad (4)$$

where

$$A(\alpha) = \int_{-\infty}^\infty f(x) \cos \alpha x \, dx \quad (5)$$

$$B(\alpha) = \int_{-\infty}^\infty f(x) \sin \alpha x \, dx. \quad (6)$$

Convergence of a Fourier Integral

Sufficient conditions under which a Fourier integral converges to $f(x)$ are similar to, but are slightly more restrictive than, the conditions for a Fourier series.

THEOREM 14.1   **Conditions for Convergence**

Let f and f' be piecewise continuous on every finite interval and let f be absolutely integrable on $(-\infty, \infty)$.* Then the Fourier integral of f on the interval converges to $f(x)$ at a point of continuity. At a point of discontinuity the Fourier integral will converge to the average

$$\frac{f(x+) + f(x-)}{2},$$

where $f(x+)$ and $f(x-)$ denote the limit of f at x from the right and from the left, respectively.

---

* This means that the integral $\int_{-\infty}^\infty |f(x)| \, dx$ converges.

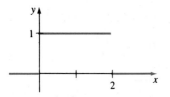

Figure 14.4

EXAMPLE I

Find the Fourier integral representation of the function

$$f(x) = \begin{cases} 0, & x < 0 \\ 1, & 0 < x < 2 \\ 0, & x > 2. \end{cases}$$

Solution The function, whose graph is shown in Figure 14.4, satisfies the hypotheses of Theorem 14.1. Hence from (5) and (6) we have at once

$$A(\alpha) = \int_{-\infty}^{\infty} f(x) \cos \alpha x \, dx$$

$$= \int_{-\infty}^{0} f(x) \cos \alpha x \, dx + \int_{0}^{2} f(x) \cos \alpha x \, dx + \int_{2}^{\infty} f(x) \cos \alpha x \, dx$$

$$= \int_{0}^{2} \cos \alpha x \, dx = \frac{\sin 2\alpha}{\alpha}$$

$$B(\alpha) = \int_{-\infty}^{\infty} f(x) \sin \alpha x \, dx = \int_{0}^{2} \sin \alpha x \, dx = \frac{1 - \cos 2\alpha}{\alpha}.$$

Substituting these coefficients in (4) then gives

$$f(x) = \frac{1}{\pi} \int_{0}^{\infty} \left[\left(\frac{\sin 2\alpha}{\alpha} \right) \cos \alpha x + \left(\frac{1 - \cos 2\alpha}{\alpha} \right) \sin \alpha x \right] d\alpha.$$

When we use trigonometric identities, the last integral simplifies to

$$f(x) = \frac{2}{\pi} \int_{0}^{\infty} \frac{\sin \alpha \cos \alpha (x - 1)}{\alpha} \, d\alpha. \tag{7}$$

∎

The Fourier integral can be used to evaluate integrals. For example, at $x = 1$ it follows from Theorem 14.1 that (7) converges to $f(1)$; that is,

$$\int_{0}^{\infty} \frac{\sin \alpha}{\alpha} \, d\alpha = \frac{\pi}{2}.$$

The latter result is worthy of special note since it cannot be obtained in the "usual" manner; the integrand $(\sin x)/x$ does not possess an antiderivative that is an elementary function.

Cosine and Sine Integrals

When f is an even function on the interval $(-\infty, \infty)$, then the product $f(x) \cos \alpha x$ is also an even function, whereas $f(x) \sin \alpha x$ is an odd function.

As a consequence of property (g) of Section 11.3, $B(\alpha) = 0$ and so (4) becomes

$$f(x) = \frac{2}{\pi} \int_0^\infty \left(\int_0^\infty f(t) \cos \alpha t \, dt \right) \cos \alpha x \, d\alpha.$$

Here we have also used property (f) of Section 11.3 to write

$$\int_{-\infty}^\infty f(t) \cos \alpha t \, dt = 2 \int_0^\infty f(t) \cos \alpha t \, dt.$$

Similarly, when f is an odd function on $(-\infty, \infty)$, the products $f(x) \cos \alpha x$ and $f(x) \sin \alpha x$ are odd and even functions, respectively. Therefore $A(\alpha) = 0$ and

$$f(x) = \frac{2}{\pi} \int_0^\infty \left(\int_0^\infty f(t) \sin \alpha t \, dt \right) \sin \alpha x \, d\alpha.$$

We summarize:

DEFINITION 14.3   **Fourier Cosine and Sine Integrals**

(i) The Fourier integral of an even function on the interval $(-\infty, \infty)$ is the **cosine integral**

$$f(x) = \frac{2}{\pi} \int_0^\infty A(\alpha) \cos \alpha x \, d\alpha, \tag{8}$$

where

$$A(\alpha) = \int_0^\infty f(x) \cos \alpha x \, dx. \tag{9}$$

(ii) The Fourier integral of an odd function on the interval $(-\infty, \infty)$ is the **sine integral**

$$f(x) = \frac{2}{\pi} \int_0^\infty B(\alpha) \sin \alpha x \, d\alpha, \tag{10}$$

where

$$B(\alpha) = \int_0^\infty f(x) \sin \alpha x \, dx. \tag{11}$$

EXAMPLE 2

Find the Fourier integral representation of the function

$$f(x) = \begin{cases} 1, & |x| < a \\ 0, & |x| > a. \end{cases}$$

Figure 14.5

Solution It is apparent from Figure 14.5 that f is an even function. Hence we represent f by the Fourier cosine integral (8). From (9) we obtain

$$A(\alpha) = \int_0^\infty f(x) \cos \alpha x \, dx$$

$$= \int_0^a f(x) \cos \alpha x \, dx + \int_a^\infty f(x) \cos \alpha x \, dx$$

$$= \int_0^a \cos \alpha x \, dx = \frac{\sin a\alpha}{\alpha}$$

and so
$$f(x) = \frac{2}{\pi} \int_0^\infty \frac{\sin a\alpha \cos \alpha x}{\alpha} \, d\alpha. \qquad \blacksquare$$

The integrals (8) and (10) can be used when f is neither odd nor even and defined only on the half-line $(0, \infty)$. In this case, (8) represents f on the interval $(0, \infty)$ and its even (but not periodic) extension to $(-\infty, 0)$, whereas (10) represents f on $(0, \infty)$ and its odd extension to the interval $(-\infty, 0)$. The next example illustrates this concept.

Figure 14.6

EXAMPLE 3

Represent $\qquad\qquad f(x) = e^{-x}, \qquad x > 0,$

(a) by a cosine integral   and   **(b)** by a sine integral.

Solution   The graph of the function is given in Figure 14.6.

(a) Using integration by parts, we find

$$A(\alpha) = \int_0^\infty e^{-x} \cos \alpha x \, dx = \frac{1}{1 + \alpha^2}.$$

Therefore the cosine integral of f is

$$f(x) = \frac{2}{\pi} \int_0^\infty \frac{\cos \alpha x}{1 + \alpha^2} \, d\alpha.$$

(b) Similarly, we have

$$B(\alpha) = \int_0^\infty e^{-x} \sin \alpha x \, dx = \frac{\alpha}{1 + \alpha^2}.$$

The sine integral of f is then

$$f(x) = \frac{2}{\pi} \int_0^\infty \frac{\alpha \sin \alpha x}{1 + \alpha^2} \, d\alpha.$$

Figure 14.7 shows the graphs of the functions and their extensions represented by the two integrals.

(a) cosine integral   **(b)** sine integral

Figure 14.7   ■

Complex Form

The Fourier integral (4) also possesses an equivalent **complex form**, or **exponential form**, that is analogous to the complex form of a Fourier series (see Problem 21, Exercises 11.2). If (5) and (6) are substituted in (4), then

$$f(x) = \frac{1}{\pi} \int_0^\infty \int_{-\infty}^\infty f(t)[\cos \alpha t \cos \alpha x + \sin \alpha t \sin \alpha x]\, dt\, d\alpha$$

$$= \frac{1}{\pi} \int_0^\infty \int_{-\infty}^\infty f(t) \cos \alpha(t - x)\, dt\, d\alpha$$

$$= \frac{1}{2\pi} \int_{-\infty}^\infty \int_{-\infty}^\infty f(t) \cos \alpha(t - x)\, dt\, d\alpha \tag{12}$$

$$= \frac{1}{2\pi} \int_{-\infty}^\infty \int_{-\infty}^\infty f(t)[\cos \alpha(t - x) + i \sin \alpha(t - x)]\, dt\, d\alpha \tag{13}$$

$$= \frac{1}{2\pi} \int_{-\infty}^\infty \int_{-\infty}^\infty f(t)e^{i\alpha(t - x)}\, dt\, d\alpha$$

$$= \frac{1}{2\pi} \int_{-\infty}^\infty \left(\int_{-\infty}^\infty f(t)e^{i\alpha t}\, dt \right)e^{-i\alpha x}\, d\alpha. \tag{14}$$

We note that (12) follows from the fact that the integrand is an even function of α. In (13) we have simply added zero to the integrand;

$$i \int_{-\infty}^\infty \int_{-\infty}^\infty f(t) \sin \alpha(t - x)\, dt\, d\alpha = 0$$

because the integrand is an odd function of α. The integral in (14) can be expressed as

$$f(x) = \frac{1}{2\pi} \int_{-\infty}^\infty C(\alpha)e^{-i\alpha x}\, d\alpha, \tag{15}$$

where

$$C(\alpha) = \int_{-\infty}^\infty f(x)e^{i\alpha x}\, dx. \tag{16}$$

This latter form of the Fourier integral will be put to use in the next section when we return to the solution of boundary-value problems.

14.3 **EXERCISES**

Answers to odd-numbered problems begin on page A–47.

In Problems 1–6 find the Fourier integral representation of the given function.

1. $f(x) = \begin{cases} 0, & x < -1 \\ -1, & -1 < x < 0 \\ 2, & 0 < x < 1 \\ 0, & x > 1 \end{cases}$

2. $f(x) = \begin{cases} 0, & x < \pi \\ 4, & \pi < x < 2\pi \\ 0, & x > 2\pi \end{cases}$

3. $f(x) = \begin{cases} 0, & x < 0 \\ x, & 0 < x < 3 \\ 0, & x > 3 \end{cases}$

4. $f(x) = \begin{cases} 0, & x < 0 \\ \sin x, & 0 \le x \le \pi \\ 0, & x > \pi \end{cases}$

5. $f(x) = \begin{cases} 0, & x < 0 \\ e^{-x}, & x > 0 \end{cases}$

6. $f(x) = \begin{cases} e^x, & |x| < 1 \\ 0, & |x| > 1 \end{cases}$

In Problems 7–12 represent the given function by an appropriate cosine or sine integral.

7. $f(x) = \begin{cases} 0, & x < -1 \\ -5, & -1 < x < 0 \\ 5, & 0 < x < 1 \\ 0, & x > 1 \end{cases}$

8. $f(x) = \begin{cases} 0, & |x| < 1 \\ \pi, & 1 < |x| < 2 \\ 0, & |x| > 2 \end{cases}$

9. $f(x) = \begin{cases} |x|, & |x| < \pi \\ 0, & |x| > \pi \end{cases}$

10. $f(x) = \begin{cases} x, & |x| < \pi \\ 0, & |x| > \pi \end{cases}$

11. $f(x) = e^{-|x|} \sin x$*

12. $f(x) = xe^{-|x|}$

In Problems 13–16 find the cosine and sine integral representations of the given function.

13. $f(x) = e^{-kx}$, $k > 0, x > 0$

14. $f(x) = e^{-x} - e^{-3x}$, $x > 0$

15. $f(x) = xe^{-2x}$, $x > 0$

16. $f(x) = e^{-x} \cos x$, $x > 0$

In Problems 17 and 18 solve the given integral equation for the function f.

17. $\displaystyle\int_0^\infty f(x) \cos \alpha x \, dx = e^{-\alpha}$

---

* The integrals in Problems 11–16 are not difficult if you remember some Laplace transform results.

18. $\displaystyle\int_0^\infty f(x)\sin\alpha x\, dx = \begin{cases} 1, & 0 < \alpha < 1 \\ 0, & \alpha > 1 \end{cases}$

19. (a) Use (7) to show that

$$\int_0^\infty \frac{\sin 2x}{x}\, dx = \frac{\pi}{2}.$$

   [*Hint:* α is a dummy variable of integration.]

(b) Show in general that, for $k > 0$,

$$\int_0^\infty \frac{\sin kx}{x}\, dx = \frac{\pi}{2}.$$

20. Use the complex form (15) to find the Fourier integral representation of $f(x) = e^{-|x|}$. Show that the result is the same as that obtained from (8).

14.4 FOURIER TRANSFORMS

Transform Pairs

Up to now we have been using a symbolic representation for the inverse of a Laplace transform: $\mathscr{L}^{-1}\{F(s)\} = f(t)$. Actually, the inverse Laplace transform is itself an integral. If

$$\mathscr{L}\{f(t)\} = \int_0^\infty e^{-st}f(t)\, dt = F(s), \tag{1}$$

then it can be proved that

$$\mathscr{L}^{-1}\{F(s)\} = \frac{1}{2\pi i}\int_{\gamma - i\infty}^{\gamma + i\infty} e^{st}F(s)\, ds = f(t). \tag{2}$$

The foregoing integral is called a **contour integral**; its evaluation requires the use of complex variables and is beyond the scope of this text. The point is this: integral transforms occur in **transform pairs**; that is, if

$$F(\alpha) = \int_a^b f(x)K(\alpha, x)\, dx \tag{3}$$

is an integral that transforms $f(x)$ into $F(\alpha)$, then the function f is recovered by means of another integral transform

$$f(x) = \int_c^d F(\alpha)H(\alpha, x)\, d\alpha, \tag{4}$$

called the **inversion integral**, or **inverse transform**. The functions K and H are called the **kernels** of transforms (3) and (4), respectively. For example, in (1) we identify the kernel of the transform as $K(s, t) = e^{-st}$, whereas in (2) we can take the kernel of the inverse transform to be $H(s, t) = e^{st}/2\pi i$.

Fourier Transforms

The Fourier integral is the source of three new integral transforms. From (16)–(15), (11)–(10), and (9)–(8) of the preceding section we are prompted to define the following transform pairs.

DEFINITION 14.4   **Fourier Transform Pairs**

(i) **Fourier transform:**
$$\mathscr{F}\{f(x)\} = \int_{-\infty}^{\infty} f(x)e^{i\alpha x}\,dx = F(\alpha) \tag{5}$$

Inverse Fourier transform:
$$\mathscr{F}^{-1}\{F(\alpha)\} = \frac{1}{2\pi}\int_{-\infty}^{\infty} F(\alpha)e^{-i\alpha x}\,d\alpha = f(x) \tag{6}$$

(ii) **Fourier sine transform:**
$$\mathscr{F}_s\{f(x)\} = \int_{0}^{\infty} f(x)\sin\alpha x\,dx = F(\alpha) \tag{7}$$

Inverse Fourier sine transform:
$$\mathscr{F}_s^{-1}\{F(\alpha)\} = \frac{2}{\pi}\int_{0}^{\infty} F(\alpha)\sin\alpha x\,d\alpha = f(x) \tag{8}$$

(iii) **Fourier cosine transform:**
$$\mathscr{F}_c\{f(x)\} = \int_{0}^{\infty} f(x)\cos\alpha x\,dx = F(\alpha) \tag{9}$$

Inverse Fourier cosine transform:
$$\mathscr{F}_c^{-1}\{F(\alpha)\} = \frac{2}{\pi}\int_{0}^{\infty} F(\alpha)\cos\alpha x\,d\alpha = f(x) \tag{10}$$

Existence

The conditions under which (5), (7), and (9) exist are more stringent than those for the Laplace transform. For example, you should verify that $\mathscr{F}\{1\}$, $\mathscr{F}_s\{1\}$, and $\mathscr{F}_c\{1\}$ do not exist. Sufficient conditions for existence are that f be absolutely integrable on the appropriate interval and that f and f' be piecewise continuous on every finite interval.

Operational Properties

Since our immediate goal is to apply these new transforms to boundary-value problems, we need to examine the transforms of derivatives.

Fourier transform   Suppose that f is continuous and absolutely integrable on the interval $(-\infty, \infty)$ and f' is piecewise continuous on every finite interval. If $f(x) \to 0$ as $x \to \pm\infty$, then integration by parts gives

$$\mathscr{F}\{f'(x)\} = \int_{-\infty}^{\infty} f'(x)e^{i\alpha x}\,dx$$

$$= f(x)e^{i\alpha x}\Big|_{-\infty}^{\infty} - i\alpha \int_{-\infty}^{\infty} f(x)e^{i\alpha x}\,dx$$

$$= -i\alpha \int_{-\infty}^{\infty} f(x)e^{i\alpha x}\,dx;$$

that is,
$$\mathscr{F}\{f'(x)\} = -i\alpha F(\alpha). \tag{11}$$

Similarly, under the added assumptions that f' is continuous on $(-\infty, \infty)$, $f''(x)$ is piecewise continuous on every finite interval, and $f'(x) \to 0$ as $x \to \pm\infty$, we have

$$\mathscr{F}\{f''(x)\} = (-i\alpha)^2 \mathscr{F}\{f(x)\} = -\alpha^2 F(\alpha). \tag{12}$$

It is important to be aware that the sine and cosine transforms are not suitable for transforming the first derivative (or for that matter, any derivative of *odd* order). It is left as an exercise to show that

$$\mathscr{F}_s\{f'(x)\} = -\alpha \mathscr{F}_c\{f(x)\} \tag{13}$$

and
$$\mathscr{F}_c\{f'(x)\} = \alpha \mathscr{F}_s\{f(x)\} - f(0). \tag{14}$$

The difficulty is apparent; the transform of $f'(x)$ is not expressed in terms of the original integral transform.

Fourier sine transform Suppose that f and f' are continuous, f is absolutely integrable on the interval $[0, \infty)$, and f'' is piecewise continuous on every finite interval. If $f \to 0$ and $f' \to 0$ as $x \to \infty$, then

$$\mathscr{F}_s\{f''(x)\} = \int_{0}^{\infty} f''(x) \sin \alpha x\,dx$$

$$= f'(x) \sin \alpha x\Big|_{0}^{\infty} - \alpha \int_{0}^{\infty} f'(x) \cos \alpha x\,dx$$

$$= -\alpha \left[f(x) \cos \alpha x\Big|_{0}^{\infty} + \alpha \int_{0}^{\infty} f(x) \sin \alpha x\,dx \right]$$

$$= \alpha f(0) - \alpha^2 \mathscr{F}_s\{f(x)\};$$

that is,
$$\mathscr{F}_s\{f''(x)\} = -\alpha^2 F(\alpha) + \alpha f(0). \tag{15}$$

Fourier cosine transform Under the same assumptions that lead to (15), we find the Fourier cosine transform of $f''(x)$ to be

$$\mathscr{F}_c\{f''(x)\} = -\alpha^2 F(\alpha) - f'(0). \tag{16}$$

A natural question is: How do we know which transform to use on a given boundary-value problem? Clearly, to use a Fourier transform the domain of the variable to be eliminated must be $(-\infty, \infty)$. To utilize a sine or cosine transform the domain of at least one of the variables in the problem must be $[0, \infty)$. But the determining factor in choosing between the sine transform and the cosine transform is the type of boundary condition specified at zero.

In the examples that follow, we shall assume without further mention that both u and $\partial u/\partial x$ (or $\partial u/\partial y$) approach zero as $x \to \pm\infty$. This is not a major restriction since these conditions hold in most applications.

EXAMPLE 1

Solve the heat equation

$$k\frac{\partial^2 u}{\partial x^2} = \frac{\partial u}{\partial t}, \qquad -\infty < x < \infty, \quad t > 0,$$

subject to $\qquad u(x, 0) = f(x) \quad$ where $\quad f(x) = \begin{cases} 1, & |x| \leq 1 \\ 0, & |x| > 1. \end{cases}$

Solution The problem can be interpreted as finding the temperature $u(x, t)$ in an infinite rod. Since the domain of x is the infinite interval $(-\infty, \infty)$, we use the Fourier transform (5) and define

$$\mathscr{F}\{u(x, t)\} = \int_{-\infty}^{\infty} u(x, t)e^{i\alpha x}\, dx = U(\alpha, t).$$

Transforming the partial differential equation and using (12) yield

$$\mathscr{F}\left\{k\frac{\partial^2 u}{\partial x^2}\right\} = \mathscr{F}\left\{\frac{\partial u}{\partial t}\right\}$$

$$-k\alpha^2 U(\alpha, t) = \frac{dU}{dt} \quad \text{or} \quad \frac{dU}{dt} + k\alpha^2 U = 0.$$

Thus $\qquad\qquad\qquad U(\alpha, t) = ce^{-k\alpha^2 t}.$ $\qquad\qquad$ **(17)**

Now the transform of the initial condition is

$$\mathscr{F}\{u(x, 0)\} = \int_{-\infty}^{\infty} f(x)e^{i\alpha x}\, dx = \int_{-1}^{1} e^{i\alpha x}\, dx$$

$$= \frac{e^{i\alpha} - e^{-i\alpha}}{i\alpha}$$

or $\qquad\qquad\qquad U(\alpha, 0) = 2\frac{\sin \alpha}{\alpha}.$

Using this last condition in (17) gives $U(\alpha, 0) = c = (2 \sin \alpha)/\alpha$, and so

$$U(\alpha, t) = 2 \frac{\sin \alpha}{\alpha} e^{-k\alpha^2 t}.$$

It then follows from the inversion integral (6) that

$$u(x, t) = \frac{1}{\pi} \int_{-\infty}^{\infty} \frac{\sin \alpha}{\alpha} e^{-k\alpha^2 t} e^{-i\alpha x} \, d\alpha.$$

The last expression can be simplified somewhat by writing $e^{-i\alpha x} = \cos \alpha x - i \sin \alpha x$ and noting $\int_{-\infty}^{\infty} \frac{\sin \alpha}{\alpha} e^{-k\alpha^2 t} \sin \alpha x \, d\alpha = 0$ since the integrand is an odd function of α. Hence we finally have

$$u(x, t) = \frac{1}{\pi} \int_{-\infty}^{\infty} \frac{\sin \alpha \cos \alpha x}{\alpha} e^{-k\alpha^2 t} \, d\alpha. \tag{18}$$

∎

EXAMPLE 2

The steady-state temperature in a semi-infinite plate is determined from

$$\frac{\partial^2 u}{\partial x^2} + \frac{\partial^2 u}{\partial y^2} = 0, \qquad 0 < x < \pi, \quad y > 0$$

$$u(0, y) = 0, \qquad u(\pi, y) = e^{-y}, \qquad y > 0$$

$$\left. \frac{\partial u}{\partial y} \right|_{y=0} = 0, \qquad 0 < x < \pi.$$

Solve for $u(x, y)$.

Solution   The domain of the variable y and the prescribed condition at $y = 0$ indicate that the Fourier cosine transform is suitable for the problem. We define

$$\mathscr{F}_c\{u(x, y)\} = \int_0^{\infty} u(x, y) \cos \alpha y \, dy = U(x, \alpha).$$

In view of (16), $\mathscr{F}_c\left\{\dfrac{\partial^2 u}{\partial x^2}\right\} + \mathscr{F}_c\left\{\dfrac{\partial^2 u}{\partial y^2}\right\} = \mathscr{F}_c\{0\}$

becomes

$$\frac{d^2 U}{dx^2} - \alpha^2 U(x, \alpha) - u_y(x, 0) = 0 \quad \text{or} \quad \frac{d^2 U}{dx^2} - \alpha^2 U = 0.$$

Since the domain of x is a finite interval, we choose to write the solution of the ordinary differential equation as

$$U(x, \alpha) = c_1 \cosh \alpha x + c_2 \sinh \alpha x. \tag{19}$$

Now $\mathscr{F}_c\{u(0, y)\} = \mathscr{F}_c\{0\}$ and $\mathscr{F}_c\{u(\pi, y)\} = \mathscr{F}_c\{e^{-y}\}$ are in turn equivalent to

$$U(0, \alpha) = 0 \quad \text{and} \quad U(\pi, \alpha) = \frac{1}{1 + \alpha^2}.$$

Applying these latter conditions, the solution (19) gives $c_1 = 0$ and $c_2 = 1/[(1 + \alpha^2) \sinh \alpha\pi]$. Therefore

$$U(x, \alpha) = \frac{\sinh \alpha x}{(1 + \alpha^2) \sinh \alpha\pi},$$

and so from (10) we arrive at

$$u(x, y) = \frac{2}{\pi} \int_0^\infty \frac{\sinh \alpha x}{(1 + \alpha^2) \sinh \alpha\pi} \cos \alpha y \, d\alpha. \tag{20}$$

∎

Had $u(x, 0)$ been given in Example 2 rather than $u_y(x, 0)$, then the sine transform would have been appropriate.

Remark   You may be wondering whether the integrals in (18) and (20) can be evaluated. In *most* cases the answer is no (see Problem 22 for an exception). One way of finding a specific value, say, of the temperature $u(1, 1)$ in (20) is to resort to numerical integration.

14.4 EXERCISES

Answers to odd-numbered problems begin on page A−47.

In Problems 1–21 use the Fourier integral transforms of this section to solve the given boundary-value problem. Make assumptions about boundedness where necessary.

1. $k\dfrac{\partial^2 u}{\partial x^2} = \dfrac{\partial u}{\partial t}, \quad -\infty < x < \infty, t > 0$

   $u(x, 0) = e^{-|x|}, \quad -\infty < x < \infty$

2. $k\dfrac{\partial^2 u}{\partial x^2} = \dfrac{\partial u}{\partial t}, \quad -\infty < x < \infty, t > 0$

   $u(x, 0) = \begin{cases} 0, & x < -1 \\ -100, & -1 < x < 0 \\ 100, & 0 < x < 1 \\ 0, & x > 1 \end{cases}$

3. Use the result $\mathscr{F}\{e^{-x^2/4p^2}\} = 2\sqrt{\pi}pe^{-p^2\alpha^2}$ to solve the boundary-value problem

   $$k\frac{\partial^2 u}{\partial x^2} = \frac{\partial u}{\partial t}, \quad -\infty < x < \infty, \quad t > 0$$

   $$u(x, 0) = e^{-x^2}, \quad -\infty < x < \infty.$$

4. **(a)** If $\mathscr{F}\{f(x)\} = F(\alpha)$ and $\mathscr{F}\{g(x)\} = G(\alpha)$, then the **convolution theorem** for the Fourier transform is given by

$$\int_{-\infty}^{\infty} f(\tau)g(x - \tau)\,d\tau = \mathscr{F}^{-1}\{F(\alpha)G(\alpha)\}.$$

Use this result and $\mathscr{F}\{e^{-x^2/4p^2}\} = 2\sqrt{\pi}pe^{-p^2\alpha^2}$ to show that a solution of the boundary-value problem

$$k\frac{\partial^2 u}{\partial x^2} = \frac{\partial u}{\partial t}, \qquad -\infty < x < \infty, \quad t > 0$$

$$u(x, 0) = f(x), \qquad -\infty < x < \infty,$$

is given by

$$u(x, t) = \frac{1}{2\sqrt{k\pi t}} \int_{-\infty}^{\infty} f(\tau)e^{-(x-\tau)^2/4kt}\,d\tau.$$

(b) By using the change of variables $u = (x - \tau)/2\sqrt{kt}$ and Problem 9 of Exercises 14.1, show that the solution of part (a) when

$$f(x) = \begin{cases} u_0, & |x| < 1 \\ 0, & |x| > 1. \end{cases}$$

is

$$u(x, t) = \frac{u_0}{2}\left[\operatorname{erf}\left(\frac{x+1}{2\sqrt{kt}}\right) - \operatorname{erf}\left(\frac{x-1}{2\sqrt{kt}}\right)\right].$$

5. Find the temperature $u(x, t)$ in a semi-infinite rod if $u(0, t) = u_0$, $t > 0$, and $u(x, 0) = 0$, $x > 0$.

6. Use the result $\displaystyle\int_0^{\infty} \frac{\sin \alpha x}{\alpha}\,d\alpha = \frac{\pi}{2}$, $x > 0$, to show that the solution of Problem 5 can be written as

$$u(x, t) = u_0 - \frac{2u_0}{\pi} \int_0^{\infty} \frac{\sin \alpha x}{\alpha} e^{-k\alpha^2 t}\,d\alpha.$$

7. Find the temperature $u(x, t)$ in a semi-infinite rod if $u(0, t) = 0$, $t > 0$, and
$$u(x, 0) = \begin{cases} 1, & 0 < x < 1 \\ 0, & x > 1. \end{cases}$$

8. Solve Problem 5 if the condition at the left boundary is

$$\frac{\partial u}{\partial x}\bigg|_{x=0} = -A, \qquad t > 0.$$

9. Solve Problem 7 if the end $x = 0$ is insulated.

10. Find the temperature $u(x, t)$ in a semi-infinite rod if $u(0, t) = 1$, $t > 0$, and $u(x, 0) = e^{-x}$, $x > 0$.

11. (a) $a^2 \dfrac{\partial^2 u}{\partial x^2} = \dfrac{\partial^2 u}{\partial t^2}, \qquad -\infty < x < \infty, \quad t > 0$

$u(x, 0) = f(x), \qquad \dfrac{\partial u}{\partial t}\bigg|_{t=0} = g(x), \qquad -\infty < x < \infty$

(b) If $g(x) = 0$, show that the solution of part (a) can be written as
$u(x, t) = \frac{1}{2}[f(x + at) + f(x - at)]$.

12. Find the displacement $u(x, t)$ of a semi-infinite string if $u(0, t) = 0, t > 0$,

$u(x, 0) = xe^{-x}, \quad \dfrac{\partial u}{\partial t}\bigg|_{t=0} = 0, x > 0$.

13. Solve the problem in Example 2 if the boundary conditions at $x = 0$ and $x = \pi$ are reversed: $u(0, y) = e^{-y}, u(\pi, y) = 0, y > 0$.

14. Solve the problem in Example 2 if the boundary condition at $y = 0$ is $u(x, 0) = 1, 0 < x < \pi$.

15. Find the steady-state temperature $u(x, y)$ in a plate defined by $x \geq 0$, $y \geq 0$ if the boundary $x = 0$ is insulated, and at $y = 0$,

$u(x, 0) = \begin{cases} 50, & 0 < x < 1 \\ 0, & x > 1. \end{cases}$

16. Solve Problem 15 if the boundary condition at $x = 0$ is $u(0, y) = 0, y > 0$.

17. $\dfrac{\partial^2 u}{\partial x^2} + \dfrac{\partial^2 u}{\partial y^2} = 0, \qquad x > 0, \quad 0 < y < 2$

$u(0, y) = 0, \qquad 0 < y < 2$
$u(x, 0) = f(x), \qquad u(x, 2) = 0, \qquad x > 0$

18. $\dfrac{\partial^2 u}{\partial x^2} + \dfrac{\partial^2 u}{\partial y^2} = 0, \qquad 0 < x < \pi, \quad y > 0$

$u(0, y) = f(y), \qquad \dfrac{\partial u}{\partial x}\bigg|_{x=\pi} = 0, \qquad y > 0$

$\dfrac{\partial u}{\partial y}\bigg|_{y=0} = 0, \qquad 0 < x < \pi$

In Problems 19 and 20 find the steady-state temperature in the plate given in the figure. [*Hint:* One way of proceeding is to express Problems 19 and 20 as two- and three-boundary-value problems, respectively. Use the superposition principle (see Section 12.5).]

19.

20.

Figure 14.8   Figure 14.9

$u = e^{-x^2}$

insulated

Figure 14.10

21. Use the transform $\mathscr{F}\{e^{-x^2/4p^2}\}$ given in Problem 3 to find the steady-state temperature in the infinite strip shown in Figure 14.10.

22. The solution of Problem 16 can be integrated. Use entries 42 and 43 of the table in Appendix II to show that

$$u(x, y) = \frac{100}{\pi}\left[\arctan\frac{x}{y} - \frac{1}{2}\arctan\frac{x+1}{y} - \frac{1}{2}\arctan\frac{x-1}{y}\right].$$

23. Derive (13).   **24.** Derive (14).

Chapter 14   *REVIEW*

The **error function** and **complementary error function**

$$\text{erf}(x) = \frac{2}{\sqrt{\pi}}\int_0^x e^{-u^2}\,du, \qquad \text{erfc}(x) = 1 - \text{erf}(x) = \frac{2}{\sqrt{\pi}}\int_x^\infty e^{-u^2}\,du$$

occur often in finding inverse Laplace transforms and in the solution of boundary-value problems.

The **Fourier integral**

$$f(x) = \frac{1}{\pi}\int_0^\infty [A(\alpha)\cos\alpha x + B(\alpha)\sin\alpha x]\,d\alpha, \tag{1}$$

where

$$A(\alpha) = \int_{-\infty}^\infty f(x)\cos\alpha x\,dx,$$

$$B(\alpha) = \int_{-\infty}^\infty f(x)\sin\alpha x\,dx,$$

is a way of representing a nonperiodic function that is defined on the infinite interval $(-\infty, \infty)$. When f is an even function, (1) becomes a **cosine integral**; when f is odd, (1) reduces to a **sine integral**.

Integral transforms occur in transform pairs: the transform itself and the inversion integral. To solve a boundary-value problem by an integral transform, we:

(**i**) transform the partial differential equation into an ordinary differential equation.

(**ii**) transform all boundary or initial conditions that depend on the transformed variable.

(**iii**) solve the ordinary differential equation subject to the conditions found in step (ii).

(**iv**) find the inverse transform of the function found in step (iii). This is the solution of the original boundary-value problem.

Chapter 14   *REVIEW EXERCISES*

Answers to odd-numbered problems begin on page A-48.

In Problems 1–13 solve the given boundary-value problem by an appropriate integral transform. Make assumptions about boundedness where necessary.

1. $\dfrac{\partial^2 u}{\partial x^2} + \dfrac{\partial^2 u}{\partial y^2} = 0, \quad x > 0, 0 < y < \pi$

$\dfrac{\partial u}{\partial x}\bigg|_{x=0} = 0, \quad 0 < y < \pi$

$u(x, 0) = 0, \quad \dfrac{\partial u}{\partial y}\bigg|_{y=\pi} = e^{-x}, \quad x > 0$

2. $\dfrac{\partial^2 u}{\partial x^2} = \dfrac{\partial u}{\partial t}, \quad 0 < x < 1, t > 0$

$u(0, t) = 0, \quad u(1, t) = 0, \quad t > 0$

$u(x, 0) = 50 \sin 2\pi x, \quad 0 < x < 1$

3. $\dfrac{\partial^2 u}{\partial x^2} - hu = \dfrac{\partial u}{\partial t}, \quad h > 0, x > 0, t > 0$

$u(0, t) = 0, \quad \lim_{x \to \infty} \dfrac{\partial u}{\partial x} = 0, \quad t > 0$

$u(x, 0) = u_0, \quad x > 0$

4. $\dfrac{\partial u}{\partial t} - \dfrac{\partial^2 u}{\partial x^2} = e^{-|x|}, \quad -\infty < x < \infty, t > 0$

$u(x, 0) = 0, \quad -\infty < x < \infty$

5. $\dfrac{\partial^2 u}{\partial x^2} = \dfrac{\partial u}{\partial t}, \quad x > 0, t > 0$

$u(0, t) = t, \quad \lim_{x \to \infty} u(x, t) = 0$

$u(x, 0) = 0, \quad x > 0 \quad [\textit{Hint:} \text{ Use Theorem 7.9.}]$

6. $\dfrac{\partial^2 u}{\partial x^2} = \dfrac{\partial^2 u}{\partial t^2}, \quad 0 < x < 1, t > 0$

$u(0, t) = 0, \quad u(1, t) = 0, \quad t > 0$

$u(x, 0) = \sin \pi x, \quad \dfrac{\partial u}{\partial t}\bigg|_{t=0} = -\sin \pi x, \quad 0 < x < 1$

7. $k\dfrac{\partial^2 u}{\partial x^2} = \dfrac{\partial u}{\partial t}, \quad -\infty < x < \infty, t > 0$

$u(x, 0) = \begin{cases} 0, & x < 0 \\ u_0, & 0 < x < \pi \\ 0, & x > \pi \end{cases}$

8. $\dfrac{\partial^2 u}{\partial x^2} + \dfrac{\partial^2 u}{\partial y^2} = 0, \quad 0 < x < \pi, y > 0$

$$u(0, y) = 0, \quad u(\pi, y) = \begin{cases} 0, & 0 < y < 1 \\ 1, & 1 < y < 2 \\ 0, & y > 2 \end{cases}$$

$$\left.\dfrac{\partial u}{\partial y}\right|_{y=0} = 0, \quad 0 < x < \pi$$

9. $\dfrac{\partial^2 u}{\partial x^2} + \dfrac{\partial^2 u}{\partial y^2} = 0, \quad x > 0, y > 0$

$$u(0, y) = \begin{cases} 50, & 0 < y < 1 \\ 0, & y > 1 \end{cases}$$

$$u(x, 0) = \begin{cases} 100, & 0 < x < 1 \\ 0, & x > 1 \end{cases}$$

10. $\dfrac{\partial^2 u}{\partial x^2} + r = \dfrac{\partial u}{\partial t}, \quad 0 < x < 1, t > 0$

$$\left.\dfrac{\partial u}{\partial x}\right|_{x=0} = 0, \quad u(1, t) = 0, \quad t > 0$$

$$u(x, 0) = 0, \quad 0 < x < 1$$

11. $\dfrac{\partial^2 u}{\partial x^2} + \dfrac{\partial^2 u}{\partial y^2} = 0, \quad x > 0, 0 < y < \pi$

$$u(0, y) = A, \quad 0 < y < \pi$$

$$\left.\dfrac{\partial u}{\partial y}\right|_{y=0} = 0, \quad \left.\dfrac{\partial u}{\partial y}\right|_{y=\pi} = Be^{-x}, \quad x > 0$$

12. $\dfrac{\partial^2 u}{\partial x^2} = \dfrac{\partial u}{\partial t}, \quad 0 < x < 1, t > 0$

$$u(0, t) = u_0, \quad u(1, t) = u_0, \quad t > 0$$

$$u(x, 0) = 0, \quad 0 < x < 1$$

[*Hint:* Use the identity $\sinh(x - y) = \sinh x \cosh y - \cosh x \sinh y$ and then use Problem 6, Exercises 14.1.]

13. $k\dfrac{\partial^2 u}{\partial x^2} = \dfrac{\partial u}{\partial t}, \quad -\infty < x < \infty, t > 0$

$$u(x, 0) = \begin{cases} 0, & x < 0 \\ e^{-x}, & x > 0 \end{cases}$$

CHAPTER 15

NUMERICAL METHODS FOR PARTIAL DIFFERENTIAL EQUATIONS

15.1 Elliptic Equations
15.2 Parabolic Equations
15.3 Hyperbolic Equations

Chapter 15 Review
Chapter 15 Review Exercises

Important Concepts

Difference equations
Mesh size
Mesh points
Dirichlet problem
Gauss-Seidel iteration
Explicit finite difference method
Implicit finite difference method
Crank-Nicholson method

In Chapters 12, 13, and 14 it may have occurred to you to ask a basic question: How do we obtain a numerical value of, say, the displacement $u(x, t)$ of a vibrating string for a specific point x along the string and time t, or the temperature $u(x, y)$ at a specified point $P(x, y)$? Of course, we could substitute values of x and t in a solution such as (5) of Section 12.4 and (5) of Section 12.5 and sum an infinite series. But analogous to the material studied in Chapter 9 there are numerical procedures for linear partial differential equations. For certain purposes it is perfectly reasonable to accept an approximation to the solution of

Continues

Continued

the heat equation, the wave equation, or Laplace's equation. Moreover, in practice it may be necessary to solve, say, Laplace's equation in a region of the *xy*-plane, such as illustrated in the preceding figure. For complicated regions such as this, the methods of separation of variables and integral transforms may not be convenient or even applicable. Indeed, a numerical procedure may be the *only* way of attacking a boundary-value problem.

In this chapter the key concept is the notion of a finite difference equation (see Section 9.8). We approximate the solution of a partial differential equation by replacing it with a difference equation.

15.1 ELLIPTIC EQUATIONS

Using the difference quotients derived in Section 9.8, we can easily replace a linear second-order partial differential equation by a difference equation. In the discussion that follows we shall confine our attention to elliptic partial differential equations such as Laplace's equation.

A Difference Equation Replacement for Laplace's Equation

If *u* is a function of two variables *x* and *y*, we can form two central differences

$$u(x + h, y) - 2u(x, y) + u(x - h, y) \quad \text{and} \quad u(x, y + h) - 2u(x, y) + u(x, y - h).$$

It then follows from (6) of Section 9.8 that finite difference approximations for the second partial derivatives u_{xx} and u_{yy} are given by

$$\frac{\partial^2 u}{\partial x^2} \approx \frac{1}{h^2} [u(x + h, y) - 2u(x, y) + u(x - h, y)] \tag{1}$$

$$\frac{\partial^2 u}{\partial y^2} \approx \frac{1}{h^2} [u(x, y + h) - 2u(x, y) + u(x, y - h)]. \tag{2}$$

Now by adding (1) and (2) we obtain a **five-point approximation** to the Laplacian:

$$\frac{\partial^2 u}{\partial x^2} + \frac{\partial^2 u}{\partial y^2} \approx \frac{1}{h^2} [u(x + h, y) + u(x, y + h) + u(x - h, y) + u(x, y - h) - 4u(x, y)].$$

Hence we can replace Laplace's equation $\dfrac{\partial^2 u}{\partial x^2} + \dfrac{\partial^2 u}{\partial y^2} = 0$ by the difference equation

$$u(x + h, y) + u(x, y + h) + u(x - h, y) + u(x, y - h) - 4u(x, y) = 0. \quad \textbf{(3)}$$

If we adopt the notation $u(x, y) = u_{ij}$ and

$$u(x + h, y) = u_{i+1, j}$$

$$u(x, y + h) = u_{i, j+1}$$

$$u(x - h, y) = u_{i-1, j}$$

$$u(x, y - h) = u_{i, j-1},$$

then (3) becomes

$$u_{i+1, j} + u_{i, j+1} + u_{i-1, j} + u_{i, j-1} - 4u_{ij} = 0. \quad \textbf{(4)}$$

To understand (4) a little better, suppose a rectangular grid consisting of horizontal lines spaced h units apart and vertical lines spaced h units apart is placed over a region R bounded by a curve C in which we are seeking a solution of Laplace's equation. The number h is called the **mesh size**. See Figure 15.1(a). The points of intersection $P_{ij} = P(ih, jh)$, i and j integers, of the lines are called **mesh points**, or **lattice points**. A mesh point is an **interior point** if its four nearest neighboring mesh points are points of R. Points in R or on C that are not interior points are called **boundary points**. For example, in Figure 15.1(a) we have

$$P_{20} = P(2h, 0), \qquad P_{11} = P(h, h), \qquad P_{21} = P(2h, h), \qquad P_{22} = P(2h, 2h),$$

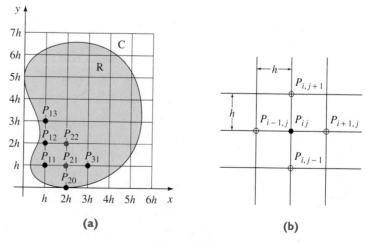

(a)   (b)

Figure 15.1

and so on. Of the points listed, P_{21} and P_{22} are interior points, P_{20} and P_{11} are boundary points. In Figure 15.1(a) interior points are the colored dots and the boundary points are the black dots. Now from (4) we see that

$$u_{ij} = \frac{1}{4}[u_{i+1,j} + u_{i,j+1} + u_{i-1,j} + u_{i,j-1}] \tag{5}$$

and so, as shown in Figure 15.1(b), the value of u_{ij} at an interior mesh point of R is the average of the values of u at four neighboring mesh points. The neighboring points $P_{i+1,j}$, $P_{i,j+1}$, $P_{i-1,j}$, and $P_{i,j-1}$ correspond, respectively, to the four points on a compass E, N, W, and S.

Dirichlet Problem

Recall, in the **Dirichlet problem** for Laplace's equation $\nabla^2 u = 0$ the values of $u(x, y)$ are prescribed on the boundary C of a region R. The basic idea is to find an approximate solution to Laplace's equation at interior mesh points by replacing the partial differential equation at these points by the difference equation (4). Hence, the approximate values of u at the mesh points, namely, the u_{ij}, are related to each other and, possibly, to known values of u if a mesh point lies on the boundary C. In this manner we obtain a system of linear algebraic equations that we solve for the unknown u_{ij}. The following example illustrates the method for a square region.

EXAMPLE I

In Problem 14 of Exercises 12.5 you were asked to solve the boundary-value problem

$$\frac{\partial^2 u}{\partial x^2} + \frac{\partial^2 u}{\partial y^2} = 0, \qquad 0 < x < 2, \quad 0 < y < 2$$

$$u(0, y) = 0, \qquad u(2, y) = y(2 - y), \qquad 0 < y < 2$$

$$u(x, 0) = 0, \qquad u(x, 2) = \begin{cases} x, & 0 < x < 1 \\ 2 - x, & 1 \leq x < 2 \end{cases}$$

utilizing the superposition principle. To apply the present numerical method let us start with a mesh size of $h = 2/3$. As we see in Figure 15.2, that choice yields four interior points and eight boundary points. The numbers listed next to the boundary points are the exact values of u obtained from the specified condition along that boundary. For example, at $P_{31} = P(3h, h) = P(2, \frac{2}{3})$ we have $x = 2$ and $y = \frac{2}{3}$ and so the condition $u(2, y)$ gives $u(2, \frac{2}{3}) = \frac{2}{3}(2 - \frac{2}{3}) = \frac{8}{9}$. Similarly, at $P_{13} = P(\frac{2}{3}, 2)$ the condition $u(x, 2)$ gives $u(\frac{2}{3}, 2) = \frac{2}{3}$. We now

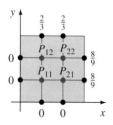

Figure 15.2

apply (4) at each interior point. For example, at P_{11} we have $i = 1$ and $j = 1$, so (4) becomes

$$u_{21} + u_{12} + u_{01} + u_{10} - 4u_{11} = 0.$$

Since $u_{01} = u(0, \frac{2}{3}) = 0$ and $u_{10} = u(\frac{2}{3}, 0) = 0$, the foregoing equation becomes $-4u_{11} + u_{21} + u_{12} = 0$. Repeating this, in turn, at P_{21}, P_{12}, and P_{22}, we get three additional equations:

$$
\begin{aligned}
-4u_{11} + u_{21} + u_{12} \qquad\qquad &= 0 \\[2mm]
u_{11} - 4u_{21} + \qquad\quad u_{22} &= -\frac{8}{9} \\[2mm]
u_{11} \qquad\quad - 4u_{12} + u_{22} &= -\frac{2}{3} \\[2mm]
u_{21} + u_{12} - 4u_{22} &= -\frac{14}{9}.
\end{aligned}
\tag{6}
$$

Using the computer algebra system Mathematica to solve the system, we find the approximate temperatures at the four interior points to be:

$$u_{11} = \frac{7}{36} = 0.1944, \qquad u_{21} = \frac{5}{12} = 0.4167,$$

$$u_{12} = \frac{13}{36} = 0.3611, \qquad u_{22} = \frac{7}{12} = 0.5833. \qquad\blacksquare$$

As in the discussion of ordinary differential equations, we expect that a smaller value of h will improve the accuracy of the approximation. However, using a smaller mesh size means, of course, that there are more interior mesh points, and correspondingly there is a much larger system of equations to be solved. For a *square* region whose length of side is L, a mesh size of $h = L/n$ will yield a total of $(n - 1)^2$ interior mesh points. In Example 1, for $n = 8$, the mesh size is a reasonable $h = 2/8 = 1/4$, but the number of interior points is $(8 - 1)^2 = 49$. Thus we have 49 equations in 49 unknowns. In the next example, we use a mesh size of $h = 1/2$.

Figure 15.3

EXAMPLE 2

As we see in Figure 15.3, with $n = 4$, a mesh size $h = 2/4 = 1/2$ for the square in Example 1 gives $3^2 = 9$ interior mesh points. Applying (4) at these points and using the indicated boundary conditions, we get nine equations in nine unknowns. So that you can verify the results, we give the system in an unsimplified form:

$$u_{21} + u_{12} + 0 + 0 - 4u_{11} = 0$$

$$u_{31} + u_{22} + u_{11} + 0 - 4u_{21} = 0$$

$$\frac{3}{4} + u_{32} + u_{21} + 0 - 4u_{31} = 0$$

$$u_{22} + u_{13} + u_{11} + 0 - 4u_{12} = 0$$

$$u_{32} + u_{23} + u_{12} + u_{21} - 4u_{22} = 0 \qquad (7)$$

$$1 + u_{33} + u_{22} + u_{31} - 4u_{32} = 0$$

$$u_{23} + \frac{1}{2} + 0 + u_{12} - 4u_{13} = 0$$

$$u_{33} + 1 + u_{13} + u_{22} - 4u_{23} = 0$$

$$\frac{3}{4} + \frac{1}{2} + u_{23} + u_{32} - 4u_{33} = 0.$$

In this case, Mathematica yields

$$u_{11} = \frac{7}{64} = 0.1094, \qquad u_{21} = \frac{51}{224} = 0.2277, \qquad u_{31} = \frac{177}{448} = 0.3951$$

$$u_{12} = \frac{47}{224} = 0.2098, \qquad u_{22} = \frac{13}{32} = 0.4063, \qquad u_{32} = \frac{135}{224} = 0.6027$$

$$u_{13} = \frac{145}{448} = 0.3237, \qquad u_{23} = \frac{131}{224} = 0.5848, \qquad u_{33} = \frac{39}{64} = 0.6094. \qquad \blacksquare$$

After we simplify (7), it is interesting to note that the 9×9 matrix of coefficients is

$$\begin{pmatrix} -4 & 1 & 0 & 1 & 0 & 0 & 0 & 0 & 0 \\ 1 & -4 & 1 & 0 & 1 & 0 & 0 & 0 & 0 \\ 0 & 1 & -4 & 0 & 0 & 1 & 0 & 0 & 0 \\ 1 & 0 & 0 & -4 & 1 & 0 & 1 & 0 & 0 \\ 0 & 1 & 0 & 1 & -4 & 1 & 0 & 1 & 0 \\ 0 & 0 & 1 & 0 & 1 & -4 & 0 & 0 & 1 \\ 0 & 0 & 0 & 1 & 0 & 0 & -4 & 1 & 0 \\ 0 & 0 & 0 & 0 & 1 & 0 & 1 & -4 & 1 \\ 0 & 0 & 0 & 0 & 0 & 1 & 0 & 1 & -4 \end{pmatrix} \qquad (8)$$

This is an example of a **sparse matrix**, in that a large percentage of the entries are zeros. The matrix (8) is also an example of a **banded matrix**. These kinds of matrices are characterized by the properties that the entries on the main diagonal and on diagonals (or bands) parallel to the main diagonal are all nonzero.

Gauss-Seidel Iteration

Problems requiring approximations to solutions of partial differential equations invariably lead to large systems of linear algebraic equations. It is not uncommon to have to solve systems involving hundreds of equations. Although a direct method of solution such as Gaussian elimination leaves unchanged the zero entries outside the bands in a matrix such as (8), it does fill in the positions between the bands with nonzeros. Since storing very large matrices uses up a large portion of computer memory, it is usual practice to solve a large system in an indirect manner. One popular indirect method is called **Gauss-Seidel iteration**.

We shall illustrate this method for the system in (6).* For the sake of simplicity we replace the double-subscripted variables u_{11}, u_{21}, u_{12}, and u_{22} by x_1, x_2, x_3, and x_4, respectively.

EXAMPLE 3

Step 1:   *Solve each equation for the variables on the main diagonal of the system.*   That is, in (6) solve the first equation for x_1, the second equation for x_2, and so on:

$$
\begin{aligned}
x_1 &= \qquad\quad 0.25x_2 + 0.25x_3 \\[4pt]
x_2 &= 0.25x_1 + \qquad\qquad\quad 0.25x_4 + 0.2222 \\[4pt]
x_3 &= 0.25x_1 + \qquad\qquad\quad 0.25x_4 + 0.1667 \qquad\qquad \textbf{(9)} \\[4pt]
x_4 &= \qquad\quad 0.25x_2 + 0.25x_3 \qquad\; + 0.3889.
\end{aligned}
$$

These equations can be obtained directly by using (5) rather than (4) at the interior points.

Step 2:   *Iterations.*   We start by making an initial guess for the values of x_1, x_2, x_3, and x_4. If this were simply a system of linear equations and we knew nothing about the solution, we could start with $x_1 = 0$, $x_2 = 0$, $x_3 = 0$, $x_4 = 0$. But since the solution of (9) represents approximations to a solution of a boundary-value problem, it would seem reasonable to use as the initial guess for the values of $x_1 = u_{11}$, $x_2 = u_{21}$, $x_3 = u_{12}$, and $x_4 = u_{22}$ the average of all the boundary conditions. In this case the average of the numbers at the

---

* In the context of approximating a solution to Laplace's equation, this iteration technique is often referred to as **Liebman's method**.

eight boundary points shown in Figure 15.2 is approximately 0.4. Thus our initial guess is $x_1 = 0.4$, $x_2 = 0.4$, $x_3 = 0.4$, and $x_4 = 0.4$. Iterations of the Gauss-Seidel method use the x values as soon as they are computed. Note that the first equation in (9) depends only on x_2 and x_3; thus, substituting $x_2 = 0.4$ and $x_3 = 0.4$ gives $x_1 = 0.2$. Since the second and third equations depend on x_1 and x_4, we use the newly calculated values $x_1 = 0.2$ and $x_4 = 0.4$ to obtain $x_2 = 0.3722$ and $x_3 = 0.3167$. The fourth equation depends on x_2 and x_3 so we use the new values $x_2 = 0.3722$ and $x_3 = 0.3167$ to get $x_4 = 0.5611$. In summary, the first iteration has given the values

$$x_1 = 0.2, \qquad x_2 = 0.3722, \qquad x_3 = 0.3167, \qquad x_4 = 0.5611.$$

Note how close these numbers are already to the actual values given at the end of Example 1.

The second iteration starts with substituting $x_2 = 0.3722$ and $x_3 = 0.3167$ into the first equation. This gives $x_1 = 0.1722$. From $x_1 = 0.1722$ and the last computed value of x_4, namely, $x_4 = 0.5611$, the second and third equations give, in turn, $x_2 = 0.4055$ and $x_3 = 0.3500$. Using these two values, we find from the fourth equation that $x_4 = 0.5678$. At the end of the second iteration we have

$$x_1 = 0.1722, \qquad x_2 = 0.4055, \qquad x_3 = 0.3500, \qquad x_4 = 0.5678.$$

The third through seventh iterations are summarized in the accompanying table.

Iteration	3rd	4th	5th	6th	7th
x_1	0.1889	0.1931	0.1941	0.1944	0.1944
x_2	0.4139	0.4160	0.4165	0.4166	0.4166
x_3	0.3584	0.3605	0.3610	0.3611	0.3611
x_4	0.5820	0.5830	0.5833	0.5833	0.5833

Computer programs for Gauss-Seidel iteration are provided in the software supplement for this text.

Note To apply Gauss-Seidel iteration to a general system of n linear equations in n unknowns, the variable x_i must actually appear in the ith equation of the system. Moreover, after solving each equation for x_i, $i = 1, 2, \ldots, n$, the resulting system has the form $\mathbf{X} = \mathbf{AX} + \mathbf{B}$, where all the entries on the main diagonal of $\mathbf{A}$ are zero.

Remarks (i) In the examples given in this section the values of u_{ij} were determined using known values of u at boundary points. But what do we do if the region is such that boundary points do not coincide with the actual boundary C of the region R? In this case, the required values can be obtained by interpolation.

(ii) It is sometimes possible to cut down the number of equations to solve by using symmetry. Consider the rectangular region shown in Figure 15.4.

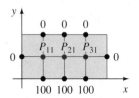

Figure 15.4

The boundary conditions are $u = 0$ along $x = 0$, $x = 2$, $y = 1$, and $u = 100$ along $y = 0$. The region is symmetric about the lines $x = 1$ and $y = 1/2$, and the interior points P_{11} and P_{31} are equidistant from the neighboring boundary points at which the specified values of u are the same. Consequently we assume that $u_{11} = u_{31}$ and so the system of three equations in three unknowns reduces to two equations in two unknowns. See Problem 2.

(iii) Although on a computer this may not be noticeable, convergence of Gauss-Seidel iteration may not be particularly fast. Also, in a more general setting, Gauss-Seidel iteration may not converge at all. If a system of n linear equations in n unknowns is written in the standard form $\mathbf{AX} = \mathbf{B}$, then a sufficient condition for convergence is that the coefficient matrix $\mathbf{A}$ be **diagonally dominant**. This means that the absolute value of each entry on the main diagonal of $\mathbf{A}$ is greater than the sum of the absolute values of the remaining entries in the same row. For example, the coefficient matrix (8) for the system (7) is clearly diagonally dominant, since in each row $|-4|$ is greater than the sum of the absolute values of the other entries. You should verify that the coefficient matrix for the system (6) has the same property. Diagonal dominance guarantees that Gauss-Seidel iteration converges for any initial values.

15.1 EXERCISES

Answers to odd-numbered problems begin on page A–48.

In Problems 1–4 use (4) to approximate the solution of Laplace's equation at the interior points of the given region. Use symmetry when possible.

1. $u(0, y) = 0$, $u(3, y) = y(2 - y)$, $0 < y < 2$
   $u(x, 0) = 0$, $u(x, 2) = x(3 - x)$, $0 < x < 3$
   mesh size: $h = 1$

2. $u(0, y) = 0$, $u(2, y) = 0$, $0 < y < 1$
   $u(x, 0) = 100$, $u(x, 1) = 0$, $0 < x < 2$
   mesh size: $h = 1/2$

3. $u(0, y) = 0$, $u(1, y) = 0$, $0 < y < 1$
   $u(x, 0) = 0$, $u(x, 1) = \sin \pi x$, $0 < x < 1$
   mesh size: $h = 1/3$

4. $u(0, y) = 108y^2(1 - y)$, $u(1, y) = 0$, $0 < y < 1$
   $u(x, 0) = 0$, $u(x, 1) = 0$, $0 < x < 1$
   mesh size: $h = 1/3$

In Problems 5 and 6 use (5) and Gauss-Seidel iteration to approximate the solution of Laplace's equation at the interior points of a unit square. Use the mesh size $h = 1/4$. In Problem 5 the boundary conditions are given; in Problem 6 the values of u at boundary points are given in Figure 15.5.

5. $u(0, y) = 0$, $u(1, y) = 100y$, $0 < y < 1$
   $u(x, 0) = 0$, $u(x, 1) = 100x$, $0 < x < 1$

6.

Figure 15.5

Figure 15.6

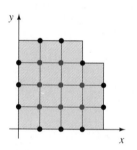

Figure 15.7

7. (a) In Problem 12 of Exercises 12.6 you solved a potential problem using a special form of Poisson's equation $\dfrac{\partial^2 u}{\partial x^2} + \dfrac{\partial^2 u}{\partial y^2} = f(x, y)$. Show that the difference equation replacement for Poisson's equation is

$$u_{i+1,j} + u_{i,j+1} + u_{i-1,j} + u_{i,j-1} - 4u_{ij} = h^2 f(x, y).$$

(b) Use the result in part (a) to approximate the solution of the Poisson equation $\dfrac{\partial^2 u}{\partial x^2} + \dfrac{\partial^2 u}{\partial y^2} = -2$ at the interior points of the region in Figure 15.6. The mesh size is $h = 1/2$, $u = 1$ at every point along $ABCD$ and $u = 0$ at every point along $DEFGA$. Use symmetry and, if necessary, Gauss-Seidel iteration.

8. Use the result in part (a) of Problem 7 to approximate the solution of the Poisson equation $\dfrac{\partial^2 u}{\partial x^2} + \dfrac{\partial^2 u}{\partial y^2} = -64$ at the interior points of the region in Figure 15.7. The mesh is $h = 1/8$ and $u = 0$ at every point on the boundary of the region. If necessary, use Gauss-Seidel iteration.

9. Use the approximation

$$\frac{\partial u}{\partial t} \approx \frac{u(x, t + k) - u(x, t)}{k}$$

to find a difference equation replacement for the heat equation $\dfrac{\partial^2 u}{\partial x^2} = \dfrac{\partial u}{\partial t}$. Note that since x and t usually represent, respectively, spatial and time variables, we do not assume that the mesh size h in the x-direction is the same as the mesh size k in the t-direction.

10. Find a difference equation replacement for the wave equation $\dfrac{\partial^2 u}{\partial x^2} = \dfrac{\partial^2 u}{\partial t^2}$.

15.2 PARABOLIC EQUATIONS

In this section we examine two methods for approximating the solution of a boundary-value problem of the type

$$c \frac{\partial^2 u}{\partial x^2} = \frac{\partial u}{\partial t}, \qquad 0 < x < a, \quad t > 0$$

$$u(0, t) = T_1, \qquad u(a, t) = T_2, \qquad t > 0 \tag{1}$$

$$u(x, 0) = f(x), \qquad 0 < x < a.$$

The equation in this problem is the one-dimensional heat equation and is, as we know from Section 12.1, a parabolic partial differential equation. The constants T_1 and T_2 represent temperatures at the endpoints of the rod. Recall, the function f can be interpreted as the initial temperature distribution in a homogeneous rod extending from $x = 0$ to $x = a$. Although we shall not prove it, the boundary-value problem has a unique solution when f is continuous on the closed interval $[0, a]$. This latter condition will be assumed and so we replace the initial condition in (1) by $u(x, 0) = f(x), 0 \le x \le a$.

A Difference Equation Replacement for the Heat Equation

To approximate the solution of (1) we first replace the heat equation by a difference equation. Using the central difference approximation (1) of Section 15.1:

$$\frac{\partial^2 u}{\partial x^2} \approx \frac{1}{h^2} \left[u(x + h, t) - 2u(x, t) + u(x - h, t) \right]$$

and the forward difference approximation (3) of Section 9.8:

$$\frac{\partial u}{\partial t} \approx \frac{1}{k} \left[u(x, t + k) - u(x, t) \right],$$

we can replace the heat equation $c \dfrac{\partial^2 u}{\partial x^2} = \dfrac{\partial u}{\partial t}$ by the following difference equation:

$$\frac{c}{h^2} \left[u(x + h, t) - 2u(x, t) + u(x - h, t) \right] = \frac{1}{k} \left[u(x, t + k) - u(x, t) \right]. \tag{2}$$

If we let $\lambda = ck/h^2$ and

$$u(x, t) = u_{ij}$$

$$u(x + h, t) = u_{i+1, j}$$

$$u(x - h, t) = u_{i-1, j}$$

$$u(x, t + k) = u_{i, j+1},$$

the (2) becomes, after simplifying,

$$u_{i,j+1} = \lambda u_{i+1,j} + (1 - 2\lambda)u_{ij} + \lambda u_{i-1,j}. \tag{3}$$

We wish to approximate a solution of a problem such as (1) on a rectangular region in the xt-plane defined by the inequalities $0 \le x \le a$, $0 \le t \le T$ for some value of time T. Over this region we place a rectangular grid consisting of vertical lines h units apart and horizontal lines k units apart. If we choose two positive integers n and m and define

$$h = \frac{a}{n} \quad \text{and} \quad k = \frac{T}{m},$$

then the vertical and horizontal grid lines are defined by

$$x_i = ih, \qquad i = 0, 1, 2, \ldots, n$$

$$t_j = jk, \qquad j = 0, 1, 2, \ldots, m.$$

See Figure 15.8.

Figure 15.8

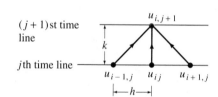

Figure 15.9

As illustrated in Figure 15.9, the idea here is to use formula (3) to estimate the values of the solution $u(x, t)$ at the points on the $(j + 1)$st time line using only values from the jth time line. For example, the values on the first time line $(j = 1)$ depend on the initial condition $u_{i,0} = u(x_i, 0) = f(x_i)$ given on the zeroth time line $(j = 0)$. This kind of numerical procedure is called an **explicit finite difference method**.

EXAMPLE 1

Consider the boundary-value problem

$$\frac{\partial^2 u}{\partial x^2} = \frac{\partial u}{\partial t}, \qquad 0 < x < 1, \quad 0 < t < 0.5$$

$$u(0, t) = 0, \qquad u(1, t) = 0, \qquad 0 \le t \le 0.5$$

$$u(x, 0) = \sin \pi x, \qquad 0 \le x \le 1.$$

First we identify $c = 1$, $a = 1$, and $T = 0.5$. If we choose, say, $n = 5$ and $m = 50$, then $h = 1/5 = 0.2$, $k = 0.5/50 = 0.01$, $\lambda = 0.25$, and

$$x_i = i\frac{1}{5}, \qquad i = 0, 1, 2, 3, 4, 5$$

$$t_j = j\frac{1}{100}, \qquad j = 0, 1, 2, \ldots, 50.$$

Thus (3) becomes

$$u_{i, j+1} = 0.25(u_{i+1, j} + 2u_{ij} + u_{i-1, j}).$$

By setting $j = 0$ in this formula, we get a formula for the approximations to the temperature u on the first time line:

$$u_{i, 1} = 0.25(u_{i+1, 0} + 2u_{i, 0} + u_{i-1, 0}).$$

If we then let $i = 1, \ldots, 4$ in the last equation, we obtain, in turn,

$$u_{11} = 0.25(u_{20} + 2u_{10} + u_{00})$$

$$u_{21} = 0.25(u_{30} + 2u_{20} + u_{10})$$

$$u_{31} = 0.25(u_{40} + 2u_{30} + u_{20})$$

$$u_{41} = 0.25(u_{50} + 2u_{40} + u_{40}).$$

The first equation in this list is interpreted as

$$u_{11} = 0.25(u(x_2, 0) + 2u(x_1, 0) + u(0, 0))$$

$$= 0.25(u(0.4, 0) + 2u(0.2, 0) + u(0, 0)).$$

From the initial condition $u(x, 0) = \sin \pi x$, the last line becomes

$$u_{11} = 0.25(0.951056516 + 2(0.587785252) + 0) = 0.531656755.$$

This number represents an approximation to the temperature $u(0.2, 0.01)$.
The remaining calculations are summarized in Table 15.1.

TABLE 15.1 Explicit Difference Equation Approximation
with $h = 0.2$, $k = 0.01$, $\lambda = 0.25$

Time	$x = 0.20$	$x = 0.40$	$x = 0.60$	$x = 0.80$
0.00	0.5878	0.9511	0.9511	0.5878
0.01	0.5317	0.8602	0.8602	0.5317
0.02	0.4809	0.7781	0.7781	0.4809
0.03	0.4350	0.7038	0.7038	0.4350
0.04	0.3934	0.6366	0.6366	0.3934
0.05	0.3559	0.5758	0.5758	0.3559
0.06	0.3219	0.5208	0.5208	0.3219

TABLE 15.1 *Continued*

Time	x = 0.20	x = 0.40	x = 0.60	x = 0.80
0.07	0.2911	0.4711	0.4711	0.2911
0.08	0.2633	0.4261	0.4261	0.2633
0.09	0.2382	0.3854	0.3854	0.2382
0.10	0.2154	0.3486	0.3486	0.2154
0.11	0.1949	0.3153	0.3153	0.1949
0.12	0.1763	0.2852	0.2852	0.1763
0.13	0.1594	0.2580	0.2580	0.1594
0.14	0.1442	0.2333	0.2333	0.1442
0.15	0.1304	0.2111	0.2111	0.1304
0.16	0.1180	0.1909	0.1909	0.1180
0.17	0.1067	0.1727	0.1727	0.1067
0.18	0.0965	0.1562	0.1562	0.0965
0.19	0.0873	0.1413	0.1413	0.0873
0.20	0.0790	0.1278	0.1278	0.0790
0.21	0.0714	0.1156	0.1156	0.0714
0.22	0.0646	0.1045	0.1045	0.0646
0.23	0.0584	0.0946	0.0946	0.0584
0.24	0.0529	0.0855	0.0855	0.0529
0.25	0.0478	0.0774	0.0774	0.0478
0.26	0.0432	0.0700	0.0700	0.0432
0.27	0.0391	0.0633	0.0633	0.0391
0.28	0.0354	0.0572	0.0572	0.0354
0.29	0.0320	0.0518	0.0518	0.0320
0.30	0.0289	0.0468	0.0468	0.0289
0.31	0.0262	0.0424	0.0424	0.0262
0.32	0.0237	0.0383	0.0383	0.0237
0.33	0.0214	0.0347	0.0347	0.0214
0.34	0.0194	0.0313	0.0313	0.0194
0.35	0.0175	0.0284	0.0284	0.0175
0.36	0.0159	0.0256	0.0256	0.0159
0.37	0.0143	0.0232	0.0232	0.0143
0.38	0.0130	0.0210	0.0210	0.0130
0.39	0.0117	0.0190	0.0190	0.0117
0.40	0.0106	0.0172	0.0172	0.0106
0.41	0.0096	0.0155	0.0155	0.0096
0.42	0.0087	0.0140	0.0140	0.0087
0.43	0.0079	0.0127	0.0127	0.0079
0.44	0.0071	0.0115	0.0115	0.0071
0.45	0.0064	0.0104	0.0104	0.0064
0.46	0.0058	0.0094	0.0094	0.0058
0.47	0.0053	0.0085	0.0085	0.0053
0.48	0.0048	0.0077	0.0077	0.0048
0.49	0.0043	0.0070	0.0070	0.0043
0.50	0.0039	0.0063	0.0063	0.0039

Computer programs for the method illustrated in Example 1 are provided in the software supplement for this text.

You should verify, using the methods of Chapter 12, that the exact solution of the boundary-value problem in Example 1 is given by $u(x, t) = e^{-\pi^2 t} \sin \pi x$. Some sample values are given here:

Exact	Approximation
$u(0.4, 0.05) = 0.5806$	$u_{25} = 0.5758$
$u(0.6, 0.06) = 0.5261$	$u_{36} = 0.5208$
$u(0.2, 0.10) = 0.2191$	$u_{1,10} = 0.2154$
$u(0.8, 0.14) = 0.1476$	$u_{4,14} = 0.1442$

These approximations are comparable to the exact values and are accurate enough for some purposes. But there is a difficulty with the foregoing method. Recall, a numerical method is **unstable** if round-off errors or any other errors grow too rapidly as the computations proceed. The numerical procedure illustrated in Example 1 can exhibit this kind of behavior. It can be proved that the procedure is stable if λ is less than or equal to 0.5 but unstable otherwise. Note that in Example 1 we chose the values of h and k so that $\lambda = 0.25$.

The next example illustrates what happens when a larger step size k is chosen so that λ is greater than 0.5.

EXAMPLE 2

Approximate the solution of the boundary-value problem

$$\frac{\partial^2 u}{\partial x^2} = \frac{\partial u}{\partial t}, \quad 0 < x < 1, \quad 0 < t < 1$$

$$u(0, t) = 0, \quad u(1, t) = 0, \quad 0 \le t \le 1$$

$$u(x, 0) = \sin \pi x, \quad 0 \le x \le 1,$$

using (3) with $n = 5$ and $m = 25$.

Solution In this case $c = 1$, $a = 1$, $T = 1$, $h = 1/5 = 0.2$, $k = 1/25 = 0.04$, and $\lambda = 1$. With the aid of a computer we obtain the results given in Table 15.2.

TABLE 15.2 Explicit Difference Equation Approximation with $h = 0.2$, $k = 0.04$, $\lambda = 1$

Time	$x = 0.20$	$x = 0.40$	$x = 0.60$	$x = 0.80$
0.00	0.5878	0.9511	0.9511	0.5878
0.04	0.3633	0.5878	0.5878	0.3633
0.08	0.2245	0.3633	0.3633	0.2245
0.12	0.1388	0.2245	0.2245	0.1388
0.16	0.0858	0.1388	0.1388	0.0858
0.20	0.0530	0.0857	0.0858	0.0530
0.24	0.0327	0.0530	0.0530	0.0328
0.28	0.0203	0.0327	0.0328	0.0202
0.32	0.0124	0.0204	0.0201	0.0126
0.36	0.0080	0.0121	0.0129	0.0075
0.40	0.0041	0.0088	0.0067	0.0054
0.44	0.0047	0.0020	0.0075	0.0012
0.48	−0.0026	0.0102	−0.0043	0.0063
0.52	0.0128	−0.0171	0.0208	−0.0106
0.56	−0.0299	0.0507	−0.0485	0.0314
0.60	0.0806	−0.1291	0.1306	−0.0799
0.64	−0.2097	0.3403	−0.3396	0.2105
0.68	0.5500	−0.8895	0.8903	−0.5500
0.72	−1.4395	2.3298	−2.3299	1.4403
0.76	3.7693	−6.0992	6.1000	−3.7702
0.80	−9.8685	15.9685	−15.9693	9.8702
0.84	25.8370	−41.8063	41.8080	−25.8395
0.88	−67.6432	109.4512	−109.4538	67.6475
0.92	177.0945	−286.5482	286.5525	−177.1013
0.96	−463.6426	750.1950	−750.2019	463.6537
1.00	1213.8374	−1964.0393	1964.0503	−1213.8555

For values of t from 0 to 0.4 the results given in Table 15.2 are believable, although the accuracy, upon closer examination, is found to be poor. The results for values of t greater than 0.5 are not believable. Because the values of $u(x, 0) = \sin \pi x$ on the interval $0 \le x \le 1$ are never negative, we expect that the solution $u(x, t)$ should never be negative in this example. Moreover, the large values near the end of the output clearly reveal the large errors produced by the instability of the procedure for this choice of h and k. Of course, $\lambda = 1$ exceeds 0.5 and so the instability was predictable. For $\lambda = 0.5$, the numerical results are satisfactory, but some increase in accuracy results from using smaller values of k with consequent smaller values of λ. The necessity of using very small step sizes in the time direction is the principal fault of this method.

Crank-Nicholson Method

There are **implicit finite difference methods** for solving parabolic partial differential equations. These methods require that we solve a system of equations to determine the approximate values of u on the $(j + 1)$st time line. However, implicit methods do not suffer from instability problems.

The algorithm introduced by J. Crank and P. Nicholson in 1947 is used mostly for solving the heat equation and consists of replacing the second partial derivative in $c \dfrac{\partial^2 u}{\partial x^2} = \dfrac{\partial u}{\partial t}$ by an average of two central difference quotients, one evaluated at t and the other at $t + k$:

$$\frac{c}{2}\left[\frac{u(x+h,t)-2u(x,t)+u(x-h,t)}{h^2} + \frac{u(x+h,t+k)-2u(x,t+k)+u(x-h,t+k)}{h^2}\right] = \frac{1}{k}[u(x,t+k)-u(x,t)]. \quad (4)$$

If we again define $\lambda = ck/h^2$, then after rearranging terms, we can write (4) as

$$-u_{i-1,j+1} + \alpha u_{i,j+1} - u_{i+1,j+1} = u_{i+1,j} - \beta u_{ij} + u_{i-1,j}, \quad (5)$$

where $\alpha = 2(1 + 1/\lambda)$ and $\beta = 2(1 - 1/\lambda)$, $j = 0, 1, \ldots, m-1$, and $i = 1, 2, \ldots, n-1$.

For each choice of j the difference equation (5) for $i = 1, 2, \ldots, n-1$ gives $n - 1$ equations in $n - 1$ unknowns $u_{i,j+1}$. Because of the prescribed boundary conditions, the values of $u_{i,j+1}$ are known for $i = 0$ and for $i = n$. For example, in the case $n = 4$ the system of equations for determining the approximate values of u on the $(j + 1)$st time line is

$$-u_{0,j+1} + \alpha u_{1,j+1} - u_{2,j+1} = u_{2,j} - \beta u_{1,j} + u_{0,j}$$

$$-u_{1,j+1} + \alpha u_{2,j+1} - u_{3,j+1} = u_{3,j} - \beta u_{2,j} + u_{1,j}$$

$$-u_{2,j+1} + \alpha u_{3,j+1} - u_{4,j+1} = u_{4,j} - \beta u_{3,j} + u_{2,j}$$

or

$$\alpha u_{1,j+1} - u_{2,j+1} \qquad\qquad = b_1$$

$$-u_{1,j+1} + \alpha u_{2,j+1} - u_{3,j+1} = b_2 \qquad (6)$$

$$-u_{2,j+1} + \alpha u_{3,j+1} = b_3,$$

where

$$b_1 = u_{2,j} - \beta u_{1,j} + u_{0,j} + u_{0,j+1}$$

$$b_2 = u_{3,j} - \beta u_{2,j} + u_{1,j}$$

$$b_3 = u_{4,j} - \beta u_{3,j} + u_{2,j} + u_{4,j+1}.$$

In general, if we use the difference equation (5) to determine values of u on the $(j + 1)$st time line, we need to solve a linear system $\mathbf{AX} = \mathbf{B}$, where the coefficient matrix $\mathbf{A}$ is a **tridiagonal matrix**:

$$\mathbf{A} = \begin{pmatrix} \alpha & -1 & 0 & 0 & 0 & \cdots & & 0 \\ -1 & \alpha & -1 & 0 & 0 & & & 0 \\ 0 & -1 & \alpha & -1 & 0 & & & 0 \\ 0 & 0 & -1 & \alpha & -1 & & & 0 \\ \vdots & & & & & \ddots & & \vdots \\ 0 & 0 & 0 & 0 & 0 & & \alpha & -1 \\ 0 & 0 & 0 & 0 & 0 & \cdots & -1 & \alpha \end{pmatrix}$$

and the entries of the column matrix **B** are

$$b_1 = u_{2,j} - \beta u_{1,j} + u_{0,j} + u_{0,j+1}$$

$$b_2 = u_{3,j} - \beta u_{2,j} + u_{1,j}$$

$$b_3 = u_{4,j} - \beta u_{3,j} + u_{2,j}$$

$$\vdots$$

$$b_{n-1} = u_{n,j} - \beta u_{n-1,j} + u_{n-2,j} + u_{n,j+1}.$$

Computer programs for the Crank-Nicholson method are provided in the software supplement for this text.

EXAMPLE 3

Use the Crank-Nicholson method to approximate the solution of the boundary-value problem

$$0.25 \frac{\partial^2 u}{\partial x^2} = \frac{\partial u}{\partial t}, \qquad 0 < x < 2, \quad 0 < t < 0.3$$

$$u(0, t) = 0, \qquad u(2, t) = 0, \qquad 0 \le t \le 0.3$$

$$u(x, 0) = \sin \pi x, \qquad 0 \le x \le 2,$$

using $n = 8$ and $m = 30$.

Solution From the identifications $a = 2$, $T = 0.3$, $h = 1/4 = 0.25$, $k = 1/100 = 0.01$, and $c = 0.25$ we get $\lambda = 0.04$. With the aid of a computer we get the results in Table 15.3.

TABLE 15.3 Crank-Nicholson Method with $h = 0.25$, $k = 0.01$, $\lambda = 0.25$

Time	$x = 0.25$	$x = 0.50$	$x = 0.75$	$x = 1.00$	$x = 1.25$	$x = 1.50$	$x = 1.75$
0.00	0.7071	1.0000	0.7071	0.0000	−0.7071	−1.0000	−0.7071
0.01	0.6907	0.9768	0.6907	0.0000	−0.6907	−0.9768	−0.6907
0.02	0.6747	0.9542	0.6747	0.0000	−0.6747	−0.9542	−0.6747
0.03	0.6591	0.9321	0.6591	0.0000	−0.6591	−0.9321	−0.6591
0.04	0.6438	0.9105	0.6438	0.0000	−0.6438	−0.9105	−0.6438
0.05	0.6289	0.8894	0.6289	0.0000	−0.6289	−0.8894	−0.6289
0.06	0.6144	0.8688	0.6144	0.0000	−0.6144	−0.8688	−0.6144
0.07	0.6001	0.8487	0.6001	0.0000	−0.6001	−0.8487	−0.6001
0.08	0.5862	0.8291	0.5862	0.0000	−0.5862	−0.8291	−0.5862
0.09	0.5727	0.8099	0.5727	0.0000	−0.5727	−0.8099	−0.5727
0.10	0.5594	0.7911	0.5594	0.0000	−0.5594	−0.7911	−0.5594
0.11	0.5464	0.7728	0.5464	0.0000	−0.5464	−0.7728	−0.5464
0.12	0.5338	0.7549	0.5338	0.0000	−0.5338	−0.7549	−0.5338
0.13	0.5214	0.7374	0.5214	0.0000	−0.5214	−0.7374	−0.5214
0.14	0.5093	0.7203	0.5093	0.0000	−0.5093	−0.7203	−0.5093
0.15	0.4975	0.7036	0.4975	0.0000	−0.4975	−0.7036	−0.4975

TABLE 15.3 *Continued*

Time	x = 0.25	x = 0.50	x = 0.75	x = 1.00	x = 1.25	x = 1.50	x = 1.75
0.16	0.4860	0.6873	0.4860	0.0000	−0.4860	−0.6873	−0.4860
0.17	0.4748	0.6714	0.4748	0.0000	−0.4748	−0.6714	−0.4748
0.18	0.4638	0.6559	0.4638	0.0000	−0.4638	−0.6559	−0.4638
0.19	0.4530	0.6407	0.4530	0.0000	−0.4530	−0.6407	−0.4530
0.20	0.4425	0.6258	0.4425	0.0000	−0.4425	−0.6258	−0.4425
0.21	0.4323	0.6114	0.4323	0.0000	−0.4323	−0.6114	−0.4323
0.22	0.4223	0.5972	0.4223	0.0000	−0.4223	−0.5972	−0.4223
0.23	0.4125	0.5834	0.4125	0.0000	−0.4125	−0.5834	−0.4125
0.24	0.4029	0.5699	0.4029	0.0000	−0.4029	−0.5699	−0.4029
0.25	0.3936	0.5567	0.3936	0.0000	−0.3936	−0.5567	−0.3936
0.26	0.3845	0.5438	0.3845	0.0000	−0.3845	−0.5438	−0.3845
0.27	0.3756	0.5312	0.3756	0.0000	−0.3756	−0.5312	−0.3756
0.28	0.3669	0.5189	0.3669	0.0000	−0.3669	−0.5189	−0.3669
0.29	0.3584	0.5068	0.3584	0.0000	−0.3584	−0.5068	−0.3584
0.30	0.3501	0.4951	0.3501	0.0000	−0.3501	−0.4951	−0.3501

■

You should verify that the function satisfying all the conditions in Example 3 is $u(x, t) = e^{-\pi^2 t/4} \sin \pi x$. The sample comparisons listed here show that the absolute errors are of the order 10^{-2} or 10^{-3}. Smaller errors can be obtained by decreasing either h or k.

Exact	Approximation
$u(0.75, 0.05) = 0.6250$	$u_{35} = 0.6289$
$u(0.50, 0.20) = 0.6105$	$u_{2,20} = 0.6259$
$u(0.25, 0.10) = 0.5525$	$u_{1,10} = 0.5594$

15.2 EXERCISES

Answers to odd-numbered problems begin on page A–48.

1. Use the difference equation (3) to approximate the solution of the boundary-value problem

$$\frac{\partial^2 u}{\partial x^2} = \frac{\partial u}{\partial t}, \quad 0 < x < 2, \quad 0 < t < 1$$

$$u(0, t) = 0, \quad u(2, t) = 0, \quad 0 \le t \le 1$$

$$u(x, 0) = \begin{cases} 1, & 0 \le x \le 1 \\ 0, & 1 < x \le 2 \end{cases}.$$

Use $n = 8$ and $m = 40$.

2. Using the Fourier series solution obtained in Problem 1, Exercises 12.3, with $L = 2$, one can sum the first 20 terms to estimate the values for $u(0.25, 0.1)$, $u(1, 0.5)$, and $u(1.5, 0.8)$ for the solution $u(x, t)$ of Problem 1 above. A student wrote a computer program to do this and obtained the results $u(0.25, 0.1) = 0.3794$, $u(1, 0.5) = 0.1854$, and $u(1.5, 0.8) = 0.0623$. Assume these results are accurate for all digits given. Compare these values with the approximations obtained in Problem 1 above. Find the absolute errors in each case.

3. Solve Problem 1 by the Crank-Nicholson method with $n = 8$ and $m = 40$. Use the values for $u(0.25, 0.1)$, $u(1, 0.5)$, and $u(1.5, 0.8)$ given in Problem 2 to compute the absolute errors.

4. Repeat Problem 1 using $n = 8$ and $m = 20$. Use the values for $u(0.25, 0.1)$, $u(1, 0.5)$, and $u(1.5, 0.8)$ given in Problem 2 to compute the absolute errors. Why are the approximations so inaccurate in this case?

5. Solve Problem 1 by the Crank-Nicholson method with $n = 8$ and $m = 20$. Use the values for $u(0.25, 0.1)$, $u(1, 0.5)$, and $u(1.5, 0.8)$ given in Problem 2 to compute the absolute errors. Compare the absolute errors with those obtained in Problem 4.

6. It was shown in Section 12.2 that if a rod of length L is made of a material with thermal conductivity K, specific heat γ, and density ρ, the temperature $u(x, t)$ satisfies the partial differential equation

$$\frac{K}{\gamma\rho}\frac{\partial^2 u}{\partial x^2} = \frac{\partial u}{\partial t}, \qquad 0 < x < L.$$

Consider the boundary-value problem consisting of the foregoing equation and conditions

$$u(0, t) = 0, \qquad u(L, t) = 0, \qquad 0 \le t \le 10$$

$$u(x, 0) = f(x), \qquad 0 \le x \le L.$$

Use the difference equation (3) in this section with $n = 10$ and $m = 10$ to approximate the solution of the boundary-value problem when:

(a) $L = 20$, $K = 0.15$, $\rho = 8.0$, $\gamma = 0.11$, $f(x) = 30$

(b) $L = 50$, $K = 0.15$, $\rho = 8.0$, $\gamma = 0.11$, $f(x) = 30$

(c) $L = 20$, $K = 1.10$, $\rho = 2.7$, $\gamma = 0.22$, $f(x) = 0.5x(20 - x)$

(d) $L = 100$, $K = 1.04$, $\rho = 10.6$, $\gamma = 0.06$,

$$f(x) = \begin{cases} 0.8x, & 0 \le x \le 50 \\ 40 - 0.8(x - 50), & 50 < x \le 100 \end{cases}$$

7. Solve Problem 6 by the Crank-Nicholson method with $n = 10$ and $m = 10$.

8. Repeat Problem 6 if the endpoint temperatures are $u(0, t) = 0$, $u(L, t) = 20$, $0 \le t \le 10$.

9. Solve Problem 8 by the Crank-Nicholson method.

10. Consider the boundary-value problem in Example 3. Assume that $n = 4$.

    (a) Find the new value of λ.

    (b) Use the Crank-Nicholson difference equation (5) to find the system of equations for u_{11}, u_{21}, and u_{31}, that is, the approximate values of u on the first time line. [*Hint:* Set $j = 0$ in (5) and let i take on the values 1, 2, 3.]

    (c) Solve the system of three equations without the aid of a computer program. Compare your results with the corresponding entries in Table 15.3.

11. Consider a rod whose length is $L = 20$ for which $K = 1.05$, $\rho = 10.6$, and $\gamma = 0.056$. Suppose

    $$u(0, t) = 20, \qquad u(20, t) = 30$$

    $$u(x, 0) = 50.$$

    (a) Use the method outlined in Section 12.6 to find the steady-state solution $\psi(x)$.

    (b) Use the Crank-Nicholson method to approximate the temperatures $u(x, t)$ for $0 \leq t \leq T_{max}$. Select T_{max} large enough to allow the temperatures to approach the steady-state values. Compare the approximations for $t = T_{max}$ with the values of $\psi(x)$ found in part (a).

15.3 HYPERBOLIC EQUATIONS

We turn now to a numerical method for approximating a solution of a boundary-value problem involving the one-dimensional wave equation:

$$c^2 \frac{\partial^2 u}{\partial x^2} = \frac{\partial^2 u}{\partial t^2}, \qquad 0 < x < a, \quad t > 0$$

$$u(0, t) = 0, \qquad u(a, t) = 0, \qquad t > 0 \tag{1}$$

$$u(x, 0) = f(x), \qquad \left. \frac{\partial u}{\partial t} \right|_{t=0} = g(x), \qquad 0 \leq x \leq a.$$

The wave equation is an example of a hyperbolic partial differential equation. This problem has a unique solution if the functions f and g have continuous second derivatives on the interval $(0, a)$ and $f(a) = f(0) = 0$.

A Difference Equation Replacement for the Wave Equation

As in the preceding two sections, our first objective is to replace the partial differential equation in (1) by a difference equation. In this case we replace the two second-order partial derivatives by the central difference approximations

$$\frac{\partial^2 u}{\partial x^2} \approx \frac{1}{h^2} [u(x + h, t) - 2u(x, t) + u(x - h, t)]$$

and

$$\frac{\partial^2 u}{\partial t^2} \approx \frac{1}{k^2} [u(x, t + k) - 2u(x, t) + u(x, t - k)].$$

Thus we replace $c^2 \dfrac{\partial^2 u}{\partial x^2} = \dfrac{\partial^2 u}{\partial t^2}$ by

$$\frac{c^2}{h^2} [u(x + h, t) - 2u(x, t) + u(x - h, t)] = \frac{1}{k^2} [u(x, t + k) - 2u(x, t) + u(x, t - k)]. \quad (2)$$

If we let $\lambda = ck/h$, then (2) can be expressed as

$$u_{i, j+1} = \lambda^2 u_{i+1, j} + 2(1 - \lambda^2) u_{ij} + \lambda^2 u_{i-1, j} - u_{i, j-1} \quad (3)$$

for $i = 1, 2, \ldots, n - 1$, and $j = 1, 2, \ldots, m - 1$.

The numerical method based on equation (3), like the first method considered in Section 15.2, is an explicit finite difference method. As before, we use the difference equation to approximate the solution $u(x, t)$ of (1) over a rectangular region in the xt-plane defined by the inequalities $0 \le x \le a$, $0 \le t \le T$. If n and m are positive integers and

$$h = \frac{a}{n} \quad \text{and} \quad k = \frac{T}{m},$$

the vertical and horizontal grid lines on this region are defined by

$$x_i = ih, \quad i = 0, 1, 2, \ldots, n$$

$$t_j = jk, \quad j = 0, 1, 2, \ldots, m.$$

As shown in Figure 15.10, (3) enables us to obtain the approximation $u_{i, j+1}$ on the $(j + 1)$st time line from the values indicated on the jth and $(j - 1)$st time lines. Moreover, we use

Figure 15.10

$$u_{0, j} = u(0, jk) = 0, \quad u_{n, j} = u(a, jk) = 0 \quad \Leftarrow \boxed{\text{boundary conditions}}$$

and

$$u_{i, 0} = u(x_i, 0) = f(x_i). \quad \Leftarrow \boxed{\text{initial condition}}$$

There is one minor problem in getting started. You can see from (3) that for $j = 1$ we need to know the values of $u_{i, 1}$, that is, the estimates of u on the first time line, in order to find $u_{i, 2}$. But from Figure 15.10, with $j = 0$, we see that the values of $u_{i, 1}$ on the first time line depend on the values of $u_{i, 0}$ on the zeroth time line and on the values of $u_{i, -1}$. To compute these latter values we make use of the initial-velocity condition $u_t(x, 0) = g(x)$. At $t = 0$ it follows from (5) of Section 9.8 that

$$g(x_i) = u_t(x_i, 0) \approx \frac{u(x_i, k) - u(x_i, -k)}{2k}. \quad (4)$$

In order to make sense of the term $u(x_i, -k) = u_{i,-1}$ in (4) we have to imagine $u(x, t)$ extended backward in time. It follows from (4) that

$$u(x_i, -k) \approx u(x_i, k) - 2kg(x_i).$$

This last result suggests that we define

$$u_{i,-1} = u_{i,1} - 2kg(x_i) \tag{5}$$

in the iteration of (3). By substituting (5) into (3) when $j = 0$ we get the special case

$$u_{i,1} = \frac{\lambda^2}{2}(u_{i+1,0} + u_{i-1,0}) + (1 - \lambda^2)u_{i,0} + kg(x_i). \tag{6}$$

Computer programs for this method are provided in the software supplement for this text.

EXAMPLE I

Approximate the solution of the boundary-value problem

$$4\frac{\partial^2 u}{\partial x^2} = \frac{\partial^2 u}{\partial t^2}, \qquad 0 < x < 1, \quad 0 < t < 1$$

$$u(0, t) = 0, \qquad u(1, t) = 0 \qquad 0 \le t \le 1$$

$$u(x, 0) = \sin \pi x, \qquad \frac{\partial u}{\partial t}\bigg|_{t=0} = 0, \qquad 0 \le x \le 1,$$

using (3) with $n = 5$ and $m = 20$.

Solution We make the identifications $c = 2$, $a = 1$, and $T = 1$. With $n = 5$ and $m = 20$, we get $h = 1/5 = 0.2$, $k = 1/20 = 0.05$, and $\lambda = 0.5$. Thus, with $g(x) = 0$, equations (6) and (3) become, respectively,

$$u_{i,1} = 0.125(u_{i+1,0} + u_{i-1,0}) + 0.75u_{i,0} \tag{7}$$

$$u_{i,j+1} = 0.25u_{i+1,j} + 1.5u_{ij} + 0.25u_{i-1,j} - u_{i,j-1}. \tag{8}$$

For $i = 1, 2, 3, 4$, equation (7) yields the following values for the $u_{i,1}$ on the first time line:

$$u_{11} = 0.125(u_{20} + u_{00}) + 0.75u_{10} = 0.55972100$$

$$u_{21} = 0.125(u_{30} + u_{10}) + 0.75u_{20} = 0.90564761$$

$$u_{31} = 0.125(u_{40} + u_{20}) + 0.75u_{30} = 0.90564761 \tag{9}$$

$$u_{41} = 0.125(u_{50} + u_{30}) + 0.75u_{40} = 0.55972100.$$

Note that the results given in (9) were obtained from the initial condition $u(x, 0) = \sin \pi x$. For example, $u_{20} = \sin(0.2\pi)$, and so on. Now $j = 1$ in (8) gives

$$u_{i,2} = 0.25u_{i+1,1} + 1.5u_{i,1} + 0.25u_{i-1,1} - u_{i,0}$$

and so for $i = 1, 2, 3, 4$, we get

$$u_{12} = 0.25u_{21} + 1.5u_{11} + 0.25u_{01} - u_{10}$$

$$u_{22} = 0.25u_{31} + 1.5u_{21} + 0.25u_{11} - u_{20}$$

$$u_{32} = 0.25u_{41} + 1.5u_{31} + 0.25u_{21} - u_{30}$$

$$u_{42} = 0.25u_{51} + 1.5u_{41} + 0.25u_{31} - u_{40}.$$

Using the boundary conditions, initial conditions, and the data obtained in (9), we get from these equations the approximations for u on the second time line. These results and the remaining calculations are summarized in Table 15.4.

TABLE 15.4 Explicit Difference Equation Approximation with $h = 0.2$, $k = 0.05$, $\lambda = 0.5$

Time	$x = 0.20$	$x = 0.40$	$x = 0.60$	$x = 0.80$
0.00	0.5878	0.9511	0.9511	0.5878
0.05	0.5597	0.9056	0.9056	0.5597
0.10	0.4782	0.7738	0.7738	0.4782
0.15	0.3510	0.5680	0.5680	0.3510
0.20	0.1903	0.3080	0.3080	0.1903
0.25	0.0115	0.0185	0.0185	0.0115
0.30	−0.1685	−0.2727	−0.2727	−0.1685
0.35	−0.3324	−0.5378	−0.5378	−0.3324
0.40	−0.4645	−0.7516	−0.7516	−0.4645
0.45	−0.5523	−0.8936	−0.8936	−0.5523
0.50	−0.5873	−0.9503	−0.9503	−0.5873
0.55	−0.5663	−0.9163	−0.9163	−0.5663
0.60	−0.4912	−0.7947	−0.7947	−0.4912
0.65	−0.3691	−0.5973	−0.5973	−0.3691
0.70	−0.2119	−0.3428	−0.3428	−0.2119
0.75	−0.0344	−0.0556	−0.0556	−0.0344
0.80	0.1464	0.2369	0.2369	0.1464
0.85	0.3132	0.5068	0.5068	0.3132
0.90	0.4501	0.7283	0.7283	0.4501
0.95	0.5440	0.8803	0.8803	0.5440
1.00	0.5860	0.9482	0.9482	0.5860

It is readily verified that the exact solution of the problem in Example 1 is $u(x, t) = \sin \pi x \cos 2\pi t$. With this function we can compare the exact results with the approximations given in Table 15.4. For example, the following are some sample values:

Exact	Approximation
$u(0.4, 0.25) = 0$	$u_{25} = 0.0185$
$u(0.6, 0.3) = -0.2939$	$u_{36} = -0.2727$
$u(0.2, 0.5) = -0.5878$	$u_{1,10} = -0.5873$
$u(0.8, 0.7) = -0.1816$	$u_{4,14} = -0.2119$

As you can see, the approximations are in the same "ball park" as the exact values but the accuracy is not particularly impressive. We can, however, obtain more accurate results. The accuracy of this algorithm varies with the choice of λ. Of course, λ is determined by the choice of the integers n and m, which in turn, determine the values of the step sizes h and k. It can be proved that the best accuracy is always obtained from this method when the ratio $\lambda = kc/h$ is equal to one, in other words, when the step in the time direction is $k = h/c$. For example, the choice $n = 8$ and $m = 16$ yields $h = 1/8, k = 1/16$, and $\lambda = 1$. The sample values listed here clearly show the improved accuracy.

Exact	Approximation
$u(0.25, 0.3125) = -0.2706$	$u_{25} = -0.2706$
$u(0.375, 0.375) = -0.6533$	$u_{36} = -0.6533$
$u(0.125, 0.625) = -0.2706$	$u_{1,10} = -0.2706$

We note in conclusion that this explicit finite difference method for the wave equation is stable when $\lambda \leq 1$ and unstable when $\lambda > 1$.

Remark It is worthwhile to spend a few moments considering the sources of error in the finite difference methods covered in the last several sections. The error in the numerical solution (that is, in the approximations) arises from two sources:

  (i) replacement of the partial differential equation by a difference equation, and

  (ii) round-off error in computer arithmetic.

The error in the values of the solution u that arises from approximation of the partial differential equation by a difference equation is an example of **discretization error**. If $u(x_i, t_j)$ denotes the exact value of the solution at the ijth grid point, and u_{ij} is the numerical estimate arising from a finite difference method, then the total error at the ijth point is

$$u(x_i, t_j) - u_{ij} = \textit{discretization error} + \textit{round-off error}.$$

One might expect the numerical estimates to improve as h and k are decreased. Eventually, however, the benefit from a decrease in the step size is offset by the increase in round-off error that results from increasing the amount of computer arithmetic required. Theoretical bounds from the discretization error can be established for some difference methods. Such bounds can sometimes be used to prove that the discretization error does decrease in magnitude as h and k are decreased. Unfortunately, such bounds are not ordinarily useful in practice for actual estimation of the total error in the numerical solution. Such an analysis of the difference method presented in this section reveals that if $\lambda = 1$, then the only errors involved are round-off errors in the computer arithmetic. This explains why the accuracy in Example 1 can be increased by taking $\lambda = 1$.

Round-off error is difficult to analyze and it is not feasible in practice to try to estimate the actual magnitude of such error. As a consequence we *do* need to be concerned with the rate of growth of round-off error as the computations proceed.

15.3 EXERCISES

Answers to odd-numbered problems begin on page A–54.

1. Use the difference equation (3) to approximate the solution of the boundary-value problem

$$c^2 \frac{\partial^2 u}{\partial x^2} = \frac{\partial^2 u}{\partial t^2}, \qquad 0 < x < a, \quad 0 < t < T$$

$$u(0, t) = 0, \qquad u(a, t) = 0, \qquad 0 \le t \le T$$

$$u(x, 0) = f(x), \qquad \left. \frac{\partial u}{\partial t} \right|_{t=0} = 0, \qquad 0 \le x \le a,$$

when:

(a) $c = 1,\ a = 1,\ T = 1,\ f(x) = x(1 - x);\quad n = 4$ and $m = 10$

(b) $c = 1,\ a = 2,\ T = 1,\ f(x) = e^{-16(x-1)^2};\quad n = 5$ and $m = 10$

(c) $c = \sqrt{2},\ a = 1,\ T = 1,\ f(x) = \begin{cases} 0, & 0 \le x \le 0.5 \\ 0.5, & 0.5 < x \le 1 \end{cases};\quad n = 10$ and $m = 25$

2. Given the boundary-value problem

$$\frac{\partial^2 u}{\partial x^2} = \frac{\partial^2 u}{\partial t^2}, \qquad 0 < x < 1, \quad 0 < t < 0.5$$

$$u(0, t) = 0, \qquad u(1, t) = 0, \qquad 0 \le t \le 0.5$$

$$u(x, 0) = \sin \pi x, \qquad \left. \frac{\partial u}{\partial t} \right|_{t=0} = 0, \qquad 0 \le x \le 1.$$

(a) Use the methods of Chapter 12 to verify that the solution of the problem is $u(x, t) = \sin \pi x \cos \pi t$.

(b) Use the method of this section to approximate the solution of the problem without the aid of a computer program. Use $n = 4$ and $m = 5$.

(c) Compute the absolute error at each interior grid point.

3. Approximate the solution of the boundary-value problem in Problem 2 using a computer program with **(a)** $n = 5$, $m = 10$ and **(b)** $n = 5$, $m = 20$.

4. Given the boundary-value problem

$$\frac{\partial^2 u}{\partial x^2} = \frac{\partial^2 u}{\partial t^2}, \qquad 0 < x < 1, \quad 0 < t < 1$$

$$u(0, t) = 0, \qquad u(1, t) = 0, \qquad 0 \le t \le 1$$

$$u(x, 0) = x(1 - x), \qquad \left.\frac{\partial u}{\partial t}\right|_{t=0} = 0, \qquad 0 \le x \le 1.$$

Use $h = k = 1/5$ in equation (6) to compute the values of $u_{i,1}$ by hand.

5. It was shown in Section 12.2 that the equation of a vibrating string is

$$\frac{T}{\rho} \frac{\partial^2 u}{\partial x^2} = \frac{\partial^2 u}{\partial t^2},$$

where T is the constant magnitude of the tension in the string and ρ is its mass per unit length. Suppose a string of length 60 cm is secured to the x-axis at its ends and is released from rest from the initial displacement

$$f(x) = \begin{cases} 0.01x, & 0 \le x \le 30 \\ 0.30 - \dfrac{x - 30}{100}, & 30 < x \le 60. \end{cases}$$

Use the difference equation (3) in this section to approximate the solution of the boundary-value problem when $h = 10$, $k = 5\sqrt{\rho/T}$ and where $\rho = 0.0225$ g/cm, $T = 1.4 \times 10^7$ dynes. Use $m = 50$.

6. Repeat Problem 5 using

$$f(x) = \begin{cases} 0.2x, & 0 \le x \le 15 \\ 0.30 - \dfrac{x - 15}{150}, & 15 < x \le 60, \end{cases}$$

and $h = 10$, $k = 2.5\sqrt{\rho/T}$. Use $m = 50$.

Chapter 15 *REVIEW*

The difference equation

$$u_{i+1, j} + u_{i, j+1} + u_{i-1, j} + u_{i, j-1} - 4u_{ij} = 0 \qquad \text{(1)}$$

gives us a method for approximating solutions of Laplace's equation in a region in the xy-plane. When we place a rectangular grid over the region,

(1) yields a system of linear equations for the u_{ij} at mesh points of the grid. If the system of equations is not too large, we can solve it either by a direct method such as Gaussian elimination or by an indirect method such as Gauss-Seidel iteration.

Solutions of the heat equation over a rectangular grid in the xt-plane can be approximated by

$$u_{i,j+1} = \lambda u_{i+1,j} + (1 - 2\lambda)u_{ij} + \lambda u_{i-1,j}, \tag{2}$$

where $\lambda = ck/h^2$. This method is stable for $\lambda \le 0.5$ and unstable for $\lambda > 0.5$. There is no problem with stability in the Crank-Nicholson method, which utilizes the difference equation

$$-u_{i-1,j+1} + \alpha u_{i,j+1} - u_{i+1,j+1} = u_{i+1,j} - \beta u_{ij} + u_{i-1,j}, \tag{3}$$

where $\alpha = 2(1 + 1/\lambda)$ and $\beta = 2(1 - 1/\lambda)$. This difference equation also yields a system of equations.

Solutions of the wave equation over a rectangular grid in the xt-plane can be approximated by

$$u_{i,j+1} = \lambda^2 u_{i+1,j} + 2(1 - \lambda^2)u_{ij} + \lambda^2 u_{i-1,j} - u_{i,j-1}, \tag{4}$$

where $\lambda = ck/h$. This finite difference method is stable for $\lambda \le 1$ and unstable for $\lambda > 1$.

Chapter 15 REVIEW EXERCISES

Answers to odd-numbered problems begin on page A–58.

1. Consider the boundary-value problem

$$\frac{\partial^2 u}{\partial x^2} + \frac{\partial^2 u}{\partial y^2} = 0, \qquad 0 < x < 2, \quad 0 < y < 1$$

$$u(0, y) = 0, \qquad u(2, y) = 50, \qquad 0 < y < 1$$

$$u(x, 0) = 0, \qquad u(x, 1) = 0, \qquad 0 < x < 2.$$

Approximate the solution of the differential equation at the interior points of the region with mesh size $h = 1/2$. Use Gaussian elimination or Gauss-Seidel iteration.

2. Solve Problem 1 using mesh size $h = 1/4$. Use Gauss-Seidel iteration.

3. Consider the boundary-value problem

$$\frac{\partial^2 u}{\partial x^2} = \frac{\partial u}{\partial t}, \qquad 0 < x < 1, \quad 0 < t < 0.05$$

$$u(0, t) = 0, \qquad u(1, t) = 0, \qquad t > 0$$

$$u(x, 0) = x, \qquad 0 < x < 1.$$

Note that the initial temperature $u(x, 0) = x$ indicates that the temperature at the right boundary $x = 1$ should be $u(1, 0) = 1$, whereas the boundary conditions imply that $u(1, 0) = 0$. The computer programs for the explicit finite difference method in the software supplement that accompanies this text were written so that the boundary conditions prevail for all times considered including $t = 0$.

(a) Use one of these programs to complete the following table.

Time	x = 0.00	x = 0.20	x = 0.40	x = 0.60	x = 0.80	x = 1.00
0.00	0.0000	0.2000	0.4000	0.6000	0.8000	0.0000
0.01	0.0000					0.0000
0.02	0.0000					0.0000
0.03	0.0000					0.0000
0.04	0.0000					0.0000
0.05	0.0000					0.0000

(b) Modify, say, the BASIC program in the software supplement so that the initial condition prevails at the boundaries at $t = 0$. Use this program to complete the following table.

Time	x = 0.00	x = 0.20	x = 0.40	x = 0.60	x = 0.80	x = 1.00
0.00	0.0000	0.2000	0.4000	0.6000	0.8000	1.0000
0.01	0.0000					0.0000
0.02	0.0000					0.0000
0.03	0.0000					0.0000
0.04	0.0000					0.0000
0.05	0.0000					0.0000

(c) Are the tables in parts (a) and (b) related in any way? Use a larger time interval if necessary.

APPENDICES

I Gamma Function
II Laplace Transforms
III Review of Determinants
IV Complex Numbers

Important Concepts

Gamma function
Generalized factorial
Minor
Cofactor
Cramer's rule
Complex number
Real part
Imaginary part
Conjugate
Modulus
Polar form
Argument

/ GAMMA FUNCTION

Euler's integral definition of the **gamma function*** is

$$\Gamma(x) = \int_0^\infty t^{x-1} e^{-t}\, dt. \qquad (1)$$

Convergence of the integral requires that $x - 1 > -1$, or $x > 0$. The recurrence relation

$$\Gamma(x + 1) = x\Gamma(x), \qquad (2)$$

which we saw in Section 6.5, can be obtained from (1) with integration by parts. Now when $x = 1$,

$$\Gamma(1) = \int_0^\infty e^{-t}\, dt = 1$$

and thus (2) gives $\qquad \Gamma(2) = 1\Gamma(1) = 1$

$$\Gamma(3) = 2\Gamma(2) = 2 \cdot 1$$

$$\Gamma(4) = 3\Gamma(3) = 3 \cdot 2 \cdot 1,$$

and so on. In this manner it is seen that when n is a positive integer,

$$\Gamma(n + 1) = n!$$

For this reason the gamma function is often called the **generalized factorial function**.

Although the integral form (1) does not converge for $x < 0$, it can be shown by means of alternative definitions that the gamma function is defined for all real and complex numbers *except* $x = -n, n = 0, 1, 2, \ldots$. As a consequence, (2) is actually valid for $x \neq -n$. Considered as a function of a real variable x, the graph of $\Gamma(x)$ is as given in Figure A.1. Observe that the nonpositive integers correspond to vertical asymptotes of the graph.

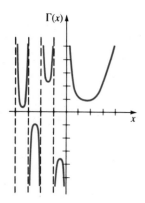

Figure A.1

---

* This function was first defined by Leonhard Euler in his text *Institutiones calculi integralis* published in 1768.

In Problems 27–33 of Exercises 6.5 we utilized the fact that $\Gamma(\frac{1}{2}) = \sqrt{\pi}$. This result can be derived from (1) by setting $x = 1/2$:

$$\Gamma(\tfrac{1}{2}) = \int_0^\infty t^{-1/2} e^{-t}\, dt. \tag{3}$$

When we let $t = u^2$, (3) can be written as

$$\Gamma(\tfrac{1}{2}) = 2 \int_0^\infty e^{-u^2}\, du.$$

But

$$\int_0^\infty e^{-u^2}\, du = \int_0^\infty e^{-v^2}\, dv,$$

and so

$$[\Gamma(\tfrac{1}{2})]^2 = \left(2 \int_0^\infty e^{-u^2}\, du\right)\left(2 \int_0^\infty e^{-v^2}\, dv\right)$$

$$= 4 \int_0^\infty \int_0^\infty e^{-(u^2+v^2)}\, du\, dv.$$

Switching to polar coordinates $u = r \cos\theta$, $v = r \sin\theta$ enables us to evaluate the double integral:

$$4 \int_0^\infty \int_0^\infty e^{-(u^2+v^2)}\, du\, dv = 4 \int_0^{\pi/2} \int_0^\infty e^{-r^2} r\, dr\, d\theta = \pi.$$

Hence

$$[\Gamma(\tfrac{1}{2})]^2 = \pi \quad \text{or} \quad \Gamma(\tfrac{1}{2}) = \sqrt{\pi}.$$

EXAMPLE I

Evaluate $\Gamma(-1/2)$.

Solution   In view of (2) it follows that, with $x = -1/2$,

$$\Gamma\left(\frac{1}{2}\right) = -\frac{1}{2} \Gamma\left(-\frac{1}{2}\right).$$

Therefore

$$\Gamma\left(-\frac{1}{2}\right) = -2\Gamma\left(\frac{1}{2}\right) = -2\sqrt{\pi}. \qquad\blacksquare$$

APPENDIX I
EXERCISES

Answers to odd-numbered problems begin on page A–58.

1. Evaluate.

   (a) $\Gamma(5)$ **(b)** $\Gamma(7)$ **(c)** $\Gamma(-3/2)$ **(d)** $\Gamma(-5/2)$

2. Use (1) and the fact that $\Gamma(6/5) = 0.92$ to evaluate $\displaystyle\int_0^\infty x^5 e^{-x^5}\,dx$. [*Hint:* Let $t = x^5$.]

3. Use (1) and the fact that $\Gamma(5/3) = 0.89$ to evaluate $\displaystyle\int_0^\infty x^4 e^{-x^3}\,dx$.

4. Evaluate $\displaystyle\int_0^1 x^3 \left(\ln\frac{1}{x}\right)^3 dx$. [*Hint:* Let $t = -\ln x$.]

5. Use the fact that $\Gamma(x) > \displaystyle\int_0^1 t^{x-1} e^{-t}\,dt$ to show that $\Gamma(x)$ is unbounded as $x \to 0^+$.

6. Use (1) to derive (2) for $x > 0$.

|| LAPLACE TRANSFORMS

$f(t)$	$\mathcal{L}\{f(t)\} = F(s)$
1. 1	$\dfrac{1}{s}$
2. t	$\dfrac{1}{s^2}$
3. t^n	$\dfrac{n!}{s^{n+1}}$, n a positive integer
4. $t^{-1/2}$	$\sqrt{\dfrac{\pi}{s}}$
5. $t^{1/2}$	$\dfrac{\sqrt{\pi}}{2s^{3/2}}$
6. t^α	$\dfrac{\Gamma(\alpha + 1)}{s^{\alpha+1}}$, $\alpha > -1$
7. $\sin kt$	$\dfrac{k}{s^2 + k^2}$
8. $\cos kt$	$\dfrac{s}{s^2 + k^2}$
9. $\sin^2 kt$	$\dfrac{2k^2}{s(s^2 + 4k^2)}$
10. $\cos^2 kt$	$\dfrac{s^2 + 2k^2}{s(s^2 + 4k^2)}$

$f(t)$	$\mathscr{L}\{f(t)\} = F(s)$
11. e^{at}	$\dfrac{1}{s-a}$
12. $\sinh kt$	$\dfrac{k}{s^2 - k^2}$
13. $\cosh kt$	$\dfrac{s}{s^2 - k^2}$
14. $\sinh^2 kt$	$\dfrac{2k^2}{s(s^2 - 4k^2)}$
15. $\cosh^2 kt$	$\dfrac{s^2 - 2k^2}{s(s^2 - 4k^2)}$
16. te^{at}	$\dfrac{1}{(s-a)^2}$
17. $t^n e^{at}$	$\dfrac{n!}{(s-a)^{n+1}}$, n a positive integer
18. $e^{at}\sin kt$	$\dfrac{k}{(s-a)^2 + k^2}$
19. $e^{at}\cos kt$	$\dfrac{s-a}{(s-a)^2 + k^2}$
20. $e^{at}\sinh kt$	$\dfrac{k}{(s-a)^2 - k^2}$
21. $e^{at}\cosh kt$	$\dfrac{s-a}{(s-a)^2 - k^2}$
22. $t\sin kt$	$\dfrac{2ks}{(s^2 + k^2)^2}$
23. $t\cos kt$	$\dfrac{s^2 - k^2}{(s^2 + k^2)^2}$
24. $\sin kt + kt\cos kt$	$\dfrac{2ks^2}{(s^2 + k^2)^2}$
25. $\sin kt - kt\cos kt$	$\dfrac{2k^3}{(s^2 + k^2)^2}$
26. $t\sinh kt$	$\dfrac{2ks}{(s^2 - k^2)^2}$
27. $t\cosh kt$	$\dfrac{s^2 + k^2}{(s^2 - k^2)^2}$

$f(t)$	$\mathcal{L}\{f(t)\} = F(s)$
28. $\dfrac{e^{at} - e^{bt}}{a - b}$	$\dfrac{1}{(s - a)(s - b)}$
29. $\dfrac{ae^{at} - be^{bt}}{a - b}$	$\dfrac{s}{(s - a)(s - b)}$
30. $1 - \cos kt$	$\dfrac{k^2}{s(s^2 + k^2)}$
31. $kt - \sin kt$	$\dfrac{k^3}{s^2(s^2 + k^2)}$
32. $\dfrac{a \sin bt - b \sin at}{ab(a^2 - b^2)}$	$\dfrac{1}{(s^2 + a^2)(s^2 + b^2)}$
33. $\dfrac{\cos bt - \cos at}{a^2 - b^2}$	$\dfrac{s}{(s^2 + a^2)(s^2 + b^2)}$
34. $\sin kt \sinh kt$	$\dfrac{2k^2 s}{s^4 + 4k^4}$
35. $\sin kt \cosh kt$	$\dfrac{k(s^2 + 2k^2)}{s^4 + 4k^4}$
36. $\cos kt \sinh kt$	$\dfrac{k(s^2 - 2k^2)}{s^4 + 4k^4}$
37. $\cos kt \cosh kt$	$\dfrac{s^3}{s^4 + 4k^4}$
38. $J_0(kt)$	$\dfrac{1}{\sqrt{s^2 + k^2}}$
39. $\dfrac{e^{bt} - e^{at}}{t}$	$\ln \dfrac{s - a}{s - b}$
40. $\dfrac{2(1 - \cos kt)}{t}$	$\ln \dfrac{s^2 + k^2}{s^2}$
41. $\dfrac{2(1 - \cosh kt)}{t}$	$\ln \dfrac{s^2 - k^2}{s^2}$
42. $\dfrac{\sin at}{t}$	$\arctan\left(\dfrac{a}{s}\right)$
43. $\dfrac{\sin at \cos bt}{t}$	$\dfrac{1}{2} \arctan \dfrac{a + b}{s} + \dfrac{1}{2} \arctan \dfrac{a - b}{s}$

$f(t)$	$\mathcal{L}\{f(t)\} = F(s)$
44. $\dfrac{1}{\sqrt{\pi t}}\, e^{-a^2/4t}$	$\dfrac{e^{-a\sqrt{s}}}{\sqrt{s}}$
45. $\dfrac{a}{2\sqrt{\pi t^3}}\, e^{-a^2/4t}$	$e^{-a\sqrt{s}}$
46. $\operatorname{erfc}\left(\dfrac{a}{2\sqrt{t}}\right)$	$\dfrac{e^{-a\sqrt{s}}}{s}$
47. $2\sqrt{\dfrac{t}{\pi}}\, e^{-a^2/4t} - a\operatorname{erfc}\left(\dfrac{a}{2\sqrt{t}}\right)$	$\dfrac{e^{-a\sqrt{s}}}{s\sqrt{s}}$
48. $e^{ab}e^{b^2t}\operatorname{erfc}\left(b\sqrt{t} + \dfrac{a}{2\sqrt{t}}\right)$	$\dfrac{e^{-a\sqrt{s}}}{\sqrt{s}(\sqrt{s}+b)}$
49. $-e^{ab}e^{b^2t}\operatorname{erfc}\left(b\sqrt{t} + \dfrac{a}{2\sqrt{t}}\right) + \operatorname{erfc}\left(\dfrac{a}{2\sqrt{t}}\right)$	$\dfrac{be^{-a\sqrt{s}}}{s(\sqrt{s}+b)}$
50. $\delta(t)$	1
51. $\delta(t - t_0)$	e^{-st_0}
52. $e^{at}f(t)$	$F(s - a)$
53. $f(t - a)\,\mathcal{U}(t - a)$	$e^{-as}F(s)$
54. $\mathcal{U}(t - a)$	$\dfrac{e^{-as}}{s}$
55. $f^{(n)}(t)$	$s^n F(s) - s^{(n-1)}f(0) - \cdots - f^{(n-1)}(0)$
56. $t^n f(t)$	$(-1)^n\dfrac{d^n}{ds^n}F(s)$
57. $\displaystyle\int_0^t f(\tau)g(t - \tau)\, d\tau$	$F(s)G(s)$

III   REVIEW OF DETERMINANTS

The determinant of a 2×2 matrix $\mathbf{A}$ is defined by

$$\begin{vmatrix} a_{11} & a_{12} \\ a_{21} & a_{22} \end{vmatrix} = a_{11}a_{22} - a_{12}a_{21}.$$

Minors and Cofactors

For an $n \times n$ matrix $\mathbf{A}$, let a_{ij} be the entry in the ith row and jth column. The **minor** M_{ij} associated with a_{ij} is the determinant of the $(n-1) \times (n-1)$ matrix obtained by deleting the ith row and the jth column of the matrix. The **cofactor** C_{ij} associated with a_{ij} is a *signed minor*, specifically

$$C_{ij} = (-1)^{i+j}M_{ij}.$$

EXAMPLE 1

The cofactors of the entries in the 3×3 matrix

$$\begin{pmatrix} 2 & 4 & 7 \\ 1 & 2 & 3 \\ 1 & 5 & 3 \end{pmatrix} \qquad \textbf{(1)}$$

are

$$C_{11} = (-1)^{1+1}M_{11} \qquad\qquad C_{12} = (-1)^{1+2}M_{12} \qquad\qquad C_{13} = (-1)^{1+3}M_{13}$$

$$= \begin{vmatrix} 2 & 3 \\ 5 & 3 \end{vmatrix} = -9 \qquad\qquad = (-1)\begin{vmatrix} 1 & 3 \\ 1 & 3 \end{vmatrix} = 0 \qquad\qquad = \begin{vmatrix} 1 & 2 \\ 1 & 5 \end{vmatrix} = 3$$

$$C_{21} = (-1)^{2+1}M_{21} \qquad\qquad C_{22} = (-1)^{2+2}M_{22} \qquad\qquad C_{23} = (-1)^{2+3}M_{23}$$

$$= (-1)\begin{vmatrix} 4 & 7 \\ 5 & 3 \end{vmatrix} = 23 \qquad\qquad = \begin{vmatrix} 2 & 7 \\ 1 & 3 \end{vmatrix} = -1 \qquad\qquad = (-1)\begin{vmatrix} 2 & 4 \\ 1 & 5 \end{vmatrix} = -6$$

$$C_{31} = (-1)^{3+1}M_{31} \qquad\qquad C_{32} = (-1)^{3+2}M_{32} \qquad\qquad C_{33} = (-1)^{3+3}M_{33}$$

$$= \begin{vmatrix} 4 & 7 \\ 2 & 3 \end{vmatrix} = -2 \qquad\qquad = (-1)\begin{vmatrix} 2 & 7 \\ 1 & 3 \end{vmatrix} = 1 \qquad\qquad = \begin{vmatrix} 2 & 4 \\ 1 & 2 \end{vmatrix} = 0. \qquad \blacksquare$$

Expansion by Cofactors

It can be proved that a determinant can be **expanded** in terms of cofactors:

Multiply the entries a_{ij} in any row (or column) by their corresponding cofactors C_{ij} and add the n products.

Thus a 3×3 determinant* can be expanded into three 2×2 determinants, a 4×4 determinant can be expanded into four 3×3 determinants, and so on.

---

* Even though a determinant of a matrix of numbers is a number, it is sometimes convenient to refer to it as if it were an array.

E X A M P L E 2

Evaluate the determinant of the matrix in (1).

Solution   Expanding by the first row gives

$$\begin{vmatrix} 2 & 4 & 7 \\ 1 & 2 & 3 \\ 1 & 5 & 3 \end{vmatrix} = 2\begin{vmatrix} 2 & 3 \\ 5 & 3 \end{vmatrix} + 4(-1)\begin{vmatrix} 1 & 3 \\ 1 & 3 \end{vmatrix} + 7\begin{vmatrix} 1 & 2 \\ 1 & 5 \end{vmatrix} = 3.$$

Alternatively, we can expand the determinant by, say, the second column:

$$\begin{vmatrix} 2 & 4 & 7 \\ 1 & 2 & 3 \\ 1 & 5 & 3 \end{vmatrix} = 4(-1)\begin{vmatrix} 1 & 3 \\ 1 & 3 \end{vmatrix} + 2\begin{vmatrix} 2 & 7 \\ 1 & 3 \end{vmatrix} + 5(-1)\begin{vmatrix} 2 & 7 \\ 1 & 3 \end{vmatrix} = 3.$$ ∎

We note that if a determinant has a row (or a column) containing many zero entries, then wisdom dictates that we expand the determinant by that row (or column).

Cramer's Rule

Determinants are sometimes useful in solving algebraic systems of n linear equations in n unknowns:

$$a_{11}x_1 + a_{12}x_2 + \cdots + a_{1n}x_n = b_1$$

$$a_{21}x_1 + a_{22}x_2 + \cdots + a_{2n}x_n = b_2$$

$$\vdots \qquad\qquad \vdots$$

$$a_{n1}x_1 + a_{n2}x_2 + \cdots + a_{nn}x_n = b_n.$$ (2)

Let **A** be the matrix of coefficients of (2) and let

$$\det \mathbf{A} = \begin{vmatrix} a_{11} & a_{12} & \cdots & a_{1n} \\ a_{21} & a_{22} & \cdots & a_{2n} \\ \vdots & & & \vdots \\ a_{n1} & a_{n2} & \cdots & a_{nn} \end{vmatrix}.$$

If

$$\det \mathbf{A}_k = \begin{vmatrix} a_{11} & a_{12} & \cdots & a_{1,k-1} & b_1 & a_{1,k+1} & \cdots & a_{1n} \\ a_{21} & a_{22} & \cdots & a_{2,k-1} & b_2 & a_{2,k+1} & \cdots & a_{2n} \\ \vdots & & & \vdots & \vdots & \vdots & & \vdots \\ a_{n1} & a_{n2} & \cdots & a_{n,k-1} & b_n & a_{n,k+1} & \cdots & a_{nn} \end{vmatrix}$$

*k*th column
↓

is the same as det $\mathbf{A}$ except that its kth column has been replaced by the column

$$
\begin{array}{c}
b_1 \\
b_2 \\
\vdots \\
b_n
\end{array}
$$

then (2) has the unique solution

$$
x_1 = \frac{\det \mathbf{A}_1}{\det \mathbf{A}}, \ x_2 = \frac{\det \mathbf{A}_2}{\det \mathbf{A}}, \ldots, x_n = \frac{\det \mathbf{A}_n}{\det \mathbf{A}}
\tag{3}
$$

whenever det $\mathbf{A} \neq 0$. This method of solving (2) by determinants is known as **Cramer's rule.***

EXAMPLE 3

Solve the system

$$
\begin{aligned}
3x + 2y + \ z &= 7 \\
x - \ y + 3z &= 3 \\
5x + 4y - 2z &= 1
\end{aligned}
$$

by Cramer's rule.

Solution   The solution requires the calculation of four determinants:

$$
\det \mathbf{A} = \begin{vmatrix} 3 & 2 & 1 \\ 1 & -1 & 3 \\ 5 & 4 & -2 \end{vmatrix} = 13, \qquad
\det \mathbf{A}_1 = \begin{vmatrix} 7 & 2 & 1 \\ 3 & -1 & 3 \\ 1 & 4 & -2 \end{vmatrix} = -39
$$

$$
\det \mathbf{A}_2 = \begin{vmatrix} 3 & 7 & 1 \\ 1 & 3 & 3 \\ 5 & 1 & -2 \end{vmatrix} = 78, \qquad
\det \mathbf{A}_3 = \begin{vmatrix} 3 & 2 & 7 \\ 1 & -1 & 3 \\ 5 & 4 & 1 \end{vmatrix} = 52.
$$

Hence (3) gives

$$
x = \frac{\det \mathbf{A}_1}{\det \mathbf{A}} = -3, \qquad y = \frac{\det \mathbf{A}_2}{\det \mathbf{A}} = 6, \qquad z = \frac{\det \mathbf{A}_3}{\det \mathbf{A}} = 4. \qquad ■
$$

---

* This rule was named after **Gabriel Cramer** (1704–1752), a Swiss mathematician who was the first to publish this result in 1750.

Homogeneous Systems

If $b_i = 0, i = 1, 2, \ldots, n$, then the system of equations (2) is said to be **homogeneous**. If at least one of the b_i is not zero, the system is **nonhomogeneous**. Now if $\det \mathbf{A} \neq 0$, (3) implies that the only solution of a homogeneous system is $x_1 = 0, x_2 = 0, \ldots, x_n = 0$. If $\det \mathbf{A} = 0$, then a homogeneous system of n linear equations in n unknowns has infinitely many solutions. These solutions can be found by solving the system through elimination. If $\det \mathbf{A} = 0$, then a nonhomogeneous system may have either infinitely many solutions or no solution at all.

APPENDIX III EXERCISES

Answers to odd-numbered problems begin on page A–58.

In Problems 1–8 evaluate the given determinant.

1. $\begin{vmatrix} 2 & 4 & 6 \\ -1 & 5 & 1 \\ 0 & 2 & -3 \end{vmatrix}$

2. $\begin{vmatrix} 1 & 4 & 2 \\ -2 & 6 & 3 \\ 9 & 8 & 4 \end{vmatrix}$

3. $\begin{vmatrix} 2 & 0 & 5 \\ 0 & 7 & 9 \\ -6 & 1 & 4 \end{vmatrix}$

4. $\begin{vmatrix} 79 & 81 & 40 \\ 22 & 16 & 59 \\ 0 & 0 & 0 \end{vmatrix}$

5. $\begin{vmatrix} 1 & 2 & 3 & 4 \\ 1 & 1 & 0 & 0 \\ 8 & 7 & 0 & 0 \\ 9 & 5 & 3 & 0 \end{vmatrix}$

6. $\begin{vmatrix} 1 & 0 & 9 & 0 & 3 \\ 2 & 1 & 7 & 0 & 0 \\ 0 & 0 & 2 & 0 & 0 \\ -1 & 1 & 5 & 2 & 2 \\ 2 & 2 & 8 & 1 & 1 \end{vmatrix}$

7. $\begin{vmatrix} e^t & e^{3t} & e^{-t} \\ e^t & 3e^{3t} & -e^{-t} \\ e^t & 9e^{3t} & e^{-t} \end{vmatrix}$

8. $\begin{vmatrix} e^{2t} & \sin t & \cos t \\ 2e^{2t} & \cos t & -\sin t \\ 4e^{2t} & -\sin t & -\cos t \end{vmatrix}$

In Problems 9–12 use Cramer's rule to solve the given system of equations.

9. $\begin{aligned} 2x + y &= 1 \\ 3x + 2y &= -2 \end{aligned}$

10. $\begin{aligned} 5x + 4y &= -1 \\ 10x - 6y &= 5 \end{aligned}$

11. $\begin{aligned} x + 2y + z &= 8 \\ 2x - 2y + 2z &= 7 \\ x - 4y + 3z &= 1 \end{aligned}$

12. $\begin{aligned} 4x + 3y + 2z &= 8 \\ -x + 2z &= 12 \\ 3x + 2y + z &= 3 \end{aligned}$

13. Given the system

$$x - y + 2z = 0$$

$$2x + y - z = 0$$

$$4x - y + 3z = 0.$$

(a) If A denotes the matrix of coefficients, show that $\det A = 0$.

(b) Show that the system has infinitely many solutions.

(c) Explain the geometric significance of the system.

14. Given the system

$$a_1x + b_1y = c_1$$

$$a_2x + b_2y = c_2.$$

Let A denote the matrix of coefficients.

(a) Explain the geometric significance of $\det A \neq 0$.

(b) Explain the geometric significance of $\det A = 0$.

IV COMPLEX NUMBERS

A **complex number** is any expression of the form

$$z = a + bi, \qquad \text{where} \quad i^2 = -1.$$

The real numbers a and b are called the **real** and **imaginary parts** of z, respectively. In practice, the symbol i is written $i = \sqrt{-1}$. The number $\bar{z} = a - bi$ is called the **conjugate** of z.

EXAMPLE I

From the properties of radicals, we have

$$\sqrt{-25} = \sqrt{25}\,\sqrt{-1} = 5i. \qquad \blacksquare$$

EXAMPLE 2

The conjugates of the complex numbers $z_1 = 4 + 5i$ and $z_2 = 3 - 2i$ are, in turn, $\bar{z}_1 = 4 - 5i$ and $\bar{z}_2 = 3 + 2i$. $\qquad \blacksquare$

Sum, Difference, and Product

The **sum**, **difference**, and **product** of two complex numbers $z_1 = a_1 + b_1 i$ and $z_2 = a_2 + b_2 i$ are defined as follows:

(i) $z_1 + z_2 = (a_1 + a_2) + (b_1 + b_2)i$

(ii) $z_1 - z_2 = (a_1 - a_2) + (b_1 - b_2)i$

(iii) $z_1 z_2 = (a_1 a_2 - b_1 b_2) + (a_1 b_2 + b_1 a_2)i.$

In other words, to add or subtract two complex numbers we simply add or subtract the corresponding real and imaginary parts. To multiply two complex numbers we use the distributive law and the fact that $i^2 = -1$.

EXAMPLE 3

If $z_1 = 4 + 5i$ and $z_2 = 3 - 2i$, then

$$z_1 + z_2 = (4 + 3) + (5 + (-2))i = 7 + 3i$$

$$z_1 - z_2 = (4 - 3) + (5 - (-2))i = 1 + 7i$$

$$z_1 z_2 = (4 + 5i)(3 - 2i)$$

$$= (4 + 5i)3 + (4 + 5i)(-2i)$$

$$= 12 + 15i - 8i - 10i^2$$

$$= (12 + 10) + (15 - 8)i = 22 + 7i. \qquad \blacksquare$$

The product of a complex number $z = a + bi$ and its conjugate $\bar{z} = a - bi$ is the real number

$$z\bar{z} = a^2 + b^2. \tag{1}$$

Quotient

The quotient of two complex numbers z_1 and z_2 is found by multiplying the numerator and denominator of z_1/z_2 by the conjugate of the denominator z_2 and using (1). The next example illustrates the procedure.

EXAMPLE 4

$$\frac{z_1}{z_2} = \frac{4 + 5i}{3 - 2i}$$

$$= \frac{4 + 5i}{3 - 2i}\frac{3 + 2i}{3 + 2i}$$

$$= \frac{12 + 15i + 8i + 10i^2}{9 + 4} = \frac{2}{13} + \frac{23}{13}i \qquad \blacksquare$$

Geometric Interpretation

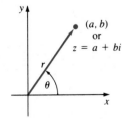

Figure A.2

By taking a and b to be the x- and y-coordinates of a point in the plane, we can interpret a complex number $z = a + bi$ as a **vector** from the origin terminating at (a, b) (see Figure A.2). The length of the vector is called the **modulus** of z and is written as r, or $|z|$. From the Pythagorean theorem it follows that

$$r = |z| = \sqrt{a^2 + b^2}.$$

If θ is an angle the vector makes with the positive x-axis, then from Figure A.2 we see that

$$a = r \cos \theta \quad \text{and} \quad b = r \sin \theta.$$

Thus

$$z = r \cos \theta + ir \sin \theta \quad \text{or} \quad z = r(\cos \theta + i \sin \theta). \tag{2}$$

This latter form is called the **polar form** of the complex number z. The angle θ is called an **argument** of z.

Euler's Formula

The power series $e^X = \sum_{n=0}^{\infty} \dfrac{X^n}{n!}$ is known to converge for all real and complex numbers. If we let $X = i\theta$, θ a real number, then

$$e^{i\theta} = \sum_{n=0}^{\infty} \frac{(i\theta)^n}{n!} = 1 + i\theta + \frac{i^2\theta^2}{2!} + \frac{i^3\theta^3}{3!} + \frac{i^4\theta^4}{4!} + \frac{i^5\theta^5}{5!} + \frac{i^6\theta^6}{6!} + \frac{i^7\theta^7}{7!} + \cdots. \tag{3}$$

Now $i^2 = -1$, $i^3 = -i$, $i^4 = 1$, $i^5 = i$, and so on. Thus (3) can be separated into real and imaginary parts:

$$e^{i\theta} = \left(1 - \frac{\theta^2}{2!} + \frac{\theta^4}{4!} - \frac{\theta^6}{6!} + \cdots\right) + i\left(\theta - \frac{\theta^3}{3!} + \frac{\theta^5}{5!} - \frac{\theta^7}{7!} + \cdots\right). \tag{4}$$

But from calculus we recall that

$$\cos \theta = \sum_{n=0}^{\infty} \frac{(-1)^n}{(2n)!} \theta^{2n} \quad \text{and} \quad \sin \theta = \sum_{n=0}^{\infty} \frac{(-1)^n}{(2n+1)!} \theta^{2n+1},$$

where each series converges for every real number θ. Hence (4) can be written as

$$e^{i\theta} = \cos \theta + i \sin \theta. \tag{5}$$

This last result is known as **Euler's formula**. Note that in view of (5), the polar form (2) of a complex number can be expressed in the compact form

$$z = re^{i\theta}. \tag{6}$$

EXAMPLE 5

Find the polar form (6) of $z = 1 - i$.

Solution The graph of the complex number is given in Figure A.3. Since $a = 1$ and $b = -1$, the modulus of z is

$$r = \sqrt{1^2 + (-1)^2} = \sqrt{2}.$$

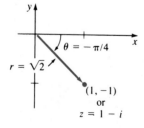

Figure A.3

As seen in the Figure A.3, $\tan\theta = -1$ and so we can take an argument of z to be $\theta = -\pi/4$. Therefore, the polar form of the number is

$$z = \sqrt{2}\,e^{-i\pi/4}.$$ ■

APPENDIX IV
EXERCISES

Answers to odd-numbered problems begin on page A–58.

In Problems 1–10 let $z_1 = 2 - i$ and $z_2 = 5 + 3i$. Perform the indicated operation.

1. $z_1 + \bar{z}_2$

2. $4z_1 + z_2$

3. $2z_1 - 3z_2$

4. $z_1 z_2$

5. $(z_1)^2$

6. $\bar{z}_1(i + z_2)$

7. z_1/z_2

8. z_2/z_1

9. $1/z_2$

10. z_1/i

In Problems 11–20 write the given complex number in the polar form (6).

11. $z = i$

12. $z = -4i$

13. $z = i^2$

14. $z = 6i^5$

15. $z = 2 + 2i$

16. $z = -\sqrt{5} - \sqrt{5}\,i$

17. $z = 6 + 6\sqrt{3}\,i$

18. $z = -10\sqrt{3} + 10i$

19. $z = i(1 - \sqrt{3}\,i)$

20. $z = -7 + 7i$

In Problems 21 and 22 express the given complex number in polar form in the form $z = a + bi$.

21. $z = 8e^{-i\pi}$

22. $z = 2e^{i7\pi/4}$

23. Prove **DeMoivre's theorem**:* For any positive integer n,

$$[r(\cos\theta + i\sin\theta)]^n = r^n[\cos n\theta + i\sin n\theta].$$

24. Use DeMoivre's theorem of Problem 23 to evaluate $(1 + i)^{10}$.

25. Use Euler's formula to show that

$$\cos\theta = \frac{e^{i\theta} + e^{-i\theta}}{2} \quad \text{and} \quad \sin\theta = \frac{e^{i\theta} - e^{-i\theta}}{2i}.$$

---

* This theorem was named after the French mathematician **Abraham DeMoivre** (1667–1754).

ANSWERS TO ODD-NUMBERED PROBLEMS

EXERCISES 1.1, Page 10

1. linear, second-order

3. nonlinear, first-order

5. linear, fourth-order

7. nonlinear, second-order

9. linear, third-order

11. $2y' + y = 2(-\frac{1}{2})e^{-x/2} + e^{-x/2} = 0$

13. $\dfrac{dy}{dx} - 2y - e^{3x} = (3e^{3x} + 20e^{2x})$
    $-2(e^{3x} + 10e^{2x}) - e^{3x} = 0$

15. $y' - 25 - y^2 = 25 \sec^2 5x - 25(1 + \tan^2 5x)$
    $= 25 \sec^2 5x - 25 \sec^2 5x = 0$

17. $y' + y - \sin x = \frac{1}{2} \cos x + \frac{1}{2} \sin x$
    $-10e^{-x} + \frac{1}{2} \sin x - \frac{1}{2} \cos x + 10e^{-x} - \sin x = 0$

19. $yx^2 = -1$ implies $d(yx^2) = 0$
    or $2yx\,dx + x^2\,dy = 0$

21. $y - 2xy' - y(y')^2 = y - 2x\dfrac{c_1}{2y} - y\dfrac{c_1^2}{4y^2}$
    $= \dfrac{y^2 - (c_1 x + (c_1^2/4))}{y}$
    $= \dfrac{y^2 - y^2}{y} = 0$

23. $y' - \dfrac{1}{x} y - 1 = 1 + \ln x - \ln x - 1 = 0$

25. $\dfrac{d}{dt} \ln \dfrac{2 - X}{1 - X} = 1$
    $\left[\dfrac{-1}{2 - X} + \dfrac{1}{1 - X} \right] \dfrac{dX}{dt} = 1$
    simplifies to $\dfrac{dX}{dt} = (2 - X)(1 - X)$

27. The differential of $c_1 = xe^{y/x}/(x + y)^2$ is
    $\{(x + y)^2[xe^{y/x}(x\,dy - y\,dx/x^2) + e^{y/x}\,dx]$
    $- xe^{y/x}2(x + y)(dx + dy)\}/(x + y)^4 = 0$.
    Multiplying by $-x^2(x + y)^3 e^{-y/x}$ and simplifying
    yield $(x^2 + y^2)\,dx + (x^2 - xy)\,dy = 0$.

29. $y'' - 6y' + 13y = 5e^{3x} \cos 2x - 12e^{3x} \sin 2x$
    $+ 12e^{3x} \sin 2x - 18e^{3x} \cos 2x + 13e^{3x} \cos 2x = 0$

31. $y'' = \cosh x + \sinh x = y$

33. $y'' + (y')^2 = -\dfrac{1}{(x + c_1)^2} + \dfrac{1}{(x + c_1)^2} = 0$

35. $x\dfrac{d^2 y}{dx^2} + 2\dfrac{dy}{dx} = x(2c_2 x^{-3}) + 2(-c_2 x^{-2}) = 0$

37. $x^2 y'' - 3xy' + 4y$
    $= x^2(5 + 2\ln x) - 3x(3x + 2x\ln x)$
    $+ 4(x^2 + x^2 \ln x) = 9x^2 - 9x^2$
    $+ 6x^2 \ln x - 6x^2 \ln x = 0$

39. $y''' - 3y'' + 3y' - y$
    $= x^2 e^x + 6xe^x + 6e^x - 3x^2 e^x - 12xe^x$
    $- 6e^x + 3x^2 e^x + 6xe^x - x^2 e^x = 0$

41. For $x < 0$,
    $xy' - 2y = x(-2x) - 2(-x^2) = 0$;
    for $x \geq 0$,
    $xy' - 2y = x(2x) - 2(x^2) = 0$.

43. $y - xy' - (y')^2 = cx + c^2 - x(c) - c^2 = 0$;
    $k = -1/4$

45. $y = -1$

47. $m = 2$ and $m = 3$

49. $m = \dfrac{1 \pm \sqrt{5}}{2}$

51. For $y = x^2$,
$$x^2 y'' - 4xy' + 6y = x^2(2) - 4x(2x) + 6x^2$$
$$= 8x^2 - 8x^2 = 0;$$
for $y = x^3$,
$$x^2 y'' - 4xy' + 6y = x^2(6x) - 4x(3x^2) + 6x^3$$
$$= 12x^3 - 12x^3 = 0;$$
yes; yes

53. (a) $y = 0$;   (b) no real solution;
(c) $y = 1$ or $y = -1$

EXERCISES 1.2, Page 26

1. $\dfrac{dv}{dt} + \dfrac{k}{m} v = g$

3. (a) $k = gR^2$;   (b) $\dfrac{d^2 r}{dt^2} - \dfrac{gR^2}{r^2} = 0$;

(c) $v \dfrac{dv}{dr} - \dfrac{gR^2}{r^2} = 0$

5. $L \dfrac{di}{dt} + Ri = E(t)$ **7.** $\dfrac{dh}{dt} = -\dfrac{\pi}{750} \sqrt{h}$

9. $\dfrac{dh}{dt} = -\dfrac{1}{30\sqrt{h}(10 - h)}$ **11.** $\dfrac{dx}{dt} + kx = r$, $k > 0$

13. $mx'' = -k \cos \theta$ $\quad my'' = -mg - k \sin \theta$

$\quad\quad = -k \cdot \dfrac{1}{v} \dfrac{dx}{dt}$ $\quad = -mg - k \cdot \dfrac{1}{v} \dfrac{dy}{dt}$

$\quad\quad = -|c| \dfrac{dx}{dt}$ $\quad = -mg - |c| \dfrac{dy}{dt}$

15. Using $\tan \phi = \dfrac{x}{y}$, $\quad \tan\left(\dfrac{\pi}{2} - \theta\right) = \dfrac{dy}{dx}$,

$\tan \theta = \dfrac{dx}{dy}$, and $\tan \phi = \tan 2\theta$

$\quad\quad\quad\quad\quad = \dfrac{2\tan\theta}{1 - \tan^2\theta}$,

we obtain $x\left(\dfrac{dx}{dy}\right)^2 + 2y \dfrac{dx}{dy} = x$.

17. By combining Newton's second law of motion with his law of gravitation, we obtain
$$m \dfrac{d^2 y}{dt^2} = -k_1 \dfrac{mM}{y^2},$$
where M is the mass of the earth and k_1 is a constant of proportionality. Dividing by m gives
$$\dfrac{d^2 y}{dt^2} = -\dfrac{k}{y^2},$$

where $k = k_1 M$. The constant k is gR^2, where R is the radius of the earth. This follows from the fact that on the surface of the earth $y = R$ so that
$$k_1 \dfrac{mM}{R^2} = mg$$
$$k_1 M = gR^2 \quad \text{or} \quad k = gR^2.$$
If $t = 0$ is the time at which burnout occurs, then
$$y(0) = R + y_B,$$
where y_B is the distance from the earth's surface to the rocket at the time of burnout, and
$$y'(0) = V_B$$
is the corresponding velocity at that time.

19. $\dfrac{dy}{dx} = -\dfrac{y}{\sqrt{s^2 - y^2}}$ **21.** $\dfrac{dA}{dt} = k(M - A)$, $k > 0$

CHAPTER 1 REVIEW EXERCISES, Page 30

1. ordinary, first-order, nonlinear

3. partial, second-order   **9.** $y = x^2$ **11.** $y = \dfrac{x^2}{2}$

13. $y = 0$, $y = e^x$ **15.** $y = 0$, $y = \cos x$, $y = \sin x$

17. $x < 0$ or $x > 1$ **19.** $\dfrac{dh}{dt} = -\dfrac{25\sqrt{2g}}{16\pi} h^{-3/2}$

EXERCISES 2.1, Page 37

1. half-planes defined by either $y > 0$ or $y < 0$

3. half-planes defined by either $x > 0$ or $x < 0$

5. the regions defined by either $y > 2$, $y < -2$, or $-2 < y < 2$

7. any region not containing $(0, 0)$

9. the entire xy-plane   **11.** $y = 0$, $y = x^3$

13. There is some interval around $x = 0$ on which the unique solution is $y = 0$.

15. $y = 0$, $y = x$. No, the given function is nondifferentiable at $x = 0$.

17. yes   **19.** no

EXERCISES 2.2, Page 44

1. $y = -\frac{1}{5} \cos 5x + c$ **3.** $y = \frac{1}{3} e^{-3x} + c$

5. $y = x + 5 \ln|x + 1| + c$ **7.** $y = cx^4$

9. $y^{-2} = 2x^{-1} + c$ **11.** $-3 + 3x \ln|x| = xy^3 + cx$

13. $-3e^{-2y} = 2e^{3x} + c$ **15.** $2 + y^2 = c(4 + x^2)$

17. $y^2 = x - \ln|x + 1| + c$

19. $\dfrac{x^3}{3} \ln x - \dfrac{1}{9} x^3 = \dfrac{y^2}{2} + 2y + \ln|y| + c$

21. $S = ce^{kr}$ **23.** $\dfrac{P}{1 - P} = ce^t$ or $P = \dfrac{ce^t}{1 + ce^t}$

25. $4 \cos y = 2x + \sin 2x + c$

27. $-2 \cos x + e^y + ye^{-y} + e^{-y} = c$

29. $(e^x + 1)^{-2} + 2(e^y + 1)^{-1} = c$

31. $(y + 1)^{-1} + \ln|y + 1| = \dfrac{1}{2} \ln \left| \dfrac{x + 1}{x - 1} \right| + c$

33. $y - 5 \ln|y + 3| = x - 5 \ln|x + 4| + c$

or $\left(\dfrac{y + 3}{x + 4} \right)^5 = c_1 e^{y - x}$

35. $-\cot y = \cos x + c$ **37.** $y = \sin \left(\dfrac{x^2}{2} + c \right)$

39. $-y^{-1} = \tan^{-1}(e^x) + c$

41. $(1 + \cos x)(1 + e^y) = 4$

43. $\sqrt{y^2 + 1} = 2x^2 + \sqrt{2}$

45. $x = \tan(4y - 3\pi/4)$ **47.** $xy = e^{-(1 + 1/x)}$

49. (a) $y = 3 \dfrac{1 - e^{6x}}{1 + e^{6x}}$;   (b) $y = 3$;

(c) $y = 3 \dfrac{2 - e^{6x - 2}}{2 + e^{6x - 2}}$

51. $y = 1$ **53.** $y = 1$ **55.** $y = 1 + \dfrac{1}{10} \tan \dfrac{x}{10}$

57. $y = -x - 1 + \tan(x + c)$

59. $2y - 2x + \sin 2(x + y) = c$

61. $4(y - 2x + 3) = (x + c)^2$

EXERCISES 2.3, Page 52

1. homogeneous of degree 3

3. homogeneous of degree 2

5. not homogeneous   **7.** homogeneous of degree 0

9. homogeneous of degree -2

11. $x \ln|x| + y = cx$

13. $(x - y) \ln|x - y| = y + c(x - y)$

15. $x + y \ln|x| = cy$

17. $\ln(x^2 + y^2) + 2 \tan^{-1}(y/x) = c$

19. $4x = y(\ln|y| - c)^2$ **21.** $y^9 = c(x^3 + y^3)^2$

23. $(y/x)^2 = 2 \ln|x| + c$ **25.** $e^{2x/y} = 8 \ln|y| + c$

27. $x \cos(y/x) = c$ **29.** $y + x = cx^2 e^{y/x}$

31. $y^3 + 3x^3 \ln|x| = 8x^3$ **33.** $y^2 = 4x(x + y)^2$

35. $\ln|x| = e^{y/x} - 1$

37. $4x \ln|y/x| + x \ln x + y - x = 0$

39. $3x^{3/2} \ln x + 3x^{1/2}y + 2y^{3/2} = 5x^{3/2}$

41. $(x + y) \ln|y| + x = 0$

43. $\ln|y| = -2(1 - x/y)^{1/2} + \sqrt{2}$

45. By homogeneity the equation can be written as

$$M(x/y, 1) \, dx + N(x/y, 1) \, dy = 0.$$

With $v = x/y$, it follows that

$$M(v, 1)(v \, dy + y \, dv) + N(v, 1) \, dy = 0$$

$$[vM(v, 1) + N(v, 1)] \, dy + yM(v, 1) \, dv = 0$$

or $\dfrac{dy}{y} + \dfrac{M(v, 1) \, dv}{vM(v, 1) + N(v, 1)} = 0.$

47. $\dfrac{dy}{dx} = -\dfrac{M(x, y)}{N(x, y)} = -\dfrac{y^n M(x/y, 1)}{y^n N(x/y, 1)}$

$= -\dfrac{M(x/y, 1)}{N(x/y, 1)} = G(x/y)$

EXERCISES 2.4, Page 60

1. $x^2 - x + \frac{3}{2}y^2 + 7y = c$ **3.** $\frac{5}{2}x^2 + 4xy - 2y^4 = c$

5. $x^2y^2 - 3x + 4y = c$

7. not exact, but is homogeneous

9. $xy^3 + y^2 \cos x - \frac{1}{2}x^2 = c$ **11.** not exact

13. $xy - 2xe^x + 2e^x - 2x^3 = c$

15. $x + y + xy - 3 \ln|xy| = c$

17. $x^3y^3 - \tan^{-1}3x = c$

19. $-\ln|\cos x| + \cos x \sin y = c$

21. $y - 2x^2y - y^2 - x^4 = c$

23. $x^4y - 5x^3 - xy + y^3 = c$

25. $\frac{1}{3}x^3 + x^2y + xy^2 - y = \frac{4}{3}$

27. $4xy + x^2 - 5x + 3y^2 - y = 8$

29. $y^2 \sin x - x^3y - x^2 + y \ln y - y = 0$

31. $k = 10$ **33.** $k = 1$

35. $M(x, y) = ye^{xy} + y^2 - (y/x^2) + h(x)$

37. $M(x, y) = 6xy^3$
$N(x, y) = 4y^3 + 9x^2y^2$
$\partial M/\partial y = 18xy^2 = \partial N/\partial x$
solution is $3x^2y^3 + y^4 = c$

39. $M(x, y) = -x^2y^2 \sin x + 2xy^2 \cos x$
$N(x, y) = 2x^2y \cos x$
$\partial M/\partial y = -2x^2y \sin x + 4xy \cos x = \partial N/\partial x$
solution is $x^2y^2 \cos x = c$

41. $M(x, y) = 2xy^2 + 3x^2$
$N(x, y) = 2x^2y$
$\partial M/\partial y = 4xy = \partial N/\partial x$
solution is $x^2y^2 + x^3 = c$

43. A separable first-order differential equation can be written $h(y)\, dy - g(x)\, dx = 0$. Identifying $M(x, y) = -g(x)$ and $N(x, y) = h(y)$, we find that $\partial M/\partial y = 0 = \partial N/\partial x$.

EXERCISES 2.5, Page 69

1. $y = ce^{5x}, \quad -\infty < x < \infty$

3. $y = \frac{1}{3} + ce^{-4x}, \quad -\infty < x < \infty$

5. $y = \frac{1}{4}e^{3x} + ce^{-x}, \quad -\infty < x < \infty$

7. $y = \frac{1}{3} + ce^{-x^3}, \quad -\infty < x < \infty$

9. $y = x^{-1}\ln x + cx^{-1}, \quad 0 < x < \infty$

11. $x = -\frac{4}{5}y^2 + cy^{-1/2}, \quad 0 < y < \infty$

13. $y = -\cos x + \dfrac{\sin x}{x} + \dfrac{c}{x}, \quad 0 < x < \infty$

15. $y = \dfrac{c}{e^x + 1}, \quad -\infty < x < \infty$

17. $y = \sin x + c \cos x, \quad -\pi/2 < x < \pi/2$

19. $y = \frac{1}{7}x^3 - \frac{1}{5}x + cx^{-4}, \quad 0 < x < \infty$

21. $y = \dfrac{1}{2x^2} e^x + \dfrac{c}{x^2} e^{-x}, \quad 0 < x < \infty$

23. $y = \sec x + c \csc x, \quad 0 < x < \pi/2$

25. $x = \dfrac{1}{2} e^y - \dfrac{1}{2y} e^y + \dfrac{1}{4y^2} e^y + \dfrac{c}{y^2} e^{-y}, \quad 0 < y < \infty$

27. $y = e^{-3x} + \dfrac{c}{x} e^{-3x}, \quad 0 < x < \infty$

29. $x = 2y^6 + cy^4, \quad 0 < y < \infty$

31. $y = e^{-x}\ln(e^x + e^{-x}) + ce^{-x}, \quad -\infty < x < \infty$

33. $x = \dfrac{1}{y} + \dfrac{c}{y} e^{-y^2}, \quad 0 < y < \infty$

35. $(\sec \theta + \tan \theta)r = \theta - \cos \theta + c, \quad -\pi/2 < \theta < \pi/2$

37. $y = \frac{5}{3}(x + 2)^{-1} + c(x + 2)^{-4}, \quad -2 < x < \infty$

39. $y = 10 + ce^{-\sinh x}, \quad -\infty < x < \infty$

41. $y = 4 - 2e^{-5x}, \quad -\infty < x < \infty$

43. $i(t) = E/R + (i_0 - E/R)e^{-Rt/L}, \quad -\infty < t < \infty$

45. $y = \sin x \cos x - \cos x, \quad -\pi/2 < x < \pi/2$

47. $T(t) = 50 + 150e^{kt}, \quad -\infty < t < \infty$

49. $(x + 1)y = x \ln x - x + 21, \quad 0 < x < \infty$

51. $y = \dfrac{2x}{x - 2}, \quad 2 < x < \infty$

53. $x = \frac{1}{2}y + 8/y, \quad 0 < y < \infty$

55. $y = \begin{cases} \frac{1}{2}(1 - e^{-2x}), & 0 \le x \le 3 \\ \frac{1}{2}(e^6 - 1)e^{-2x}, & x > 3 \end{cases}$

57. $y = \begin{cases} \frac{1}{2} + \frac{3}{2}e^{-x^2}, & 0 \le x < 1 \\ (\frac{1}{2}e + \frac{3}{2})e^{-x^2}, & x \ge 1 \end{cases}$

EXERCISES 2.6, Page 75

1. $y^3 = 1 + cx^{-3}$ **3.** $y^{-3} = x + \frac{1}{3} + ce^{3x}$

5. $e^{x/y} = cx$ **7.** $y^{-3} = -\frac{9}{5}x^{-1} + \frac{49}{5}x^{-6}$

9. $x^{-1} = 2 - y^2 - e^{-y^2/2}$, the equation is Bernoulli in the variable x.

11. $y = 2 + \dfrac{1}{ce^{-3x} - 1/3}$ **13.** $y = \dfrac{2}{x} + \dfrac{1}{cx^{-3} - x/4}$

15. $y = -e^x + \dfrac{1}{ce^{-x} - 1}$ **17.** $y = -2 + \dfrac{1}{ce^{-x} - 1}$

19. $y = cx + 1 - \ln c; \ y = 2 + \ln x$

21. $y = cx - c^3; \ 27y^2 = 4x^3$

23. $y = cx - e^c; \ y = x \ln x - x$

EXERCISES 2.7, Page 79

1. $x^2e^{2y} = 2x \ln x - 2x + c$

3. $e^{-x} = y \ln|y| + cy$ **5.** $-e^{-y/x^4} = x^2 + c$

7. $x^2 + y^2 = x - 1 + ce^{-x}$ **9.** $\ln(\tan y) = x + cx^{-1}$

11. $x^3y^3 = 2x^3 - 9 \ln|x| + c$

13. $e^y = -e^{-x} \cos x + ce^{-x}$

15. $y^2 \ln x = ye^y - e^y + c$

17. $y = \ln|\cos(c_1 - x)| + c_2$

19. $y = -\dfrac{1}{c_1}(1 - c_1^2x^2)^{1/2} + c_2$

CHAPTER 2

ANSWERS TO ODD-NUMBERED PROBLEMS

21. The given equation is a Clairaut equation in $u = y'$. The solution is $y = c_1 x^2/2 + x + c_1^3 x + c_2$.

23. $y = c_1 + c_2 x^2$ **25.** $\frac{1}{3}y^3 - c_1 y = x + c_2$

27. $y = -\sqrt{1 - x^2}$

EXERCISES 2.8, Page 82

1. $y_1(x) = 1 - x$

$y_2(x) = 1 - \dfrac{x}{1!} + \dfrac{x^2}{2!}$

$y_3(x) = 1 - \dfrac{x}{1!} + \dfrac{x^2}{2!} - \dfrac{x^3}{3!}$

$y_4(x) = 1 - \dfrac{x}{1!} + \dfrac{x^2}{2!} - \dfrac{x^3}{3!} + \dfrac{x^4}{4!}$;

$y_n(x) \to e^{-x}$ as $n \to \infty$

3. $y_1(x) = 1 + x^2$

$y_2(x) = 1 + \dfrac{x^2}{1!} + \dfrac{x^4}{2!}$

$y_3(x) = 1 + \dfrac{x^2}{1!} + \dfrac{x^4}{2!} + \dfrac{x^6}{3!}$

$y_4(x) = 1 + \dfrac{x^2}{1!} + \dfrac{x^4}{2!} + \dfrac{x^6}{3!} + \dfrac{x^8}{4!}$

$y_n(x) \to e^{x^2}$ as $n \to \infty$

5. $y_1(x) = y_2(x) = y_3(x) = y_4(x) = 0$; $y_n(x) \to 0$ as $n \to \infty$

7. **(a)** $y_1(x) = x$

$y_2(x) = x + \frac{1}{3}x^3$

$y_3(x) = x + \frac{1}{3}x^3 + \frac{2}{15}x^5 + \frac{1}{63}x^7$

(b) $y = \tan x$

(c) The Maclaurin series expansion of $\tan x$ is $x + \frac{1}{3}x^3 + \frac{2}{15}x^5 + \frac{17}{315}x^7 + \cdots, |x| < \pi/2.$

CHAPTER 2 REVIEW EXERCISES, Page 84

1. the regions defined by $x^2 + y^2 > 25$ and $x^2 + y^2 < 25$

3. false

5. **(a)** linear in x

(b) homogeneous, exact, linear in y

(c) Clairaut

(d) Bernoulli in x

(e) separable

(f) separable, Ricatti

(g) linear in x

(h) homogeneous

(i) Bernoulli

(j) homogeneous, exact, Bernoulli

(k) separable, homogeneous, exact, linear in x and in y

(l) exact, linear in y

(m) homogeneous

(n) separable

(o) Clairaut

(p) Ricatti

7. $2y^2 \ln y - y^2 = 4xe^x - 4e^x - 1$

9. $2y^2 + x^2 = 9x^6$ **11.** $e^{xy} - 4y^3 = 5$

13. $y = \frac{1}{4} - 320(x^2 + 4)^{-4}$

15. $y = \dfrac{1}{x^4 - x^4 \ln|x|}$ **17.** $x^2 - \sin \dfrac{1}{y^2} = c$

19. $y_1(x) = 1 + x + \frac{1}{3}x^3$

$y_2(x) = 1 + x + x^2 + \frac{2}{3}x^3 + \frac{1}{6}x^4 + \frac{2}{15}x^5 + \frac{1}{63}x^7$

EXERCISES 3.1, Page 92

1. $x^2 + y^2 = c_2^2$ **3.** $2y^2 + x^2 = c_2$

5. $2 \ln|y| = x^2 + y^2 + c_2$ **7.** $y^2 = 2x + c_2$

9. $2x^2 + 3y^2 = c_2$ **11.** $x^3 + y^3 = c_2$

13. $y^2 \ln|y| + x^2 = c_2 y^2$ **15.** $y^2 - x^2 = c_2 x$

17. $2y^2 = 2 \ln|x| + x^2 + c_2$

19. $y = \frac{1}{4} - \frac{1}{6}x^2 + c_2 x^{-4}$ **21.** $2y^3 = 3x^2 + c_2$

23. $2 \ln(\cosh y) + x^2 = c_2$ **25.** $y^{5/3} = x^{5/3} + c_2$

27. $y = 2 - x + 3e^{-x}$ **29.** $r = c_2 \sin \theta$

31. $r^2 = c_2 \cos 2\theta$ **33.** $r = c_2 \csc \theta$

35. Let β be the angle of inclination, measured from the positive x-axis, of the tangent line to a member of the given family, and ϕ the angle of inclination of the

(a)

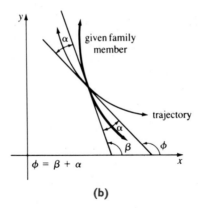

$\phi = \beta + \alpha$

(b)

tangent to a trajectory. At the point where the curves intersect, the angle between the tangents is α. From the accompanying figures we conclude that there exist two possible cases and that $\phi = \beta \pm \alpha$. Thus the slope of the tangent line to a trajectory is

$$\frac{dy}{dx} = \tan \phi = \tan(\beta \pm \alpha)$$

$$= \frac{\tan \beta \pm \tan \alpha}{1 \mp \tan \beta \tan \alpha}$$

$$= \frac{f(x, y) \pm \tan \alpha}{1 \mp f(x, y) \tan \alpha}.$$

37. $\mp \dfrac{2}{\sqrt{3}} \tan^{-1}\left(\dfrac{y}{x}\right) + \ln c_2(x^2 + y^2) = 0$

39. Since the given equation is quadratic in c_1, it follows from the quadratic formula that

$$c_1 = -x \pm \sqrt{x^2 + y^2}.$$

Differentiating this last expression and solving for dy/dx give

$$\frac{dy}{dx} = \frac{-x + \sqrt{x^2 + y^2}}{y}$$

and

$$\frac{dy}{dx} = \frac{-x - \sqrt{x^2 + y^2}}{y}.$$

These two equations correspond to choosing $c_1 > 0$ and $c_1 < 0$ in the given family, respectively. Forming the product of these derivatives yields

$$\left(\frac{dy}{dx}\right)_{(1)} \cdot \left(\frac{dy}{dx}\right)_{(2)} = \frac{x^2 - x^2 - y^2}{y^2} = -1.$$

This shows that the family is self-orthogonal.

41. The differential equation of the orthogonal family is $(x - y)\, dx + (x + y)\, dy = 0$. The verification follows by substituting $x = c_2 e^{-t} \cos t$ and $y = c_2 e^{-t} \sin t$ into the equation.

EXERCISES 3.2, Page 103

1. 7.9 years; 10 years   **3.** 760   **5.** 11 hours

7. 136.5 hours

9. $I(15) = 0.000098 I_0$, or $I(15)$ is approximately 0.1% of I_0.

11. 15,600 years

13. $T(1) = 36.67$ degrees; approximately 3.06 minutes

15. $i(t) = \frac{3}{5} - \frac{3}{5}e^{-500t}$; $i \to \frac{3}{5}$ as $t \to \infty$

17. $q(t) = \frac{1}{100} - \frac{1}{100}e^{-50t}$; $i(t) = \frac{1}{2}e^{-50t}$

19. $i(t) = \begin{cases} 60 - 60e^{-t/10}, & 0 \le t \le 20 \\ 60(e^2 - 1)e^{-t/10}, & t > 20 \end{cases}$

21. $A(t) = 200 - 170e^{-t/50}$

23. $A(t) = 1000 - 1000e^{-t/100}$ **25.** 64.38 lb

27. **(a)** $v(t) = \dfrac{mg}{k} + \left(v_0 - \dfrac{mg}{k}\right)e^{-kt/m}$

(b) $v \to \dfrac{mg}{k}$ as $t \to \infty$

(c) $s(t) = \dfrac{mg}{k}t - \dfrac{m}{k}\left(v_0 - \dfrac{mg}{k}\right)e^{-kt/m}$

$$+ \frac{m}{k}\left(v_0 - \frac{mg}{k}\right) + s_0$$

29. $E(t) = E_0 e^{-(t-t_1)/RC}$

31. **(a)** $P(t) = P_0 e^{(k_1 - k_2)t}$

(b) $k_1 > k_2$, births surpass deaths so population increases.

$k_1 = k_2$, a constant population since the number of births equals the number of deaths.

$k_1 < k_2$, deaths surpass births so population decreases.

33. From $r^2\, d\theta = \dfrac{L}{M}\, dt$ we get

$$A = \frac{1}{2}\int_{\theta_1}^{\theta_2} r^2\, d\theta = \frac{1}{2}\frac{L}{M}\int_a^b dt = \frac{1}{2}\frac{L}{M}(b - a).$$

EXERCISES 3.3, Page 116

1. 1834; 2000   **3.** 1,000,000; 52.9 months

5. (a) Separating variables gives

$$\frac{dP}{P(a - b \ln P)} = dt,$$

so that

$$-(1/b) \ln|a - b \ln P| = t + c_1$$

$$a - b \ln P = c_2 e^{-bt} \qquad (e^{-bc_1} = c_2)$$

$$\ln P = (a/b) - ce^{-bt} \qquad (c_2/b = c)$$

$$P(t) = e^{a/b} \cdot e^{-ce^{-bt}}.$$

(b) If $P(0) = P_0$, then

$$P_0 = e^{a/b} e^{-c} = e^{a/b - c}$$

and so $\quad \ln P_0 = (a/b) - c$

$$c = (a/b) - \ln P_0.$$

7. 29.3 grams; $X \to 60$ as $t \to \infty$; 0 grams of A and 30 grams of B.

9. For $\alpha \neq \beta$ the differential equation separates as

$$\frac{1}{\alpha - \beta}\left[-\frac{1}{\alpha - X} + \frac{1}{\beta - X}\right] dx = k\, dt.$$

It follows immediately that

$$\frac{1}{\alpha - \beta}[\ln|\alpha - X| - \ln|\beta - X|] = kt + c$$

or $\qquad \dfrac{1}{\alpha - \beta} \ln\left|\dfrac{\alpha - X}{\beta - X}\right| = kt + c.$

For $\alpha = \beta$ the equation can be written as

$$(\alpha - X)^{-2} dX = k\, dt.$$

It follows that $(\alpha - X)^{-1} = kt + c$ or

$$X = \alpha - \frac{1}{kt + c}.$$

11. (a) $v^2 = (2gR^2/y) + v_0^2 - 2gR.$

We note that as y increases, v decreases. In particular, if $v_0^2 - 2gR < 0$, then there must be some value of y for $v = 0$; the rocket stops and returns to earth under the influence of gravity. However, if $v_0^2 - 2gR \geq 0$, then $v > 0$ for all values of y. Hence we should have $v_0 \geq \sqrt{2gR}$.

(b) Using the values $R = 4000$ miles, $g = 32$ ft/s^2, 1 ft $= 1/5280$ mi, and 1 s $= 1/3600$ h, it follows that $v_0 \geq 25{,}067$ mi/h.

13. Using the condition $y'(1) = 0$, we find

$$\frac{dy}{dx} = \frac{1}{2}[x^{v_1/v_2} - x^{-v_1/v_2}].$$

Now if $v_1 = v_2$, then $y = \frac{1}{4}x^2 - \frac{1}{2}\ln x - \frac{1}{4}$; if $v_1 \neq v_2$, then

$$y = \frac{1}{2}\left[\frac{x^{1+(v_1/v_2)}}{1 + \dfrac{v_1}{v_2}} - \frac{x^{1-(v_1/v_2)}}{1 - \dfrac{v_1}{v_2}}\right] + \frac{v_1 v_2}{v_2^2 - v_1^2}.$$

15. $2h^{1/2} = -\frac{1}{25}t + 2\sqrt{20};\ t = 50\sqrt{20}\,\mathrm{s}$

17. To evaluate the indefinite integral of the left side of

$$\frac{\sqrt{100 - y^2}}{y}\,dy = -dx$$

we use the substitution $y = 10 \cos \theta$. It follows that

$$x = 10 \ln\left(\frac{10 + \sqrt{100 - y^2}}{y}\right) - \sqrt{100 - y^2}.$$

19. Under the substitution $w = x^2$, the differential equation becomes

$$w = y\frac{dw}{dy} + \frac{1}{4}\left(\frac{dw}{dy}\right)^2,$$

which is Clairaut's equation. The solution is

$$x^2 = cy + \frac{c^2}{4}.$$

If $2c_1 = c$, then we recognize

$$x^2 = 2c_1 y + c_1^2$$

as describing a family of parabolas.

21. $-\gamma \ln y + \delta y = \alpha \ln x - \beta x + c$

23. (a) The equation $2\dfrac{d^2\theta}{dt^2}\dfrac{d\theta}{dt} + 2\dfrac{g}{l}\sin\theta\dfrac{d\theta}{dt} = 0$ is the same as

$$\frac{d}{dt}\left(\frac{d\theta}{dt}\right)^2 + 2\frac{g}{l}\sin\theta\frac{d\theta}{dt} = 0.$$

Integrating this last equation with respect to t and using the initial conditions give the result.

(b) From (a),

$$dt = \sqrt{\frac{l}{2g}} \, \frac{d\theta}{\sqrt{\cos\theta - \cos\theta_0}}.$$

Integrating this last equation gives the time for the pendulum to move from $\theta = \theta_0$ to $\theta = 0$:

$$t = \sqrt{\frac{l}{2g}} \int_0^{\theta_0} \frac{d\theta}{\sqrt{\cos\theta - \cos\theta_0}}.$$

The period is the total time T to go from $\theta = \theta_0$ to $\theta = -\theta_0$ and back again to $\theta = \theta_0$. This is

$$T = 4\sqrt{\frac{l}{2g}} \int_0^{\theta_0} \frac{d\theta}{\sqrt{\cos\theta - \cos\theta_0}}$$

$$= 2\sqrt{\frac{2l}{g}} \int_0^{\theta_0} \frac{d\theta}{\sqrt{\cos\theta - \cos\theta_0}}.$$

CHAPTER 3 REVIEW EXERCISES, Page 121

1. $y^3 + 3/x = c_2$ **3.** $2(y-2)^2 + (x-1)^2 = c_2^2$
5. $P(45) = 8.99$ billion

7. $x(t) = \dfrac{\alpha c_1 e^{\alpha k_1 t}}{1 + c_1 e^{\alpha k_1 t}}$, $y(t) = c_2(1 + c_1 e^{\alpha k_1 t})^{k_2/k_1}$

9. (a) $T(t) = \dfrac{T_2 + BT_1}{1 + B} + \dfrac{T_1 - T_2}{1 + B} e^{k(1+B)t}$;

  (b) $\dfrac{T_2 + BT_1}{1 + B}$;   **(c)** $\dfrac{T_2 + BT_1}{1 + B}$

EXERCISES 4.1, Page 149

1. $y = \frac{1}{2}e^x - \frac{1}{2}e^{-x}$ **3.** $y = \frac{3}{5}e^{4x} + \frac{2}{5}e^{-x}$
5. $y = 3x - 4x \ln x$ **7.** $y = 0, y = x^2$
9. (a) $y = e^x \cos x - e^x \sin x$
  (b) no solution
  (c) $y = e^x \cos x + e^{-\pi/2} e^x \sin x$
  (d) $y = c_2 e^x \sin x$, where c_2 is arbitrary
11. $(-\infty, 2)$ **13.** $\lambda = n, n = 1, 2, 3, \ldots$
15. dependent   **17.** dependent   **19.** dependent
21. independent
23. $W(x^{1/2}, x^2) = \frac{3}{2}x^{3/2} \neq 0$ on $(0, \infty)$
25. $W(\sin x, \csc x) = -2 \cot x$.
  $W = 0$ only at $x = \pi/2$ in the interval.

27. $W(e^x, e^{-x}, e^{4x}) = -30e^{4x} \neq 0$ on $(-\infty, \infty)$
29. no
31. (a) $y'' - 2y^3 = \dfrac{2}{x^3} - 2\left(\dfrac{1}{x}\right)^3 = 0$

  (b) $y'' - 2y^3 = \dfrac{2c}{x^2} - 2\dfrac{c^3}{x^3} = \dfrac{2}{x^3}c(1 - c^2) \neq 0$
  for $c \neq 0, \pm 1$

33. The functions satisfy the differential equation and are linearly independent on the interval since
  $W(e^{-3x}, e^{4x}) = 7e^x \neq 0$; $y = c_1 e^{-3x} + c_2 e^{4x}$.
35. The functions satisfy the differential equation and are linearly independent on the interval since
  $W(e^x \cos 2x, e^x \sin 2x) = 2e^{2x} \neq 0$;
  $y = c_1 e^x \cos 2x + c_2 e^x \sin 2x$.
37. The functions satisfy the differential equation and are linearly independent on the interval since
  $W(x^3, x^4) = x^6 \neq 0$; $y = c_1 x^3 + c_2 x^4$.
39. The functions satisfy the differential equation and are linearly independent on the interval since
  $W(x, x^{-2}, x^{-2} \ln x) = 9x^{-6} \neq 0$;
  $y = c_1 x + c_2 x^{-2} + c_3 x^{-2} \ln x$.
41. e^{2x} and e^{5x} form a fundamental set of solutions of the homogeneous equation; $6e^x$ is a particular solution of the nonhomogeneous equation.
43. e^{2x} and xe^{2x} form a fundamental set of solutions of the homogeneous equation; $x^2 e^{2x} + x - 2$ is a particular solution of the nonhomogeneous equation.
45. (a) The accompanying graphs show that y_1 and y_2 are not multiples of each other.
  Also,

$$x^2 y_1'' - 4xy_1' + 6y_1 = x^2(6x) - 4x(3x^2) + 6x^3$$
$$= 12x^3 - 12x^3 = 0.$$

For $x \geq 0$ the demonstration that y_2 is a solution of the equation is exactly as given above for y_1.

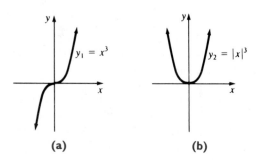

(a)   (b)

For $x < 0$, $y_2 = -x^3$ and so

$$x^2 y_2'' - 4xy_2' + 6y_2 = x^2(-6x)$$
$$-4x(-3x^2) + 6(-x^3)$$
$$= -12x^3 + 12x^3 = 0.$$

(b) For $x \geq 0$,

$$W(y_1, y_2) = \begin{vmatrix} x^3 & x^3 \\ 3x^2 & 3x^2 \end{vmatrix} = 3x^5 - 3x^5 = 0.$$

For $x < 0$,

$$W(y_1, y_2) = \begin{vmatrix} x^3 & -x^3 \\ 3x^2 & -3x^2 \end{vmatrix} = -3x^5 + 3x^5 = 0.$$

Thus $W(y_1, y_2) = 0$ for every real value of x.
(c) No, $a_2(x) = x^2$ is zero at $x = 0$.
(d) Since $Y_1 = y_1$, we need only show

$$x^2 Y_2'' - 4x Y_2' + 6Y_2 = x^2(2) - 4x(2x) + 6x^2$$
$$= 8x^2 - 8x^2 = 0$$

and $W(x^3, x^2) = -x^4$. Thus Y_1 and Y_2 are linearly independent solutions on the interval.
(e) $Y_1 = x^3$, $Y_2 = x^2$, or $y_2 = |x|^3$
(f) Neither; we form a general solution on an interval for which $a_2(x) \neq 0$ for every x in the interval. The linear combination

$$y = c_1 Y_1 + c_2 Y_2$$

would be a general solution of the equation on, say, the interval $(0, \infty)$.

47. (a) Since y_1 and y_2 are solutions of the given differential equation we have

$$a_2(x)y_1'' + a_1(x)y_1' + a_0(x)y_1 = 0$$

and $\quad a_2(x)y_2'' + a_1(x)y_2' + a_0(x)y_2 = 0.$

Now multiply the first equation by y_2 and the second by y_1 and subtract the first from the second:

$$a_2(x)[y_1 y_2'' - y_2 y_1''] + a_1(x)[y_1 y_2' - y_2 y_1'] = 0.$$

Now it is easily verified that

$$\frac{dW}{dx} = \frac{d}{dx}(y_1 y_2' - y_2 y_1') = y_1 y_2'' - y_2 y_1'',$$

and so it follows that

$$a_2(x)\frac{dW}{dx} + a_1(x)W = 0.$$

(b) Since this last equation is a linear first-order differential equation, the integrating factor is

$$e^{\int [a_1(x)/a_2(x)]\,dx}.$$

Therefore, from

$$\frac{d}{dx}\left[e^{\int [a_1(x)/a_2(x)]\,dx} W \right] = 0$$

we obtain $\quad W = c e^{-\int [a_1(x)/a_2(x)]\,dx}.$
(c) Substituting $x = x_0$ in the given result, we find $c = W(x_0)$.
(d) Since an exponential function is never zero when $W(x_0) \neq 0$, it follows from part (c) that $W \neq 0$. On the other hand, if $W(x_0) = 0$, we have immediately that $W = 0$.

49. From part (c) of Problem 47 we have

$$W(y_1, y_2) = W(y_1(x_0), y_2(x_0)) e^{-\int_{x_0}^x dt/t}$$
$$= \begin{vmatrix} k_1 & k_3 \\ k_2 & k_4 \end{vmatrix} e^{-\ln(x/x_0)}$$
$$= (k_1 k_4 - k_3 k_2)\left(\frac{x_0}{x}\right).$$

EXERCISES 4.2, Page 158

1. $y_2 = e^{-5x}$ **3.** $y_2 = xe^{2x}$ **5.** $y_2 = \sin 4x$
7. $y_2 = \sinh x$ **9.** $y_2 = xe^{2x/3}$ **11.** $y_2 = x^4 \ln|x|$
13. $y_2 = 1$ **15.** $y_2 = x^2 + x + 2$
17. $y_2 = x\cos(\ln x)$ **19.** $y_2 = x$ **21.** $y_2 = x\ln x$
23. $y_2 = x^3$ **25.** $y_2 = x^2$ **27.** $y_2 = 3x + 2$
29. $y_2 = \frac{1}{2}[\tan x \sec x + \ln|\sec x + \tan x|]$
31. $y_2 = e^{2x},\ y_p = -\frac{1}{2}$ **33.** $y_2 = e^{2x},\ y_p = \frac{5}{2}e^{3x}$

EXERCISES 4.3, Page 167

1. $y = c_1 + c_2 e^{-x/4}$ **3.** $y = c_1 e^{-6x} + c_2 e^{6x}$
5. $y = c_1 \cos 3x + c_2 \sin 3x$ **7.** $y = c_1 e^{3x} + c_2 e^{-2x}$
9. $y = c_1 e^{-4x} + c_2 xe^{-4x}$
11. $y = c_1 e^{(-3+\sqrt{29})x/2} + c_2 e^{(-3-\sqrt{29})x/2}$
13. $y = c_1 e^{2x/3} + c_2 e^{-x/4}$
15. $y = e^{2x}(c_1 \cos x + c_2 \sin x)$

17. $y = e^{-x/3}\left(c_1 \cos \dfrac{\sqrt{2}}{3}x + c_2 \sin \dfrac{\sqrt{2}}{3}x\right)$

19. $y = c_1 + c_2 e^{-x} + c_3 e^{5x}$

21. $y = c_1 e^x + e^{-x/2}\left(c_2 \cos \dfrac{\sqrt{3}}{2}x + c_3 \sin \dfrac{\sqrt{3}}{2}x\right)$

23. $y = c_1 e^{-x} + c_2 e^{3x} + c_3 x e^{3x}$

25. $y = c_1 e^x + e^{-x}(c_2 \cos x + c_3 \sin x)$

27. $y = c_1 e^{-x} + c_2 x e^{-x} + c_3 x^2 e^{-x}$

29. $y = c_1 + c_2 x + e^{-x/2}\left(c_3 \cos \dfrac{\sqrt{3}}{2}x + c_4 \sin \dfrac{\sqrt{3}}{2}x\right)$

31. $y = c_1 \cos \dfrac{\sqrt{3}}{2}x + c_2 \sin \dfrac{\sqrt{3}}{2}x$

$\qquad + c_3 x \cos \dfrac{\sqrt{3}}{2}x + c_4 x \sin \dfrac{\sqrt{3}}{2}x$

33. $y = c_1 + c_2 e^{-2x} + c_3 e^{2x} + c_4 \cos 2x + c_5 \sin 2x$

35. $y = c_1 e^x + c_2 x e^x + c_3 e^{-x} + c_4 x e^{-x} + c_5 e^{-5x}$

37. $y = 2 \cos 4x - \tfrac{1}{2} \sin 4x$

39. $y = -\tfrac{3}{4} e^{-5x} + \tfrac{3}{4} e^{-x}$

41. $y = -e^{x/2} \cos(x/2) + e^{x/2} \sin(x/2)$ **43.** $y = 0$

45. $y = e^{2(x-1)} - e^{x-1}$

47. $y = \tfrac{5}{36} - \tfrac{5}{36} e^{-6x} + \tfrac{1}{6} x e^{-6x}$

49. $y = -\dfrac{1}{6} e^{2x} + \dfrac{1}{6} e^{-x} \cos \sqrt{3}x - \dfrac{\sqrt{3}}{6} e^{-x} \sin \sqrt{3}x$

51. $y = 2 - 2e^x + 2xe^x - \tfrac{1}{2} x^2 e^x$ **53.** $y = e^{5x} - xe^{5x}$

55. $y = -2 \cos x$

57. $\dfrac{d^3 y}{dx^3} + 6\dfrac{d^2 y}{dx^2} - 15\dfrac{dy}{dx} - 100y = 0$

59. $y = c_1 e^x + e^{4x}(c_2 \cos x + c_3 \sin x)$

61. $y'' - 3y' - 18y = 0$ **63.** $y''' - 7y'' = 0$

65. $y = e^{-\sqrt{2}x/2}\left(c_1 \cos \dfrac{\sqrt{2}}{2}x + c_2 \sin \dfrac{\sqrt{2}}{2}x\right)$

$\qquad + e^{\sqrt{2}x/2}\left(c_3 \cos \dfrac{\sqrt{2}}{2}x + c_4 \sin \dfrac{\sqrt{2}}{2}x\right)$

EXERCISES 4.4, Page 179

1. $y = c_1 e^{-x} + c_2 e^{-2x} + 3$

3. $y = c_1 e^{5x} + c_2 x e^{5x} + \tfrac{6}{5}x + \tfrac{3}{5}$

5. $y = c_1 e^{-2x} + c_2 x e^{-2x} + x^2 - 4x + \tfrac{7}{2}$

7. $y = c_1 \cos \sqrt{3}x + c_2 \sin \sqrt{3}x$
$\qquad + (-4x^2 + 4x - \tfrac{4}{3})e^{3x}$

9. $y = c_1 + c_2 e^x + 3x$

11. $y = c_1 e^{x/2} + c_2 x e^{x/2} + 12 + \tfrac{1}{2} x^2 e^{x/2}$

13. $y = c_1 \cos 2x + c_2 \sin 2x - \tfrac{3}{4} x \cos 2x$

15. $y = c_1 \cos x + c_2 \sin x - \tfrac{1}{2} x^2 \cos x + \tfrac{1}{2} x \sin x$

17. $y = c_1 e^x \cos 2x + c_2 e^x \sin 2x + \tfrac{1}{4} x e^x \sin 2x$

19. $y = c_1 e^{-x} + c_2 x e^{-x} - \tfrac{1}{2} \cos x + \tfrac{12}{25} \sin 2x - \tfrac{9}{25} \cos 2x$

21. $y = c_1 + c_2 x + c_3 e^{6x} - \tfrac{1}{4} x^2 - \tfrac{6}{37} \cos x + \tfrac{1}{37} \sin x$

23. $y = c_1 e^x + c_2 x e^x + c_3 x^2 e^x - x - 3 - \tfrac{2}{3} x^3 e^x$

25. $y = c_1 \cos x + c_2 \sin x + c_3 x \cos x$
$\qquad + c_4 x \sin x + x^2 - 2x - 3$

27. $y_p = 4 + \tfrac{4}{3} \cos 2x$ **29.** $y = \sqrt{2} \sin 2x - \tfrac{1}{2}$

31. $y = -200 + 200 e^{-x/5} - 3x^2 + 30x$

33. $y = -10 e^{-2x} \cos x + 9 e^{-2x} \sin x + 7 e^{-4x}$

35. $x = \dfrac{F_0}{2\omega^2} \sin \omega t - \dfrac{F_0}{2\omega} t \cos \omega t$

37. $y = -\dfrac{1}{6} \cos x - \dfrac{\pi}{4} \sin x + \dfrac{1}{2} x \sin x + \dfrac{1}{3} \sin 2x$

39. $y = 11 - 11 e^x + 9x e^x + 2x - 12x^2 e^x + \tfrac{1}{2} e^{5x}$

41. $y = 6 \cos x - 6(\cot 1) \sin x + x^2 - 1$

43. $y = \begin{cases} \cos 2x + \tfrac{5}{6} \sin 2x + \tfrac{1}{3} \sin x, & 0 \le x \le \pi/2 \\ \tfrac{2}{3} \cos 2x + \tfrac{5}{6} \sin 2x, & x > \pi/2 \end{cases}$

EXERCISES 4.5, Page 186

1. $(D + 5)y = 9 \sin x$ **3.** $(3D^2 - 5D + 1)y = e^x$

5. $(D^3 - 4D^2 + 5D)y = 4x$

7. $(3D - 2)(3D + 2)$ **9.** $(D - 6)(D + 2)$

11. $D(D + 5)^2$ **13.** $(D - 1)(D - 2)(D + 5)$

15. $D(D + 2)(D^2 - 2D + 4)$ **21.** D^4

23. $D(D - 2)$ **25.** $D^2 + 4$ **27.** $D^3(D^2 + 16)$

29. $(D + 1)(D - 1)^3$ **31.** $D(D^2 - 2D + 5)$

33. $1, x, x^2, x^3, x^4$ **35.** $e^{6x}, e^{-3x/2}$

37. $\cos \sqrt{5}x, \sin \sqrt{5}x$ **39.** $1, e^{5x}, xe^{5x}$

EXERCISES 4.6, Page 193

1. $y = c_1 e^{-3x} + c_2 e^{3x} - 6$ **3.** $y = c_1 + c_2 e^{-x} + 3x$

5. $y = c_1 e^{-2x} + c_2 x e^{-2x} + \tfrac{1}{2} x + 1$

7. $y = c_1 + c_2 x + c_3 e^{-x} + \tfrac{2}{3} x^4 - \tfrac{8}{3} x^3 + 8x^2$

9. $y = c_1 e^{-3x} + c_2 e^{4x} + \frac{1}{7} x e^{4x}$

11. $y = c_1 e^{-x} + c_2 e^{3x} - e^x + 3$

13. $y = c_1 \cos 5x + c_2 \sin 5x + \frac{1}{4} \sin x$

15. $y = c_1 e^{-3x} + c_2 x e^{-3x} - \frac{1}{49} x e^{4x} + \frac{2}{343} e^{4x}$

17. $y = c_1 e^{-x} + c_2 e^x + \frac{1}{6} x^3 e^x - \frac{1}{4} x^2 e^x + \frac{1}{4} x e^x - 5$

19. $y = e^x (c_1 \cos 2x + c_2 \sin 2x) + \frac{1}{3} e^x \sin x$

21. $y = c_1 \cos 5x + c_2 \sin 5x - 2x \cos 5x$

23. $y = e^{-x/2} \left(c_1 \cos \dfrac{\sqrt{3}}{2} x + c_2 \sin \dfrac{\sqrt{3}}{2} x \right)$

$+ \sin x + 2 \cos x - x \cos x$

25. $y = c_1 + c_2 x + c_3 e^{-8x} + \frac{11}{256} x^2 + \frac{7}{32} x^3 - \frac{1}{16} x^4$

27. $y = c_1 e^x + c_2 x e^x + c_3 x^2 e^x + \frac{1}{6} x^3 e^x + x - 13$

29. $y = c_1 + c_2 x + c_3 e^x + c_4 x e^x + \frac{1}{2} x^2 e^x + \frac{1}{2} x^2$

31. $y = c_1 e^{x/2} + c_2 e^{-x/2} + c_3 \cos \dfrac{x}{2} + c_4 \sin \dfrac{x}{2} + \frac{1}{8} x e^{x/2}$

33. $y = \frac{5}{8} e^{-8x} + \frac{5}{8} e^{8x} - \frac{1}{4}$

35. $y = -\frac{41}{125} + \frac{41}{125} e^{5x} - \frac{1}{10} x^2 + \frac{9}{25} x$

37. $y = -\pi \cos x - \frac{11}{3} \sin x - \frac{8}{3} \cos 2x + 2x \cos x$

39. $y = 2e^{2x} \cos 2x - \frac{3}{64} e^{2x} \sin 2x + \frac{1}{8} x^3 + \frac{3}{16} x^2 + \frac{3}{32} x$

41. $y_p = Ax e^x + B e^x \cos 2x + C e^x \sin 2x$
$+ Ex e^x \cos 2x + Fx e^x \sin 2x$

EXERCISES 4.7, Page 201

1. $y = c_1 \cos x + c_2 \sin x + x \sin x$
$+ \cos x \ln|\cos x|; \quad (-\pi/2, \pi/2)$

3. $y = c_1 \cos x + c_2 \sin x + \frac{1}{2} \sin x - \frac{1}{2} x \cos x$
$= c_1 \cos x + c_3 \sin x - \frac{1}{2} x \cos x; \quad (-\infty, \infty)$

5. $y = c_1 \cos x + c_2 \sin x + \frac{1}{2} - \frac{1}{6} \cos 2x; \quad (-\infty, \infty)$

7. $y = c_1 e^x + c_2 e^{-x} + \frac{1}{4} x e^x - \frac{1}{4} x e^{-x}$
$= c_1 e^x + c_2 e^{-x} + \frac{1}{2} x \sinh x; \quad (-\infty, \infty)$

9. $y = c_1 e^{2x} + c_2 e^{-2x}$

$+ \dfrac{1}{4} \left(e^{2x} \ln|x| - e^{-2x} \displaystyle\int_{x_0}^{x} \dfrac{e^{4t}}{t} \, dt \right),$

$x_0 > 0; (0, \infty)$

11. $y = c_1 e^{-x} + c_2 e^{-2x} + (e^{-x} + e^{-2x})$
$\times \ln(1 + e^x); \quad (-\infty, \infty)$

13. $y = c_1 e^{-2x} + c_2 e^{-x} - e^{-2x} \sin e^x; \quad (-\infty, \infty)$

15. $y = c_1 e^x + c_2 x e^x - \frac{1}{2} e^x \ln(1 + x^2)$
$+ x e^x \tan^{-1} x; \quad (-\infty, \infty)$

17. $y = c_1 e^{-x} + c_2 x e^{-x} + \frac{1}{2} x^2 e^{-x} \ln x$
$- \frac{3}{4} x^2 e^{-x}; \quad (0, \infty)$

19. $y = c_1 e^x \cos 3x + c_2 e^x \sin x$
$- \frac{1}{27} e^x \cos 3x \ln|\sec 3x + \tan 3x|;$
$(-\pi/6, \pi/6)$

21. $y = c_1 + c_2 \cos x + c_3 \sin x - \ln|\cos x|$
$- \sin x \ln|\sec x + \tan x|; \quad (-\pi/2, \pi/2)$

23. $y = c_1 e^x + c_2 e^{2x} + c_3 e^{-x} + \frac{1}{8} e^{3x}; \quad (-\infty, \infty)$

25. $y = \frac{1}{4} e^{-x/2} + \frac{3}{4} e^{x/2} + \frac{1}{8} x^2 e^{x/2} - \frac{1}{4} x e^{x/2}$

27. $y = \frac{4}{9} e^{-4x} + \frac{25}{36} e^{2x} - \frac{1}{4} e^{-2x} + \frac{1}{9} e^{-x}$

29. $y = c_1 x + c_2 x \ln x + \frac{2}{3} x (\ln x)^3$

31. $y = c_1 x^{-1/2} \cos x + c_2 x^{-1/2} \sin x + x^{-1/2}$

33. (a) $y_{p_1} = 4x^2 - 16x + 21$ **(b)** $y_{p_2} = x e^{-x} \ln x$
(c) $y_p = 4x^2 - 16x + 21 + x e^{-x} \ln x$

CHAPTER 4 REVIEW EXERCISES, Page 204

1. $y = 0$

3. False, the functions $f_1(x) = 0$ and $f_2(x) = e^x$ are linearly dependent on $(-\infty, \infty)$ but f_2 is not a constant multiple of f_1.

5. $(-\infty, 0); (0, \infty)$ **7.** false   **9.** $y_p = A + Bx e^x$

11. $y_2 = \sin 2x$ **13.** $y = c_1 e^{(1 + \sqrt{3})x} + c_2 e^{(1 - \sqrt{3})x}$

15. $y = c_1 + c_2 e^{-5x} + c_3 x e^{-5x}$

17. $y = c_1 e^{-x/3} + e^{-3x/2} \left(c_2 \cos \dfrac{\sqrt{7}}{2} x + c_3 \sin \dfrac{\sqrt{7}}{2} x \right)$

19. $y = e^{3x/2} \left(c_1 \cos \dfrac{\sqrt{11}}{2} x + c_2 \sin \dfrac{\sqrt{11}}{2} x \right)$

$+ \frac{4}{5} x^3 + \frac{36}{25} x^2 + \frac{46}{125} x - \frac{222}{625}$

21. $y = c_1 + c_2 e^{2x} + c_3 e^{3x} + \frac{1}{5} \sin x - \frac{1}{5} \cos x + \frac{4}{3} x$

23. $y = e^{x - \pi} \cos x$

25. $y = e^x (c_1 \cos x + c_2 \sin x)$
$- e^x \cos x \ln|\sec x + \tan x|$

27. $y = \frac{2}{5} e^{x/2} - \frac{2}{5} e^{3x} + x e^{3x} - 4$

EXERCISES 5.1, Page 219

1. A weight of 4 lb ($\frac{1}{8}$ slug), attached to a spring, is released from a point 3 units above the equilibrium position with an initial upward velocity of 2 ft/s. The spring constant is 3 lb/ft.

3. $x(t) = 2\sqrt{2}\sin\left(5t - \dfrac{\pi}{4}\right)$

5. $x(t) = \sqrt{5}\sin(\sqrt{2}\,t + 3.6052)$

7. $x(t) = \dfrac{\sqrt{101}}{10}\sin(10t + 1.4711)$

9. 8 lb

11. $\sqrt{2}\,\pi/8$ **13.** $x(t) = -\frac14\cos 4\sqrt{6}\,t$

15. (a) $x(\pi/12) = -1/4;\quad x(\pi/8) = -1/2;$
$\quad\quad x(\pi/6) = -1/4;\quad x(\pi/4) = 1/2;$
$\quad\quad x(9\pi/32) = \sqrt{2}/4$
   (b) 4 ft/s; downward
   (c) $t = (2n+1)\pi/16,\quad n = 0, 1, 2, \ldots$

17. (a) the 20-kg mass
   (b) the 20-kg mass; the 50-kg mass
   (c) $t = n\pi, n = 0, 1, 2, \ldots$; at the equilibrium position; the 50-kg mass is moving upward whereas the 20-kg mass is moving upward when n is even and downward when n is odd.

19. $x(t) = \frac12\cos 2t + \frac34\sin 2t$
$\quad\quad = \dfrac{\sqrt{13}}{4}\sin(2t + 0.5880)$

21. (a) $x(t) = -\frac23\cos 10t + \frac12\sin 10t$
$\quad\quad\quad = \frac56\sin(10t - 0.927)$
   (b) 5/6 ft; $\pi/5$
   (c) 15 cycles
   (d) 0.721 s
   (e) $(2n+1)\pi/20 + 0.0927, n = 0, 1, 2, \ldots$
   (f) $x(3) = -0.597$ ft
   (g) $x'(3) = -5.814$ ft/s
   (h) $x''(3) = 59.702$ ft/s^2
   (i) $\pm 8\frac13$ ft/s
   (j) $0.1451 + n\pi/5; .0.3545 + n\pi/5, n = 0, 1, 2, \ldots$
   (k) $0.3545 + n\pi/5,\quad n = 0, 1, 2, \ldots$

23. 120 lb/ft; $x(t) = \dfrac{\sqrt{3}}{12}\sin 8\sqrt{3}\,t$

25. Using $x(t) = c_1\cos \omega t + c_2\sin \omega t$, $x(0) = x_0$, and $x'(0) = v_0$, we find $c_1 = x_0$ and $c_2 = v_0/\omega$. The result follows from $A = \sqrt{c_1^2 + c_2^2}$.

27. $x(t) = 2\sqrt{2}\cos\left(5t + \dfrac{5\pi}{4}\right)$

29. When $\omega t + \phi = (2m+1)\pi/2$, $|x''| = A\omega^2$. But $T = 2\pi/\omega$ implies $\omega = 2\pi/T$ and $\omega^2 = 4\pi^2/T^2$. Therefore, the magnitude of the acceleration is $|x''| = 4\pi^2 A/T^2$.

EXERCISES 5.2, Page 229

1. A 2-lb weight is attached to a spring whose constant is 1 lb/ft. The system is damped with a resisting force numerically equal to 2 times the instantaneous velocity. The weight starts from the equilibrium position with an upward velocity of 1.5 ft/s.

3. (a) above;   **(b)** heading upward

5. (a) below;   **(b)** heading upward

7. $\frac14 s; \frac12 s, x(\frac12) = e^{-2}$; that is, the weight is approximately 0.14 ft below the equilibrium position.

9. (a) $x(t) = \frac43 e^{-2t} - \frac13 e^{-8t}$
   (b) $x(t) = -\frac23 e^{-2t} + \frac53 e^{-8t}$

11. (a) $x(t) = e^{-2t}[-\cos 4t - \frac12\sin 4t]$
   (b) $x(t) = \dfrac{\sqrt{5}}{2}\,e^{-2t}\sin(4t + 4.249)$
   (c) $t = 1.294$ s

13. (a) $\beta > 5/2;$ **(b)** $\beta = 5/2;$ **(c)** $0 < \beta < 5/2$

15. $x(t) = \frac27 e^{-7t}\sin 7t$ **17.** $v_0 > 2$ ft/s

19. Suppose $\gamma = \sqrt{\omega^2 - \lambda^2}$. Then the derivative of $x(t) = Ae^{-\lambda t}\sin(\gamma t + \phi)$ is

$$x'(t) = Ae^{-\lambda t}[\gamma\cos(\gamma t + \phi) - \lambda\sin(\gamma t + \phi)].$$

So $x'(t) = 0$ implies

$$\tan(\gamma t + \phi) = \frac{\gamma}{\lambda},$$

from which it follows that

$$t = \frac1\gamma\left[\tan^{-1}\frac{\gamma}{\lambda} + k\pi - \phi\right].$$

The difference between the t values between two successive maxima (or minima) is then

$$t_{k+2} - t_k = (k+2)(\pi/\gamma) - k(\pi/\gamma) = 2\pi/\gamma.$$

21. $t_{k+1}^* - t_k^* = \dfrac{(2k+3)\pi/2 - \phi}{\sqrt{\omega^2 - \lambda^2}} - \dfrac{(2k+1)\pi/2 - \phi}{\sqrt{\omega^2 - \lambda^2}}$
$\quad\quad = \dfrac{\pi}{\sqrt{\omega^2 - \lambda^2}}$

23. Let the quasi-period $2\pi/\sqrt{\omega^2 - \lambda^2}$ be denoted by T_q. From equation (15) we find

$$\frac{x_n}{x_{n+2}} = \frac{x(t)}{x(t + T_q)}$$

$$= \frac{e^{-\lambda t}\sin(\sqrt{\omega^2 - \lambda^2}\,t + \phi)}{e^{-\lambda(t + T_q)}\sin(\sqrt{\omega^2 - \lambda^2}(t + T_q) + \phi)}$$

$$= e^{\lambda T_q}$$

since

$$\sin(\sqrt{\omega^2 - \lambda^2}\,t + \phi) = \sin(\sqrt{\omega^2 - \lambda^2}(t + T_q) + \phi).$$

Therefore

$$\ln\left(\frac{x_n}{x_{n+2}}\right) = \lambda T_q = \frac{2\pi\lambda}{\sqrt{\omega^2 - \lambda^2}}.$$

EXERCISES 5.3, Page 239

1. $x(t) = e^{-t/2}\left(-\frac{4}{3}\cos\frac{\sqrt{47}}{2}t - \frac{64}{3\sqrt{47}}\sin\frac{\sqrt{47}}{2}t\right)$

$\qquad + \frac{10}{3}(\cos 3t + \sin 3t)$

3. $x(t) = \frac{1}{4}e^{-4t} + te^{-4t} - \frac{1}{4}\cos 4t$

5. $x(t) = -\frac{1}{2}\cos 4t$

$\qquad + \frac{9}{4}\sin 4t + \frac{1}{2}e^{-2t}\cos 4t - 2e^{-2t}\sin 4t$

7. $m\dfrac{d^2x}{dt^2} = -k(x - h) - \beta\dfrac{dx}{dt}$ or

$\dfrac{d^2x}{dt^2} + 2\lambda\dfrac{dx}{dt} + \omega^2 x = \omega^2 h(t)$, where

$2\lambda = \beta/m$ and $\omega^2 = k/m$.

9. (a) $x(t) = \frac{2}{3}\sin 4t - \frac{1}{3}\sin 8t$

$\quad$ **(b)** $t = n\pi/4$, $\quad n = 0, 1, 2, \ldots$

$\quad$ **(c)** $t = \pi/6 + n\pi/2$, $\quad n = 0, 1, 2, \ldots,$

$\qquad$ and $t = \pi/3 + n\pi/2$, $\quad n = 0, 1, 2, \ldots$

$\quad$ **(d)** $\sqrt{3}/2$ cm, $-\sqrt{3}/2$ cm

$\quad$ **(e)**

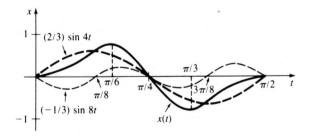

11. (a) $g'(\gamma) = 0$ implies $\gamma(\gamma^2 - \omega^2 + 2\lambda^2) = 0$ so that either $\gamma = 0$ or $\gamma = \sqrt{\omega^2 - 2\lambda^2}$. The first deriv-

ative test can be used to verify that $g(\gamma)$ is a maximum at the latter value.

$\quad$ **(b)** $g(\sqrt{\omega^2 - 2\lambda^2}) = F_0/2\lambda\sqrt{\omega^2 - \lambda^2}$

13. $x_p = -5\cos 2t + 5\sin 2t$

$\qquad = 5\sqrt{2}\sin\left(2t - \frac{\pi}{4}\right)$

15. (a) $x(t) = x_c + x_p$

$\qquad = c_1\cos\omega t + c_2\sin\omega t + \dfrac{F_0}{\omega^2 - \gamma^2}\cos\gamma t,$

$\quad$ where the initial conditions imply that
$c_1 = -F_0/(\omega^2 - \gamma^2)$ and $c_2 = 0$.

$\quad$ **(b)** By L'Hôpital's rule the given limit is the same as

$$\lim_{\gamma \to \omega}\frac{F_0(-t\sin\gamma t)}{-2\gamma} = \frac{F_0}{2\omega}t\sin\omega t.$$

17. $x(t) = -\cos 2t - \frac{1}{8}\sin 2t + \frac{3}{4}t\sin 2t + \frac{5}{4}t\cos 2t$

19. (a) Recall that

$$\cos(u - v) = \cos u\cos v + \sin u\sin v$$

$$\cos(u + v) = \cos u\cos v - \sin u\sin v.$$

$\quad$ Subtracting gives

$$\sin u\sin v = \frac{1}{2}[\cos(u - v) - \cos(u + v)].$$

$\quad$ Setting $u = \frac{1}{2}(\gamma - \omega)t$ and $v = \frac{1}{2}(\gamma + \omega)t$ then gives

$$\sin\tfrac{1}{2}(\gamma - \omega)t\,\sin\tfrac{1}{2}(\gamma + \omega)t = \tfrac{1}{2}[\cos\omega t - \cos\gamma t],$$

$\quad$ from which the result follows.

$\quad$ **(b)** For small ε, $\gamma \approx \omega$ so $\gamma + \omega \approx 2\gamma$ and therefore

$$\frac{-2F_0}{(\omega + \gamma)(\omega - \gamma)}\sin\frac{1}{2}(\gamma - \omega)t\,\sin\frac{1}{2}(\gamma + \omega)t$$

$$\approx \frac{F_0}{2\gamma\varepsilon}\sin\varepsilon t\,\sin\frac{1}{2}(2\gamma)t.$$

$\quad$ **(c)** By L'Hôpital's rule the given limit is the same as

$$\lim_{\varepsilon \to 0}\frac{F_0 t\cos\varepsilon t\,\sin\gamma t}{2\gamma} = \frac{F_0}{2\gamma}t\sin\gamma t = \frac{F_0}{2\omega}t\sin\omega t.$$

EXERCISES 5.4, Page 246

1. $q(t) = -\frac{15}{4}\cos 4t + \frac{15}{4}$; $i(t) = 15\sin 4t$

3. underdamped $\qquad$ **5.** 4.568 coulombs; 0.0509 s

7. $q(t) = 10 - 10e^{-3t}(\cos 3t + \sin 3t)$;
$i(t) = 60e^{-3t} \sin 3t$; 10.432 coulombs

9. $q_p = \frac{100}{13} \sin t + \frac{150}{13} \cos t$
$i_p = \frac{100}{13} \cos t - \frac{150}{13} \sin t$

13. $q(t) = -\frac{1}{2}e^{-10t}(\cos 10t + \sin 10t) + \frac{3}{2}$; $\frac{3}{2}$ coulombs

15. Show that $dZ/dC = 0$ when $C = 1/L\gamma^2$. At this value, Z is a minimum and, correspondingly, the amplitude E_0/Z is a maximum.

17. $q(t) = \left(q_0 - \dfrac{E_0 C}{1 - \gamma^2 LC} \right) \cos \dfrac{t}{\sqrt{LC}}$

$+ \sqrt{LC}\, i_0 \sin \dfrac{t}{\sqrt{LC}} + \dfrac{E_0 C}{1 - \gamma^2 LC} \cos \gamma t$;

$i(t) = i_0 \cos \dfrac{t}{\sqrt{LC}}$

$- \dfrac{1}{\sqrt{LC}} \left(q_0 - \dfrac{E_0 C}{1 - \gamma^2 LC} \right) \sin \dfrac{t}{\sqrt{LC}}$

$- \dfrac{E_0 C\gamma}{1 - \gamma^2 LC} \sin \gamma t$

19. $\theta(t) = \dfrac{1}{2} \cos 4t + \dfrac{\sqrt{3}}{2} \sin 4t$; 1; $\pi/2$; $2/\pi$

CHAPTER 5 REVIEW EXERCISES, Page 249

1. 8 ft   **3.** 5/4 m

5. False; there could be an impressed force driving the system.

7. overdamped   **9.** 9/2 lb/ft

11. $x(t) = -\frac{2}{3}e^{-2t} + \frac{1}{3}e^{-4t}$ **13.** $0 < m \le 2$

15. $\gamma = 8\sqrt{3}/3$

17. $x(t) = e^{-4t}$

$\times \left(\dfrac{26}{17} \cos 2\sqrt{2}t + \dfrac{28\sqrt{2}}{17} \sin 2\sqrt{2}t \right) + \dfrac{8}{17} e^{-t}$

19. (a) $q(t) = -\dfrac{1}{150} \sin 100t + \dfrac{1}{75} \sin 50t$

(b) $i(t) = -\dfrac{2}{3} \cos 100t + \dfrac{2}{3} \cos 50t$

(c) $t = n\pi/50$, $n = 0, 1, 2, \ldots$

EXERCISES 6.1, Page 265

1. $y = c_1 x^{-1} + c_2 x^2$ **3.** $y = c_1 + c_2 \ln x$

5. $y = c_1 \cos(2 \ln x) + c_2 \sin(2 \ln x)$

7. $y = c_1 x^{(2-\sqrt{6})} + c_2 x^{(2+\sqrt{6})}$

9. $y_1 = c_1 \cos(\frac{1}{5} \ln x) + c_2 \sin(\frac{1}{5} \ln x)$

11. $y = c_1 x^{-2} + c_2 x^{-2} \ln x$

13. $y = x[c_1 \cos(\ln x) + c_2 \sin(\ln x)]$

15. $y = x^{-1/2} \left[c_1 \cos\left(\dfrac{\sqrt{3}}{6} \ln x \right) + c_2 \sin\left(\dfrac{\sqrt{3}}{6} \ln x \right) \right]$

17. $y = c_1 x^3 + c_2 \cos(\sqrt{2} \ln x) + c_3 \sin(\sqrt{2} \ln x)$

19. $y = c_1 x^{-1} + c_2 x^2 + c_3 x^4$

21. $y = c_1 + c_2 x + c_3 x^2 + c_4 x^{-3}$

23. $y = 2 - 2x^{-2}$

25. $y = \cos(\ln x) + 2 \sin(\ln x)$

27. $y = 2(-x)^{1/2} - 5(-x)^{1/2} \ln(-x)$

29. $y = c_1 + c_2 \ln x + \dfrac{x^2}{4}$

31. $y = c_1 x^{-1/2} + c_2 x^{-1} + \frac{1}{15}x^2 - \frac{1}{6}x$

33. $y = c_1 x + c_2 x \ln x + x(\ln x)^2$

35. $y = c_1 x^{-1} + c_2 x^{-8} + \frac{1}{30}x^2$

37. $y = x^2[c_1 \cos(3 \ln x) + c_2 \sin(3 \ln x)] + \frac{4}{13} + \frac{3}{10}x$

39. $y = c_1 x^2 + c_2 x^{-10} - \frac{1}{7}x^{-3}$

41. $u(r) = \left(\dfrac{u_0 - u_1}{b - a} \right)\dfrac{ab}{r} + \dfrac{u_1 b - u_0 a}{b - a}$

43. $y = c_1(x - 1)^{-1} + c_2(x - 1)^4$

45. $y = c_1 \cos(\ln(x + 2)) + c_2 \sin(\ln(x + 2))$

EXERCISES 6.2, Page 276

1. $(-1, 1]$ **3.** $[-\frac{1}{2}, \frac{1}{2})$ **5.** $[2, 4]$

7. $(-5, 15)$ **9.** $\{0\}$

11. $x + x^2 + \frac{1}{3}x^3 - \frac{1}{30}x^5 + \cdots$

13. $x - \frac{2}{3}x^3 + \frac{2}{15}x^5 - \frac{4}{315}x^7 + \cdots$

15. $x^2 - \frac{2}{3}x^4 + \frac{23}{45}x^6 - \frac{44}{105}x^8 + \cdots$

17. $x + \frac{1}{3}x^3 + \frac{2}{15}x^5 + \frac{17}{315}x^7 + \cdots$

19. $1 + \frac{1}{2}x^2 - \frac{1}{12}x^4 + \frac{1}{24}x^6 - \cdots$

21. $y = ce^{-x}$; $y = c_0 \displaystyle\sum_{n=0}^{\infty} \dfrac{(-1)^n}{n!} x^n$

23. $y = ce^{x^3/3}$; $y = c_0 \displaystyle\sum_{n=0}^{\infty} \dfrac{1}{n!} \left(\dfrac{x^3}{3} \right)^n$

25. $y = c/(1 - x)$; $y = c_0 \displaystyle\sum_{n=0}^{\infty} x^n$

27. $y = C_1 \cos x + C_2 \sin x$;

$$y = c_0 \sum_{n=0}^{\infty} \frac{(-1)^n}{(2n)!} x^{2n} + c_1 \sum_{n=0}^{\infty} \frac{(-1)^n}{(2n+1)!} x^{2n+1}$$

29. $y = C_1 + C_2 e^x$;

$$y = c_0 + c_1 \sum_{n=1}^{\infty} \frac{x^n}{n!} = c_0 - c_1 + c_1 \sum_{n=0}^{\infty} \frac{x^n}{n!}$$

$$= c_0 - c_1 + c_1 e^x$$

EXERCISES 6.3, Page 285

1. $y_1(x) = c_0 \left[1 + \dfrac{1}{3 \cdot 2} x^3 + \dfrac{1}{6 \cdot 5 \cdot 3 \cdot 2} x^6 \right.$

$$\left. + \dfrac{1}{9 \cdot 8 \cdot 6 \cdot 5 \cdot 3 \cdot 2} x^9 + \cdots \right]$$

$y_2(x) = c_1 \left[x + \dfrac{1}{4 \cdot 3} x^4 + \dfrac{1}{7 \cdot 6 \cdot 4 \cdot 3} x^7 \right.$

$$\left. + \dfrac{1}{10 \cdot 9 \cdot 7 \cdot 6 \cdot 4 \cdot 3} x^{10} + \cdots \right]$$

3. $y_1(x) = c_0 \left[1 - \dfrac{1}{2!} x^2 - \dfrac{3}{4!} x^4 - \dfrac{21}{6!} x^6 - \cdots \right]$

$y_2(x) = c_1 \left[x + \dfrac{1}{3!} x^3 + \dfrac{5}{5!} x^5 + \dfrac{45}{7!} x^7 + \cdots \right]$

5. $y_1(x) = c_0 \left[1 - \dfrac{1}{3!} x^3 + \dfrac{4^2}{6!} x^6 - \dfrac{7^2 \cdot 4^2}{9!} x^9 + \cdots \right]$

$y_2(x) = c_1 \left[x - \dfrac{2^2}{4!} x^4 + \dfrac{5^2 \cdot 2^2}{7!} x^7 \right.$

$$\left. - \dfrac{8^2 \cdot 5^2 \cdot 2^2}{10!} x^{10} + \cdots \right]$$

7. $y_1(x) = c_0$; $\quad y_2(x) = c_1 \sum_{n=1}^{\infty} \dfrac{1}{n} x^n$

9. $y_1(x) = c_0 \sum_{n=0}^{\infty} x^{2n}$; $\quad y_2(x) = c_1 \sum_{n=0}^{\infty} x^{2n+1}$

11. $y_1(x) = c_0 \left[1 + \dfrac{1}{4} x^2 - \dfrac{7}{4 \cdot 4!} x^4 + \dfrac{23 \cdot 7}{8 \cdot 6!} x^6 - \cdots \right]$

$y_2(x) = c_1 \left[x - \dfrac{1}{6} x^3 + \dfrac{14}{2 \cdot 5!} x^5 - \dfrac{34 \cdot 14}{4 \cdot 7!} x^7 - \cdots \right]$

13. $y_1(x) = c_0 [1 + \frac{1}{2} x^2 + \frac{1}{6} x^3 + \frac{1}{6} x^4 + \cdots]$
$y_2(x) = c_1 [x + \frac{1}{2} x^2 + \frac{1}{2} x^3 + \frac{1}{4} x^4 + \cdots]$

15. $y(x) = -2[1 + \frac{1}{2!} x^2 + \frac{1}{3!} x^3 + \frac{1}{4!} x^4 + \cdots] + 6x$
$= 8x - 2e^x$

17. $y(x) = 3 - 12x^2 + 4x^4$

19. $y_1(x) = c_0 [1 - \frac{1}{6} x^3 + \frac{1}{120} x^5 + \cdots]$
$y_2(x) = c_1 [x - \frac{1}{12} x^4 + \frac{1}{180} x^6 + \cdots]$

21. $y_1(x) = c_0 [1 - \frac{1}{2} x^2 + \frac{1}{6} x^3 - \frac{1}{40} x^5 + \cdots]$
$y_2(x) = c_1 [x - \frac{1}{6} x^3 + \frac{1}{12} x^4 - \frac{1}{60} x^5 + \cdots]$

23. $y_1(x) = c_0 \left[1 + \dfrac{1}{3!} x^3 + \dfrac{4}{6!} x^6 + \dfrac{7 \cdot 4}{9!} x^9 + \cdots \right.$

$$\left. + c_1 \left[x + \dfrac{2}{4!} x^4 + \dfrac{5 \cdot 2}{7!} x^7 + \dfrac{8 \cdot 5 \cdot 2}{10!} x^{10} + \cdots \right] \right.$$

$$\left. + \dfrac{1}{2!} x^2 + \dfrac{3}{5!} x^5 + \dfrac{6 \cdot 3}{8!} x^8 + \dfrac{9 \cdot 6 \cdot 3}{11!} x^{11} + \cdots \right]$$

25. For $n = 1$: $y = x$; for $n = 2$: $y = 1 - 2x^2$.

EXERCISES 6.4, Page 304

1. $x = 0$, irregular singular point

3. $x = -3$, regular singular point; $x = 3$, irregular singular point

5. $x = 0, 2i, -2i$, regular singular points

7. $x = -3, 2$, regular singular points

9. $x = 0$, irregular singular point; $x = -5, 5, 2$, regular singular points

11. $r_1 = \frac{3}{2}, r_2 = 0$;

$$y(x) = C_1 x^{3/2} \left[1 - \dfrac{2}{5} x + \dfrac{2^2}{7 \cdot 5 \cdot 2} x^2 \right.$$

$$\left. - \dfrac{2^3}{9 \cdot 7 \cdot 5 \cdot 3!} x^3 + \cdots \right]$$

$$+ C_2 \left[1 + 2x - 2x^2 + \dfrac{2^3}{3 \cdot 3!} x^3 - \cdots \right]$$

13. $r_1 = \frac{7}{8}, r_2 = 0$;

$$y(x) = C_1 x^{7/8} \left[1 - \dfrac{2}{15} x + \dfrac{2^2}{23 \cdot 15 \cdot 2} x^2 \right.$$

$$\left. - \dfrac{2^3}{31 \cdot 23 \cdot 15 \cdot 3!} x^3 + \cdots \right]$$

$$+ C_2 \left[1 - 2x + \dfrac{2^2}{9 \cdot 2} x^2 - \dfrac{2^3}{17 \cdot 9 \cdot 3!} x^3 + \cdots \right]$$

15. $r_1 = \frac{1}{3}, r_2 = 0$;

$$y(x) = C_1 x^{1/3} \left[1 + \dfrac{1}{3} x + \dfrac{1}{3^2 \cdot 2} x^2 + \dfrac{1}{3^3 \cdot 3!} x^3 + \cdots \right]$$

$$+ C_2 \left[1 + \dfrac{1}{2} x + \dfrac{1}{5 \cdot 2} x^2 + \dfrac{1}{8 \cdot 5 \cdot 2} x^3 + \cdots \right]$$

17. $r_1 = \frac{5}{2}, r_2 = 0$;

$$y(x) = C_1 x^{5/2}\left[1 + \frac{2 \cdot 2}{7}x + \frac{2^2 \cdot 3}{9 \cdot 7}x^2 \right.$$
$$\left. + \frac{2^3 \cdot 4}{11 \cdot 9 \cdot 7}x^3 + \cdots\right]$$
$$+ C_2\left[1 + \frac{1}{3}x - \frac{1}{6}x^2 - \frac{1}{6}x^3 - \cdots\right]$$

19. $r_1 = \frac{2}{3}, r_2 = \frac{1}{3}$;
$$y(x) = C_1 x^{2/3}[1 - \frac{1}{2}x + \frac{5}{28}x^2 - \frac{1}{21}x^3 + \cdots]$$
$$+ C_2 x^{1/3}[1 - \frac{1}{2}x + \frac{1}{5}x^2 - \frac{7}{120}x^3 + \cdots]$$

21. $r_1 = 1, r_2 = -\frac{1}{2}$;
$$y(x) = C_1 x\left[1 + \frac{1}{5}x + \frac{1}{5 \cdot 7}x^2 + \frac{1}{5 \cdot 7 \cdot 9}x^3 + \cdots\right]$$
$$+ C_2 x^{-1/2}\left[1 + \frac{1}{2}x + \frac{1}{2 \cdot 4}x^2 \right.$$
$$\left. + \frac{1}{2 \cdot 4 \cdot 6}x^3 + \cdots\right]$$

23. $r_1 = 0, r_2 = -1$;

$$y(x) = C_1 x^{-1}\sum_{n=0}^{\infty}\frac{1}{(2n)!}x^{2n}$$
$$+ C_2 x^{-1}\sum_{n=0}^{\infty}\frac{1}{(2n+1)!}x^{2n+1}$$
$$= \frac{1}{x}[C_1 \cosh x + C_2 \sinh x]$$

25. $r_1 = 4, r_2 = 0$;

$$y(x) = C_1\left[1 + \frac{2}{3}x + \frac{1}{3}x^2\right] + C_2\sum_{n=0}^{\infty}(n+1)x^{n+4}$$

27. $r_1 = r_2 = 0$;

$$y(x) = C_1 y_1(x) + C_2\left[y_1(x)\ln x + y_1(x)\right.$$
$$\left. \times\left(-x + \frac{1}{4}x^2 - \frac{1}{3 \cdot 3!}x^3 + \frac{1}{4 \cdot 4!}x^4 - \cdots\right)\right],$$

where $y_1(x) = \sum_{n=0}^{\infty}\frac{1}{n!}x^n = e^x$

29. $r_1 = r_2 = 0$;
$$y(x) = C_1 y_1(x) + C_2[y_1(x)\ln x + y_1(x)$$
$$\times(2x + \frac{5}{4}x^2 + \frac{23}{27}x^3 + \cdots)],$$

where $y_1(x) = \sum_{n=0}^{\infty}\frac{(-1)^n}{(n!)^{2n}}x^n$

31. $r_1 = r_2 = 1$;
$$y(x) = C_1 xe^{-x} + C_2 xe^{-x}$$
$$\times\left[\ln x + x + \frac{1}{4}x^2 + \frac{1}{3 \cdot 3!}x^3 + \cdots\right]$$

33. $r_1 = 2, r_2 = 0$;

$$y(x) = C_1 x^2 + C_2\left[\frac{1}{2}x^2\ln x - \frac{1}{2} + x - \frac{1}{3!}x^3 + \cdots\right]$$

35. The method of Frobenius yields only the trivial solution $y(x) = 0$.

37. The assumption $y = \sum_{n=0}^{\infty}c_n x^{n+r}$ leads to

$$c_n[(n+r)(n+r+2) - 8] = 0$$

for $n \geq 0$. For $n = 0$ and $c_0 \neq 0$ we have $r^2 + 2r - 8 = 0$ and so $r_1 = 2, r_2 = -4$. For these values we are forced to take $c_n = 0$ for $n > 0$. Hence a solution exists of the form $y = c_0 x^r$. It follows that the general solution on $0 < x < \infty$ is $y = C_1 x^2 + C_2 x^{-4}$.

39. $r(r-1) + \frac{5}{3}r - \frac{1}{3} = 0$; $r_1 = \frac{1}{3}, r_2 = -1$

EXERCISES 6.5, Page 318

1. $y = c_1 J_{1/3}(x) + c_2 J_{-1/3}(x)$

3. $y = c_1 J_{5/2}(x) + c_2 J_{-5/2}(x)$

5. $y = c_1 J_0(x) + c_2 Y_0(x)$

7. $y = c_1 J_2(3x) + c_2 Y_2(3x)$

9. After we use the change of variables, the differential equation becomes

$$x^2 v'' + xv' + (\lambda^2 x^2 - \frac{1}{4})v = 0.$$

Since the solution of the last equation is

$$v = c_1 J_{1/2}(\lambda x) + c_2 J_{-1/2}(\lambda x),$$

we find

$$y = c_1 x^{-1/2}J_{1/2}(\lambda x) + c_2 x^{-1/2}J_{-1/2}(\lambda x).$$

11. After substituting into the differential equation, we find

$$xy'' + (1 + 2n)y' + xy$$
$$= x^{-n-1}[x^2 J_n'' + xJ_n' + (x^2 - n^2)J_n]$$
$$= x^{-n-1} \cdot 0 = 0.$$

13. From Problem 10 with $n = \frac{1}{2}$ we find $y = x^{1/2}J_{1/2}(x)$; from Problem 11 with $n = -\frac{1}{2}$ we find $y = x^{1/2}J_{-1/2}(x)$.

15. From Problem 10 with $n = -1$ we find $y = x^{-1}J_{-1}(x)$; from Problem 11 with $n = 1$ we find $y = x^{-1}J_1(x)$ but since $J_{-1}(x) = -J_1(x)$, no new solution results.

17. From Problem 12 with $\lambda = 1$ and $v = \pm\frac{3}{2}$ we find $y = \sqrt{x}J_{3/2}(x)$ and $y = \sqrt{x}J_{-3/2}(x)$.

19. Using the hint, we can write

$$xJ_v'(x) = -v\sum_{n=0}^{\infty} \frac{(-1)^n}{n!\Gamma(1+v+n)}\left(\frac{x}{2}\right)^{2n+v}$$

$$+ 2\sum_{n=0}^{\infty} \frac{(-1)^n(n+v)}{n!(n+v)\Gamma(n+v)}\left(\frac{x}{2}\right)^{2n+v}$$

$$= -v\sum_{n=0}^{\infty} \frac{(-1)^n}{n!\Gamma(1+v+n)}\left(\frac{x}{2}\right)^{2n+v}$$

$$+ x\sum_{n=0}^{\infty} \frac{(-1)^n}{n!\Gamma(n+v)}\left(\frac{x}{2}\right)^{2n+v-1}$$

$$= -vJ_v(x) + xJ_{v-1}(x).$$

21. Subtracting the equations

$$xJ_v'(x) = vJ_v(x) - xJ_{v+1}(x)$$

$$xJ_v'(x) = -vJ_v(x) + xJ_{v-1}(x)$$

gives $\quad 2vJ_v(x) = xJ_{v+1}(x) + xJ_{v-1}(x).$

23. From Problem 20, $\dfrac{d}{dr}[rJ_1(r)] = rJ_0(r)$. Therefore

$$\int_0^x rJ_0(r)\,dr = \int_0^x \frac{d}{dr}[rJ_1(r)]\,dr = rJ_1(r)\Big|_0^x$$

$$= xJ_1(x).$$

25. Start with

$$x^nJ_0 = x^{n-1}xJ_0 = x^{n-1}\frac{d}{dx}(xJ_1)$$

and then integrate by parts.

27. $J_{-1/2}(x) = \sqrt{\dfrac{2}{\pi x}}\cos x$

29. $J_{-3/2}(x) = \sqrt{\dfrac{2}{\pi x}}\left[-\sin x - \dfrac{\cos x}{x}\right]$

31. $J_{-5/2}(x) = \sqrt{\dfrac{2}{\pi x}}\left[\dfrac{3}{x}\sin x + \left(\dfrac{3}{x^2} - 1\right)\cos x\right]$

33. $J_{-7/2}(x) = \sqrt{\dfrac{2}{\pi x}}\left[\left(1 - \dfrac{15}{x^2}\right)\sin x + \left(\dfrac{6}{x} - \dfrac{15}{x^3}\right)\cos x\right]$

35. $y = c_1 I_v(x) + c_2 I_{-v}(x), \quad v \neq$ integer

37. Since $1/\Gamma(1 - m + n) = 0$ when $n \le m - 1$, m a positive integer,

$$J_{-m}(x) = \sum_{n=0}^{\infty} \frac{(-1)^n}{n!\Gamma(1-m+n)}\left(\frac{x}{2}\right)^{2n-m}$$

$$= \sum_{n=m}^{\infty} \frac{(-1)^n}{n!\Gamma(1-m+n)}\left(\frac{x}{2}\right)^{2n-m}$$

$$= \sum_{k=0}^{\infty} \frac{(-1)^{k+m}}{(k+m)!\Gamma(1+k)}\left(\frac{x}{2}\right)^{2k+m} \quad (n = k+m)$$

$$= (-1)^m \sum_{k=0}^{\infty} \frac{(-1)^k}{\Gamma(1+k+m)k!}\left(\frac{x}{2}\right)^{2k+m}$$

$$= (-1)^m J_m(x).$$

39. (a) $P_6(x) = \frac{1}{16}(231x^6 - 315x^4 + 105x^2 - 5)$

$P_7(x) = \frac{1}{16}[429x^7 - 693x^5 + 315x^3 - 35x]$

(b) $y = P_6(x)$ satisfies $(1 - x^2)y'' - 2xy' + 42y = 0.$

$y = P_7(x)$ satisfies $(1 - x^2)y'' - 2xy' + 56y = 0.$

41. If $x = \cos\theta$, then $\dfrac{dy}{d\theta} = \dfrac{dy}{dx}\dfrac{dx}{d\theta} = -\sin\theta\dfrac{dy}{dx}$ and

$\dfrac{d^2y}{d\theta^2} = \sin^2\theta\dfrac{d^2y}{dx^2} - \cos\theta\dfrac{dy}{dx}$. Now the original equation can be written as

$$\frac{d^2y}{d\theta^2} + \frac{\cos\theta}{\sin\theta}\frac{dy}{d\theta} + n(n+1)y = 0,$$

and so

$$\sin^2\theta\frac{d^2y}{dx^2} - 2\cos\theta\frac{dy}{dx} + n(n+1)y = 0.$$

Since $x = \cos\theta$ and $\sin^2\theta = 1 - \cos^2\theta = 1 - x^2$, we obtain

$$(1 - x^2)\frac{d^2y}{dx^2} - 2x\frac{dy}{dx} + n(n+1)y = 0.$$

43. Using binomial series we have formally

$$(1 - 2xt + t^2)^{-1/2} = 1 + \frac{1}{2}(2xt - t^2) + \frac{1\cdot3}{2^2 2!}$$

$$\times (2xt - t^2)^2 + \cdots.$$

Grouping by powers of t, we then find

$$(1 - 2xt + t^2)^{-1/2} = 1 \cdot t^0 + x \cdot t + \tfrac{1}{2}(3x^2 - 1)t^2 + \cdots$$
$$= P_0(x)t^0 + P_1(x)t + P_2(x)t^2 + \cdots.$$

45. For $k = 1$, $P_2(x) = \tfrac{1}{2}[3xP_1(x) - P_0(x)]$
$$= \tfrac{1}{2}(3x^2 - 1).$$
For $k = 2$, $P_3(x) = \tfrac{1}{3}[5xP_2(x) - 2P_1(x)]$
$$= \tfrac{1}{2}(5x^3 - 3x).$$
For $k = 3$, $P_4(x) = \tfrac{1}{4}[7xP_3(x) - 3P_2(x)]$
$$= \tfrac{1}{8}(35x^4 - 30x^2 + 3).$$
For $k = 4$, $P_5(x) = \tfrac{1}{5}[9xP_4(x) - 4P_3(x)]$
$$= \tfrac{1}{8}(63x^5 - 70x^3 + 15x).$$
For $k = 5$, $P_6(x) = \tfrac{1}{6}[11xP_5(x) - 5P_4(x)]$
$$= \tfrac{1}{16}(231x^6 - 315x^4 + 105x^2 - 5)$$

47. For $n = 0, 1, 2, 3$, the values of the integrals are $2, \tfrac{2}{3}$, $\tfrac{2}{5}$, and $\tfrac{2}{7}$, respectively. In general,

$$\int_{-1}^{1} P_n^2(x)\, dx = \frac{2}{2n + 1}, \quad n = 0, 1, 2, \ldots.$$

49. y_2 is obtained from (4) of Section 4.2.

CHAPTER 6 REVIEW EXERCISES, Page 322

1. $y = c_1 x^{-1/3} + c_2 x^{1/2}$

3. $y(x) = c_1 x^2 + c_2 x^3 + x^4 - x^2 \ln x$

5. The singular points are $x = 0$, $x = -1 + \sqrt{3}i$, $x = -1 - \sqrt{3}i$. All other finite values of x, real or complex, are ordinary points.

7. regular singular point $x = 0$; irregular singular point $x = 5$

9. regular singular point; $x = -3$, $x = 3$; irregular singular point $x = 0$

11. $|x| < \infty$

13. $y_1(x) = c_0\left[1 - \dfrac{1}{3 \cdot 2}x^3 + \dfrac{1}{6 \cdot 5 \cdot 3 \cdot 2}x^6\right.$
$$\left. - \dfrac{1}{9 \cdot 8 \cdot 6 \cdot 5 \cdot 3 \cdot 2}x^9 + \cdots\right]$$
$$y_2(x) = c_1\left[x - \dfrac{1}{4 \cdot 3}x^4 + \dfrac{1}{7 \cdot 6 \cdot 4 \cdot 3}x^7\right.$$
$$\left. - \dfrac{1}{10 \cdot 9 \cdot 7 \cdot 6 \cdot 4 \cdot 3}x^{10} + \cdots\right]$$

15. $y_1(x) = c_0[1 + \tfrac{3}{2}x^2 + \tfrac{1}{2}x^3 + \tfrac{5}{8}x^4 + \cdots]$
$$y_2(x) = c_1[x + \tfrac{1}{2}x^3 + \tfrac{1}{4}x^4 + \cdots]$$

17. $r_1 = 1$, $r_2 = -\tfrac{1}{2}$;

$$y(x) = C_1 x\left[1 + \frac{1}{5}x + \frac{1}{7 \cdot 5 \cdot 2}x^2\right.$$
$$\left. + \frac{1}{9 \cdot 7 \cdot 5 \cdot 3 \cdot 2}x^3 + \cdots\right]$$
$$+ C_2 x^{-1/2}\left[1 - x - \frac{1}{2}x^2 - \frac{1}{3^2 \cdot 2}x^3 - \cdots\right]$$

19. $r_1 = 3$, $r_2 = 0$;

$$y_1(x) = C_3\left[x^3 + \frac{5}{4}x^4 + \frac{11}{8}x^5 + \cdots\right]$$
$$y(x) = C_1 y_1(x) + C_2\left[-\frac{1}{36}y_1(x)\ln x + y_1(x)\right.$$
$$\left. \times \left(-\frac{1}{3}\frac{1}{x^3} + \frac{1}{4}\frac{1}{x^2} + \frac{1}{16}\frac{1}{x} + \cdots\right)\right]$$

21. $r_1 = r_2 = 0$; $\quad y(x) = C_1 e^x + C_2 e^x \ln x$

23. $y(x) = c_0\left[1 - \frac{1}{2^2}x^2 + \frac{1}{2^4(1 \cdot 2)^2}x^4\right.$
$$\left. - \frac{1}{2^6(1 \cdot 2 \cdot 3)^2}x^6 + \cdots\right]$$

EXERCISES 7.1, Page 333

1. $\dfrac{2}{s}e^{-s} - \dfrac{1}{s}$ **3.** $\dfrac{1}{s^2} - \dfrac{1}{s^2}e^{-s}$ **5.** $\dfrac{1 + e^{-s\pi}}{s^2 + 1}$

7. $\dfrac{e^{-s}}{s} + \dfrac{e^{-s}}{s^2}$ **9.** $\dfrac{1}{s} - \dfrac{1}{s^2} + \dfrac{e^{-s}}{s^2}$ **11.** $\dfrac{e^7}{s - 1}$

13. $\dfrac{1}{(s - 4)^2}$ **15.** $\dfrac{1}{s^2 + 2s + 2}$ **17.** $\dfrac{s^2 - 1}{(s^2 + 1)^2}$

19. $\dfrac{48}{s^5}$ **21.** $\dfrac{4}{s^2} - \dfrac{10}{s}$ **23.** $\dfrac{2}{s^3} + \dfrac{6}{s^2} - \dfrac{3}{s}$

25. $\dfrac{6}{s^4} + \dfrac{6}{s^3} + \dfrac{3}{s^2} + \dfrac{1}{s}$ **27.** $\dfrac{1}{s} + \dfrac{1}{s - 4}$

29. $\dfrac{1}{s} + \dfrac{2}{s - 2} + \dfrac{1}{s - 4}$ **31.** $\dfrac{8}{s^3} - \dfrac{15}{s^2 + 9}$

33. Use $\sinh kt = \dfrac{e^{kt} - e^{-kt}}{2}$ to show that

$$\mathcal{L}\{\sinh kt\} = \frac{k}{s^2 - k^2}.$$

35. $\dfrac{1}{2(s - 2)} - \dfrac{1}{2s}$ **37.** $\dfrac{2}{s^2 + 16}$

39. $\dfrac{1}{2}\left(\dfrac{s}{s^2+9}+\dfrac{s}{s^2+1}\right)$ **41.** $\dfrac{1}{2}\left(\dfrac{3}{s^2+9}-\dfrac{1}{s^2+1}\right)$

43. The result follows by letting $u=st$ in
$$\mathscr{L}\{t^\alpha\}=\int_0^\infty t^\alpha e^{-st}\,dt.$$

45. $\dfrac{\frac{1}{2}\Gamma(\frac{1}{2})}{s^{3/2}}=\dfrac{\sqrt{\pi}}{2s^{3/2}}$

47. On $0\le t\le 1,\ e^{-st}\ge e^{-s}\ (s>0)$. Therefore,
$$\int_0^1 e^{-st}\frac{1}{t^2}\,dt\ge e^{-s}\int_0^1\frac{1}{t^2}\,dt.$$

The latter integral diverges.

EXERCISES 7.2, Page 341

1. $\frac{1}{2}t^2$ **3.** $t-2t^4$ **5.** $1+3t+\frac{3}{2}t^2+\frac{1}{6}t^3$

7. $t-1+e^{2t}$ **9.** $\frac{1}{4}e^{-t/4}$ **11.** $\frac{5}{7}\sin 7t$

13. $\cos(t/2)$ **15.** $\frac{1}{4}\sinh 4t$ **17.** $2\cos 3t-2\sin 3t$

19. $\frac{1}{3}-\frac{1}{3}e^{-3t}$ **21.** $\frac{3}{4}e^{-3t}+\frac{1}{4}e^t$

23. $0.3e^{0.1t}+0.6e^{-0.2t}$ **25.** $\frac{1}{2}e^{2t}-e^{3t}+\frac{1}{2}e^{6t}$

27. $-\frac{1}{3}e^{-t}+\frac{8}{15}e^{2t}-\frac{1}{5}e^{-3t}$ **29.** $\frac{1}{4}t-\frac{1}{8}\sin 2t$

31. $-\frac{1}{4}e^{-2t}+\frac{1}{4}\cos 2t+\frac{1}{4}\sin 2t$

33. $\frac{1}{3}\sin t-\frac{1}{6}\sin 2t$ **35.** $1/s$

EXERCISES 7.3, Page 352

1. $\dfrac{1}{(s-10)^2}$ **3.** $\dfrac{6}{(s+2)^4}$ **5.** $\dfrac{3}{(s-1)^2+9}$

7. $\dfrac{3}{(s-5)^2-9}$ **9.** $\dfrac{1}{(s-2)^2}+\dfrac{2}{(s-3)^2}+\dfrac{1}{(s-4)^2}$

11. $\dfrac{1}{2}\left[\dfrac{1}{s+1}-\dfrac{s+1}{(s+1)^2+4}\right]$ **13.** $\frac{1}{2}t^2e^{-2t}$

15. $e^{3t}\sin t$ **17.** $e^{-2t}\cos t-2e^{-2t}\sin t$

19. $e^{-t}-te^{-t}$ **21.** $5-t-5e^{-t}-4te^{-t}-\frac{3}{2}t^2e^{-t}$

23. $\dfrac{e^{-s}}{s^2}$ **25.** $\dfrac{e^{-2s}}{s^2}+2\dfrac{e^{-2s}}{s}$ **27.** $\dfrac{s}{s^2+4}e^{-\pi s}$

29. $\dfrac{6e^{-s}}{(s-1)^4}$ **31.** $\frac{1}{2}(t-2)^2\mathscr{U}(t-2)$

33. $-\sin t\,\mathscr{U}(t-\pi)$ **35.** $\mathscr{U}(t-1)-e^{-(t-1)}\mathscr{U}(t-1)$

37. $\dfrac{s^2-4}{(s^2+4)^2}$ **39.** $\dfrac{6s^2+2}{(s^2-1)^3}$ **41.** $\dfrac{12s-24}{[(s-2)^2+36]^2}$

43. $\frac{1}{2}t\sin t$ **45.** (c)   **47.** (f)   **49.** (a)

51. $f(t)=2-4\mathscr{U}(t-3)$;
$$\mathscr{L}\{f(t)\}=\frac{2}{s}-\frac{4}{s}e^{-3s}$$

53. $f(t)=t^2\mathscr{U}(t-1)$
$$=(t-1)^2\mathscr{U}(t-1)+2(t-1)\mathscr{U}(t-1)+\mathscr{U}(t-1);$$
$$\mathscr{L}\{f(t)\}=2\frac{e^{-s}}{s^3}+2\frac{e^{-s}}{s^2}+\frac{e^{-s}}{s}$$

55. $f(t)=t-t\mathscr{U}(t-2)$
$$=t-(t-2)\mathscr{U}(t-2)-2\mathscr{U}(t-2);$$
$$\mathscr{L}\{f(t)\}=\frac{1}{s^2}-\frac{e^{-2s}}{s^2}-2\frac{e^{-2s}}{s}$$

57. $f(t)=\mathscr{U}(t-a)-\mathscr{U}(t-b)$;
$$\mathscr{L}\{f(t)\}=\frac{e^{-as}}{s}-\frac{e^{-bs}}{s}$$

59. $f(t)$

61. $\dfrac{e^{-t}-e^{3t}}{t}$ **63.** $\dfrac{\sin 2t}{t}$

EXERCISES 7.4, Page 361

1. Since $f'(t)=e^t,\ f(0)=1$, it follows from (1) that $\mathscr{L}\{e^t\}=s\mathscr{L}\{e^t\}-1$. Solving gives $\mathscr{L}\{e^t\}=1/(s-1)$.

3. $(s^2+3s)F(s)-s-2$ **5.** $F(s)=\dfrac{2s-1}{(s-1)^2}$

7. $\dfrac{1}{s(s-1)}$ **9.** $\dfrac{s+1}{s[(s+1)^2+1]}$ **11.** $\dfrac{1}{s^2(s-1)}$

13. $\dfrac{3s^2+1}{s^2(s^2+1)^2}$ **15.** $\dfrac{6}{s^5}$ **17.** $\dfrac{48}{s^8}$

19. $\dfrac{s-1}{(s+1)[(s-1)^2+1]}$ **21.** $\displaystyle\int_0^t f(\tau)e^{-5(t-\tau)}\,d\tau$

23. $1-e^{-t}$ **25.** $-\frac{1}{3}e^{-t}+\frac{1}{3}e^{2t}$ **27.** $\frac{1}{4}t\sin 2t$

29. The result follows from letting $u=t-\tau$ in the first integral.

31. $\dfrac{(1-e^{-as})^2}{s(1-e^{-2as})}=\dfrac{1-e^{-as}}{s(1+e^{-as})}$ **33.** $\dfrac{a}{s}\left(\dfrac{1}{bs}-\dfrac{1}{e^{bs}-1}\right)$

35. $\dfrac{\coth(\pi s/2)}{s^2+1}$ **37.** $\dfrac{1}{s^2+1}$

EXERCISES 7.5, Page 374

1. $y = -1 + e^t$

3. $y = te^{-4t} + 2e^{-4t}$

5. $y = \frac{4}{3}e^{-t} - \frac{1}{3}e^{-4t}$

7. $y = \frac{1}{9}t + \frac{2}{27} - \frac{2}{27}e^{3t} + \frac{10}{9}te^{3t}$

9. $y = \frac{1}{20}t^5e^{2t}$

11. $y = \cos t - \frac{1}{2}\sin t - \frac{1}{2}t\cos t$

13. $y = \frac{1}{2} - \frac{1}{2}e^t\cos t + \frac{1}{2}e^t\sin t$

15. $y = -\frac{8}{9}e^{-t/2} + \frac{1}{9}e^{-2t} + \frac{5}{18}e^t + \frac{1}{2}e^{-t}$

17. $y = \cos t$ **19.** $y = [5 - 5e^{-(t-1)}]\mathcal{U}(t-1)$

21. $y = -\frac{1}{4} + \frac{1}{2}t + \frac{1}{4}e^{-2t} - \frac{1}{4}\mathcal{U}(t-1)$
$\quad -\frac{1}{2}(t-1)\mathcal{U}(t-1)$
$\quad +\frac{1}{4}e^{-2(t-1)}\mathcal{U}(t-1)$

23. $y = \cos 2t - \frac{1}{6}\sin 2(t-2\pi)\mathcal{U}(t-2\pi)$
$\quad +\frac{1}{3}\sin(t-2\pi)\mathcal{U}(t-2\pi)$

25. $y = \sin t + [1 - \cos(t-\pi)]\mathcal{U}(t-\pi)$
$\quad - [1 - \cos(t-2\pi)]\mathcal{U}(t-2\pi)$

27. $y = (e+1)te^{-t} + (e-1)e^{-t}$ **29.** $f(t) = \sin t$

31. $f(t) = -\frac{1}{8}e^{-t} + \frac{1}{8}e^t + \frac{3}{4}te^t + \frac{1}{4}t^2e^t$

33. $f(t) = e^{-t}$

35. $f(t) = \frac{3}{8}e^{2t} + \frac{1}{8}e^{-2t} + \frac{1}{2}\cos 2t + \frac{1}{4}\sin 2t$

37. $y = \sin t - \frac{1}{2}t\sin t$

39. $i(t) = 20,000[te^{-100t} - (t-1)e^{-100(t-1)}\mathcal{U}(t-1)]$

41. $q(t) = \dfrac{E_0C}{1 - kRC}(e^{-kt} - e^{-t/RC})$;

$\quad q(t) = \dfrac{E_0}{R}te^{-t/RC}$

43. $q(t) = \frac{2}{5}\mathcal{U}(t-3) - \frac{2}{5}e^{-5(t-3)}\mathcal{U}(t-3)$

45. $i(t) = \dfrac{1}{101}e^{-10t} - \dfrac{1}{101}\cos t + \dfrac{10}{101}\sin t$

$\quad -\dfrac{10}{101}e^{-10(t-3\pi/2)}\mathcal{U}\left(t - \dfrac{3\pi}{2}\right)$

$\quad +\dfrac{10}{101}\cos\left(t - \dfrac{3\pi}{2}\right)\mathcal{U}\left(t - \dfrac{3\pi}{2}\right)$

$\quad +\dfrac{1}{101}\sin\left(t - \dfrac{3\pi}{2}\right)\mathcal{U}\left(t - \dfrac{3\pi}{2}\right)$

47. $i(t) = \dfrac{t}{R} + \dfrac{L}{R^2}(e^{-Rt/L} - 1)$

$\quad +\dfrac{1}{R}\sum_{n=1}^{\infty}(e^{-R(t-n)/L} - 1)\mathcal{U}(t-n)$

For $0 \le t < 2$,

$$i(t) = \begin{cases} \dfrac{t}{R} + \dfrac{L}{R^2}(e^{-Rt/L} - 1), & 0 \le t < 1 \\[2mm] \dfrac{t}{R} + \dfrac{L}{R^2}(e^{-Rt/L} - 1) \\[2mm] \quad +\dfrac{1}{R}(e^{-R(t-1)/L} - 1), & 1 \le t < 2 \end{cases}$$

49. $q(t) = \frac{3}{5}e^{-10t} + 6te^{-10t} - \frac{3}{5}\cos 10t$;
$\quad i(t) = -60te^{-10t} + 6\sin 10t$;
$\quad$ steady-state current is $6\sin 10t$

51. $q(t) = \dfrac{E_0}{L\left(k^2 + \dfrac{1}{LC}\right)}[e^{-kt} - \cos(t/\sqrt{LC})]$

$\quad +\dfrac{kE_0\sqrt{C/L}}{k^2 + \dfrac{1}{LC}}\sin(t/\sqrt{LC})$

53. $x(t) = -\dfrac{3}{2}e^{-7t/2}\cos\dfrac{\sqrt{15}}{2}t - \dfrac{7\sqrt{15}}{10}e^{-7t/2}\sin\dfrac{\sqrt{15}}{2}t$

55. $y(x) = \dfrac{w_0}{EI}\left(\dfrac{L^2}{4}x^2 - \dfrac{L}{6}x^3 + \dfrac{1}{24}x^4\right)$;

$\quad \dfrac{17w_0L^4}{384EI}; \dfrac{w_0L^4}{8EI}$

57. $y(x) = \dfrac{w_0L^2}{16EI}x^2 - \dfrac{w_0L}{12EI}x^3 + \dfrac{w_0}{24EI}x^4$

$\quad -\dfrac{w_0}{24EI}\left(x - \dfrac{L}{2}\right)^4\mathcal{U}\left(x - \dfrac{L}{2}\right)$

59. $y = \frac{1}{3}t^3 + \frac{1}{2}ct^2$

EXERCISES 7.6, Page 383

1. $y = e^{3(t-2)}\mathcal{U}(t-2)$ **3.** $y = \sin t + \sin t\,\mathcal{U}(t-2\pi)$

5. $y = -\cos t\,\mathcal{U}\left(t - \dfrac{\pi}{2}\right) + \cos t\,\mathcal{U}\left(t - \dfrac{3\pi}{2}\right)$

7. $y = \frac{1}{2} - \frac{1}{2}e^{-2t} + [\frac{1}{2} - \frac{1}{2}e^{-2(t-1)}]\mathcal{U}(t-1)$

9. $y = e^{-2(t-2\pi)}\sin t\,\mathcal{U}(t-2\pi)$

11. $y = e^{-2t}\cos 3t + \frac{2}{3}e^{-2t}\sin 3t$
$\quad +\frac{1}{3}e^{-2(t-\pi)}\sin 3(t-\pi)\mathcal{U}(t-\pi)$
$\quad +\frac{1}{3}e^{-2(t-3\pi)}\sin 3(t-3\pi)\mathcal{U}(t-3\pi)$

13. $y(x) = \begin{cases} \dfrac{P_0}{EI}\left(\dfrac{L}{4}x^2 - \dfrac{1}{6}x^3\right), & 0 \le x < L/2 \\[2mm] \dfrac{P_0L^2}{4EI}\left(\dfrac{1}{2}x - \dfrac{L}{12}\right), & L/2 \le x \le L \end{cases}$

15. From (7) with $f(t) = e^{-st}$, we have

$$\mathcal{L}\{\delta(t - t_0)\} = \int_0^\infty e^{-st} \delta(t - t_0)\, dt = e^{-st_0}.$$

17. $y = e^{-t} \cos t + e^{-(t-3\pi)} \sin t \,\mathcal{U}(t - 3\pi)$

19. $i(t) = \dfrac{1}{L} e^{-Rt/L}$; no

CHAPTER 7 REVIEW EXERCISES, Page 385

1. $\dfrac{1}{s^2} - \dfrac{2}{s^2} e^{-s}$ **3.** false   **5.** true   **7.** $\dfrac{1}{s+7}$

9. $\dfrac{2}{s^2 + 4}$ **11.** $\dfrac{4s}{(s^2 + 4)^2}$ **13.** $\frac{1}{6}t^5$ **15.** $\frac{1}{2}t^2 e^{5t}$

17. $e^{5t} \cos 2t + \frac{5}{2} e^{5t} \sin 2t$

19. $\cos \pi(t - 1)\mathcal{U}(t - 1) + \sin \pi(t - 1)\mathcal{U}(t - 1)$

21. -5 **23.** $e^{-ks}F(s - a)$

25. (a) $f(t) = t - (t - 1)\mathcal{U}(t - 1) - \mathcal{U}(t - 4)$

   (b) $\mathcal{L}\{f(t)\} = \dfrac{1}{s^2} - \dfrac{1}{s^2} e^{-s} - \dfrac{1}{s} e^{-4s}$

   (c) $\mathcal{L}\{e^t f(t)\} = \dfrac{1}{(s-1)^2} - \dfrac{1}{(s-1)^2} e^{-(s-1)}$
$$- \dfrac{1}{s-1} e^{-4(s-1)}$$

27. (a) $f(t) = 2 + (t - 2)\mathcal{U}(t - 2)$

   (b) $\mathcal{L}\{f(t)\} = \dfrac{2}{s} + \dfrac{1}{s^2} e^{-2s}$

   (c) $\mathcal{L}\{e^t f(t)\} = \dfrac{2}{s-1} + \dfrac{1}{(s-1)^2} e^{-2(s-1)}$

29. $y = 5te^t + \frac{1}{2}t^2 e^t$

31. $y = 5\mathcal{U}(t - \pi) - 5e^{2(t-\pi)} \cos \sqrt{2}(t - \pi)\mathcal{U}(t - \pi)$
$$+ 5\sqrt{2}e^{2(t-\pi)} \sin \sqrt{2}(t - \pi)\mathcal{U}(t - \pi)$$

33. $y = -\frac{2}{125} - \frac{2}{25}t - \frac{1}{5}t^2 + \frac{127}{125}e^{5t}$
$$- \left[-\frac{37}{125} - \frac{12}{25}(t - 1) - \frac{1}{5}(t - 1)^2 \right.$$
$$\left. + \frac{37}{125}e^{5(t-1)} \right]\mathcal{U}(t - 1)$$

35. $y = 1 + t + \frac{1}{2}t^2$ **37.** $i(t) = -9 + 2t + 9e^{-t/5}$

39. $y(x) = \dfrac{w_0}{12EIL} \left[-\dfrac{1}{5}x^5 + \dfrac{L}{2}x^4 - \dfrac{L^2}{2}x^3 + \dfrac{L^3}{4}x^2 \right.$
$$\left. + \dfrac{1}{5}\left(x - \dfrac{L}{2}\right)^5 \mathcal{U}\left(x - \dfrac{L}{2}\right) \right]$$

EXERCISES 8.1, Page 397

1. $x = c_1 e^t + c_2 t e^t$
$y = (c_1 - c_2)e^t + c_2 t e^t$

3. $x = c_1 \cos t + c_2 \sin t + t + 1$
$y = c_1 \sin t - c_2 \cos t + t - 1$

5. $x = \frac{1}{2}c_1 \sin t + \frac{1}{2}c_2 \cos t - 2c_3 \sin \sqrt{6}t$
$\qquad - 2c_4 \cos \sqrt{6}t$
$y = c_1 \sin t + c_2 \cos t + c_3 \sin \sqrt{6}t + c_4 \cos \sqrt{6}t$

7. $x = c_1 e^{2t} + c_2 e^{-2t} + c_3 \sin 2t + c_4 \cos 2t + \frac{1}{5}e^t$
$y = c_1 e^{2t} + c_2 e^{-2t} - c_3 \sin 2t - c_4 \cos 2t - \frac{1}{5}e^t$

9. $x = c_1 - c_2 \cos t + c_3 \sin t + \frac{17}{15}e^{3t}$
$y = c_1 + c_2 \sin t + c_3 \cos t - \frac{4}{15}e^{3t}$

11. $x = c_1 e^t + c_2 e^{-t/2} \cos \dfrac{\sqrt{3}}{2}t + c_3 e^{-t/2} \sin \dfrac{\sqrt{3}}{2}t$

$y = \left(-\dfrac{3}{2}c_2 - \dfrac{\sqrt{3}}{2}c_3 \right)e^{-t/2} \cos \dfrac{\sqrt{3}}{2}t$

$\qquad + \left(\dfrac{\sqrt{3}}{2}c_2 - \dfrac{3}{2}c_3 \right)e^{-t/2} \sin \dfrac{\sqrt{3}}{2}t$

13. $x = c_1 e^{4t} + \frac{4}{3}e^t$
$y = -\frac{3}{4}c_1 e^{4t} + c_2 + 5e^t$

15. $x = c_1 + c_2 t + c_3 e^t + c_4 e^{-t} - \frac{1}{2}t^2$
$y = (c_1 - c_2 + 2) + (c_2 + 1)t + c_4 e^{-t} - \frac{1}{2}t^2$

17. $x = c_1 e^t + c_2 e^{-t/2} \sin \dfrac{\sqrt{3}}{2}t + c_3 e^{-t/2} \cos \dfrac{\sqrt{3}}{2}t$

$y = c_1 e^t + \left(-\dfrac{1}{2}c_2 - \dfrac{\sqrt{3}}{2}c_3 \right)e^{-t/2} \sin \dfrac{\sqrt{3}}{2}t$

$\qquad + \left(\dfrac{\sqrt{3}}{2}c_2 - \dfrac{1}{2}c_3 \right)e^{-t/2} \cos \dfrac{\sqrt{3}}{2}t$

$z = c_1 e^t + \left(-\dfrac{1}{2}c_2 + \dfrac{\sqrt{3}}{2}c_3 \right)e^{-t/2} \sin \dfrac{\sqrt{3}}{2}t$

$\qquad + \left(-\dfrac{\sqrt{3}}{2}c_2 - \dfrac{1}{2}c_3 \right)e^{-t/2} \cos \dfrac{\sqrt{3}}{2}t$

19. $x = -6c_1 e^{-t} - 3c_2 e^{-2t} + 2c_3 e^{3t}$
$y = c_1 e^{-t} + c_2 e^{-2t} + c_3 e^{3t}$
$z = 5c_1 e^{-t} + c_2 e^{-2t} + c_3 e^{3t}$

21. $x = -c_1 e^{-t} + c_2 + \frac{1}{3}t^3 - 2t^2 + 5t$
$y = c_1 e^{-t} + 2t^2 - 5t + 5$

23. $x = e^{-3t+3} - te^{-3t+3}$
$y = -e^{-3t+3} + 2te^{-3t+3}$

25. $Dx - Dy = 0$
$(D - 1)x - y = 0$

EXERCISES 8.2, Page 405

1. $x = -\frac{1}{3}e^{-2t} + \frac{1}{3}e^{t}$ **3.** $x = -\cos 3t - \frac{5}{3}\sin 3t$
$y = \frac{1}{3}e^{-2t} + \frac{2}{3}e^{t}$ $y = 2\cos 3t - \frac{7}{3}\sin 3t$

5. $x = -2e^{3t} + \frac{5}{2}e^{2t} - \frac{1}{2}$
$y = \frac{8}{3}e^{3t} - \frac{5}{2}e^{2t} - \frac{1}{6}$

7. $x = -\frac{1}{2}t - \frac{3}{4}\sqrt{2}\sin\sqrt{2}t$
$y = -\frac{1}{2}t + \frac{3}{4}\sqrt{2}\sin\sqrt{2}t$

9. $x = 8 + \dfrac{2}{3!}t^3 + \dfrac{1}{4!}t^4$

$y = -\dfrac{2}{3!}t^3 + \dfrac{1}{4!}t^4$

11. $x = \frac{1}{2}t^2 + t + 1 - e^{-t}$
$y = -\frac{1}{3} + \frac{1}{3}e^{-t} + \frac{1}{3}te^{-t}$

13. $x_1 = \dfrac{1}{5}\sin t + \dfrac{2\sqrt{6}}{15}\sin\sqrt{6}t + \dfrac{2}{5}\cos t - \dfrac{2}{5}\cos\sqrt{6}t$

$x_2 = \dfrac{2}{5}\sin t - \dfrac{\sqrt{6}}{15}\sin\sqrt{6}t + \dfrac{4}{5}\cos t + \dfrac{1}{5}\cos\sqrt{6}t$

15. **(b)** $i_2 = \frac{100}{9} - \frac{100}{9}e^{-900t}$
$i_3 = \frac{80}{9} - \frac{80}{9}e^{-900t}$
(c) $i_1 = 20 - 20e^{-900t}$

17. $i_2 = -\frac{20}{13}e^{-2t} + \frac{375}{1469}e^{-15t} + \frac{145}{113}\cos t + \frac{85}{113}\sin t$
$i_3 = \frac{30}{13}e^{-2t} + \frac{250}{1469}e^{-15t} - \frac{280}{113}\cos t + \frac{810}{113}\sin t$

19. $i_1 = \frac{6}{5} - \frac{6}{5}e^{-100t}\cos 100t$
$i_2 = \frac{6}{5} - \frac{6}{5}e^{-100t}\cos 100t - \frac{6}{5}e^{-100t}\sin 100t$

21. **(b)** $q = 50e^{-t}\sin(t - 1)\mathcal{U}(t - 1)$

EXERCISES 8.3, Page 412

1. $x_1' = x_2$
$x_2' = -4x_1 + 3x_2 + \sin 3t$

3. $x_1' = x_2$
$x_2' = x_3$
$x_3' = 10x_1 - 6x_2 + 3x_3 + t^2 + 1$

5. $x_1' = x_2$
$x_2' = x_3$
$x_3' = x_4$
$x_4' = -x_1 - 4x_2 + 2x_3 + t$

7. $x_1' = x_2$

$x_2' = \dfrac{t}{t + 1}x_1$

9. $x' = -2x + y + 5t$
$y' = 2x + y - 2t$

11. $Dx = t^2 + 5t - 2$
$Dy = -x + 5t - 2$

13. The system is degenerate.

15. $Dx = u$
$Dy = v$
$Du = w$
$Dv = 10t^2 - 4u + 3v$
$Dw = 4x + 4v - 3w$

17. $x_1 = \frac{25}{2}e^{-t/25} + \frac{25}{2}e^{-3t/25}$
$x_2 = \frac{25}{4}e^{-t/25} - \frac{25}{4}e^{-3t/25}$

19. $x_1' = \frac{1}{50}x_2 - \frac{3}{50}x_1$
$x_2' = \frac{3}{50}x_1 - \frac{7}{100}x_2 + \frac{1}{100}x_3$
$x_3' = \frac{1}{20}x_2 - \frac{1}{20}x_3$

EXERCISES 8.4, Page 432

1. **(a)** $\begin{pmatrix} 2 & 11 \\ 2 & -1 \end{pmatrix}$ **(b)** $\begin{pmatrix} -6 & 1 \\ 14 & -19 \end{pmatrix}$

(c) $\begin{pmatrix} 2 & 28 \\ 12 & -12 \end{pmatrix}$

3. **(a)** $\begin{pmatrix} -11 & 6 \\ 17 & -22 \end{pmatrix}$ **(b)** $\begin{pmatrix} -32 & 27 \\ -4 & -1 \end{pmatrix}$

(c) $\begin{pmatrix} 19 & -18 \\ -30 & 31 \end{pmatrix}$ **(d)** $\begin{pmatrix} 19 & 6 \\ 3 & 22 \end{pmatrix}$

5. **(a)** $\begin{pmatrix} 9 & 24 \\ 3 & 8 \end{pmatrix}$ **(b)** $\begin{pmatrix} 3 & 8 \\ -6 & -16 \end{pmatrix}$

(c) $\begin{pmatrix} 0 & 0 \\ 0 & 0 \end{pmatrix}$ **(d)** $\begin{pmatrix} -4 & -5 \\ 8 & 10 \end{pmatrix}$

7. **(a)** 180 **(b)** $\begin{pmatrix} 4 & 8 & 10 \\ 8 & 16 & 20 \\ 10 & 20 & 25 \end{pmatrix}$ **(c)** $\begin{pmatrix} 6 \\ 12 \\ -5 \end{pmatrix}$

9. **(a)** $\begin{pmatrix} 7 & 38 \\ 10 & 75 \end{pmatrix}$ **(b)** $\begin{pmatrix} 7 & 38 \\ 10 & 75 \end{pmatrix}$

11. $\begin{pmatrix} -14 \\ 1 \end{pmatrix}$ **13.** $\begin{pmatrix} -38 \\ -2 \end{pmatrix}$ **15.** singular

17. nonsingular; $\mathbf{A}^{-1} = \dfrac{1}{4}\begin{pmatrix} -5 & -8 \\ 3 & 4 \end{pmatrix}$

19. nonsingular; $\mathbf{A}^{-1} = \dfrac{1}{2}\begin{pmatrix} 0 & -1 & 1 \\ 2 & 2 & -2 \\ -4 & -3 & 5 \end{pmatrix}$

21. nonsingular; $\mathbf{A}^{-1} = -\dfrac{1}{9}\begin{pmatrix} -2 & -2 & -1 \\ -13 & 5 & 7 \\ 8 & -1 & -5 \end{pmatrix}$

23. $\det \mathbf{A}(t) = 2e^{3t} \neq 0$ for every value of t;

$$\mathbf{A}^{-1}(t) = \dfrac{1}{2e^{3t}}\begin{pmatrix} 3e^{4t} & -e^{4t} \\ -4e^{-t} & 2e^{-t} \end{pmatrix}$$

25. $\dfrac{d\mathbf{X}}{dt} = \begin{pmatrix} -5e^{-t} \\ -2e^{-t} \\ 7e^{-t} \end{pmatrix}$

27. $\dfrac{d\mathbf{X}}{dt} = 4\begin{pmatrix} 1 \\ -1 \end{pmatrix}e^{2t} - 12\begin{pmatrix} 2 \\ 1 \end{pmatrix}e^{-3t}$

29. (a) $\begin{pmatrix} 4e^{4t} & -\pi \sin \pi t \\ 2 & 6t \end{pmatrix}$ (b) $\begin{pmatrix} \frac{1}{4}e^8 - \frac{1}{4} & 0 \\ 4 & 6 \end{pmatrix}$

(c) $\begin{pmatrix} \frac{1}{4}e^{4t} - \frac{1}{4} & (1/\pi)\sin \pi t \\ t^2 & t^3 - t \end{pmatrix}$

31. $x = 3,\ y = 1,\ z = -5$

33. $x = 2 + 4t,\ y = -5 - t,\ z = t$

35. $x = -\frac{1}{2},\ y = \frac{3}{2},\ z = \frac{7}{2}$

37. $x_1 = 1,\ x_2 = 0,\ x_3 = 2,\ x_4 = 0$

41. $\lambda_1 = 6,\ \lambda_2 = 1,\ \mathbf{K}_1 = \begin{pmatrix} 2 \\ 7 \end{pmatrix},\ \mathbf{K}_2 = \begin{pmatrix} 1 \\ 1 \end{pmatrix}$

43. $\lambda_1 = \lambda_2 = -4,\ \mathbf{K}_1 = \begin{pmatrix} 1 \\ -4 \end{pmatrix}$

45. $\lambda_1 = 0,\ \lambda_2 = 4,\ \lambda_3 = -4,$

$\mathbf{K}_1 = \begin{pmatrix} 9 \\ 45 \\ 25 \end{pmatrix},\ \mathbf{K}_2 = \begin{pmatrix} 1 \\ 1 \\ 1 \end{pmatrix},\ \mathbf{K}_3 = \begin{pmatrix} 1 \\ 9 \\ 1 \end{pmatrix}$

47. $\lambda_1 = \lambda_2 = \lambda_3 = -2,$

$\mathbf{K}_1 = \begin{pmatrix} 2 \\ -1 \\ 0 \end{pmatrix},\ \mathbf{K}_2 = \begin{pmatrix} 0 \\ 0 \\ 1 \end{pmatrix}$

49. $\lambda_1 = 3i,\ \lambda_2 = -3i,$

$\mathbf{K}_1 = \begin{pmatrix} 1 - 3i \\ 5 \end{pmatrix},\ \mathbf{K}_2 = \begin{pmatrix} 1 + 3i \\ 5 \end{pmatrix}$

51. $\dfrac{d}{dt}\begin{pmatrix} a_{11}(t) & a_{12}(t) \\ a_{21}(t) & a_{22}(t) \end{pmatrix}\begin{pmatrix} x_1(t) \\ x_2(t) \end{pmatrix}$

$= \dfrac{d}{dt}\begin{pmatrix} a_{11}(t)x_1(t) + a_{12}(t)x_2(t) \\ a_{21}(t)x_1(t) + a_{22}(t)x_2(t) \end{pmatrix}$

$= \begin{pmatrix} a_{11}(t)x_1'(t) + a_{11}'(t)x_1(t) + a_{12}(t)x_2'(t) + a_{12}'(t)x_2(t) \\ a_{21}(t)x_1'(t) + a_{21}'(t)x_1(t) + a_{22}(t)x_2'(t) + a_{22}'(t)x_2(t) \end{pmatrix}$

$= \begin{pmatrix} a_{11}(t)x_1'(t) + a_{12}(t)x_2'(t) + a_{11}'(t)x_1(t) + a_{12}'(t)x_2(t) \\ a_{21}(t)x_1'(t) + a_{22}(t)x_2'(t) + a_{21}'(t)x_1(t) + a_{22}'(t)x_2(t) \end{pmatrix}$

$= \begin{pmatrix} a_{11}(t) & a_{12}(t) \\ a_{21}(t) & a_{22}(t) \end{pmatrix}\begin{pmatrix} x_1'(t) \\ x_2'(t) \end{pmatrix} + \begin{pmatrix} a_{11}'(t) & a_{12}'(t) \\ a_{21}'(t) & a_{22}'(t) \end{pmatrix}\begin{pmatrix} x_1(t) \\ x_2(t) \end{pmatrix}$

$= \mathbf{A}(t)\mathbf{X}'(t) + \mathbf{A}'(t)\mathbf{X}(t)$

53. Since $\mathbf{A}^{-1}$ exists, $\mathbf{AB} = \mathbf{AC}$ implies $\mathbf{A}^{-1}(\mathbf{AB}) = \mathbf{A}^{-1}(\mathbf{AC}),\ (\mathbf{A}^{-1}\mathbf{A})\mathbf{B} = (\mathbf{A}^{-1}\mathbf{A})\mathbf{C},\ \mathbf{IB} = \mathbf{IC},$ or $\mathbf{B} = \mathbf{C}$.

55. No, since in general $\mathbf{AB} \neq \mathbf{BA}$.

EXERCISES 8.5, Page 450

1. $\mathbf{X}' = \begin{pmatrix} 3 & -5 \\ 4 & 8 \end{pmatrix}\mathbf{X}$, where $\mathbf{X} = \begin{pmatrix} x \\ y \end{pmatrix}$

3. $\mathbf{X}' = \begin{pmatrix} -3 & 4 & -9 \\ 6 & -1 & 0 \\ 10 & 4 & 3 \end{pmatrix}\mathbf{X}$, where $\mathbf{X} = \begin{pmatrix} x \\ y \\ z \end{pmatrix}$

5. $\mathbf{X}' = \begin{pmatrix} 1 & -1 & 1 \\ 2 & 1 & -1 \\ 1 & 1 & 1 \end{pmatrix}\mathbf{X} + \begin{pmatrix} 0 \\ -3t^2 \\ t^2 \end{pmatrix} + \begin{pmatrix} t \\ 0 \\ -t \end{pmatrix}$

$+ \begin{pmatrix} -1 \\ 0 \\ 2 \end{pmatrix}$, where $\mathbf{X} = \begin{pmatrix} x \\ y \\ z \end{pmatrix}$

7. $\dfrac{dx}{dt} = 4x + 2y + e^t$

$\dfrac{dy}{dt} = -x + 3y - e^t$

9. $\dfrac{dx}{dt} = x - y + 2z + e^{-t} - 3t$

$\dfrac{dy}{dt} = 3x - 4y + z + 2e^{-t} + t$

$\dfrac{dz}{dt} = -2x + 5y + 6z + 2e^{-t} - t$

11. $X' = \begin{pmatrix} -5e^{-5t} \\ -10e^{-5t} \end{pmatrix}$

$\begin{pmatrix} 3 & -4 \\ 4 & -7 \end{pmatrix} X = \begin{pmatrix} 3-8 \\ 4-14 \end{pmatrix} e^{-5t} = \begin{pmatrix} -5 \\ -10 \end{pmatrix} e^{-5t} = X'$

13. $X' = \begin{pmatrix} \frac{3}{2} \\ -3 \end{pmatrix}$

$\begin{pmatrix} -1 & \frac{1}{4} \\ 1 & -1 \end{pmatrix} X = \begin{pmatrix} 1+\frac{1}{2} \\ -1-2 \end{pmatrix} e^{-3t/2} = \begin{pmatrix} \frac{3}{2} \\ -3 \end{pmatrix} e^{-3t/2} = X'$

15. $\dfrac{dX}{dt} = \begin{pmatrix} 0 \\ 0 \\ 0 \end{pmatrix}$

$\begin{pmatrix} 1 & 2 & 1 \\ 6 & -1 & 0 \\ -1 & -2 & -1 \end{pmatrix} X = \begin{pmatrix} 1+12-13 \\ 6-6 \\ -1-12+13 \end{pmatrix} = \begin{pmatrix} 0 \\ 0 \\ 0 \end{pmatrix} = \dfrac{dX}{dt}$

17. Yes; $W(X_1, X_2) = -2e^{-8t} \neq 0$ implies X_1 and X_2 are linearly independent on $(-\infty, \infty)$.

19. No; $W(X_1, X_2, X_3) =$

$\begin{vmatrix} 1+t & 1 & 3+2t \\ -2+2t & -2 & -6+4t \\ 4+2t & 4 & 12+4t \end{vmatrix} = 0$ for every t.

The solution vectors are linearly dependent on $(-\infty, \infty)$. Note that $X_3 = 2X_1 + X_2$.

21. $\dfrac{dX_p}{dt} = \begin{pmatrix} 2 \\ -1 \end{pmatrix}$

$\begin{pmatrix} 1 & 4 \\ 3 & 2 \end{pmatrix} X_p + \begin{pmatrix} 2 \\ -4 \end{pmatrix} t - \begin{pmatrix} 7 \\ 18 \end{pmatrix}$

$= \begin{pmatrix} (2-4)t + 9 + 2t - 7 \\ (6-2)t + 17 - 4t - 18 \end{pmatrix} = \begin{pmatrix} 2 \\ -1 \end{pmatrix} = \dfrac{dX_p}{dt}$

23. $X_p' = \begin{pmatrix} 2e^t + te^t \\ -te^t \end{pmatrix}$

$\begin{pmatrix} 2 & 1 \\ 3 & 4 \end{pmatrix} X_p - \begin{pmatrix} 1 \\ 7 \end{pmatrix} e^t = \begin{pmatrix} 3e^t + te^t - e^t \\ 7e^t - te^t - 7e^t \end{pmatrix}$

$= \begin{pmatrix} 2e^t + te^t \\ -te^t \end{pmatrix} = X_p'$

25. Let $X_1 = \begin{pmatrix} 6 \\ -1 \\ -5 \end{pmatrix} e^{-t}$, $X_2 = \begin{pmatrix} -3 \\ 1 \\ 1 \end{pmatrix} e^{-2t}$,

$X_3 = \begin{pmatrix} 2 \\ 1 \\ 1 \end{pmatrix} e^{3t}$, and $A = \begin{pmatrix} 0 & 6 & 0 \\ 1 & 0 & 1 \\ 1 & 1 & 0 \end{pmatrix}$. Then

$AX_1 = \begin{pmatrix} -6 \\ 1 \\ 5 \end{pmatrix} e^{-t} = X_1'$,

$AX_2 = \begin{pmatrix} 6 \\ -2 \\ -2 \end{pmatrix} e^{-2t} = X_2'$,

$AX_3 = \begin{pmatrix} 6 \\ 3 \\ 3 \end{pmatrix} e^{3t} = X_3'$, and

$W(X_1, X_2, X_3) = \begin{vmatrix} 6e^{-t} & -3e^{-2t} & 2e^{3t} \\ -e^{-t} & e^{-2t} & e^{3t} \\ -5e^{-t} & e^{-2t} & e^{3t} \end{vmatrix}$

$= 20 \neq 0$.

Therefore, X_1, X_2, X_3 form a fundamental set of solutions of $X' = AX$ on $(-\infty, \infty)$. By definition

$$X = c_1 X_1 + c_2 X_2 + c_3 X_3$$

is the general solution.

27. $\Phi(t) = \begin{pmatrix} e^{2t} & e^{7t} \\ -2e^{2t} & 3e^{7t} \end{pmatrix}$

$\Phi^{-1}(t) = \dfrac{1}{5e^{9t}} \begin{pmatrix} 3e^{7t} & -e^{7t} \\ 2e^{2t} & e^{2t} \end{pmatrix}$

29. $\Phi(t) = \begin{pmatrix} -e^t & -te^t \\ 3e^t & 3te^t - e^t \end{pmatrix}$

$\Phi^{-1}(t) = \dfrac{1}{e^{2t}} \begin{pmatrix} 3te^t - e^t & te^t \\ -3e^t & -e^t \end{pmatrix}$

31. $\Psi(t) = \begin{pmatrix} \frac{3}{5}e^{2t} + \frac{2}{5}e^{7t} & -\frac{1}{5}e^{2t} + \frac{1}{5}e^{7t} \\ -\frac{6}{5}e^{2t} + \frac{6}{5}e^{7t} & \frac{2}{5}e^{2t} + \frac{3}{5}e^{7t} \end{pmatrix}$

33. $\Psi(t) = \begin{pmatrix} 3te^t + e^t & te^t \\ -9te^t & -3te^t + e^t \end{pmatrix}$

35. $X(t_0) = \Phi(t_0)C$ implies $C = \Phi^{-1}(t_0)X(t_0)$. Substituting in $X = \Phi(t)C$ gives $X = \Phi(t)\Phi^{-1}(t_0)X_0$.

37. Comparing $X = \Phi(t)\Phi^{-1}(t_0)X_0$ and $X = \Psi(t)X_0$ implies $[\Psi(t) - \Phi(t)\Phi^{-1}(t_0)]X_0 = 0$. Since this last equation is to hold for any X_0, we conclude $\Psi(t) = \Phi(t)\Phi^{-1}(t_0)$.

EXERCISES 8.6, Page 466

1. $X = c_1 \begin{pmatrix} 1 \\ 2 \end{pmatrix} e^{5t} + c_2 \begin{pmatrix} 1 \\ -1 \end{pmatrix} e^{-t}$

3. $X = c_1 \begin{pmatrix} 2 \\ 1 \end{pmatrix} e^{-3t} + c_2 \begin{pmatrix} 2 \\ 5 \end{pmatrix} e^{t}$

5. $X = c_1 \begin{pmatrix} 5 \\ 2 \end{pmatrix} e^{8t} + c_2 \begin{pmatrix} 1 \\ 4 \end{pmatrix} e^{-10t}$

7. $X = c_1 \begin{pmatrix} 1 \\ 0 \\ 0 \end{pmatrix} e^{t} + c_2 \begin{pmatrix} 2 \\ 3 \\ 1 \end{pmatrix} e^{2t} + c_3 \begin{pmatrix} 1 \\ 0 \\ 2 \end{pmatrix} e^{-t}$

9. $X = c_1 \begin{pmatrix} -1 \\ 0 \\ 1 \end{pmatrix} e^{-t} + c_2 \begin{pmatrix} 1 \\ 4 \\ 3 \end{pmatrix} e^{3t} + c_3 \begin{pmatrix} 1 \\ -1 \\ 3 \end{pmatrix} e^{-2t}$

11. $X = c_1 \begin{pmatrix} 4 \\ 0 \\ -1 \end{pmatrix} e^{-t} + c_2 \begin{pmatrix} -12 \\ 6 \\ 5 \end{pmatrix} e^{-t/2} + c_3 \begin{pmatrix} 4 \\ 2 \\ -1 \end{pmatrix} e^{-3t/2}$

13. $X = 3 \begin{pmatrix} 1 \\ 1 \end{pmatrix} e^{t/2} + 2 \begin{pmatrix} 0 \\ 1 \end{pmatrix} e^{-t/2}$

15. $X = c_1 \begin{pmatrix} \cos t \\ 2 \cos t + \sin t \end{pmatrix} e^{4t} + c_2 \begin{pmatrix} \sin t \\ 2 \sin t - \cos t \end{pmatrix} e^{4t}$

17. $X = c_1 \begin{pmatrix} \cos t \\ -\cos t - \sin t \end{pmatrix} e^{4t} + c_2 \begin{pmatrix} \sin t \\ -\sin t + \cos t \end{pmatrix} e^{4t}$

19. $X = c_1 \begin{pmatrix} 5 \cos 3t \\ 4 \cos 3t + 3 \sin 3t \end{pmatrix} + c_2 \begin{pmatrix} 5 \sin 3t \\ 4 \sin 3t - 3 \cos 3t \end{pmatrix}$

21. $X = c_1 \begin{pmatrix} 1 \\ 0 \\ 0 \end{pmatrix} + c_2 \begin{pmatrix} -\cos t \\ \cos t \\ \sin t \end{pmatrix} + c_3 \begin{pmatrix} \sin t \\ -\sin t \\ \cos t \end{pmatrix}$

23. $X = c_1 \begin{pmatrix} 0 \\ 2 \\ 1 \end{pmatrix} e^{t} + c_2 \begin{pmatrix} \sin t \\ \cos t \\ \cos t \end{pmatrix} e^{t} + c_3 \begin{pmatrix} \cos t \\ -\sin t \\ -\sin t \end{pmatrix} e^{t}$

25. $X = \begin{pmatrix} 28 \\ -5 \\ 25 \end{pmatrix} e^{2t} + c_2 \begin{pmatrix} 5 \cos 3t \\ -4 \cos 3t - 3 \sin 3t \\ 0 \end{pmatrix} e^{-2t}$
$\quad + c_3 \begin{pmatrix} 5 \sin 3t \\ -4 \sin 3t + 3 \cos 3t \\ 0 \end{pmatrix} e^{-2t}$

27. $X = -\begin{pmatrix} 25 \\ -7 \\ 6 \end{pmatrix} e^{t} - \begin{pmatrix} \cos 5t - 5 \sin 5t \\ \cos 5t \\ \cos 5t \end{pmatrix}$
$\quad + 6 \begin{pmatrix} 5 \cos 5t + \sin 5t \\ \sin 5t \\ \sin 5t \end{pmatrix}$

29. $X = c_1 \begin{pmatrix} 1 \\ 3 \end{pmatrix} + c_2 \left\{ \begin{pmatrix} 1 \\ 3 \end{pmatrix} t + \begin{pmatrix} \frac{1}{4} \\ -\frac{1}{4} \end{pmatrix} \right\}$

31. $X = c_1 \begin{pmatrix} 1 \\ 1 \end{pmatrix} e^{2t} + c_2 \left\{ \begin{pmatrix} 1 \\ 1 \end{pmatrix} te^{2t} + \begin{pmatrix} -\frac{1}{3} \\ 0 \end{pmatrix} e^{2t} \right\}$

33. $X = c_1 \begin{pmatrix} 1 \\ 1 \\ 1 \end{pmatrix} e^{t} + c_2 \begin{pmatrix} 1 \\ 1 \\ 0 \end{pmatrix} e^{2t} + c_3 \begin{pmatrix} 1 \\ 0 \\ 1 \end{pmatrix} e^{2t}$

35. $X = c_1 \begin{pmatrix} -4 \\ -5 \\ 2 \end{pmatrix} + c_2 \begin{pmatrix} 2 \\ 0 \\ -1 \end{pmatrix} e^{5t}$
$\quad + c_3 \left\{ \begin{pmatrix} 2 \\ 0 \\ -1 \end{pmatrix} te^{5t} + \begin{pmatrix} -\frac{1}{2} \\ -\frac{1}{2} \\ -1 \end{pmatrix} e^{5t} \right\}$

37. $X = c_1 \begin{pmatrix} 0 \\ 1 \\ 1 \end{pmatrix} e^{t} + c_2 \left\{ \begin{pmatrix} 0 \\ 1 \\ 1 \end{pmatrix} te^{t} + \begin{pmatrix} 0 \\ 1 \\ 0 \end{pmatrix} e^{t} \right\}$
$\quad + c_3 \left\{ \begin{pmatrix} 0 \\ 1 \\ 1 \end{pmatrix} \frac{t^2}{2} e^{t} + \begin{pmatrix} 0 \\ 1 \\ 0 \end{pmatrix} te^{t} + \begin{pmatrix} \frac{1}{2} \\ 0 \\ 0 \end{pmatrix} e^{t} \right\}$

39. $X = -7 \begin{pmatrix} 2 \\ 1 \end{pmatrix} e^{4t} + 13 \begin{pmatrix} 2t + 1 \\ t + 1 \end{pmatrix} e^{4t}$

41. $X = \begin{pmatrix} \frac{6}{5} e^{5t} - \frac{1}{5} e^{-5t} \\ \frac{2}{5} e^{5t} + \frac{3}{5} e^{-5t} \end{pmatrix}$

43. $X = c_1 t^2 \begin{pmatrix} 3 \\ 1 \end{pmatrix} + c_2 t^4 \begin{pmatrix} 1 \\ 1 \end{pmatrix}$

EXERCISES 8.7, Page 472

1. $X = c_1 \begin{pmatrix} -1 \\ 1 \end{pmatrix} e^{-t} + c_2 \begin{pmatrix} -3 \\ 1 \end{pmatrix} e^{t} + \begin{pmatrix} -1 \\ 3 \end{pmatrix}$

3. $X = c_1 \begin{pmatrix} 1 \\ -1 \end{pmatrix} e^{-2t} + c_2 \begin{pmatrix} 1 \\ 1 \end{pmatrix} e^{4t} + \begin{pmatrix} -\frac{1}{4} \\ \frac{3}{4} \end{pmatrix} t^2$
$\quad + \begin{pmatrix} \frac{1}{4} \\ -\frac{1}{4} \end{pmatrix} t + \begin{pmatrix} -2 \\ \frac{3}{4} \end{pmatrix}$

5. $X = c_1 \begin{pmatrix} 1 \\ -3 \end{pmatrix} e^{3t} + c_2 \begin{pmatrix} 1 \\ 9 \end{pmatrix} e^{7t} + \begin{pmatrix} \frac{55}{36} \\ -\frac{19}{4} \end{pmatrix} e^{t}$

7. $X = c_1 \begin{pmatrix} 1 \\ 0 \\ 0 \end{pmatrix} e^{t} + c_2 \begin{pmatrix} 1 \\ 1 \\ 0 \end{pmatrix} e^{2t} + c_3 \begin{pmatrix} 1 \\ 2 \\ 2 \end{pmatrix} e^{5t} - \begin{pmatrix} \frac{3}{2} \\ \frac{7}{2} \\ 2 \end{pmatrix} e^{4t}$

9. $X = 13\begin{pmatrix} 1 \\ -1 \end{pmatrix} e^t + 2\begin{pmatrix} -4 \\ 6 \end{pmatrix} e^{2t} + \begin{pmatrix} -9 \\ 6 \end{pmatrix}$

11. $X = c_1 \begin{pmatrix} 1 \\ 1 \end{pmatrix} + c_2 \begin{pmatrix} 1 \\ -1 \end{pmatrix} e^{2t} + \begin{pmatrix} -\frac{3}{2} \\ -\frac{3}{2} \end{pmatrix} t + \begin{pmatrix} -\frac{5}{2} \\ 1 \end{pmatrix}$

EXERCISES 8.8, Page 476

1. $X = c_1 \begin{pmatrix} 1 \\ 1 \end{pmatrix} + c_2 \begin{pmatrix} 3 \\ 2 \end{pmatrix} e^t - \begin{pmatrix} 11 \\ 11 \end{pmatrix} t - \begin{pmatrix} 15 \\ 10 \end{pmatrix}$

3. $X = c_1 \begin{pmatrix} 2 \\ 1 \end{pmatrix} e^{t/2} + c_2 \begin{pmatrix} 10 \\ 3 \end{pmatrix} e^{3t/2}$

$- \begin{pmatrix} \frac{13}{2} \\ \frac{13}{4} \end{pmatrix} t e^{t/2} - \begin{pmatrix} \frac{15}{2} \\ \frac{9}{4} \end{pmatrix} e^{t/2}$

5. $X = c_1 \begin{pmatrix} 2 \\ 1 \end{pmatrix} e^t + c_2 \begin{pmatrix} 1 \\ 1 \end{pmatrix} e^{2t} + \begin{pmatrix} 3 \\ 3 \end{pmatrix} e^t + \begin{pmatrix} 4 \\ 2 \end{pmatrix} t e^t$

7. $X = c_1 \begin{pmatrix} 4 \\ 1 \end{pmatrix} e^{3t} + c_2 \begin{pmatrix} -2 \\ 1 \end{pmatrix} e^{-3t} + \begin{pmatrix} -12 \\ 0 \end{pmatrix} t - \begin{pmatrix} \frac{4}{3} \\ \frac{4}{3} \end{pmatrix}$

9. $X = c_1 \begin{pmatrix} 1 \\ -1 \end{pmatrix} e^t + c_2 \begin{pmatrix} -t \\ \frac{1}{2} - t \end{pmatrix} e^t + \begin{pmatrix} \frac{1}{2} \\ -2 \end{pmatrix} e^{-t}$

11. $X = c_1 \begin{pmatrix} \cos t \\ \sin t \end{pmatrix} + c_2 \begin{pmatrix} \sin t \\ -\cos t \end{pmatrix}$

$+ \begin{pmatrix} \cos t \\ \sin t \end{pmatrix} t + \begin{pmatrix} -\sin t \\ \cos t \end{pmatrix} \ln|\cos t|$

13. $X = c_1 \begin{pmatrix} \cos t \\ \sin t \end{pmatrix} e^t + c_2 \begin{pmatrix} \sin t \\ -\cos t \end{pmatrix} e^t + \begin{pmatrix} \cos t \\ \sin t \end{pmatrix} t e^t$

15. $X = c_1 \begin{pmatrix} \cos t \\ -\sin t \end{pmatrix} + c_2 \begin{pmatrix} \sin t \\ \cos t \end{pmatrix} + \begin{pmatrix} \cos t \\ -\sin t \end{pmatrix} t$

$+ \begin{pmatrix} -\sin t \\ \sin t \tan t \end{pmatrix} - \begin{pmatrix} \sin t \\ \cos t \end{pmatrix} \ln|\cos t|$

17. $X = c_1 \begin{pmatrix} 2\sin t \\ \cos t \end{pmatrix} e^t + c_2 \begin{pmatrix} 2\cos t \\ -\sin t \end{pmatrix} e^t + \begin{pmatrix} 3\sin t \\ \frac{3}{2}\cos t \end{pmatrix} t e^t$

$+ \begin{pmatrix} \cos t \\ -\frac{1}{2}\sin t \end{pmatrix} e^t \ln|\sin t| + \begin{pmatrix} 2\cos t \\ -\sin t \end{pmatrix} e^t \ln|\cos t|$

19. $X = c_1 \begin{pmatrix} 1 \\ -1 \\ 0 \end{pmatrix} + c_2 \begin{pmatrix} 1 \\ 1 \\ 0 \end{pmatrix} e^{2t} + c_3 \begin{pmatrix} 0 \\ 0 \\ 1 \end{pmatrix} e^{3t}$

$+ \begin{pmatrix} -\frac{1}{4}e^{2t} + \frac{1}{2}te^{2t} \\ -e^t + \frac{1}{4}e^{2t} + \frac{1}{2}te^{2t} \\ \frac{1}{2}t^2 e^{3t} \end{pmatrix}$

21. $X = \begin{pmatrix} 2 \\ 2 \end{pmatrix} te^{2t} + \begin{pmatrix} -1 \\ 1 \end{pmatrix} e^{2t} + \begin{pmatrix} -2 \\ 2 \end{pmatrix} te^{4t} + \begin{pmatrix} 2 \\ 0 \end{pmatrix} e^{4t}$

23. $X = \begin{pmatrix} -2 \\ 4 \end{pmatrix} e^{2t} + \begin{pmatrix} 7 \\ -9 \end{pmatrix} e^{7t} + \begin{pmatrix} 20 \\ 60 \end{pmatrix} te^{7t}$

25. (b) $\begin{pmatrix} i_1 \\ i_2 \end{pmatrix} = 2\begin{pmatrix} 1 \\ 3 \end{pmatrix} e^{-2t} + \frac{6}{29} \begin{pmatrix} 3 \\ -1 \end{pmatrix} e^{-12t}$

$+ \begin{pmatrix} \frac{332}{29} \\ \frac{276}{29} \end{pmatrix} \sin t - \begin{pmatrix} \frac{76}{29} \\ \frac{168}{29} \end{pmatrix} \cos t$

EXERCISES 8.9, Page 480

1. $\begin{pmatrix} \cosh t & \sinh t \\ \sinh t & \cosh t \end{pmatrix}$

3. $X = \begin{pmatrix} \cosh t & \sinh t \\ \sinh t & \cosh t \end{pmatrix} \begin{pmatrix} c_1 \\ c_2 \end{pmatrix} = c_1 \begin{pmatrix} \cosh t \\ \sinh t \end{pmatrix} + c_2 \begin{pmatrix} \sinh t \\ \cosh t \end{pmatrix}$

5. $X = c_1 \begin{pmatrix} \cosh t \\ \sinh t \end{pmatrix} + c_2 \begin{pmatrix} \sinh t \\ \cosh t \end{pmatrix} - \begin{pmatrix} 1 \\ 1 \end{pmatrix}$

7. $X = c_1 \begin{pmatrix} 1 \\ 0 \end{pmatrix} e^t + c_2 \begin{pmatrix} 0 \\ 1 \end{pmatrix} e^{2t} + \begin{pmatrix} -t - 1 \\ \frac{1}{2}e^{4t} \end{pmatrix}$

9. $P = \begin{pmatrix} 1 & 1 \\ 1 & 3 \end{pmatrix}, P^{-1} = \begin{pmatrix} \frac{3}{2} & -\frac{1}{2} \\ -\frac{1}{2} & \frac{1}{2} \end{pmatrix},$

$D = \begin{pmatrix} 3 & 0 \\ 0 & 5 \end{pmatrix}, PDP^{-1} = \begin{pmatrix} 2 & 1 \\ -3 & 6 \end{pmatrix}$

11. $e^{tA} = e^{PDP^{-1}}$

$= I + tPDP^{-1} + \frac{t^2}{2!}(PDP^{-1})^2 + \cdots$

$= PP^{-1} + tPDP^{-1} + \frac{t^2}{2!}PD^2P^{-1} + \cdots$

$= P\left[I + tD + \frac{t^2}{2!}D^2 + \cdots\right]P^{-1}$

$= Pe^{tD}P^{-1}$

13. $X = \begin{pmatrix} \frac{3}{2}e^{3t} - \frac{1}{2}e^{5t} & -\frac{1}{2}e^{3t} + \frac{1}{2}e^{5t} \\ \frac{3}{2}e^{3t} - \frac{3}{2}e^{5t} & -\frac{1}{2}e^{3t} + \frac{3}{2}e^{5t} \end{pmatrix} \begin{pmatrix} c_1 \\ c_2 \end{pmatrix}$

CHAPTER 8 REVIEW EXERCISES, Page 482

1. true   3. $\begin{pmatrix} -2 & 1 \\ \frac{3}{2} & -\frac{1}{2} \end{pmatrix}$ 5. true   7. false

9. true   11. false

13. $x = -c_1 e^t - \frac{3}{2}c_2 e^{2t} + \frac{5}{2}$
$y = c_1 e^t + c_2 e^{2t} - 3$

15. $x = c_1 e^t + c_2 e^{5t} + t e^t$
   $y = -c_1 e^t + 3c_2 e^{5t} - t e^t + 2 e^t$

17. $x = -\frac{1}{4} + \frac{9}{8} e^{-2t} + \frac{1}{8} e^{2t}$
   $y = t + \frac{9}{4} e^{-2t} - \frac{1}{4} e^{2t}$

19. (a) $\begin{pmatrix} t^3 + 3t^2 + 5t - 2 \\ -t^3 - t + 2 \\ 4t^3 + 12t^2 + 8t + 1 \end{pmatrix}$; **(b)** $\begin{pmatrix} 3t^2 + 6t + 5 \\ -3t^2 - 1 \\ 12t^2 + 24t + 8 \end{pmatrix}$

21. $Dx = u$
   $Dy = v$
   $Du = -2u + v - 2x - \ln t + 10t - 4$
   $Dv = -u - x + 5t - 2$

23. $\mathbf{X} = c_1 \begin{pmatrix} 1 \\ -1 \end{pmatrix} e^t + c_2 \left\{ \begin{pmatrix} 1 \\ -1 \end{pmatrix} t e^t + \begin{pmatrix} 0 \\ 1 \end{pmatrix} e^t \right\}$

25. $\mathbf{X} = c_1 \begin{pmatrix} \cos 2t \\ -\sin 2t \end{pmatrix} e^t + c_2 \begin{pmatrix} \sin 2t \\ \cos 2t \end{pmatrix} e^t$

27. $\mathbf{X} = c_1 \begin{pmatrix} -1 \\ 1 \\ 0 \end{pmatrix} + c_2 \begin{pmatrix} -1 \\ 0 \\ 1 \end{pmatrix} + c_3 \begin{pmatrix} 1 \\ 1 \\ 1 \end{pmatrix} e^{3t}$

29. $\mathbf{X} = c_1 \begin{pmatrix} 1 \\ 0 \end{pmatrix} e^{2t} + c_2 \begin{pmatrix} 4 \\ 1 \end{pmatrix} e^{4t} + \begin{pmatrix} 16 \\ -4 \end{pmatrix} t + \begin{pmatrix} 11 \\ -1 \end{pmatrix}$

31. $\mathbf{X} = c_1 \begin{pmatrix} \cos t \\ \cos t - \sin t \end{pmatrix} + c_2 \begin{pmatrix} \sin t \\ \sin t + \cos t \end{pmatrix} - \begin{pmatrix} 1 \\ 1 \end{pmatrix}$

$\quad + \begin{pmatrix} \sin t \\ \sin t + \cos t \end{pmatrix} \ln|\csc t - \cot t|$

EXERCISES 9.1, Page 492

1.

3.

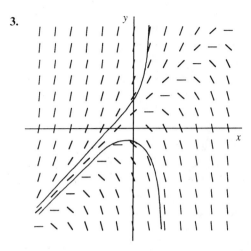

5. a family of vertical lines $x = c - 4$

7. a family of hyperbolas $x^2 - y^2 = c$

9. a family of circles $x^2 + (y + 1)^2 = c^2$ with center at $(0, -1)$

11. a family of hyperbolas $xy + y^2 = c$

13. a family of straight lines $y = c(x - 2) + 1$ passing through $(2, 1)$

15.

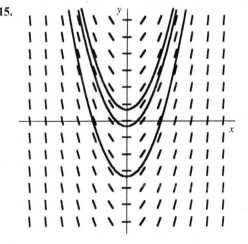

17.

19.

21.

23. $y = \dfrac{\alpha - c\gamma}{c\delta - \beta}\, x$

25. $3x + 2y = -\dfrac{3}{2}$

27. $y = \pm\sqrt{2}\,x$

29. $y = 4x;\ y = -x$

EXERCISES 9.2, Page 502

1. $y = 1 - x + \tan(x + \pi/4)$

3. (a)

x_n	y_n
1.00	5.0000
1.10	3.8000
1.20	2.9800
1.30	2.4260
1.40	2.0582
1.50	1.8207

(b)

x_n	y_n
1.00	5.0000
1.05	4.4000
1.10	3.8950
1.15	3.4707
1.20	3.1151
1.25	2.8179
1.30	2.5702
1.35	2.3647
1.40	2.1950
1.45	2.0557
1.50	1.9424

5. (a)

x_n	y_n
0.00	0.0000
0.10	0.1000
0.20	0.2010
0.30	0.3050
0.40	0.4143
0.50	0.5315

(b)

x_n	y_n
0.00	0.0000
0.05	0.0500
0.10	0.1001
0.15	0.1506
0.20	0.2018
0.25	0.2538
0.30	0.3070
0.35	0.3617
0.40	0.4183
0.45	0.4770
0.50	0.5384

7. (a)

x_n	y_n
0.00	0.0000
0.10	0.1000
0.20	0.1905
0.30	0.2731
0.40	0.3492
0.50	0.4198

(b)

x_n	y_n
0.00	0.0000
0.05	0.0500
0.10	0.0976
0.15	0.1429
0.20	0.1863
0.25	0.2278
0.30	0.2676
0.35	0.3058
0.40	0.3427
0.45	0.3782
0.50	0.4124

9. (a)

x_n	y_n
0.00	0.5000
0.10	0.5250
0.20	0.5431
0.30	0.5548
0.40	0.5613
0.50	0.5639

(b)

x_n	y_n
0.00	0.5000
0.05	0.5125
0.10	0.5232
0.15	0.5322
0.20	0.5395
0.25	0.5452
0.30	0.5496
0.35	0.5527
0.40	0.5547
0.45	0.5559
0.50	0.5565

11. (a)

x_n	y_n
1.00	1.0000
1.10	1.0000
1.20	1.0191
1.30	1.0588
1.40	1.1231
1.50	1.2194

(b)

x_n	y_n
1.00	1.0000
1.05	1.0000
1.10	1.0049
1.15	1.0147
1.20	1.0298
1.25	1.0506
1.30	1.0775
1.35	1.1115
1.40	1.1538
1.45	1.2057
1.50	1.2696

13. (a) $h = 0.1$ $h = 0.05$

x_n	y_n
1.00	5.0000
1.10	3.9900
1.20	3.2545
1.30	2.7236
1.40	2.3451
1.50	2.0801

x_n	y_n
1.00	5.0000
1.05	4.4475
1.10	3.9763
1.15	3.5751
1.20	3.2342
1.25	2.9452
1.30	2.7009
1.35	2.4952
1.40	2.3226
1.45	2.1786
1.50	2.0592

(b) $h = 0.1$ $h = 0.05$

x_n	y_n
0.00	0.0000
0.10	0.1005
0.20	0.2030
0.30	0.3098
0.40	0.4234
0.50	0.5470

x_n	y_n
0.00	0.0000
0.05	0.0501
0.10	0.1004
0.15	0.1512
0.20	0.2028
0.25	0.2554
0.30	0.3095
0.35	0.3652
0.40	0.4230
0.45	0.4832
0.50	0.5465

(c) $h = 0.1$ $h = 0.05$

x_n	y_n
0.00	0.0000
0.10	0.0952
0.20	0.1822
0.30	0.2622
0.40	0.3363
0.50	0.4053

x_n	y_n
0.00	0.0000
0.05	0.0488
0.10	0.0953
0.15	0.1397
0.20	0.1823
0.25	0.2231
0.30	0.2623
0.35	0.3001
0.40	0.3364
0.45	0.3715
0.50	0.4054

(e) $h = 0.1$ $h = 0.05$

x_n	y_n
1.00	1.0000
1.10	1.0095
1.20	1.0404
1.30	1.0967
1.40	1.1866
1.50	1.3260

x_n	y_n
1.00	1.0000
1.05	1.0024
1.10	1.0100
1.15	1.0228
1.20	1.0414
1.25	1.0663
1.30	1.0984
1.35	1.1389
1.40	1.1895
1.45	1.2526
1.50	1.3315

(d) $h = 0.1$ $h = 0.05$

x_n	y_n
0.00	0.5000
0.10	0.5215
0.20	0.5362
0.30	0.5449
0.40	0.5490
0.50	0.5503

x_n	y_n
0.00	0.5000
0.05	0.5116
0.10	0.5214
0.15	0.5294
0.20	0.5359
0.25	0.5408
0.30	0.5444
0.35	0.5469
0.40	0.5484
0.45	0.5492
0.50	0.5495

15.

x_n	Euler	Improved Euler
1.0	1.0000	1.0000
1.1	1.2000	1.2469
1.2	1.4938	1.6668
1.3	1.9711	2.6427
1.4	2.9060	8.7989

17. Using $y' = f(x, y)$ gives

$$\int_{x_n}^{x_{n+1}} y' \, dx = \int_{x_n}^{x_{n+1}} f(x, y) \, dx$$

$$y(x_{n+1}) - y(x_n) \approx f(x_n, y_n)(x_{n+1} - x_n)$$

$$= hf(x_n, y_n)$$

$$y(x_{n+1}) \approx y(x_n) + hf(x_n, y_n).$$

We write this as

$$y_{n+1} = y_n + hf(x_n, y_n).$$

EXERCISES 9.3, Page 506

1. (a)

x_n	y_n
1.00	5.0000
1.10	3.9900
1.20	3.2545
1.30	2.7236
1.40	2.3451
1.50	2.0801

(b)

x_n	y_n
1.00	5.0000
1.05	4.4475
1.10	3.9763
1.15	3.5751
1.20	3.2342
1.25	2.9452
1.30	2.7009
1.35	2.4952
1.40	2.3226
1.45	2.1786
1.50	2.0592

3. (a)

x_n	y_n
0.00	0.0000
0.10	0.1000
0.20	0.2020
0.30	0.3082
0.40	0.4211
0.50	0.5438

(b)

x_n	y_n
0.00	0.0000
0.05	0.0500
0.10	0.1003
0.15	0.1510
0.20	0.2025
0.25	0.2551
0.30	0.3090
0.35	0.3647
0.40	0.4223
0.45	0.4825
0.50	0.5456

5. (a)

x_n	y_n
0.00	0.0000
0.10	0.0950
0.20	0.1818
0.30	0.2617
0.40	0.3357
0.50	0.4046

(b)

x_n	y_n
0.00	0.0000
0.05	0.0488
0.10	0.0952
0.15	0.1397
0.20	0.1822
0.25	0.2230
0.30	0.2622
0.35	0.2999
0.40	0.3363
0.45	0.3714
0.50	0.4053

7. (a)

x_n	y_n
0.00	0.5000
0.10	0.5213
0.20	0.5355
0.30	0.5438
0.40	0.5475
0.50	0.5482

(b)

x_n	y_n
0.00	0.5000
0.05	0.5116
0.10	0.5213
0.15	0.5293
0.20	0.5357
0.25	0.5406
0.30	0.5441
0.35	0.5466
0.40	0.5480
0.45	0.5487
0.50	0.5490

9. (a)

x_n	y_n
1.00	1.0000
1.10	1.0100
1.20	1.0410
1.30	1.0969
1.40	1.1857
1.50	1.3226

(b)

x_n	y_n
1.00	1.0000
1.05	1.0025
1.10	1.0101
1.15	1.0229
1.20	1.0415
1.25	1.0663
1.30	1.0983
1.35	1.1387
1.40	1.1891
1.45	1.2518
1.50	1.3301

11.

x_n	Euler	Improved Euler	Three-Term Taylor
1.0	1.0000	1.0000	1.0000
1.1	1.2000	1.2469	1.2400
1.2	1.4938	1.6668	1.6345
1.3	1.9711	2.6427	2.4600
1.4	2.9060	8.7988	5.6353

13.

x_n	Improved Euler	Three-Term Taylor	True Value
1.00	5.0000	5.0000	5.0000
1.10	5.5300	5.5300	5.5310
1.20	6.1262	6.1262	6.1284
1.30	6.7954	6.7954	6.7992
1.40	7.5454	7.5454	7.5510
1.50	8.3847	8.3847	8.3923

EXERCISES 9.4, Page 512

1.

x_n	y_n
1.00	5.0000
1.10	3.9724
1.20	3.2284
1.30	2.6945
1.40	2.3163
1.50	2.0533

3.

x_n	y_n
0.00	0.0000
0.10	0.1003
0.20	0.2027
0.30	0.3093
0.40	0.4228
0.50	0.5463

5.

x_n	y_n
0.00	0.0000
0.10	0.0953
0.20	0.1823
0.30	0.2624
0.40	0.3365
0.50	0.4055

7.

x_n	y_n
0.00	0.5000
0.10	0.5213
0.20	0.5358
0.30	0.5443
0.40	0.5482
0.50	0.5493

9.

x_n	y_n
1.00	1.0000
1.10	1.0101
1.20	1.0417
1.30	1.0989
1.40	1.1905
1.50	1.3333

11. $v(5) \approx 35.7678$

13.

1.93	12.50	36.46	47.23	49.00

15.

x_n	y_n
1.00	1.0000
1.10	1.2511
1.20	1.6934
1.30	2.9425
1.40	903.0283

EXERCISES 9.5, Page 516

1. $y(x) = -x + e^x$; $y(0.2) = 1.0214$, $y(0.4) = 1.0918$, $y(0.6) = 1.2221$, $y(0.8) = 1.4255$

3.

x_n	y_n
0.00	1.0000
0.20	0.7328
0.40	0.6461
0.60	0.6585
0.80	0.7232

5.

x_n	y_n	x_n	y_n
0.00	0.0000	0.00	0.0000
0.20	0.2027	0.10	0.1003
0.40	0.4228	0.20	0.2027
0.60	0.6841	0.30	0.3093
0.80	1.0297	0.40	0.4228
1.00	1.5569	0.50	0.5463
		0.60	0.6842
		0.70	0.8423
		0.80	1.0297
		0.90	1.2603
		1.00	1.5576

7. x_n	y_n	x_n	y_n
0.00	0.0000	0.00	0.0000
0.20	0.0026	0.10	0.0003
0.40	0.0201	0.20	0.0026
0.60	0.0630	0.30	0.0087
0.80	0.1360	0.40	0.0200
1.00	0.2385	0.50	0.0379
		0.60	0.0629
		0.70	0.0956
		0.80	0.1360
		0.90	0.1837
		1.00	0.2384

9. x_n	y_n
0.00	1.0000
0.10	1.0052
0.20	1.0214
0.30	1.0499
0.40	1.0918

EXERCISES 9.6, Page 522

1. The improved Euler method is a second-order Runge-Kutta method, so letting $k = 3$, $a = x_n$, $x = x_{n+1} = x_n + h$, we get

$$y(x_{n+1}) = y(x_n) + y'(x_n)h + y''(x_n)\frac{h^2}{2} + y^{(3)}(c)\frac{h^3}{3!}.$$

The derivation of the Runge-Kutta method requires that the formula agree with the Taylor series through the h^2 term, so the local truncation error is $y^{(3)}(c)(h^3/3!)$.

3. For the fourth-order Runge-Kutta method, we let $k = 4$, $a = x_n$, $x = x_{n+1} = x_n + h$ and get

$$y(x_{n+1}) = y(x_n) + y'(x_n)h + \cdots + y^{(4)}(x_n)\frac{h^4}{4!} + y^{(5)}(c)\frac{h^5}{5!}.$$

The derivation of this Runge-Kutta method requires that the formula agree with the Taylor series

through the h^4 term, so the local truncation error is $y^{(5)}(c)(h^5/5!)$.

5. (a) $y_1 = 1.2200$

(b) $y^{(3)}(c)\dfrac{h^3}{3!} = 8e^{2c}\dfrac{h^3}{3!} \le 8e^{2(0.1)}\dfrac{(0.1)^3}{3!} = 0.001629$

(c) Actual value is $y(0.1) = e^{2(0.1)} = 1.221403$.
Error is $1.221403 - 1.22 = 0.001403 \le 0.001629$.

(d) If $h = 0.05$, $y_2 = 1.221025$.

(e) Error with $h = 0.1$ is 0.001403.
Error with $h = 0.05$ is
$1.221403 - 1.221025 = 0.000378$.

7. (a) $y_1 = 1.221400$.

(b) $y^{(5)}(c)\dfrac{h^5}{5!} = 32e^{2c}\dfrac{h^5}{5!} \le 32e^{2(0.1)}\dfrac{(0.1)^5}{5!}$

$= 3.257 \times 10^{-6}$

(c) Actual value is $y(0.1) = e^{2(0.1)} = 1.221402758$.
Error is $1.221402758 - 1.2214 = 2.758 \times 10^{-6}$
$\le 3.257 \times 10^{-6}$.

(d) If $h = 0.05$, $y_2 = 1.22140257$.

(e) Error with $h = 0.1$ is 2.758×10^{-6}.
Error with $h = 0.05$ is 1.873×10^{-7}.

9. (a) $y_1 = 0.8250$

(b) $y^{(3)}(c)\dfrac{h^3}{3!} = 10e^{-2c}\dfrac{h^3}{3!} \le 10e^{-2(0)}\dfrac{(0.1)^3}{3!}$

$= 0.001667$

(c) Actual value is $y(0.1) = 0.8234$.
Error is $0.8250 - 0.8234 = 0.001600 \le 0.001667$.

(d) If $h = 0.05$, $y_2 = 0.8238$.

(e) Error with $h = 0.1$ is 0.001600.
Error with $h = 0.05$ is
$0.008238 - 0.008234 = 0.0004$.

11. (a) $y_1 = 0.82341667$

(b) $y^{(5)}(c)\dfrac{h^5}{5!} = 40e^{-2c}\dfrac{h^5}{5!} \le 40e^{2(0)}\dfrac{(0.1)^5}{5!}$

$= 3.333 \times 10^{-6}$

(c) Actual value is $y(0.1) = 0.8234134413$.
Error is $3.225 \times 10^{-6} \le 3.333 \times 10^{-6}$.

(d) If $h = 0.05$, $y_2 = 0.82341363$.

(e) Error with $h = 0.1$ is 3.225×10^{-6}.
Error with $h = 0.05$ is 1.854×10^{-7}.

13. (a) $\left| y^{(3)}(c)\dfrac{h^3}{3!} \right| = 114e^{-3(c-1)}\dfrac{h^3}{3!}$

(b) $114e^{-3(c-1)}\dfrac{h^3}{3!} \le 114e^{-3(1-1)}\dfrac{(0.1)^3}{3!} = 0.019$

(c) From Problem 3 in Exercises 9.2 with $h = 0.1$, $y_5 = 2.0801$.
From Problem 3 in Exercises 9.2 with $h = 0.05$, $y_{10} = 2.0592$.

(d) The error with $h = 0.1$ is 0.0269.
The error with $h = 0.05$ is 0.0060.

15. (a) $\left| y^{(5)}(c) \dfrac{h^5}{5!} \right| = 1026e^{-3(c-1)} \dfrac{h^5}{5!}$

(b) $1026e^{-3(c-1)} \dfrac{h^5}{5!} \leq 1026^{-3(1-1)} \dfrac{(0.1)^5}{5!}$

$= 8.55 \times 10^{-5}$

(c) From calculation with $h = 0.1$, $y_5 = 2.05333883$.
From calculation with $h = 0.05$, $y_{10} = 2.05322299$.

(d) The error with $h = 0.1$ is 0.00012259.
The error with $h = 0.05$ is 0.00006757.

17. (a) $y^{(3)}(c) \dfrac{h^3}{3!} = \dfrac{2}{(c+1)^3} \dfrac{h^3}{3!}$

(b) $\dfrac{2}{(c+1)^3} \dfrac{h^3}{3!} \leq 2 \dfrac{(0.1)^3}{3!} = 0.0003333$

(c) From Problem 13(c) in Exercises 9.2 with $h = 0.1$, $y_5 = 0.40528104$.
From Problem 13(c) in Exercises 9.2 with $h = 0.05$, $y_{10} = 0.40541888$.

(d) The error with $h = 0.1$ is 0.0001840683428.
The error with $h = 0.05$ is 0.00004622850912.

19. (a) $y^{(5)}(c) \dfrac{h^5}{5!} = \dfrac{24}{(c+1)^5} \dfrac{h^5}{5!}$

(b) $\dfrac{24}{(c+1)^5} \dfrac{h^5}{5!} \leq 24 \dfrac{(0.1)^3}{5!} = 2.0000 \times 10^{-6}$

(c) From calculation with $h = 0.1$, $y_5 = 0.40546517$.
From calculation with $h = 0.05$, $y_{10} = 0.40546511$.

(d) The error with $h = 0.1$ is $5.978114648 \times 10^{-8}$.
The error with $h = 0.05$ is $3.388367642 \times 10^{-9}$.

EXERCISES 9.7, Page 527

1. $y(x) = -2e^{2x} + 5xe^{2x}$; $y(0.2) = -1.4918$, $y_2 = -1.6800$

3. $y_1 = -1.4928$, $y_2 = -1.4919$

5. $y_1 = 1.4640$, $y_2 = 1.4640$

7. $x_1 = 8.3055$, $y_1 = 3.4199$;
$x_2 = 8.3055$, $y_2 = 3.4199$

9. $x_1 = -3.9123$, $y_1 = 4.2857$;
$x_2 = -3.9123$, $y_2 = 4.2857$

EXERCISES 9.8, Page 532

1. $y_1 = -5.6774$
$y_2 = -2.5807$
$y_3 = 6.3226$

3. $y_1 = -0.2259$
$y_2 = -0.3356$
$y_3 = -0.3308$
$y_4 = -0.2167$

5. $y_1 = 3.3751$
$y_2 = 3.6306$
$y_3 = 3.6448$
$y_4 = 3.2355$
$y_5 = 2.1411$

7. $y_1 = 3.8842$
$y_2 = 2.9640$
$y_3 = 2.2064$
$y_4 = 1.5826$
$y_5 = 1.0681$
$y_6 = 0.6430$
$y_7 = 0.2913$

9. $y_1 = 0.2660$
$y_2 = 0.5097$
$y_3 = 0.7357$
$y_4 = 0.9471$
$y_5 = 1.1465$
$y_6 = 1.3353$
$y_7 = 1.5149$
$y_8 = 1.6855$
$y_9 = 1.8474$

11. $y_1 = 0.3492$
$y_2 = 0.7202$
$y_3 = 1.1363$
$y_4 = 1.6233$
$y_5 = 2.2118$
$y_6 = 2.9386$
$y_7 = 3.8490$

13. (c) $y_0 = -2.2755$
$y_1 = -2.0755$
$y_2 = -1.8589$
$y_3 = -1.6126$
$y_4 = -1.3275$

CHAPTER 9 REVIEW EXERCISES, Page 534

1. All isoclines $y = cx$ are solutions of the differential equation.

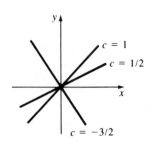

3. Comparison of Numerical Methods with $h = 0.1$

x_n	Euler	Improved Euler	Three-Term Taylor	Runge-Kutta
1.00	2.0000	2.0000	2.0000	2.0000
1.10	2.1386	2.1549	2.1556	2.1556
1.20	2.3097	2.3439	2.3453	2.3454
1.30	2.5136	2.5672	2.5694	2.5695
1.40	2.7504	2.8246	2.8277	2.8278
1.50	3.0201	3.1157	3.1198	3.1197

Comparison of Numerical Methods with $h = 0.05$

x_n	Euler	Improved Euler	Three-Term Taylor	Runge-Kutta
1.00	2.0000	2.0000	2.0000	2.0000
1.05	2.0693	2.0735	2.0735	2.0736
1.10	2.1469	2.1554	2.1556	2.1556
1.15	2.2329	2.2459	2.2462	2.2462
1.20	2.3272	2.3450	2.3454	2.3454
1.25	2.4299	2.4527	2.4532	2.4532
1.30	2.5410	2.5689	2.5695	2.5695
1.35	2.6604	2.6937	2.6944	2.6944
1.40	2.7883	2.8269	2.8278	2.8278
1.45	2.9245	2.9686	2.9696	2.9696
1.50	3.0690	3.1187	3.1198	3.1197

5. Comparison of Numerical Methods with $h = 0.1$

x_n	Euler	Improved Euler	Three-Term Taylor	Runge-Kutta
0.50	0.5000	0.5000	0.5000	0.5000
0.60	0.6000	0.6048	0.6050	0.6049
0.70	0.7095	0.7191	0.7195	0.7194
0.80	0.8283	0.8427	0.8433	0.8431
0.90	0.9559	0.9752	0.9759	0.9757
1.00	1.0921	1.1163	1.1172	1.1169

Comparison of Numerical Methods with $h = 0.05$

x_n	Euler	Improved Euler	Three-Term Taylor	Runge-Kutta
0.50	0.5000	0.5000	0.5000	0.5000
0.55	0.5500	0.5512	0.5512	0.5512
0.60	0.6024	0.6049	0.6049	0.6049
0.65	0.6573	0.6609	0.6610	0.6610
0.70	0.7144	0.7193	0.7194	0.7194
0.75	0.7739	0.7800	0.7802	0.7801
0.80	0.8356	0.8430	0.8431	0.8431
0.85	0.8996	0.9082	0.9083	0.9083
0.90	0.9657	0.9755	0.9757	0.9757
0.95	1.0340	1.0451	1.0453	1.0452
1.00	1.1044	1.1168	1.1170	1.1169

7. $h = 0.2$: $y(0.2) \approx 3.2$
$h = 0.1$: $y(0.2) \approx 3.23$

9. $x(0.2) \approx 1.62$, $y(0.2) \approx 1.84$

EXERCISES 10.1, Page 552

1. $x' = y$
$y' = -9 \sin x$; critical points at $(\pm n\pi, 0)$

3. $x' = y$
$y' = x^2 + y(x^3 - 1)$; critical point at $(0, 0)$

5. $x' = y$
$y' = \varepsilon x^3 - x$,

critical points at $(0, 0), \left(\dfrac{1}{\sqrt{\varepsilon}}, 0\right), \left(-\dfrac{1}{\sqrt{\varepsilon}}, 0\right)$

7. $(0, 0)$ and $(-1, -1)$

9. $(0, 0)$ and $(\frac{4}{3}, \frac{4}{3})$

11. $(0, 0), (10, 0), (0, 16)$, and $(4, 12)$

13. $(0, y)$, y arbitrary

15. $(0, 0), (0, 1), (0, -1), (1, 0), (-1, 0)$

17. **(a)** $x = c_1 e^{5t} - c_2 e^{-t}$ **(b)** $x = -2e^{-t}$
$y = 2c_1 e^{5t} + c_2 e^{-t}$ $y = 2e^{-t}$

(c)

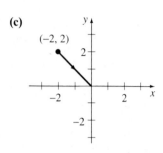

19. (a) $x = c_1(4 \cos 3t - 3 \sin 3t) + c_2(4 \sin 3t + 3 \cos 3t)$
$y = c_1(5 \cos 3t) + c_2(5 \sin 3t)$

(b) $x = 4 \cos 3t - 3 \sin 3t$
$y = 5 \cos 3t$

(c)

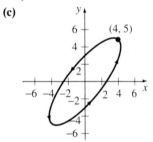

21. (a) $x = c_1(\sin t - \cos t)e^{4t} + c_2(-\sin t - \cos t)e^{4t}$
$y = 2c_1(\cos t)e^{4t} + 2c_2(\sin t)e^{4t}$

(b) $x = (\sin t - \cos t)e^{4t}$
$y = 2(\cos t)e^{4t}$

(c)

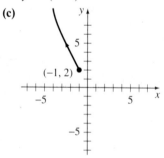

23. $r = \dfrac{1}{\sqrt[4]{4t + c_1}}$, $\theta = t + c_2$; $r = 4\dfrac{1}{\sqrt[4]{1024t + 1}}$, $\theta = t$;
the solution spirals toward the origin as t increases.

25. $r = \dfrac{1}{\sqrt{1 + c_1 e^{-2t}}}$, $\theta = t + c_2$; $r = 1$, $\theta = t$ (or $x = \cos t$ and $y = \sin t$) is the solution that satisfies
$X(0) = (1, 0)$; $r = \dfrac{1}{\sqrt{1 - \frac{3}{4}e^{-2t}}}$, $\theta = t$ is the solution

that satisfies $X(0) = (2, 0)$. This solution spirals toward the circle $r = 1$ as t increases.

27. If $X(t) = (x(t), y(t))$ is a solution, $(d/dt)f(x(t), y(t)) = (\partial f/\partial x)x'(t) + (\partial f/\partial y)y'(t) = QP - PQ = 0$. Therefore $f(x(t), y(t)) = c$ for some constant c.

EXERCISES 10.2, Page 563

1. (a) If $X(0) = X_0$ lies on the line $y = 2x$, then $X(t)$ approaches $(0, 0)$ along this line. For all other initial conditions, $X(t)$ approaches $(0, 0)$ from the direction determined by the line $y = -x/2$.

(b)

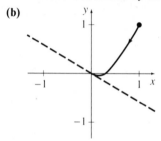

3. (a) All solutions are unstable spirals that become unbounded as t increases.

(b)

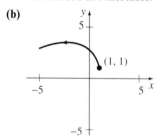

5. (a) All solutions approach $(0, 0)$ from the direction specified by the line $y = x$.

(b)

7. (a) If $\mathbf{X}(0) = \mathbf{X}_0$ lies on the line $y = 3x$, then $\mathbf{X}(t)$ approaches $(0, 0)$ along this line. For all other initial conditions, $\mathbf{X}(t)$ becomes unbounded and $y = x$ serves as the asymptote.

(b)

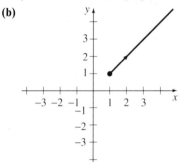

9. saddle point   **11.** saddle point

13. degenerate stable node   **15.** stable spiral

17. $|\mu| < 1$

19. $\mu < -1$ for a saddle point; $-1 < \mu < 3$ for an unstable spiral point

21. $\mathbf{X}(t) = \mathbf{X}_c(t) - \mathbf{A}^{-1}\mathbf{F}$ and $\mathbf{X}_c(t)$ approaches $(0, 0)$ as t increases.

EXERCISES 10.3, Page 575

1. $r = r_0 e^{\alpha t}$

3. $x = 0$ is unstable; $x = n + 1$ is asymptotically stable.

5. $T = T_0$ is unstable.

7. $x = \alpha$ is unstable; $x = \beta$ is asymptotically stable.

9. $P = a/b$ is asymptotically stable; $P = c$ is unstable.

11. $(\frac{1}{2}, 1)$ is a stable spiral point.

13. $(\sqrt{2}, 0)$ and $(-\sqrt{2}, 0)$ are saddle points; $(\frac{1}{2}, -\frac{7}{4})$ is a stable spiral point.

15. $(1, 1)$ is a stable node; $(1, -1)$ is a saddle point; $(2, 2)$ is a saddle point; $(2, -2)$ is an unstable spiral point.

17. $(0, -1)$ is a saddle point; $(0, 0)$ is unclassified; $(0, 1)$ is stable but we are unable to classify further.

19. $(0, 0)$ is an unstable node; $(10, 0)$ is a saddle point; $(0, 16)$ is a saddle point; $(4, 12)$ is a stable node.

21. $\theta = 0$ is a saddle point. It is not possible to classify either $\theta = \pi/3$ or $\theta = -\pi/3$.

23. It is not possible to classify $x = 0$.

25. It is not possible to classify $x = 0$, but $x = 1/\sqrt{\varepsilon}$ and $x = -1/\sqrt{\varepsilon}$ are each saddle points.

27. $\tau = 0$ and $\Delta = \beta$. Therefore $(0, 0)$ is a saddle point if $\beta < 0$.

29. (a) $(0, 0)$ is a stable spiral point.
   (b) For $\mathbf{X}_1, \Delta = -6.07298$.

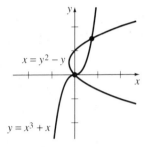

31. The solution satisfying $\mathbf{X}(0) = (x_0, 0)$ with $x_0 > 0$ is $y^2 = x_0^4 - x^4$, and if $-x_0 < x < x_0$, there are two corresponding values of y.

33. (a) At both $(1, 0)$ and $(-1, 0)$, $\tau = 0$ and $\Delta = 4$ and so no conclusions can be drawn.
   (b) The family of circles $(x - c)^2 + y^2 = c^2 - 1$ is shown in the figure.

35. $|v_0| < \frac{1}{2}\sqrt{2}$

37. If $\beta > 0$, $(0, 0)$ is the only critical point and is stable. If $\beta < 0$, $(0, 0)$, $(\hat{x}, 0)$, and $(-\hat{x}, 0)$, where $\hat{x}^2 = -\alpha/\beta$, are critical points. $(0, 0)$ is stable, while $(\hat{x}, 0)$ and $(-\hat{x}, 0)$ are each saddle points.

EXERCISES 10.4, Page 589

1. $|\omega_0| < \sqrt{\dfrac{3g}{L}}$

5. (a) First show that $y^2 = v_0^2 + g \ln\left(\dfrac{1 + x^2}{1 + x_0^2}\right)$.

7. If $x_m < x_1 < x_M$, then $F(x_1) > F(x_m) = F(x_M)$. Letting $x = x_1$, we get

$$G(y) = \frac{c_0}{F(x_1)} = \frac{F(x_m)G(a/b)}{F(x_1)} < G(a/b)$$

and from Figure 10.28(b), there are two solutions y_1 and y_2 to $G(y) = c_0/F(x_1)$ that satisfy $y_1 < a/b < y_2$.

9. (a) The new critical point is $(d/c - \varepsilon_2/c, a/b + \varepsilon_1/b)$. Note that the new differential equation is also a Lotka-Volterra predator-prey differential equation, and so the new critical point is a center.

(b) yes

11. If we let $c = 1 - (K_1/K_2)\alpha_{21}$, show that $\tau = -r_1 + r_2 c$, $\Delta = -r_1 r_2 c$, and $\tau^2 - 4\Delta = (cr_2 + r_1)^2$.

13. Show that $\alpha_{12}\alpha_{21} > 1$ and conclude that $\Delta < 0$.

15. (a) $(0, 0)$ is the only critical point.

(b) $\tau = 0$, $\Delta = k/m$, and so $(0, 0)$ is a center, a stable spiral point, or an unstable spiral point. Since $(0, 0)$ must be asymptotically stable (from physical considerations), $(0, 0)$ must be a stable spiral point.

17. If $(0, 0)$ were a stable spiral point, there would exist an x with more than two corresponding values of y. This is impossible since $y^2 + 2F(x) = c$ has at most two solutions for a given value of x.

19. The critical points are $(0, \beta)$ and $(\hat{x}, \hat{y})$, where $\hat{y} = 1/(\alpha - 1)$ and $\hat{x} = \alpha(\beta - \hat{y})$. The critical point $(\hat{x}, \hat{y})$ will lie in the first quadrant provided $\alpha > 1$ and $\beta - \hat{y} > 0$; that is, $\beta > 1/(\alpha - 1)$. Show that $\Delta = ((\alpha - 1)^2/\alpha^2)\hat{x}$, $\tau = -(\Delta + 1)$, and $\tau^2 - 4\Delta = (\Delta - 1)^2$ to conclude that $(\hat{x}, \hat{y})$ is a stable node.

CHAPTER 10 REVIEW EXERCISES, Page 594

1. true   **3.** a center or a saddle point

5. false   **7.** false

9. $r = 1/\sqrt[3]{3t + 1}$, $\theta = t$. The solution curve spirals toward the origin.

11. (a) center;   **(b)** degenerate stable node

13. stable node for $\mu < -2$; stable spiral point for $-2 < \mu < 0$; unstable spiral point for $0 < \mu < 2$; unstable node for $\mu > 2$.

15. Show that $y^2 = (1 + x_0^2 - x^2)^2 - 1$.

17. (a) $(0, 0)$ is a stable spiral point in case $\beta < 2ml\sqrt{g/l - \omega^2}$.

(b) $(\hat{x}, 0)$ and $(-\hat{x}, 0)$ are stable spiral points in case $\beta < 2ml\sqrt{\omega^2 - g^2/(\omega^2 l^2)}$.

EXERCISES 11.1, Page 603

1. $\displaystyle\int_{-2}^{2} x \cdot x^2 \, dx = \frac{x^4}{4}\Big|_{-2}^{2} = 4 - 4 = 0$

3. $\displaystyle\int_{0}^{2} e^x(xe^{-x} - e^{-x}) \, dx = \int_{0}^{2} (x - 1) \, dx$

$$= \frac{x^2}{2} - x \Big|_{0}^{2}$$

$$= 0$$

5. $\displaystyle\int_{-\pi/2}^{\pi/2} x \cos 2x \, dx = [\tfrac{1}{2}x \sin 2x + \tfrac{1}{4}\cos 2x]_{-\pi/2}^{\pi/2}$

$$= -\tfrac{1}{4} + \tfrac{1}{4} = 0$$

7. $\displaystyle\int_{0}^{\pi/2} \sin(2m + 1)x \sin(2n + 1)x \, dx$

$$= \frac{1}{2}\int_{0}^{\pi/2} [\cos 2(m - n)x$$

$$- \cos 2(m + n + 1)x] \, dx$$

$$= \frac{1}{4}\left[\frac{\sin 2(m - n)x}{m - n} - \frac{\sin 2(m + n + 1)x}{m + n + 1}\right]_{0}^{\pi/2}$$

$$= 0, \, m \neq n; \, \sqrt{\pi/2}$$

9. $\displaystyle\int_{0}^{\pi} \sin mx \sin nx \, dx$

$$= \frac{1}{2}\int_{0}^{\pi} [\cos(m - n)x - \cos(m + n)x] \, dx$$

$$= \frac{1}{2}\left[\frac{\sin(m - n)x}{m - n} - \frac{\sin(m + n)x}{m + n}\right]_{0}^{\pi}$$

$$= 0, \, m \neq n; \, \sqrt{\pi/2}$$

11. $\displaystyle\int_{0}^{p} \cos\frac{n\pi}{p} x \, dx = \frac{p}{n\pi}\sin\frac{n\pi}{p}x\Big|_{0}^{p} = 0, \, n \neq 0;$

$$\int_{0}^{p} \cos\frac{m\pi}{p} x \cos\frac{n\pi}{p} x \, dx$$

$$= \frac{1}{2}\int_{0}^{p}\left[\cos\frac{(m - n)\pi}{p}x + \cos\frac{(m + n)\pi}{p}x\right] dx$$

$$= \frac{p}{2\pi}\left[\frac{\sin\dfrac{(m - n)\pi}{p}x}{m - n} + \frac{\sin\dfrac{(m + n)\pi}{p}x}{m + n}\right]_{0}^{p}$$

$$= 0, \, m \neq n; \, \|1\| = \sqrt{p}, \, \left\|\cos\frac{n\pi}{p}x\right\| = \sqrt{p/2}$$

13. For example

$$\int_{-\infty}^{\infty} e^{-x^2} H_0(x) H_1(x)\, dx = \int_{-\infty}^{\infty} e^{-x^2}(2x)\, dx$$

$$= -\int_{-\infty}^{0} e^{-x^2}(-2x\, dx) - \int_{0}^{\infty} e^{-x^2}(-2x\, dx)$$

$$= -e^{-x^2}\Big|_{-\infty}^{0} - e^{-x^2}\Big|_{0}^{\infty} = -1 - (-1) = 0.$$

The results $\displaystyle\int_{-\infty}^{\infty} e^{-x^2} H_0(x) H_2(x)\, dx = 0$

and $\displaystyle\int_{-\infty}^{\infty} e^{-x^2} H_1(x) H_2(x)\, dx = 0$

follow from integration by parts.

15. $\displaystyle\int_{a}^{b} \phi_n(x)\, dx = \int_{a}^{b} \phi_0(x)\phi_n(x)\, dx$

$$= 0 \quad \text{for } n = 1, 2, 3, \dots$$

17. $\|\phi_m(x) + \phi_n(x)\|^2$

$$= \int_{a}^{b} [\phi_m(x) + \phi_n(x)]^2\, dx$$

$$= \int_{a}^{b} \phi_m^2(x)\, dx + 2 \underbrace{\int_{a}^{b} \phi_m(x)\phi_n(x)\, dx}_{\text{zero by orthogonality}}$$

$$+ \int_{a}^{b} \phi_n^2(x)\, dx$$

$$= \int_{a}^{b} \phi_m^2(x)\, dx + \int_{a}^{b} \phi_n^2(x)\, dx$$

$$= \|\phi_m(x)\|^2 + \|\phi_n(x)\|^2$$

19. The nontrivial function $f(x) = x^2$ is orthogonal to each function in the orthogonal set.

EXERCISES 11.2, Page 609

1. $f(x) = \dfrac{1}{2} + \dfrac{1}{\pi} \displaystyle\sum_{n=1}^{\infty} \dfrac{1 - (-1)^n}{n} \sin nx$

3. $f(x) = \dfrac{3}{4} + \displaystyle\sum_{n=1}^{\infty} \left\{ \dfrac{(-1)^n - 1}{n^2 \pi^2} \cos n\pi x - \dfrac{1}{n\pi} \sin n\pi x \right\}$

5. $f(x) = \dfrac{\pi^2}{6} + \displaystyle\sum_{n=1}^{\infty} \left\{ \dfrac{2(-1)^n}{n^2} \cos nx \right.$

$$\left. + \left(\dfrac{(-1)^{n+1}\pi}{n} + \dfrac{2}{\pi n^3}[(-1)^n - 1] \right) \sin nx \right\}$$

7. $f(x) = \pi + 2 \displaystyle\sum_{n=1}^{\infty} \dfrac{(-1)^{n+1}}{n} \sin nx$

9. $f(x) = \dfrac{1}{\pi} + \dfrac{1}{2} \sin x + \dfrac{1}{\pi} \displaystyle\sum_{n=2}^{\infty} \dfrac{(-1)^n + 1}{1 - n^2} \cos nx$

11. $f(x) = -\dfrac{1}{4} + \dfrac{1}{\pi} \displaystyle\sum_{n=1}^{\infty} \left\{ -\dfrac{1}{n} \sin \dfrac{n\pi}{2} \cos \dfrac{n\pi}{2} x \right.$

$$\left. + \dfrac{3}{n} \left(1 - \cos \dfrac{n\pi}{2} \right) \sin \dfrac{n\pi}{2} x \right\}$$

13. $f(x) = \dfrac{9}{4} + 5 \displaystyle\sum_{n=1}^{\infty} \left\{ \dfrac{(-1)^n - 1}{n^2 \pi^2} \cos \dfrac{n\pi}{5} x \right.$

$$\left. + \dfrac{(-1)^{n+1}}{n\pi} \sin \dfrac{n\pi}{5} x \right\}$$

15. $f(x) = \dfrac{2 \sinh \pi}{\pi} \left[\dfrac{1}{2} + \displaystyle\sum_{n=1}^{\infty} \dfrac{(-1)^n}{1 + n^2} (\cos nx - n \sin nx) \right]$

17. At the endpoint $x = \pi$, the series will converge to

$$\dfrac{f(\pi-) + f(-\pi+)}{2} = \dfrac{\pi^2}{2}.$$

Substituting $x = \pi$ into the series gives

$$\dfrac{\pi^2}{6} + 2 \displaystyle\sum_{n=1}^{\infty} \dfrac{1}{n^2}.$$

Equating the two results then yields

$$\dfrac{\pi^2}{6} = \displaystyle\sum_{n=1}^{\infty} \dfrac{1}{n^2} = 1 + \dfrac{1}{2^2} + \dfrac{1}{3^2} + \dfrac{1}{4^2} + \cdots.$$

Now at $x = 0$ the series converges to

$$f(0) = 0 = \dfrac{\pi^2}{6} + \displaystyle\sum_{n=1}^{\infty} \dfrac{2(-1)^n}{n^2}.$$

This implies

$$\dfrac{\pi^2}{12} = \displaystyle\sum_{n=1}^{\infty} \dfrac{(-1)^{n+1}}{n^2} = 1 - \dfrac{1}{2^2} + \dfrac{1}{3^2} - \dfrac{1}{4^2} + \cdots.$$

19. Set $x = \pi/2$.

EXERCISES 11.3, Page 619

1. odd **3.** neither even nor odd **5.** even

7. odd **9.** neither even nor odd

11. $f(x) = \dfrac{2}{\pi} \displaystyle\sum_{n=1}^{\infty} \dfrac{1 - (-1)^n}{n} \sin nx$

13. $f(x) = \dfrac{\pi}{2} + \dfrac{2}{\pi} \displaystyle\sum_{n=1}^{\infty} \dfrac{(-1)^n - 1}{n^2} \cos nx$

15. $f(x) = \dfrac{1}{3} + \dfrac{4}{\pi^2} \displaystyle\sum_{n=1}^{\infty} \dfrac{(-1)^n}{n^2} \cos n\pi x$

17. $f(x) = \dfrac{2\pi^2}{3} + 4\sum\limits_{n=1}^{\infty} \dfrac{(-1)^{n+1}}{n^2}\cos nx$

19. $f(x) = \dfrac{2}{\pi}\sum\limits_{n=1}^{\infty} \dfrac{1-(-1)^n(1+\pi)}{n}\sin nx$

21. $f(x) = \dfrac{3}{4} + \dfrac{4}{\pi^2}\sum\limits_{n=1}^{\infty} \dfrac{\cos\frac{n\pi}{2}-1}{n^2}\cos\frac{n\pi}{2}x$

23. $f(x) = \dfrac{2}{\pi} + \dfrac{2}{\pi}\sum\limits_{n=2}^{\infty} \dfrac{1+(-1)^n}{1-n^2}\cos nx$

25. $f(x) = \dfrac{1}{2} + \dfrac{2}{\pi}\sum\limits_{n=1}^{\infty} \dfrac{\sin\frac{n\pi}{2}}{n}\cos n\pi x$

$f(x) = \dfrac{2}{\pi}\sum\limits_{n=1}^{\infty} \dfrac{1-\cos\frac{n\pi}{2}}{n}\sin n\pi x$

27. $f(x) = \dfrac{2}{\pi} + \dfrac{4}{\pi}\sum\limits_{n=1}^{\infty} \dfrac{(-1)^n}{1-4n^2}\cos 2nx$

$f(x) = \dfrac{8}{\pi}\sum\limits_{n=1}^{\infty} \dfrac{n}{4n^2-1}\sin 2nx$

29. $f(x) = \dfrac{\pi}{4} + \dfrac{2}{\pi}\sum\limits_{n=1}^{\infty} \dfrac{2\cos\frac{n\pi}{2}-(-1)^n-1}{n^2}\cos nx$

$f(x) = \dfrac{4}{\pi}\sum\limits_{n=1}^{\infty} \dfrac{\sin\frac{n\pi}{2}}{n^2}\sin nx$

31. $f(x) = \dfrac{3}{4} + \dfrac{4}{\pi^2}\sum\limits_{n=1}^{\infty} \dfrac{\cos\frac{n\pi}{2}-1}{n^2}\cos\frac{n\pi}{2}x$

$f(x) = \sum\limits_{n=1}^{\infty} \left\{\dfrac{4}{n^2\pi^2}\sin\dfrac{n\pi}{2} - \dfrac{2}{n\pi}(-1)^n\right\}\sin\dfrac{n\pi}{2}x$

33. $f(x) = \dfrac{5}{6} + \dfrac{2}{\pi^2}\sum\limits_{n=1}^{\infty} \dfrac{3(-1)^n-1}{n^2}\cos n\pi x$

$f(x) = 4\sum\limits_{n=1}^{\infty} \left\{\dfrac{(-1)^{n+1}}{n\pi} + \dfrac{(-1)^n-1}{n^3\pi^3}\right\}\sin n\pi x$

35. $f(x) = \dfrac{4\pi^2}{3} + 4\sum\limits_{n=1}^{\infty} \left\{\dfrac{1}{n^2}\cos nx - \dfrac{\pi}{n}\sin nx\right\}$

37. $f(x) = \dfrac{3}{2} - \dfrac{1}{\pi}\sum\limits_{n=1}^{\infty} \dfrac{1}{n}\sin 2n\pi x$

39. $x_p(t) = \dfrac{10}{\pi}\sum\limits_{n=1}^{\infty} \dfrac{1-(-1)^n}{n(10-n^2)}\sin nt$

41. $x_p(t) = \dfrac{\pi^2}{18} + 16\sum\limits_{n=1}^{\infty} \dfrac{1}{n^2(n^2-48)}\cos nt$

43. (b) $y(x) = \dfrac{2w_0L^4}{EI\pi^5}\sum\limits_{n=1}^{\infty} \dfrac{(-1)^{n+1}}{n^5}\sin\dfrac{n\pi}{L}x$

45. Let f and g be even functions. Define $F(x) = f(x)g(x)$. Then

$$F(-x) = f(-x)g(-x) = f(x)g(x) = F(x).$$

47. Let f and g be even functions. Define $F(x) = f(x) + g(x)$. Then

$$F(-x) = f(-x) + g(-x)$$
$$= f(x) + g(x) = F(x).$$

49. Let f be an odd function. Then

$$\int_{-a}^{a} f(x)\,dx = \int_{-a}^{0} f(x)\,dx + \int_{0}^{a} f(x)\,dx = I_1 + I_2.$$

In I_1 let $-x = t$ and $-dx = dt$ so that

$$I_1 = \int_{a}^{0} f(-t)(-dt) = \int_{0}^{a} f(-t)\,dt$$

$$= -\int_{0}^{a} f(t)\,dt = -I_2.$$

Therefore $I_1 + I_2 = 0$.

51. Adding the results of Problems 13 and 14 and dividing by 2 give:

$$\dfrac{\pi}{4} + \sum\limits_{n=1}^{\infty} \left\{\dfrac{(-1)^n-1}{\pi n^2}\cos nx + \dfrac{(-1)^{n+1}}{n}\sin nx\right\}.$$

53. $f(x, y) = \dfrac{1}{4} + \dfrac{1}{\pi^2}\sum\limits_{m=1}^{\infty} \dfrac{(-1)^m-1}{m^2}\cos m\pi x$

$+ \dfrac{1}{\pi^2}\sum\limits_{n=1}^{\infty} \dfrac{(-1)^n-1}{n^2}\cos n\pi y$

$+ \dfrac{4}{\pi^4}\sum\limits_{m=1}^{\infty}\sum\limits_{n=1}^{\infty} \dfrac{[(-1)^m-1][(-1)^n-1]}{m^2n^2}$

$\times \cos m\pi x \cos n\pi y$

EXERCISES 11.4, Page 630

1. $\lambda = n^2, n = 1, 2, 3, \ldots$;
$y = \sin nx$

3. $\lambda = (2n-1)^2\pi^2/4L^2, n = 1, 2, 3, \ldots$;
$y = \cos((2n-1)\pi x/2L)$

5. $\lambda = n^2, n = 0, 1, 2, \ldots$;
$y = \cos nx$

7. $\lambda_n = x_n^2$, where the x_n, $n = 1, 2, 3, \ldots$, are the consecutive positive roots of $\cot \sqrt{\lambda} = \sqrt{\lambda}$; $y = \cos \sqrt{\lambda_n} x$

9. $\lambda = n^2\pi^2/25, n = 1, 2, 3, \ldots$;
$y = e^{-x} \sin(n\pi x/5)$

11. $\lambda = n\pi/L, n = 1, 2, 3, \ldots$;
$y = \sin(n\pi x/L)$

15. $\lambda = n^2, n = 1, 2, 3, \ldots$;
$y = \sin(n \ln x)$;

$$\frac{d}{dx}[xy'] + \lambda \frac{1}{x} y = 0;$$

$$\int_1^{e^\pi} \frac{1}{x} \sin(m \ln x) \sin(n \ln x)\, dx = 0, m \neq n$$

17. $\frac{1}{2}[1 + \sin^2 \sqrt{\lambda_n}]$

19. $\frac{d}{dx}[xe^{-x}y'] + ne^{-x}y = 0$;

$$\int_0^\infty e^{-x} L_m(x) L_n(x)\, dx = 0, m \neq n$$

21. $\lambda_1 = 4.1159, \lambda_2 = 24.1393$,
$\lambda_3 = 63.6592, \lambda_4 = 122.8883$

23. Since $r(a) = r(b)$ and

$$y_n(a) = y_n(b), \qquad y_m(a) = y_m(b)$$

$$y_n'(a) = y_n'(b), \qquad y_m'(a) = y_m'(b)$$

the right-hand member of (8) becomes

$$r(a)[y_m(a)y_n'(a) - y_n(a)y_m'(a)$$
$$- y_m(a)y_n'(a) + y_n(a)y_m'(a)] = 0.$$

EXERCISES 11.5, Page 637

1. 1.277, 2.339, 3.391, 4.441

3. $f(x) = \displaystyle\sum_{i=1}^\infty \frac{1}{\lambda_i J_1(2\lambda_i)} J_0(\lambda_i x)$

5. $f(x) = 4 \displaystyle\sum_{i=1}^\infty \frac{\lambda_i J_1(2\lambda_i)}{(4\lambda_i^2 + 1)J_0^2(2\lambda_i)} J_0(\lambda_i x)$

7. $f(x) = 20 \displaystyle\sum_{i=1}^\infty \frac{\lambda_i J_2(4\lambda_i)}{(2\lambda_i^2 + 1)J_1^2(4\lambda_i)} J_1(\lambda_i x)$

9. $f(x) = \dfrac{9}{2} - 4 \displaystyle\sum_{i=1}^\infty \frac{J_2(3\lambda_i)}{\lambda_i^2 J_0^2(3\lambda_i)} J_0(\lambda_i x)$

11. $f(x) = \frac{1}{3}P_0(x) + \frac{2}{3}P_2(x)$

13. $f(x) = \frac{1}{4}P_0(x) + \frac{1}{2}P_1(x) + \frac{5}{16}P_2(x) - \frac{3}{32}P_4(x) + \cdots$.

15. Use $\cos 2\theta = 2\cos^2\theta - 1$.

19. $f(x) = \frac{1}{2}P_0(x) + \frac{5}{8}P_2(x) - \frac{3}{16}P_4(x) + \cdots$;
$f(x) = |x|$ on $(-1, 1)$

CHAPTER 11 REVIEW EXERCISES, Page 640

1. true.   **3.** cosine   **5.** $\frac{3}{2}$ **7.** false

9. $\dfrac{1}{\sqrt{1 - x^2}}, -1 \leq x \leq 1$

11. $\displaystyle\int_0^L \sin\frac{(2m + 1)\pi}{2L} x \sin\frac{(2n + 1)\pi}{2L} x\, dx$

$= \dfrac{1}{2}\displaystyle\int_0^L \left[\cos\frac{(m - n)\pi}{L} x - \cos\frac{(m + n + 1)\pi}{L} x \right] dx$

$= \dfrac{L}{2\pi}\left[\dfrac{\sin\frac{(m - n)\pi}{L} x}{m - n} - \dfrac{\sin\frac{(m + n + 1)\pi}{L} x}{m + n + 1} \right]_0^L$

$= 0, m \neq n$

13. $f(x) = \dfrac{1}{2} + \dfrac{2}{\pi}\displaystyle\sum_{n=1}^\infty \left\{ \dfrac{1}{n^2\pi}[(-1)^n - 1] \cos n\pi x \right.$

$\left. + \dfrac{2}{n}(-1)^n \sin n\pi x \right\}$

15. $f(x) = 1 - e^{-1} + 2 \displaystyle\sum_{n=1}^\infty \dfrac{1 - (-1)^n e^{-1}}{1 + n^2\pi^2} \cos n\pi x$

17. $\lambda = (2n - 1)^2\pi^2/36, n = 1, 2, 3, \ldots$;

$y = \cos\left(\dfrac{2n - 1}{2} \pi \ln x\right)$

19. $f(x) = \dfrac{1}{4}\displaystyle\sum_{i=1}^\infty \dfrac{J_1(2\lambda_i)}{\lambda_i J_1^2(4\lambda_i)} J_0(\lambda_i x)$

EXERCISES 12.1, Page 648

1. The possible cases can be summarized in one form $u = c_1 e^{c_2(x + y)}$, where c_1 and c_2 are constants.

3. $u = c_1 e^{y + c_2(x - y)}$ **5.** $u = c_1(xy)^{c_2}$

7. not separable

9. $u = e^{-t}(A_1 e^{k\lambda^2 t} \cosh \lambda x + B_1 e^{k\lambda^2 t} \sinh \lambda x)$
$u = e^{-t}(A_2 e^{-k\lambda^2 t} \cos \lambda x + B_2 e^{-k\lambda^2 t} \sin \lambda x)$
$u = (c_7 x + c_8)c_9 e^{-t}$

11. $u = (c_1 \cosh \lambda x + c_2 \sinh \lambda x)(c_3 \cosh \lambda at$
$\qquad + c_4 \sinh \lambda at)$
$\quad u = (c_5 \cos \lambda x + c_6 \sin \lambda x)(c_7 \cos \lambda at + c_8 \sin \lambda at)$
$\quad u = (c_9 x + c_{10})(c_{11} t + c_{12})$

13. $u = (c_1 \cosh \lambda x + c_2 \sinh \lambda x)(c_3 \cos \lambda y + c_4 \sin \lambda y)$
$\quad u = (c_5 \cos \lambda x + c_6 \sin \lambda x)(c_7 \cosh \lambda y + c_8 \sinh \lambda y)$
$\quad u = (c_9 x + c_{10})(c_{11} y + c_{12})$

15. For $\lambda^2 > 0$ there are three possibilities:

$$u = (c_1 \cosh \lambda x$$
$$\quad + c_2 \sinh \lambda x)(c_3 \cosh \sqrt{1 - \lambda^2}\, y$$
$$\quad + c_4 \sinh \sqrt{1 - \lambda^2}\, y), \quad \lambda^2 < 1$$

$$u = (c_1 \cosh \lambda x$$
$$\quad + c_2 \sinh \lambda x)(c_3 \cos \sqrt{\lambda^2 - 1}\, y$$
$$\quad + c_4 \sin \sqrt{\lambda^2 - 1}\, y), \quad \lambda^2 > 1$$

$$u = (c_1 \cosh x + c_2 \sinh x)$$
$$\quad \times (c_3 y + c_4), \quad \lambda^2 = 1$$

The results for the case $-\lambda^2 < 0$ are similar. For $\lambda^2 = 0$ we have

$$u = (c_1 x + c_2)(c_3 \cosh y + c_4 \sinh y).$$

17. elliptic   **19.** parabolic   **21.** hyperbolic

23. parabolic   **25.** hyperbolic

27. Using $-\lambda^2$ as a separation constant, we obtain

$$T' + k\lambda^2 T = 0$$

$$rR'' + R' + \lambda^2 r R = 0.$$

This last equation can be written as

$$r^2 R'' + r R' + \lambda^2 r^2 R = 0,$$

which we recognize as Bessel's equation with $v = 0$. The solutions of the respective equations are as indicated in the problem.

29. $u = e^{n(-3x+y)}$, $u = e^{n(2x+y)}$

31. The equation $x^2 + 4y^2 = 4$ defines an ellipse. The partial differential equation is hyperbolic outside the ellipse, parabolic on the ellipse, and elliptic inside the ellipse.

EXERCISES 12.2, Page 656

1. $k \dfrac{\partial^2 u}{\partial x^2} = \dfrac{\partial u}{\partial t}$, $\quad 0 < x < L, t > 0$

$\quad u(0, t) = 0, \quad \left.\dfrac{\partial u}{\partial x}\right|_{x=L} = 0, \quad t > 0$

$\quad u(x, 0) = f(x), \quad 0 < x < L$

3. $k \dfrac{\partial^2 u}{\partial x^2} = \dfrac{\partial u}{\partial t}$, $\quad 0 < x < L, t > 0$

$\quad u(0, t) = 100, \quad \left.\dfrac{\partial u}{\partial x}\right|_{x=L} = -hu(L, t), \quad t > 0$

$\quad u(x, 0) = f(x), \quad 0 < x < L$

5. $a^2 \dfrac{\partial^2 u}{\partial x^2} = \dfrac{\partial^2 u}{\partial t^2}$, $\quad 0 < x < L, t > 0$

$\quad u(0, t) = 0, \quad u(L, t) = 0, \quad t > 0$

$\quad u(x, 0) = x(L - x), \quad \left.\dfrac{\partial u}{\partial t}\right|_{t=0} = 0, \quad 0 < x < L$

7. $a^2 \dfrac{\partial^2 u}{\partial x^2} - 2\beta \dfrac{\partial u}{\partial t} = \dfrac{\partial^2 u}{\partial t^2}$, $\quad 0 < x < L, t > 0$

$\quad u(0, t) = 0, \quad u(L, t) = \sin \pi t, \quad t > 0$

$\quad u(x, 0) = f(x), \quad \left.\dfrac{\partial u}{\partial t}\right|_{t=0} = 0, \quad 0 < x < L$

9. $\dfrac{\partial^2 u}{\partial x^2} + \dfrac{\partial^2 u}{\partial y^2} = 0$, $\quad 0 < x < 4, \quad 0 < y < 2$

$\quad \left.\dfrac{\partial u}{\partial x}\right|_{x=0} = 0, \quad u(4, y) = f(y), \quad 0 < y < 2$

$\quad \left.\dfrac{\partial u}{\partial y}\right|_{y=0} = 0, \quad u(x, 2) = 0, \quad 0 < x < 4$

EXERCISES 12.3, Page 660

1. $u(x, t) = \dfrac{2}{\pi} \displaystyle\sum_{n=1}^{\infty} \left(\dfrac{-\cos \dfrac{n\pi}{2} + 1}{n} \right) e^{-k(n^2\pi^2/L^2)t} \sin \dfrac{n\pi}{L} x$

3. $u(x, t) = \dfrac{1}{L} \displaystyle\int_0^L f(x)\, dx$
$\qquad + \dfrac{2}{L} \displaystyle\sum_{n=1}^{\infty} \left(\displaystyle\int_0^L f(x) \cos \dfrac{n\pi}{L} x\, dx \right)$
$\qquad \times e^{-k(n^2\pi^2/L^2)t} \cos \dfrac{n\pi}{L} x$

5. $u(x, t) = e^{-ht} \left[\dfrac{1}{L} \displaystyle\int_0^L f(x)\, dx \right.$

$\left. + \dfrac{2}{L} \displaystyle\sum_{n=1}^{\infty} \left(\int_0^L f(x) \cos \dfrac{n\pi}{L} x\, dx \right) \right.$

$\left. \times\, e^{-k(n^2\pi^2/L^2)t} \cos \dfrac{n\pi}{L} x \right]$

EXERCISES 12.4, Page 663

1. $u(x, t) = \dfrac{L^2}{\pi^3} \displaystyle\sum_{n=1}^{\infty} \dfrac{1 - (-1)^n}{n^3} \cos \dfrac{n\pi a}{L} t \sin \dfrac{n\pi}{L} x$

3. $u(x, t) = \dfrac{6\sqrt{3}}{\pi^2} \left(\cos \dfrac{\pi a}{L} t \sin \dfrac{\pi}{L} x \right.$

$\left. - \dfrac{1}{5^2} \cos \dfrac{5\pi a}{L} t \sin \dfrac{5\pi}{L} x \right.$

$\left. + \dfrac{1}{7^2} \cos \dfrac{7\pi a}{L} t \sin \dfrac{7\pi}{L} x - \cdots \right)$

5. $u(x, t) = \dfrac{1}{a} \sin at \sin x$

7. $u(x, t) = \dfrac{8h}{\pi^2} \displaystyle\sum_{n=1}^{\infty} \dfrac{\sin \dfrac{n\pi}{2}}{n^2} \cos \dfrac{n\pi a}{L} t \sin \dfrac{n\pi}{L} x$

9. $u(x, t) = e^{-\beta t} \displaystyle\sum_{n=1}^{\infty} A_n\{\cos q_n t + \dfrac{\beta}{q_n} \sin q_n t\} \sin nx,$

where $A_n = \dfrac{2}{\pi} \displaystyle\int_0^\pi f(x) \sin nx\, dx$ and $q_n = \sqrt{n^2 - \beta^2}$

11. $u(x, t) = \displaystyle\sum_{n=1}^{\infty} \left(A_n \cos \dfrac{n^2\pi^2}{L^2} at + B_n \sin \dfrac{n^2\pi^2}{L^2} at \right)$

$\times \sin \dfrac{n\pi}{L} x,$

where $A_n = \dfrac{2}{L} \displaystyle\int_0^L f(x) \sin \dfrac{n\pi}{L} x\, dx$

$B_n = \dfrac{2L}{n^2\pi^2 a} \displaystyle\int_0^L g(x) \sin \dfrac{n\pi}{L} x\, dx$

15. $u(x, t) = t + \sin x \cos 2at$

17. $u(x, t) = \dfrac{1}{2a} \sin 2x \sin 2at$

EXERCISES 12.5, Page 669

1. $u(x, y) = \dfrac{2}{a} \displaystyle\sum_{n=1}^{\infty} \left(\dfrac{1}{\sinh \dfrac{n\pi}{a} b} \int_0^a f(x) \sin \dfrac{n\pi}{a} x\, dx \right)$

$\times \sinh \dfrac{n\pi}{a} y \sin \dfrac{n\pi}{a} x$

3. $u(x, y) = \dfrac{2}{a} \displaystyle\sum_{n=1}^{\infty} \left(\dfrac{1}{\sinh \dfrac{n\pi}{a} b} \int_0^a f(x) \sin \dfrac{n\pi}{a} x\, dx \right)$

$\times \sinh \dfrac{n\pi}{a} (b - y) \sin \dfrac{n\pi}{a} x$

5. $u(x, y) = \dfrac{1}{2} x + \dfrac{2}{\pi^2} \displaystyle\sum_{n=1}^{\infty} \dfrac{1 - (-1)^n}{n^2 \sinh n\pi} \sinh n\pi x \cos n\pi y$

7. $u(x, y) = \dfrac{2}{\pi} \displaystyle\sum_{n=1}^{\infty} \dfrac{[1 - (-1)^n]}{n}$

$\times \dfrac{n \cosh nx + \sinh nx}{n \cosh n\pi + \sinh n\pi} \sin ny$

9. $u(x, y) = \dfrac{2}{\pi} \displaystyle\sum_{n=1}^{\infty} \left(\int_0^\pi f(x) \sin nx\, dx \right) e^{-ny} \sin nx$

11. $u(x, y) = \displaystyle\sum_{n=1}^{\infty} \left(A_n \cosh \dfrac{n\pi}{a} y + B_n \sinh \dfrac{n\pi}{a} y \right) \sin \dfrac{n\pi}{a} x,$

where $A_n = \dfrac{2}{a} \displaystyle\int_0^a f(x) \sin \dfrac{n\pi}{a} x\, dx$

$B_n = \dfrac{1}{\sinh \dfrac{n\pi}{a} b} \left(\dfrac{2}{a} \int_0^a g(x) \sin \dfrac{n\pi}{a} x\, dx \right.$

$\left. - A_n \cosh \dfrac{n\pi}{a} b \right)$

13. $u = u_1 + u_2,$ where

$u_1(x, y) = \dfrac{2}{\pi} \displaystyle\sum_{n=1}^{\infty} \dfrac{1 - (-1)^n}{n \sinh n\pi} \sinh ny \sin nx$

$u_2(x, y) = \dfrac{2}{\pi} \displaystyle\sum_{n=1}^{\infty} \dfrac{[1 - (-1)^n]}{n}$

$\times \dfrac{\sinh nx + \sinh n(\pi - x)}{\sinh n\pi} \sin ny$

EXERCISES 12.6, Page 673

1. $u(x, t) = 100 + \dfrac{200}{\pi} \displaystyle\sum_{n=1}^{\infty} \dfrac{(-1)^n - 1}{n} e^{-kn^2\pi^2 t} \sin n\pi x$

3. $u(x, t) = u_0 - \dfrac{r}{2k} x(x-1) + 2 \sum\limits_{n=1}^{\infty} \left[\dfrac{u_0}{n\pi} + \dfrac{r}{kn^3\pi^3} \right]$

$\times \, [(-1)^n - 1]e^{-kn^2\pi^2 t} \sin n\pi x$

5. $u(x, t) = \psi(x) + \sum\limits_{n=1}^{\infty} A_n e^{-kn^2\pi^2 t} \sin n\pi x$, where

$$\psi(x) = \dfrac{A}{k\beta^2} \left[-e^{-\beta x} + (e^{-\beta} - 1)x + 1 \right]$$

$$A_n = 2 \int_0^1 [f(x) - \psi(x)] \sin n\pi x \, dx$$

7. $\psi(x) = u_0 \left[1 - \dfrac{\sinh \sqrt{h/k}\, x}{\sinh \sqrt{h/k}} \right]$

9. $u(x, t) = \dfrac{A}{6a^2}(x - x^3)$

$+ \dfrac{2A}{a^2\pi^3} \sum\limits_{n=1}^{\infty} \dfrac{(-1)^n}{n^3} \cos n\pi a t \sin n\pi x$

11. $u(x, y) = (u_0 - u_1)y + u_1$

$+ \dfrac{2}{\pi} \sum\limits_{n=1}^{\infty} \dfrac{u_0(-1)^n - u_1}{n} e^{-n\pi x} \sin n\pi y$

EXERCISES 12.7, Page 678

1. $u(x, t) = 2h \sum\limits_{n=1}^{\infty} \dfrac{\sin \lambda_n}{\lambda_n[h + \sin^2 \lambda_n]} e^{-k\lambda_n^2 t} \cos \lambda_n x$,

where the λ_n are the consecutive positive roots of $\cot \lambda = \lambda/h$

3. $u(x, y) = \sum\limits_{n=1}^{\infty} A_n \sinh \lambda_n y \sin \lambda_n x$,

where $A_n = \dfrac{2h}{\sinh \lambda_n b[ah + \cos^2 \lambda_n a]} \int_0^a f(x) \sin \lambda_n x \, dx$

and the λ_n are the consecutive positive roots of $\tan \lambda a = -\lambda/h$

5. $u(x, t) = \sum\limits_{n=1}^{\infty} A_n e^{-k(2n-1)^2\pi^2 t/4L^2} \sin\left(\dfrac{2n-1}{2L} \right)\pi x$,

where $A_n = \dfrac{2}{L} \int_0^L f(x) \sin\left(\dfrac{2n-1}{2L} \right)\pi x \, dx$

7. $u(x, y) = \dfrac{4u_0}{\pi} \sum\limits_{n=1}^{\infty} \dfrac{1}{(2n-1) \cosh\left(\dfrac{2n-1}{2} \right)\pi}$

$\times \cosh\left(\dfrac{2n-1}{2} \right)\pi x \sin\left(\dfrac{2n-1}{2} \right)\pi y$

9. (b) 1.8751, 4.6941

EXERCISES 12.8, Page 682

1. $u(x, y, t) = \sum\limits_{m=1}^{\infty} \sum\limits_{n=1}^{\infty} A_{mn} e^{-k(m^2+n^2)t} \sin mx \sin ny$,

where $A_{mn} = \dfrac{4u_0}{mn\pi^2} [1 - (-1)^m][1 - (-1)^n]$

3. $u(x, y, t) = \sum\limits_{m=1}^{\infty} \sum\limits_{n=1}^{\infty} A_{mn} \sin mx \sin ny \cos a\sqrt{m^2+n^2}\, t$,

where $A_{mn} = \dfrac{16}{m^3 n^3 \pi^2} [(-1)^m - 1][(-1)^n - 1]$

5. $u(x, y, z) = \sum\limits_{m=1}^{\infty} \sum\limits_{n=1}^{\infty} A_{mn} \sinh \omega_{mn} z \sin \dfrac{m\pi}{a} x \sin \dfrac{n\pi}{b} y$,

where $\omega_{mn} = \sqrt{(m\pi/a)^2 + (n\pi/b)^2}$ and

$$A_{mn} = \dfrac{4}{ab \sinh(c\omega_{mn})} \int_0^b \int_0^a f(x, y)$$

$$\times \sin \dfrac{m\pi}{a} x \sin \dfrac{n\pi}{b} y \, dx \, dy$$

CHAPTER 12 REVIEW EXERCISES, Page 683

1. $u = c_1 e^{(c_2 x + y/c_2)}$ **3.** $\psi(x) = u_0 + \dfrac{(u_1 - u_0)}{1 + \pi} x$

5. $u(x, t) = \dfrac{2h}{\pi^2 a} \sum\limits_{n=1}^{\infty} \dfrac{\cos \dfrac{n\pi}{4} - \cos \dfrac{3n\pi}{4}}{n^2}$

$\times \sin n\pi a t \sin n\pi x$

7. $u(x, y) = \dfrac{100}{\pi} \sum\limits_{n=1}^{\infty} \dfrac{1 - (-1)^n}{n \sinh n\pi} \sinh nx \sin ny$

9. $u(x, y) = \dfrac{100}{\pi} \sum\limits_{n=1}^{\infty} \dfrac{1 - (-1)^n}{n} e^{-nx} \sin ny$

11. $u(x, t) = e^{-t} \sin x$

13. $u(x, t) = e^{-(x+t)} \sum\limits_{n=1}^{\infty} A_n[\sqrt{n^2 + 1} \cos \sqrt{n^2 + 1}\, t$

$+ \sin \sqrt{n^2 + 1}\, t] \sin nx$

EXERCISES 13.1, Page 698

1. $u(r, \theta) = \dfrac{u_0}{2} + \dfrac{u_0}{\pi} \sum\limits_{n=1}^{\infty} \dfrac{1 - (-1)^n}{n} r^n \sin n\theta$

3. $u(r, \theta) = \dfrac{2\pi^2}{3} - 4 \sum\limits_{n=1}^{\infty} \dfrac{r^n}{n^2} \cos n\theta$

5. $u(r, \theta) = A_0 + \sum\limits_{n=1}^{\infty} r^{-n}(A_n \cos n\theta + B_n \sin n\theta)$, where

$$A_0 = \frac{1}{2\pi} \int_0^{2\pi} f(\theta) \, d\theta$$

$$A_n = \frac{c^n}{\pi} \int_0^{2\pi} f(\theta) \cos n\theta \, d\theta$$

$$B_n = \frac{c^n}{\pi} \int_0^{2\pi} f(\theta) \sin n\theta \, d\theta$$

7. $u(r, \theta) = \dfrac{1}{2} + \dfrac{2}{\pi} \sum\limits_{n=1}^{\infty} \dfrac{\sin \frac{n\pi}{2}}{n} \left(\dfrac{r}{c}\right)^{2n} \cos 2n\theta$

9. $u(r, \theta) = A_0 \ln\left(\dfrac{r}{b}\right) + \sum\limits_{n=1}^{\infty} \left[\left(\dfrac{b}{r}\right)^n - \left(\dfrac{r}{b}\right)^n\right]$

$\qquad \times [A_n \cos n\theta + B_n \sin n\theta]$, where

$$A_0 \ln\left(\frac{a}{b}\right) = \frac{1}{2\pi} \int_0^{2\pi} f(\theta) \, d\theta,$$

$$\left[\left(\frac{b}{a}\right)^n - \left(\frac{a}{b}\right)^n\right] A_n = \frac{1}{\pi} \int_0^{2\pi} f(\theta) \cos n\theta \, d\theta,$$

$$\left[\left(\frac{b}{a}\right)^n - \left(\frac{a}{b}\right)^n\right] B_n = \frac{1}{\pi} \int_0^{2\pi} f(\theta) \sin n\theta \, d\theta$$

11. $u(r, \theta) = \dfrac{4}{\pi} \sum\limits_{n=1}^{\infty} \dfrac{1 - (-1)^n}{n^3} \dfrac{r^{2n} - b^{2n}}{a^{2n} - b^{2n}} \left(\dfrac{a}{r}\right)^n \sin n\theta$

13. $u(r, \theta) = \dfrac{u_0}{2} + \dfrac{2u_0}{\pi} \sum\limits_{n=1}^{\infty} \dfrac{\sin \frac{n\pi}{2}}{n} \left(\dfrac{r}{2}\right)^n \cos n\theta$

EXERCISES 13.2, Page 704

1. $u(r, t) = \dfrac{2}{ac} \sum\limits_{n=1}^{\infty} \dfrac{\sin \lambda_n at \, J_0(\lambda_n r)}{\lambda_n^2 J_1(\lambda_n c)}$

3. $u(r, z) = u_0 \sum\limits_{n=1}^{\infty} \dfrac{\sinh \lambda_n(4 - z) \, J_0(\lambda_n r)}{\lambda_n \sinh 4\lambda_n \, J_1(2\lambda_n)}$

5. $u(r, t) = \sum\limits_{n=1}^{\infty} A_n J_0(\lambda_n r) e^{-k\lambda_n^2 t}$, where

$$A_n = \frac{2}{c^2 J_1^2(\lambda_n c)} \int_0^c r J_0(\lambda_n r) f(r) \, dr$$

7. $u(r, t) = \sum\limits_{n=1}^{\infty} A_n J_0(\lambda_n r) e^{-k\lambda_n^2 t}$, where

$$A_n = \frac{2\lambda_n^2}{(\lambda_n^2 + h^2) J_0^2(\lambda_n)} \int_0^1 r J_0(\lambda_n r) f(r) \, dr$$

9. $u(r, t) = 100 + 50 \sum\limits_{n=1}^{\infty} \dfrac{J_1(\lambda_n) J_0(\lambda_n r)}{\lambda_n J_1^2(2\lambda_n)} e^{-\lambda_n^2 t}$

11. (b) $u(x, t) = \sum\limits_{n=1}^{\infty} A_n \cos(\lambda_n \sqrt{g} \, t) J_0(2\lambda_n \sqrt{x})$, where

$$A_n = \frac{2}{LJ_1^2(2\lambda_n \sqrt{L})} \int_0^{\sqrt{L}} v J_0(2\lambda_n v) f(v^2) \, dv$$

13. (a) $\lambda_1 = 0.2405, \lambda_2 = 0.5520, \lambda_3 = 0.8654$;
$\qquad A_1 = 0.7966, A_2 = 0.0687, A_3 = 0.0535$;
$\qquad S_3 = 0.7966 \cos(0.2405t) J_0(0.2405r)$
$\qquad\quad + 0.0687 \cos(0.5520t) J_0(0.5520r)$
$\qquad\quad + 0.0535 \cos(0.8654t) J_0(0.8654r)$

(b)

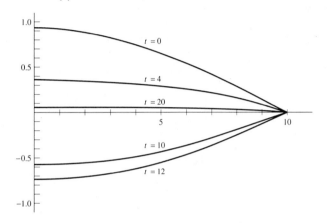

EXERCISES 13.3, Page 708

1. $u(r, \theta) = 50\left[\dfrac{1}{2} P_0(\cos \theta) + \dfrac{3}{4}\left(\dfrac{r}{c}\right) P_1(\cos \theta)\right.$

$\qquad \left. - \dfrac{7}{16}\left(\dfrac{r}{c}\right)^3 P_3(\cos \theta) + \dfrac{11}{32}\left(\dfrac{r}{c}\right)^5 P_5(\cos \theta) + \cdots\right]$

3. $u(r, \theta) = \dfrac{r}{c} \cos \theta$

5. $u(r, \theta) = \sum\limits_{n=0}^{\infty} A_n \dfrac{b^{2n+1} - r^{2n+1}}{b^{2n+1} r^{n+1}} P_n(\cos \theta)$, where

$$\frac{b^{2n+1} - a^{2n+1}}{b^{2n+1} a^{n+1}} A_n = \frac{2n+1}{2} \int_0^{\pi} f(\theta) P_n(\cos \theta) \sin \theta \, d\theta$$

7. $u(r, \theta) = \sum\limits_{n=0}^{\infty} A_n r^{2n} P_{2n}(\cos \theta)$, where

$$A_n = \frac{(4n+1)}{c^{2n}} \int_0^{\pi/2} f(\theta) P_{2n}(\cos \theta) \sin \theta \, d\theta$$

9. $u(r, t) = 100 + \dfrac{200}{\pi r} \sum\limits_{n=1}^{\infty} \dfrac{(-1)^n}{n} e^{-n^2\pi^2 t} \sin n\pi r$

11. $u(r, t) = \sum_{n=1}^{\infty} \left(A_n \cos \dfrac{n\pi a}{c} t + B_n \sin \dfrac{n\pi a}{c} t \right) \sin \dfrac{n\pi}{c} r,$

where $A_n = \dfrac{2}{c} \displaystyle\int_0^c r f(r) \sin \dfrac{n\pi}{c} r \, dr,$

$B_n = \dfrac{2}{n\pi a} \displaystyle\int_0^c r g(r) \sin \dfrac{n\pi}{c} r \, dr$

CHAPTER 13 REVIEW EXERCISES, Page 710

1. $u(r, \theta) = \dfrac{2u_0}{\pi} \displaystyle\sum_{n=1}^{\infty} \dfrac{1 - (-1)^n}{n} \left(\dfrac{r}{c} \right)^n \sin n\theta$

3. $u(r, \theta) = \dfrac{4u_0}{\pi} \displaystyle\sum_{n=1}^{\infty} \dfrac{1 - (-1)^n}{n^3} \sin n\theta$

5. $u(r, \theta) = \dfrac{2u_0}{\pi} \displaystyle\sum_{n=1}^{\infty} \dfrac{r^{4n} + r^{-4n}}{2^{4n} + 2^{-4n}} \dfrac{1 - (-1)^n}{n} \sin 4n\theta$

7. $u(r, t) = 2e^{-ht} \displaystyle\sum_{n=1}^{\infty} \dfrac{J_0(\lambda_n r)}{\lambda_n J_1(\lambda_n)} e^{-\lambda_n^2 t}$

9. $u(r, z) = 50 \displaystyle\sum_{n=1}^{\infty} \dfrac{\cosh \lambda_n z \, J_0(\lambda_n r)}{\lambda_n \cosh 4\lambda_n \, J_1(2\lambda_n)}$

11. $u(r, \theta) = 100 \left[\dfrac{3}{2} r P_1(\cos \theta) - \dfrac{7}{8} r^3 P_3(\cos \theta) \right.$

$\left. + \dfrac{11}{16} r^5 P_5(\cos \theta) + \cdots \right]$

EXERCISES 14.1, Page 717

1. (a) Let $\tau = u^2$ in the integral $\operatorname{erf}(\sqrt{t})$.

(b) $\mathscr{L}\{\operatorname{erf}(\sqrt{t})\} = \dfrac{1}{\sqrt{\pi}} \mathscr{L}\{1\} \mathscr{L}\{t^{-1/2} e^{-t}\}$

$= \left. \dfrac{1}{\sqrt{\pi}} \dfrac{1}{s} \left(\dfrac{\sqrt{\pi}}{s^{1/2}} \right) \right|_{s \to s+1}$

$= \dfrac{1}{s\sqrt{s+1}}$

3. $\mathscr{L}\{e^t \operatorname{erf}(\sqrt{t})\} = \mathscr{L}\{\operatorname{erf}(\sqrt{t})\}|_{s \to s-1}$

$= \left. \dfrac{1}{s\sqrt{s+1}} \right|_{s \to s-1}$

$= \dfrac{1}{(s-1)\sqrt{s}}$

5. $\mathscr{L}\left\{ e^{-Gt/C} \operatorname{erf}\left(\dfrac{x}{2} \sqrt{\dfrac{RC}{t}} \right) \right\}$

$= \mathscr{L}\left\{ e^{-Gt/C} \left[1 - \operatorname{erfc}\left(\dfrac{x}{2} \sqrt{\dfrac{RC}{t}} \right) \right] \right\}$

$= \left. \dfrac{1}{s + G/C} - \dfrac{e^{-x\sqrt{RC}\sqrt{s}}}{s} \right|_{s \to s + G/C}$

$= \dfrac{1}{s + G/C} - \dfrac{e^{-x\sqrt{RC}\sqrt{s + G/C}}}{s + G/C}$

$= \dfrac{C}{Cs + G} \{ 1 - e^{-x\sqrt{RCs + RG}} \}$

7. $y(t) = e^{\pi t} \operatorname{erfc}(\sqrt{\pi t})$

9. Use the property $\displaystyle\int_0^b - \int_0^a = \int_0^b + \int_a^0$

EXERCISES 14.2, Page 723

1. $u(x, t) = A \cos \dfrac{a\pi}{L} t \sin \dfrac{\pi}{L} x$

3. $u(x, t) = f\left(t - \dfrac{x}{a} \right) \mathscr{U}\left(t - \dfrac{x}{a} \right)$

5. $u(x, t) = \left[\dfrac{1}{2} g\left(t - \dfrac{x}{a} \right)^2 + A \sin \omega\left(t - \dfrac{x}{a} \right) \right]$

$\times \mathscr{U}\left(t - \dfrac{x}{a} \right) - \dfrac{1}{2} g t^2$

7. $u(x, t) = a \dfrac{F_0}{E} \displaystyle\sum_{n=0}^{\infty} (-1)^n \left\{ \left(t - \dfrac{2nL + L - x}{a} \right) \right.$

$\times \mathscr{U}\left(t - \dfrac{2nL + L - x}{a} \right)$

$\left. - \left(t - \dfrac{2nL + L + x}{a} \right) \mathscr{U}\left(t - \dfrac{2nL + L + x}{a} \right) \right\}$

9. $u(x, t) = (t - x) \sinh(t - x) \mathscr{U}(t - x)$
$+ x e^{-x} \cosh t - e^{-x} t \sinh t$

11. $u(x, t) = u_0 \operatorname{erfc}\left(\dfrac{x}{2\sqrt{kt}} \right)$

13. $u(x, t) = u_1 + (u_0 - u_1) \operatorname{erfc}\left(\dfrac{x}{2\sqrt{t}} \right)$

15. $u(x, t) = u_0 \left[1 - \left\{ \operatorname{erfc}\left(\dfrac{x}{2\sqrt{t}} \right) \right.\right.$

$\left.\left. - e^{x+t} \operatorname{erfc}\left(\sqrt{t} + \dfrac{x}{2\sqrt{t}} \right) \right\} \right]$

17. $u(x, t) = \dfrac{x}{2\sqrt{\pi}} \displaystyle\int_0^t \dfrac{f(t - \tau)}{\tau^{3/2}} e^{-x^2/4\tau} \, d\tau$

19. $u(x, t) = 60 + 40 \, \text{erfc}\left(\dfrac{x}{2\sqrt{t - 2}}\right) \mathscr{U}(t - 2)$

21. $u(x, t) = 100\left[-e^{1-x+t} \, \text{erfc}\left(\sqrt{t} + \dfrac{1 - x}{2\sqrt{t}}\right) \right.$

$\left. + \text{erfc}\left(\dfrac{1 - x}{2\sqrt{t}}\right) \right]$

23. $u(x, t) = u_0 + u_0 e^{-(\pi^2/L^2)t} \sin\left(\dfrac{\pi}{L} x\right)$

25. $u(x, t) = u_0 - u_0 \displaystyle\sum_{n=0}^{\infty} (-1)^n \left[\text{erfc}\left(\dfrac{2n + 1 - x}{2\sqrt{kt}}\right) \right.$

$\left. + \text{erfc}\left(\dfrac{2n + 1 + x}{2\sqrt{kt}}\right) \right]$

27. $u(x, t) = u_0 e^{-Gt/C} \, \text{erf}\left(\dfrac{x}{2}\sqrt{\dfrac{RC}{t}}\right)$

29. $u(x, t) = A\sqrt{\dfrac{k}{\pi t}} \, e^{-x^2/4kt}$; an impulse of heat, or flash burn, takes place at $x = 0$

EXERCISES 14.3, Page 734

1. $f(x) = \dfrac{1}{\pi} \displaystyle\int_0^{\infty} \dfrac{\sin \alpha \cos \alpha x + 3(1 - \cos \alpha) \sin \alpha x}{\alpha} \, d\alpha$

3. $f(x) = \dfrac{1}{\pi} \displaystyle\int_0^{\infty} [A(\alpha) \cos \alpha x + B(\alpha) \sin \alpha x] \, d\alpha$,

where $A(\alpha) = (3\alpha \sin 3\alpha + \cos 3\alpha - 1)/\alpha^2$
$B(\alpha) = (\sin 3\alpha - 3\alpha \cos 3\alpha)/\alpha^2$

5. $f(x) = \dfrac{1}{\pi} \displaystyle\int_0^{\infty} \dfrac{\cos \alpha x + \alpha \sin \alpha x}{1 + \alpha^2} \, d\alpha$

7. $f(x) = \dfrac{10}{\pi} \displaystyle\int_0^{\infty} \dfrac{(1 - \cos \alpha) \sin \alpha x}{\alpha} \, d\alpha$

9. $f(x) = \dfrac{2}{\pi} \displaystyle\int_0^{\infty} \dfrac{(\pi\alpha \sin \pi\alpha + \cos \pi\alpha - 1) \cos \alpha x}{\alpha^2} \, d\alpha$

11. $f(x) = \dfrac{4}{\pi} \displaystyle\int_0^{\infty} \dfrac{\alpha \sin \alpha x}{4 + \alpha^4} \, d\alpha$

13. $f(x) = \dfrac{2k}{\pi} \displaystyle\int_0^{\infty} \dfrac{\cos \alpha x}{k^2 + \alpha^2} \, d\alpha$

$f(x) = \dfrac{2}{\pi} \displaystyle\int_0^{\infty} \dfrac{\alpha \sin \alpha x}{k^2 + \alpha^2} \, d\alpha$

15. $f(x) = \dfrac{2}{\pi} \displaystyle\int_0^{\infty} \dfrac{(4 - \alpha^2) \cos \alpha x}{(4 + \alpha^2)^2} \, d\alpha$

$f(x) = \dfrac{8}{\pi} \displaystyle\int_0^{\infty} \dfrac{\alpha \sin \alpha x}{(4 + \alpha^2)^2} \, d\alpha$

17. $f(x) = \dfrac{2}{\pi} \dfrac{1}{1 + x^2}, \; x > 0$

19. Let $x = 2$ in (7). Use a trigonometric identity and replace α by x. In part (b) make the change of variable $2x = kt$.

EXERCISES 14.4, Page 740

1. $u(x, t) = \dfrac{1}{\pi} \displaystyle\int_{-\infty}^{\infty} \dfrac{e^{-k\alpha^2 t}}{1 + \alpha^2} e^{-i\alpha x} \, d\alpha$

$= \dfrac{1}{\pi} \displaystyle\int_{-\infty}^{\infty} \dfrac{\cos \alpha x}{1 + \alpha^2} e^{-k\alpha^2 t} \, d\alpha$

3. $u(x, t) = \dfrac{1}{\sqrt{1 + 4kt}} e^{-x^2/(1 + 4kt)}$

5. $u(x, t) = \dfrac{2u_0}{\pi} \displaystyle\int_0^{\infty} \dfrac{1 - e^{-k\alpha^2 t}}{\alpha} \sin \alpha x \, d\alpha$

7. $u(x, t) = \dfrac{2}{\pi} \displaystyle\int_0^{\infty} \dfrac{1 - \cos \alpha}{\alpha} e^{-k\alpha^2 t} \sin \alpha x \, d\alpha$

9. $u(x, t) = \dfrac{2}{\pi} \displaystyle\int_0^{\infty} \dfrac{\sin \alpha}{\alpha} e^{-k\alpha^2 t} \cos \alpha x \, d\alpha$

11. **(a)** $u(x, t) = \dfrac{1}{2\pi} \displaystyle\int_{-\infty}^{\infty} \left(F(\alpha) \cos \alpha a t \right.$

$\left. + G(\alpha) \dfrac{\sin \alpha a t}{\alpha a} \right) e^{-i\alpha x} \, d\alpha$

13. $u(x, y) = \dfrac{2}{\pi} \displaystyle\int_0^{\infty} \dfrac{\sinh \alpha(\pi - x)}{(1 + \alpha^2) \sinh \alpha\pi} \cos \alpha y \, d\alpha$

15. $u(x, y) = \dfrac{100}{\pi} \displaystyle\int_0^{\infty} \dfrac{\sin \alpha}{\alpha} e^{-\alpha y} \cos \alpha x \, d\alpha$

17. $u(x, y) = \dfrac{2}{\pi} \displaystyle\int_0^{\infty} F(\alpha) \dfrac{\sinh \alpha(2 - y)}{\sinh 2\alpha} \sin \alpha x \, d\alpha$

19. $u(x, y) = \dfrac{2}{\pi} \displaystyle\int_0^{\infty} \dfrac{\alpha}{1 + \alpha^2} [e^{-\alpha x} \sin \alpha y + e^{-\alpha y} \sin \alpha x] \, d\alpha$

21. $u(x, y) = \dfrac{1}{2\sqrt{\pi}} \displaystyle\int_{-\infty}^{\infty} \dfrac{e^{-\alpha^2/4} \cosh \alpha y}{\cosh \alpha} e^{-i\alpha x} \, d\alpha$

$= \dfrac{1}{2\sqrt{\pi}} \displaystyle\int_{-\infty}^{\infty} \dfrac{e^{-\alpha^2/4} \cosh \alpha y}{\cosh \alpha} \cos \alpha x \, d\alpha$

CHAPTER 14 REVIEW EXERCISES, Page 744

1. $u(x, y) = \dfrac{2}{\pi} \displaystyle\int_0^\infty \dfrac{\sinh \alpha y}{\alpha(1 + \alpha^2) \cosh \alpha\pi} \cos \alpha x \, d\alpha$

3. $u(x, t) = u_0 e^{-ht} \, \mathrm{erf}\left(\dfrac{x}{2\sqrt{t}}\right)$

5. $u(x, t) = \displaystyle\int_0^t \mathrm{erfc}\left(\dfrac{x}{2\sqrt{\tau}}\right) d\tau$

7. $u(x, t) = \dfrac{u_0}{2\pi} \displaystyle\int_{-\infty}^\infty \dfrac{\sin \alpha(\pi - x) + \sin \alpha x}{\alpha} e^{-k\alpha^2 t} \, d\alpha$

9. $u(x, y) = \dfrac{100}{\pi} \displaystyle\int_0^\infty \left(\dfrac{1 - \cos \alpha}{\alpha}\right)[e^{-\alpha x} \sin \alpha y$
$\qquad\qquad + 2e^{-\alpha y} \sin \alpha x] \, d\alpha$

11. $u(x, y) = \dfrac{2}{\pi} \displaystyle\int_0^\infty \left(\dfrac{B \cosh \alpha y}{(1 + \alpha^2) \sinh \alpha\pi} + \dfrac{A}{\alpha}\right) \sin \alpha x \, d\alpha$

13. $u(x, y) = \dfrac{1}{2\pi} \displaystyle\int_{-\infty}^\infty \dfrac{\cos \alpha x + \alpha \sin \alpha x}{1 + \alpha^2} e^{-k\alpha^2 t} \, d\alpha$

EXERCISES 15.1, Page 755

1. $u_{11} = 11/15$, $u_{21} = 14/15$

3. $u_{11} = u_{21} = \sqrt{3}/16$, $u_{22} = u_{12} = 3\sqrt{3}/16$

5. $u_{21} = u_{12} = 12.50$, $u_{31} = u_{13} = 18.75$, $u_{32} = u_{23} = 37.50$
$u_{11} = 6.25$, $u_{22} = 25.00$, $u_{33} = 56.25$

7. (b) $u_{14} = u_{41} = 0.5427$, $u_{24} = u_{42} = 0.6707$,
$\qquad u_{34} = u_{43} = 0.6402$
$\qquad u_{33} = 0.9451$, $u_{44} = 0.4451$

9. $u_{i, j+1} = (1 - 2\lambda)u_{ij} + \lambda(u_{i+1, j} + u_{i-1, j})$, where
$\lambda = k/h^2$

EXERCISES 15.2, Page 765

1.

Time	x = 0.25	x = 0.50	x = 0.75	x = 1.00	x = 1.25	x = 1.50	x = 1.75
0.000	1.0000	1.0000	1.0000	1.0000	0.0000	0.0000	0.0000
0.025	0.6000	1.0000	1.0000	0.6000	0.4000	0.0000	0.0000
0.050	0.5200	0.8400	0.8400	0.6800	0.3200	0.1600	0.0000
0.075	0.4400	0.7120	0.7760	0.6000	0.4000	0.1600	0.0640
0.100	0.3728	0.6288	0.6800	0.5904	0.3840	0.2176	0.0768
0.125	0.3261	0.5469	0.6237	0.5437	0.4000	0.2278	0.1024
0.150	0.2840	0.4893	0.5610	0.5182	0.3886	0.2465	0.1116
0.175	0.2525	0.4358	0.5152	0.4835	0.3836	0.2494	0.1209
0.200	0.2248	0.3942	0.4708	0.4562	0.3699	0.2517	0.1239
0.225	0.2027	0.3571	0.4343	0.4275	0.3571	0.2479	0.1255
0.250	0.1834	0.3262	0.4007	0.4021	0.3416	0.2426	0.1242
0.275	0.1672	0.2989	0.3715	0.3773	0.3262	0.2348	0.1219
0.300	0.1530	0.2752	0.3448	0.3545	0.3101	0.2262	0.1183
0.325	0.1407	0.2541	0.3209	0.3329	0.2943	0.2166	0.1141
0.350	0.1298	0.2354	0.2990	0.3126	0.2787	0.2067	0.1095
0.375	0.1201	0.2186	0.2790	0.2936	0.2635	0.1966	0.1046
0.400	0.1115	0.2034	0.2607	0.2757	0.2488	0.1865	0.0996
0.425	0.1036	0.1895	0.2438	0.2589	0.2347	0.1766	0.0945
0.450	0.0965	0.1769	0.2281	0.2432	0.2211	0.1670	0.0896
0.475	0.0901	0.1652	0.2136	0.2283	0.2083	0.1577	0.0847
0.500	0.0841	0.1545	0.2002	0.2144	0.1961	0.1487	0.0800
0.525	0.0786	0.1446	0.1876	0.2014	0.1845	0.1402	0.0755
0.550	0.0736	0.1354	0.1759	0.1891	0.1735	0.1320	0.0712
0.575	0.0689	0.1269	0.1650	0.1776	0.1632	0.1243	0.0670
0.600	0.0645	0.1189	0.1548	0.1668	0.1534	0.1169	0.0631

Time	$x = 0.25$	$x = 0.50$	$x = 0.75$	$x = 1.00$	$x = 1.25$	$x = 1.50$	$x = 1.75$
0.625	0.0605	0.1115	0.1452	0.1566	0.1442	0.1100	0.0594
0.650	0.0567	0.1046	0.1363	0.1471	0.1355	0.1034	0.0559
0.675	0.0532	0.0981	0.1279	0.1381	0.1273	0.0972	0.0525
0.700	0.0499	0.0921	0.1201	0.1297	0.1196	0.0914	0.0494
0.725	0.0468	0.0864	0.1127	0.1218	0.1124	0.0859	0.0464
0.750	0.0439	0.0811	0.1058	0.1144	0.1056	0.0807	0.0436
0.775	0.0412	0.0761	0.0994	0.1074	0.0992	0.0758	0.0410
0.800	0.0387	0.0715	0.0933	0.1009	0.0931	0.0712	0.0385
0.825	0.0363	0.0671	0.0876	0.0948	0.0875	0.0669	0.0362
0.850	0.0341	0.0630	0.0823	0.0890	0.0822	0.0628	0.0340
0.875	0.0320	0.0591	0.0772	0.0836	0.0772	0.0590	0.0319
0.900	0.0301	0.0555	0.0725	0.0785	0.0725	0.0554	0.0300
0.925	0.0282	0.0521	0.0681	0.0737	0.0681	0.0521	0.0282
0.950	0.0265	0.0490	0.0640	0.0692	0.0639	0.0489	0.0265
0.975	0.0249	0.0460	0.0601	0.0650	0.0600	0.0459	0.0249
1.000	0.0234	0.0432	0.0564	0.0610	0.0564	0.0431	0.0233

3.

Time	$x = 0.25$	$x = 0.50$	$x = 0.75$	$x = 1.00$	$x = 1.25$	$x = 1.50$	$x = 1.75$
0.000	1.0000	1.0000	1.0000	1.0000	0.0000	0.0000	0.0000
0.025	0.7074	0.9520	0.9566	0.7444	0.2545	0.0371	0.0053
0.050	0.5606	0.8499	0.8685	0.6633	0.3303	0.1034	0.0223
0.075	0.4684	0.7473	0.7836	0.6191	0.3614	0.1529	0.0462
0.100	0.4015	0.6577	0.7084	0.5837	0.3753	0.1871	0.0684
0.125	0.3492	0.5821	0.6428	0.5510	0.3797	0.2101	0.0861
0.150	0.3069	0.5187	0.5857	0.5199	0.3778	0.2247	0.0990
0.175	0.2721	0.4652	0.5359	0.4901	0.3716	0.2329	0.1078
0.200	0.2430	0.4198	0.4921	0.4617	0.3622	0.2362	0.1132
0.225	0.2186	0.3809	0.4533	0.4348	0.3507	0.2358	0.1160
0.250	0.1977	0.3473	0.4189	0.4093	0.3378	0.2327	0.1166
0.275	0.1798	0.3181	0.3881	0.3853	0.3240	0.2275	0.1157
0.300	0.1643	0.2924	0.3604	0.3626	0.3097	0.2208	0.1136
0.325	0.1507	0.2697	0.3353	0.3412	0.2953	0.2131	0.1107
0.350	0.1387	0.2495	0.3125	0.3211	0.2808	0.2047	0.1071
0.375	0.1281	0.2313	0.2916	0.3021	0.2666	0.1960	0.1032
0.400	0.1187	0.2150	0.2725	0.2843	0.2528	0.1871	0.0989
0.425	0.1102	0.2002	0.2549	0.2675	0.2393	0.1781	0.0946
0.450	0.1025	0.1867	0.2387	0.2517	0.2263	0.1692	0.0902
0.475	0.0955	0.1743	0.2236	0.2368	0.2139	0.1606	0.0858
0.500	0.0891	0.1630	0.2097	0.2228	0.2020	0.1521	0.0814
0.525	0.0833	0.1525	0.1967	0.2096	0.1906	0.1439	0.0772
0.550	0.0779	0.1429	0.1846	0.1973	0.1798	0.1361	0.0731
0.575	0.0729	0.1339	0.1734	0.1856	0.1696	0.1285	0.0691
0.600	0.0683	0.1256	0.1628	0.1746	0.1598	0.1214	0.0653

Time	x = 0.25	x = 0.50	x = 0.75	x = 1.00	x = 1.25	x = 1.50	x = 1.75
0.625	0.0641	0.1179	0.1530	0.1643	0.1506	0.1145	0.0617
0.650	0.0601	0.1106	0.1438	0.1546	0.1419	0.1080	0.0582
0.675	0.0564	0.1039	0.1351	0.1455	0.1336	0.1018	0.0549
0.700	0.0530	0.0976	0.1270	0.1369	0.1259	0.0959	0.0518
0.725	0.0497	0.0917	0.1194	0.1288	0.1185	0.0904	0.0488
0.750	0.0467	0.0862	0.1123	0.1212	0.1116	0.0852	0.0460
0.775	0.0439	0.0810	0.1056	0.1140	0.1050	0.0802	0.0433
0.800	0.0413	0.0762	0.0993	0.1073	0.0989	0.0755	0.0408
0.825	0.0388	0.0716	0.0934	0.1009	0.0931	0.0711	0.0384
0.850	0.0365	0.0674	0.0879	0.0950	0.0876	0.0669	0.0362
0.875	0.0343	0.0633	0.0827	0.0894	0.0824	0.0630	0.0341
0.900	0.0323	0.0596	0.0778	0.0841	0.0776	0.0593	0.0341
0.925	0.0303	0.0560	0.0732	0.0791	0.0730	0.0558	0.0302
0.950	0.0285	0.0527	0.0688	0.0744	0.0687	0.0526	0.0284
0.975	0.0268	0.0496	0.0647	0.0700	0.0647	0.0495	0.0268
1.000	0.0253	0.0466	0.0609	0.0659	0.0608	0.0465	0.0252

Absolute errors are approximately 2.2×10^{-2}, 3.7×10^{-2}, 1.3×10^{-2}.

5.

Time	x = 0.25	x = 0.50	x = 0.75	x = 1.00	x = 1.25	x = 1.50	x = 1.75
0.00	1.0000	1.0000	1.0000	1.0000	0.0000	0.0000	0.0000
0.05	0.5265	0.8693	0.8852	0.6141	0.3783	0.0884	0.0197
0.10	0.3972	0.6551	0.7043	0.5883	0.3723	0.1955	0.0653
0.15	0.3042	0.5150	0.5844	0.5192	0.3812	0.2261	0.1010
0.20	0.2409	0.4171	0.4901	0.4620	0.3636	0.2385	0.1145
0.25	0.1962	0.3452	0.4174	0.4092	0.3391	0.2343	0.1178
0.30	0.1631	0.2908	0.3592	0.3624	0.3105	0.2220	0.1145
0.35	0.1379	0.2482	0.3115	0.3208	0.2813	0.2056	0.1077
0.40	0.1181	0.2141	0.2718	0.2840	0.2530	0.1876	0.0993
0.45	0.1020	0.1860	0.2381	0.2514	0.2265	0.1696	0.0904
0.50	0.0888	0.1625	0.2092	0.2226	0.2020	0.1523	0.0816
0.55	0.0776	0.1425	0.1842	0.1970	0.1798	0.1361	0.0732
0.60	0.0681	0.1253	0.1625	0.1744	0.1597	0.1214	0.0654
0.65	0.0599	0.1104	0.1435	0.1544	0.1418	0.1079	0.0582
0.70	0.0528	0.0974	0.1268	0.1366	0.1257	0.0959	0.0518
0.75	0.0466	0.0860	0.1121	0.1210	0.1114	0.0851	0.0460
0.80	0.0412	0.0760	0.0991	0.1071	0.0987	0.0754	0.0408
0.85	0.0364	0.0672	0.0877	0.0948	0.0874	0.0668	0.0361
0.90	0.0322	0.0594	0.0776	0.0839	0.0774	0.0592	0.0320
0.95	0.0285	0.0526	0.0687	0.0743	0.0686	0.0524	0.0284
1.00	0.0252	0.0465	0.0608	0.0657	0.0607	0.0464	0.0251

Absolute errors are approximately 1.8×10^{-2}, 3.7×10^{-2}, 1.3×10^{-2}.

7. (a)

Time	$x = 2.00$	$x = 4.00$	$x = 6.00$	$x = 8.00$	$x = 10.00$	$x = 12.00$	$x = 14.00$	$x = 16.00$	$x = 18.00$
0.00	30.0000	30.0000	30.0000	30.0000	30.0000	30.0000	30.0000	30.0000	30.0000
1.00	28.7733	29.9749	29.9995	30.0000	30.0000	30.0000	29.9995	29.9749	28.7733
2.00	27.6450	29.9037	29.9970	29.9999	30.0000	29.9999	29.9970	29.9037	27.6450
3.00	26.6051	29.7938	29.9911	29.9997	30.0000	29.9997	29.9911	29.7938	26.6051
4.00	25.6452	29.6517	29.9805	29.9991	29.9999	29.9991	29.9805	29.6517	25.6452
5.00	24.7573	29.4829	29.9643	29.9981	29.9998	29.9981	29.9643	29.4829	24.7573
6.00	23.9347	29.2922	29.9421	29.9963	29.9996	29.9963	29.9421	29.2922	23.9347
7.00	23.1711	29.0836	29.9134	29.9936	29.9992	29.9936	29.9134	29.0836	23.1711
8.00	22.4612	28.8606	29.8782	29.9898	29.9986	29.9898	29.8782	28.8606	22.4612
9.00	21.7999	28.6263	29.8362	29.9848	29.9977	29.9848	29.8362	28.6263	21.7999
10.00	21.1829	28.3831	29.7878	29.9782	29.9964	29.9782	29.7878	28.3831	21.1829

(b)

Time	$x = 5.00$	$x = 10.00$	$x = 15.00$	$x = 20.00$	$x = 25.00$	$x = 30.00$	$x = 35.00$	$x = 40.00$	$x = 45.00$
0.00	30.0000	30.0000	30.0000	30.0000	30.0000	30.0000	30.0000	30.0000	30.0000
1.00	29.7968	29.9993	30.0000	30.0000	30.0000	30.0000	30.0000	29.9993	29.7968
2.00	29.5964	29.9973	30.0000	30.0000	30.0000	30.0000	30.0000	29.9973	29.5964
3.00	29.3987	29.9939	30.0000	30.0000	30.0000	30.0000	30.0000	29.9939	29.3987
4.00	29.2036	29.9893	29.9999	30.0000	30.0000	30.0000	29.9999	29.9893	29.2036
5.00	29.0112	29.9834	29.9998	30.0000	30.0000	30.0000	29.9998	29.9834	29.0112
6.00	28.8212	29.9762	29.9997	30.0000	30.0000	30.0000	29.9997	29.9762	28.8213
7.00	28.6339	29.9679	29.9995	30.0000	30.0000	30.0000	29.9995	29.9679	28.6339
8.00	28.4490	29.9585	29.9992	30.0000	30.0000	30.0000	29.9993	29.9585	28.4490
9.00	28.2665	29.9479	29.9989	30.0000	30.0000	30.0000	29.9989	29.9479	28.2665
10.00	28.0864	29.9363	29.9986	30.0000	30.0000	30.0000	29.9986	29.9363	28.0864

(c)

Time	$x = 2.00$	$x = 4.00$	$x = 6.00$	$x = 8.00$	$x = 10.00$	$x = 12.00$	$x = 14.00$	$x = 16.00$	$x = 18.00$
0.00	18.0000	32.0000	42.0000	48.0000	50.0000	48.0000	42.0000	32.0000	18.0000
1.00	16.4489	30.1970	40.1561	46.1495	48.1486	46.1495	40.1561	30.1970	16.4489
2.00	15.3312	28.5348	38.3465	44.3067	46.3001	44.3067	38.3465	28.5348	15.3312
3.00	14.4216	27.0416	36.6031	42.4847	44.4619	42.4847	36.6031	27.0416	14.4216
4.00	13.6371	25.6867	34.9416	40.6988	42.6453	40.6988	34.9416	25.6867	13.6371
5.00	12.9378	24.4419	33.3628	38.9611	40.8634	38.9611	33.3628	24.4419	12.9378
6.00	12.3012	23.2863	31.8624	37.2794	39.1273	37.2794	31.8624	23.2863	12.3012
7.00	11.7137	22.2051	30.4350	35.6578	37.4446	35.6578	30.4350	22.2051	11.7137
8.00	11.1659	21.1877	29.0757	34.0984	35.8202	34.0984	29.0757	21.1877	11.1659
9.00	10.6517	20.2261	27.7799	32.6014	34.2567	32.6014	27.7799	20.2261	10.6517
10.00	10.1665	19.3143	26.5439	31.1662	32.7549	31.1662	26.5439	19.3143	10.1665

(d)

Time	$x = 10.00$	$x = 20.00$	$x = 30.00$	$x = 40.00$	$x = 50.00$	$x = 60.00$	$x = 70.00$	$x = 80.00$	$x = 90.00$
0.00	8.0000	16.0000	24.0000	32.0000	40.0000	32.0000	24.0000	16.0000	8.0000
1.00	8.0000	16.0000	24.0000	31.9979	39.7425	31.9979	24.0000	16.0000	8.0000
2.00	8.0000	16.0000	23.9999	31.9918	39.4932	31.9918	23.9999	16.0000	8.0000
3.00	8.0000	16.0000	23.9997	31.9820	39.2517	31.9820	23.9997	16.0000	8.0000
4.00	8.0000	16.0000	23.9993	31.9686	39.0175	31.9686	23.9993	16.0000	8.0000
5.00	8.0000	16.0000	23.9987	31.9520	38.7905	31.9520	23.9987	16.0000	8.0000
6.00	8.0000	15.9999	23.9978	31.9323	38.5701	31.9323	23.9978	15.9999	8.0000
7.00	8.0000	15.9999	23.9966	31.9097	38.3561	31.9097	23.9966	15.9999	8.0000
8.00	8.0000	15.9998	23.9950	31.8844	38.1483	31.8844	23.9950	15.9998	8.0000
9.00	8.0000	15.9997	23.9931	31.8566	37.9463	31.8566	23.9931	15.9997	8.0000
10.00	8.0000	15.9996	23.9908	31.8265	37.7498	31.8265	23.9908	15.9996	8.0000

9. (a)

Time	$x = 2.00$	$x = 4.00$	$x = 6.00$	$x = 8.00$	$x = 10.00$	$x = 12.00$	$x = 14.00$	$x = 16.00$	$x = 18.00$
0.00	30.0000	30.0000	30.0000	30.0000	30.0000	30.0000	30.0000	30.0000	30.0000
1.00	28.7733	29.9749	29.9995	30.0000	30.0000	30.0000	29.9998	29.9916	29.5911
2.00	27.6450	29.9037	29.9970	29.9999	30.0000	30.0000	29.9990	29.9679	29.2150
3.00	26.6051	29.7938	29.9911	29.9997	30.0000	29.9999	29.9970	29.9313	28.8684
4.00	25.6452	29.6517	29.9805	29.9991	30.0000	29.9997	29.9935	29.8839	28.5484
5.00	24.7573	29.4829	29.9643	29.9981	29.9999	29.9994	29.9881	29.8276	28.2524
6.00	23.9347	29.2922	29.9421	29.9963	29.9997	29.9988	29.9807	29.7641	27.9782
7.00	23.1711	29.0836	29.9134	29.9936	29.9995	29.9979	29.9711	29.6945	27.7237
8.00	22.4612	28.8606	29.8782	29.9899	29.9991	29.9966	29.9594	29.6202	27.4870
9.00	21.7999	28.6263	29.8362	29.9848	29.9985	29.9949	29.9454	29.5421	27.2666
10.00	21.1829	28.3831	29.7878	29.9783	29.9976	29.9927	29.9293	29.4610	27.0610

(b)

Time	$x = 5.00$	$x = 10.00$	$x = 15.00$	$x = 20.00$	$x = 25.00$	$x = 30.00$	$x = 35.00$	$x = 40.00$	$x = 45.00$
0.00	30.0000	30.0000	30.0000	30.0000	30.0000	30.0000	30.0000	30.0000	30.0000
1.00	29.7968	29.9993	30.0000	30.0000	30.0000	30.0000	30.0000	29.9998	29.9323
2.00	29.5964	29.9973	30.0000	30.0000	30.0000	30.0000	30.0000	29.9991	29.8655
3.00	29.3987	29.9939	30.0000	30.0000	30.0000	30.0000	30.0000	29.9980	29.7996
4.00	29.2036	29.9893	29.9999	30.0000	30.0000	30.0000	30.0000	29.9964	29.7345
5.00	29.0112	29.9834	29.9998	30.0000	30.0000	30.0000	29.9999	29.9945	29.6704
6.00	28.8212	29.9762	29.9997	30.0000	30.0000	30.0000	29.9999	29.9921	29.6071
7.00	28.6339	29.9679	29.9995	30.0000	30.0000	30.0000	29.9998	29.9893	29.5446
8.00	28.4490	29.9585	29.9992	30.0000	30.0000	30.0000	29.9997	29.9862	29.4830
9.00	28.2665	29.9479	29.9989	30.0000	30.0000	30.0000	29.9996	29.9827	29.4222
10.00	28.0864	29.9363	29.9986	30.0000	30.0000	30.0000	29.9995	29.9788	29.3621

(c)

Time	x = 2.00	x = 4.00	x = 6.00	x = 8.00	x = 10.00	x = 12.00	x = 14.00	x = 16.00	x = 18.00
0.00	18.0000	32.0000	42.0000	48.0000	50.0000	48.0000	42.0000	32.0000	18.0000
1.00	16.4489	30.1970	40.1562	46.1502	48.1531	46.1773	40.3274	31.2520	22.9449
2.00	15.3312	28.5350	38.3477	44.3130	46.3327	44.4671	39.0872	31.5755	24.6930
3.00	14.4219	27.0429	36.6090	42.5113	44.5759	42.9362	38.1976	31.7478	25.4131
4.00	13.6381	25.6913	34.9606	40.7728	42.9127	41.5716	37.4340	31.7086	25.6986
5.00	12.9409	24.4545	33.4091	39.1182	41.3519	40.3240	36.7033	31.5136	25.7663
6.00	12.3088	23.3146	31.9546	37.5566	39.8880	39.1565	35.9745	31.2134	25.7128
7.00	11.7294	22.2589	30.5939	36.0884	38.5109	38.0470	35.2407	30.8434	25.5871
8.00	11.1946	21.2785	29.3217	34.7092	37.2109	36.9834	34.5032	30.4279	25.4167
9.00	10.6987	20.3660	28.1318	33.4130	35.9801	35.9591	33.7660	29.9836	25.2181
10.00	10.2377	19.5150	27.0178	32.1929	34.8117	34.9710	33.0338	29.5224	25.0019

(d)

Time	x = 10.00	x = 20.00	x = 30.00	x = 40.00	x = 50.00	x = 60.00	x = 70.00	x = 80.00	x = 90.00
0.00	8.0000	16.0000	24.0000	32.0000	40.0000	32.0000	24.0000	16.0000	8.0000
1.00	8.0000	16.0000	24.0000	31.9979	39.7425	31.9979	24.0000	16.0026	8.3218
2.00	8.0000	16.0000	23.9999	31.9918	39.4932	31.9918	24.0001	16.0225	8.9350
3.00	8.0000	16.0000	23.9997	31.9820	39.2517	31.9820	24.0002	16.0391	9.2272
4.00	8.0000	16.0000	23.9993	31.9686	39.0175	31.9687	24.0002	16.0391	9.2272
5.00	8.0000	16.0000	23.9987	31.9520	38.7905	31.9520	24.0003	16.0599	9.5103
6.00	8.0000	15.9999	23.9978	31.9323	38.5701	31.9324	24.0005	16.0845	9.7846
7.00	8.0000	15.9999	23.9966	31.9097	38.3561	31.9098	24.0008	16.1126	10.0506
8.00	8.0000	15.9998	23.9950	31.8844	38.1483	31.8846	24.0012	16.1441	10.3084
9.00	8.0000	15.9997	23.9931	31.8566	37.9463	31.8569	24.0017	16.1786	10.5585
10.00	8.0000	15.9996	23.9908	31.8265	37.7499	31.8269	24.0023	16.2160	10.8012

11. (a) $\psi(x) = \frac{1}{2}x + 20$

(b)

Time	x = 4.00	x = 8.00	x = 12.00	x = 16.00
0.00	50.0000	50.0000	50.0000	50.0000
10.00	32.7433	44.2679	45.4228	38.2971
20.00	29.9946	36.2354	38.3148	35.8160
30.00	26.9487	32.1409	34.0874	32.9644
40.00	25.2691	29.2562	31.2704	31.2580
50.00	24.1178	27.4348	29.4296	30.1207
60.00	23.3821	26.2339	28.2356	29.3810
70.00	22.8995	25.4560	27.4554	28.8998
80.00	22.5861	24.9481	26.9482	28.5859
90.00	22.3817	24.6176	26.6175	28.3817
100.00	22.2486	24.4022	26.4023	28.2486
110.00	22.1619	24.2620	26.2620	28.1619
120.00	22.1055	24.1707	26.1707	28.1055
130.00	22.0687	24.1112	26.1112	28.0687
140.00	22.0447	24.0724	26.0724	28.0447
150.00	22.0291	24.0472	26.0472	28.0291
160.00	22.0190	24.0307	26.0307	28.0190
170.00	22.0124	24.0200	26.0200	28.0124
180.00	22.0081	24.0130	26.0130	28.0081
190.00	22.0052	24.0085	26.0085	28.0052
200.00	22.0034	24.0055	26.0055	28.0034
210.00	22.0022	24.0036	26.0036	28.0022
220.00	22.0015	24.0023	26.0023	28.0015
230.00	22.0009	24.0015	26.0015	28.0009
240.00	22.0006	24.0010	26.0010	28.0006
250.00	22.0004	24.0007	26.0007	28.0004
260.00	22.0003	24.0004	26.0004	28.0003
270.00	22.0002	24.0003	26.0003	28.0002
280.00	22.0001	24.0002	26.0002	28.0001
290.00	22.0001	24.0001	26.0001	28.0001
300.00	22.0000	24.0001	26.0001	28.0000
310.00	22.0000	24.0001	26.0001	28.0000
320.00	22.0000	24.0000	26.0000	28.0000
330.00	22.0000	24.0000	26.0000	28.0000
340.00	22.0000	24.0000	26.0000	28.0000
350.00	22.0000	24.0000	26.0000	28.0000

EXERCISES 15.3, Page 772

1. (a)

Time	x = 0.25	x = 0.50	x = 0.75
0.00	0.1875	0.2500	0.1875
0.10	0.1775	0.2400	0.1775
0.20	0.1491	0.2100	0.1491
0.30	0.1066	0.1605	0.1066
0.40	0.0556	0.0938	0.0556
0.50	0.0019	0.0148	0.0019
0.60	−0.0501	−0.0682	−0.0501
0.70	−0.0970	−0.1455	−0.0970
0.80	−0.1361	−0.2072	−0.1361
0.90	−0.1648	−0.2462	−0.1648
1.00	−0.1802	−0.2591	−0.1802

(b)

Time	x = 0.4	x = 0.8	x = 1.2	x = 1.
0.00	0.0032	0.5273	0.5273	0.003
0.10	0.0194	0.5109	0.5109	0.019
0.20	0.0652	0.4638	0.4638	0.065
0.30	0.1318	0.3918	0.3918	0.131
0.40	0.2065	0.3035	0.3035	0.206
0.50	0.2743	0.2092	0.2092	0.274
0.60	0.3208	0.1190	0.1190	0.320
0.70	0.3348	0.0413	0.0413	0.334
0.80	0.3094	−0.0180	−0.0180	0.309
0.90	0.2443	−0.0568	−0.0568	0.244
1.00	0.1450	−0.0768	−0.0768	0.145

(c) Time	x = 0.1	x = 0.2	x = 0.3	x = 0.4	x = 0.5	x = 0.6	x = 0.7	x = 0.8	x = 0.9
0.00	0.0000	0.0000	0.0000	0.0000	0.0000	0.5000	0.5000	0.5000	0.5000
0.04	0.0000	0.0000	0.0000	0.0000	0.0800	0.4200	0.5000	0.5000	0.4200
0.08	0.0000	0.0000	0.0000	0.0256	0.2432	0.2568	0.4744	0.4744	0.2312
0.12	0.0000	0.0000	0.0082	0.1126	0.3411	0.1589	0.3792	0.3710	0.0462
0.16	0.0000	0.0026	0.0472	0.2394	0.3076	0.1898	0.2108	0.1663	−0.0496
0.20	0.0008	0.0187	0.1334	0.3264	0.2146	0.2651	0.0215	−0.0933	−0.0605
0.24	0.0071	0.0657	0.2447	0.3159	0.1735	0.2463	−0.1266	−0.3056	−0.0625
0.28	0.0299	0.1513	0.3215	0.2371	0.2013	0.0849	−0.2127	−0.3829	−0.1223
0.32	0.0819	0.2525	0.3168	0.1737	0.2033	−0.1345	−0.2580	−0.3223	−0.2264
0.36	0.1623	0.3197	0.2458	0.1657	0.0877	−0.2853	−0.2843	−0.2104	−0.2887
0.40	0.2412	0.3129	0.1727	0.1583	−0.1223	−0.3164	−0.2874	−0.1473	−0.2336
0.44	0.2657	0.2383	0.1399	0.0658	−0.3046	−0.2761	−0.2549	−0.1565	−0.0761
0.48	0.1965	0.1410	0.1149	−0.1216	−0.3593	−0.2381	−0.1977	−0.1715	0.0800
0.52	0.0466	0.0531	0.0225	−0.3093	−0.2992	−0.2260	−0.1451	−0.1144	0.1300
0.56	−0.1161	−0.0466	−0.1662	−0.3876	−0.2188	−0.2114	−0.1085	0.0111	0.0602
0.60	−0.2194	−0.2069	−0.3875	−0.3411	−0.1901	−0.1662	−0.0666	0.1140	−0.0446
0.64	−0.2485	−0.4290	−0.5362	−0.2611	−0.2021	−0.0969	0.0012	0.1084	−0.0843
0.68	−0.2559	−0.6276	−0.5625	−0.2503	−0.1993	−0.0298	0.0720	0.0068	−0.0354
0.72	−0.3003	−0.6865	−0.5097	−0.3230	−0.1585	0.0156	0.0893	−0.0874	0.0384
0.76	−0.3722	−0.5652	−0.4538	−0.4029	−0.1147	0.0289	0.0265	−0.0849	0.0596
0.80	−0.3867	−0.3464	−0.4172	−0.4068	−0.1172	−0.0046	−0.0712	−0.0005	0.0155
0.84	−0.2647	−0.1633	−0.3546	−0.3214	−0.1763	−0.0954	−0.1249	0.0665	−0.0386
0.88	−0.0254	−0.0738	−0.2202	−0.2002	−0.2559	−0.2215	−0.1079	0.0385	−0.0468
0.92	0.2064	−0.0157	−0.0325	−0.1032	−0.3067	−0.3223	−0.0804	−0.0636	−0.0127
0.96	0.3012	0.1081	0.1380	−0.0487	−0.2974	−0.3407	−0.1250	−0.1548	0.0092
1.00	0.2378	0.3032	0.2392	−0.0141	−0.2223	−0.2762	−0.2481	−0.1840	−0.0244

3. (a)

Time	x = 0.2	x = 0.4	x = 0.6	x = 0.8
0.00	0.5878	0.9511	0.9511	0.5878
0.05	0.5808	0.9397	0.9397	0.5808
0.10	0.5599	0.9059	0.9059	0.5599
0.15	0.5256	0.8505	0.8505	0.5256
0.20	0.4788	0.7748	0.7748	0.4788
0.25	0.4206	0.6806	0.6806	0.4206
0.30	0.3524	0.5701	0.5701	0.3524
0.35	0.2757	0.4460	0.4460	0.2757
0.40	0.1924	0.3113	0.3113	0.1924
0.45	0.1046	0.1692	0.1692	0.1046
0.50	0.0142	0.0230	0.0230	0.0142

(b)

Time	x = 0.2	x = 0.4	x = 0.6	x = 0.8
0.00	0.5878	0.9511	0.9511	0.5878
0.03	0.5860	0.9482	0.9482	0.5860
0.05	0.5808	0.9397	0.9397	0.5808
0.08	0.5721	0.9256	0.9256	0.5721
0.10	0.5599	0.9060	0.9060	0.5599
0.13	0.5445	0.8809	0.8809	0.5445
0.15	0.5257	0.8507	0.8507	0.5257
0.18	0.5039	0.8153	0.8153	0.5039
0.20	0.4790	0.7750	0.7750	0.4790
0.23	0.4513	0.7302	0.7302	0.4513
0.25	0.4209	0.6810	0.6810	0.4209
0.28	0.3879	0.6277	0.6277	0.3879
0.30	0.3527	0.5706	0.5706	0.3527
0.33	0.3153	0.5102	0.5102	0.3153
0.35	0.2761	0.4467	0.4467	0.2761
0.38	0.2352	0.3806	0.3806	0.2352
0.40	0.1929	0.3122	0.3122	0.1929
0.43	0.1495	0.2419	0.2419	0.1495
0.45	0.1052	0.1701	0.1701	0.1052
0.48	0.0602	0.0974	0.0974	0.0602
0.50	0.0149	0.0241	0.0241	0.0149

5.

Time	x = 10	x = 20	x = 30	x = 40	x = 50
0.00000	0.1000	0.2000	0.3000	0.2000	0.1000
0.20045	0.1000	0.2000	0.2750	0.2000	0.1000
0.40089	0.1000	0.1938	0.2125	0.1938	0.1000
0.60134	0.0984	0.1688	0.1406	0.1688	0.0984
0.80178	0.0898	0.1191	0.0828	0.1191	0.0898
1.00223	0.0661	0.0531	0.0432	0.0531	0.0661
1.20268	0.0226	−0.0121	0.0085	−0.0121	0.0226
1.40312	−0.0352	−0.0635	−0.0365	−0.0635	−0.0352
1.60357	−0.0913	−0.1011	−0.0950	−0.1011	−0.0913
1.80401	−0.1271	−0.1347	−0.1566	−0.1347	−0.1271
2.00446	−0.1329	−0.1719	−0.2072	−0.1719	−0.1329
2.20491	−0.1153	−0.2081	−0.2402	−0.2081	−0.1153
2.40535	−0.0920	−0.2292	−0.2571	−0.2292	−0.0920
2.60580	−0.0801	−0.2230	−0.2601	−0.2230	−0.0801
2.80624	−0.0838	−0.1903	−0.2445	−0.1903	−0.0838
3.00669	−0.0932	−0.1445	−0.2018	−0.1445	−0.0932
3.20713	−0.0921	−0.1003	−0.1305	−0.1003	−0.0921
3.40758	−0.0701	−0.0615	−0.0440	−0.0615	−0.0701
3.60803	−0.0284	−0.0205	0.0336	−0.0205	−0.0284

Time	x = 10	x = 20	x = 30	x = 40	x = 50
3.80847	0.0224	0.0321	0.0842	0.0321	0.0224
4.00892	0.0700	0.0953	0.1087	0.0953	0.0700
4.20936	0.1064	0.1555	0.1265	0.1555	0.1064
4.40981	0.1285	0.1962	0.1588	0.1962	0.1285
4.61026	0.1354	0.2106	0.2098	0.2106	0.1354
4.81070	0.1273	0.2060	0.2612	0.2060	0.1273
5.01115	0.1070	0.1955	0.2851	0.1955	0.1070
5.21159	0.0821	0.1853	0.2641	0.1853	0.0821
5.41204	0.0625	0.1689	0.2038	0.1689	0.0625
5.61249	0.0539	0.1347	0.1260	0.1347	0.0539
5.81293	0.0520	0.0781	0.0526	0.0781	0.0520
6.01338	0.0436	0.0086	−0.0080	0.0086	0.0436
6.21382	0.0156	−0.0564	−0.0604	−0.0564	0.0156
6.41427	−0.0343	−0.1043	−0.1107	−0.1043	−0.0343
6.61472	−0.0931	−0.1364	−0.1578	−0.1364	−0.0931
6.81516	−0.1395	−0.1630	−0.1942	−0.1630	−0.1395
7.01561	−0.1568	−0.1915	−0.2150	−0.1915	−0.1568
7.21605	−0.1436	−0.2173	−0.2240	−0.2173	−0.1436
7.41650	−0.1129	−0.2263	−0.2297	−0.2263	−0.1129
7.61695	−0.0824	−0.2078	−0.2336	−0.2078	−0.0824
7.81739	−0.0625	−0.1644	−0.2247	−0.1644	−0.0625
8.01784	−0.0526	−0.1106	−0.1856	−0.1106	−0.0526
8.21828	−0.0440	−0.0611	−0.1091	−0.0611	−0.0440
8.41873	−0.0287	−0.0192	−0.0085	−0.0192	−0.0287
8.61918	−0.0038	0.0229	0.0867	0.0229	−0.0038
8.81962	0.0287	0.0743	0.1500	0.0743	0.0287
9.02007	0.0654	0.1332	0.1755	0.1332	0.0654
9.22051	0.1027	0.1858	0.1799	0.1858	0.1027
9.42096	0.1352	0.2160	0.1872	0.2160	0.1352
9.62140	0.1540	0.2189	0.2089	0.2189	0.1540
9.82185	0.1506	0.2030	0.2356	0.2030	0.1506
10.02230	0.1226	0.1822	0.2461	0.1822	0.1226

Note: Time is expressed in milliseconds.

CHAPTER 15 REVIEW EXERCISES, Page 774

1. $u_{11} = 0.8929$, $u_{21} = 3.5714$, $u_{31} = 13.3929$

3. (a)

Time	x = 0.00	x = 0.20	x = 0.40	x = 0.60	x = 0.80	x = 1.00
0.00	0.0000	0.2000	0.4000	0.6000	0.8000	0.0000
0.01	0.0000	0.2000	0.4000	0.6000	0.5500	0.0000
0.02	0.0000	0.2000	0.4000	0.5375	0.4250	0.0000
0.03	0.0000	0.2000	0.3844	0.4750	0.3469	0.0000
0.04	0.0000	0.1961	0.3609	0.4203	0.2922	0.0000
0.05	0.0000	0.1883	0.3346	0.3734	0.2512	0.0000

(b)

Time	x = 0.00	x = 0.20	x = 0.40	x = 0.60	x = 0.80	x = 1.00
0.00	0.0000	0.2000	0.4000	0.6000	0.8000	1.0000
0.01	0.0000	0.2000	0.4000	0.6000	0.8000	0.0000
0.02	0.0000	0.2000	0.4000	0.6000	0.5500	0.0000
0.03	0.0000	0.2000	0.4000	0.5375	0.4250	0.0000
0.04	0.0000	0.2000	0.3844	0.4750	0.3469	0.0000
0.05	0.0000	0.1961	0.3609	0.4203	0.2922	0.0000

(c) Yes; the table in part (b) is the table in part (a) shifted downward.

APPENDIX I EXERCISES, Page APP–3

1. (a) 24   **(b)** 720   **(c)** $4\sqrt{\pi}/3$ **(d)** $-8\sqrt{\pi}/15$

3. 0.297

5. $\Gamma(x) > \displaystyle\int_0^1 t^{x-1}e^{-t}\,dt > e^{-1}\int_0^1 t^{x-1}\,dt = \dfrac{1}{xe}$
for $x > 0$. As $x \to 0^+$, $1/x \to +\infty$.

APPENDIX III EXERCISES, Page APP–11

1. -58 **3.** 248   **5.** 12   **7.** $16e^{3t}$

9. $x = 4$, $y = -7$ **11.** $x = 4$, $y = 3/2$, $z = 1$

13. Let $z = t$, t any real number. The solution is $x = -t/3$, $y = 5t/3$, $z = t$; the system represents the line of intersection of two planes.

APPENDIX IV EXERCISES, Page APP–15

1. $7 - 4i$ **3.** $-11 - 11i$ **5.** $3 - 4i$

7. $\dfrac{7}{34} - \dfrac{11}{34}i$ **9.** $\dfrac{5}{34} - \dfrac{3}{34}i$ **11.** $z = e^{i\pi/2}$

13. $z = e^{i\pi}$ **15.** $z = 2\sqrt{2}e^{i\pi/4}$ **17.** $z = 12e^{i\pi/3}$

19. $z = 2e^{i\pi/6}$ **21.** $z = -8$

INDEX

Abel, Niels Henrik, 153
Abel's formula, 153
Absolute converge of a power series, 268
Absolute error, 496
Action potential, 537
Adams-Bashforth/Adams-Moulton method, 513
Amplitude of free vibrations, 217
Analytic at a point, 269
Annihilator differential operator, 183
Applications of differential equations:
  atomic physics, 96–98
  chemistry, 111–114
  cooling, 19, 98
  continuous compound interest, 24–25
  deflection of beams, 22–23, 372–374
  discharge through an orifice, 21
  falling body, 13
  pendulum motion, 15
  population growth, 23–24, 94, 108
  memory, 29, 107
  mixtures, 101
  networks, 403
  rotating string, 16–17
  series circuits, 18–19, 99–100, 242
  spring-mass systems, 14, 212, 400
  suspended wire, 20–21
  temperature distributions, 650, 653
Arc, 547
Argument of a complex number, 790
Arithmetic of power series, 269
Asymptotically stable critical point, 566
Augmented matrix, 423
  elementary row operations on, 424
Autonomous systems, 545
  applications of, 578
Auxiliary equation, 160, 259

Axis of symmetry, 22
Axon, 537

Backward difference, 529
Beams:
  cantilever, 22, 679
  embedded, 23, 373
  simply supported, 664
  static deflection of, 22–23, 372
  vibrating, 664, 679
Beats, 241
Bernoulli, Jakob, 71
Bernoulli's differential equation, 72
Bessel's differential equation, 306
  parametric, 310
  solution of, 307
Bessel, Friedrich Wilhelm, 306
Bessel functions:
  differential recursions relations for, 312–313
  of first kind, 308–309
  graphs of, 310
  modified, 319
  numerical values of, 311
  properties of, 310–311
  of second kind, 309
  spherical, 313
Boundary conditions, 132, 653
Boundary points, 749
Boundary-value problem, 132, 655
Branch point, 403
Buckling of thin column, 286

Calculation of order α, 518
Canonical form of a system of linear differential equations, 408
Carbon dating, 97
Catenary, 21
Cauchy, Augustin-Louis, 259

Cauchy-Euler differential equation, 258
    method of solution, 259
Center, 559
Central difference, 529
Chaos, 207
Characteristic equation, 160, 429
Characteristic functions, 624
Characteristic values, 428
Characteristic vectors, 428
Chebyshev's differential equation,
    640
Chemical reactions, 111–112
Circuits, differential equations of, 18,
    99–100
Clairaut, Alexis Claude, 71
Clairaut's differential equation, 74
Cofactor, 420
Competition model, 586
Complementary error function, 716
Complementary function, 147
Complete set of functions, 603
Complete solution, 9
Complex number, APP-12
    argument of, APP-14
    conjugate of, APP-12
    difference of, APP-12
    imaginary part of, APP-12
    modulus of, APP-13–APP-14
    polar form of, APP-14
    product of, APP-12
    quotient of, APP-13
    real part of, APP-12
    sum of, APP-12
    vector interpretation, APP-13
Continuing method, 514
Continuous compound interest,
    24–26
Convergence of a Fourier series, 608
Convolution integral, 356
Convolution theorem, 357
Cooling, Newton's law of, 19, 98
Coupled springs, 389, 400–401
Cover-up method, 341
Cramer, Gabriel, APP-10
Cramer's rule, APP-9–APP-10
Crank-Nicholson method, 762–763
Critical damping, 223
Critical point, 547
    asymptotically stable, 566
    center, 559

degenerate nodes, 557–558
locally stable, 554, 566
saddle, 556
stable, 566
stable node, 555, 557–558
stable spiral point, 560
unstable, 554, 567
unstable node, 555, 558
unstable spiral point, 560
Critical loads, 630
Critical speeds, 630
Critically damped circuit, 243
Curvature, 23
Cycle, 547

D'Alembert, Jean le Rond, 665
D'Alembert's solution, 665
Damped amplitude, 227
Damped motion, 222
Damping constant, 222
Damping factor, 223
Decay, radioactive, 94, 96
Deflection curve, 17, 22
Degenerate node, 557
    stable, 557–558
    unstable, 558
Degenerate system of differential
    equations, 410
DeMoivre, Abraham, APP-15
DeMoivre's theorem, APP-15
Determinants, APP-7
    expansion by cofactors, 418,
    APP-8
Diagonally dominant matrix, 755
Difference equation, 530, 749
Difference quotients, 529
Differential equation, 2
    linear, 3
    nonlinear, 4
    ordinary, 2
    partial, 2
Differential equation of a family of
    curves, 88
Differential operator, 181
Diffusion equation, 656
Dirac delta function, 379–380
    Laplace transform of, 380
Dirac, Paul Adrian Maurice, 380
Direction field, 488

Dirichlet, Peter G. L., 654
Dirichlet condition, 653
Dirichlet problem, 667, 750
    superposition principle for, 668
Discharge through an orifice,
    differential equation of, 21–22
Discretization error, 517
Distributions, 382
Double eigenvalues, 696
Double cosine series, 621
Double sine series, 621, 682

Effective spring constant, 221
Eigenfunctions, 624, 658
Eigenvalues:
    double, 696
    of a matrix, 428–429
    of a boundary-value problem, 624,
    658
Eigenvectors of a matrix, 428
Elastic curve, 22
Electrical networks, 403
Electrical vibrations, 243
Electromotive force, 19
Elementary row operations, 424
    notation for, 426
Elimination method, 390, 424
Elliptic partial differential equation,
    647
Emf, 19
Epidemics, 24, 108, 110–111
Equation of motion, 214–215
Equidimensional equation, 258
Equilibrium position, 14, 213
Error function, 716
Errors:
    discretization, 517
    formula, 517
    global truncation, 519
    local truncation, 517
    round off, 517
Escape velocity, 114–115, 117
Essential parameters, 9
Euler, Leonhard, 161
Euler load, 630
Euler-Mascheroni constant, 311
Euler's formula, 161
    derivation of, APP-14
Euler's methods, 494–495, 499

Even function, 610
Exact differential, 54
    criterion for, 55
Exact differential equation, 54
Excitation function, 149
Existence and uniqueness of a
    solution, 35, 131, 439
Expansion of a function:
    in a cosine series, 612
    in a Fourier-Bessel series, 634
    in a Fourier-Legendre series,
    636–637
    in a Fourier series, 606
    in a series of orthogonal functions,
    601–602
    in a sine series, 612
Explicit finite-difference method, 758
Explicit solution, 6
Exponential growth, 94–96, 123
Exponential order, 327
Exponents of a singularity, 292
Extreme displacement, 215

Family of solutions, 9
Fick's law, 107
Finite difference equation, 529–530
Finite differences, 529
Finite elements methods, 540
First buckling mode, 630
First harmonic, 663
First-order chemical reactions,
    111–112
First-order differential equations:
    applications of, 88, 94, 108
    solutions of, 38, 46, 53, 62, 71
First translation theorem, 343
Fitzhugh-Nagumo equation, 538
Five-point approximation of
    Laplacian, 748
Flow of heat, 650–651
Forced motion, 233
Forcing function, 149
Forward difference, 529
Fourier, Jean-Baptiste Joseph, 606
Fourier-Bessel series, 634
    convergence of, 635
Fourier-Legendre series, 636
    convergence of, 636
Fourier coefficients, 606

Fourier cosine transform, 736
  existence of, 736
  operational properties of, 737
Fourier integral, 729
  complex form, 733
  convergence of, 729
  cosine form, 731
  sine form, 731
Fourier series, 606
  complex, 610
  convergence of, 607–608
  cosine, 612
  generalized, 603
  sine, 612
  in two variables, 621, 682
Fourier sine transform, 736
  existence of, 736
  operational properties of, 737
Fourier transform, 736
  existence of, 736
  operational properties of, 736–737
Free motion:
  damped, 222
  undamped, 213–214
Frequency, 214
Frobenius, Ferdinand Georg, 289
Frobenius, method of, 289–290
Fundamental set of solutions:
  of a linear differential equation,
    142, 144
  of a system of linear differential
    equations, 443
Fundamental matrix, 447

Galloping Gertie, 253
Gamma function, 308, 334, APP-2
Gauss, Karl Friedrich, 424
Gaussian elimination, 424
Gauss-Jordan elimination, 424
Gauss-Seidel iteration, 753
General solution:
  of a differential equation, 9
  of a linear differential equation, 65,
    144, 147
  of a system of linear differential
    equations, 443, 445
Generalized factorial function, APP-2
Generalized Fourier series, 601–602
Generalized functions, 382

Generating function, 317
Gibbs, Josiah Willard, 614
Gibbs phenomenon, 614
Global truncation error, 519
Gompertz curve, 111
Growth and decay, 94

Half-life, 96
Half-range expansions, 614–615
Heat equation:
  derivation of one-dimensional, 651
  difference equation replacement
    for, 757–758, 763
  one-dimensional, 650
  solution of, 657
  two-dimensional, 680
Hermite, Charles, 286, 631
Hermite's differential equation, 286,
  631
Heun's formula, 499
Hodgkin-Huxley equation, 538
Homogeneous differential equation:
  ordinary, 48, 139
  partial, 644
Homogeneous functions of degree n, 46
Homogeneous systems:
  of algebraic equations, 429
  of linear differential equations, 408
Hooke, Robert, 212
Hooke's law, 14, 212
Hyperbolic partial differential
  equation, 647

Identity matrix, 417
Impedance, 245
Implicit finite difference method, 762
Implicit solution, 6
Impressed force, 232
Improved Euler method, 499
Indicial equation, 292-293
Indicial roots, 292
Initial conditions, 34, 130, 653
Initial-value problem:
  for a linear differential equation,
    34, 130
  for a system of linear differential
    equations, 438
Inner product, 598

Input function, 149
Insulated boundaries, 654
Integral equation, 369
Integral transform, 325, 735
Integrating factor, 59, 63
Integrodifferential equation, 370
Interest compounded continuously, 24–25
Interior mesh points, 529, 749
Inverse Fourier cosine transform, 736
Inverse Fourier sine transform, 736
Inverse Fourier transform, 736
Inverse integral transform, 735
Inverse Laplace transform, 335, 735
Inverse matrix, 419–420
Irregular singular point, 287
  at ∞, 304
Isoclines, 488
Isogonal trajectories, 93

Jacobian matrix, 569–570
Jordan, Wilhelm, 424

Kepler's second law of planetary motion, 107
Kernel of a transform, 735
Kirchhoff's laws, 18, 242
Kutta, Martin W., 508

Laguerre, Edmond, 631
Laguerre's differential equation, 631
Laplace, Pierre Simon Marquis de, 326
Laplace's equation, 650, 653
  difference equation replacement for, 748–749
  solution of, 666
Laplace transform:
  convolution theorem for, 357
  definition of, 325
  of derivatives, 356
  differentiation of, 351
  of Dirac delta function, 380
  existence of, 328
  of integrals, 357
  inverse of, 335
  linearity of, 326–327
  of periodic functions, 359

tables of, APP-4–APP-7
translation theorems of, 343, 347–348
Laplacian, 650
  five-point approximation for, 748
  in polar coordinates, 694
  in cylindrical coordinates, 702
  in spherical coordinates, 707
Lattice points, 749
Law of Malthus, 121
Law of mass action, 114
Lazer-McKenna model, 254–255
Legendre, Adrien Marie, 306
Legendre's differential equation, 306
  solution of, 314
Legendre polynomials, 316
  generating function for, 317
  graphs of, 316
  properties of, 316–318
  recurrence relation for, 317
Libby, Willard, 97
Liebman's method, 753
Lineal elements, 488
Linear dependence:
  of functions, 134
  of solution vectors, 441
Linear independence:
  of eigenvectors, 431
  of functions, 134
  of solutions, 141–142
  of solution vectors, 440–442
Linear ordinary differential equations, 3
  applications of, 94, 211
  complementary function for, 147
  first-order, 62
  general solution of, 65, 144, 147
  higher-order, 129
  homogeneous, 139
  nonhomogeneous, 139
  particular solution of, 145
  superposition principle(s) for, 140, 148
Linear operator, 181–182, 327
Linear partial differential equations, 644
  homogeneous, 644
  nonhomogeneous, 644
  superposition principle for, 646
Linear property, 324–325

Linear system, 149
Linear systems of algebraic equations, 423
Linear systems of differential equations, 408
    stability criteria for, 563
Linear transform, 327
Linearization, 568–569
Local truncation error, 517
Logarithmic decrement, 231
Logistic equation, 24, 108
Logistic function, 108–109
Logistic growth model, 125
Lotka, A. J., 119
Lotka and Volterra, equations of:
    competition model, 586–589
    predator-prey model, 119, 583–586
L-R series circuit, differential equation of, 99
L-R-C series circuit, differential equation of, 19, 242

Malthus, law of, 121
Malthus, Thomas R., 123–124
Mathematical model, 13, 87
Matrices:
    addition of, 415
    augmented, 423
    associative law of, 418
    banded, 752
    characteristic equation of, 429
    column, 414
    definition of, 414
    derivative of, 422
    determinant of, 418
    diagonally dominant, 755
    difference of, 416
    distributive law of, 418
    eigenvalues of, 428
    equality of, 414
    exponential, 478
    fundamental, 447
    integral of, 422
    inverse of, 420
    Jacobian, 569–570
    multiples of, 414
    multiplicative identity, 417
    multiplicative inverse, 419

nonsingular, 419
product of, 416
singular, 419
size, 414
sparse, 752
square, 414
sum of, 415
transpose of, 419
tridiagonal, 763
vector, 414
Wronski, 448
zero, 418
Matrix form of a linear system, 436
Meander function, 362
Mechanical resonance, 237
Mesh points, 529, 749
Mesh size, 749
Method of Frobenius, 290
Middle C, 687
Milne's method, 515
Minor, 420
Mixtures, 101
Mode of vibration, 662, 688
Modified Bessel function of the first kind, 319
Modified logistic equation, 116
Modulus of a complex number, APP-13–APP-14
Multistep method, 514
Multiplicity of eigenvalues, 462

Natural frequency, 214
Natural period, 214
Nerve impulse models, 537
Neumann, C. G., 310, 654
Neumann's condition, 653
Neumann's function, 310
Newton's law of cooling, 19, 98
Newton's second law of motion, 14, 213
Nodal line, 702
Node:
    degenerate, 557–558
    stable, 555
    unstable, 555
Nodes, 662
Nonhomogeneous linear differential equation:
    ordinary, 139
    partial, 644

Nonhomogeneous systems of linear differential equations, 408
Nonlinear differential equation, 4
Nonsingular matrix, 419
Normal form, 408
Normal modes, 662
Norm, 599
Normalized set of functions, 601
Numerical methods:
　Adams-Bashforth/Adams-Moulton, 513–514
　Crank-Nicholson method, 762–764
　Euler's method, 495
　Errors in, 517
　Finite difference methods, 528–530
　Gauss-Seidel iteration, 753
　Heun's formula, 499
　Improved Euler method, 499
　Milne's method, 515
　Runge-Kutta methods, 508–509
　Stability of, 521
　Three-term Taylor method, 504

Octave, 687
Odd function, 610
Operator, differential, 181
Order of a differential equation, 3
Ordinary differential equation, definition of, 2
Ordinary point, 277
Orthogonal curves, 88
Orthogonal functions, 598
Orthogonal set, 599
　with respect to a weight function, 602
Orthogonal trajectories, 89
Orthonormal set of functions, 599
Output, 149
Overdamped circuit, 243
Overdamped motion, 223, 591
Overtones, 663

Parabolic partial differential equation, 647
Parametric Bessel equation, 310
Partial differential equation, 2
　homogeneous, 644
　linear, 644

nonhomogeneous, 644
separable, 644
Particular solution, 9, 145
Peano, Guiseppe, 37
Pendulum:
　nonlinear, 579
　simple, 15–16
Percentage relative error, 496–497
Period, 214
Periodic boundary conditions, 631
Periodic extension, 608
Periodic functions, Laplace transform of, 359
Periodic solution, 547
Phase angle, 217, 227
Phase plane method, 574
Physical pendulum, 15
Picard, Charles Émile, 35
Picard's method of successive approximations, 80–81
Piecewise continuous functions, 327
Plane autonomous system, 545, 554
　stability criteria for, 563
Plucked string, 653, 663
Poisson, Simeon-Denis, 674
Poisson's equation, 674, 756
　difference equation replacement for, 756
Poisson's integral formula for a circle, 699
Polar form of a complex number, APP-14
Population dynamics, 123
Power series, review of, 267–269
Power series solutions, 272
Population growth, 94, 108, 123
Predator-prey model, 119, 583
Predictor-corrector methods, 513–515
Pure resonance, 236

Quasi frequency, 227
Quasi period, 227

Radial symmetry, 699
Radial vibrations, 699
Radioactive decay, 94, 96
Radius of convergence, 268

Rational roots of a polynomial
    equation, 165
R-C series circuit, differential equation
    of, 100
Reactance, 245
Reactions, chemical, 111–114
Rectangular pulse, 354
Rectified sine wave, 363
Recurrence relation, 273
Reduced row-echelon form, 425
Reduction of order, 154
Regular singular point, 287
    at ∞, 304
Regular Sturm-Liouville problem, 625
Relative error, 496
Resonance:
    mechanical, 237
    electrical, 244
Resonance curve, 236–237
Response of a system, 99, 149
Ricatti, Jacobo Francesco, 71
Ricatti's differential equation, 73
Robin, Victor G., 654
Robin condition, 654
Rodrigues' formula, 320
Rodrigues, Olinde, 320
Rotating string, 16–18
Round-off error, 517
Row-echelon form, 425
Runge, Carl D. T., 508
Runge-Kutta methods, 508

Saddle point, 556
Sawtooth function, 362
Schwartz, Laurent, 382
Second-order chemical reactions, 112
Second translation theorem, 348
Self-adjoint form, 628
Self-orthogonal family of curves, 93
Separable differential equations:
    ordinary, 39
    partial, 644
Separation constant, 645
Series:
    Fourier, 606, 612
    Fourier-Bessel, 634
    Fourier-Lengendre, 636
    power, 267
    solutions of ordinary differential
        equations, 272, 277, 286

Sifting property, 383
Simple harmonic electrical vibrations,
    243
Simple harmonic motion, 214
Simple pendulum, 15–16, 246
Single-step method, 514
Singular point of a differential equation,
    277
    irregular, 287
    regular, 287
    at ∞, 304
Singular matrix, 419
Singular solution, 9
Sliding bead, 120, 580
Slope field, 488
Solution of a differential equation, 4
    complete, 9
    explicit, 6
    general, 9, 65, 144, 147
    implicit, 6
    n-parameter family of, 9
    number of, 7
    particular, 9
    singular, 9
    trivial, 5
Solution of a system of differential
        equations, 390, 437
    general, 443, 445
    particular, 444
Spherical Bessel functions, 313
Spiral points, 559–560
Spread of a disease, 24, 108
Square norm, 599
Square wave, 362
Stability of a linear system, 554
    criteria for, 563
Stability of a numerical method, 521
Staircase function, 354
Standing waves, 662, 688, 701
Starting methods, 514
State of system, 13, 149, 545
Stationary point, 547
Steady-state:
    current, 100, 244
    solution, 234, 244, 672
    temperature, 650, 653
Stefan's law of radiation, 118
Sturm-Liouville problem:
    properties of, 626
    regular, 625
    singular, 627

Substitutions, 76
Superposition principle:
   for Dirichlet's problem, 668
   for homogeneous linear equations, 140
   for homogeneous linear systems, 439
   for nonhomogeneous linear equations, 148
   for homogeneous linear partial differential equations, 646
Synthetic division, 165
Systems of linear differential equations, solutions of:
   Laplace transform method, 398
   by matrices, 453
   operator method, 390

Tacoma Narrows bridge collapse, 253
Tangent lines, method of, 494
Telegraph equation, 656
Thermal diffusivity, 651
Third-order chemical reactions, 117
Three-term Taylor method, 503–504
Total differential, 54
Tractrix, 28, 118
Trajectories, orthogonal, 89
Transient:
   solution, 234, 244, 672
   term, 100, 234
Translation theorems for Laplace transform, 343, 348
Transpose of a matrix, 419
Traveling waves, 665
Triangular wave, 362
Tridiagonal matrix, 763
Trivial solution, 5
Truncation errors, 517
Twisted shaft, 245, 676
Two-point boundary-value problem, 132

Underdamped circuit, 243
Underdamped motion, 224, 591
Undetermined coefficients:
   for linear differential equations, 169, 187

for systems of linear differential equations, 469
Uniqueness theorems, 35, 131, 439
Units, 15, 19
Unit impulse, 379
Unit step function, 345
   Laplace transform of, 348
Unstable critical point, 554
Unstable numerical method, 521, 761

Variables, separable, 39
Variation of parameters:
   for linear differential equations, 195
   for systems of linear differential equations, 473
Vectors, definition of, 414
   solutions of systems of linear differential equations, 437
Verhulst, P. F., 108, 125
Vibrations, spring-mass systems, 14, 212–239, 400–403
Volterra integral equation, 369
Volterra, Vito, 119
Volterra's principle, 590

Wave-equation:
   derivation of one-dimensional, 651–653
   difference equation replacement for, 768
   one-dimensional, 650
   solution of, 660
   two-dimensional, 680
Weight, 15
Weight function, orthogonality with respect to, 602
Wronski, Josef Maria Hoëne, 137
Wronski matrix, 448
Wronskian, 136–137, 442
   Abel's formula for, 153

Zero matrix, 418

TABLE OF INTEGRALS

1. $\int u\,dv = uv - \int v\,du$

2. $\int u^n\,du = \dfrac{1}{n+1}u^{n+1} + C, n \neq -1$

3. $\int \dfrac{du}{u} = \ln|u| + C$

4. $\int e^u\,du = e^u + C$

5. $\int a^u\,du = \dfrac{1}{\ln a}a^u + C$

6. $\int \sin u\,du = -\cos u + C$

7. $\int \cos u\,du = \sin u + C$

8. $\int \sec^2 u\,du = \tan u + C$

9. $\int \csc^2 u\,du = -\cot u + C$

10. $\int \sec u \tan u\,du = \sec u + C$

11. $\int \csc u \cot u\,du = -\csc u + C$

12. $\int \tan u\,du = -\ln|\cos u| + C$

13. $\int \cot u\,du = \ln|\sin u| + C$

14. $\int \sec u\,du = \ln|\sec u + \tan u| + C$

15. $\int \csc u\,du = \ln|\csc u - \cot u| + C$

16. $\int \dfrac{du}{\sqrt{a^2 - u^2}} = \sin^{-1}\dfrac{u}{a} + C$

17. $\int \dfrac{du}{a^2 + u^2} = \dfrac{1}{a}\tan^{-1}\dfrac{u}{a} + C$

18. $\int \dfrac{du}{u\sqrt{u^2 - a^2}} = \dfrac{1}{a}\sec^{-1}\dfrac{u}{a} + C$

19. $\int \dfrac{du}{a^2 - u^2} = \dfrac{1}{2a}\ln\left|\dfrac{u+a}{u-a}\right| + C$

20. $\int \dfrac{du}{u^2 - a^2} = \dfrac{1}{2a}\ln\left|\dfrac{u-a}{u+a}\right| + C$

21. $\int \sin^2 u\,du = \tfrac{1}{2}u - \tfrac{1}{4}\sin 2u + C$

22. $\int \cos^2 u\,du = \tfrac{1}{2}u + \tfrac{1}{4}\sin 2u + C$

23. $\int \tan^2 u\,du = \tan u - u + C$

24. $\int \cot^2 u\,du = -\cot u - u + C$

25. $\int \sin^3 u\,du = -\tfrac{1}{3}(2 + \sin^2 u)\cos u + C$

26. $\int \cos^3 u\,du = \tfrac{1}{3}(2 + \cos^2 u)\sin u + C$

27. $\int \tan^3 u\,du = \tfrac{1}{2}\tan^2 u + \ln|\cos u| + C$

28. $\int \cot^3 u\,du = -\tfrac{1}{2}\cot^2 u - \ln|\sin u| + C$

29. $\int \sec^3 u\,du = \tfrac{1}{2}\sec u \tan u + \tfrac{1}{2}\ln|\sec u + \tan u| + C$

30. $\int \csc^3 u\,du = -\tfrac{1}{2}\csc u \cot u + \tfrac{1}{2}\ln|\csc u - \cot u| + C$

31. $\int \sin^n u\,du = -\dfrac{1}{n}\sin^{n-1}u \cos u + \dfrac{n-1}{n}\int \sin^{n-2}u\,du$

32. $\int \cos^n u\,du = \dfrac{1}{n}\cos^{n-1}u \sin u + \dfrac{n-1}{n}\int \cos^{n-2}u\,du$

33. $\int \tan^n u\,du = \dfrac{1}{n-1}\tan^{n-1}u - \int \tan^{n-2}u\,du$

34. $\int \cot^n u\,du = \dfrac{-1}{n-1}\cot^{n-1}u - \int \cot^{n-2}u\,du$

35. $\int \sec^n u\,du = \dfrac{1}{n-1}\tan u \sec^{n-2}u + \dfrac{n-2}{n-1}\int \sec^{n-2}u\,du$

36. $\int \csc^n u\,du = \dfrac{-1}{n-1}\cot u \csc^{n-2}u + \dfrac{n-2}{n-1}\int \csc^{n-2}u\,du$

37. $\int \sin au \sin bu\,du = \dfrac{\sin(a-b)u}{2(a-b)} - \dfrac{\sin(a+b)u}{2(a+b)} + C$

38. $\int \cos au \cos bu\,du = \dfrac{\sin(a-b)u}{2(a-b)} + \dfrac{\sin(a+b)u}{2(a+b)} + C$

39. $\int \sin au \cos bu\,du = -\dfrac{\cos(a-b)u}{2(a-b)} - \dfrac{\cos(a+b)u}{2(a+b)} + C$

40. $\int u \sin u\,du = \sin u - u \cos u + C$

41. $\int u \cos u\,du = \cos u + u \sin u + C$

42. $\int u^n \sin u\,du = -u^n \cos u + n \int u^{n-1} \cos u\,du$

43. $\int u^n \cos u\,du = u^n \sin u - n \int u^{n-1} \sin u\,du$

44. $\int \sin^n u \cos^m u\,du = -\dfrac{\sin^{n-1}u \cos^{m+1}u}{n+m} + \dfrac{n-1}{n+m}\int \sin^{n-2}u \cos^m u\,du = \dfrac{\sin^{n+1}u \cos^{m-1}u}{n+m} + \dfrac{m-1}{n+m}\int \sin^n u \cos^{m-2}u\,du$